T0210734

Lecture Notes of the Institute for Computer Sciences, Social Informatics and Telecommunications Engineering 262

More information about this series at http://www.springer.com/series/8197

Xingang Liu · Dai Cheng
Lai Jinfeng (Eds.)

Communications
and Networking

13th EAI International Conference, ChinaCom 2018
Chengdu, China, October 23–25, 2018
Proceedings

 Springer

Editors
Xingang Liu (iD)
School of Electronic Engineering
University of Electronic Science
and Technology of China
Chengdu, Sichuan, China

Lai Jinfeng (iD)
University of Electronic Science
and Technology of China
Chengdu, China

Dai Cheng
University of Electronic Science
and Technology of China
Chengdu, China

ISSN 1867-8211   ISSN 1867-822X (electronic)
Lecture Notes of the Institute for Computer Sciences, Social Informatics
and Telecommunications Engineering
ISBN 978-3-030-06160-9   ISBN 978-3-030-06161-6 (eBook)
https://doi.org/10.1007/978-3-030-06161-6

Library of Congress Control Number: 2018964610

This Springer imprint is published by the registered company Springer Nature Switzerland AG
The registered company address is: Gewerbestrasse 11, 6330 Cham, Switzerland

Preface

We are delighted to introduce the proceedings of the 2018 European Alliance for Innovation (EAI) International Conference on Communications and Networking in China (ChinaCom 2018). This conference has brought together Chinese and international researchers and practitioners in networking and communications under one roof, building a showcase of these fields in China.

The technical program of ChinaCom 2018 consisted of 22 tracks: Wireless Communications and Networking, Next-Generation WLAN, Big Data Networks, Cloud Communications and Networking, Ad Hoc and Sensor Networks, Satellite and Space Communications and Networking, Optical Communications and Networking, Information and Coding Theory, Multimedia Communications and Smart Networking, Green Communications and Computing, Signal Processing for Communications, Network and Information Security, Machine-to-Machine and Internet of Things, Communication QoS, Reliability and Modeling, Cognitive Radio and Networks, Smart Internet of Things Modeling, Pattern Recognition and Image Signal Processing, Digital Audio and Video Signal Processing, Antenna and Microwave Communications, Radar Imaging and Target Recognition, and Video Coding and Image Signal Processing.

Following the great success of the past ChinaCom events held during 2006–2017, ChinaCom 2018 received more than 114 submitted papers, from which 71 papers were selected for presentation. The Technical Program Committee (TPC) did an outstanding job in organizing and invited three keynote speakers: Dr. Michael Pecht from Maryland University, UK, Stephen Weinstein from Communication Theory and Technology Consulting LLC, UK, and Han-Chieh Chao from National Dongwa University, Hualien, China.

Coordination with the steering chairs, Imrich Chlamtac and Bruno Kessler were, was essential for the success of the conference. We sincerely appreciate their constant support and guidance. It was also a great pleasure to work with such an excellent Organizing Committee and we thank them for their hard work in organizing and supporting the conference. In particular, we thank the TPC, led by our TPC co-chairs, Dr. Xingang Liu, Dr. Supeng Leng, and Dr. Jinfeng Lai, who completed the peer-review process of the technical papers and compiled a high-quality technical program. We are also grateful to the conference manager, Kristina Lappyova, for her support, and all the authors who submitted their papers to the ChinaCom 2018 conference.

We strongly believe that ChinaCom 2018 provided a good forum for all researchers, developers, and practitioners to discuss all science and technology aspects that are relevant to smart grids. We also expect that future ChinaCom conferences will be as successful and stimulating as the present one.

November 2018

Hsiao-Hwa Chen
Gharavi Hamid
Xingang Liu

Organization

Steering Committee

Imrich Chlamtac   EAI/Create-Net, Italy
Bruno Kessler   University of Trento, Italy

Organizing Committee

General Chair

Hsiao-Hwa Chen   National Cheng Kung University, China Taiwan

General Co-chairs

Yiming Pi   UESTC, China
Hamid Gharavi   National Institute of Standards and Technology, USA

TPC Chairs and Co-chairs

Xingang Liu   UESTC, China
Supeng Leng   UESTC, China
Jinfeng Lai   UESTC, China

Sponsorship and Exhibit Chair

Local Chairs

Zongjie Cao   UESTC, China
Jinfeng Hu   UESTC, China

Workshops Chair

Bo Yan   UESTC, China

Publicity and Social Media Chair

Publications Chair

Cheng Dai   UESTC, China

Web Chair

Cheng Dai   UESTC, China

Posters and PhD Track Chair

Panels Chair

Wang Jin   Yangzhou University, China

Demos Chair

Tutorials Chair

Jian Shen   Nanjing University of Information Science and Technology,
                             China

Technical Program Committee

Xiang Chen   Sun Yat-sen University, China
Xiaoge Huang   Chongqing University of Posts and Telecommunications,
                          China
Gongpu Wang   Beijing Jiaotong University, China
Haixia Zhang   School of Information, Science and Engineering,
                          Shandong University, China
Guang Tan   SIAT Chinese Academy of Science, China
Yongming Huang   Southeast University, China
Xiang Chen   Sun Yat-sen University, China
Zhiyuan Ren   Xidian University, China
Jinglun Shi   South China University of Technology, China

Contents

Main Track

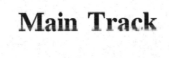

Main Track

Joint QoS-Aware Downlink and Resource Allocation for Throughput Maximization in Narrow-Band IoT with NOMA

Wei Chen[✉], Heli Zhang, Hong Ji, and Xi Li

Key Laboratory of Universal Wireless Communications, Ministry of Education,
Beijing University of Posts and Telecommunications,
Beijing, People's Republic of China
{weic,zhangheli,jihong,lixi}@bupt.edu.cn

Abstract. Narrow-Band Internet of Things (NB-IoT) is 3GPPs cellular technology designed for Low-Power Wide Area Network (LPWN) and it is a promising approach that NB-IoT combines with NOMA which is designed for accommodating more devices in the 5G era. Previous works mainly focus on uplink channel resource allocation to achieve connectivity maximization in NB-IoT with NOMA; however, few articles consider NB-IoT downside issues and downlink resource allocation problem to achieve maximum system throughput has not been studied in NB-IoT with NOMA. Thus, in this paper to provide a reliable and seamless service for NB-IoT users (NUs) and maximizing network downlink throughput, we propose a resource allocation algorithm for joint equipment QoS requirements and resource allocation fairness. In this scheme, we design algorithm to implement the mapping between NUs and subchannels for suboptimal system throughput. Then we convert the power allocation problem of the NUs on the same subchannel into a DC problem and we design algorithm to solve it to get suboptimal solution. Numerical results show that the proposed scheme achieves a better performance compared with exiting schemes in terms of the system throughput.

Keywords: Narrow band internet of things · Downlink resource allocation algorithm · Quality of service (QoS) · Throughput maximization

1 Introduction

Internet of Things (IoT) has been adopted to incorporate the digital information and the real world of devices. By 2020, it is expected that the IoT connections in the world will reach tens of billions level, far exceeding the number of concurrent personal computers and mobile phones; moving forward to 2024, the overall IoT industry is expected to generate a revenue of 4.3 trillion dollars which come

X. Liu et al. (Eds.): ChinaCom 2018, LNICST 262, pp. 3–13, 2019.
https://doi.org/10.1007/978-3-030-06161-6_1

from different sectors such as device connectivity, manufacturing, and other value added services [1]. Low-Power Wide Area Network (LPWN) represents a novel technology to enable a much wider range of IoT applications and complete short-range wireless technologies [2]. The most viable LPWN technology for tomorrow's need is the capability to download data to devices, in future even more so than today [3]. Narrow-Band Internet of Things (NB-IoT) is 3GPP's cellular technology designed for Low-Power Wide Area Network (LPWN). NB-IOT has been designed with the following performance objectives: extended coverage, support for massive devices, support delivery of IP and non-IP data, low cost of device, low deployment cost, low power consumption [4].

However, a single NB-IoT carrier spans one PRB in uplink and downlink. Considering the large-scale of NB-IoT users (NUs) requesting resources, it may degrade the network performance. Lots of works have been done for better network performance. In [5], the author proposes an interference aware resource allocation for NB-IoT by formulating the rate maximization problem considering the overhead of control channels, time offset, and repetition factor. The authors in [6] put forward a classification Back-off method to classify different types of devices in the Back-off mechanism to improve system capacity. In [7] a low area interleaver which is an important component of turbo coding is implemented to achieve the required error correction with 5% area saving by sharing resources. However, the above work just improved the performance of the NB-IoT network from the implementation, but it does not fundamentally solve the problem of insufficient bandwidth resources. In [8], the author proposes a powerdomain uplink non-orthogonal multiple access (NOMA) scheme which allows multiple MTCEs to share the same sub-carrier to achieve large-scale equipment access through allocating sub-carriers. However, the author did not consider the downlink of the NB-IoT network and the impact of NOMA on the network downlink whose base station has the same transmit power. And to some extent, it aggravates the downlink resource allocation problem which is carried out in [9]. Therefore, these aspects will be included in our work of downlink resource allocation in NB-IoT.

In this paper, we focus on the resource allocation in downlink NB-IoT with NOMA for throughput maximization. We propose joint QoS-aware downlink and resource allocation algorithm consisting of subchannels allocation algorithm and subchannels power allocation algorithm that satisfy NUs QoS requirements and ensure the fairness. In our algorithm, we firstly assign subchannels to NUs from a global optimal point. Then we convert the power allocation problem to DC representation and get suboptimal solution. Through the simulation, we can get that our algorithm has better performance than the traditional.

The remainder of our work is organized as follows. Section 2 gives the system model and problem formulation. In Sect. 3, the proposed optimization algorithm is developed. Simulation results and discussions are given in Sect. 4. Finally, we conclude this paper in Sect. 5.

2 System Model

Consider the downlink of a multi-user communication NB-IoT network including single base station (BS) as shown. The SIC is applied at the received of NUs and active NUs share a system bandwidth of one PRB for downlink data transmissions (Fig. 1).

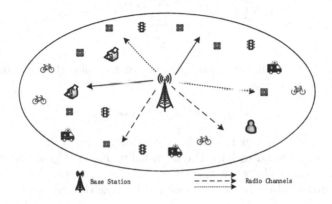

Fig. 1. System model of NB-IoT network with NOMA

We denote u as index for uth NU in the set of U NUs, which is expressed as $u \in \{1, 2, ..., U\}$. Each NU has a downlink rate demand r_u which is no more than r_{max}. However, r_{max} is equal to 0 for NUs without rate requirements. And we denote n as index for nth subchannel in the set of N subchannels, which is expressed as $n \in \{1, 2, ...N\}$. Moreover U NUs are uniformly distributed in a circular region D. According to NB-IoT standard we can get that the bandwidth of the system is BW equaling 180 kHz and the constraint of the bandwidth of subchannels B is expressed as $B - 15$ kHz, so we can get $N = BW/B$. Moreover, each NU can only assign one subchannel. Let $K_n \in \{K_1, K_2, ..., K_N\}$ be the set of NUs allocated on the subchannel n, so we can get $K_n = \{k_{1,n}, k_{2,n}, ..., k_{m_n,n}\}$ where m_n means the number of NUs allocated to subchannel n and the $k_{i,n} \in \{1, 2, ...U\}$ means the ith NUs on the subchannel n. And the power which allocated to NU u on subchannel n is expressed as $p_{u,n}$. We denote P and p_n as the total transmitted power of the BS and the total power allocated to subchannel n.

2.1 Communication Model

For the purpose of throughput-maximum, we consider incorporate NOMA technology into the NB-IoT downlink network. Moreover, successive interference cancellation (SIC) precess is implemented at NU receiver to reduce the interference form other NUs on the same subchannel. We assume that the blocking fading of the channel is taken into consideration with the assumptions that it remains the same within a subchannel and it varies independently across different subchannel.

According to decoding order, the NU $k_{i,n}$ on the subchannel n can be successfully decoded and remove the interference symbols from NUs $k_{j,n}$ with $j > i$. And the interference symbol from NUs $k_{j,n}$ with $j < i$ will be treated as noise by NU $k_{i,n}$. Therefore the SINR of User $k_{i,n}$ with SIC at receiver is denoted as

$$SINR_{k_{i,n},n} = \frac{p_{k_{i,n},n}H_{k_{i,n},n}}{1 + \sum_{j=1}^{i-1} p_{k_{j,n},n}H_{k_{i,n},n}}. \tag{1}$$

where $H_{u,n}$ is the channel response normalized by noise (CRNN) of NU u on subchannel n expressed as

$$H_{u,n} = |h_{u,n}|^2 / noise_n. \tag{2}$$

where $h_{u,n} = g_n.PL^{-1}(d)$ is the coefficient of subchannel n from the BS to NU u in which g_n is assumed to have Rayleigh fading channel gain and $PL^{-1}(d)$ is the path loss function between the BS and NU u at distance d, $noise_n = \sigma_n^2$ means the additive white Gaussian noise (AWGN) with zero mean and variance σ_n^2.

Consequently, the sum rate $R_{k_{i,n},n}$ of NU $k_{i,n}$ on subchannel n through the allocated resources can be expressed as

$$R_{k_{i,n},n} = B \log_2 \left(1 + SINR_{k_{i,n},n}\right). \tag{3}$$

And the achievable sum rate R_n on subchannel n can be expressed in terms of the aggregate rate of NUs allocated the same subchannel n. It is

$$R_n = \sum_{i=1}^{m_n} R_{k_{i,n},n}. \tag{4}$$

In the rest of paper, we simply represent downlink NB-IoT network using NOMA technology with SC and SIC to maximize system throughput in the event of NU's downlink demand.

2.2   Problem Formulation

The implementation complexity of SIC at the receiver increases with the maximum number of the NUs allocated on the same subchannel. In order to keep the receiver complexity comparatively low and restrict the error propagation, we consider the simple case where no more than two users can be allocated on the same subchannel, so the sum rate of subchannel can be expressed as

$$R_n = B \log_2 \left(1 + p_n\alpha_n H_{k_{1,n},n}\right) +$$
$$B \log_2 \left(1 + \frac{p_n(1 - \alpha_n)H_{k_{2,n},n}}{1 + p_n\alpha_n H_{k_{2,n},n}}\right), \alpha_n \in (0, 1]. \tag{5}$$

where α_n is the proportional factor to allocate power among the two NUs on subchannel n.

In this paper, the objective is to maximize the system throughput that meet the QoS requirements of NU. For above purpose, we formulate it as an optimization problem which consist of channel assignment and power allocation. We formulate the throughput problem as

$$\max \sum_{i=1}^{N} \sum_{j=1}^{m_i} R_{k_{j,i},i}$$

$$\text{s.t.} \quad C1 : k_{j,i} \in \{1, 2, ..., U\} \quad \forall j, i$$
$$C2 : R_{k_{j,i},i} \geq r_{k_{j,i}} \quad \forall j, i \tag{6}$$
$$C3 : m_i \in \{1, 2\} \quad \forall i$$
$$C4 : \sum_{i \in K_n} p_{i,n} \leq p_n, \sum_{n=1}^{N} p_n \leq P.$$

The problem 6 is a mixed-integer non-liner programming problem, which is extremely difficult to derive a globally optimal solution with low computation complexity. Because of the above, we assume that equal power is allocated to subchannels and propose a novel resource allocation algorithm to solve it which is show below.

3   Proposed Algorithm

As aforementioned, to obtain the optimal solution of 6, we propose a joint QoS-Aware downlink and resource allocation algorithm. Considering the limited resources of the system, we will select $U = 2N$ downlink users randomly from downlink users. Taking into account the fairness of equipment resource allocation, We set a priority parameter i_u for each NU u which increases as the number of resource allocation failures increases and is used for channel resource allocation. Based on the priority of the device we update the CRNN parameter, so we can get

$$H'_{u,n} = \left(\frac{i_u U}{\sum_{u \in U} i_u} \right)^{\beta} H_{u,n}, \quad \alpha \geq 0. \tag{7}$$

where β is used to adjust the balance between fairness and system throughput.

The design idea of the algorithm is to allocate the device to specific subchannels and calculate optional proportional factor (α_n) in the event of NU's downlink rate requirements. We assume that the system allocates equal power to each subchannel so $p_n = P/N$. Moreover, we consider that all subchannels of NB-IoT network is available, but it is also feasible in portion of the system resources available scenes.

Algorithm 1 Dynamic Suboptimal Subchannel Allocation Algorithm

1: **Input:** BW, B, P, U, N, K_N, $H_{u,n}$, $H'_{u,n}$, Av_u, p_n, r_u, $\forall n \in \{1, 2, 3, ..., N\}, \forall u \in \{1, 2, 3, ..., U\}$

2: **Initialization:**
   (a)Initialize BW, P, B, U, r_u , and let $K_n = \varnothing$;
   (b)Initialize the set of NUs Re_{un} to record NUs who have not been allocated;
   (c)Calculate p_n, N, $H_{u,n}$, $H'_{u,n}$, Av_n;

3: **while** Re_{un} is not empty **do**
4:    **for** each $u \in Re_{un}$ **do**
5:       **if** $|Av_u| = 0$ **then**
6:          Remove the NU u from Re_{un}; continue;
7:       **end if**
8:       Get $n = arg \max\limits_{H_{u,n} \in Av_u} H_{u,n}$;
9:       **if** $|K_n| = 0$ **then**
10:         Remove the $H_{u,n}$ from Av_u and the u from Re_{un};
11:         Add the NU u to K_n; continue;
12:       **end if**
13:       **if** $|K_n| > 0$ **then**
14:         step1: Remove the $H_{u,n}$ from Av_u;
15:         step2: Find out all the combinations that have one or two NUs as a set. Then calculate the most appropriate power factor α_n for each collection by the algorithm 2;
16:         step3: According to the parameters α_n calculated in step2, we recalculate the downlink rate obtained for each combination where parameter $H'_{u,n}$ is used instead of $H_{u,n}$. Then get combination K_{result} satisfying maximum subchannel n downlink rate from all above updated downlink rates of combinations;
17:         step4: Get K_n according the combination choose in step3. Remove the allocated users from Re_{un} and the unallocated users to Re_{un}.
18:       **end if**
19:    **end for**
20: **end while**
21: Calculate the system throughput R_{all} according to 6.
22: **Output** R_{all}.

3.1 Subchannels Allocation Algorithm

Assuming no more than two NUs can share the same subchannel due to the complexity of decoding and NU's downlink QoS requirements. We denote the subchannels which have not been paired with the NU u as Av_u and whose initial value can be expressed as

$$Av_u = \{H_{u,1}, H_{u,2}, ..., H_{u,N}\}, \ \forall u \in \{1, 2, ..., U\}. \tag{8}$$

Based on the perfect channel state information, we can easily get that the NU u can get better performance on subchannel i than on subchannel j if $H_{u,i} > H_{u,j}$. In the rest of the subsection, we propose a suboptimal algorithm named as dynamic suboptimal subchannel Allocation Algorithm (DSSA) for the problem above, as shown in Algorithm 1.

Algorithm 1 describes the proposed DSSA to maximize the system's throughput in the case of ensuring the fairness of resource allocation and satisfying the QoS requirements of the NUs. In lines 1–2, we have initialized the algorithm's input variables according to NB-IoT standards and actual scene requirements. In lines 3–7, we guarantee that the algorithm will be terminated if all devices have already been allocated to the channel or if the device does not find the desired subchannel meeting the requirements. In lines 9–11 it considers how to handle in the absence of equipment allocated to the subchannel. And the other conditions considered in lines 12–17 where step 4 ensures Fairness by using updated $H'_{u,n}$.

3.2 Subchannel Power Allocation Algorithm

The main idea of this subsection is to design power allocation algorithm for maximizing downlink subchannel rate. We propose a algorithm named as QoS-Aware power allocation algorithm (QAPA) as shown in Algorithm 2 to solve problem above.

To realize the maximum throughput of the subchannel without considering the QoS requirement of the NU. Let's assume that there are two NUs $k_{1,n}$ and $k_{2,n}$ allocated on the subchannel n. The problem above of finding α_n to maximize throughput of subchannel n can be restated as

$$\max B \log_2 \left(1 + p_n \alpha_n H_{k_{1,n},n}\right) + $$
$$B \log_2 \left(1 + \frac{p_n(1 - \alpha_n) H_{k_{2,n},n}}{1 + p_n \alpha_n H_{k_{2,n},n}}\right), \alpha_n \in (0, 1]. \tag{9}$$

Algorithm 2 QoS-Aware power allocation algorithm

1: **Input:** denote c_{in} as the input NUs combination, denote $user$ as NUs in c_{in}, p_n
2: **if** $|c_{in}| == 1$ **then**
3:     Calculate the downlink rate of NU $user1$ r'_{user1}, when $\alpha_n = 1$;
4:     **if** $r'_{user1} \geq r_{user1}$ **then**
5:         Set $\alpha_n = 1$;
6:     **end if**
7: **end if**
8: **if** $|c_{in}| == 2$ **then**
9:     Calculate the optimal solution α_n through algorithm 3;
10:     Calculate the downlink rate of two NUs r'_{user1} and r'_{user2} allocated subchannel;

11:     **if** $r'_{user1} \geq r_{user1} \&\& r'_{user2} \geq r_{user2}$ **then**
12:         Set α the value calculated through algorithm 3;
13:     **end if**
14: **end if**
15: **Output** α_n.

The problem 9 is solved through DC planning approach. First We transform the formula 9 into a formula for the general definition of DC planning problem that is denoted as

$$\min \quad f(\alpha_n) = m(\alpha_n) - g(\alpha_n), \alpha_n \in (0,1]. \tag{10}$$

where $m(\alpha_n) = -B\log_2\left(1 + p_n\alpha_n H_{k_{1,n},n}\right)$ and $g(\alpha_n) = B\log_2\left(1 + \frac{p_n(1-\alpha_n)H_{k_{2,n},n}}{1+p_n\alpha_n H_{k_{2,n},n}}\right)$. We can easily get

$$\nabla^2 m(\alpha_n) > 0 \quad and \quad \nabla^2 g(\alpha_n) > 0, \tag{11}$$

so we can prove that 10 can be solved using DC planning approach.

We convert the above-mentioned constrained DC planning problem into an unconstrained DC programming problem. We denote the representative function of α_n as

$$I(\alpha_n) = \begin{cases} 0, & \alpha_n \in (0,1] \\ +\infty, & \alpha_n \notin (0,1], \end{cases} \tag{12}$$

so the problem can be expressed as

$$F(\alpha_n) = f(\alpha_n) + I(\alpha_n) = M(\alpha_n) + G(\alpha_n). \tag{13}$$

where $M(\alpha_n) = m(\alpha_n) - I(\alpha_n)$ and $G(\alpha_n) = g(\alpha_n)$. Then we can denote conjugate function of function M and G as

$$M^*(\beta_n) = sup\{\beta_n^T \alpha_n - M(\alpha_n)|\alpha_n \in (0,1]\}, \tag{14}$$

$$G^*(\beta_n) = sup\{\beta_n^T \alpha_n - G(\alpha_n)|\alpha_n \in (0,1]\}. \tag{15}$$

Finally, We use difference of convex functions algorithm (DCA) to solve 13. The detail steps about DCA algorithm is shown in Algorithm 1, through which we can acquire α_n.

Algorithm 3 Solution for Subchannel Power Allocation

1: **Initialization:**
   Initialize $\alpha_n^{(0)}$, $\beta_n^{(0)} \in \partial G(\alpha_n^{(0)})$, $k = 0$, $\varepsilon > 0$;
2: **while** $F(\alpha_n^{(k+1)}) - F(\alpha_n^{(k)}) \le \varepsilon$ $\quad or \quad$ $\|\alpha_n^{(k+1)} - \alpha_n^{(k)}\| \le \varepsilon$ **do**
3:    Define convex approximation as
4:    $P(\alpha_n) = M(\alpha_n) - G(\alpha_n^{(k)}) - (\alpha_n - \alpha_n^{(k)}, \beta_n^{(k)}), \alpha_n \in (0,1]$ and
5:    $D(\beta_n) = G^*(\beta_n) - M^*(\beta_n^k) - (\alpha_n^{(k+1)}, \beta_n - \beta_n^{(k)}), \alpha_n \in (0,1]$;
6:    Solve the convex problem. We can get $\alpha_n^{(k+1)} = arg\min P(\alpha_n)$ and $\beta_n^{(k+1)} = arg\min D(\beta_n)$;
7:    $k \leftarrow k + 1$;
8: **end while**
9: **Output** α_n.

---

4   Simulation Results and Discussion

In this section, some simulation results are presented to illustrate the performance of the proposed algorithm. The simulation assumptions that nearly follows 3GPP standards [10]. The NUs are distributed uniformly in the area of $1000 \times 1000 \, \text{m}^2$. The full $180 \, \text{kHz}$ bandwidth, i.e. 12 subchannels at $15 \, \text{kHz}$ subchannels spacing in downlink, is used for analysis. The wireless channel is modeled as Rayleigh fading channel including pathloss, where the channel coefficient is $h_i^2 = h_0^2 L_i^{-\kappa}$, where L_i^2 is the distance between AP and user i. The pathloss exponent $\kappa = 4$ and h_0 is the complex Gaussian channel coefficient. Moreover, we assume that BS transmit power is $32 \, \text{dBm}$ and BS cable loss is $3 \, \text{dB}$. We consider additive white Gaussian noise with power spectral density $-174 \, \text{dBm/Hz}$ and noise figure of $5 \, \text{dB}$. In addition, we can easily get that most NUs in a NB-IoT network are rate insensitive devices so we set two-thirds of the NUs does not have a rate requirement and the other have the downlink rate requirement. We compare the performance of the following algorithms in this simulation:

- Proposed scheme: The resource allocation that jointly considering NU's QoS requirement and the fairness with for NB-IoT NOMA.
- OFDMA scheme: Each subchannel can only be allocated one NU and the power allocation is equal on each subchannel [11].
- NOMA-EQ: The subchannel allocation is same with the proposed scheme but equal power allocation scheme replaces our proposed QoS-Aware power allocation scheme.

Figure 2 illustrates the performance (System Throughput) versus the number of NUs. In the simulation the number of NUs is from 0 to 32 in one base station (BS) and the downlink rate requirement is assigned from [80, 120] kbps. As indicated in Fig. 2, we can get that the proposed scheme improves system throughput by nearly 20% compared to NOMA-EQ scheme and compared to OFDMA scheme the system throughput is improved by nearly 65%. First, the system's throughput increases with the number of devices because system resources are not saturated. Then, Three schemes eventually converge because of limited system resources. Moreover, Algorithm 3 converges earlier than other algorithms because NOMA is not used resulting in supporting fewer NUs.

Figure 3 illustrate the number of successfully served NUs versus total users. In the simulation the number of NUs is from 0 to 32 in one base station (BS) and the downlink rate requirement is assigned from [80, 120] kbps. And we can find that compared with other schemes the number of devices that the system can successfully serve the most under the scenario where proposed scheme is used. In Fig. 3, it can be observed that the proposed scheme and NOMA-EQ scheme perform much better than OFDMA scheme. Moreover, the proposed scheme improves successfully served NUs by nearly 10% compared to NOMA-EQ scheme, because the system cannot meet the downlink demand of some NUs when the power is evenly distributed.

Figure 4 illustrates the impact of the NUs QoS requirement on the system throughput. Here, 32 NUs are deployed in the range and we

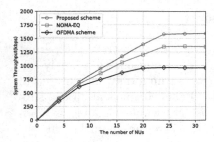

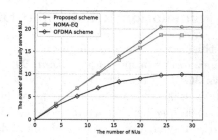

Fig. 2. The system throughput versus the number of NUs.

Fig. 3. The number of successfully served NUs versus total users.

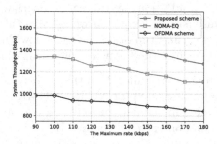

Fig. 4. The system throughput versus the NU QoS requirement

consider that the maximum rate takes the values $(90, 100, 110, 120, 130, 140, 150, 160, 170, 180)$ kbps. In Fig. 4, as the maximum rate continues to increase system throughput continues to decline. However, it can be observed that the proposed scheme still performs better than NOMA-EQ scheme and OFDMA scheme.

5   Conclusion

In this paper, we focus on the issue of downlink in NB-IoT network with NOMA. In order to maximize downlink throughput of NB-IoT network with NOMA, we propose a scheme with the consideration of NU's QoS and the fairness of resource allocation. First, we propose an algorithm called Dynamic Suboptimal subchannel Allocation Algorithm (DSSA) to find the best user-channel matching relationship. Furthermore, to satisfy NU's QoS requirement in DSSA, we propose a QoS-Aware power allocation algorithm (QAPA). Last, in QAPA we transform the power allocation problem between NUs of the same subchannel into a DC problem and the difference of convex functions algorithm (DCA) is utilized to find successive convex approximation. Simulation results prove the system performance has been improved significantly. For future work, to improve the system performance, We will consider joint uplink and downlink system resource allocation in the NB-IoT network with NOMA.

Acknowledgements. This paper is sponsored by National Natural Science Foundation of China (Grant 61671088 and 61771070).

References

1. Ericsson, O.: Technical report. White Paper, Cellular networks for massive IoT: enabling low power wide area applications (2016). https://www.ericsson.com/res/docs/whitepapers/wp_iot.pdf
2. Poursafar, N., Alahi, M.E.E., Mukhopadhyay, S.: Long-range wireless technologies for IoT applications: a review. In: 2017 Eleventh International Conference on Sensing Technology (ICST), Dec 2017, pp. 1–6 (2017)
3. White Paper, O.: Narrowband IoT groundbreaking in the internet of things (2016). https://iot.t-mobile.nl/downloads/NB-IoTDTAG.pdf
4. Beyene, Y.D., Jantti, R., Ruttik, K., Iraji, S.: On the performance of narrowband internet of things (NB-IoT). In: 2017 IEEE Wireless Communications and Networking Conference (WCNC), March 2017, pp. 1–6 (2017)
5. Malik, H., Pervaiz, H., Alam, M.M., Moullec, Y.L., Kuusik, A., Imran, M.A.: Radio resource management scheme in NB-IoT systems. IEEE Access **6**, 15051–15064 (2018)
6. Zhao, Y., Liu, K., Yan, H., Huang, L.: A classification back-off method for capacity optimization in NB-IoT random access. In: 2017 11th IEEE International Conference on Anti-counterfeiting, Security, and Identification (ASID), Oct 2017, pp. 104–108 (2017)
7. Abdelbaky, A., Mostafa, H.: New low area NB-IoT turbo encoder interleaver by sharing resources. In: 2017 29th International Conference on Microelectronics (ICM), Dec 2017, pp. 1–4 (2017)
8. Mostafa, A.E., Zhou, Y., Wong, Y.W.S.: Connectivity maximization for narrowband IoT systems with NOMA. In: 2017 IEEE International Conference on Communications (ICC), May 2017, pp. 1–6 (2017)
9. Boisguene, R., Tseng, S.C., Huang, C.W., Lin, P.: A survey on NB-IoT downlink scheduling: issues and potential solutions. In: 2017 13th International Wireless Communications and Mobile Computing Conference (IWCMC), June 2017, pp. 547–551 (2017)
10. Cellular system support for ultra-low complexity and low throughput internet of things (CIoT). 3GPP Technical report (TR), v 13.1.0 Release 13 (2016)
11. Li, B., Qu, Q., Yan, Z., Yang, M.: Survey on OFDMA based MAC protocols for the next generation WLAN. In: 2015 IEEE Wireless Communications and Networking Conference Workshops (WCNCW), March 2015, pp. 131–135 (2015)

MIMO-UFMC Transceiver Schemes for Millimeter Wave Wireless Communications

Stefano Buzzi[1]([⊠]) [iD], Carmen D'Andrea[1] [iD], Dejian Li[2], and Shulan Feng[2]

[1] University of Cassino and Southern Lazio, Cassino, Italy
{buzzi,carmen.dandrea}@unicas.it
[2] Hisilicon Technologies Co., Ltd, Beijing, People's Republic of China
{lidejian,Shulan.Feng}@hisilicon.com

Abstract. This paper provides results on the use of UFMC modulation scheme in MIMO wireless links operating at mmWave frequencies. First of all, full mathematical details on the processing needed to realize a MIMO-UFMC transceiver at mmWave, taking into account also the hybrid analog/digital nature of the beamformers, are given. Then, we propose several reception structures, considering also the case of continuous packet transmission with no guard intervals among the packets. In particular, an adaptive low complexity MMSE receiver is proposed that is shown to achieve very satisfactory performance. A channel independent transmit beamformer is also considered, as as to avoid the need for channel state information at the transmitter. Numerical results show that the proposed transceiver schemes are effective, as well as that the continuous packet transmission scheme, despite increased interference, attains the highest values of system throughput.

Keywords: MIMO-UFMC · Millimeter waves · 5G Networks

1 Introduction

The Universal Filtered MultiCarrier (UFMC), one of the modulations that is considered as an alternative to orthogonal frequency division multiplexing (OFDM) for future wireless systems [2], is an intermediate scheme between filtered-OFDM and Filter Bank MultiCarrier (FBMC). Indeed, while in filtered OFDM the whole OFDM signal is filtered to reduce OOB emissions and achieve better spectral containment [1], and while in FBMC each subcarrier is individually filtered [5], in UFMC the subcarriers are grouped in contiguous, non-overlapping blocks, called *subbands*, and each subband is individually filtered [7].

This paper has been supported by Huawei through HIRP OPEN Agreement No. HO2016050002BM.

© ICST Institute for Computer Sciences, Social Informatics and Telecommunications Engineering 2019
Published by Springer Nature Switzerland AG 2019. All Rights Reserved
X. Liu et al. (Eds.): ChinaCom 2018, LNICST 262, pp. 14–23, 2019.
https://doi.org/10.1007/978-3-030-06161-6_2

UFMC has been received an increasing attention is the last few years; however, despite the relevance of multi-antenna processing to future fifth-generation (5G) wireless systems, its use in conjunction with a multiple-input-multiple-output (MIMO) configuration has not been fully addressed in the open literature; similarly, the potentialities coming from the use of UFMC modulation at millimeter wave (mmWave) frequencies have not yet been properly investigated.

This paper proposes transceiver schemes for the MIMO-UFMC modulation operating at mmWave frequencies. The contributions of this paper may be summarized as follows: (a) we provide the full mathematical model of a MIMO-UFMC scheme, taking into account the presence of hybrid analog/digital beamformers; (b) we propose several reception structures and, among these, an adaptive linear minimum mean square error (MMSE) receiver that can be implemented without resorting to an explicit channel estimation phase; (c) we propose a modified UFMC transmission scheme wherein contiguous data packets are transmitted with no guard intervals; and, finally, (d) we propose and evaluate the performance of a channel-independent pre-coding beamforming structure at the transmitter, so as to avoid the need for channel state information at the transmitter. Numerical results will show the effectiveness of the proposed transceiver algorithms and of the continuous packet transmission scheme with no guard intervals.

This paper is organized as follows. Next section contains the description of the mathematical model and of the transceiver processing for an UFMC system with multiple antennas. Section 3 contains the derivation of the linear MMSE receiver for MIMO-UFMC systems, including the definition of a channel-independent beamformer, while Sect. 4 contains the discussion of the numerical results. Finally, concluding remarks are given in Sect. 5.

2   MIMO-UFMC Transceiver Processing

In this section we generalize to the MIMO case the UFMC single-packet modulation scheme detailed in [7], presenting three different receiving structures. We will refer to the scheme reported in Fig. 1. We will denote by M the multiplexing order, by N_T and N_R the number of transmit and receive antennas, respectively, and by N_T^{RF} and N_R^{RF} the number of RF chains at the transmitter and at the receiver, respectively. We assume that the k subcarriers are split in B subbands of D subcarriers each (thus implying that $k = BD$). Assume that a sequence of Mk data symbols is to be transmitted; these symbols are arranged into an $(M \times k)$-dimensional matrix, that we denote by $\mathbf{S}$. The columns of $\mathbf{S}$ undergo a digital precoding transformation; in particular, denoting by $\mathbf{Q}_{BB}(n)$ the $(N_T^{\mathrm{RF}} \times M)$-dimensional matrix representing the digital precoder for the n-th column of $\mathbf{S}$, the useful data at the output of the digital precoding stage can be represented by the $(N_T^{\mathrm{RF}} \times k)$-dimensional matrix $\mathbf{X}$, whose n-th column, $\mathbf{X}(:,n)$ say, is expressed as $\mathbf{X}(:,n) = \mathbf{Q}_{BB}(n)\mathbf{S}(:,n)$. After digital precoding, each of the N_T^{RF} rows of the matrix $\mathbf{X}$ goes through an

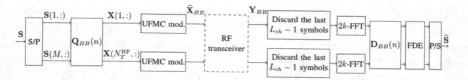

Fig. 1. Block scheme of the UFMC multi antenna transceiver. The dashed box "RF transceiver" contains the cascade of a bank of N_T^{RF} transmit shaping filters, a bank of N_T^{RF} power amplifiers, the analog RF precoding matrix $\mathbf{Q}_{\mathrm{RF}}$, the N_T transmit antennas, the $(N_R \times N_T)$-dimensional matrix-valued MIMO channel impulse response, the N_R receive antennas, the analog RF post-coding matrix $\mathbf{D}_{\mathrm{RF}}$, and a bank of N_R^{RF} receive shaping filters

UFMC modulator; the outputs of the N_T^{RF} parallel UFMC modulators can be grouped in the matrix $\widetilde{\mathbf{X}}_{BB}$ of dimension $[N_T^{\mathrm{RF}} \times (k + L - 1)]$. Each UFMC modulator uses a finite impulse response (FIR) passband filter in order to improve the frequency localization property of the input signal. Denoting by $\mathbf{g} \triangleq [g_0, g_1, \ldots, g_{L-1}]^T$ the L-dimensional vector representing the prototype filter, the FIR filter used in the i-th subband to process the vector in input at the generic UFMC modulator is denoted by $\mathbf{g}_i$ and its entries $g_{i,0}, g_{i,1}, \ldots, g_{i,L-1}$ are defined as $g_{i,\ell} = g_\ell e^{j2\pi \frac{F_i \ell}{k}}, i = 0, \ldots, B - 1, \quad \ell = 0, \ldots, L - 1$, with $F_i \triangleq \frac{D-1}{2} + iD$ the normalized frequency shift of the filter tuned to the i-th subband. Denoting by $\mathbf{X}(\ell, :)$ the ℓ-th row of the matrix $\mathbf{X}$, the ℓ-th row of the output matrix $\widetilde{\mathbf{X}}_{BB}$ is written as

$$\widetilde{\mathbf{X}}_{BB}(\ell, :)^T = \sum_{i=0}^{B-1} \mathbf{G}_i \mathbf{W}_{k-IFFT} \mathbf{P}_i \mathbf{X}(\ell, :)^T, \tag{1}$$

where $\mathbf{G}_i$ is the Toeplitz $[(k + L - 1) \times k]$-dimensional matrix describing the discrete convolution operation with the filter $\mathbf{g}_i$, the matrix $\mathbf{W}_{k,IFFT}$ denotes the $(k \times k)$-dimensional matrix representing the isometric IFFT transformation and the $(k \times k)$-dimensional matrix $\mathbf{P}_i = \mathrm{diag}\left([\; \underbrace{0 \ldots 0}_{iD} \underbrace{1 \ldots 1}_{D} \underbrace{0 \ldots 0}_{k-(i+1)D} \;] \right)$, for $i = 0, \ldots, B - 1$. Equation (1) can be compactly written in matrix notations as $\widetilde{\mathbf{X}}_{BB} = \mathbf{X} \left(\sum_{i=0}^{B-1} \mathbf{P}_i^T \mathbf{W}_{k-IFFT}^T \mathbf{G}_i^T \right)$. The columns of $\widetilde{\mathbf{X}}_{BB}$ are then fed to the MIMO RF transceiver scheme, that is made of the receive and transmit shaping filters, the analog precoding and post-coding matrices $\mathbf{Q}_{\mathrm{RF}}$ (of dimension $N_T \times N_T^{\mathrm{RF}}$) and $\mathbf{D}_{\mathrm{RF}}$ (of dimension $N_R \times N_R^{\mathrm{RF}}$), respectively, and of the MIMO channel impulse response. Assuming that the power amplifiers operate in the linear regime, the RF transceiver block can be modeled as an LTI filter with $(N_R^{\mathrm{RF}} \times N_T^{\mathrm{RF}})$-dimensional matrix-valued impulse response $\mathbf{L}(\ell) = \sqrt{P_T} \mathbf{D}_{\mathrm{RF}}^H \widetilde{\mathbf{H}}(\ell) \mathbf{Q}_{\mathrm{RF}}$, wherein P_T is the transmitted power, $\widetilde{\mathbf{H}}(\ell)$, with $\ell = 0, \ldots, L_{ch} - 1$, is the matrix-valued $(N_R \times N_T)$-dimensional millimeter wave (mmWave) channel impulse response including also the transmit and receive rectangular shaping filters [4], with L_{ch} the length of the channel impulse response (in discrete samples). The output of the RF transceiver can be represented through a matrix $\mathbf{Y}_{BB}$ of dimension $[N_R^{\mathrm{RF}} \times (k + L + L_{ch} - 2)]$. The m-th column of $\mathbf{Y}_{BB}$ is easily seen to be expressed as

$$\mathbf{Y}_{BB}(:,m) = \mathbf{D}_{\mathrm{RF}}^H \left[\sum_{\ell=0}^{L_{ch}-1} \sqrt{P_T} \widetilde{\mathbf{H}}(\ell) \mathbf{Q}_{\mathrm{RF}} \widetilde{\mathbf{X}}_{BB}(:,m-\ell) + \mathbf{w}(m) \right], \qquad (2)$$

where we have assumed that $\widetilde{\mathbf{X}}(:,m)$ is zero for $m \le 0$. The vector $\mathbf{w}(n)$ represents the additive thermal noise contribution. At this point, following the usual UFMC processing, the last $L_{ch} - 1$ columns of the matrix $\mathbf{Y}_{BB}$ are discarded, and each row of the resulting matrix, say $\widetilde{\mathbf{Y}}_{BB}$, is passed through an FFT on $2k$ points. The output of the FFT is downsampled by a factor of 2, so that we get a matrix of dimension $N_R^{\mathrm{RF}} \times k$, and finally, digital postcoding is applied. Denoting by $\mathbf{D}_{BB}(n)$ the $(N_R^{\mathrm{RF}} \times M)$-dimensional matrix representing the digital postcoder for the n-th column of the data matrix, we finally get a $(M \times k)$-dimensional matrix $\mathbf{Y}_{dec}$, whose n-th column can be shown to be approximately expressed as

$$\mathbf{Y}_{dec}(:,n+1) \approx \frac{2k}{\sqrt{2}} \sqrt{P_T} \mathbf{D}_{BB}^H(n) \mathbf{D}_{\mathrm{RF}}^H \overline{\mathcal{H}}(2n) \times$$

$$\mathbf{Q}_{\mathrm{RF}} \mathcal{G}_{\lfloor n/D \rfloor}(2n) \mathbf{Q}_{BB}(n) \mathbf{S}(:,n+1) + \mathbf{D}_{BB}^H(n) \mathbf{D}_{\mathrm{RF}}^H \mathcal{W}(2n) , \quad (3)$$

with $n = 0, \ldots, k - 1$. In the above equation, $\mathcal{H}(2n)$ is an $N_R \times N_T$ matrix whose (p,q)-th entry is the $2n$-th coefficient of the isometric $2k$-point FFT of the sequence $\widetilde{\mathbf{H}}_{p,q}(0), \ldots, \widetilde{\mathbf{H}}_{p,q}(L_{ch}-1)$; similarly, $\mathcal{G}_i(2n)$ denotes the $2n$-th coefficient of the isometric $2k$-points FFT of the i-th subband filter $\mathbf{g}_i$, and $\mathcal{W}(2n)$ is an N_R-dimensional vector whose ℓ-th entry is the $2n$-th coefficient of the isometric $2k$-points FFT of the noise sequence $w_\ell(0), \ldots w_\ell(K + L + L_{ch} - 3), \forall \ell = 1, \ldots N_R$. We remark that (3) holds approximately and not with a perfect equality since we have discarded the last $L_{ch} - 1$ symbols. Now, given (3), an estimate of the n-th column of the data symbols matrix $\mathbf{S}$ can be simply obtained as

$$\widehat{\mathbf{S}}_{\mathrm{id}}(:,n+1) = \mathbf{E}(n+1)\mathbf{Y}_{dec}(:,n+1) , \qquad (4)$$

where $\mathbf{E}(n+1) = \left[\frac{2k}{\sqrt{2}} \sqrt{P_T} \mathbf{D}_{BB}^H(n) \mathbf{D}_{\mathrm{RF}}^H \overline{\mathcal{H}}(2n) \mathbf{Q}_{\mathrm{RF}} \mathcal{G}_{\lfloor \frac{n}{D} \rfloor}(2n) \mathbf{Q}_{BB}(n) \right]^+$, and $(\cdot)^+$ denotes Moore–Penrose generalized inverse.

A different processing can be obtained by avoiding the use of the approximate relation (3). We thus consider the received matrix $\mathbf{Y}_{BB}$ in Eq. (2), discard the last $L_{ch} - 1$ columns of the matrix, compute the FFT on $2k$ points, downsample the result by a factor of 2, and finally, apply digital postcoding, namely:

$$\mathbf{Y}_{\mathrm{dis}} = \mathbf{Y}_{BB} \mathbf{D}_{L_{ch}-1} \mathbf{W}_{2k,FFT}(1 : k + L - 1, :) , \qquad (5)$$

where the matrix $\mathbf{W}_{2k,FFT}$ denotes the $(2k \times 2k)$-dimensional matrix representing the isometric FFT transformation and $\mathbf{D}_{L_{ch}-1} = \left[\mathbf{I}_{k+L-1} \ \mathbf{0}_{L_{ch}-1 \times k+L-1} \right]^T$

is the $[(k + L + L_{ch} - 2) \times (k + L - 1)]$-dimensional matrix used for discarding the last $L_{ch} - 1$ columns of the matrix $\mathbf{Y}_{BB}$. An estimate of the m-th column of the data symbols matrix $\mathbf{S}$ can be thus obtained as

$$\widehat{\mathbf{S}}_{\text{dis}}(:, m) = \mathbf{E}(m)\mathbf{Y}_{\text{dis}}(:, m) . \tag{6}$$

We can also avoid discarding the last $L_{ch} - 1$ columns of $\mathbf{Y}_{BB}$; in this case we obtain the processing

$$\widehat{\mathbf{S}}_{\text{no dis}}(:, m) = \mathbf{E}(m)\mathbf{Y}_{BB}\mathbf{W}_{2k,FFT}(1 : k + L + L_{ch} - 2, m) . \tag{7}$$

Equations (6) and (7), have to be computed for $m = 1, \ldots, k$.

2.1   Channel Dependent (CD) Beamformer Design

We now address the beamformers choice by referring to (3), which shows that the precoding matrices multiply by the right the FFT channel coefficient $\overline{\mathcal{H}}(2n)$, while the postcoding matrices multiply this same coefficient by the left. Denoting by $\mathbf{Q}^{\text{opt}}(n)$ and $\mathbf{D}^{\text{opt}}(n)$ the "optimal" precoding and postcoding matrices[1] for the transmission and detection of the n-th column of $\mathbf{S}$, it is seen from (3) that, upon letting $\overline{\mathcal{H}}(2n) = \overline{\mathbf{U}}(2n)\overline{\mathbf{\Lambda}}(2n)\overline{\mathbf{V}}^H(2n)$ be the singular-value-decomposition of $\overline{\mathcal{H}}(2n)$, the matrix $\mathbf{Q}^{\text{opt}}(n)$ should contain on its columns the M columns of $\overline{\mathbf{V}}(2n)$ associated with the largest eigenvalues of $\overline{\mathcal{H}}(2n)$, and, similarly, the matrix $\mathbf{D}^{\text{opt}}(n)$ should contain on its columns the M columns of $\overline{\mathbf{U}}(2n)$ associated with the largest eigenvalues of $\overline{\mathcal{H}}(2n)$. Given the matrices $\mathbf{Q}^{\text{opt}}(n)$ and $\mathbf{D}^{\text{opt}}(n)$, the beamformers $\mathbf{Q}_{\text{BB}}(n)$, $\mathbf{Q}_{\text{RF}}$, $\mathbf{D}_{\text{BB}}(n)$, and $\mathbf{D}_{\text{RF}}$ are obtained following the approximation algorithm reported in [6].

3   MIMO-UFMC Scheme with Linear MMSE Equalization at the Receiver

The transceiver processing described in the previous section requires the knowledge of the channel impulse response and is suited for a single packet transmission, i.e. for the case in which a single isolated block of kM symbols is transmitted. In practice, however, several blocks are to be continuously transmitted. In this case, consecutive UFMC blocks are usually spaced in discrete-time by a number of intervals equal to $L-1$ [7]; since the channel is time-dispersive, at the receiver there will be inter-block interference (IBI): in particular, the first $L_{ch} - 1$ samples of the received signal $\mathbf{Y}_{BB}$ will be corrupted by the tail of the preceding block of data symbols. In this case, the single packet processing described in the previous section is suboptimal and alternative interference-suppressing schemes are to be envisaged. We now describe a linear MMSE-based processing operating directly on the matrix $\mathbf{Y}_{BB}$ reported in (2). The receiver processing is

---

[1] By the adjective "optimal" we mean here the beamforming matrices that we would use in the case in which the number of RF coincides with the number of antennas.

adaptive and so it, based on a known training sequence, automatically learns the interference-suppressing detection matrix; as a consequence, the detection strategy that we are going to illustrate can be used also in the case in which multiple packets are continuously transmitted, either with a guard-time between them, as recommended in [7], or with no guard-time at all. Starting from the matrix $\mathbf{Y}_{BB}$, we denote by $\mathbf{Z}_{BB} = \mathbf{Y}_{BB}\mathbf{W}_{2k,FFT}(1:k+L+L_{ch}-2,:)$ the $\left(N_R^{RF} \times 2K\right)$-dimensional matrix contains the $2k$-points FFT of the matrix $\mathbf{Y}_{BB}$ in Eq. (2). We denote by J the number of columns of the matrix $\mathbf{Z}_{BB}$ that we use to decode the symbol transmitted on the generic subcarrier, i.e., to limit system complexity, we use a window of data of dimension JN_R^{RF}. In order to estimate the symbol transmitted on the n-th subcarrier, we consider the $\left(N_R^{RF} \times J\right)$-dimensional matrix $\mathbf{Z}_{BB}^{(n)}$. We denote as $\mathbf{z}_{BB}^{(n)}$ the vector-stacked version of $\mathbf{Z}_{BB}^{(n)}$, i.e. $\mathbf{z}_{BB}^{(n)} = \text{vec}\left(\mathbf{Z}_{BB}^{(n)}\right)$, and we consider the linear processing

$$\widehat{\mathbf{S}}_{\mathrm{mmse}}(:,n) = \mathbf{d}_{\mathrm{mmse}}(n)^H \mathbf{z}_{BB}^{(n)} . \tag{8}$$

The detection vector can be shown to be expressed as $\mathbf{d}_{\mathrm{mmse}}(n) = \mathbf{R}_{z,n}^{-1}\mathbf{R}_{zs,n}$. In order to specify the expression of the matrices $\mathbf{R}_{z,n}$ and $\mathbf{R}_{zs,n}$, we start by considering the matrices $\widetilde{\mathbf{R}}_z$ and $\widetilde{\mathbf{R}}_{zs}$ that are obtained through time-averages approximating the matrices $\mathbf{R}_z = \mathbb{E}\left[\mathbf{z}_{BB}\mathbf{z}_{BB}^H\right]$ and $\mathbf{R}_{zs}^{(n)} = \mathbb{E}\left[\mathbf{z}_{BB}\mathbf{S}(:,n)^H\right]$, where $\mathbf{z}_{BB} = \text{vec}\left(\mathbf{Z}_{BB}\right)$, as follows:

$$\mathbf{R}_z \approx \widetilde{\mathbf{R}}_z = \frac{1}{N_{\mathrm{cov}}}\sum_{\ell=1}^{N_{\mathrm{cov}}} \mathbf{z}_{BB,\ell}\mathbf{z}_{BB,\ell}^H, \quad \mathbf{R}_{zs}^{(n)} \approx \widetilde{\mathbf{R}}_{zs}^{(n)} = \frac{1}{N_{\mathrm{cov}}}\sum_{\ell=1}^{N_{\mathrm{cov}}} \mathbf{z}_{BB,\ell}\mathbf{S}_\ell(:,n)^H . \tag{9}$$

In the above equations, N_{cov} is the number of samples used to compute the time averages, and the temporal index "ℓ" has been introduced in order to denote data coming from the ℓ-th transmitted packet. Given (9), the selection of the quantities $\mathbf{R}_{z,n}$, $\mathbf{R}_{zs,n}$ and $\mathbf{z}_{BB}^{(n)}$, depends on the subcarrier index n, and the choice is made in order to select, for each n, the columns of $\mathbf{Z}_{BB}$ that are most relevant for detecting the n-th column of the data matrix $\mathbf{S}$. For the sake of brevity, the full details are omitted and the main steps are summarized in Algorithm 1. The proposed procedure has a computational cost proportional to $k\left(N_R^{RF}J\right)^3$, whereas the complexity of the full linear MMSE receiver, that elaborates the full matrix $\mathbf{Z}_{BB}$ and corresponding to the choice $J = 2k$, the complexity climbs up to $\left[N_R^{RF}\left(k+L+L_{ch}-2\right)\right]^3$.

3.1   Channel Independet (CI) Beamforming

The outlined MMSE processing relies on a sequence of pilot symbols, and does not require any prior channel estimation. Nonetheless, implementing the beamformers as outlined in Sect. 2.1 still requires knowledge of the channel state. In order to come up with a transceiver processing that does not rely on prior channel estimation, we propose a possible channel-independent beamforming scheme that can be easily implemented through the use of 0-1 switches.

Algorithm 1 Procedure for the selection of the quantities $\widetilde{\mathbf{R}}_{z,n}$, $\widetilde{\mathbf{R}}_{zs,n}$ and $\mathbf{Z}_{\mathrm{BB},n}$. The notation $\widetilde{\mathbf{R}}_z(a:b)$ denotes selection of a submatrix of $\widetilde{\mathbf{R}}_z$ containing the entries whose column and row coordinates are in the range $(a:b)$.

1: **if** $n == 1$ **or** $n == k$ **then**
2:   **if** $n == 1$ **then**
3:     $I_{\min,1} = 1$, $I_{\max,1} = J - 2$, $I_{\min,2} = 2k - 1$, $I_{\max,2} = 2k$
4:   **else if** $n == k$ **then**
5:     $I_{\min,1} = 1$, $I_{\max,1} = 2$, $I_{\min,2} = 2k - J + 3$, $I_{\max,2} = 2k$
6:   **end if**
7:   $\mathbf{Z}_{\mathrm{BB}}^{(n)} = [\mathbf{Z}_{\mathrm{BB}}(:, I_{\min,1} : I_{\max,1}), \mathbf{Z}_{\mathrm{BB}}(:, I_{\min,2} : I_{\max,2})]$.
8:   $\mathbf{R}_{z,n} = \left[\widetilde{\mathbf{R}}_z\left(N_R^{\mathrm{RF}}(I_{\min,1} - 1) + 1 : N_R^{\mathrm{RF}} I_{\max,1}\right),\right.$
9:   $\left.\widetilde{\mathbf{R}}_z\left(N_R^{\mathrm{RF}}(I_{\min,2} - 1) + 1 : N_R^{\mathrm{RF}} I_{\max,2}\right)\right]$.
10:   $\mathbf{R}_{zs,n} = \left[\widetilde{\mathbf{R}}_{zs}^{(n)}\left(N_R^{\mathrm{RF}}(I_{\min,1} - 1) + 1 : N_R^{\mathrm{RF}} I_{\max,1}, :\right),\right.$
11:   $\left.\widetilde{\mathbf{R}}_{zs}^{(n)}\left(N_R^{\mathrm{RF}}(I_{\min,2} - 1) + 1 : N_R^{\mathrm{RF}} I_{\max,2}, :\right)\right]$.
12: **else**
13:   **if** $2n - \frac{J}{2} \geq 1$ **and** $2n + \frac{J}{2} - 1 \leq 2k$ **then**
14:     $I_{\min} = 2n - \frac{J}{2}$, $I_{\max} = 2n + \frac{J}{2} - 1$
15:   **else if** $2n - \frac{J}{2} < 1$ **then**
16:     $I_{\min} = 1$, $I_{\max} = J$
17:   **else if** $2n + \frac{J}{2} - 1 > 2k$ **then**
18:     $I_{\min} = 2k - J + 1$, $I_{\max} = 2k$
19:   **end if**
20:   $\mathbf{Z}_{\mathrm{BB}}^{(n)} = \mathbf{Z}_{\mathrm{BB}}\left(:, N_R^{\mathrm{RF}}(I_{\min} - 1) + 1 : N_R^{\mathrm{RF}} I_{\max}\right)$.
21:   $\mathbf{R}_{z,n} = \widetilde{\mathbf{R}}_z\left(N_R^{\mathrm{RF}}(I_{\min} - 1) + 1 : N_R^{\mathrm{RF}} I_{\max}\right)$.
22:   $\mathbf{R}_{zs,n} = \widetilde{\mathbf{R}}_{zs}^{(n)}\left(N_R^{\mathrm{RF}}(I_{\min} - 1) + 1 : N_R^{\mathrm{RF}} I_{\max}, :\right)$.
23: **end if**

The digital precoding $(N_T^{\mathrm{RF}} \times M)$-dimensional matrices are $\mathbf{Q}_{\mathrm{BB}}^{\mathrm{CI}}(n) = \mathbf{I}_M \otimes \mathbf{1}_{N_T^{\mathrm{RF}}/M}$ $\forall n = 1, \ldots, k$, where $\mathbf{I}_M$ is the $(M \times M)$-dimensional identity matrix and $\mathbf{1}_{N_T^{\mathrm{RF}}/M}$ is the $\frac{N_T^{\mathrm{RF}}}{M}$-dimensional vector whose entries are all equal to 1, and $\otimes$ denotes the Kronecker product. Notice also that the above defined digital precoding matrices are no longer dependent on the subcarrier index. The analog precoding $(N_T \times N_T^{\mathrm{RF}})$- dimensional matrix is $\mathbf{Q}_{\mathrm{RF}}^{\mathrm{CI}} = \mathbf{I}_{N_T^{\mathrm{RF}}} \otimes \mathbf{1}_{N_T/N_T^{\mathrm{RF}}}$, and the analog postcoding $(N_R \times N_R^{\mathrm{RF}})$- dimensional matrix is $\mathbf{D}_{\mathrm{RF}}^{\mathrm{CI}} = \mathbf{I}_{N_R^{\mathrm{RF}}} \otimes \mathbf{1}_{N_R/N_R^{\mathrm{RF}}}$.

4   Performance Measures and Numerical Results

In order to evaluate the performance of the transceiver architectures proposed in the paper, we will three different figures of merit. The first one is the root mean square error (RMSE) defined as $\mathrm{RMSE} = \mathbb{E}\left[\frac{|s - \hat{s}|^2}{|s|^2}\right]$, where s and $\hat{s}$ are the generic symbol transmitted and estimated, respectively.

The second one is the bit-error-rate (BER), while, finally, the third one is the throughput, that is measured in bit/s, and depends on the system BER

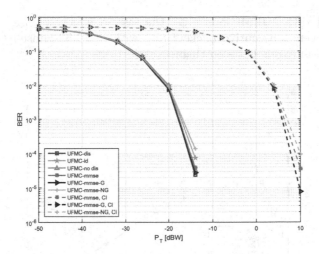

Fig. 2. BER versus transmit power with $M = 1$, performance of MIMO-UFMC transceiver architectures with CD and CI beamformers

and on the cardinality of the used modulation. Denoting by T_s the signaling time, i.e. assuming that the modulator transmits a data-symbol of cardinality $\mathcal{M}$ every T_s seconds, kM symbols are transmitted in $(k + L - 1)T_s$ seconds, in the case in which we consider the guard intervals in the packet transmission, and in kT_s seconds, in the case in which we do not consider the guard intervals between the consecutive blocks. Denoting by W the communication bandwidth, the throughputs T_G and T_{NG} are expressed as

$$T_G = \frac{W \log_2(\mathcal{M})\, kM}{k + L - 1}(1 - \text{BER})\,, \qquad T_{NG} = W \log_2(\mathcal{M})\, M(1 - \text{BER}) \quad [\text{bit/s}]\,. \quad (10)$$

when we consider the guard interval between packets or not, respectively.

In our simulation setup, we consider a communication bandwidth of $W = 500\,\text{MHz}$ centered over a mmWave carrier frequency. The MIMO propagation channel has been generated according to the statistical procedure described in [3,4]. We assume a distance between transmitter and receiver of 50 m. The additive thermal noise is assumed to have a power spectral density of $-174\,\text{dBm/Hz}$, while the front-end receiver is assumed to have a noise figure of 3 dB. For the prototype filter in the UFMC modulators we use a Dolph–Chebyshev filter with length $L = 16$ and side-lobe attenuation with respect to the peak of the main lobe equal to 100 dB. We use $k = 128$ subcarriers, $B = 8$ subbands (which leads to $D = 16$ subcarriers in each subband), and we assume 4-QAM modulation. We consider the antenna configuration $N_R \times N_T = 16 \times 64$, and we assume hybrid beamforming with $N_T^{\text{RF}} = 16$ and $N_R^{\text{RF}} = 4$. In the figures we denote as "UFMC-id" the case in which the estimate of the n-th column of the data symbols matrix $\mathbf{S}$ is expressed as (4), as "UFMC-dis" the case in which we use (6), as "UFMC-no dis" the case in which we use (7), as "UFMC-mmse" the case in which we use (8). With regard to the continuous packet transmission, we label with "UFMC-mmse-G" the case in which we use $L - 1$ guard intervals between consecutive packets and with "UFMC-mmse-NG" the case without guard intervals between them.

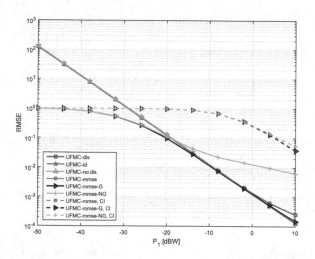

Fig. 3. RMSE versus transmit power with $M = 1$, performance of MIMO-UFMC transceiver architectures with CD and CI beamformers.

In Figs. 2, 3, and 4 we show the BER, the RMSE and the throughput versus the transmit power, for the case $M = 1$, considering the hybrid implementation of CD and CI beamformers. The obtained results show that the performances of the "UFMC-mmse" are superior to the ones offered by the standard MIMO-UFMC, even using beamformers based on the ideal received channel model in Eq. (3) and described in Sect. 2.1. This behaviour can be justified by the fact that the MMSE-based receiver described here performs an online MMSE detection of the data symbols, automatically rejecting the interference contribution. We can see that in terms of BER and RMSE the performances of "UFMC-mmse-NG" are worse than the ones obtained with "UFMC-mmse-G", because of the increased overlap (i.e., interference) of the data corresponding to consecutive blocks. Nevertheless, the throughput of "UFMC-mmse-NG" is larger than that obtained with the "UFMC-mmse-G", i.e., the increase in the system BER due to increased interference is compensated by the increased efficiency due to the fact that there are no guard intervals. For the case in which transmitter and receiver have no CSI, the CI beamformers of Sect. 3.1 are employed.

5   Conclusions

This paper presented several signal processing schemes have been developed for data detection in MIMO-UFMC systems, taking into account also the hybrid nature of the beamformers. In particular, an adaptive MMSE receiver has been

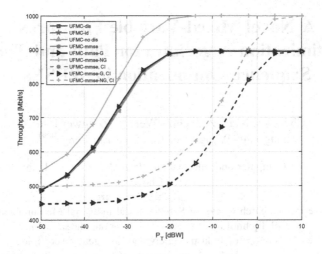

Fig. 4. Throughput versus transmit power with $M = 1$, performance of MIMO-UFMC transceiver architectures with CD and CI beamformers.

proposed that, in conjunction with CI beamforming, does not require prior channel estimation. This receiver, for the case in which no guard-time is inserted between consecutive packets at the transmitter, has been shown to be able to achieve increased performance in terms of system throughput.

References

1. Abdoli, J., Jia, M., Ma, J.: Filtered OFDM: a new waveform for future wireless systems. In: 2015 IEEE 16th International Workshop on Signal Processing Advances in Wireless Communications (SPAWC), pp. 66–70 (2015). https://doi.org/10.1109/SPAWC.2015.7227001
2. Banelli, P., Buzzi, S., Colavolpe, G., Modenini, A., Rusek, F., Ugolini, A.: Modulation formats and waveforms for 5G networks: who will be the heir of OFDM? IEEE Signal Process. Mag. **31**(6), 80–93 (2014)
3. Buzzi, S., D'Andrea, C.: On clustered statistical MIMO millimeter wave channel simulation (2016). https://arxiv.org/abs/1604.00648
4. Buzzi, S., D'Andrea, C., Foggi, T., Ugolini, A., Colavolpe, G.: Single-carrier modulation versus OFDM for millimeter-wave wireless MIMO. IEEE Trans. Commun. **66**, 1335–1348 (2018). https://doi.org/10.1109/TCOMM.2017.2771334
5. Farhang-Boroujeny, B.: OFDM versus filter bank multicarrier. IEEE Signal Process. Mag. **28**(3), 92–112 (2011)
6. Ghauch, H., Kim, T., Bengtsson, M., Skoglund, M.: Subspace estimation and decomposition for large millimeter-wave MIMO systems. IEEE J. Sel. Top. Signal Process. **10**(3), 528–542 (2016). https://doi.org/10.1109/JSTSP.2016.2538178
7. Vakilian, V., Wild, T., Schaich, F., ten Brink, S., Frigon, J.F.: Universal-filtered multi-carrier technique for wireless systems beyond LTE. In: 2013 IEEE Globecom Workshops (GC Wkshps), pp. 223–228 (2013). https://doi.org/10.1109/GLOCOMW.2013.6824990

A Novel Mixed-Variable Fireworks Optimization Algorithm for Path and Time Sequence Optimization in WRSNs

Chengkai Xia[1], Zhenchun Wei[1,2,3], Zengwei Lyu[1],
Liangliang Wang[1], Fei Liu[1], and Lin Feng[1(✉)]

[1] School of Computer and Information, Hefei University of Technology,
Hefei, China
fenglin@hfut.edu.cn
[2] Engineering Research Center of Safety Critical Industrial Measurement and
Control Technology, Ministry of Education, Hefei, China
[3] Key Laboratory of Industry Safety and Emergency Technology,
Hefei, Anhui Province, China

Abstract. To prolong the lifespan of the network, the auxiliary charging equipment is introduced into the traditional Wireless Sensor Networks (WSNs), known as Wireless Rechargeable Sensor Networks (WRSNs). Different from existing researches, in this paper, a periodic charging and data collecting model in WRSNs is proposed to keep the network working perpetually and improve data collection ratio. Meanwhile, the Wireless Charging Vehicle (WCV) has more working patterns, charging, waiting, and collecting data when staying at the sensor nodes. Then, the simultaneous optimization for the traveling path and time sequence is formulated to be a mixed-variable optimization problem. A novel Mixed-Variable Fireworks Optimization Algorithm (MVFOA) is proposed to solve it. A large number of experiments show the feasibility of the MVFOA, and MVFOA is superior to the Greedy Algorithm.

Keywords: Wireless rechargeable sensor networks · Mixed-variable optimization Fireworks algorithm

1 Introduction

To extend the lifespan of the network, the auxiliary charging equipment is introduced into the traditional Wireless Sensor Networks (WSNs), known as Wireless Rechargeable Sensor Networks (WRSNs). The wireless power transfer [1] is a promising technology which can be used in WRSN. Xie et al. use a Wireless Charging Vehicle (WCV) to periodically provide the whole network with perpetual energy [2]. According to this charging method, the WCV starts from a service station to charge all the sensor nodes in network. To enhance the performance of the network, especially the data transmission, the WCV takes more functions such as data colleting in [3–6]. The joint charging and data collecting methods have attracted more focus. However, in existing literature, when it comes to the traveling path of the WCV and the time sequence of data collecting, they are optimized respectively. In fact, the traveling path

X. Liu et al. (Eds.): ChinaCom 2018, LNICST 262, pp. 24–34, 2019.
https://doi.org/10.1007/978-3-030-06161-6_3

of the WCV and staying time at sensor nodes for charging or data collecting act and react upon one another. To obtain the traveling path of WCV, it is inevitable to schedule the order of sensor nodes visited by WCV, which is a combinatorial problem. To achieve the most suitable time sequence when WCV stays at each sensor node, time, as the continuous variable, should be calculated. If we take them into considerations at the same time, optimizing the traveling path and time sequence simultaneously, it is a mixed-variables optimization problem [7]. Furthermore, the traveling path of the WCV in WRSN can be treated a Traveling Salesman Problem (TSP), a typical NP complete combinatorial optimum problem. Especially, the order of sensor nodes visited by WCV also has effect on the staying time at each sensor node to improve some performance of network such as data collection ratio. Therefore, it is extremely difficult to solve this sort of mixed-variables optimization problem. Fireworks algorithm [8], proposed by Tan, is a novel swarm intelligence algorithm recently, which shows great edge in solving continuous variable problem [9] and discrete combinatorial problem [10].

In this paper, a periodic charging and data collecting planning, different from the non-periodic charging planning which fails to ensure permanent operation of the whole network, is formulated with optimizing the traveling path and the time sequence simultaneously to achieve higher data collection ratio in a cycle. Inspired by the framework of fireworks algorithm, a novel Mixed-Variable Fireworks Optimization Algorithm (MVFOA) for path and time sequence optimization in WRSNs is proposed. The main contributions are as follows:

- It is the first attempt to formulate and solve the path and time sequence mixed-variables optimization problem in WRSNs.
- A novel MVFOA is proposed and the encoding of the firework is designed because of the adaption of this mixed-variable optimization problem.

2 Modeling and Problem Statement

In a periodic charging and data collecting planning, the WCV undertakes two tasks, charging the sensor nodes and collecting data. To keep the network working perpetually, the WCV can travel through the whole network periodically [1]. That is to say that the WCV arrives at and leaves each sensor node at the same time in different cycles and the energy of each node varies regularly. How to schedule a best traveling path of the WCV to make all the sensor nodes periodically visited by the WCV once and make for more data collection ratio? At the same time, how to arrange the time sequence when the WCV stays at each sensor node to keep the network working perpetually and collect more data in a cycle?

2.1 Network Model

N sensor nodes are deployed in a 2-D area. The set of the sensor nodes is denoted as $V_{Sensor} = \{v_1, v_2 \ldots v_N\}$. All the positions of the sensor nodes are fixed and they are powered by the same type of battery. The battery capacity of each sensor node is $E_{\max}$.

The energy consumption power of sensor node v_i is P_i. A wireless charging vehicle (WCV) armed with data collection module is deployed in this network to provide energy for sensor node one-to-one before the energy of sensor node is lower than $E_{\min}$ and collect data generated by sensor nodes around it when staying at some sensor node. The charging power of the WCV is P_c. It is assumed that sensor node v_i sends the generated data by frequency R_i ($R_i = kP_i, k > 0$ and k is a constant) to the WCV when the WCV stays at some sensor node and sensor node v_i is within the communication radius of the WCV. The WCV will return to the Service Station (SS) to get maintenance and upload the data through the base station. It assumed that this network is a delay-tolerant network.

2.2   Construction of Working Cycle

According to the above-mentioned network model, the construction of working cycle will be shown as follows. It is assumed that the initial energy of each sensor node is $E_{\max}$. To keep the energy of each sensor node varying successively, the working cycle consists of two parts, the ordinary working cycle and the adjustment working cycle.

Ordinary Working Cycle.
The WCV starts from SS, visits all the sensor nodes once to replenish energy for sensor nodes and collect data, and then returns to SS, which is denoted as a working cycle T.

$P = (\pi_0, \pi_1, \pi_2, \ldots \pi_N, \pi_0)$ is denoted as the traveling path of the WCV, where π_0 is SS. Apparently, the traveling time between two neighbor nodes is $D_{\pi_i,\pi_{i+1}}/V, i = 1, 2 \cdots N$, where V is the traveling speed of the WCV, π_i is the ith sensor node visited by the WCV in the traveling path and $D_{\pi_i,\pi_{i+1}}$ is the distance between two neighbor nodes. Therefore, in a working cycle T, the total traveling time is $T_t = \frac{D_{\pi_N,\pi_0}}{V} + \sum_{i=1}^{N} \frac{D_{\pi_{i-1},\pi_i}}{V}, i = 1, 2 \cdots N$.

When the WCV arrives at sensor node v_i, it is possible that the remaining energy of sensor node v_i is still higher than $E_{\min}$. Thus, the WCV can wait a waiting interval Δt_i without charging and then spend time tc_i replenishing energy up to $E_{\max}$ for the sensor node with the precondition being guaranteed that the remaining energy of sensor node v_i is still higher than $E_{\min}$. Therefore, the total staying time at sensor node v_i is $(tc_i + \Delta t_i)$, in which the WCV can collect data with its communication radius. Compared with arriving at sensor node and charging immediately and then leaving, the WCV can collect more data because of the waiting interval Δt_i. Thus, the total waiting interval is $\sum_{i=1}^{N} \Delta t_i$ and the total charging time is $\sum_{i=1}^{N} tc_i$ in a working cycle T. To sum up, a working cycle T is as follows.

$$T = T_t + \sum_{i=1}^{N} \Delta t_i + \sum_{i=1}^{N} tc_i \tag{1}$$

The energy varying of sensor node v_i in each working cycle is shown in Fig. 1. a_i is the time when the WCV just arrives at sensor node v_i in the first ordinary working cycle.

To keep sensor node v_i working perpetually, it needs to be replenished energy period-ically. Thus, the energy consumption of sensor node v_i must equals the replenishment energy from the WCV to keep the energy varying regularly. For sensor node v_i,

$$T \cdot P_i = tc_i \cdot P_c \qquad (2)$$

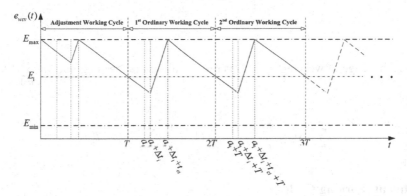

Fig. 1. The energy varying of sensor node with time in each working cycle

Therefore, the charging time for sensor node v_i is $tc_i = \frac{T \cdot P_i}{P_c}$. To avoid the sensor node death, it is essential that the remaining energy of sensor node v_i must be higher than E_{min} when the WCV just starts to charge. Then, the following formula must be satisfied,

$$e_i(a_i + \Delta t_i) = E_i - (a_i - T) \cdot P_i - \Delta t_i \cdot P_i \geq E_{min} \qquad (3)$$

where E_i is the remaining energy of sensor node v_i at the end of each working cycle. Because of the periodic varying, the energy of sensor node in the same time in different working cycles has the same energy level. Thus,

$$E_i = e_i(2T) = e_i(a_i + \Delta t_i + tc_i) - (2T - a_i - \Delta t_i - tc_i) \cdot P_i$$
$$= E_{max} - (2T - a_i - \Delta t_i - tc_i) \cdot P_i \qquad (4)$$

Furthermore, combining (3) with (4), (5) must be satisfied.

$$e_i(a_i + \Delta t_i) = E_{max} - (2T - a_i - \Delta t_i - tc_i) \cdot P_i - (a_i - T) \cdot P_i - \Delta t_i \cdot P_i \geq E_{min}$$
$$= E_{max} - (T - tc_i) \cdot P_i \geq E_{min} \qquad (5)$$

To avoid the sensor node death, the working cycle T should be restricted. For sensor node v_i, combining (2) with (5), the following formulate must be satisfied.

$$T \leq \frac{E_{max} - E_{min}}{P_i \cdot \left(1 - \frac{P_i}{P_c}\right)} \qquad (6)$$

If the WCV charges all the sensor nodes immediately when arriving at them, there are no waiting intervals and then $\sum_{i=1}^{N} \Delta t_i = 0$. Thus, combining (1) and (2), T must satisfy,

$$T \geq \frac{T_t}{(1 - \frac{1}{P_c} \cdot \sum_{i=1}^{N} P_i)}. \tag{7}$$

Once the range of working cycle T is restricted, the total waiting interval $\sum_{i=1}^{N} \Delta t_i$ will be given.

$$\sum_{i=1}^{N} \Delta t_i \leq \frac{E_{\max} - E_{\min}}{P_i \cdot (1 - \frac{P_i}{P_c})} (1 - \frac{1}{P_c} \cdot \sum_{i=1}^{N} P_i) - T_t \tag{8}$$

Adjustment Working Cycle.
As shown in Fig. 1, the initial energy of sensor node v_i is $E_{\max}$. To make the energy vary continuously with the next working cycle, the charging time tc'_i in the adjustment working cycle should be adjusted. $E_{\max} - E_i = (2T - a_i - \Delta t_i - tc_i) \cdot P_i = T \cdot P_i - tc'_i \cdot P_c$, so we have

$$tc'_i = \frac{(a_i - T + \Delta t_i + tc_i) \cdot P_i}{P_c}. \tag{9}$$

Furthermore, the arrival time and departure time of WCV in the adjustment working cycle must be consistent with those in ordinary working cycles. Thus, the total staying time at sensor node v_i in the adjustment working cycle is also $(tc_i + \Delta t_i)$. Therefore, the waiting interval $\Delta t'_i$ in the adjustment working cycle is as follows.

$$\Delta t'_i = \Delta t_i + tc_i - \frac{(a_i - T + \Delta t_i + tc_i) \cdot P_i}{P_c} \tag{10}$$

2.3   Path and Time Sequence Mixed-Variable Optimization Problem

It is assumed that sensor node v_i sends the generated data by frequency R_i to the WCV when the WCV stays at some sensor node and sensor node v_i is within the communication radius of the WCV. The WCV is also treated as a mobile sink to collect data. To pay more attention to the performance of data collection, data collection ratio $(\sum_{i=1}^{N} C_i)/T$ in a working cycle is proposed, where C_i is the data collection times when the WCV

$$C_i = \sum_{j=1}^{M_i} R_i \lfloor \Delta t_i + tc_i \rfloor, j = 1, 2 \ldots M_i. \tag{11}$$

stays at sensor node v_i. M_i is the number of the sensor nodes within the communication radius of the WCV staying at sensor node v_i. Then, the mixed-variable optimization problem can be formulated as follows,

$$Obj : \max \frac{\sum_{i=1}^{N} C_i}{T}, i = 1, 2 \cdots N. \tag{12}$$

$$s.t. : (1) - (11)$$

where $T, \Delta t_i, tc_i$ and the traveling path of the WCV are variables and $P_i, P_c, E_{\max}$ and $E_{\min}$ are constants.

3 Mixed-Variable Fireworks Optimization Algorithm

3.1 Encoding Design of MVFOA

Inspired by the explosion of fireworks, Tan et al. proposed [8, 9] Fireworks Algorithm (FA), which has great advantages in solving continuous variable problem [9] and discrete combinatorial problem [10], respectively. In this paper, to solve this mixed-variable optimization problem, the encoding of firework is designed, shown in Fig. 2, because the traveling path of the WCV and staying time at sensor nodes for charging or data collecting act and react upon one another. The upper sequence is the traveling path with its corresponding waiting interval in the lower sequence. Thus, the suitable waiting interval can be bonded with the related sensor nodes with the 2-OPT explosion [10]. Because the time is continuous variable, waiting intervals in the lower time sequence need to be given precision.

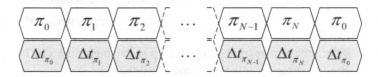

Fig. 2. Encoding of firework in MVFOA

3.2 Steps of MVFOA

Inspired by the framework of FA, the 2-OPT explosion is used to create sparks in view of optimizing the traveling path. When it comes to the time sequence, each waiting interval is treated as a dimension, similar to the explosion and mutation in FA. To decrease the computation complexity and the amount of computation, an initial solution optimization algorithm based on FA is shown in Algorithm 1.

Algorithm 1 *Initial Solution Optimization Algorithm*

Input: The number of initial firework M , m=0;
Output: M' fireworks which satisfy the constraint condition;
1: Initialization: Generate M fireworks randomly. Set the time sequence as 0;
2: **while** m< M' **do**
3:    Calculate the fitness of fireworks(total traveling time) and record the worst;
4:    Calculate the number of sparks of each firework;
5:    Generate sparks using 2-OPT and create mutation sparks using 2h-OPT ;
6:    Select the best spark or firework into next generation;
7:    Select (M-1) sparks or fireworks randomly into next generation;
8:    m=m+1 when the best spark or firework satisfies the constraint condition;
9: **end while**

Algorithm 2: *Mixed-Variable Fireworks Optimization Algorithm*

Input: M' fireworks which satisfy the constraint condition, iteration I
Output: The best group of traveling path and time sequence of the WCV ;
1: Initialization: i=0. Fix the traveling path of M' fireworks, calculate the range of the total waiting interval, and distribute a random total waiting interval according to the proportion of power of the sensor nodes in the traveling path;
2: **while** i<I **do**
3:    Calculate the fitness of fireworks(optimization objective) and record the worst;
4:    Calculate the number of sparks of each firework;
5:    **for** 1: M'
6:        Generate sparks using 2-OPT and create mutation sparks using 2h-OPT;
7:        Delete the spark if the traveling path fails to satisfy the constraint condition;
8:        Modify the time sequence if the total waiting interval is over the boundary;
9:        **for** each spark or firework
10:           Calculate the fitness f and the margin total waiting interval Tp;
11:           **for** each sensor node in the traveling path of corresponding spark
12:               Update the waiting intervals of random Z sensor nodes to make
                   $\Delta t_i' \sim U(\Delta t_i - \frac{Tp}{Z}, \Delta t_i + \frac{Tp}{Z})$. Modify the time sequence if the total
                   waiting interval is over the boundary and calculate the fitness;
13:           **end for**
14:           **if** the fitness of spark $> f$, increase Tp at the proportion k1(k1>1)
15:           **else** decrease Tp at the proportion k2(0<k2<1);
16:           Save the spark with the best fitness into next iteration;
17:        **end for**
18:    **end for**
19:    Select the best spark or firework into next generation;
20:    Select ($M'-1$) sparks or fireworks randomly into next generation;
21:    i1=i1+1;
22: **end while**

The steps of MVFOA is shown in Algorithm 2. Because the traveling path has great effect on the time sequence, it is important to take the path optimization into the prior consideration, and then the corresponding time sequence will be optimized. Benefit from the special structure of coding the firework, the explosion of firework works successfully both in continues variables and combinational variables.

4 Simulation

To get the best parameters of the MVFOA, the number of fireworks and spark coefficient which are two important parameters of the MVFOA, show the influence of different communication radius with the traveling path and time sequence, and demonstrate the performance of MVFOA, several groups of experiments are executed. Sensor nodes are distributed randomly in a 1000 m × 1000 m area. The Service Station is located at coordinate (0,0). $E_{max} = 10800J$. $E_{min} = 540J$. $P_c = 2.5W$. P_i is set from 0.01 W to 0.1 W randomly. The speed of the WCV is 5 m/s. Firstly, under the same network setting where 20 sensor nodes are deployed in the network and the communication radius of the WCV is 100 m, the average convergence iterations under different number of fireworks and spark coefficient are shown in Fig. 3. Each situation is executed repeatedly for 20 times. As shown in Fig. 3, 5 fireworks and spark coefficient 60 have advantage in less convergence iterations, which is used in the following experiments.

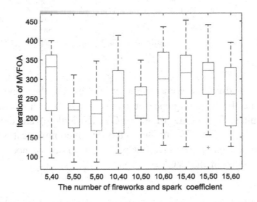

Fig. 3. Influence of the number of fireworks and spark coefficient on the convergence iterations

Secondly, to show the influence of different communication radius with the traveling path and time sequence, three groups of experiments are executed in a 40 sensor nodes network with 50 m, 100 m, 150 m communication radius of the WCV, respectively. To collect more data, the WCV must spend less time on moving. Therefore, there is no the crossover of route as shown in Fig. 4. When communication radius of the WCV is 50 m, the WCV can usually collect data from only one sensor node, the one the WCV visits. Thus, the distance of the traveling path is the shortest. As the increase of the communication radius, the traveling path becomes more and more intricate contributing to collecting more data. That is because the sensor node will communicate with WCV more times with the reciprocating and intricate path. As for the time sequence, the proportion of waiting time at each sensor node is similar to the proportion of the power the corresponding sensor node when the communication radius is 50 m because the high power of sensor node has high frequency to send data. As the increase of the radius, the WCV will wait for more time at several sensor nodes because there are more sensor nodes within its communication radius. With the reciprocating and intricate path, some sensor nodes even can communicate with the WCV many times. Therefore, the time sequence is not only related to the power of sensor nodes but also to the reciprocating path.

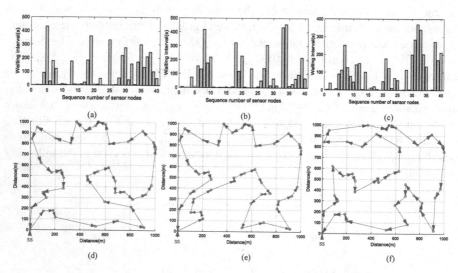

Fig. 4. The influence of different communication radius on the traveling path and time sequence. (a) and (d) are the time sequence and traveling path in a 40 sensor nodes network within the 50 m communication radius of the WCV. (b) and (e) are the time sequence and traveling path in a 40 sensor nodes network within the 100 m communication radius of the WCV. (c) and (f) are the time sequence and traveling path in a 40 sensor nodes network within the 150 m communication radius of the WCV.

In the end, MVFOA is compared with the Greedy Algorithm with the same parameter except for the number of sensor nodes. Greedy Algorithm is that the WCV will travel along the shortest path and the distribution of the time sequence is the proportion of the power of sensor nodes. As shown in Fig. 5, the MVFOA achieves higher optimization objective value than Greedy Algorithm with the increase of the number of sensor nodes.

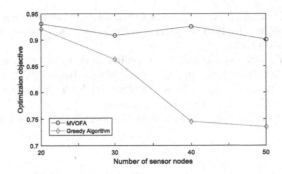

Fig. 5. Comparison between MVFOA with the Greedy Algorithm under different number of sensor nodes

5   Conclusion

In this paper, a periodic charging and data collecting model in WRSNs is proposed. The WCV has more working patterns, charging, waiting, and collecting data when staying at the sensor nodes. The optimization for the traveling path and time sequence is formulated to be a mixed-variable optimization problem. Inspired by the framework of FA, MVFOA is proposed and the coding of the firework is redesigned because of the adaption of this mixed-variable optimization problem. Simulations show the feasibility of the MVFOA, and MVFOA is superior to the Greedy Algorithm in solving proposed mixed-variable optimization problem.

References

1. Kurs, A., Karalis, A., Moffatt, R., Joannopoulos, J.D., Fisher, P., Soljai, M.: Wireless power transfer via strongly coupled magnetic resonances. Science **317**(5834), 83–86 (2007)
2. Xie, L., Shi, Y., Hou, Y.T., Sherali, H.D.: Making sensor networks immortal: an energy-renewal approach with wireless power transfer. IEEE/ACM Trans. Netw. **20**(6), 1748–1761 (2012)
3. Xie, L., Shi, Y., Hou, Y.T., Lou, W., Sherali, H.D., Zhou, H.: A mobile platform for wireless charging and data collection in sensor networks. IEEE J. Sel. Areas Commun. **33**(8), 1521–1533 (2015)
4. Guo, S., Wang, C., Yang, Y.: Joint mobile data gathering and energy provisioning in wireless rechargeable sensor networks. IEEE Trans. Mob. Comput. **13**(12), 2836–2852 (2014)
5. Wang, C., Li, J., Yang, Y.: Low-latency mobile data collection for Wireless Rechargeable Sensor Networks. In: International Conference on Communications, pp. 6524–6529. IEEE, London (2015)
6. Zhong, P., Li, Y.T., Liu, W.R., Duan, G.H., Chen, Y.W., Xiong, N.: Joint mobile data collection and wireless energy transfer in wireless rechargeable sensor networks. Sensors **17**(8), 1–23 (2017)

7. Lin, Y., Du, W., Liao, T., Stützle, T.: Three L-SHADE based algorithms on mixed-variables optimization problems. In: IEEE Congress on Evolutionary Computation, pp. 2274–2281. IEEE, San Sebastian (2017)
8. Tan, Y., Zhu, Y.: Fireworks algorithm for optimization. In: International Conference on Advances in Swarm Intelligence, pp. 355–364. Springer, Berlin (2010)
9. Li, J., Zheng, S., Tan, Y.: Adaptive fireworks algorithm. In: IEEE Congress on Evolutionary Computation, pp. 3214–3221. IEEE, Beijing (2014)
10. Tan, Y.: Fireworks Algorithm: A Novel Swarm Intelligence Optimization Method. Springer, Berlin (2015)

Aggregating Multidimensional Wireless Link Information for Device-Free Localization

Dongping Yu⬤, Yan Guo$^{(\boxtimes)}$, Ning Li, and Sixing Yang

College of Communications Engineering, Army Engineering University of PLA,
Nanjing 210007, China
guoyan_1029@sina.com

Abstract. Device-free localization (DFL) is an emerging and promising technique, which can realize target localization without the requirement of attaching any wireless devices to targets. By analyzing the shadowing loss caused by targets on wireless links, we can estimate the target locations. However, for existing DFL approaches, a large number of wireless links is required to guarantee a certain localization precision, which may lead to high hardware cost. In this paper, we propose a novel multi-target device-free localization method with multidimensional wireless link information (MDMI). Unlike previous works that measure RSS only on a single transmission power level, MDMI collects RSS measurements from multiple transmission power levels to enrich the measurement information. Furthermore, the compressive sensing (CS) theory is applied by exploiting the inherent spatial sparsity of DFL. We model the DFL problem as a joint sparse recovery problem and adopt the multiple sparse Bayesian learning (M-SBL) algorithm to reconstruct the sparse vectors of different transmission power levels. Numerical simulation results demonstrate the outstanding performance of the proposed method.

Keywords: Device-free localization · Wireless sensor network ·
Compressive sensing · Sparse Bayesian learning

1 Introduction

In the last decade, target localization has grasped great attention since it is pivotal in many location-based services (LBS). To address the localization problem of multiple targets, an intense research work has been carried out by the scientific community [1]. With the widespread usage of wireless networks, target location estimation can be realized by analyzing the target-induced perturbations in the radio frequency (RF) field. Based on this insight, the device-free localization (DFL) [2,3] has been proposed, which do not require targets to carry any wireless devices, nor to participate actively in the localization process. It is attractive

© ICST Institute for Computer Sciences, Social Informatics and Telecommunications Engineering 2019
Published by Springer Nature Switzerland AG 2019. All Rights Reserved
X. Liu et al. (Eds.): ChinaCom 2018, LNICST 262, pp. 35–45, 2019.
https://doi.org/10.1007/978-3-030-06161-6_4

and promising for a wide number of applications, such as intrusion detection, emergency rescue, healthcare, and smart spaces, etc. [4].

The DFL technique can enable existing wireless infrastructures (e.g., WiFi, WSNs, Bluetooth, etc.) to have the ability of location awareness while, at the same time, do not disturb the normal communication tasks. Received signal strength (RSS) is a common signature of target location. In the literature, many RSS-based multi-target DFL approaches have been developed. Based on how to utilize the RSS measurements, there are three types of DFL approaches, including geometry-based approaches, fingerprinting-based approaches, and radio tomographic imaging (RTI)-based approaches. The geometry-based approaches exploit the geometry information of shadowed links to locate targets [5]. However, they need a prior knowledge of the deployment of wireless nodes, and suffer from low localization accuracy. The fingerprinting-based DFL approaches can achieve an improved accuracy [6], whereas a labor-intensive and time-consuming training process is required to build and update the radio map. The RTI-based approaches [7] infer the target positions according to the principle of computed tomography (CT). They use an empirical model to quantify the relationship between RSS variations and target locations. Unfortunately, a sufficient number of wireless links is required to cover the area of interest.

As a new and promising technique, the compressive sensing (CS) [8] theory asserts that a small number of measurements (undersampled) will suffice for recovering signals that are compressible or sparse under a certain basis. Recent works have shown the potential of applying the CS theory in multi-target DFL. Compared to traditional DFL approaches, the CS-based DFL method demands much less number of wireless links (or measurements). As a representative CS-based DFL method, LCS [9] has proven that the product of the dictionary obeys restricted isometry property (RIP) with high probability. Different with LCS, E-HIPA [10] does not require a prior knowledge of target number. It adopts an adaptive orthogonal matching pursuit algorithm to reconstruct the sparse location vector. Moreover, in order to adapt to the changes in radio environments, DR-DFL [11] presents a dictionary refinement algorithm.

However, existing CS-based multi-target DFL approaches collect RSS measurements from just one transmission power level. It is assumed that each wireless link can only provide one reading of the RSS. To enrich the measurement information, MDMI proposes to collect RSS measurements from multiple transmission power levels. By doing so, the performance of multi-target DFL can be further improved with the assistance of power diversity. Hence, better localization accuracy can be achieved without increasing the number of wireless links. To leverage the advantage of CS in sparse recovery, we model the multi-target DFL with multiple transmission power levels as a joint sparse recovery problem. The sparse vectors corresponding to different transmission power levels share a common sparsity pattern. We reconstruct them by using the multiple sparse Bayesian learning (M-SBL) algorithm [12], and estimate the number and locations of multiple targets according to the reconstructed sparse vectors. The rest of the paper is organized as follows: In Sect. 2, we give the signal model and model the DFL problem as a joint sparse recovery problem. The design and implementation of the proposed MDMI method are illustrated in Sect. 3, and the simulation results are shown in Sect. 4. Finally, conclusions are given in Sect. 5.

2   Problem Statement and Motivation

2.1   Problem Statement

Fig. 1. Illustration of multi-target device-free localization.

Suppose a wireless network is deployed in the area of interest. When multiple targets entering into the area, some wireless links will be shadowed. As a consequence, the RSS readings on these shadowed links may be different from the measurements when no target is present. Our MDMI method attempt to leverage the changes of RSS to realize target localization. For simplicity, an illustration of the CS-based multi-target DFL is shown in Fig. 1. The wireless nodes are uniformly deployed around the perimeter of the monitoring area $\mathcal{A}$, and K targets are randomly distributed in it. We divide $\mathcal{A}$ into N equal-sized grids, thus the target locations can be represented as

$$\boldsymbol{\theta} = [\theta_1, \theta_2, ..., \theta_n, ..., \theta_N]^T \qquad (1)$$

where $\boldsymbol{\theta} \in \mathbb{R}^{N \times 1}$ denotes the location vector, $\theta_n \in \{0, 1\}$ denotes the n-th entry of $\boldsymbol{\theta}$. If there is a target in grid n, we set $\theta_n = 1$; otherwise $\theta_n = 0$. In this sense, K also represents the sparsity level of $\boldsymbol{\theta}$. The aim of the CS-based DFL is equivalent to reconstruct $\boldsymbol{\theta}$ by exploiting RSS measurements.

According to the shadowing model, the RSS measurement of link m with transmission power level e can be given as

$$R(m, e) = G(m, e) + P(m, e) - L(m, e) - 10\beta \lg d_m - S(m, e) + \epsilon(m, e) \quad (2)$$

where $G(m, e)$ denotes the receiver gain (dB), $P(m, e)$ is the transmission power, $L(m, e)$ is the signal attenuation power of unit distance, β is the path-loss exponent, and d_m represents the length of link m. The above-mentioned parameters are constant with time. On the contrary, $S(m, e)$ and $\epsilon(m, e)$ are time-variant parameters. $S(m, e)$ denotes the shadowing loss, which is caused by the targets that attenuate radio signals. $\epsilon(m, e)$ is the measurement noise. We denote $R_0(m, e)$ as the RSS measurement when $\mathcal{A}$ is vacant. Then, the change of RSS corresponding to link m and transmission power level e can be written as

$$\Delta R(m, e) = R(m, e) - R_0(m, e) \approx -S(m, e) + \epsilon(m, e) - \epsilon_0(m, e) \quad (3)$$

As mentioned earlier, $\mathcal{A}$ is divided into multiple grids. Hence, we approximate $S(m, e)$ by the summation of attenuation that occurs in each grid, i.e.,

$$\Delta R(m, e) = \sum_{n=1}^{N} \Delta p_n(m, e) \cdot \theta_n + \Delta \epsilon(m, e) \quad (4)$$

where $\Delta p_n(m, e)$ denotes the shadowing loss on link m that contributed by a target in grid n. $\Delta \epsilon(m, e)$ is the change of measurement noise. Based on the saddle surface (SaS) model [13], $\Delta p_n(m, e)$ can be calculated as

$$\Delta p_n(m, e) = \left(\frac{1 - \rho}{\lambda_1^2} U_{m,n}^2 + \rho \cdot \left(1 - \frac{V_{m,n}^2}{\lambda_2^2} \right) \right) \cdot \gamma^e \quad (5)$$

where $(U_{m,n}, V_{m,n})$ is the coordinate of grid n. According to the SaS model, only the grids in the elliptical spatial impact area of link m will have a nonzero $\Delta p_n(m, e)$, and $\Delta p_n(m, e)$ is very different at different locations within the spatial impact area. In this model, ρ represents the shadow rate, which is defined as the normalized shadowing effect in the midpoint of the line-of-sight (LOS) path. γ^e denotes the maximum shadowing effect corresponding to power level e. Based on (4), the RSS variations on M links can be expressed as

$$\mathbf{y}^e = \mathbf{\Phi}^e \, \boldsymbol{\theta} + \boldsymbol{\epsilon}^e \quad (6)$$

where $\mathbf{y}^e \in \mathbb{R}^{M \times 1}$ is the measurement vector corresponding to power level e, $\mathbf{\Phi}^e \in \mathbb{R}^{M \times N}$ is a dictionary, and $\boldsymbol{\epsilon}^e \in \mathbb{R}^{M \times 1}$ is the noise vector. We denote $\phi_{m,n}^e$ as the (m, n)-th element of $\mathbf{\Phi}^e$, which is equal to $\Delta p_n(m, e)$.

2.2   Motivation

In fact, $\Delta p_n(m, e)$ can be decomposed as $\Delta p_n(m, e) = (\phi_{m,n}^e / \gamma^e) \cdot \gamma^e$. We define $\mathbf{w}^e = \gamma^e \cdot \boldsymbol{\theta}$ and $\phi_{m,n} = (\phi_{m,n}^e / \gamma^e)$. It is assumed that we can collect RSS measurements form E different transmission power levels. Thus, the CS-based DFL can be reformulated as a joint sparse recovery problem as follows:

$$\mathbf{Y} = \mathbf{\Phi} \mathbf{W} + \mathbf{\Xi} \quad (7)$$

where $\mathbf{Y} \in \mathbb{R}^{M \times E}$ is the measurement matrix, and $\mathbf{Y} = [\mathbf{y}^1, ..., \mathbf{y}^E]$. $\mathbf{\Xi} \in \mathbb{R}^{M \times E}$ is the noise matrix, and $\mathbf{\Xi} = [\boldsymbol{\epsilon}^1, ..., \boldsymbol{\epsilon}^E]$. $\mathbf{\Phi} \in \mathbb{R}^{M \times N}$ is a dictionary, whose

(m,n)-th element is $\phi_{m,n}$. $\mathbf{W} \in \mathbb{R}^{N \times E}$ is a coefficient matrix, whose e-th component $\mathbf{w}^e$ is a K-sparse coefficient vector. $\mathbf{W}$ satisfies $rd(\mathbf{W}) = K$, where

$$rd(\mathbf{W}) \triangleq \sum_{n=1}^{N} \mathcal{I}\left[\|W_{n\cdot}\| > 0\right] \tag{8}$$

$rd(\cdot)$ represents a row-diversity measure, which counts the number of rows that have nonzero values. $\mathcal{I}[\cdot]$ denotes the indicator function. $\|\cdot\|$ is an arbitrary vector norm, and $W_{n\cdot}$ is the n-th row of $\mathbf{W}$. To reconstruct $\mathbf{W}$, we formulate the following relaxed optimization problem:

$$\hat{\mathbf{W}} = \arg\min_{\mathbf{W}}(\|\mathbf{Y} - \mathbf{\Phi}\mathbf{W}\|_{\mathcal{F}}^2 + \ell \cdot rd(\mathbf{W})) \tag{9}$$

where $\|\cdot\|_{\mathcal{F}}$ denotes the Frobenius norm, ℓ is a tradeoff parameter, and $rd(\mathbf{W})$ is the regularization term. However, directly solving (9) is NP-hard, and the optimal value of ℓ is generally not available.

3   Joint Sparse Recovery

To bypass the requirement of estimating ℓ, we resort to a Bayesian probabilistic approach. By applying an $\exp[-(\cdot)]$ transformation, the optimization problem in (9) can be viewed as a maximum a posterior probability (MAP) estimation task, which is summarized as follows:

$$\hat{\mathbf{W}} = \arg\max_{\mathbf{W}} p(\mathbf{Y}|\mathbf{W}) \cdot p(\mathbf{W}) = \arg\max_{\mathbf{W}} p(\mathbf{W}|\mathbf{Y}) \tag{10}$$

To solve the above problem, we resort to the M-SBL algorithm. Firstly, a Gaussian distribution is imposed on the likelihood function for each $\mathbf{y}^e$ and $\mathbf{w}^e$, i.e.,

$$p(\mathbf{y}^e|\mathbf{w}^e,\sigma) = \left(2\pi\sigma^2\right)^{-\frac{N}{2}} \exp\left(-\frac{\|\mathbf{y}^e - \mathbf{\Phi}\mathbf{w}^e\|_2^2}{2\sigma^2}\right) \tag{11}$$

where σ^2 denotes the noise variance. Secondly, to induce the sparsity of $\mathbf{w}^e$, a Gaussian prior is imposed on the n-th row of $\mathbf{W}$, i.e.,

$$p(W_{n\cdot};\alpha_n) = \mathcal{N}(0,\alpha_n\boldsymbol{I}) \tag{12}$$

where α_n denotes the common variance of the elements in $W_{n\cdot}$. Here, $\{\alpha_1,...,\alpha_N\}$ is used for encouraging the joint sparsity of $\{\mathbf{w}^1,...,\mathbf{w}^E\}$. Based on (12), the prior distribution of $\mathbf{W}$ can be given as

$$p(\mathbf{W};\boldsymbol{\alpha}) = \prod_{n=1}^{N} p(W_{n\cdot};\alpha_n) \tag{13}$$

where $\boldsymbol{\alpha} = [\alpha_1,...,\alpha_N]^T$. Based on the likelihood function and the prior distributions, the posterior of $\mathbf{w}^e$ can be written as

$$p(\mathbf{w}^e|\mathbf{y}^e;\alpha_n) = \frac{p(\mathbf{w}^e,\mathbf{y}^e;\boldsymbol{\alpha})}{\int p(\mathbf{w}^e,\mathbf{y}^e;\boldsymbol{\alpha})d\mathbf{w}^e} = \mathcal{N}(\boldsymbol{\mu}_e,\boldsymbol{\Sigma}) \tag{14}$$

where Σ denotes the covariance matrix. It can be given as

$$\Sigma = \mathrm{Cov}\left[\mathbf{w}^e \,|\mathbf{y}^e; \boldsymbol{\alpha}\right] = \boldsymbol{\Gamma} - \boldsymbol{\Gamma}\boldsymbol{\Phi}^T\boldsymbol{\Theta}^{-1}\boldsymbol{\Phi}\boldsymbol{\Gamma} \tag{15}$$

where $\boldsymbol{\Gamma} = \mathrm{diag}\left(\boldsymbol{\alpha}\right)$ and $\boldsymbol{\Theta} = \sigma^2\,\boldsymbol{I} + \boldsymbol{\Phi}\boldsymbol{\Gamma}\boldsymbol{\Phi}^T$. The mean of $\mathbf{W}$ is

$$\boldsymbol{\Pi} = [\boldsymbol{\mu}_1, ..., \boldsymbol{\mu}_E] = \mathrm{E}\left[\mathbf{W}\,|\mathbf{Y}; \boldsymbol{\alpha}\right] = \boldsymbol{\Gamma}\boldsymbol{\Phi}^T\boldsymbol{\Theta}^{-1}\mathbf{Y} \tag{16}$$

The e-th column of $\boldsymbol{\Pi}$ represents the mean vector of $\mathbf{w}^e$. To find the optimal value of $\boldsymbol{\alpha}$, we maximize the marginal likelihood with respect to $\boldsymbol{\alpha}$. Based on it, the cost function can be expressed as

$$\mathcal{L}\left(\boldsymbol{\alpha}\right) = -2\log\int p\left(\mathbf{Y}, \mathbf{W}\right)p\left(\mathbf{W}\right)d\mathbf{W} = E \cdot \log|\boldsymbol{\Theta}| + \sum_{e=1}^{E}(\mathbf{y}^e)^T\boldsymbol{\Theta}^{-1}\mathbf{y}^e \tag{17}$$

From (17), the update rule of α_n can be given as

$$\alpha_n^* = \frac{1}{E}\left\|\Pi_{n\cdot}\right\|_2^2 + \Sigma_{nn}, \quad \forall n = 1, ..., N \tag{18}$$

In the same way, σ^2 can be updated as

$$\left(\sigma^2\right)^* = \frac{\frac{1}{E}\left\|\mathbf{Y} - \boldsymbol{\Phi}\boldsymbol{\Pi}\right\|_{\mathcal{F}}^2}{M - N + \sum_{n=1}^{N}\frac{\Sigma_{nn}}{\alpha_n}} \tag{19}$$

We estimate the posterior of $\mathbf{W}$ and the parameters $\boldsymbol{\alpha}$ and σ^2 by maximizing a marginal likelihood function via an iterative algorithm, which is summarized in Algorithm 1. To estimate target locations, a sparse vector $\Pi_{\cdot\hat{e}}$ is chosen from $\boldsymbol{\Pi}$. In step 10, a sparsity threshold η_{th} is adopted to filter out the negligible but nonzero coefficients of $\Pi_{\cdot\hat{e}}$. Consequently, we can calculate the estimated coordinates of targets based on $\hat{\boldsymbol{\theta}}$, and estimate the target number as $\hat{K} = \|\hat{\boldsymbol{\theta}}\|_0$.

Algorithm 1 Location Vector Estimation

1: **Initialization:**
2: $\gamma_{\mathrm{th}} = 10^{-3}$, $\tau_{\max} = 10^3$, $\eta_{\mathrm{th}} = -10\mathrm{dB}$, $\gamma = \tau = 0$.
3: **while** ($\gamma \geqslant \gamma_{\mathrm{th}}$ or $\tau \leqslant \tau_{\max}$) **do**
4:    Calculate Σ and $\boldsymbol{\Pi}$ by using (15)–(16).
5:    Update $\boldsymbol{\alpha}$ and σ^2 by using (18)–(19).
6:    $\gamma \leftarrow \|\mathbf{Y} - \boldsymbol{\Phi}\boldsymbol{\Pi}\|, \tau \leftarrow \tau + 1$.
7: **end while**
8: Choose $\hat{e}$ that minimizes $\|\mathbf{y}^e - \boldsymbol{\Phi}\Pi_{\cdot e}\|$.
9: If $20\lg(\Pi_{n\hat{e}}/\max_i|\Pi_{i\hat{e}}|) < \eta_{\mathrm{th}}$, set $\Pi_{n\hat{e}} = 0$ for all n.
10: Let the estimated location vector $\hat{\boldsymbol{\theta}} = \Pi_{\cdot\hat{e}}$.

---

4   Numerical Results

4.1   Simulation Setup

In this section, we conduct numerical simulations to demonstrate the superior performance of MDMI. For a typical scenario of CS-based multi-target DFL, the monitoring area $\mathcal{A}$ is set as a $14\,\mathrm{m} \times 14\,\mathrm{m}$ square region. $\mathcal{A}$ is divided into $N = 784$ equal-sized grids, and the side length of each grid is $0.5\,\mathrm{m}$. To locate the targets, a wireless network with $M = 28$ wireless links is deployed in $\mathcal{A}$. The signal-to-noise ratio (SNR) is defined as $\mathrm{SNR(dB)} \triangleq 10\lg(\|\boldsymbol{\Phi\theta}\|_2^2/M\sigma^2)$.

To evaluate the localization and counting performance, we define the following two metrics: (1) Average localization error ($AvgErr$), which denotes the average Euclidean distance between the true and estimated target locations; (2) Correct counting rate ($CoCoun$), which represents the probability of correctly estimating the target number (i.e., $\hat{K} = K$). In our simulations, we compare the localization performance of MDMI with the CS-based multi-target DFL approaches that adopt the following sparse recovery algorithms: orthogonal matching pursuit (OMP) [10], basis pursuit (BP) [8], greedy matching pursuit algorithm (GMP) [9], Bayesian compressive sensing (BCS) [14], and variational EM algorithm [11].

4.2   Impact of the Number of Iterations

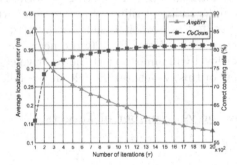

Fig. 2. The performance of MDMI when τ varies from 10^2 to 2×10^3.

In the first simulation, we investigate the impact of the number of iterations on the performance of MDMI. In Sect. 3, an iterative two-step procedure is adopted to estimate the posterior of $\mathbf{w}^e$ and the parameters α and σ^2. Intuitively, the estimation accuracy is closely related to the iteration number τ. To verify this, we test the performance of MDMI when τ varies from 10^2 to 2×10^3. As can be seen from Fig. 2, $AvgErr$ decreases rapidly as the increasing of τ. At the same time, $CoCoun$ is increased. The simulation results confirm our analysis. Although we can achieve a better performance with a larger τ, it may result in heavy computational load. For this reason, we set $\tau_{\max} = 10^3$ as a tradeoff between accuracy and complexity.

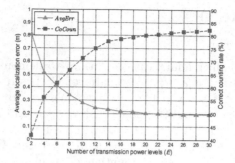

Fig. 3. The performance of MDMI when E varies from 2 to 30.

4.3   Impact of the Number of Transmission Power Levels

In the second simulation, we test the effect of the number of transmission power levels on the target localization and counting performance. The key novelty of the proposed MDMI is the utilization of the RSS measurements that collected from multiple transmission power levels. If we increase the number of transmission power level E, the power diversity of RSS measurements will be improved, and more useful information will be provided. To validate the effectiveness of MDMI, we conduct a quantitative analysis to investigate how the number of transmission power levels affects the localization and counting performance. Figure 3 shows $AvgErr$ and $CoCoun$ under different values of E. The simulation results confirm the effectiveness of MDMI. However, it is noteworthy that $AvgErr$ decreases very slowly when E exceeds 20. When $E > 20$, the negative effect of increasing the transmission power level will almost offset the positive effect contributed by power diversity. In view of this, we choose $E = 20$ in the following simulations.

4.4   Localization Error vs. Number of Targets

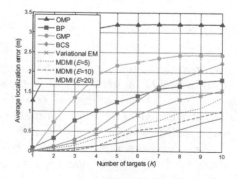

Fig. 4. Impact of the number of targets on average localization error.

In the third simulation, we turn our attention to the impact of the number of targets on localization accuracy. Figure 4 shows the performances of multiple DFL approaches under different numbers of targets. When K increases from 1 to 10, the $AvgErr$ for all approaches increase dramatically. It should be pointed out that, with the increase in K, the joint sparsity level of $\{\mathbf{w}^e\}_{e=1}^E$ will decease accordingly. In this case, the reconstruction accuracy of the location vector will be degraded according to the principle of CS. Furthermore, owing to the aggregating of multidimensional measurement information, MDMI can achieve the lowest $AvgErr$ among all approaches.

4.5   Localization Error vs. SNR

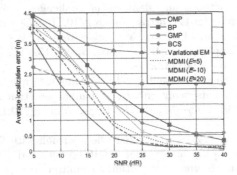

Fig. 5. Impact of the signal-to-noise ratio on average localization error.

In the last simulation, the localization performances of DFL approaches under different SNR is demonstrated. Figure 5 shows the results of the simulation. As SNR increases from 5 to 40 dB, the $AvgErr$ of all DFL methods experience a greatly drop. We observe that the MDMI($E = 20$) outperforms other DFL methods in most cases (SNR > 9 dB). In addition, when SNR < 30 dB, the difference in $AvgErr$ among MDMI($E = 5$), MDMI($E = 10$) and MDMI($E = 20$) is relatively high. This implies that we can mitigate the influence of measurement noise by increasing the power diversity of RSS measurements.

5   Conclusion

In this paper, a novel CS-based multi-target DFL method (MDMI) is developed to reduce the number of wireless links that required for multi-target DFL. Unlike existing CS-based DFL methods for multiple targets which collect measurements from just one transmission power level, MDMI proposes to exploit multidimensional wireless link information from multiple transmission power levels. It models the CS-based multi-target DFL problem as a joint sparse recovery problem, and adopts the multiple sparse Bayesian learning (M-SBL) algorithm to reconstruct the sparse vectors of different transmission power levels. To validate the merits of MDMI, we perform an extensive simulation study compared with the state-of-the-art CS-based multi-target DFL approaches. Simulation results confirm the effectiveness of the proposed method.

Acknowledgment. This work was supported in part by the National Natural Science Foundation of China under grant 61871400, and 61571463; the Natural Science Foundation of Jiangsu Province under grant BK20171401.

References

1. Khalajmehrabadi, A., Gatsis, N., Akopian, D.: Modern WLAN fingerprinting indoor positioning methods and deployment challenges. IEEE Commun. Surv. Tuts. **19**(3), 1974–2002 (2017). https://doi.org/10.1109/COMST.2017.2671454
2. Wang, J., Gao, Q., Pan, M., Fang, Y.: Device-free wireless sensing: challenges, opportunities, and applications. IEEE Netw. **32**(2), 132–137 (2018). https://doi.org/10.1109/mnet.2017.1700133
3. Lei, Q., Zhang, H., Sun, H., Tang, L.: Fingerprint-based device-free localization in changing environments using enhanced channel selection and logistic regression. IEEE Access **66**, 2569–2577 (2018). https://doi.org/10.1109/ACCESS.2017.2784387
4. Zhou, Z., Wu, C., Yang, Z., Liu, Y.: Sensorless sensing with WiFi. Tsinghua Sci. Technol. **20**(1), 1–6 (2015). https://doi.org/10.1109/tst.2015.7040509
5. Zhang, D., et al.: Fine-grained localization for multiple transceiver-free objects by using RF-based technologies. IEEE Trans. Parallel Distrib. Syst. **25**(6), 1464–1475 (2014). https://doi.org/10.1109/tpds.2013.243
6. Zhang, D., Liu, Y., Guo, X., Ni, L.: RASS: a real-time, accurate, and scalable system for tracking transceiver-free objects. IEEE Trans. Parallel Distrib. Syst. **24**(5), 996–1008 (2013). https://doi.org/10.1109/tpds.2012.134
7. Wang, Q., Yigitler, H., Jantti, R., Huang, X.: Localizing multiple objects using radio tomographic imaging technology. IEEE Trans. Veh. Technol. **65**(5), 3641–3656 (2016). https://doi.org/10.1109/tvt.2015.2432038
8. Candes, E., Wakin, M.: An introduction to compressive sampling. IEEE Signal Process. Mag. **25**(2), 21–30 (2008). https://doi.org/10.1109/msp.2007.914731
9. Wang, J., Fang, D., Chen, X., Yang, Z., Xing, T., Cai, L.: LCS: compressive sensing based device-free localization for multiple targets in sensor networks. In: IEEE INFOCOM 2013, Turin, Italy, pp. 14–19 (2013). https://doi.org/10.1109/infcom.2013.6566752
10. Wang, J., et al.: E-HIPA: an energy-efficient framework for high-precision multi-target-adaptive device-free localization. IEEE Trans. Mob. Comput. **16**(3), 716–729 (2017). https://doi.org/10.1109/tmc.2016.2567396
11. Yu, D., Guo, Y., Li, N., Fang, D.: Dictionary refinement for compressive sensing based device-free localization via the variational EM algorithm. IEEE Access **4**, 9743–9757 (2016). https://doi.org/10.1109/access.2017.2649540
12. Wipf, D., Rao, B.: An empirical Bayesian strategy for solving the simultaneous sparse approximation problem. IEEE Trans. Signal Process. **55**(7), 3704–3716 (2007). https://doi.org/10.1109/TSP.2007.894265

13. Wang, J., Gao, Q., Pan, M., Zhang, X., Yu, Y., Wang, H.: Towards accurate device-free wireless localization with a saddle surface model. IEEE Trans. Veh. Technol. **65**(8), 6665–6677 (2016). https://doi.org/10.1109/tvt.2015.2476495
14. Ji, S., Xue, Y., Carin, L.: Bayesian compressive sensing. IEEE Trans. Signal Process. **56**(6), 2346–2356 (2008). https://doi.org/10.1109/TSP.2007.914345

Software Defined Industrial Network: Architecture and Edge Offloading Strategy

Fangmin Xu[1], Huanyu Ye[1(✉)], Shaohua Cui[2], Chenglin Zhao[1],
and Haipeng Yao[1]

[1] Key Laboratory of Universal Wireless Communications, Ministry of Education,
Beijing University of Posts and Telecommunications, Beijing 100876, China
huanyuyebupt@gmail.com
[2] China Petroleum Technology and Development Corporation (CPTDC), Beijing
100028, China
shcui@163.com

Abstract. The integration of the internet and the traditional manufacturing industry has identified the "Industrial Internet of Things" (IIoT) as a popular research topic. However, traditional industrial networks continue to face challenges of resource management and limited raw data storage and computation capacity. In this paper, we propose a Software Defined Industrial Network (SDIN) architecture to address the existing drawbacks in IIoT such as resource utilization, data processing and system compatibility. The architecture is developed based on the Software Defined Network (SDN) architecture, combining hierarchical cloud and edge computing technologies. Based on the SDIN architecture, a novel centralized computation offloading strategy in industrial application is proposed. The simulation results confirm that the SDIN architecture is feasible and effective in the application of edge computing.

Keywords: Software defined industrial network · Industrial internet of things · Edge computing · Computing offloading · Time delay

1 Introduction

Intelligent manufacturing (IM), which has been driven by Information and communication technologies (ICT), greatly improves the automation level, production quality and efficiency of manufacturing industry. These information

Supported by Key Program of the National Natural Science Foundation of China (Grant No 61431008) and Project of intelligent manufacturing integrated standardization and new model application.

X. Liu et al. (Eds.): ChinaCom 2018, LNICST 262, pp. 46–56, 2019.
https://doi.org/10.1007/978-3-030-06161-6_5

and communication technologies (ICTs) provide reduced Capital Expenditure (CAPEX) and Operating Expense (OPEX) with higher efficiency and effectiveness. However, owing to the inherent limitations and complex network protocols in traditional industrial networks, traditional industrial Ethernets cannot manage distributed resources flexibly. Moreover, collected raw data are becoming increasingly more granular and voluminous.To address these drawbacks, bringing the cloud computing resources nearer to the underlying networks is attractive and promising.

In [1], the authors proposed a low latency mobile edge computing (MEC) framework based on the SDN architecture. Security and privacy are two of the main challenges to the IoT; hence, [2] proposed solutions and models for securing IoT devices and communications using the SDN architecture. Meanwhile, the performance of SDIN using in data offloading and edge computing in several scenarios including cloud server and mobile tasks was discussed in [3–5]. Computation offloading technology as discussed in [6,7]. In [8] and [9], mobile devices can extend the standby time by computation offloading.

Other previous research addressed special purposes such as energy saving and real-time communication; however, they did not focus on the details of the entire system architecture and operation. Therefore, we present a new software defined industrial network (SDIN) architecture that is the combination of Software Defined Networking (SDN) and IIoT. Industry network intelligence and control are logically centralized in SDIN to provide greater processing performance and avoid the majority of the aforementioned drawbacks. Particularly, our contributions are as follows:

- We propose a new SDIN architecture, and analyze the control and management process.
- Based on the SDIN architecture and the characteristics of industrial computing tasks, we propose a hybrid centralized edge computing offloading strategy.

The remainder of this paper is organized as follows. Section 2 identifies the architecture of SDIN. Section 3 analyses the application of SDIN in solving the computing offloading problem and the unique features. The system model, problem formulation, and solutions are provided in Sect. 4. We present a performance simulation in Sect. 5. Finally, Sect. 6 concludes the work.

2 Software Defined Industrial Network

We propose an SDIN architecture as displayed in Fig. 1. The SDIN architecture contains four layers: field devices layer, data transport layer, distributed control layer and cloud platform. Field devices include the basic infrastructures such as robot arms, conveyor belts, lathes and deployed sensors, etc.

(1) Data Transport Layer

This layer is composed of SDN switches, wireless access points (APs). The APs emphasize authentication where IoT devices access the network and for data

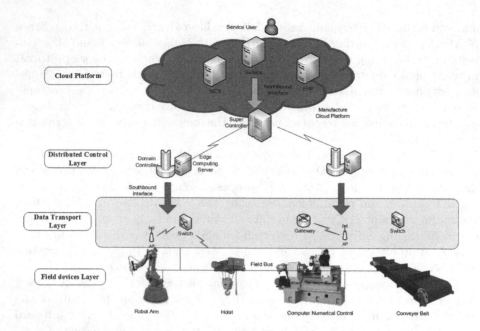

Fig. 1. Proposed software defined industrial network architecture

transporting. Data plane devices receive commands from control plane through southbound interface, such as computing decisions.

(2) Distributed Control Layer

The distributed SDN controllers are responsible for the centralized management of the edge computation servers and the authorization of multiuser access. Controllers receive requests from devices in the data transport plane and execute the offloading decision algorithm considering both the mission requirements and status of the edge computation servers. As displayed in Fig. 1, two-tier heterogeneous controller structure (domain controller and super controller) is one of the typical deployment solutions in large scale manufacture enterprise.

(3) Cloud Platform

The cloud platform includes a series of cloud service applications composed of industry application systems (Such as MES (Manufacturing Execution System), EPR (Enterprise Resources Planning)). The super controller places emphasis on resource management and the authorization of the distributed SDN controllers.

Besides, edge computing servers are more powerful than the local computing nodes in field devices. Due to the edge servers are located near the factories and production lines, it can provide lower and more stable latency than cloud computing server.

3 Computing Task and Offload in Industrial Scenario

In the concept of the intelligent factory, there are many computing tasks during the production process, such as Automatic Guided Vehicle (AGV) navigation, operation control of mechanical arms, and the identification of product imperfection. These services and applications can require significant computation resources and constrained time delay. However, the computational capabilities of the field devices are limited owing cost and size limitations.

3.1 Industrial Computing Task

After investigation, the industrial computing task has following unique features:

(1) Stringent computing delay tolerant
In industrial internet, the distributed sensors, actuators, machines, and other computing devices need to collaborate together to achieve real-time operation or complete the production tasks. In order to minimize the influence to the production line, the latency-sensitive industrial applications require delay from tens of milliseconds to hundreds of milliseconds.

(2) Diverse computing factors
As mentioned before, typical computing tasks in modern factory could be classified into following types with diverse characteristics and QoS requirements.

- Image or video recognition (such as quality inspection).
  Characteristic: huge amounts of raw input data, small size of computing result data.
- Localization and mapping (such as welding robot positioning guide).
  Characteristic: high computing accuracy, huge computing resource.
- Production planning and scheduling.
  Characteristic: multiple data sources, complex computing, low frequency.

To study the effects of the computation task characteristics on the design of offloading schemes, we classify the tasks into I types. Each type of computing task has various QoS requirements. According to computation task types, the field devices can be correspondingly classified into I types. Let us assume that one field device only generates one type of computing task.

(3) Regular task pattern
Generally, the computing tasks originate from the production line which has a fixed production cycle. For instance, the cycle of one product line is five products per minute, therefore the cycle of product imperfection identify task is also five times per minute.

To simplify the analysis, the arrival of single industrial computing task is modeled as regular arrival. However, considering the asynchronism of different field devices and production lines, the arrival of tasks at the edge server could be treated as poisson flow.

3.2   Computing Offload Procedure Based on the SDIN Architecture

The decision of whether local computing or offloading the computing task to the edge server is an important and difficult procedure. Figure 2 displays the computing decision process. The normal working process includes two stages: maintenance and update. In the maintenance stage, domain controllers broadcast the domain offloading strategy (the offloading probability of each computing task type, denoted as ξ_i, $i \in I = [1, 2, ...I]$) to field devices (① in Fig. 2). Field devices generate a random number between 0 and 1. If this number is less than ξ_i, then the computing will be offloaded to the edge server. Otherwise, the computing will be implemented in local computing unit. Update stage is trigged by the change of manufacturing environment and other factors, such as the increasing of computing task frequency, adjust of production scheduling. It contains three parts: requests collection, mode decision, computing, as follows:

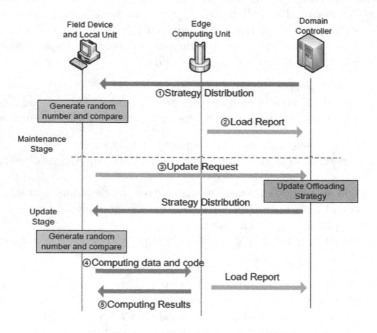

Fig. 2. The computing offloading decision process

(1) Update Request
The devices collect update data (information of computing tasks) and send it to the data transport plane through the APs. Then the update request message is send to its domain SDIN controller by southbound interface (③ in Fig. 2).

(2) Mode Decision
The domain controller exacts the computing capability parameters from the latest update request message and edge computing server (from the load report

message, as ② in Fig. 2), then it will execute the decision algorithm described in Sect. 4 and return the domain offloading strategy (the offloading probability of each computing task type, ξ_i).

(3) Computing

Similar with the maintenance stage, field devices decide whether offloading the task to the edge server or not after received the domain offloading strategy. For instance, if the device chooses the remote computing, field device will upload the necessary data and code to the edge server (④ in Fig. 2), and receive the computing result from the server afterward (⑤ in Fig. 2).

4 System Model and Offloading Algorithm

Considering most of computing devices in current factory are powered with electricity instead of batteries, the energy consumption of computing and data transfer is not a great issue. Therefore, the computing offload policy only considers the goal of minimizing computing delay.

The computing latency is divided into the following five aspects for decision analysis: local computing delay D_{Local}, data transmission delay from the local to the edge server T_{Trans}, task queuing delay for the edge computing server D_{queue}, edge computing server calculation delay D_{Remote}, and computing result return delay from the edge server to the local devices D_{Result}.

We denote the computing tasks of type i by $I_i = (D_i, C_i, T_i)$, where D_i denotes data size, C_i represents the size of computation data involved in the number of CPU cycles required to complete the type-i task, and T_i represents the maximum delay tolerance of the type-i task.

According to their computation task types, the proportion of the field devices with type-i tasks is given by π_i, where $i \in I$, and $\sum_{i \in I} \pi_i = 1$.

4.1 System Model

The local computing unit capability is defined as f_i^l, and f_i^r represents the CPU computation cycles per second that the edge server can provide. We assume that the transmission bandwidth is not constrained. Infinite buffer exists in the edge servers, and at most one computing task is served by the edge server simultaneously. Then the latency components could be calculated as (1)–(4).

$$D_{Local}(i) = \frac{C_i}{f_i^l} \tag{1}$$

$$D_{Remote}(i) = \frac{C_i}{f_r^l} \tag{2}$$

$$T_{Trans}(i) = \frac{D_i}{r_i} \tag{3}$$

$$D_{Result}(i) = \frac{D_i^r}{r_i} = \frac{K_i C_i}{r_i} \tag{4}$$

where r_i represents the transmission rate between the local node and the edge server node (either uplink or downlink), the unit is Kbps; D_i^r represents the size of computation result for type-i task. In general, its size is proportional to the amount of computation data C_i, the proportion factor is a constant K_i.

The queue delay D_{queue} is estimated by multiple class M/D/1 queue theory. Here we consider two typical cases: FIFO with equal priority (EP in short) and Non-preemptive priority queue (NPP in short). Denoted the arrival rate of all type-i tasks by λ_i, which is usually a known parameter at the domain controller. Therefore, the real arrival rate of type-i tasks at the edge server is $\xi_i \lambda_i$.

The mean service time of type-i $E(S_i)$ is deterministic, and equals to $D_{queue}(i)$. We define the probability that the server is busy and busy with a type-i task as ρ and ρ_i respectively. Obviously

$$\rho_i = \xi_i \lambda_i E(S_i) = \frac{\xi_i \lambda_i C_i}{f_i^r} \tag{5}$$

$$\rho = \sum_{i \in I} \rho_i = \sum_{i \in I} \frac{\xi_i \lambda_i C_i}{f_i^r} \tag{6}$$

$$E(S) = \sum_{i \in I} \frac{\xi_i \lambda_i}{\sum_{i \in I} \xi_i \lambda_i} E(S_i) = \sum_{i \in I} \frac{\xi_i \lambda_i}{\sum_{i \in I} \xi_i \lambda_i} \frac{C_i}{f_i^r} \tag{7}$$

Case A: Equal Priority (EP):

Based on Little theory and PASTA (Poisson arrivals see time averages) property, the average queueing latency is equal to the average queueing latency of each class, which could be estimated by

$$E(D_{queue}) = \frac{\sum_{i \in I} \rho_i \frac{E(S_i)}{2}}{1 - \rho} = \frac{\sum_{i \in I} \frac{\xi_i \lambda_i (C_i)^2}{2(f_i^r)^2}}{1 - \sum_{i \in I} \frac{\xi_i \lambda_i C_i}{f_i^r}} \tag{8}$$

Case B: Non-preemptive priority (NPP)

If type 1 has non-preemptive priority over type 2, then a type 2 task cannot be preempted once it enters service. Type 1 task still have priority over any type 2 task that are waiting but not being served. Let us assume that if $i < j$, then type i has non-preemptive priority over type j.

For type 1 task,

$$E(D_{Queue(1)}) = \frac{\sum_{i \in I} \rho_i \frac{E(S_i)}{2}}{1 - \rho_1} = \frac{\sum_{i \in I} \frac{\xi_i \lambda_i (C_i)^2}{2*(f_i^r)^2}}{1 - \frac{\xi_1 \lambda_1 C_1}{f_1^r}} \tag{9}$$

For type $i > 1$, again using little and PASTA,

$$E(D_{Queue}(i)) = \frac{\sum_{i \in I} \rho_i \frac{E(S_i)}{2}}{(1 - \sum_{k=1}^{i-1} \rho_k)(1 - \sum_{k=1}^{i} \rho_k)}$$

$$= \frac{\sum_{i \in I} \rho_i \frac{E(S_i)}{2}}{(1 - \sum_{k=1}^{i-1} \frac{\xi_k \lambda_k C_k}{f_k^r})(1 - \sum_{k=1}^{i} \frac{\xi_k \lambda_k C_k}{f_k^r})} \tag{10}$$

Therefore, the average computing latency for each type of task could be obtained as:

$$E(L_i) = (1 - \xi_i)D_{Local}(i) +$$
$$\xi_i(D_{Remote}(i) + T_{Data}(i) + E(D_{Queue}(i)) + T_{Result}(i)) \quad (11)$$

4.2 Offloading Policy

The optimization goal is minimizing the total computing latency of all computing units, including the edge offloading units and local computing units under the constraints of allowed maximum delay tolerant of each unit T_i, the optimization problem is mathematically modeled as:

$$\min_{\xi_i} \sum_{i \in I} \pi_i E(L_i) \quad (12)$$
$$\text{s.t.} E(L_i) \leq T_i, 0 \leq \xi_i \leq 1, i \in I$$

It is known from [8] that the optimization problem above is convex optimization. Thus, one can use the block coordinate descent (BCD) approach to deal with it as in the following iterative algorithm.

Algorithm 1 Proposed iterative algorithm based on BCD

1: Initiate : random choose $(\xi_i, i \in I)$
2: Repeat
3:    for $i \in I$
4:    update ξ_i with all ξ_j (for all $j \neq i$) fixed by
5:        $\xi_i = \xi_i + \nabla_i \sum_{i \in I} \pi_i E(L_i)$
6: Until $|\sum_{\xi_i, i \in I} \pi_i E(L_i) - \sum_{\xi_i, i \in I} \pi_i E(L_i)| \leq \varepsilon$, or maximum number of iterations is reached.
7: End Repeat
8: Return $(\xi_i, i \in I)$

---

5 Simulation and Result Analysis

In this section, we use MATLAB simulation to evaluate the performance of proposed edge computation offload scheme. The computation tasks of field devices are classified into four types with the probabilities $\pi : \{0.1, 0.3, 0.4, 0.2\}$.

The incoming computing flow of each type obeys poisson distribution of parameter λ. Other parameters are listed in Table 1. In addition, the computational capability $f_i^l = 2\,\text{GHz}$, $f_i^r = 10\,\text{GHz}$, $r_i = 20\,\text{Mbps}$, $K_i = 10^{-5}$. The link bandwidth bottleneck and transmission error are ignored.

Table 1. Parameters of various computation tasks

Parameter	Value	Unit
D	$\{0.2,\ 0.5,\ 3,\ 6\}$	Mbits
C	$\{10^8,\ 2*10^8,\ 3*10^8,\ 5*10^8\}$	Cycles
T	$\{0.05,\ 0.1,\ 0.2,\ 0.4\}$	Seconds

Figure 3 evaluates the percentage of various types of the tasks that are off-loaded under different computing task density λ in equal priority case. Because the tasks of higher priority are more sensitive to the delay constraints, the equal priority case cannot improve the probability of processing for high-priority tasks. We found that the offloading percentage of the type-1 and the type-4 tasks is nearly 0% and 100% in all the λ values, respectively. With the increase of task density, the offloading percentage decrease due to higher queuing latency at the edge server.

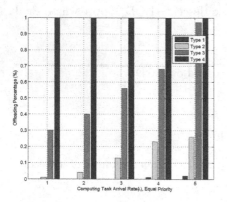

Fig. 3. EP offloading percentage   **Fig. 4.** NPP offloading percentage

Figure 4 shows the offload probability for each type of tasks under different computing task density λ in non-preemptive priority case. Compared with EP case, type-2 tasks will increase the offload probability slightly due to higher priority in high load region. Priority is given to high-priority tasks, which have less impact on the delay of subsequent tasks and easier to be flexibly chore-ographed.So Non-preemptive priority case can improve the service rate of high-priority tasks and further reduce average delay overall service.

Figure 5 depicts the average delay of proposed central offloading scheme, All-local computing and All-remote computing scheme in EP case. Figure 6 shows the probability of outage (The probability that the computing latency larger than the maximal allowed latency) of those schemes in EP case. The delay and outage probability prove the feasibility of proposed offloading scheme. With the growing

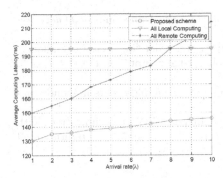

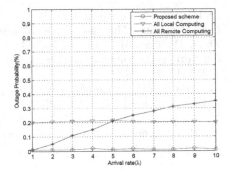

Fig. 5. Average delay   **Fig. 6.** Outage probability

of computing load, the performance of All-remote computing solution will grow worse due to the increasing queuing delay. The offloading scheme proposed in the paper can greatly reduce the computing latency and improve the computing QoS for different users.

6 Conclusion

In this paper, we propose a new SDIN architecture and a kind of centralized computing offloading strategies based on our SDIN architecture. The simulation results have indicated that the proposed SDIN architecture is feasible and effective in computing offloading. And to a certain extent, our architecture can provide traditional industries a better resource management solution and more flexible production scheme which means the production efficiency can possibly be improved.

References

1. Schweissguth, E., Danielis, P., Niemann, C., Timmermann, D.: Application-aware industrial ethernet based on an SDN-supported TDMA approach. In: 2016 IEEE World Conference on Factory Communication Systems (WFCS), Aveiro, Portugal (2016)
2. Aggarwal, C., Srivastava, K.: Securing IoT devices using SDN and edge computing. In: 2016 2nd International Conference on Next Generation Computing Technologies (NGCT), Dehradun, India (2016)
3. Sun, X., Ansari, N.: EdgeIoT: mobile edge computing for the internet of things. IEEE Commun. Mag. **54**(12), 22–29 (2016)
4. Dama, S., Pasca, T.V., Sathya, V.: A feasible cellular internet of things enabling edge computing and the IoT in dense futuristic cellular networks. IEEE Consum. Electron. Mag. **6**(1), 66–72 (2017)
5. Pengfei, H., Ning, H., Qiu, T.: Fog computing based face identification and resolution scheme in internet of things. IEEE Trans. Ind. Inf. **13**(4), 1910–1920 (2017)

6. Li, D., Zhou, M.-T., Zeng, P.: Green and reliable software defined industrial network. IEEE Commun. Mag. **54**(10), 30–37 (2016)
7. Zhao, P., Tian, H., Qin, C., Nie, G.: Energy-saving offloading by jointly allocating radio and computational resources for mobile edge computing. IEEE Access **5**, 11255–11268 (2017)
8. Miettinen, A.P., Nurminen, J.K.: Energy efficiency of mobile clients in cloud computing. HotCloud **10**, 4–4 (2010)
9. Li, M., Richard Yu, F., Si, P., Yao, H.: Energy-efficient M2M communications with mobile edge computing in virtualized cellular networks. In: 2017 IEEE International Conference on Communications (ICC), Paris, France (2017)

Multi-agent Deep Reinforcement Learning Based Adaptive User Association in Heterogeneous Networks

Weiwen Yi[✉], Xing Zhang, Wenbo Wang, and Jing Li

Wireless Signal Processing and Network Laboratory, Beijing University of Posts and Telecommunications, Beijing 100876, People's Republic of China
yww2013@bupt.edu.cn

Abstract. Nowadays, lots of technical challenges emerge focusing on user association in ever-increasingly complicated 5G heterogeneous networks. With distributed multiple attribute decision making (MADM) algorithm, users tend to maximize their utilities selfishly for lack of cooperation, leading to congestion. Therefore, it is efficient to apply artificial intelligence to deal with these emerging problems, which enables users to learn with incomplete environment information. In this paper, we propose an adaptive user association approach based on multi-agent deep reinforcement learning (RL), considering various user equipment types and femtocell access mechanisms. It aims to achieve a desirable trade-off between Quality of Experience (QoE) and load balancing. We formulate user association as a Markov Decision Process. And a deep RL approach, semi-distributed deep Q-network (DQN), is exploited to get the optimal strategy. Individual reward is defined as a function of transmission rate and base station load, which are adaptively balanced by a designed weight. Simulation results reveal that DQN with adaptive weight achieves the highest average reward compared with DQN with fixed weight and MADM, which indicates it obtains the best trade-off between QoE and load balancing. Compared with MADM, our approach improves by 4% ~ 11%, 32% ~ 40%, 99% in terms of QoE, load balancing and blocking probability, respectively. Furthermore, semi-distributed framework reduces computational complexity.

Keywords: Heterogeneous networks · User association · Multi-agent Deep Q-network

1 Introduction

In order to meet the demand of surging traffic, 5G heterogeneous networks (HetNets) have emerged as an essential solution, especially through the deployment of lower-power small cell base stations (BSs). Compared with traditional cellular networks, HetNets differ primarily in maximum transmit power, coverage

© ICST Institute for Computer Sciences, Social Informatics and Telecommunications Engineering 2019
Published by Springer Nature Switzerland AG 2019. All Rights Reserved
X. Liu et al. (Eds.): ChinaCom 2018, LNICST 262, pp. 57–67, 2019.
https://doi.org/10.1007/978-3-030-06161-6_6

area and spatial density. A survey demonstrates serious penetration losses of the buildings degrade quality of service (QoE) [1]. Hence, femtocells with different access mechanisms have been proposed, where subscribers of femtocells are the users registered in it and nonsubscribers are the users not registered in it [2].

- Closed access: Closed access femtocells only provide services for subscribers, which guarantee privacy and security.
- Hybrid access: Resources of hybrid access femtocells are reserved for subscribers, who may get higher rate than nonsubscribers.
- Open access: Open access femtocells are available to all users.

It is hard to cope with user association because of network heterogeneity and limited resources, which leads to user competitions and network congestion [3]. Due to incomplete information interactions and dynamic environment changes, emerging artificial intelligence method turns into an efficient tool for user association. A network-assisted approach was proposed with Q-learning to derive network information and satisfaction-based multi-criteria decision-making method was used to guide user behavior [4]. In [5], context-aware multiple radio access technology (multi-RAT) was studied. It made double decision on which exact RAT and access point to occupy with ant colony algorithm. However, complicated centralized algorithms have high requirements for the central controller's computational ability. In [3], the evolutionary game and Q-learning were implemented to help distributed individuals make decisions independently. It pursues high QoE without taking load balancing into consideration, which may bring about congestion. Moreover, users tend to maximize their utilities selfishly for lack of cooperation, such as distributed multiple attribute decision making (MADM), which results in the one-sidedness of user decisions [6]. The above related works didn't take into account QoE, load balancing and computational complexity simultaneously when dealing with user association. Therefore, one of the crucial goals for user association in HetNets is to achieve a desirable tradeoff between QoE and load balancing with an appropriate user association algorithm.

In [7], a deep RL method, termed a deep Q-network (DQN), was proposed. In complex and dynamic HetNets, users can learn optimal strategy from high-dimensional state and action space using DQN. In this paper, we propose an adaptive user association approach based on multi-agent DQN. The main contributions include:

- Our approach aims to obtain the desirable trade-off between QoE and load balancing. Considering user equipment (UE) types and femtocell access mechanisms, we exploit semi-distributed multi-agent DQN framework to achieve the optimal strategy. It can transfer the main calculations from central controller to UEs and reduce computational complexity.
- We formulate user association as a Markov decision process (MDP). And we define the individual reward as a weighted function of transmission rate and BS load. The weight is designed into the action. Such reward provides evaluative feedback for each user to make decision adaptively.

– Simulation results show that the proposed approach converges well and achieves the best trade-off between QoE and load balancing. It yields gains in terms of QoE and load balancing and significantly decreases the blocking probability compared with MADM.

2   System Model

We focus on the downlink (DL) transmission scenario of two-tier HetNet. The system model, including information sharing and distributed association scheme, is shown in Fig. 1. We consider a macrocell and N femtocells. The set of users is denoted as $\mathcal{U} = \{u|u = 1, 2, \ldots, K\}$. And the set of BSs is denoted as $\Phi = \{m|m = 0, 1, \ldots, N\}$, where macrocell is indexed by 0.

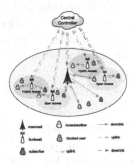

Fig. 1. System model.

The two-tier HetNet uses orthogonal spectrum with an assumption of co-tier interference [2,8]. Every femtocell is equipped with open access or hybrid access signed by 0 and 1 respectively. Therefore, the set of access mechanisms for BSs is $\mathcal{X} = \{0, 1\}$. Each BS consists of M sub-bands with bandwidth b, which are referred to time-frequency radio blocks (RBs). Hence the total bandwidth for BS is denoted as $W = Mb$. Besides, transmission power is uniformly allocated to each sub-band [8].

The spatial distribution of femtocells and users is modeled by homogeneous Poisson Point Process (PPP) with density λ_f and λ_u respectively [9]. Each user can be associated with one BS simultaneously. UE type includes registration attribute and service type. The registration attribute set is $\mathcal{A} = \{0, 1\}$, where subscribers are marked by 0 and nonsubscribers by 1. We consider two kinds of service types as $\mathcal{V} = \{0, 1\}$, where data traffic is indexed by 0 and voice calls by 1. The set of required RBs for different service types is denoted as $\mathcal{B} = \{\beta_s | s \in \mathcal{V}\}$. Therefore, the bandwidth that BS m allocates to each user u with service type s can be denoted as $\varphi_{m,u} = \eta(x, y)b\beta_s$, where $\eta(x, y) \in (0, 1]$. $\eta(x, y)$ is the match factor between registration attribute x and access mechanism y, and $x \in \mathcal{A}$ and $y \in \mathcal{X}$. If nonsubscribers associate with hybrid access femtocells, resources allocated to them will be reduced by $\eta(x, y) < 1$.

Load factor v_m is defined as the ratio of the allocated bands to the total bandwidth in Eq. (1), which indicates the BS load. $\mathcal{I}_m$ is the initial resource utilization of BS m. BS is overloaded when $v_m \geq 1$ and under-loaded when $v_m < 1$. When BS is overloaded, it will randomly block some users until it is under-loaded. Such users are regarded as blocked users, marked by set $\mathcal{O}$.

$$v_m = \frac{\mathcal{I}_m + \sum\limits_{u \in \mathcal{U}} \varphi_{m,u}}{W}. \tag{1}$$

The received signal-to-noise-plus-interference-ratio (SINR) is formulated as

$$\gamma_{m,u} = \frac{\frac{\varphi_{m,u}}{W} P_m g_{m,u} |x_{m,u}|^{-\alpha}}{\varphi_{m,u} N_0 + I_{\bar{\Phi}_u^f}}, \tag{2}$$

where $g_{m,u}$ is the exponentially distributed channel power with unit mean. $|x_{m,u}|$ indicates the distance from BS m to user u. P_m is the transmit power of BS m, α denotes the path loss exponent, and N_0 is regarded as the power spectral density of white Gaussian noise. The interference of user u is

$$I_{\bar{\Phi}_u^f} = \sum_{n \in \Phi_u^f \backslash m} \delta_n \frac{\varphi_{n,u}}{W} P_n g_{n,u} |x_{n,u}|^{-\alpha}. \tag{3}$$

When $I_{\bar{\Phi}_u^f} = 0$, the SINR degenerates into signal-to-noise-ratio (SNR). And the feasible BS set of user u is $\Phi_u^f = \{m | SNR_{m,u} \geq \gamma_{th}\}$, where γ_{th} is the SNR threshold. The interference probability of BS n detected by an arbitrary user is scaled by a thinning factor $\delta_n = \min\left(\frac{l_n}{W}, 1\right)$, where l_n is the resource utilization of BS n [9]. δ_n indicates that the interference probability is related with the sub-bands occupied. That is, if sub-bands are fully occupied, $\delta_n = 1$, and the interference from BS becomes larger than that of $\delta_n < 1$.

3 Adaptive User Association Based on Multi-agent DQN

In this section, we first formulate the problem as a MDP and elaborate the state, action and reward. Next, we review the basic conception of DQN adopted in this paper. Finally, we show the semi-distributed multi-agent DQN framework, then we get the optimal strategy using our proposed approach.

3.1 Problem Formulation

The BS environment consists of macrocell and femtocells in HetNet. In our proposed approach, users play the role of agents and interact with the BS environment. The parameters are defined as follows.

State. $\mathbf{s}_u$ indicates the state of agent (user) u with BS m selected, which is defined as $\mathbf{s}_u = (w_u, g_{m,u}, \varphi_{m,u}, v_m)$. $w_u \in \Omega$ is the weight of transmission rate discretized into F levels. $\Omega = \{\omega | \omega = 1\Delta, 2\Delta, \ldots, (F-1)\Delta\}$ is the set of weight and $\Delta = \frac{1}{F}$. And the state profile can be formulated as $\mathbf{s} = (\mathbf{s}_1, \mathbf{s}_2, \ldots, \mathbf{s}_K)$.

Action. Due to the indeterminacy of the weight, $w_u \in \Omega$ has been designed into the action. Current action of agent u can be denoted as $\mathbf{a}_u = (c_u, w_u)$, where $c_u \in \Phi_u^f$ and $w_u \in \Omega$. The action profile can be formulated as $\mathbf{a} = (\mathbf{a}_1, \mathbf{a}_2, \ldots, \mathbf{a}_K)$.

Reward. $R_u(\mathbf{s}_u, \mathbf{s}'_u, \mathbf{a}_u)$ indicates the feedback received when agent u takes the action $\mathbf{a}_u$ and turns out to be state $\mathbf{s}'_u$ from $\mathbf{s}_u$ [10]. The transmission rate of agent u refers to Shannon formula, which is formulated in Eq. (4).

$$U_u(\mathbf{s}_u, \mathbf{s}'_u, \mathbf{a}_u) = \begin{cases} \varphi_{m,u}(\mathbf{s}'_u)log(1 + \gamma_{m,u}(\mathbf{s}'_u)), u \notin \mathcal{O}(\mathbf{s}'_u) \\ 0, u \in \mathcal{O}(\mathbf{s}'_u) \end{cases}. \tag{4}$$

Conclusions as a result, we draw the following reward as a function of transmission rate and BS load as shown in Eq. (5).

$$R_u(\mathbf{s}_u, \mathbf{s}'_u, \mathbf{a}_u) = w_u(\mathbf{s}'_u) \frac{U_u(\mathbf{s}_u, \mathbf{s}'_u, \mathbf{a}_u)}{\sum_{u \in \mathcal{U}} U_u(\mathbf{s}_u, \mathbf{s}'_u, \mathbf{a}_u)} + (1 - w_u(\mathbf{s}'_u))(1 - v_m(\mathbf{s}'_u)). \tag{5}$$

There is a trade-off problem between transmission rate and BS load, which are balanced by the designed weight w_u. To seek high transmission rate, agent sets large w_u, which negatively affects BS load. Therefore, by such reward, each agent can discover actions in a more effective way, in order to contribute to the trade-off between QoE and load balancing.

3.2 Deep Q-Network

The main modification to online Q-learning in DQN module is to use a separate target network $\hat{\mathcal{Q}}_u$ with weight θ_u^- for generating the target action-value in learning update [7]. Evaluation network $\mathcal{Q}_u$ with weight θ_u is updated every step while $\hat{\mathcal{Q}}_u$ is assigned by θ_u every H step. DeepMind has proposed the DQN with the temporal-difference goal

$$y_u^t = R_u(\mathbf{s}_u, \mathbf{s}'_u, \mathbf{a}_u) + \tau \max_{\mathbf{a}'_u} \hat{\mathcal{Q}}_u(\mathbf{s}'_u, \mathbf{a}'_u; \theta_u^-), \tag{6}$$

where agent takes action $\mathbf{a}'_u$ in the next step. t indicates current training step and τ is a discounted factor. Therefore the update of θ_u can be formulated as

$$\theta_u^{t+1} = \theta_u^t + \rho\{y_u^t - \mathcal{Q}_u(\mathbf{s}_u, \mathbf{a}_u; \theta_u)\}\nabla \mathcal{Q}_u(\mathbf{s}_u, \mathbf{a}_u; \theta_u), \tag{7}$$

where ρ is the learning rate.

3.3 Proposed Algorithm

The proposed semi-distributed multi-agent DQN framework is illustrated in Fig. 2. This figure shows the interactions between agents and BS environment. After agents take actions, the information sharing scheme is executed. Then agents transform to next states, get the reward feedbacks and perform updates.

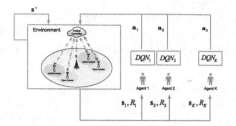

Fig. 2. Semi-distributed multi-agent DQN framework.

Algorithm 1 Multi-agent DQN Based Adaptive User Association

Initialize:
  τ, ρ, ε, K, $\mathbf{D}$ with capacity M for every agent, replace iter H, training steps T, initial state profile $\mathbf{s}$, $\mathbf{Q}$ with random weights $\boldsymbol{\theta}$, $\hat{\mathbf{Q}}$ with weights $\boldsymbol{\theta}^- = \boldsymbol{\theta}$

Output:
  Optimal strategy $\boldsymbol{\pi}_{opt}$

1: **for** $t = 1$ **to** T **do**
2: **for** $u = 1$ **to** K **do**
3:    Observe state $\mathbf{s}_u$
4:    **if** $rand() < \varepsilon$ **then**
5:      Select a random action $\mathbf{a}_u$
6:    **else**
7:      Select $\mathbf{a}_u = \arg\max_{\mathbf{a}_u} \mathcal{Q}_u(\mathbf{s}_u, \mathbf{a}_u; \theta_u)$

8:    **end if**
9:    Execute $\mathbf{a}_u$, share $\mathbf{\Gamma}_u^{UL}$ and acquire $\mathbf{\Gamma}_u^{DL}$
10:   Share U_u and acquire $\mathbf{U}_u^{DL}$
11:   Observe $\mathbf{s}_u'$ and acquire $R_u(\mathbf{s}_u, \mathbf{s}_u', \mathbf{a}_u)$
12:   Store transition $(\mathbf{s}_u, \mathbf{a}_u, \mathcal{R}_u, \mathbf{s}_u')$ in $\mathcal{D}_u$, then sample minibatch from $\mathcal{D}_u$
13:   Set y_u^t according to Eq. (6) and perform a gradient descent on $(y_u^t - \mathcal{Q}_u(\mathbf{s}_u, \mathbf{a}_u; \theta_u))^2$ with respect to θ_u according to Eq. (7)
14:   Set $\mathbf{s}_u = \mathbf{s}_u'$ and reset $\hat{\mathcal{Q}}_u = \mathcal{Q}_u$ every H step
15: **end for**
16: Decrease ε
17: **end for**
  Make a final optimal strategy $\boldsymbol{\pi}_{opt} = \mathbf{a}$

The pseudo-code of multi-agent DQN based adaptive user association algorithm is shown in Algorithm 1. $\mathbf{D} = (\mathcal{D}_u, u \in \mathcal{U})$ are the replay memories for users. $\mathbf{Q} = (\mathcal{Q}_u, u \in \mathcal{U})$ with weights $\boldsymbol{\theta} = (\theta_u, u \in \mathcal{U})$ are evaluation networks for users. And target networks for users are $\hat{\mathbf{Q}} = (\hat{\mathcal{Q}}_u, u \in \mathcal{U})$ with weights $\boldsymbol{\theta}^- = (\theta_u^-, u \in \mathcal{U})$.

At decision epochs, after current state $\mathbf{s}_u$ observed, every agent takes action $\mathbf{a}_u$, by exploration or exploitation (Line 3–8). In exploration mode, agent takes action randomly with probability ε (Line 4–5). However, in exploitation mode, agent takes action by maximum Q-value based on the previously learned $\mathcal{Q}_u$ (Line 6–7). Once agents take actions, they share $\mathbf{\Gamma}_u^{UL} = (c_u, \varphi_{m,u})$ on the UL and acquire others' information $\mathbf{\Gamma}_u^{DL} = (\mathbf{\Gamma}_i^{UL}, i \in \bar{\mathcal{U}}_u)$ on the DL, where $\bar{\mathcal{U}}_u$ indicates users except for current agent u (Line 9). Next, agents share U_u on the UL and acquire others' transmission rates $\mathbf{U}_u^{DL} = (U_i, i \in \bar{\mathcal{U}}_u)$ on the DL (Line 10). After that, agent u transforms to next state $\mathbf{s}_u'$ and gets evaluation feedback $R_u(\mathbf{s}_u, \mathbf{s}_u', \mathbf{a}_u)$ to drive the next more correct decision (Line 11).

By experience replay, we store the agent experiences, $(\mathbf{s}_u, \mathbf{a}_u, \mathcal{R}_u, \mathbf{s}_u')$ transition, into memory $\mathcal{D}_u$ with finite capacity M. If the memory buffer of $\mathcal{D}_u$ is full, we overwrite with recent transitions. Next, with full replay memory, sample uniformly minibatch from $\mathcal{D}_u$ (Line 12). Then, with temporal-difference goal y_u^t, perform a gradient descent step on evaluation network $\mathcal{Q}_u$ by RMSProp algorithm (Line 13). It's important to copy $\mathcal{Q}_u$ to target network $\hat{\mathcal{Q}}_u$ every H step. $\hat{\mathcal{Q}}_u$ is used for calculating y_u^t for the following H steps (Line 14). The policy during training is ε-greedy with ε annealed linearly. ε decreases with training steps until there is no exploration process (Line 16). Finally, after each agent repeats the above procedures T times, we get the optimal strategy π_{opt} for all users.

4   Performance Evaluation

Simulation results are presented in this section. The details of parameter setting are shown in Table 1. The access mechanisms of femtocells, registration attributes and service types of users are assigned randomly. If x=1 and y=1, match factor $\eta(x, y) = 0.6$, otherwise $\eta(x, y) - 1$. We consider MADM as baseline approach. Its utility function is formulated in Eq. (8) with fixed weight and users take actions by maximum $R_{u,m}$.

$$R_{u,m} = w_u \frac{U_{u,m}}{\sum\limits_{j \in \Phi_u^f} U_{u,j}} + (1 - w_u)(1 - v_m). \tag{8}$$

All results are averaged with P Monto Carlo simulation epochs and evaluated by four metrics. They are average reward, average transmission rate, standard deviation of resource utilization rate and blocking probability, respectively. And we consider fixed weights, $w_{1,u} = 0.2, 0.5, 0.8$, in order to investigate the effects of adaptive weight.

Figure 3a plots the convergency under user density $\lambda_u = 6 \times 10^{-6}$, 9×10^{-6} and 1.3×10^{-5}. It shows the average reward varying with the training steps. Fluctuation of average reward indicates that the exploration probability ε works. When ε decreases with training steps, reward tends to rise first and then converges to a relative stable value within a certain range. It suggests that our proposed approach converges well.

Figure 3b plots average reward varying against user density. The trade-off performance is evaluated by average reward. When user density increases, average reward decreases because of higher blocking probability. As seen in this figure, the proposed approach achieves the highest average reward compared with any other approach, which indicates it obtains the desirable tradeoff between

Table 1. Parameter setup.

Parameter	Value
Area radius	500 m
Bandwidth W	20 MHz
Transmit power of two-tier HetNet	{46, 20} dBm
Power spectral density of white Gaussian noise N_0	−174 dBm
Path loss α	4
Femtocell density λ_f	4×10^{-6}
Initial resource utilization of network	Uniform distribution
Location of N femtocells	PPP
Location of K users	PPP
Sub-band bandwidth b	180 kHz
Required RBs $\mathcal{B}$	$\mathcal{B} = \{10, 20\}$
Weight discretized level F	5
Monte Carlo simulation epochs P	300
Training steps T	20 K
SNR threshold γ_{th}	9.56 dB
Discounted factor τ	0.9
Learning rate ρ	0.05
Replace iter H	200
Exploration probability ε	0.2
Capacity M of replay memory $\mathcal{D}_u$	2000

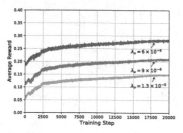

(a) Average reward vs. training step under DQN with adaptive weight.

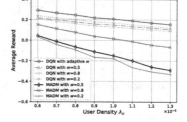

(b) Average reward vs. user density under different approaches.

Fig. 3. Average reward.

QoE and load balancing. DQN approaches have better performance than MADM approaches. It shows that by information sharing and learning, users make better decisions. MADM gets optimal strategy according to current network situation without cooperation, which leads users to simultaneously select low-load BSs that can provide high transmission rate.

Figure 4 shows the comparison of QoE and load balancing. Figure 4a plots average transmission rate against user density, which reflects QoE of users. Figure 4b investigates the standard deviation of resource utilization rate among BSs, which reflects load balancing of network. In Fig. 4a, QoE decreases when user density increases. It is due to limited resources BSs can offer and larger interference probability from other BSs. This figure shows that DQN with $w = 0.8$ gains the best QoE because of large weight of transmission rate. Our approach gains the second best QoE, by $4 \sim 11\%$ improvement than MADM. In Fig. 4b, as the user density rises, the standard deviation decreases among DQN approaches while increases slowly among MADM approaches. Lower standard deviation represents better load balancing. DQN with adaptive weight outperforms MADM approaches from the perspective of load balancing by $32 \sim 40\%$ improvement.

Comparing Fig. 4a with Fig. 4b, for DQN with fixed weight, weight can control the optimal strategy to focus more on QoE or load balancing. It can be seen that DQN with $w = 0.2$ gets the worst QoE in Fig. 4a. However, In Fig. 4b, DQN with $w = 0.2$ has the lowest standard deviation, which suggests that it performs well in load balancing because of large weight of the load. For DQN with $w = 0.8$, we observe that seeking high QoE has a negative impact on load balancing. Thus, we can infer that DQN with adaptive weight intelligently selects the appropriate weight and gets a desirable trade-off strategy. Moreover, the QoE of MADM with $w = 0.2$ decreases with user density more slowly than MADM with $w = 0.5$ and $w = 0.8$. And for DQN approaches, the gap of QoE is decreasing with user density. It shows that we urgently need to consider load balancing in the case of high user density, in order to maintain the QoE level.

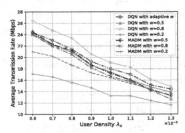

(a) Average transmission rate vs. user density.

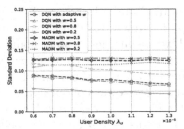

(b) Standard deviation of resource utilization rate among B-Ss vs. user density.

Fig. 4. Comparison of QoE and load balancing under different approaches.

In Fig. 5, the ordinate axis is logarithmic. As user density increases, blocking probability rises because BSs with limited resources could not accept more requests from users. MADM approaches get the worse blocking probability owing to its decision way, while DQN approaches improve by 99% compared with MADM.

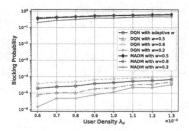

Fig. 5. Blocking probability vs. user density under different approaches.

The computational complexity of our approach depends on the number of state, action of each UE and the amount of information sharing, while by centralized algorithm it depends on the number of the cartesian product of state and action among users. It offloads the main calculations to UEs, which reduces computational complexity.

5 Conclusion

In this paper, we have studied user association problem in HetNets, considering femtocell access mechanisms and UE types. We have proposed multi-agent DQN based adaptive user association approach, aiming to jointly solve the trade-off problem from the perspective of QoE and load balancing. We formulate the problem as a MDP and adopt semi-distributed multi-agent DQN to get the optimal strategy. The reward is defined as a weighted function of transmission rate and BS load, which enables users to maintain QoE and contribute to load balancing. Therefore, by our approach, users can set their weights adaptively and select BSs intelligently to obtain the desirable trade-off strategy. Simulation results verify that the average reward of our approach outperforms DQN with fixed weight and MADM, which indicates it obtains the best trade-off between QoE and load balancing. In terms of QoE, load balancing and blocking probability, our approach improves by 4% ∼ 11%, 32% ∼ 40%, 99% respectively, compared with MADM. This is because our approach addresses user association adaptively by cooperation. The computational complexity depends on the number of state, action of each UE and the amount of information sharing. It is a relatively significantly improvement over centralized algorithms.

Acknowledgements. This work is supported by the National Science Foundation of China (NSFC) under grant 61771065, 61571054 and 61631005.

References

1. Chandrasekhar, V., Andrews, J.G., Gatherer, A.: Femtocell networks: a survey. IEEE Commun. Mag. **46**(9), 59–67 (2008)
2. De La Roche, G., Valcarce, A., López-Pérez, D., Zhang, J.: Access control mechanisms for femtocells. IEEE Commun. Mag. **48**(1), 33–39 (2010)
3. Feng, Z., Song, L., Han, Z., Zhao, X., et al.: Cell selection in two-tier femtocell networks with open/closed access using evolutionary game. In: Wireless Communications and Networking Conference (WCNC), pp. 860–865. IEEE (2013)
4. El Helou, M., Ibrahim, M., Lahoud, S., Khawam, K., Mezher, D., Cousin, B.: A network-assisted approach for rat selection in heterogeneous cellular networks. IEEE J. Sel. Areas Commun. **33**(6), 1055–1067 (2015)
5. Li, J., Zhang, X., Wang, S., Wang, W.: Context-aware multi-rat connection with bi-level decision in 5g heterogeneous networks. In: 2017 IEEE/CIC International Conference on Communications in China (ICCC), pp. 1–6 (2017). https://doi.org/10.1109/ICCChina.2017.8330398
6. Wang, L., Kuo, G.S.G.: Mathematical modeling for network selection in heterogeneous wireless networks - a tutorial. IEEE Commun. Surv. Tutor. **15**(1), 271–292 (2013)
7. Mnih, V., Kavukcuoglu, K., Silver, D., Rusu, A.A., Veness, J., Bellemare, M.G., Graves, A., Riedmiller, M., Fidjeland, A.K., Ostrovski, G., et al.: Human-level control through deep reinforcement learning. Nature **518**(7540), 529 (2015)
8. Yan, M., Feng, G., Qin, S.: Multi-rat access based on multi-agent reinforcement learning. In: GLOBECOM 2017–2017 IEEE Global Communications Conference, pp. 1–6. IEEE (2017)
9. Chae, S.H., Hong, J.P., Choi, W.: Optimal access in ofdma multi-rat cellular networks with stochastic geometry: can a single rat be better? IEEE Trans. Wirel. Commun. **15**(7), 4778–4789 (2016)
10. Liu, Y.J., Cheng, S.M., Hsueh, Y.L.: enb selection for machine type communications using reinforcement learning based markov decision process. IEEE Trans. Veh. Technol. **66**(12), 11330–11338 (2017)

A Novel Double Modulation Technique with High Spectrum Efficiency for TDCS

Bo Zheng[1,2(✉)], Heng-Yang Zhang[1], Le Sun[3], Hua-Xin Wu[4],
and Wei-Lun Liu[1]

[1] Information and Navigation Institute, Air Force Engineering University, Xi'an
710077, China
zbkgd@163.com
[2] College of Electronics and Information, Northwestern Polytechnical
University, Xi'an 710129, China
[3] PLA Unit of 94719, Ji'an 343706, China
[4] Radar NCO School, Air Force Early Warning Academy, Wuhan 430019,
China

Abstract. The modulation techniques in traditional transform domain communication system (TDCS) exist some drawbacks, such as low transmission rate and low spectrum efficiency. We propose a novel double modulation technique with high spectrum efficiency for TDCS in this paper. First, we divide the basis function averagely into several orthogonal modules, and conduct the CSK modulation. Then, the double modulation signal waveform can be obtained by employing bipolar modulation for different module combination. Furthermore, we propose two demodulation schemes for the proposed modulation technique, namely the cyclic shift keying (CSK)-bipolar and bipolar-CSK demodulation. We also derive the mathematical expressions of their bit error rate (BER) performance. Simulation results show that for different signal-to-noise ratio (SNR), the two demodulation schemes can both achieve reliable performance, satisfactory anti-interference capabilities and effectively improve spectrum efficiency. In addition, it can be verified that CSK-bipolar demodulation can achieve the same BER with less SNR compared with bipolar-CSK demodulation.

Keywords: TDCS · CSK · Bipolar modulation · Demodulation
Spectrum efficiency

1 Introduction

Transform domain communication system (TDCS) was proposed by the U.S. Air Force Institute of Technology (AFIT) in 1990s. With its unique anti-interference theory, low probability intercept (LPI) and low probability detection (LPD) performance, TDCS has attracted widespread attention in many fields, such as in aeronautical communications and satellite communications [1, 2], etc. A huge number of researches have already been carried out in this field. For instance, the flexible spectrum access and multiple access of TDCS are analyzed in [3, 4] and [5], providing theoretical bases for applications. For the problem that the peak-to-average ratio of the basis function is

X. Liu et al. (Eds.): ChinaCom 2018, LNICST 262, pp. 68–81, 2019.
https://doi.org/10.1007/978-3-030-06161-6_7

large, schemes are proposed for improvement in [6] and [7], promoting the system LPI and LPD effectively. The accurate receiving strategy in TDCS which increases the bit error rate (BER) performance and reduces the system complexity is studied in [8].

With the development of research, traditional modulation and demodulation techniques with low spectrum efficiency can no longer meet the real-time requirement for information transmission in modern communication systems. Thus, it is increasingly urgent to design an efficient and reliable modulation technique. In TDCS, the basis function is used as the modulating waveform [9]. At present, the main modulation techniques in TDCS are bipolar modulation and cyclic shift keying (CSK). As a simple modulation scheme, bipolar modulation flips the basis function, and employs different code elements to represent positive and negative energy respectively. The technique has a simple demodulation procedure and good BER performance. However, in bipolar modulation every sending waveform can only transmit a bit of binary information, leading to extremely low transmission efficiency. CSK is developed from cyclic code shift keying (CCSK) [10]. Its sending waveform set is produced by different shifts of the basis function. It improves low spectrum efficiency to a certain extent, yet its information bits are exponential to demodulation complexity. Demodulation efficiency is getting lower with the increase in the amount of information.

For problems existing in bipolar modulation and CSK, some studies on modulation technique with high spectrum efficiency for TDCS have emerged recently. In [11], a sending waveform set including more waveforms is acquired through permutation and combination of the waveforms in the orthonormal waveform set of CSK. Although the spectrum efficiency is improved, it is difficult to be used in engineering due to huge demodulation cost. In [12], a modulation technique based on cluster is presented. After spectrum sensing, the entire unoccupied spectrum are averagely divided into several clusters, and orthogonal modulating waveforms are generated. The technique also improves spectrum efficiency, yet the allocation principle of random allocation modulation scheme is not described in detail, which makes it less applicable. The above techniques only detect maximum correlation value of real part in received waveform, and discard the imaginary part directly in demodulation. For this problem, a joint modulation method of real and imaginary part of the modulating waveform is proposed in [13]. Spectrum efficiency is doubled by the method, but it reduces the orthogonality of waveforms and leads to BER increase.

Due to the drawbacks exist in the above modulation and demodulation schemes, we are motivated to design a novel double modulation technique with high spectrum efficiency for TDCS. The double modulation technique includes two stages, namely the modular CSK and modular bipolar sequently. Furthermore, two demodulation methods are presented, and performance for the methods is simulated and analyzed under different signal-to-noise ratio (SNR) and interference-to-noise ratio (INR). Results verify the effectiveness and reliability of the modulation technique on information transmission with high spectrum efficiency. The contribution of the paper can be summarized as follows. On one hand, a novel double modulation technique with high spectrum efficiency for TDCS is proposed, and the principle and process of the technique is described in detail. On the other hand, two demodulation schemes for the proposed modulation technique are presented, and their BER performance is analyzed respectively.

The remaining of the paper is organized as the following. In Sect. 2, we give a review on the principle of TDCS. In Sect. 3, the double modulation technique is described in detail, and its spectrum efficiency is analyzed. In Sect. 4, we provide two corresponding demodulation methods. Simulations are performed to verify the effectiveness of the proposed method in Sect. 5. Finally, we conclude the paper in Sect. 6.

2 Review on TDCS

TDCS is a broadband communication system. Its principle can be summarized as: the interference spectrum is eliminated in the transform domain, and a noise-like basis function is generated and used to modulate the information bits in order to achieve the goals of anti-interference, LPI and LPD. The main principle of TDCS is shown in Fig. 1.

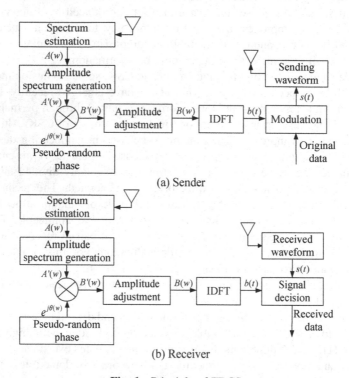

Fig. 1. Principle of TDCS.

A basis function is employed to modulate information in TDCS. When subcarrier number is N, the discrete basis function in time domain can be expressed as

$$b(n) = \frac{1}{N} \sum_{k=1}^{N-1} CA_k \, e^{j\theta_k}, \, e^{j2\pi kn/N} \tag{1}$$

Where C is the adjusting factor of amplitude, A_k is the amplitude spectrum vector, $e^{j\theta_k}$ is the pseudo-random phase, and $e^{j2\pi kn/N}$ is the coefficient of inverse discrete Fourier transform (IDFT). The basis function in frequency domain is derived through the tagged amplitudes in frequency domain mapping to random phases distributed averagely on $[0, 2\pi]$. And the basis function in time domain is the inverse transform of that in frequency domain. Thus, it can be regarded as a noise-like sequence with N points, and has a good correlation performance. Its correlation function can be expressed as

$$
\begin{aligned}
R(m) &= \sum_{m=-(N-1)}^{N-1} b(n+m)b^*(n) \\
&= \sum_{m=-(N-1)}^{N-1} \frac{C^2}{N^2} \sum_{k=1}^{N-1} A_k^2 e^{j(\theta_k - \theta_k')} e^{j\left(\frac{2\pi kn}{N} - \frac{2\pi k(n+m)}{N}\right)}
\end{aligned}
\tag{2}
$$

When $m = 0$, the auto-correlation function reaches the maximum value. When $N = 512$, the correlation performance of the basis function is shown in Fig. 2.

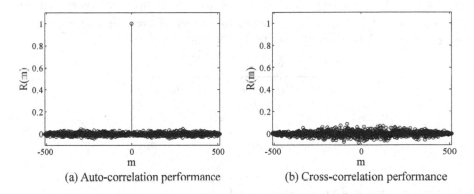

(a) Auto-correlation performance   (b) Cross-correlation performance

Fig. 2. Correlation performance of the basis function in TDCS.

3 Double Modulation Technique

3.1 Modular CSK

Since the basis function has a good correlation property, its waveforms through different time shifts have strict orthogonality. M-CSK is the cyclic shift of the basis function with the same step-length, and can be represented as

$$
\begin{aligned}
s_i(n) &= b\left(n - \frac{(i-1)T}{M}\right)_T , \quad i \in (1, M) \\
&= \frac{1}{N} \sum_{k=1}^{N-1} CA_k e^{j\theta_k} e^{j2\pi kn/N} e^{-\frac{jnS_i k}{M}}
\end{aligned}
\tag{3}
$$

From another point of view, the basis function with length N can be divided averagely into M modules before modulation, denoted as b_1, b_2,..., b_M respectively. The length of every module is N/M. The same waveform set as that in M-CSK can be derived by the cyclic shifts of the M modules in sequence. At this time every waveform can represent k bits information ($M = 2^k$). The process of the modular CSK is shown in Fig. 3.

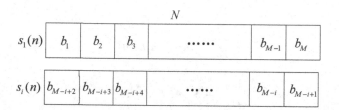

Fig. 3. Process of the modular CSK.

After the cyclic shift, the waveforms are mutually orthogonal in the waveform set. If the energy of the sending waveform is $\sqrt{\varepsilon}$, the energy distance between different waveforms is $\sqrt{2\varepsilon}$. Let ε_1, ε_2,..., ε_M denote the energy of every module respectively, and the set of modular CSK waveforms can be represented as

$$
\begin{aligned}
s_1 &= \overbrace{(\sqrt{\varepsilon},0,\cdots,0)}^{N} = \overbrace{(\varepsilon_1,\varepsilon_2,\cdots,\varepsilon_M)}^{N} \\
s_2 &= (0,\sqrt{\varepsilon},\cdots,0) = (\varepsilon_M,\varepsilon_1,\cdots,\varepsilon_{M-1}) \\
&\quad \vdots \qquad\qquad\qquad \vdots \\
s_M &= (0,\cdots,0,\sqrt{\varepsilon}) = (\varepsilon_2,\cdots,\varepsilon_M,\varepsilon_1)
\end{aligned}
\tag{4}
$$

3.2 Modular Bipolar Modulation

From (4), in the waveform set derived from the modular CSK, modules corresponding to any two waveforms are all mutually orthogonal. Hence, we can take full use of the orthogonality to achieve the second modulation. Every successive ξ modules in M modules of the CSK waveform are recombined to derive γ new modules. Thus, $\gamma = M/\xi$. Let $s_{i1}(n)$, $s_{i2}(n)$,..., $s_{i\gamma}(n)$ denote the new modules respectively. The length of every new module is $N\xi/M$, and any two modules are mutually orthogonal. When $\xi = 2$, the new modules are shown in Fig. 4.

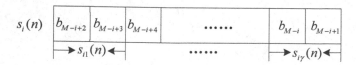

Fig.4 Combination of modules when $\xi = 2$.

We modulate every new module with bipolar modulation, and denote different information bits by its positive and negative energy, namely,

$$s_{ij}(n) = \begin{cases} s_{ij}(n), & \text{if information is } 0 \\ -s_{ij}(n), & \text{if information is } 1 \end{cases}, \qquad (5)$$

Through the modular bipolar modulation, the sending waveform set can be expressed as

$$
\begin{aligned}
s_1 &= \overbrace{((-1)^{s^1}(\varepsilon_1+\varepsilon_2), (-1)^{s^2}(\varepsilon_3+\varepsilon_4), \cdots, (-1)^{s^j}(\varepsilon_{M-1}+\varepsilon_M))}^{N} \\
s_2 &= ((-1)^{s^1}(\varepsilon_M+\varepsilon_1), (-1)^{s^2}(\varepsilon_2+\varepsilon_3), \cdots, (-1)^{s^j}(\varepsilon_{M-2}+\varepsilon_{M-1})) \\
&\ \ \vdots \\
s_M &= ((-1)^{s^1}(\varepsilon_2+\varepsilon_3), \cdots, (-1)^{s^{j-1}}(\varepsilon_{M-2}+\varepsilon_{M-1}), (-1)^{s^j}(\varepsilon_M+\varepsilon_1))
\end{aligned}, \qquad (6)
$$

Where s^1, s^2,..., s^j denote the information bit 0 or 1 after the modular bipolar modulation of the sending waveform. 0 represents the positive energy of the module, and 1 represents the negative energy of the module. $\varepsilon_1+\varepsilon_2$, $\varepsilon_2+\varepsilon_3$,... denote the energy of every module through recombination, and its energy is $1/\gamma$ of the waveform energy.

3.3 Spectrum Efficiency

For the modular CSK, every waveform can denote k bits information. For the communication system with symbol rate R_s, the bit transmission rate is

$$R_b = R_s \log_2 M = \frac{\log_2 M}{T_s}, \qquad (7)$$

Where T_s is the symbol period. Then, γ bits information is modulated with bipolar modulation with γ modules. After modulation, the bit transmission rate can be expressed as

$$R_b = R_s(\log_2 M + \gamma) = \frac{\log_2 M + \gamma}{T_s}, \qquad (8)$$

The symbol period is $T_s = 1/\Delta f$, and K_{used} is the number of available subcarrier. Thus, the signal bandwidth is

$$W_{used} = K_{used} \cdot \Delta f, \qquad (9)$$

According to the definition in 9, the spectrum efficiency of the modulation in this paper can be expressed as

$$\eta = \frac{R_b}{W_{used}} = \frac{R_s \cdot (\log_2 M + \gamma)}{K_{used} \cdot \Delta f} = \frac{\log_2 M + \gamma}{K_{used}}, \tag{10}$$

Therefore, compared with CSK, the spectrum efficiency has been improved $\frac{\gamma}{K_{used}}$.

4   Double Modulation Technique

Through two times modulations, the shift property and local turnover property of the modulating waveform have both been changed. For the demodulation of the sending waveform, both of the properties have interacted with each other. The turnover property is based on the shift property, and the shift property can be extracted simultaneously when the local turnover is exact. In this section, we proposed two demodulation schemes, and analyze their BER performance.

4.1   Csk-Bipolar Demodulation

Double modulating waveform is the combination of modular cyclic shift of the basis function and bipolar modulation. Firstly the order of the cyclic shift is demodulated, and then the sending information is recovered through modular bipolar demodulation. The demodulation flow is shown in Fig. 5.

For the sending waveform s_i, the demodulation model of the modular CSK order for the received waveform r can be expressed as

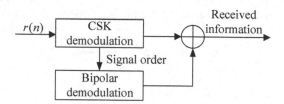

Fig. 5. Flow of CSK-bipolar demodulation.

$$a = \max \left\{ r \cdot \begin{bmatrix} s_{1,1}^* & s_{1,2}^* & \cdots & s_{1,2^\gamma}^* \\ s_{2,1}^* & \ddots & & s_{2,2^\gamma}^* \\ \vdots & & \ddots & \vdots \\ s_{M,1}^* & s_{M,2}^* & \cdots & s_{M,2^\gamma}^* \end{bmatrix} \right\}, \tag{11}$$

Where a is the order when the correlation demodulation of every waveform in the waveform set is maximum. The row number of the matrix represents the dimension M of the modular CSK orthogonal waveform set, and the column number of the matrix represents 2^γ kinds of waveform when the energy of γ modules are positive or negative values respectively in the same modulating waveform. When the received waveform is demodulated using (11), every waveform needs $M \cdot 2^\gamma$ times waveform correlation operations. The computing costs too much for engineering applications.

From (6), we can observe that modular bipolar modulation of the sending waveform does not change the orthogonality between each module. Thus, in the modular CSK demodulation of signal, positive or negative state of every module can be ignored. Thereby, the demodulation of the sending waveform can be simplified to a pure CSK demodulation. Assuming the sending signal is s_1, the received signal can be expressed as

$$r = (\sqrt{\varepsilon} + n_1, n_2, \cdots, n_M), \tag{12}$$

Where $n_1, n_2, \ldots, n_M$ are Gaussian white noise with mean value 0 and variance $\frac{N_0}{2}$. (11) can be simplified as in [14]

$$a = \max(r \cdot s_i^*) = \max \left\{ r \cdot \begin{bmatrix} \sqrt{\varepsilon}, 0, \cdots, 0 \\ 0, \sqrt{\varepsilon}, \cdots, 0 \\ \vdots \\ 0, \cdots, 0, \sqrt{\varepsilon} \end{bmatrix} \right\},$$

$$= \sum \max \left\{ \begin{array}{c} \varepsilon + \sqrt{\varepsilon} n_1 \\ \cdots \\ \sqrt{\varepsilon} \cdot n_M \end{array} \right\} \tag{13}$$

Let $z_i = r \cdot s_i^*$, and the probability that the waveform is received correctly can be expressed as in [14]

$$
\begin{aligned}
P_a &= P(z_1 > z_2, z_1 > z_3, \cdots, z_1 > z_M | s_1 send) \\
&= P(\sqrt{\varepsilon} + n_1 > n_2, \sqrt{\varepsilon} + n_1 > n_3, \cdots, \sqrt{\varepsilon} + n_1 > n_M | s_1 send) \\
&= P(\sqrt{\varepsilon} + n > n_2, \sqrt{\varepsilon} + n > n_3, \cdots, \sqrt{\varepsilon} + n > n_M | s_1 send, n_1 = n), \\
&= \int_{-\infty}^{\infty} (P(\sqrt{\varepsilon} + n > n_2 | s_1 send, n_1 = n))^{M-1} \cdot p_{n_1}(n) dn
\end{aligned} \tag{14}
$$

Where $P(\sqrt{\varepsilon} + n > n_2 | s_1 send, n_1 = n) = 1 - Q(\frac{n + \sqrt{\varepsilon}}{\sqrt{N_0/2}})$, and $p_{n_1}(n) = \frac{1}{\sqrt{\pi N_0}} e^{-\frac{n^2}{N_0}}$.

Assuming through the modular bipolar modulation the probability of information bit 0 is equal to that of 1, the decision threshold in bipolar demodulation can be set to 0. The probability of a correct decision in a module can be expressed as

$$P_b = 1 - Q\left(\sqrt{2\sqrt{\varepsilon}/\gamma N_0}\right), \tag{15}$$

Therefore, the probability that the sending waveform is correctly received absolutely is

$$P = \frac{k}{k+\gamma}P_a + \frac{\gamma}{k+\gamma}P_a \cdot (P_b)^{\gamma},\tag{16}$$

4.2 Bipolar-CSK Demodulation

For the sending waveform s_i, if every module s_{ij} can be received correctly, s_i will also inevitably be received correctly. Therefore, firstly the γ modules in the received waveform can be bipolar-based demodulated respectively, and count the modulation order of every module simultaneously. The maximum order in statistic result will be regarded as modular CSK order. The demodulation flow is shown in Fig. 6.

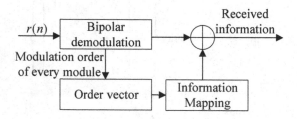

Fig. 6. Flow of bipolar-CSK demodulation.

In the modular bipolar demodulation of any received waveform r, every module is demodulated by the correlation demodulation of the corresponding module in M waveforms. Let r_1 denote the first module of the received waveform, and the demodulation model can be expressed as

$$[a,b] = \max\left\{ r_1 \cdot \begin{bmatrix} s_{1,1}^* & -s_{1,1}^* \\ s_{2,1}^* & -s_{2,1}^* \\ \vdots & \vdots \\ s_{M,1}^* & -s_{M,1}^* \end{bmatrix} \right\},\tag{17}$$

Where $a \in (1, 2, \cdots, M)$ and $b \in (1, 2)$ represent the maximum dimension number and column number of the demodulation matrix, respectively. $s_{i,1}^*$ is the conjugate of the first module in the sending waveform set. As only the maximum real parts of the correlation receiver are detected in demodulation, and the real parts of $r_1 s_{i1}^*$ and $-r_1 s_{i1}^*$ are the opposite of each other, (17) can be expressed as

$$a = \max\left[\left[\left|r_1s_{1,1}^*\right|\quad\left|r_1s_{2,1}^*\right|\quad\cdots\quad\left|r_1s_{M,1}^*\right|\right]^T\bigcap b\right.$$
$$= \max\left[(r_1s_{1,1}^*) > (-r_1s_{1,1}^*)\right] \tag{18}$$

Only when a and b are both solved correctly, the first module of the received waveform will be demodulated correctly. From (6), we can get that $s_{1,1}^*, s_{2,1}^*, \ldots, s_{M,1}^*$ are M orthogonal waveform vectors. Therefore, the model for solving a can be changed to

$$a = \max\left[r_1 \cdot \begin{pmatrix} \sqrt{\varepsilon/\gamma}, 0, \cdots, 0 \\ 0, \sqrt{\varepsilon/\gamma}, \cdots, 0 \\ \vdots \\ 0, \cdots, 0, \sqrt{\varepsilon/\gamma} \end{pmatrix}\right], \tag{19}$$

Where $\sqrt{\varepsilon/\gamma}$ is the energy of every module. Thus, the probability that a is judged correctly is

$$P_a = \int_{-\infty}^{\infty} \frac{1}{\sqrt{\pi N_0}} \left[1 - Q\left(\frac{\gamma n + \sqrt{\varepsilon}}{\gamma\sqrt{N_0/2}}\right)\right]^{M-1} e^{-\frac{n^2}{N_0}} dn, \tag{20}$$

Assuming through the modular bipolar modulation the probability of information bit 0 is equal to that of 1, the decision threshold for solving b can be set to 0. The probability for correct decision is

$$P_b = 1 - Q\left(\sqrt{2\sqrt{\varepsilon}/\gamma N_0}\right), \tag{21}$$

The demodulation of every module in the same received waveform is independent. The probability for correct decision on the bipolar modulation of any module is

$$P_{one} = P_a P_b, \tag{22}$$

For the sending waveform s_1, the order decision result of any module is $a \in (1, 2, \cdots, M)$. For a single module, the probability of correct decision on orders is P_a. The possibility number of error decision is $M - 1$, and thus the probability that the order is misjudged as i can be expressed as $\frac{1-P_a}{M-1}$. The decisions on the CSK order are statistics of the decision on each module order, and the decisions of different modules are independent and have equal probability. When the correct decision probability P_a of a single module order is larger than any error decision probability $\frac{1-P_a}{M-1}$, the decision on CSK order is correct, namely

$$P_{csk} = P(P_a > \frac{1-P_a}{M-1}), \tag{23}$$

In summary, the probability that the sending waveform is received completely correctly is

$$P = \frac{k}{k+\gamma} P_{csk} + \frac{\gamma}{k+\gamma} (P_a P_b)^{\gamma}, \tag{24}$$

5 Simulations

In simulations, the system bandwidth is 10MHz, the subcarrier number is 512, and the amount of information is 10^8 bits.

5.1 BER

We denote the CSK with $k = 4$ as CSK-4, where k represents the admissible number of information bits in CSK, and $M = 2^k$. The CSK-bipolar demodulation with $k = 4$ and $\gamma = 2$, and the bipolar-CSK demodulation with $k = 4$ and $\gamma = 2$, are represented by CSK-bipolar-4-2 and bipolar-CSK-4-2, respectively. When k is 4, 5, 6, γ is 2, 4, 8, and

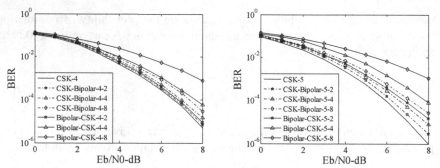

(a) BER performance when $k = 4$ and different values of γ (b) BER performance when $k = 5$ and different values of γ

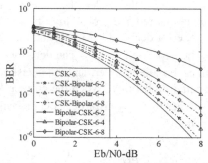

(c) BER performance when $k = 5$ and different values of γ

Fig. 7. BER under different SNR.

the jamming-to-signal power ratio (JSR) is 5 dB, the BER performance under different SNR is shown in Fig. 7.

From Fig. 7, we can acquire that when the received waveform is CSK-bipolar based demodulated, the BER decreases with the increase of k, and increases with the increase of γ. The reason lies in the error accumulation of different modules in the modular bipolar demodulation. Compared with CSK-bipolar demodulation, the BER in bipolar-CSK demodulation is higher, and its growth rate increases with the increase of γ. Since the demodulation of received waveform order is based on the correct decision on every module order, BER rises with the increase of k. Compared with CSK of the same k, BER performance for both of the two demodulation schemes decrease with the increase of γ.

When $k = 5$, γ is 2, 4, and 8 respectively, and SNR is 5 dB, BER with different JSR is shown in Fig. 8.

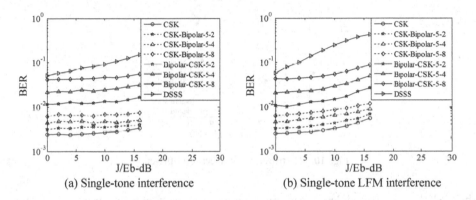

(a) Single-tone interference   (b) Single-tone LFM interference

Fig. 8. BER under different JSR

From Fig. 8, it can be observed that with the increase of JSR, BER for direct sequence spread spectrum (DSSS) system decreases sharply. Nevertheless, TDCS eliminates interference spectrum in transform domain, and thus has a good anti-interference ability. Its BER increases slowly with the increase of JSR. Under the same simulation conditions, the ability of rejecting single-tone interference is better than that of rejecting LFM interference.

5.2   Spectrum Efficiency

We estimate the spectrum of single-tone interference by FFT, and its normalized power spectrum density (PSD) is shown in Fig. 9.

When the threshold is set to the peak value of 40%, the number of available subcarrier is 511, and the BER is 10^{-4}, the spectrum efficiency and required SNR is shown in Fig. 10.

From Fig. 10, we can know that compared to CSK, the spectrum efficiency of the double modulation technique is increasing continuously with the increase of γ. Using

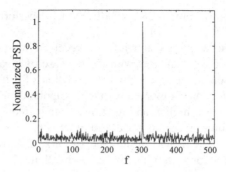

Fig. 9. Normalized PSD of the single-tone interference

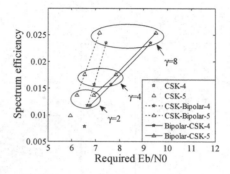

Fig. 10. Distribution of spectrum efficiency

the CSK-bipolar demodulation can bring a great improvement to spectrum efficiency at the cost of a little SNR performance. When adopting the bipolar-CSK demodulation, the improvement of spectrum efficiency costs more SNR.

6   Conclusions

In this paper, we proposed a novel double modulation technique with high spectrum efficiency for TDCS. We firstly analyzed the performance of the modulating waveform, and described the modulation flow in detail. Then we proposed two demodulation techniques for this modulation, and derived its mathematical expressions for its performance. Finally, we simulated the technique and analyzed its BER performance and spectrum efficiency with different SNR and JSR, verifying the reliability of information transmission when the two demodulation techniques cope with noises and interferences. Results show that CSK-bipolar demodulation can improve spectrum efficiency greatly at the cost of only a little SNR, while the same spectrum efficiency can be achieved at the cost of more SNR in bipolar-CSK demodulation.

Acknowledgement. This work was supported in part by the Aeronautical Science Foundation of China under Grant 20161996010.

References

1. Hu, S., Bi, G.A., Guan, Y.L., Li, S.Q.: TDCS-based cognitive radio networks with multiuser interference avoidance. IEEE Trans. Commun. **61**(12), 4828–4835 (2013)
2. Sharma, M., Gupta, R.: Comparative analysis of various communication systems for intelligent sensing of spectrum. In: Proceedings of the IEEE International Conference on Advances in Computing, Communications and Informatics (ICACCI), pp. 902–908 (2014)
3. Yang, Z.Y., Tao, R., Wang, Y., et al.: A novel multi-carrier order division multi-access communication system based on TDCS with fractional Fourier transform scheme. Wirel. Pers. Commun. **79**(2), 1301–1320 (2014)
4. Sun, H.X., Cao, F.C., Qin, H.W.: Multiple access applications of transform domain communication system based on phase coding. In: Proceedings of the 5th IEEE International Conference on Big Data and Cloud Computing (BDCloud), pp. 217–222 (2015)
5. Sharma, M., Gupta, R.: Basis function and PN phase generation in TDCS and WDCS towards dynamic spectrum access. In: Proceedings of the 5th IEEE International Conference on Communication Systems and Network Technologies (CSNT), pp. 364–368 (2015)
6. Richard, K.M., Marshall, H.: Reduction of peak-to-average power ratio in transform domain communication systems. IEEE Trans. Wirel. Commun. **8**(9), 4400–4405 (2009)
7. Wang, S., Da, X.Y., Chu, Z.Y., et al.: Magnitude weighting selection: a method for peak-to-average power ratio reduction in transform domain communication system. IET Commun. **9**(15), 1894–1901 (2015)
8. Wu, G., Hu, S.S., Li, S.Q.: Low complexity time-frequency synchronization for transform domain communications systems. In: Proceedings of the IEEE China Summit and International Conference on Signal and Information Processing (ChinaSIP), pp. 1002–1006 (2015)
9. Fumat, G., Charge, P., Zoubir, A., Fournier-Prunaret, D.: Transform domain communication systems from a multidimensional perspective, impacts on bit error rate and spectrum efficiency. IET Commun. **5**(4), 476–483 (2011)
10. Dillard, G.M., Reuter, M., Zeidler, J., Zeidler, B.: Cyclic code shift keying: a low probability of intercept communication technique. IEEE Trans. Aerosp. Electron. Syst. **39**(3), 786–798 (2003)
11. Charge, P., Zoubir, A., Fournier-Prunaret, D.: Enhancing spectral efficiency of transform domain communication systems by using a multidimensional modulation. In: Proceedings of the IEEE Conference on Cognitive Radio Oriented Wireless Networks and Communications (CROWNCOM), pp. 131–135 (2011)
12. Hu, S., Guan, Y.L., Bi, G.A.: Cluster-based transform domain communication systems for high spectrum efficiency. IET Commun. **6**(16), 2734–2739 (2012)
13. Hu, S., Bi, G.A., Guan, Y.L., Li, S.Q.: Spectrum efficiency transform domain communication systems with quadrature cyclic code shift keying. IET Commun. **7**(4), 382–390 (2013)
14. Proakis, J.G., Salehi, M.: Digital Communications, 5th edn. McGraw-Hill, New York (2008). http://www.springer.com/lncs. Accessed 21 Nov 2016

Quality of Experience Prediction of HTTP Video Streaming in Mobile Network with Random Forest

Yue Yu[✉], Yu Liu, and Yumei Wang

School of Information and Communication Engineering, Beijing University of Posts and Telecommunications, Beijing 100876, China
{yuy,liuy,ymwang}@bupt.edu.cn

Abstract. As video is witnessing a rapid growth in mobile networks, it is crucial for network service operators to understand if and how Quality of Service (QoS) metrics affect user engagement and how to optimize users' Quality of Experience (QoE). Our aim in this paper is to infer the QoE from the observable QoS metrics using machine learning techniques. For this purpose, Random Forest is applied to predict three objective QoE metrics, i.e., rebuffering frequency, mean bitrate and bitrate switch frequency, with the initial information of each video session. In our simulation, QoE of four different video streamings are analyzed with eight different system loads. Results show that sufficient prediction accuracy can be achieved for all QoE metrics with the attributes we adopted, especially with low and middle system loads. In terms of type of streamings, the prediction of all metrics for static users performs better than mobile users. Feature selection is also implemented under the highest load to examine the effect of different attributes on each QoE metric and the correlation among attributes.

Keywords: HTTP video streaming · Quality of experience · Random forest Mobile networks

1 Introduction

Video streaming is becoming more and more important in recent years. According to Cisco's forecast [1], video traffic will account for 78% of Internet traffic by 2021. HTTP video streaming is widely used in delivering on-demand multimedia content, with retransmission applied to guarantee data correctness. At the server side, single or several encoded versions are stored, where video files are divided into several chunks (segments). After being downloaded, the chunk is stored in the player's buffer for playback. Before the buffer becomes empty, users can proceed on video playing; otherwise, the video will suffer a rebuffering event.

The QoE concept has emerged mainly with the basic motivation that QoS is not powerful enough to fully express everything nowadays involved in a communication service, which is a multi-dimensional construct and consists of subjective and objective parameters [2]. When it comes to the QoE of HTTP streaming users, according to [3], it highly depends on two crucial factors: (1) the visual quality and its variation and (2) the

X. Liu et al. (Eds.): ChinaCom 2018, LNICST 262, pp. 82–91, 2019.
https://doi.org/10.1007/978-3-030-06161-6_8

frequency and duration of rebuffering events. Different from the Peak Signal to Noise Ratio (PSNR), rebuffering events cannot be directly measured but only predicted from classic QoS metrics [4]. This allows to infer QoE metrics by still relying on QoS monitoring systems. Nevertheless, it is highly complex to map between QoS and QoE metrics, as they often lay in high dimensional spaces and are subject to noise. As a consequence, it is not practical to get a closed form modeling and its experimental validation. Therefore, machine learning techniques are applied to derive the complex relationships between QoS and QoE metrics. In the context of mobile networks, it is challenging for operators to correlate the cell-related parameters like channel state information (CSI) and existing users number to QoE metrics of video consumers, due to the system complexity and difficulty in obtaining the cross-layer information. To overcome this difficulty, we have established a cross-layer simulation program that simulates the behaviors of HTTP video streamings in mobile networks as well as buffer information in user side. Thus we can access all cross-layer information for correlating the QoS parameters in data link or physical layer and the user QoE.

When video streaming service is offered over wireless networks, there are two variability time scales in QoE metrics: flow level (tens of seconds) driven by the departures/arrivals of calls, and wireless channel variability time scale (milliseconds) driven by the fast fading [5]. The analytical results in [5] demonstrate that the flow dynamics have dominant influence on QoE metrics compared to the jittering in the throughput due to the fast fading. Therefore, we model the radio access network in flow level and focus on the video flow behaviors such as arrival, departure, mobilty and rebuffering while reducing the complexity involved by packet level protocols [6]. In this paper, a flow refers to a video streaming session.

The rest of paper is organized as follows. Section 2 discusses relevant related work. In Sect. 3, we introduce the mobile network and QoE metrics. Prediction performance of four different types of video streaming is shown in Sect. 4. Section 5 concludes the paper and discusses the future works.

2   Related Work

QoE has recently gained momentum as a way to assess the perceived quality of users during videos watching. Authors of [7] studied the QoE with TCP information. Authors of [8,9] utilized flow-level model to investigate the video performance metrics, where the correlation between video rebuffering and the proposed performance metrics is not clear. Machine learning has been widely used to study both subjective and objective QoE to deal with the complexity of finding correlation between the parameters. Authors of [10] used machine learning to study the correlation between users' engagement and application metrics, such as buffer times. In [11], the cell-related parameters were first used as the research focus, but they just researched whether rebuffering occurred. Studies like the one presented in [12] proposed a QoE predicting module for adaptive HTTP streaming, without taking the traditional bitrate-constant streaming into account. Although many services have already made the migration towards adaptive streaming, their platforms continue to maintain backward compatibility with traditional bitrate-constant streaming. The investigation performed in [13] predicted QoE factors focusing on the hidden and context information, while consideration of up to 50 associated variables may increase the complexity of attribute extraction and the construction of the predictive model.

The authors of [11] used cell-related parameters (e.g., physical throughput and number of active flows) with Support Vector Machine (SVM) to predict whether a flow will encounter a rebuffering event. We consider this work as a starting point for our research and present two further contributions: (1) Instead of merely focusing on rebuffering/non-rebuffering, we bring insight into the relationship between cell-related QoS metrics and three main QoE metrics, namely rebuffering frequency, the video quality, and its variation. (2) In terms of machine learning tools, the Random Forest algorithm is adopted, which outperformes SVM in multiple classification problems and supports feature (attribute) selection analysis.

3 System Description

In this section, model of radio access network based on the flow-level concept are presented firstly. Then we show four types of HTTP streamings in our simulation. At last, we introduce the recorded attributes and QoE metrics.

3.1 Radio Access Network

Based on the concept of flow-level model in paper [8], a cell is modeled by a set of K capacity regions denoted as $R = \{R_1, \ldots, R_K\}$. In each region, physical throughputs are supposed to be homogeneous and thus, on the downlink, users are served with the same physical throughputs. Users in a cellular network are classified into static users and mobile users. The physical throughput of static users is assumed to be constant, and that of mobile users may randomly vary with time when a mobility envent occurs.

As for traffic characteristics, we follow the classical assumption that streaming flows with beginning physical throughput R_k arrive as a Poisson process with rate $\lambda_k = p_k \lambda$, where λ is the overall flow arrival rate in the cell and p_k stands for the traffic proportion with physical throughput R_k, where $\sum_k p_k = 1$. With the stability condition in paper [8], the maximum flow arrival rate, λ_{max}, guaranteeing the system stability, can be obtained. In our simulation, eight flow arrival rates normalized by the maximum value λ_{max} were demonstrated, since traffic arrival rate, λ, varies along hours in the real network. For each λ, simulator generates $m = 10^6$ streaming arrivals for the training of the Random Forest.

3.2 HTTP Video Streaming

Generally speaking, the video streaming can be categorized into two types

- Fixed bitrate streaming (also called progressive download). This is the original implementation of the HTTP video streaming and maintains a fixed bitrate for each chunk during the whole video downloading process.
- Adaptive streaming. Adaptive video streaming can switch among several optional bitrates according to the measured throughput, γ. Given the preset discrete set

$V = \{v_1, \ldots, v_M\}$, where $v_M > \ldots > v_1$, users select a video bitrate, v, for the next chunk as below, where $i = 1, \cdots, M - 1$.

$$v = \begin{cases} v_M, & \gamma \geq v_M \\ v_i, & v_i \leq \gamma < v_{i+1} \end{cases} \tag{1}$$

In order to provide a solution which will be compatible with current and previous video streaming technologies, four types of streamings are simulated. Table 1 lists the four types of streamings in our simulation.

Table 1. Types of streamings.

Type	Description
Type I	Static and adaptive streaming
Type II	Static and fixed bitrate streaming
Type III	Mobile and adaptive streaming
Type IV	Mobile and fixed bitrate streaming

3.3   Recorded Attributes

We aim to take a step closer to exploring the correlation of each user's initial QoS metrics and user's QoE by recording complete buffer statistics. Therefore, we develop an simulator that simulates the actual behavior (e.g., playback, rebuffering, and mobility) and buffer state of each user in a radio access network, driven by some flow-events. In Fig. 1, we present an illustration of a video session life time in the event-driven simulatorm, where the buffer state will switch as the corresponding flow-event occurs and the chunk events mean downloading of a new chunk. Fine-grained information about the video session in our simulation program is recorded, including the bitrate of each video segment, the bitrate switching between adjacent video segments, and the number of rebuffering events during video downloading.

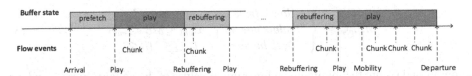

Fig. 1. An illustration of a video session life time in the event-driven simulator.

Table 2 presents all the data output by our simulator for j-th user, which can be summarized into two sets: (1) attributes set: the initial attributes recorded when user j arrives and (2) targets set: the total number of rebuffering events and the set of selected bitrates recorded during departure of user j.

Table 2. Parameters generated for j-th user in our simulations.

Set	Symbol	Description	Unit
Attributes	R_j	Physical throughput recorded at arrival	Mbps
	T_j	Video duration	S
	F_j	Numbers of flows in cell of each region	Vector
	$|F_j|$	Total number of flows in cell	Null
	F_j^r	Numbers of flows in rebuffering of each region	Vector
	$|F_j^r|$	Total number of flows in rebuffering	Null
Targets	N_j	Number of rebuffering events encountered	Null
	S_j	Set of bitrates selected	Vector

3.4 The QoE Metrics

In this subsection, we present three main QoE metrics reflecting the perceived video quality of users and the discretization for classification

- Rebuffering frequency (RF): The ratio of the number of rebuffering events to the duration of the session.
- Mean bitrate (MB, only for adaptive streaming) : The average of the bitrates weighted by the duration each bitrate is played for.
- Bitrate switch frequency (SF, only for adaptive streaming): The ratio of the number of bitrate switches to the duration of the session.

These metrics are difficult to be predicted in its raw continuous form. To simplify the classification and create a predictive model, we have further processed the metrics by labeling the data as shown in Eqs. (2)–(4).

$$RF_{label} = \begin{cases} "no\,rebuffering", & RF = 0 \\ "mild\,rebuffering", & 0 < RF < L_{rf} \\ "severe\,rebuffering", & L_{rf} \le RF \end{cases} \qquad (2)$$

where we adopt $L_{rf} = 0.1$, since [14] showed that with rebuffering ratio over 0.1, most of users abandon the video because of the quality degradation.

$$MB_{label} = \begin{cases} "low\,bitrate", & v_1 \le MB < L_{mb1} \\ "middle\,bitrate", & L_{mb1} \le MB < L_{mb2} \\ "high\,bitrate", & L_{mb2} \le MB \le v_M \end{cases} \qquad (3)$$

where v_1 and v_M are the minimum and maximum values of the optional bitrates and we set L_{mb1} as 1.5, L_{mb2} as 2, the medians of the optional bitrates.

$$SF_{label} = \begin{cases} "no\,switch", & SF = 0 \\ "mild\,switch", & 0 < SF < L_{sf} \\ "severe\,switch", & L_{sf} \le SF \end{cases} \qquad (4)$$

where L_{sf} is set to 0.3, which distinguishes mild and severe switch in this paper.

4 Simulation Analysis

In this section, we analyze the prediction performance of machine learning among different types of HTTP streaming with recorded attributes. We adopt the simulation configuration in [11] and set the optional bitrates as 1, 1.5, 2, 2.5 Mbps.

WEKA [15], one of the most popular open-source machine learning libraries, is adopted to implement the Random Forest algorithm and to investigate the prediction performance. In classification for each QoE metric, the datasets consist of instance-label pairs (X_j, Y_j), where $j = 1, \cdots, m$. X_j consists of all attributes of user j, and Y_j corresponds to each category label. For example, the prediction of the rebuffering frequency can be expressed as a three-class classification problem with instance-label pairs $(X_j, RF_{label,j})$. With the feature selection algorithms, Random Forest evaluates the predictive power of each attribute and its redundancy with each other, and tends to select attributes that have a high correlation with the target but have a low correlation with each other. Effective feature selection can significantly reduce the difficulty of attribute extraction and the complexity of the predictive model. In addition, the Random Forest algorithm can evaluate the information gain which represents the worth of each attribute in the construction of the predictive model.

In our simulation, eight flow arrival rates normalized by the maximum value λ_{max} are demonstrated to show the performance at each load. Under each load, we present the respective prediction performance for four different HTTP video streamings, as shown in Fig. 2, 3, and 4. In general, when load increases, prediction performance decreases due to the increase of uncertainty.

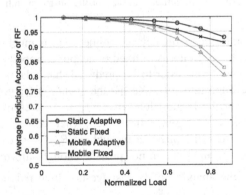

Fig. 2. The average prediction accuracy of the rebuffering frequency.

Figure 2 shows the average prediction accuracy of the rebuffering frequency. In general, sufficient accuracy can be achieved especially when the load is low. With respect to mobility of streamings, the simulation results show that static users can achieve more than 90% of accuracy even in large load, which is a significant improvement over previous approaches [16] where the achieved accuracy was approximately 84% for a binary classification and the severity of rebuffering was

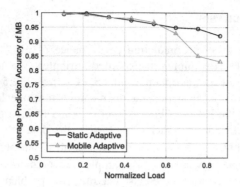

Fig. 3. The average prediction accuracy of the mean bitrate.

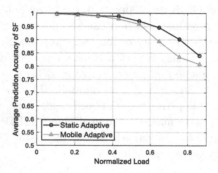

Fig. 4. The average prediction accuracy of the bitrate switch frequency.

unclear. However, rebuffering frequency of mobile users is much more difficult to be predicted when load is large. In terms of fixed or adaptive property, there is no general rule saying that fixed bitrate is easier to be predicted than adaptive streaming, where mobility plays a more important role. We list the results of feature selection for each type of streamings under the highest load in Table 3. As presented in Table 2, $F_{j, k}$ means the number of flows of region k in the cell, where $k = 1, ..., K$.

Firstly, for mobile users, the physical throughput R_j is not selected, which means that the initial physical throughput can not provide enough information to predict the rebuffering. Secondly, the abandon of $\left|F_j^r\right|$ suggests a high redundancy between $\left|F_j\right|$ and $\left|F_j^r\right|$, which may be good news for operators that they do not need to know more application information from users' side. Further, experiments show that, using the remaining attributes can achieve almost the same accuracy as overall attributes, but with reduced feature extraction overhead.

Figure 3 shows the average prediction accuracy of the mean bitrate. As mentioned earlier, the mean bitrate is only meaningful for adaptive video streaming. In terms of adaptive streaming alone, overall, over almost 85% accuracy is achieved even at high loads. Similarly, the prediction accuracy of static users can still reach more than 90%

Table 3. Attributes selected and respective information gain for RF.

Static adaptive		Static fixed		Mobile adaptive		Mobile fixed	
Attribute	Gain	Attribute	Gain	Attribute	Gain	Attribute	Gain
R_j	0.382	R_j	0.344	T_j	0.141	T_j	0.086
T_j	0.023	T_j	0.03	$|F_j|$	0.301	$|F_j|$	0.444
$|F_j|$	0.144	$|F_j|$	0.214	$F_{j,1}$	0.082	$F_{j,1}$	0.110
$F_{j,3}^r$	0.086			$F_{j,2}$	0.10	$F_{j,2}$	0.138
				$F_{j,3}$	0.15	$F_{j,3}$	0.219
				$F_{j,4}$	0.164	$F_{j,4}$	0.257
				$F_{j,5}$	0.121	$F_{j,5}$	0.179

even in high load and the impairment of mobility on predictions reduces the prediction performance for mobile users. Table 4 presents the results of feature selection for adaptive users under the highest load.

Table 4. Attributes selected and respective information gain for MB.

Static adaptive		Mobile adaptive	
Attribute	Gain	Attribute	Gain
R_j	0.254	T_j	0.044
T_j	0.002	$|F_j|$	0.57
$|F_j|$	0.308	$F_{j,3}$	0.294
$F_{j,3}$	0.195	$F_{j,4}$	0.324
$F_{j,4}$	0.219	$|F_j^r|$	0.461

Table 5 presents the confusion matrix for MB for static adaptive users under the highest load. The confusion matrix provides specific prediction accuracy of each class. We can see that the classification errors occur between instances *"Low"* and those with *"Middle"*, also between *"Middle"* and *"High"*, however, significantly fewer mis-classifications between *"Low"* and *"High"*. Possible reasons include the classifier's inability to correctly identify marginal cases which are close to the MB thresholds, and the subtle differences between instances of different classes.

Figure 4 shows the average prediction accuracy of the bitrate switch frequency. In general, the accuracy of predicting for all loads exceeding 80% can be achieved, and when the load is not so high, the accuracy is above 90%. In addition, higher prediction accuracy for static users can be achieved.

Table 6 presents the results of feature selection for adaptive users under the highest load. The information gain of T_j shows the importance of T_j for predicting SF.

Table 5. Confusion matrix for MB of static adaptive users.

Actual label	Predicted label		
	"Low"	*"Middle"*	*"High"*
"Low"	97.2%	2.2%	0.6%
"Middle"	15.7%	73%	11.3%
"High"	2.8%	9.2%	88%

Table 6. Attributes selected and respective information gain for SF.

Static Adaptive		Mobile adaptive	
Attribute	Gain	Attribute	Gain
R_j	0.050	T_j	0.381
T_j	0.126	$|F_j|$	0.083
$|F_j|$	0.066		

5 Conclusions and Feature Works

In this paper, we aim to infer the QoE metrics from the observable QoS metrics with machine learning techniques. Based on the concept of flow-level dynamics, we develop an event-driven simulator to generate datasets, by which we correlate the cell-parameters and users' QoE. We examined the prediction performance of three QoE metrics for different HTTP video streamings along different loads. Then the machine learning technique, i.e., Random Forest, is used to obtain our predictive model along the system loads. Simulation results show that, with the initial information of each video session such as number of flows and radio conditions, sufficient accuracy can be achieved. In terms of type of streamings, the prediction of all metrics for static users performs better than mobile users, due to the increase of uncertainty from mobility, which calls for more information for prediction. We also perform feature selection with the highest load as an example to examine the effect of different attributes on each QoE metric and the correlation among attributes.

Future works will consider more attributes to improve the prediction accuracy in high loads, especially for mobile users. More QoE metrics like start-up delay will be researched to completely study the perceived quality by HTTP video streaming. The application of other machine learning models such as Neural Networks may improve the prediction accuracy.

Acknowledgements. This work has been sponsored by Huawei Research Fund (grant No. YBN2016110032) and National Science Foundation of China (No. 61201149). The authors would also like to thank the reviewers for their constructive comments.

References

1. Cisco visual networking index: Global mobile data traffic forecast update 2016–2021 white paper. https://www.cisco.com/c/en/us/solutions/service-provider/visual-networking-index-vni/vni-infographic.html. Accessed 06 June 2018
2. Patrick Le Callet, S.M., Perkis, A.: Qualinet White Paper on Definitions of Quality of Experience (2012). http://www.qualinet.eu/index.php. Accessed 06 June 2018
3. Yin, X., Jindal, A., Sekar, V., Sinopoli, B.: A control-theoretic approach for dynamic adaptive video streaming over HTTP. In: Proceedings of the 2015 ACM Conference on Special Interest Group on Data, pp. 325–338, London (2015)
4. Dimopoulos, G., Leontiadis, I., Barlet-Ros, P., Papagiannaki, K.: Measuring video QoE from encrypted traffic. In: Proceedings of the 2016 Internet Measurement Conference, pp. 513–526, Santa Monica (2016)
5. Xu, Y., Elayoubi, S., Altman, E., El-Azouzi, R.: Impact of flowlevel dynamics on qoe of video streaming in wireless networks. In: 2013 Proceedings IEEE INFOCOM, pp. 2715–2723, Turin (2013)
6. Bonald, T., Proutiere, A.: A queueing analysis of data networks. Queueing Netw. **154**, 729–765 (2011)
7. Singh, K.D., Aoul, Y.H., Rubino, G.: Quality of experience estimation for adaptive HTTP/TCP video streaming using H.264/AVC. In: 2012 IEEE Consumer Communications and Networking Conference (CCNC), pp. 127–131, Las Vegas (2012)
8. Bonald, T., Elayoubi, S., Lin, Y.-T.: A flow-level performance model for mobile networks carrying adaptive streaming traffic. In: IEEE Globecom, San Diego (2015)
9. Lin, Y.-T., Bonald, T., Elayoubi, S.: Impact of chunk duration on adaptive streaming performance in mobile networks. In: 2016 IEEE Wireless Communications and Networking Conference, Doha (2016)
10. Balachandran, A., Sekar, V., Akella, A., Seshan, S., Stoica, L., Zhang, H.: Developing a predictive model of quality of experience for internet video. In: Proceedings of the ACM SIGCOMM 2013 Conference on SIGCOMM, pp. 339–350, New York (2013)
11. Lin, Y.-T., Oliveira, E.M.R., Jemaa, S.B., Elayoubi, S.E.: Machine learning for predicting QoE of video streaming in mobile networks. In: 2017 IEEE International Conference on Communications (ICC), pp. 1–6. IEEE, Paris (2017)
12. Chen, Z., Liao, N., Gu, X., Wu, F., Shi, G.: Hybrid distortion ranking tuned bitstream-layer video quality assessment. In: IEEE Trans. Circuits Syst. Video Technol. **26**(6), 1029–1043 (2016)
13. Vasilev, V., Leguay, J., Paris, S., Maggi, L., Debbah, M.: Predicting QoE factors with machine learning. In: IEEE International Conference on Communications (ICC) 2018, Kansas City (2018)
14. Krishnan, S., et al.: Video stream quality impacts viewer behavior: inferring causality using quasi-experimental designs. IEEE/ACM Trans. Netw. **21**(6), 2001–2014 (2013)
15. WEKA: Data Mining Software in Java. https://www.cs.waikato.ac.nz/ml/weka. Accessed 26 May 2018
16. Aggarwal, V., et al.: Prometheus: toward quality-of-experience estimation for mobile apps from passive network measurements. In: Proceedings of the 15th Workshop on Mobile Computing Systems and Applications, p. 18. ACM, Santa Barbara (2014)

Performance of Linearly Modulated SIMO High Mobility Systems with Channel Estimation Errors

Mahamuda Alhaji Mahamadu[1,2(✉)] and Zheng Ma[1]

[1] School of Information Science and Technology, Southwest Jiaotong University,
Chengdu 610031, China
mamahamadu@my.swjtu.edu.cn, zma@swjtu.edu.cn
[2] Council for Scientific and Industrial Research, Accra, Ghana

Abstract. This paper studies the error performance of linearly modulated single-input multiple-output (SIMO) high mobility communication systems with channel estimation errors. Channel estimation errors are unavoidable in high mobility systems, due to the rapid time-varying fading of the channel caused by severe Doppler effects, and this might have non-negligible adverse impacts on system performance. However, in high mobility communications, rapid time-varying fading channels induce Doppler diversity which can be exploited to improve system performance. Based on the statistical attributes of minimum mean square error (MMSE) channel estimation, a new optimum diversity receiver for MASK, MPSK and MQAM SIMO high mobility systems with channel estimation errors is proposed. The exact analytical error probability expressions of MPSK, MASK, and MQAM of the SIMO diversity receiver are identified and expressed as a unified expression. It quantifies the impacts of both Doppler diversity and channel estimation errors. The result is expressed as an explicit function of the channel temporal correlation, pilot and data signal-to-noise ratios (SNRs). Simulations results are used to validated analytical results. Simulation results show that MPSK, MASK, and MQAM systems have the same Doppler diversity order even though they differ in symbol error rates(SERs). Moreover, simulation results show that MQAM systems achieve better spectral efficiency than its MPSK and MASK counterparts.

Keywords: Single-input multiple-output (SIMO) systems · High mobility wireless communications · Doppler diversity · MASK · MPSK · MQAM · Channel estimation · Minimum mean square error (MMSE) · Channel estimation errors

This work was supported by The National Natural Science Foundation of China (NSFC) Project (No. 61571373), The Huawei HIRP Project (No.YB201504), The Key International Cooperation Project of Sichuan Province (No.2017HH0002), The NSFC China-Swedish Project (No.6161101297), The 111 Project (No.111-2-14), and The Young Innovative Research Team of Sichuan Province (2011JTD0007).

1   Introduction

With the ever increasing in demands for broadband wireless communications on high-speed trains and aircraft, broadband wireless communications have attracted considerable research recently. In high mobility systems, signals could encounter large Doppler spreads of the order of kilohertz [1], yet most conventional wireless communication systems are designed to operate with a Doppler spread of at most a couple of hundreds of Hertz. Large Doppler spread results in rapid time-varying fading, which is one of the principal difficulties faced in the design of reliable broadband high mobility wireless communications systems. It is not easy to estimate and track the rapid time-varying fading channel coefficients accurately. Hence channel estimation errors are bound to be present in high mobility systems, and this might have significant adverse impacts on system performance. Consequently, traditional techniques developed under the assumption of perfect channel state information (CSI) are no longer valid for high mobility systems. However, rapid time-varying fading caused by large Doppler spreads in high mobility system induces Doppler diversity which can be exploited to improve system performance. There have been some works in the literature mainly for optimizing the performance of systems with Doppler diversity [2–4]. However, all the above works are performed under the assumption of perfect CSI.

The optimum designs of Doppler diversity systems with imperfect CSI are studied in [5–7]. In [5,7], the fundamental tradeoff between imperfect CSI and Doppler diversity are analytically identified through asymptotic analysis, where the maximum Doppler diversity order with imperfect CSI is developed with the aid of a repetition code. The maximum Doppler diversity order is gotten at the price of low spectral efficiency. Spectral and Energy efficient Doppler diversity receiver are proposed in [6], and it gives a balanced tradeoff between spectral and energy efficiencies in high mobility systems. All the above works are for single-input-single-output (SISO) systems.

The design of a single-input-multiple-output (SIMO) system with imperfect CSI is discussed in [8]. It is attested that the traditional maximal ratio combining (MRC) receiver is no longer optimum in the presence of imperfect CSI. A new diversity receiver is designed by utilizing the statistics of the channel estimation errors. The results of [8] are not applicable to studies in high mobility environment because quasi-static channels are employed in the study. All the above works did not consider MASK and MQAM modulation.

In this paper, we investigate the error performance of linearly modulated single-input-multiple-output (SIMO) high mobility communication systems with channel estimation errors and Doppler diversity. For a high mobility SIMO system, there is both space diversity and Doppler diversity. The objective is to develop an optimum receiver that can effectively harvest both space diversity and Doppler diversity inherent in the system. Such a diversity receiver is designed in this paper by analyzing the statistical properties of the channel coefficients estimated by using pilot assisted MMSE channel estimation. It is different from conventional diversity receivers because the statistics of channel estimation errors are incorporated in the receiver. The analytical error probability of the proposed receiver is obtained, and it is expressed as a function of the maximum Doppler spread, the signal-to-noise ratios (SNR) of pilot and data symbols, and the temporal correlation of the channel, etc. The analytical results quantify the impact of both channel estimation errors and Doppler diversity on system performance.

To ensure that, maximum Doppler diversity is embedded in the system, a simple repetition coding is employed at the transmitter. Maximum Doppler diversity is gotten at the price of low spectral efficiency. To improve the spectral efficiency of the systems, the study seeks to adopt a spectrally efficient modulation scheme amongst the linear modulation schemes employed in the study.

The rest of the paper is organized as follows. The system model and MMSE channel estimation are given in Section II. The optimum diversity receiver with imperfect CSI is developed in Section III, where the analytical performance of the receiver is also studied. Section IV presents Numerical results, and the paper concluded in Section V.

2   System Model

We consider a SIMO system with one transmit antenna and N_R receive antennas operating in a high mobility environment. Pilot-assisted channel estimation is used to estimate and track the fast time-varying channels.

2.1   Pilot Assisted Transmission

The data symbols to be transmitted from the transmitter are divided into slots. As depicted in Fig. 1, each slot contains K unique modulated data symbols $\mathbf{s} = [s_1, \cdots, s_K]^T \in \mathcal{S}^{K \times 1}$, where $\mathcal{S}$ is the modulation alphabet set, and the superscript $(\cdot)^T$ represents the matrix transpose. To identify the maximum Doppler diversity embedded in the system, we adopt a simple repetition code, where each modulated data symbol is repeated N times. Such a repetition precoding scheme ensures that there is maximum Doppler diversity at the price of lower spectral efficiency. The error probability performance with the repetition code can serve as a lower bound for systems employing spectral-efficient precoding schemes [5]. Equally-spaced pilots have inserted among the data symbols after precoding.

The signals in a slot can be denoted as $\mathbf{x} = [\mathbf{s}, p_1, \mathbf{s}, p_2, \cdots, \mathbf{s}, p_N]^T$, where p_k, for $k = 1, \ldots, N$, are N pilot symbols, and the data symbol vector $\mathbf{s}$ is repeated N times. Without loss of generality, it is assumed that the pilot symbols are from constant amplitude modulation, such as M-ary phase shift keying (MPSK) and data symbols are equally probable from a constellation set composed of MASK symbols for MASK systems and equally probable from a constellation set composed of MQAM symbols for MQAM systems. There are totally $N_{\text{sym}} = (K + 1)N$ symbols in one slot. With such a slot structure, the time duration between two adjacent pilot symbols is $T_p = (K + 1)T_s$, where T_s is the symbol period. Thus the pilot symbols sample the channel at a rate $R_p = \frac{1}{(K+1)T_s}$.

$$N_{\text{sym}}$$

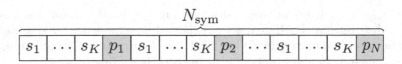

Fig. 1. The structure of the slot after precoding and insertion of pilots.

Denote the energy for each pilot and coded data symbol as E_p and E_c, respectively. The entire energy in one slot is thus $E_pN + E_cKN$, and the energy per uncoded information bit can be computed as $E_b = \frac{E_pN + E_cKN}{K\log_2 M}$, where $M = |\mathcal{S}|$ is the cardinality of the modulation constellation set. It is assumed the channels between the transmitter and each of the N_R diversity receivers are identically independently distributed (i.i.d.). For i.i.d channels, the m-th fading branch h_m and the n-th fading branch h_n have the same statistical attributes [9]. Based on this assumption, the received signals at each of the N_R receive antennas can be interpreted as a stack of N_T copies.

The coded data and pilot symbols are transmitted over the time-varying fading channel with additive white Gaussian noise (AWGN). The index of the k-th pilot symbol is denoted as $i_k = k(K+1)$, where $k = 1, \cdots, N$. Then the pilot symbols observed at the r-th receive antenna can be represented by

$$\mathbf{y}_{r,p} = \sqrt{E_p}\mathbf{X}_p\mathbf{h}_{r,p} + \mathbf{z}_{r,p}, \tag{1}$$

where $\mathbf{y}_{r,p} = [y_r(i_1), \cdots, y_r(i_N)]^T \subset \mathcal{C}^{N\times 1}$ and $\mathbf{z}_{r,p} = [z_r(i_1), \cdots, z_r(i_N)]^T \in \mathcal{C}^{N\times 1}$ are the additive white Gaussian noise (AWGN) vector and the received pilot vector respectively, with $\mathcal{C}$ denoting the set of complex numbers, $\mathbf{X}_p = \text{diag}([p_1, \cdots, p_N])$ is a diagonal matrix with the N pilot symbols on its main diagonal and $\mathbf{h}_{r,p} = [h_r(i_1), \ldots, h_r(i_N)]^T \in \mathcal{C}^{N\times 1}$ is the discrete-time channel fading vector sampled at the pilot locations for the r-th antenna. The AWGN vector is a zero-mean symmetric complex Gaussian random vector (CGRV) with covariance matrix $\sigma_z^2\mathbf{I}_N$, where σ_z^2 is the noise variance and $\mathbf{I}_N$ is a size N identity matrix. With the repetition code, each modulated data symbol is transmitted N times. The k-th data symbol s_k is transmitted over symbol indices $k_n = (n-1)(K+1) + k$, for $n = 1, \cdots, N$ in a slot. The received sample vector corresponding to the k-th data symbol s_k at the r-th antenna can then be expressed as

$$\mathbf{y}_{r,k} = \sqrt{E_c}\mathbf{h}_{r,k}s_k + \mathbf{z}_{r,k}, \tag{2}$$

where $\mathbf{y}_{r,k} = [y_r(k_1), \cdots, y_r(k_N)]^T$, $\mathbf{h}_{r,k} = [h_r(k_1), \cdots, h_r(k_N)]^T$ and $\mathbf{z}_{r,k} = [z_r(k_1), \cdots, z_r(k_N)]^T$, are length-$N$ vectors of received samples, fading coefficients, and AWGN, respectively.

Stacking up $\mathbf{y}_{r,k}$ into a column vector, we have

$$\mathbf{y}_k = \sqrt{E_c}\mathbf{h}_ks_k + \mathbf{z}_k, \tag{3}$$

where $\mathbf{y}_k = [\mathbf{y}_{1,k}^T, \cdots, \mathbf{y}_{N_R,k}^T]^T \in \mathcal{C}^{N\times 1}$, $\mathbf{h}_k = [\mathbf{h}_{1,k}^T, \cdots, \mathbf{h}_{N_R,k}^T]^T \in \mathcal{C}^{N\times 1}$ and $\mathbf{z}_k = [\mathbf{z}_{1,k}^T, \cdots, \mathbf{z}_{N_R,k}^T]^T \in \mathcal{C}^{N\times 1}$ are length N_RN vectors.

The N_R channels on different antennas are independent, and they follow Rayleigh distribution. Each channel is expected to experience wide sense stationary uncorrelated scattering (WSSUS). Hence $h_r(n)$ is a zero-mean symmetric complex Gaussian random process with the covariance function

$$\mathbb{E}[h_r(m)h_t^*(n)] = \begin{cases} J_0(2\pi f_D|m-n|T_s), & r = t \\ 0, & r \neq t \end{cases} \tag{4}$$

where $\mathbb{E}(\cdot)$ is the mathematical expectation operator, the superscript $(\cdot)^*$ denotes complex conjugate, f_D is the maximum Doppler spread of the fading channel, $J_0(x)$ is the zero-order Bessel function of the first kind and T_s is the symbol period.

2.2 Channel Estimation

The statistical attributes of the CSI estimated by using the pilot symbols are discussed in this section, and the results are utilized to design the optimum SIMO receiver for high mobility systems operating with imperfect CSI.

Since the channels observed by different antennas are independent, they can be estimated separately. The channel coefficients of the coded data symbols can be estimated using the pilot symbols at the receiver by MMSE estimation due to the temporal channel correlation, The linear MMSE estimation of the channel coefficients corresponding to the k-th data symbol at the r-th receive antenna is [7]

$$\hat{\mathbf{h}}_{r,k} = \mathbf{W}_k^H \mathbf{y}_{r,p}, \tag{5}$$

where $\mathbf{W}_k \in \mathcal{C}^{N \times N}$ is the MMSE estimation matrix. It can be represented as

$$\mathbf{W}_k^H = \sqrt{E_p}\mathbf{R}_{kp}\mathbf{X}_p^H(E_p\mathbf{X}_p\mathbf{R}_{pp}\mathbf{X}_p^H + \sigma_z^2\mathbf{I}_{N_R N})^{-1}, \tag{6}$$

where $\mathbf{R}_{kp} = \mathbb{E}[\mathbf{h}_{r,k}\mathbf{h}_{r,p}]$ and $\mathbf{R}_{pp} = \mathbb{E}[\mathbf{h}_{r,p}\mathbf{h}_{r,p}]$.

Since the channels at different antennas are identically distributed, the covariance matrices $\mathbf{R}_{kp}$ and $\mathbf{R}_{pp}$ are independent of the antenna index p. Based on (4), the matrix $\mathbf{R}_{pp}$ is a symmetric Toeplitz matrix with the first row and column being $[\rho_0, \rho_1, \cdots, \rho_{N-1}]^T$, where $\rho_n = J_0(2\pi f_D|n|T_p)$. The matrix $\mathbf{R}_{kp}$ is a Toeplitz matrix, with the first row being $[\tau_{-1}, \tau_{-2}, \cdots, \tau_{-N}]$, and the first column is $[\tau_{-1}, \tau_0, \cdots, \tau_{N-2}]^T$, where $\tau_u = J_0(2\pi f_D(uT_p + kT_s))$. The error covariance matrix, $\mathbf{R}_{ee} = \mathbb{E}[(\mathbf{h}_{r,k} - \hat{\mathbf{h}}_{r,k})(\mathbf{h}_{r,k} - \hat{\mathbf{h}}_{r,k})^H]$ is [7]

$$\mathbf{R}_{ee} = \mathbf{R}_{kk} - \mathbf{R}_{kp}\left(\mathbf{R}_{pp} + \frac{1}{\gamma_p}\mathbf{I}_N\right)^{-1}\mathbf{R}_{kp}^H \tag{7}$$

where $\mathbf{R}_{kk} = \mathbb{E}[\mathbf{h}_{r,k}\mathbf{h}_{r,k}]$.

The estimated channel vector $\hat{\mathbf{h}}_{r,k}$ is zero-mean Gaussian distributed with covariance matrix $\mathbf{R}_{\hat{k}\hat{k}} = \mathbf{R}_{kk} - \mathbf{R}_{ee}$.

In addition, conditioned on $\hat{\mathbf{h}}_{r,k}$, $\mathbf{h}_{r,k}$ is Gaussian distributed with mean $\mathbb{E}[\mathbf{h}_{r,k}|\hat{\mathbf{h}}_{r,k}] = \hat{\mathbf{h}}_{r,k}$ and covariance matrix $\mathbf{R}_{k|\hat{k}} = \mathbf{R}_{ee}$ [7].

Define $\hat{\mathbf{h}}_k = [\hat{\mathbf{h}}_{1,k}^T, \cdots, \hat{\mathbf{h}}_{N_R,k}^T]$, then $\hat{\mathbf{h}}_k$ is zero mean Gaussian distributed with covariance matrix $C_{\hat{k}\hat{k}} = \mathbf{I}_{N_R} \otimes \mathbf{R}_{\hat{k}\hat{k}}$, that is $\hat{\mathbf{h}}_k \sim \mathcal{N}(\mathbf{0}, \mathbf{I}_{N_R} \otimes \mathbf{R}_{\hat{k}\hat{k}})$. The error covariance matrix of $\hat{\mathbf{h}}_k$ is $\mathbf{C}_{ee} = \mathbf{I}_{N_R} \otimes \mathbf{R}_{ee}$.

Thus $\mathbf{h}_k$ conditioned on $\hat{\mathbf{h}}_k$ is Gaussian distributed with conditional mean and covarance matrix as

$$\mathbf{u}_{k|\hat{k}} = \mathbb{E}[\mathbf{h}_k | \hat{\mathbf{h}}_k] = \hat{\mathbf{h}}_k \tag{8}$$

$$\mathbf{C}_{kk|\hat{k}} = \mathbb{E}\left[\left(\mathbf{h}_k - \mathbf{u}_{k|\hat{k}}\right) \left(\mathbf{h}_k - \mathbf{u}_{k|\hat{k}}\right)^H \middle| \hat{\mathbf{h}}_k \right] = \mathbf{C}_{ee} \tag{9}$$

3 Optimum Diversity Receiver with Channel Estimation Errors

In this section, an optimum diversity receiver for the SIMO system operating with channel estimation errors in a high mobility environment is designed by incorporating the statistics of the channel estimation errors.

3.1 Optimum Diversity Receiver

The receiver detects s_k in (3) based on knowledge of the received data vector $\mathbf{y}_k$ and the estimated CSI vector $\hat{\mathbf{h}}_k$. We have the following proposition regarding the optimum decision rule for the SIMO system with imperfect CSI.

Proposition 1. *Consider the SIMO system defined in* (3) *with estimated CSI* $\hat{h}_k$. *If the transmitted symbols are modulated with MPSK, MQAM, MASK, and they are equiprobable, then the optimum decision rule that minimises the system error probability is*

$$\hat{s}_k = \arg\min_{s_k \in \mathcal{S}} \left\{ |\alpha_k - s_k|^2 \right\}, \tag{10}$$

where $\mathcal{S}$ *is the modulation alphabet set with Cardinality* $\mathcal{M}$, *and* α_k *is the decision variable given as*

$$\alpha_k = \sqrt{E_c} \hat{\mathbf{h}}_k^H \left(E_c \mathbf{D}_{ee} + \sigma_z^2 \mathbf{I}_{N_R N} \right)^{-1} \mathbf{y}_k. \tag{11}$$

Proof. For system with equiprobable symbols, the error probability can be minimized by using the maximum likelihood (ML) detection rule. Conditioned on the transmitted symbol s_k and the estimated CSI vector $\hat{\mathbf{h}}_k$, the received data vector $\mathbf{y}_k$ is complex Gaussian distributed. Based on (8) and (9), the conditional mean vector and covariance matrix of $\mathbf{y}_k | (s_k, \hat{\mathbf{h}}_k)$ is

$$\mathbf{u}_{y|\hat{h}} = \sqrt{E_c} \hat{\mathbf{h}}_k s_k, \tag{12a}$$

$$\mathbf{D}_{y|\hat{h}} = E_c \mathbf{D}_{ee} + \sigma_z^2 \mathbf{I}_{N_R N} \tag{12b}$$

The Maximum Likelihood (ML) detection can then be expressed as

$$\hat{s}_k = \arg\min_{s_k \in \mathcal{S}} \left\{ \left(\mathbf{y_k} - \hat{\mathbf{h}}_k s_k \right) \mathbf{D}_{y|\hat{h}}^{-1} \left(\mathbf{y_k} - \hat{\mathbf{h}}_k s_k \right)^H \right\} \tag{13}$$

Simplifying the above equation with the fact that $|s_k| = 1$ leads to (10).

3.2   Unified Error Probability with Channel Estimation Errors

The unified error probability of the optimum diversity receiver with imperfect CSI and linear modulations is derived in this section.

Proposition 2. *For linearly modulated SIMO systems with optimum diversity receiver given in* (10), *the unified symbol error rates with channel estimation errors is*

$$P(E) = \sum_{i=1}^{2} \frac{\beta_i}{\pi} \int_0^{\psi_i} \left[\det \left(\mathbf{I}_N + \frac{\xi}{\sin^2(\phi)} \boldsymbol{\Lambda} \right) \right]^{-N_R} d\phi, \tag{14}$$

where

$$\boldsymbol{\Lambda} = \mathbf{R}_{kp} \left(\mathbf{R}_{pp} + \frac{1}{\gamma_p} \mathbf{I}_N \right)^{-1} \mathbf{R}_{kp}^H \times$$

$$\left[\mathbf{R}_{pp} - \mathbf{R}_{kp} \left(\mathbf{R}_{pp} + \frac{1}{\gamma_p} \mathbf{I}_N \right)^{-1} \mathbf{R}_{kp}^H + \frac{1}{\gamma_c} \mathbf{I}_N \right]^{-1}. \tag{15}$$

and the values for ξ, β_i and ψ_i for the modulation schemes are listed in Table 1.

Proof. Based on [10–12]

$$P(E) = \sum_{i=1}^{2} \frac{\beta_i}{\pi} \int_0^{\psi_i} \left[\det \left(\mathbf{I}_{N_R N} + \frac{\xi}{\sin^2(\phi)} \boldsymbol{\Delta} \right) \right]^{-N_R} d\phi, \tag{16}$$

where

$$\boldsymbol{\Delta} = \mathbf{D}_{\hat{k}\hat{k}} \left(\mathbf{D}_{ee} + \frac{1}{\gamma_c} \mathbf{I}_{N_R N} \right)^{-1} \tag{17}$$

$$= \mathbf{I}_{N_R} \otimes \mathbf{R}_{\hat{k}\hat{k}} \left[\mathbf{I}_{N_R} \otimes \left(\mathbf{R}_{ee} + \frac{1}{\gamma_c} \mathbf{I}_N \right) \right]^{-1} \tag{18}$$

where the identity $\mathbf{A} \otimes \mathbf{B} + \mathbf{A} \otimes \mathbf{C} = \mathbf{A} \otimes (\mathbf{B} + \mathbf{C})$ is used in the second equation.

   Based on the fact that $(\mathbf{A} \otimes \mathbf{B})^{-1} = \mathbf{A}^{-1} \otimes \mathbf{B}^{-1}$ and $(\mathbf{A} \otimes \mathbf{B})(\mathbf{C} \otimes \mathbf{D}) = (\mathbf{AC}) \otimes (\mathbf{BD})$, we have

$$\boldsymbol{\Delta} = (\mathbf{I}_{N_R} \otimes \mathbf{R}_{\hat{k}\hat{k}}) \left[\mathbf{I}_{N_R} \otimes \left(\mathbf{R}_{ee} + \frac{1}{\gamma_c} \mathbf{I}_N \right)^{-1} \right] \tag{19}$$

$$= \mathbf{I}_{N_R} \otimes \left[\mathbf{R}_{\hat{k}\hat{k}} \left(\mathbf{R}_{ee} + \frac{1}{\gamma_c} \mathbf{I}_N \right)^{-1} \right] \tag{20}$$

$$= \mathbf{I}_{N_R} \otimes \boldsymbol{\Lambda} \tag{21}$$

Thus

$$\det\left(\mathbf{I}_{N_R N} + a\boldsymbol{\Delta}\right) = \det\left(\mathbf{I}_{N_R} \otimes (\mathbf{I}_N + a\boldsymbol{\Lambda})\right) \tag{22}$$

$$= \left[\det\left(\mathbf{I}_N + b\boldsymbol{\Lambda}\right)\right]^{N_R} \tag{23}$$

where $b = \frac{\xi}{\sin^2(\phi)}$, and the second equality is based on the fact that $\det(bf A \otimes \mathbf{B}) = \det(\mathbf{A})^{N_B} \det(\mathbf{B})^{N_A}$, where N_A and N_B are the dimension of the square matrices $\mathbf{A}$ and $\mathbf{B}$, respectively.

Combining (16) with (23) completes the proof.

Table 1. Parameters of the unified error-probability expressions

	ξ	β_1	β_2	ψ_1	ψ_2
MASK	$\frac{3}{M^2-1}$	$\left(2 - \frac{2}{M}\right)$	0	$\frac{\pi}{2}$	0
MQAM	$\frac{3}{M-1}$	$\left(4 - \frac{4}{\sqrt{M}}\right)$	$\left(2 - \frac{2}{\sqrt{M}}\right)^2$	$\frac{\pi}{2}$	$\frac{\pi}{4}$
MPSK	$\sin^2 \frac{\pi}{M}$	1	0	$\left(\pi - \frac{\pi}{M}\right)$	0

3.3 Normalized Diversity Order

Since one pilot symbol is transmitted for every K data symbols, the effective energy devoted for the transmission of one coded data symbol is

$$E_0 = \frac{E_p + KE_c}{K} = \frac{1}{K}E_p + E_c. \tag{24}$$

Define the effective SNR of one coded symbol as $\gamma_0 = \frac{E_0}{\sigma_z^2} = \frac{1}{K}\gamma_p + \gamma_c$. The Doppler diversity order for N repetitions of a repetition code can then be computed as [15]

$$D_N = -\lim_{\gamma_0 \to \infty} \frac{\log P(E)}{\log \gamma_0}. \tag{25}$$

For a Doppler diversity system with a codeword that covers the time duration NT_p, from (25), define the normalized Doppler diversity order as [7,12,13].

$$D = -\lim_{\substack{\gamma_0 \to \infty \\ N \to \infty}} \frac{1}{NT_p} \frac{\log P(E)}{\log \gamma_0} \tag{26}$$

In (26), the diversity order is defined as the negative slope of the error probability in log scale when γ_0 is large. The diversity order depends only on the slope instead of the actual values of error probability [12].

3.4   Spectral Efficiency

In this section, the expression for computing the spectral efficiencies of the modulation schemes in order to determine the most spectral efficient is given and verified through simulation in the next section.

The spectral efficiency of transmission of each of the linear modulation schemes is given as

$$\eta = C(1 - BER) \tag{27}$$

where $C = \frac{K}{T_s(K+1)} \log_2 M$ is the total number of data bits transmitted, M is the modulation level, T_s is the symbol period, K is the number of unique data symbols and BER is the bit error rate.

4   Numerical Results

Simulation and Analytical results are given in this part to investigate the trade-off between channel estimation errors and Doppler diversity to illustrate the error performances of the linear modulation schemes and to validate the theoretical results with the simulation results. We consider a high mobility system operating at 1.9 GHz with a symbol rate of 100 ksym/s. The range of Doppler spread is between 200 Hz ($f_D T_s = 2 \times 10^{-3}$) to 1 KHz ($f_D T_s = 10^{-2}$), which correspond to mobile speeds between 113.6 km/hr and 568.4 km/hr, The value of K is chosen as $K = 19$, so that the pilot symbols sample the channel at a rate $R_p = 5$ KHz, which is well above the Nyquist rates of the channel.

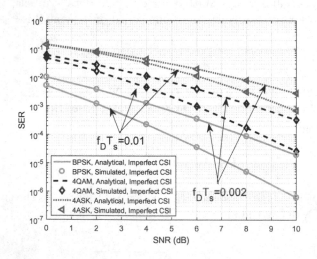

Fig. 2. The analytical and simulated SER as a function of SNR.

Figure 2 depicts the analytical and simulated SER for different values of $f_D T_s$. The parameters are $N = 5$ ($N_{\text{sym}} = 100$), $\gamma_p = \sqrt{K}\gamma_c$, and $N_R = 2$. BPSK, 4QAM and 4ASK modulation schemes are employed in the system.

The analytical results are obtained from (14), and the simulation results are obtained through Monte Carlo simulations. Each point in the simulation results is obtained by averaging over 1,000 trials. The analytical SERs of each of the modulation schemes are expressed as a unified expression in (14). At Doppler spread $f_D T_s = 0.1$, analytical the SERs can serve as lower bounds for Doppler diversity systems with channel estimation errors at lower Doppler spread (i.e $f_D T_s < 0.1$). There is excellent agreement between simulation results and analytical results. For a given Doppler spread $f_D T_s$, the slopes of the SER curves of systems with imperfect CSI are the same, which implies they have the same diversity order. As expected, the diversity order increases with $f_D T_s$.

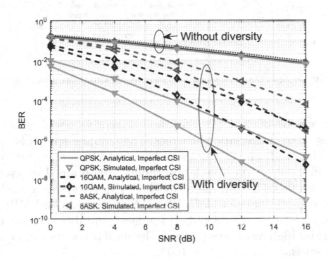

Fig. 3. BER as a function of γ_0 for different modulation schemes

Results similar to those in Fig. 2 are given in Fig. 3, with higher order of modulation together with systems without diversity, where analytical SER results for systems with Doppler diversity are obtained from (14) and dividing the SER results by $\log_2 M$ yields the analytical BER results. Then, the BER results are plotted as a function of γ_0 for different values of the modulation schemes and $N_R = 2$. Simulation parameters are $N = 5$ ($N_{sym} = 100$), $\gamma_p = \sqrt{K}\gamma_c$ [13], and modulation schemes are QPSK, 16QAM and 8PSK. The BER curves of systems with diversity have same slope for a given modulation scheme, which implies they have the identical Doppler diversity order. The results for systems without Doppler diversity are obtained from [14]. Excellent agreement is observed between simulation results and their analytical counterparts for both systems with diversity and systems without diversity. The error performance in this case of systems without diversity, is dominated by channel estimation errors. The results of Figs. 2 and 3 also show that, Doppler diversity order is independent of the modulation scheme employed in the system since the Doppler diversity is always the same irrespective of the modulation scheme employed.

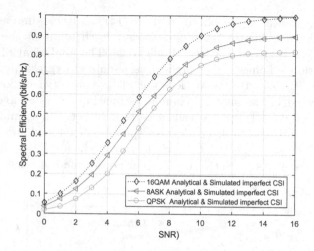

Fig. 4. Spectral efficiency for various Modulation schemes

Figure 4 shows the spectral efficiency for different values of γ_0 for $N_R = 2$ for different values of the modulation schemes. The Doppler spread is $f_D T_s = 0.01$. Analytical results are from (27) and simulation results are Monte carlo simulations where each point in the simulation results are obtained by averaging over 1,000 trials. It can be seen from the figure that, MQAM systems exhibit the best spectral efficient performance followed by MASK systems and MPSK systems. MQAM can therefore be considered as the most suitable spectral modulation schemes for SIMO high mobility systems. The similar results can be obtained at other Doppler spreads i.e. at $f_D T_s = 0.002$.

5   Conclusion

The error performance analysis of linearly modulated single-input-multiple-output (SIMO) high mobility communication systems with channel estimation errors has been investigated in this paper. An optimal diversity receiver for MPSK, MQAM and MASK SIMO systems with MMSE channel estimation errors has been derived. The optimum receiver was designed through the analysis of the statistical attributes of the estimated channel coefficients. Exact unified error probability expression of the optimal receiver has been obtained. It quantifies the impacts of both Doppler diversity and channel estimation errors. Simulation results show that MPSK, MASK, and MQAM systems have the same Doppler diversity order even though they differ in error probability performances. Moreover, simulation results also show that MQAM systems achieve better spectral efficiency than its MPSK and MASK counterparts.

References

1. Wu, J., Fan, P.: A survey on high mobility wireless communications: Challenges, opportunities and solutions. IEEE Access **4**, 450–476 (2016)
2. Sayeed, A., Aazhang, B.: Joint multipath-Doppler diversity in mobile wireless communications. IEEE Trans. Commun. **47**(1), 123–132 (1999)
3. Wu, J.: Exploring maximum Doppler diversity by Doppler domain multiplexing. In: Proceedings of IEEE Global Telecommunication Conference, GLOBECOM 2006, pp. 1–5 (2006)
4. Ma, X., Giannakis, G.: Maximum-diversity transmissions over doubly selective wireless channels. IEEE Trans. Inf. Theory **49**(7), 1832–1840 (2003)
5. Zhou, W., Wu, J., Fan, P.: Maximizing Doppler diversity for high mobility systems with imperfecr channel state information. In: Proceedings of IEEE ICC, pp. 5920–5925 (2014)
6. Zhou, W., Wu, J., Fan, P.: Energy and spectral efficient Doppler diversity transmission in high mobility systems with imperfect channel estimation. EURASIP J. Wireless Commun. Netw. **140** (2015). http://jwcn.eurasipjournals.com/content/2015/1/140
7. Zhou, W., Wu, J., Fan, P.: On the maximum Doppler diversity of high mobility systems with imperfect channel state information. In: Proceeding of IEEE International Conference Communication ICC 2015 (2015)
8. Wu, J., Xiao, C.: Optimal diversity combining based on linear estimation of Rician fading channels. IEEE Trans. Commun. **56**(10), 1612–1615 (2008)
9. Wu, J., Xiao, C.: Optimal diversity combining based on linear estimation of Rician fading channels. IEEE Int. Conf. Commun. **56**(10), 3999–4004 (2007)
10. Wu, J., Xiao, C.: Performance analysis of wireless systems with doubly selective rayloegh fading. IEEE Trans. Vech. Technol. **56**(2), 721–730 (2007)
11. Mahamadu, M.A., Wu, J., Ma, Z., Zhou, W., Fan, P.: Maximum diversity order of SIMO high mobility systems with imperfect channel state information. In: Proceedings of IEEE International Conference Communication Technology (ICCT) (2017)
12. Mahamadu, M.A., Wu, J., Ma, Z., Zhou, W., Tang, Y., Fan, P.: Fundamental tradeoff between doppler diversity and channel estimation errors in simo high mobility systems. IEEE Access **5**, 21867–21878 (2018)
13. Zhou, W., Wu, J., Fan, P.: High mobility wireless communications with doppler diversity: performance limit. IEEE Trans. Wireless Commun. **14**(12), 6981–8992 (2015)
14. Sun, N., Wu, J.: Maximizing spectral efficiency for high mobility systems with imperfect channel state information. IEEE Trans. Wireless Commun. **13**(3), 1462–1470 (2014)
15. Zheng, Z., Tse, D.N.C.: Diversity and multiplexing: a fundamental tradeoff in multiple-antenna channels. IEEE Trans. Inf. Theory **49**(5), 1073–106 (2003)

Minimum Cost Offloading Decision Strategy for Collaborative Task Execution of Platooning Assisted by MEC

Taiping Cui[✉], Xiayan Fan, Chunyan Cao, and Qianbin Chen

School of Communication and Information Engineering, Chongqing University of Posts and Telecommunications, Chongqing 400065, China
{cuitp, chenqb}@cqupt.edu.cn,
Fanxiayan, Caocyan@yeah.net

Abstract. In this paper, we study the offloading decision of collaborative task execution between platoon and MEC (Mobile Edge Computing) server. The mobile application is represented by a series of fine-grained tasks that form a linear topology, each of which is either executed on a local vehicle, offloaded to other members of the platoon, or offloaded to a MEC server. The objective of the design is to minimize the cost of task offloading and meet the deadline of tasks execution. We transform the cost minimized task decision problem into the shortest path problem, which is limited by the deadline of the tasks on a directed acyclic graph. The classical LARAC algorithm is used to solve the problem approximately. Numerical analysis shows that the scheduling method of the tasks decision can be well applied to the platoon scenario and execute the task in cooperation with the MEC server. In addition, compared with different execution models, the optimal offloading decision for collaborative task execution can significantly reduce the cost of task execution and meet lower deadlines.

Keywords: Platooning · Mobile edge computing · Offloading decision

1 Introduction

Platooning realizes the reduction of fuel consumption and gas emission, as well as safe and efficient transportation in the context of intelligent transportation system (ITS). Generally, the platoon is consisted of two parts: one is the leader and the second are the members of the platoon (including the tail vehicle, the relay vehicle and the controller). The higher the frequency of the vehicle information exchange in the platoon, the faster the mobile response of the members in the platoon, the more the status of the platoon instability can be avoided.

This work was supported in part by the National Natural Science Foundation of China (61401053, 6157073), and the Doctor Start-up Funding of CQUPT (A2016-83).

X. Liu et al. (Eds.): ChinaCom 2018, LNICST 262, pp. 104–115, 2019.
https://doi.org/10.1007/978-3-030-06161-6_10

Vehicle terminals have been widely deployed in the automotive industry. More novel and attractive vehicle services have attracted more and more people to use them. In the future, cars will be equipped with AR (Augmented Reality) applications that allow drivers to observe the surroundings of vehicles in windowless vehicles [1]. These types of vehicle applications are typical computing resource-hungry services, requiring high density computing resources and computing costs. Therefore, the tension between computing resource-hungry applications and vehicle terminals with limited computing resources poses a major challenge to the development of mobile platforms [2].

MEC provides a promising solution to this challenge and extends the capabilities of vehicle terminals, by providing additional computing, storage, and bandwidth resources in an on-demand manner. For example, there is a paper on energy consumption [3], which proposes collaborative execution between the end-user and the cloud, and accomplishes the task with minimum energy consumption within the task deadline. In addition, some people consider offloading decision from the point of view of the servers [4], set the price for the unit resource provided to each vehicle user, and use the gain function of task offloaded to maximize the profit of the server.

In this paper, we consider the platooning scenario with MEC server to complete the task offloading. Specifically, within the deadline of the task, the resource cost price is used to determine whether the tasks are offloaded to the other members of the platoon or the MEC server, or not. We aim to find an optimal collaborative offloading decision between each member of the platoon and the MEC with minimum resource cost price within the deadline. Mathematically, we model a minimum cost offloading problem as a constrained shortest path problem on a directed acyclic graph. Then we use the classical Lagrangian Relaxation Based Aggregated Cost (LARAC) algorithm to obtain a suitable result of the constrained optimization problem.

The rest of the paper is organized as follows. We give the system model in Sect. 2. In Sect. 3, delay constrained offloading decision is modeled as a limited shortest path problem. Then we can get a suitable strategy of optimal offloading decision for collaborative task execution in Sect. 4. Section 5 shows the offloading decision procedure numerical analysis, and Sect. 6 concludes the paper.

2 System Model

Suggest that the MEC server coexists with the base station (BS). Because the platoon controller controls and manages the whole platoon [5], when the platoon members communicate with the BS, they need to transmit the messages to the BS through the controller, and the members in a platoon can communicate with others directly [6]. Position of the controller in the platoon is not clearly defined. For the sake of simplification, the current researches directly select the leader as platoon controller. Also in this paper, we do not discuss in detail how to choose.

2.1   Task Model

Figure 1 illustrates the task model in a linear topology. Each task is executed in sequence, the output of the previous task is the input of the later task, and the whole application has a completion deadline T_d. Figure 1 shows that there are $n+1$ tasks in an application. Define the needed computation cycles of the task k as ω_k, $k = 0, 1, 2, ..., n+1$.

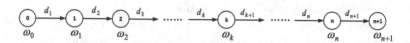

Fig. 1. Task model in a linear topology

2.2   Path Loss Model

When one vehicle is served by another, the path loss model [7] between the two vehicles is defined as $PL_v(l_{x,y}) = 63.3 + 17.7 log_{10}(l_{x,y})$. And the path loss model [8] between vehicle and BS is defined as $PL_{MEC}(l_{x,y}) = 128.1 + 37.5 log_{10}(l_{x,y})$. $l_{x,y}$ is distance in kilometers. Here, we do not consider fast fading and shadow fading. The carrier frequency used in V2I (Vehicle to Infrastructure) communication is 2 GHz and V2V communication is 5.9 GHz, so there is no interference between them, and we think of the whole channel as an ideal channel.

The MEC server numbered 0, a leader vehicle numbered 1, and the sequence number of vehicles behind it increases in turn. There are a total of m vehicles in the paltoon. $l_{x,y}$ is the transmission distance from x to y, $x \in \{0, 1, 2, ..., m\}$, $y \in \{0, 1, 2, ..., m\}$, assuming the antenna position of each vehicle is the same and that the distance between vehicles is fixed in the platoon, so $l_{x,y} = l_{y,x}$.

Considering that under ideal conditions all the vehicles in the platoon are of the same length and the same spacing so the distance between the signal receiver and the transmitter is the length of the vehicle plus the vehicle spacing. So we can get $l_{v_1,v_2} = \mu |v_1 - v_2|$, where $v_1 \in \{1, 2, ..., m\}$ is for the current vehicle number about task $k-1$ and $v_2 \in \{1, 2, ..., m\}$ is for the destination vehicle of offloading decision about task k. Specially, consider offloading tasks to the MEC server, and the distance between the leader and the BS is set to η, so we obtain $l_{1,0} = \eta$. When the task is offloaded to the MEC server, the change of distance between the leader and the BS has little effect on the path loss, so we set $l_{1,0} = l_{0,1}$. Only the leader can communicate with the BS, and the platoon members communicate with the BS through the leader.

2.3   Task Execution Model

Platoon execution. Assume that the tasks is initiated by any member of the platoon. The task execution of the platoon includes the local task execution and the tasks platoon members execution. If the task k is offloaded to execute by a member of the platoon, the completion time of the task k is defined as $t_k^{v_2} = \frac{\omega_k}{f_{v_2}}$, and the cost it has to pay is $b_k^{v_2} = \alpha_v f_{v_2}$, f_{v_2} is the computation resource that the vehicle v_2 can provide, and α_v is the computation resource cost of each unit provided by the platoon members. Suppose that the unit price of resources provided by each member is the same. It is assumed that the f_{v_2} is fixed during the execution of the task.

MEC server execution. If the task k is executed in the MEC server, the platoon members are idle during the execution of the task. Define the computing completion time of the task k offloaded to the MEC server as $t_k^0 = \frac{\omega_k}{f_0}$, and the cost of offloading the task is $b_k^0 = \alpha_{MEC} f_0$. f_0 is the computation resource that the MEC server can provide, and the computation resource cost of each unit provided by the MEC server is α_{MEC}. Note $f_0 > f_1, f_0 > f_2, ..., f_0 > f_m$, speculate that the CPU speed of the MEC server is faster than that of any platoon member.

Platoon data transmission. If the task k is executed inside the platoon, the data needs to be offloaded to the destination vehicle before it is executed. If the task is calculated on the current vehicle, there is no need to transmit the data. Define the time of data transmission that offloads the task k from vehicle v_1 to vehicle v_2 as $t_{v_1,v_2}^k = \frac{d_k}{R_{v_1,v_2}^k}$, R_{v_1,v_2}^k is the data transmission rate from vehicle v_1 to vehicle v_2. We have $R_{v_1,v_2}^k = B_1 log_2(1 + \frac{\zeta_v PL_v(l_{v_1,v_2})}{N_0})$. ζ_v is the signal transmission power of the vehicle, N_0 is the noise power and B_1 is the bandwidth used for V2V (Vehicle to Vehicle) communication. If $v_1 = v_2$, $t_{v_1,v_2} = 0$.

MEC server data transmission. If the task k needs to be executed on the MEC server, the time that the data is transferred from the vehicle v_1 to the MEC server and from the MEC server to the vehicle v_2 are $t_{v_1,0}^k = \frac{d_k}{R_{v_1,1}^k} + \frac{d_k}{R_{1,0}^k}$ and $t_{0,v_2}^k = \frac{d_k}{R_{1,v_2}^k} + \frac{d_k}{R_{0,1}^k}$ respectively. $R_{1,0}^k$ and $R_{0,1}^k$ also indicating the transmission rate of the data from the leader to the MEC server and from MEC server to leader respectively. We have $R_{1,0}^k = B_2 log_2(1 + \frac{\zeta_v PL_{MEC}(l_{1,0})}{N_0})$ and $R_{0,1}^k = B_2 log_2(1 + \frac{\zeta_{MEC} PL_{MEC}(l_{0,1})}{N_0})$. ζ_{MEC} is the signal transmission power of the BS and B_2 is the bandwidth used for V2I communication. The connection between BS and MEC is wired, and the time of wired transmission is ignored here. The same as before, $t_{0,0}^k = 0$.

In this paper, the following assumptions are used to model the practical problem of collaborative task execution. First, the tasks of the application needs to have been replicated on the other platoon members or the MEC server before it is executed. Second, the first and last task must be executed in the same vehicle in the platoon.

3   Delay Constrained Offloading

The tasks of an application is modeled by the directed acyclic graph $G = (V, C)$ between the platoon members and the MEC server. V represents the finite node set, and C represents the edge set. The task execution flow shown in Fig. 2. shows that S is the starting point of the task (Assume that the vehicle m is the requestor of the task) and D is the task destination node. An application contains $n + 2$ tasks. Except for the initial task and the last task, all the platoon members have the opportunity to execute the other n tasks. The weights of the edges of nodes x and y are a non-negative value, that is $c_{x,y}^k$, and each edge corresponds to a task offloading decision s_k. This weight value includes the cost of offloading the task and the time it takes to execute after offloading the task. Specifically, if the weight value of the edge is considered to be a cost price and the task k needs to be offloaded from x to y to be executed, we obtain the weight $c_{x,y}^k = b_k^y$. And if the weight value is considered to be a time delay, we obtain the weight value $c_{x,y}^k = t_k^y + t_{x,y}^k$ that is the sum of the computation time and the transmission time of the task. So the two directed acyclic graphs with respect to the cost price and time delay can be obtained. s_k is defined as a decision variable for state k, indicating the choice of which vehicle the task k should be offloaded to. s_0 denotes the task initiation decision and s_{n+1} denote the end decision of the task result, $s_0 = s_{n+1}$. This goal is to find an optimal offloading decision strategy $S^* = \{s_0, s_1, ..., s_{n+1}\}$.

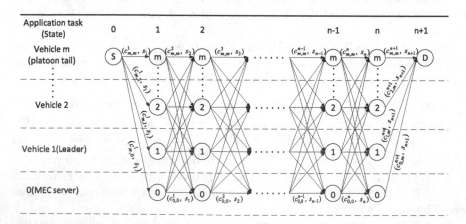

Fig. 2. The task execution flow and state transition procedure of collaborative offloading decision

Under this framework, we can transform the task optimal offloading decision into a shortest path problem to find the path with minimum cost between nodes

S and D. And it is constrained by the deadline time of the task, and the time delay path of the task can only be less than or equal to T_d. If the task time delay of a path p satisfies the constrained condition, p is an appropriate path. A path p^* is an optimal path with the minimum path cost of all appropriate paths. We can Mathematically formulate this problem as a constrained shortest path problem:

$$\min_{p\in P} \quad b(p) = \sum_{k=1}^{n} b_{x,y}^k \tag{1}$$

$$\text{s.t.} \quad (C1): \quad d(p) = \sum_{k=1}^{n+1} (t_{x,y}^k + t_k^y) \le T_d$$

$$(C2): \quad f_{v_1} < f_0, \ f_{v_2} < f_0, \quad v_1 \in x, \ v_2 \in y$$

$$(C3): \quad \alpha_v f_{v_2} < \alpha_{MEC} f_0$$

p is a path from node S to D and P is a set of all possible paths. Since each task has $m+1$ offloading decisions options, there are $(m+1)^n$ possible options for the strategy. The constrained optimization problem has been proved to be NP-complete [9].

4 Optimal Offloading Decision for Collaborative Task Execution

4.1 Based on LARAC algorithm

This constrained shortest path problem can be solved by LARAC algorithm [12]. we first define a LARAC function $L(\lambda) = \min_{p\in P}[b_\lambda(p)] - \lambda T_d$ where $b_\lambda(p) = b(p) + \lambda d(p)$ and the Lagrangian multiplier is λ. By using the Lagrangian duality principle, we can obtain the proof of $L(\lambda) \le b(p^*)$.

Then, we use algorithm 1 to find the path of smallest b_λ between S and D. In algorithm 1, $PathAlgorithm$ is a procedure of finding a shortest path of the cost C. If we find the minimum cost path within the deadline, this path is the offloading strategy, or we update p_b and p_d repeatedly to get an optimal λ. Although the algorithm cannot guarantee to find the optimal path, it can obtain a lower bound of the optimal solution. And its running time is shown to be polynomial [10].

Algorithm 1 Finding minimum cost path of b_λ for collaborative task execution

1: **Input:** S, D, T_d
2: $p_b \leftarrow PathAlgorithm(S, D, b)$
3: **if** $d(p_b) \leq T_d$ **then**
4:     *return p_b*
5: **end if**
6: $p_d \leftarrow PathAlgorithm(S, D, d)$
7: **if** $d(p_d) > T_d$ **then**
8:     *return "There is no solution"*
9: **end if**
10: **while** *true* **do**
11:     $\lambda \leftarrow \frac{c(p_b) - c(p_d)}{d(p_d) - d(p_b)}$
12:     $p_\lambda \leftarrow PathAlgorithm(S, D, b_\lambda)$
13:     **if** $b_\lambda(p_\lambda) = b_\lambda(p_d)$ **then**
14:         *return p_d*
15:     **else**
16:         **if** $d(p_\lambda \leq T_d)$ **then**
17:             $p_d \leftarrow p_\lambda$
18:         **else**
19:             $p_b \leftarrow p_\lambda$
20:         **end if**
21:     **end if**
22: **end while**
23: **Output:** p_λ^*

4.2 Dynamic Programming Algorithm for Optimal Offloading Decision

In order to apply the Algorithm 1 to the optimal offloading decision, we need to find the shortest path according to the task execution cost, execution time and aggregation cost. Specifically, we view all tasks as a multistep process with chain structure, that is, a multistep decision process.

The state transition process for the optimal offloading decision of collaborative task execution is shown in Fig. 4. State 0 and state $n + 1$ represent the start and end of the whole application execution respectively. State k represents that the task k has been executed, and $k = 0, 1, 2, ..., n + 1$. We define r_k as the location identifier of the task k that has been executed, so $r_k = 1$ indicates that the task k is executed at the position of the vehicle 1. Particularly, r_k keeps tracking the position of the task. Task $n + 1$ is considered to be the output result of the application. Since the output result need to be sent back to the starting point after the completion of task n, it is assumed that the task is initiated from vehicle m, so it can be obtained that $r_0 = m$ and $r_{n+1} = m$ are the start and end point of the application task. The output result does not need to be calculated, and the user does not need to buy the computing resource so we have $b_{n+1}^m = 0$.

In the next part, we will find the minimum cost, time delay, and aggregation cost by using the established iterative equations, respectively.

First, we define $G_{k-1}(r_{k-1})$ as the minimum cost of task $k-1$ to task $n+1$. $G_0(r_0)$ is the minimum cost for all tasks of the application, given $G_{n+1}(r_{n+1}) = 0$. On this basis, we can establish an iterative formula of the latter term value for the minimum cost. Knowing $G_k(r_k)$ at state k for location r_k, we can obtain every decision at state $k-1$ so that the cost from state $k-1$ to state $n+1$ is minimized. The backward value iteration equation of minimum purchase cost is shown as follows:

$$G_{k-1}(r_{k-1}) = \min_{s_k}[b_k(r_{k-1}, s_k) + G_k(r_k)]$$
$$G_{n+1}(r_{n+1}) = 0 \tag{2}$$

$b_k(r_{k-1}, s_k)$ refers to the cost to be paid for making the offloading decision s_k of task k after task $k-1$ at position r_{k-1} is completed. Both $G_k(r_k)$ and $b_k(r_{k-1}, s_k)$ are known values. The value of latter one can be obtained when the task is offloaded, $b_k(r_{k-1}, s_k) = b_k^y$, and $s_k = y$. The system state starts from the state n, and the value of the previous state $G_n(r_n)$ can be obtained from the initial condition $G_{n+1}(r_{n+1}) = 0$ given in the state $n+1$. Then repeat this procedure for numerical iteration, and you can find the optimal objective function value, optimal decision, and optimal path for the entire multistep decision problem in reverse order.

Second, we define $H_{k-1}(r_{k-1})$ as the minimum completion time of task $k-1$ to task $n+1$. $G_0(r_0)$ is for the minimum task completion time of all tasks, given $H_{n+1}(r_{n+1}) = 0$. The backward value iteration equation of minimum task completion time is shown as follows:

$$H_{k-1}(r_{k-1}) = \min_{s_k}[t_k(r_{k-1}, s_k) + H_k(r_k)]$$
$$H_{n+1}(r_{n+1}) = 0 \tag{3}$$

Third, we define $J_{k-1}(r_{k-1})$ as the minimum aggregated cost of task $k-1$ to task $n+1$. $J_0(r_0)$ is the minimum aggregated cost for all tasks of the application, given $J_{n+1}(r_{n+1}) = 0$. The backward value iteration equation of minimum task aggregated cost is shown as follows:

$$J_{k-1}(r_{k-1}) = \min_{s_k}[b_k(r_{k-1}, s_k) + \lambda d_k(r_{k-1}, s_k) + J_k(r_k)]$$
$$J_{n+1}(r_{n+1}) = 0 \tag{4}$$

We can use the iterative equations (2), (3), and (4) to implement the processes of $PathAlgorithm(S, D, b)$, $PathAlgorithm(S, D, d)$, and $PathAlgorithm(S, D, b_\lambda)$ to find the minimum cost, time delay and aggregated cost, respectively. Finally, the minimum cost offloading decision strategy is obtained in Algorithm 1.

5 Numerical Analysis

In this section, we evaluate the performance of task decision strategies for collaborative task execution. We use some of the parameters in [8,11] and communication between vehicles consider the use of carrier aggregation technology

[12] to increase bandwidth, as shown below: $B_1 = 20\,\text{MHz}$, $B_2 = 100\,\text{MHz}$, the thermal noise density $-174\,\text{dbm/Hz}$, $p_v = 23\,\text{dBm}$, $p_{MEC} = 46\,\text{dBm}$. Assume that the members of the platoon consist of nine vehicles, $\{f_v\} = \{3000\ 100\ 620\ 660\ 600\ 900\ 700\ 630\ 550\ 800\}$ MHz, $\alpha_v = 1$, $\alpha_{MEC} = 0.9$, $\mu = 0.008$, $\eta = 0.5$.

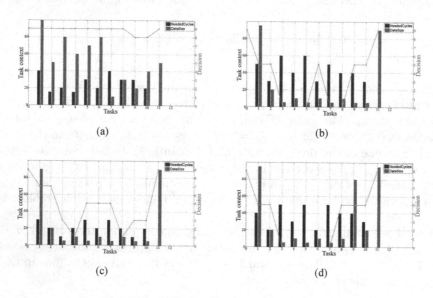

Fig. 3. Task offloading decision

We consider a mobile application that consists of 12 tasks, the application deadline is 0.4s, and the offloading decision and details are shown in Fig. 3. In Fig. 3a, the decision strategy is $S^* = \{9, 9, 9, 9, 9, 9, 9, 9, 9, 8, 8, 9\}$ and the amount of computation cycles required for each task of the application is small, but relatively large transmission data size of the task limits the execution of the task on the MEC server, because the transmission data rate between vehicles is much faster than that between vehicles and BS. If the task is to be offloaded to the MEC server, it will lead to a large transmission delay. In Fig. 3b, the decision strategy is $S^* = \{9, 5, 5, 0, 0, 0, 5, 0, 0, 5, 5, 9\}$ and the size of the transmitted data is small, and the computation cycles of each task requirement is increased. In order to satisfy the time constrained requirement of the task, a task whose computation cycles is too high is offloaded to the MEC server. The reason why the first task is not offloaded to the MEC server is that an excessively high transmission delay will be caused. In Fig. 3c, the decision strategy is $S^* = \{9, 7, 7, 3, 1, 5, 5, 5, 1, 3, 3, 9\}$ and each task requires a small amount of computation cycles, so the MEC server doesn't need to provide computational assistance to the task execution. In Fig. 3d, the decision strategy is $S^* = \{9, 5, 5, 0, 0, 0, 0, 0, 5, 5, 5, 9\}$ and the last four tasks are executed in platoon to avoid high data transmission delays from the BS back to the vehicle. These four cases show that the MEC server and the platooning can complete the task cooperatively. According to the different parameters of the task, the tasks can be offloaded reasonably, and the deadline of the task can be satisfied. In addition, there is a ratio between the computation cycles and the data size of a task, which should be greater than a threshold when a task needs to be offloaded to a MEC server; similarly, when a task needs to be offloaded to a vehicle in the platoon, the ratio of the computation cycles and the size of the data should also be greater than a threshold. We will find this threshold in future research.

In Fig. 4, we compare the cost under three execution modes and CBF (the Constrained Bellman-Ford [13]) algorithm, that is, platoon execution, collaborative execution, and the results obtained using the CBF algorithm. The parameters are set to: $\{d_k\} = \{100\ 40\ 1\ 2\ 1\ 2\ 1\ 2\ 1\ 1\ 100\}$ kb, $\{w_k\} = \{40\ 20\ 50\ 30\ 50\ 20\ 40\ 30\ 30\ 20\}$ $Mcycles$. First, compared with MEC server execution, collaborative execution can greatly reduce the cost when the task deadline is large. In most deadlines, collaborative task execution reduces task execution cost by more than four times. Second, collaborative task execution is more flexible than MEC server execution. That's because of the low transmission rate between the vehicle and the BS, the high transmission delay will be caused by the MEC server execution, so the tasks with a large amount of data transmission can not be completed within the deadline. Third, only collaborative task execution can complete the application within 0.25 s of the deadline, and it costs less than the MEC server to execute. Fourth, due to the increase of the deadline, the cost of platoon task execution and collaborative task execution is the same when the deadline is 0.45s, because after this deadline, the task execution does not require the participation of MEC server. The computational resources provided in the platoon are sufficient to enable the task to be completed within the deadline. The results of the local execution of the task were not drawn, because 3.3s is the minimum deadline, far from meeting the requirements of the task time constrained. Fifth, we can see that the collaborative task execution is very similar to the CBF algorithm, so the algorithm of collaborative task execution can be well applied in this scenario.

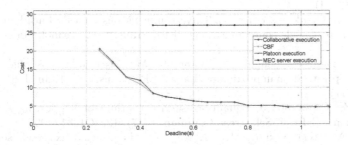

Fig. 4. Cost vs deadline

6   Conclusion

In this paper, the procedure of platoon and MEC server executing the task cooperatively within the task deadline is studied. We transform the task decision problem into the shortest path problem in a directed acyclic graph. We use the "LARAC" algorithm to obtain the optimal decision strategy for the tasks. Our research shows that there are more than one migration between the platoon members and the MEC service, and all the members have the opportunity to participate in the tasks execution. In addition, collaborative task execution can greatly reduce task execution cost and execution time.

In future research, the topological model of the tasks can be extended to various graphs (such as grid, tree, etc.). According to these structural characteristics, we will establish an optimal task decision strategy.

References

1. Shah, S.A.A., Ahmed, E., Imran, M., Zeadally, S.: 5G for Vehicular Communications. IEEE Commun. Mag. **56**(1), 111–117 (2018)
2. Mao, Y., You, C., Zhang, J., Huang, K., Letaief, K.B.: A survey on mobile edge computing: the communication perspective. IEEE Commun. Surv. Tutor. **19**(4), 2322–2358 (2017). Fourthquarter
3. Zhang, W., Wen, Y., Wu, D.O.: Energy-efficient scheduling policy for collaborative execution in mobile cloud computing. In: Proceedings IEEE INFOCOM, pp. 190–194. Turin (2013)
4. Zhang, K., Mao, Y., Leng, S., Maharjan, S., Zhang, Y.: Optimal delay constrained offloading for vehicular edge computing networks. In: 2017 IEEE International Conference on Communications (ICC), pp. 1 6. Paris (2017)
5. Khaksari, M., Fischione, C.: Performance analysis and optimization of the joining protocol for a platoon of vehicles. In: 5th International Symposium on Communications. Control and Signal Processing, pp. 1–6. Rome (2012)
6. Shao, C., Leng, S., Zhang, Y., Vinel, A., Jonsson, M.: Analysis of connectivity probability in platoon-based Vehicular Ad Hoc Networks. In: International Wireless Communications and Mobile Computing Conference (IWCMC), pp. 706–711. Nicosia (2014)
7. Karedal, J., Czink, N., Paier, A., Tufvesson, F., Molisch, A.F.: Path loss modeling for vehicle-to-vehicle communications. IEEE Trans. Veh. Technol. **60**(1), 323C328 (2011)
8. Lyu, X., Tian, H., Sengul, C., Zhang, P.: Multiuser joint task offloading and resource optimization in proximate clouds. IEEE Trans. Veh. Technol. **66**(4), 3435–3447 (2017)
9. Wang, Z., Crowcroft, J.: Quality-of-service routing for supporting multimedia applications. IEEE J. Sel. Areas Commun. **14**(7), 1228C1234 (1996)
10. Juttner, A., Szviatovski, B., Mecs, I., Rajk o, Z.: Lagrange relaxation based method for the qos routing problem. In: Proceedings of IEEE INFOCOM, vol. 2, pp. 859C868 (2001)
11. Miettinen, A.P., Nurminen, J.K.: Energy efficiency of mobile clients in cloud computing. In: Proceedings of 2nd USENIX Conference Hot Topics Cloud Computer, p. 4 (2010)

12. TD-LTE Carrier Aggregation White Paper. http://lte-tdd.org/. Accessed 12 Jun 2018
13. Widyono, R.: The design and evaluation of routing algorithms for realtime channels. Technical Report TR-94-024, University of California at Berkeley, June 1994

Cluster-Based Caching Strategy with Limited Storage in Ultra Dense Networks

Chengjia Hu$^{(\boxtimes)}$, Xi Li, Hong Ji, and Heli Zhang

Key Laboratory of Universal Wireless Communications, Ministry of Education,
Beijing University of Posts and Telecommunications,
Beijing, People's Republic of China
{hcj, lixi, jihong, zhangheli}@bupt.edu.cn

Abstract. Ultra dense network (UDN) is considered as one of the key techniques to boost the network capacity in 5G. In order to reduce the huge backhaul cost and end-to-end transmission delay, caching the popular content at the edge of UDNs is an inspiring approach. Considering that the storage capacity of a single small base station (SBS) in UDNs is usually limited, SBSs cooperation to store respective file fragments is an interesting approach that needs further investigation. In this paper, we propose a cluster-based caching strategy (CBCS) for limited storage SBSs in UDNs. A novel clustering scheme based on SBSs's load capacity and location is designed with consideration on files fragments and SBSs cooperation. We target the minimum average download delay under the constraint of the number of SBSs in a cluster. The simulation results show that the proposed algorithm could achieve a better hit ratio and has a lower average download delay.

Keywords: Ultra dense network · Clustering · Caching · Download delay · Hit ratio

1 Introduction

With the rapid deployment of intelligent terminals and the exponential growth in network traffic volume, the traditional techniques have proved incapable of satisfying the increasing traffic demand. Ultra dense network (UDN) is considered to be one of the most important techniques to improve capacity and meet the users experience. UDNs generally consists of numerous low power small base stations (SBSs) and deployment density of SBSs is much higher than Long Term Evolution (LTE) networks [1]. However, user will experience unbearable long delay in getting content due to the congestion in backhaul links, it can lead to significant decline in service quality, backhaul capability may become the bottleneck for UDNs. Recent years, caching at the network edge has been paid much attention and become an important way to tackle the backhaul limitation bottleneck

Published by Springer Nature Switzerland AG 2019. All Rights Reserved
X. Liu et al. (Eds.): ChinaCom 2018, LNICST 262, pp. 116–126, 2019.
https://doi.org/10.1007/978-3-030-06161-6_11

and reduce the latency of services [2]. Nevertheless, considering the deployment costs, the storage capacity of a single SBS in UDNs is very limited. So how to effectively cache files and transmit is an open issue.

To meet the cache limited conditions in UDNs, each SBS stores file fragments instead of full file so each SBS could cache portions of multiple files. Taking into account the densification of UDNs, cluster-based SBSs cooperation cache could further enhance the network performance. There are many existing work related to the clustering and caching issues. In [3], a cluster-based resource allocation strategy is proposed to effectively mitigate the interference and boost energy efficiency (EE) of the UDNs. In [4], in order to maximize the EE of wireless small cell networks, a locally group based on location and traffic load is proposed to couple SBSs into clusters. In [5], the adjacent SBSs are divided into disjoint clusters by using a hexagonal mesh model with distance between cluster centers is 2Rh, and they don't consider the case of empty clusters. In [6], a distributed, cached and segmented video download algorithm based on D2D communication is introduced to effectively improve the throughput in cellular networks. Li et al. [7] proposed a socially aware caching strategy for UDN to improve the network throughput. In [8], a tri-stage fairness algorithm is proposed to solve the resource allocation problem in UDNs. In [9], to resolve the interference and backhaul issues, a backhaul limited cache transmission scheme based on the linear capacity scaling law is proposed in UDN.

So far, there is few well recognized caching strategy for UNDs with limited storage. In this paper, considering the UDNs densification and limited cache capacity, we propose a clustered-based caching strategy (CBCS) to reduce download delay and improve hit ratio. A novel clustering scheme based on SBSs's load capacity and location is introduced with the same SBSs number in each cluster. Sorting the contents according to their popularity, and caching different fragments of the most popular contents in different SBSs whose in the same cluster. By changing the number of SBSs in a cluster, the minimum download delay could be found. Simulation results show that the proposed algorithm could achieve a better hit ratio and has a lower average download delay.

The structure of this paper as follows. The system model and problem formulation is written in Section II. We propose clustering and fragment-based caching strategy in Section III. In Section IV, we describes in detail the simulation results and discussions. Finally, in Section V, we conclude this paper.

2   System Model and Problem Formulation

2.1   Network Model

In Fig. 1, we describe the edge caching network in UDNs in this paper. The network contains S SBSs with U mobile users and one MBS. The users connect to SBSs and SBSs connect to MBS via wireless links. The MBS connection to the core network with wire links. We assume each SBS caches files according to the cache scheme and the size of different files are same.

2.2 SBS Clustering Model

Identifying similarities between SBSs is important step in cluster strategy. The location of SBSs and their load capacity are two important ones of many similarity features. The SBS locations reflect the capability of coordination among nearby SBSs while the SBS load capacity defines mutual interference and the willingness to cooperate [4]. Next, similarities between SBSs based on location and load capacity can be calculated as follows.

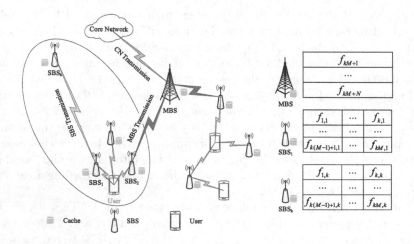

Fig. 1. Edge caching network considered in UDNs

Firstly, based on the distance between SBSs S and S_1, we calculate the Gaussian similarity as follows:

$$X_{SS_1} = \begin{cases} exp(\frac{-||X_S - X_{S_1}||^2}{2\sigma_X^2}) & \text{if } ||X_S - X_{S_1}|| \leq r, \\ 0 & \text{otherwise.} \end{cases} \quad (1)$$

Where σ_X is a constant and X_S is the coordinates of the SBS S in the Euclidean space. The r denote maximum neighborhood distances between SBSs. Furthermore, the value of X_{SS_1} is used as the (S,S_1)-th entry of the distance-based similarity matrix $\mathbb{S}$. As these SBSs get closer, the similarity will increase and SBSs are more likely to cooperate with each other.

Secondly, for the sake of simplified, we assume that the frequency of the user request file is the same. So SBS's load capacity simplified as the maximum number of users which SBS can service. So the load-based dissimilarity between SBSs S and S_1 can be calculated as follows:

$$N_{SS_1} = exp(\frac{-||N_S - N_{S_1}||^2}{2\sigma_N^2}) \quad (2)$$

Where σ_N controls the range of the dissimilarity and N_S is the maximum number of users which SBS S can service. In addition, the load-based dissimilarity matrix $\mathbb{N}$ is formed using N_{SS_1} as the (S,S_1)-th entry. In general, the SBSs with different load capacity yield more benefit by offloading traffic to one another.

Thirdly, the joint similarity matrix $\mathbb{W}$ with W_{SS_1} as the (S,S_1)-th element which combine distance-based similarity with load-based dissimilarity is formulated as follows:

$$W_{SS_1} = (X_{SS_1})^\theta \cdot (N_{SS_1})^{(1-\theta)}, \quad \theta \in [0,1] \tag{3}$$

Where θ controls the extent to which the distance and load affect the joint similarity.

Finally, we propose a clustering mechanisms based on a similarity matrix $\mathbb{W}$. A SBS was randomly selected as the center of the initial cluster and it find the most similar k the SBSs form a cluster according to the similarity matrix $\mathbb{W}$. Choose remainder SBS which closest to the previous cluster center as the center of the new cluster. It will select SBS which already within the cluster to ensure every cluster has the same number of SBSs. Once clusters are formed, the SBS which load capacity is best within the cluster is selected as the cluster head. The cluster head is used to allocating resources within a cluster and coordinate the transmissions between the cluster members.

2.3   Caching and Transmission Model

In Fig.1, we also describe the specific caching scheme and transmission model. We consider a finite and sorted by popularity content library $\mathbb{F} = \{f_1, f_2, \cdots, f_i, \cdots, f_I\}$ in UDNs, where f_i is the i-th most popular file and the size of each file are F. Each SBS can store up to M files and MBS can be cached $N (M < N < T)$ files. A cluster can cache kM files with k cooperative SBSs. We choose kM of the most popular file and each file (e.g. video) is divided into k fragments and caching in the different SBS. Moreover, files f_i with popularity order $kM < i \leq kM + N$ are cached in MBS and the remaining files are not cached.

When a user requests file from $\mathbb{F}$, he or she can obtain file mainly through the following three ways.

(1) SBS Transmission: When the file with popularity order $1 \leq i \leq kM$, k SBSs in the cluster has different fragments of this file. On the one hand, fragments are transmitted to the user directly by the SBS (e.g. SBS_1 and SBS_2) which cover the user. On the other hand, cluster head scheduling remaining SBSs (just like SBS_k) by relay sending fragments to the user.

(2) MBS Transmission: When the file with popularity order $kM < i \leq kM + N$ which caching in MBS, the SBS needs to get this file from MBS and send it to the user.

(3) Core Network Transmission: When the file not cached, the SBS needs to get this file from the core network and send it to the user.

2.4 Average Hit Ratio

We assume each local user issues an independent request for a file whose probability is related to Zipf-like popularity distribution [10]. A user has the probability p_i of requesting the i-th file. We can get p_i through

$$p_i = \frac{1/i^\alpha}{\sum\limits_{j=1}^{I} (1/j^\alpha)} \tag{4}$$

Where α is a decay constant. So the SBS expected hit ratio is calculated as

$$P_{hit}^S = \sum_{i=1}^{kM} p_i \tag{5}$$

It is obvious that the cache hit ratio will increase if the number of SBSs in a cluster increase.

2.5 Average User Download Delay

Let σ^2 be the noise power. The $h_{s,u}$, $h_{M,s}$ means the channel gain between SBS s to user u and the MBS to SBS s, respectively. Denote the transmission power of the SBSs, the MBS as P_S and P_M respectively.

In order to simplify the calculation, we assume that MBS and SBSs not reuse spectrum resources and the MBS has the bandwidth ω_M. The SBSs which in a cluster allocate orthogonal frequency band ω_S. So the signal-to-noise ratio $SNR_{s,u}$ between SBS s and user u and the $SNR_{M,s}$ between MBS and SBS s could be denoted as $SNR_{s,u} = \frac{P_S \cdot h_{s,u}}{\sigma^2}$ and $SNR_{M,s} = \frac{P_M \cdot h_{M,s}}{\sigma^2}$. Thus we would get the user u download $Delay_1^u$ of SBS transmission and the $Delay_2^u$ of MBS transmission. The expression of $Delay_1^u$ is

$$Delay_1^u = \arg\max_{s=1,2,\cdots,t} \frac{N_s F/k}{\frac{\omega_S}{k} \cdot \log_2(1 + SNR_{s,u})} \tag{6}$$

Where N_s is the number of fragments which SBS s required to transmit. Denote the number of SBSs which directly linked to the user as t.

The $Delay_2^u$ contains the delay of SBS receiving file from MBS and SBS sending the file to the user. We can get $Delay_2^u$ through

$$Delay_2^u = \frac{F}{\omega_M \cdot \log_2(1 + SNR_{M,s})} + \frac{F}{\omega_S \cdot \log_2(1 + SNR_{s,u})} \tag{7}$$

When a user requests for the files which are not cached. The user download delay contains $Delay_2^u$ and the $delay_{c,M}$ of core network sending file to MBS. To simplify, we denoted $delay_{c,M}$ as C which is a constant. Therefore, the $Delay_3^u$ is calculated as

$$Delay_3^u = Delay_2^u + C \tag{8}$$

The user obtain files probability P_1^u through SBS transmission, P_2^u through MBS transmission and P_3^u through core network transmission could be denoted as $P_1^u = P_{hit}^S$, $P_2^u = \sum_{i=kM+1}^{kM+N} p_i$ and $P_3^u = 1 - P_1^u - P_2^u$ respectively. The theoretical expression of user u average download $Delay^u$ is

$$Delay^u = P_1^u \cdot Delay_1^u + P_2^u \cdot Delay_2^u + P_3^u \cdot Delay_3^u \tag{9}$$

3   The Proposed CBCS Algorithm

In this section we rearrange the CBCS algorithm ideas. For SBSs, we propose a clustering mechanisms based on SBSs's location and load capacity in which every cluster has the same number of SBSs. For users, we count and sort the popular files that they might need. Furthermore, we can calculate each requiring probability of the files by (4). We would give the maximum number K of SBSs in a cluster so as to find the minimum value of average download delay conveniently.

After all the SBSs have been successfully clustered, and popular files have been cached in the appropriate location according to the cache strategy, then the user starts to access the SBS and requests files. Firstly, the connected user check whether required files already cached in SBSs. User can obtain required files according to SBSs transmission if required files already cached in SBSs. It follows MBS transmission if this file cached in MBS. Otherwise, it follows core network transmission. Then, the average download delay can be calculated according to (9). Our goal is to minimize the average download delay under the constraint of the number of SBSs in a cluster. The optimization problem is accordingly formulated as:

$$\begin{aligned} \underset{k}{\text{minimize}} \quad & Delay^u \\ \text{subject to:} \quad & k \in \{1, 2, \cdots, K\} \end{aligned} \tag{10}$$

Where K is the maximum number of SBSs in a cluster. Change the number of SBSs in a cluster, the minimum average download delay could be found. In Algorithm 1, we summarize the process of the CBCS algorithm as follows.

4   Simulation Results And Discussions

In this section, we evaluate the performance of the CBCS algorithm and analyze the impact of important parameters through simulation. For comparison, we simulated the "most popular" placement (MPP) scheme which each SBS cache the M most popular files mentioned in [11].

In the simulation, we let $F = 10M$ bits, $\alpha = 0.8$, $I = 200$, $M = 10$, $N = 100$, and $\beta = 4$ if there are no special instructions. Simulation area size of $200\,\text{m}*200\,\text{m}$ and we put the MBS in the center. 50 SBSs are random distribution and coverage area radius r is $50\,\text{m}$. The values of P_S and P_M are $100\,\text{mW}$ and 20W respectively, and $L_0 = -30\,\text{dB}$. The value of σ^2 is $-100\,\text{dBm}$ and bandwidth $\omega_M = \omega_S = 1\,\text{MHz}$ [12]. We assume the value of $delay_{c,M}$ as C is $1\,\text{s}$.

We can get the relationship between average hit ratio and the number of SBSs in a cluster from Fig. 2. As can be seen, increasing the value of k is beneficial to increase the cache hit ratio, and the growth rate becomes smaller. The larger the value of k is, the more files are stored in a cluster and therefore the SBS hit ratio would increase. However, file transmission collaboration between SBSs becomes

---

Algorithm 1 Minimizing Average Download Delay in UDNs via CBCS

1: Initialization:
2: (a) Set S SBSs coordinates X_S and load capacity N_S subject to random distribution. Set constant $\sigma_X, \sigma_N, r, \theta$;
3: (b) SBS storage capacity M, MBS storage capacity N, number of all files I, size of the files F;
4: (c) SBS transmission power P_S, transmission power P_M, noise power σ^2, channel gain relevant parameters L_0 and β, SBSs cluster bandwidth ω_S and MBS bandwidth ω_M, set K.
5: Calculate matrix $\mathbb{W}$ according to (1),(2),(3) and select initial SBS.
6: **for** $k = 2, \cdots, K$ **do**
7:     **while** the number of remainder SBSs $\neq 0$ **do**
8:         Find the number n of remainder SBSs which the similarity with cluster center is greater than 0.
9:         **if** $n > k - 1$ **then**
10:             Find most similar k SBSs form a cluster.
11:         **else**
12:             Select SBS already within the cluster to ensure every cluster have the k number of SBSs.
13:         The file placement strategy shown in Fig.1
14:         The users connect to the SBSs.
15:         **for** $u = 1, \cdots, U$ **do**
16:             User U_u begin to require files.
17:             **if** this file is stored in SBSs **then**
18:                 Calculate the download $Delay_1^u$.
19:             **else**
20:                 **if** this file is stored in MBS **then**
21:                     Calculate the download $Delay_2^u$.
22:                 **else**
23:                     Calculate the download $Delay_3^u$.
24:             Calculate the user u average download $Delay^u$.
25:         Calculate the average download $Delay$.
26:         Calculate the SBS expected hit ratio P_{hit}^S.
27: Output the $Delay_{min}$ and P_{hit}^S.

---

complicated. Thus, the higher average download delay appears. Further, the smaller the value of I is, the higher the hit ratio.

Figure 3 indicates the changes of average download delay with different number of SBSs in a cluster. In Fig. 3, from 1 to 6 the number of SBSs in a cluster, the user download delay firstly decreased and then increased. The reason for firstly increased is the probability of SBSs transmission increasing whose delay smaller than MBS transmission. Along with the increase of k, SBS cooperative transmission is more and more complex, finally the transmission delay over MBS transmission. Therefore, the user download delay would increase. Furthermore, when $S = 50$, combined with Fig. 2 compare MPP and CBCS ($k = 4$), we can see that average hit ratio increases from 0.35 to 0.6 while average download delay no change. The simulation results verify that our algorithm can bring better cache hit ratio in UDNs.

In Fig. 4, we compare the average download delay performance of CBCS and MPP for different total SBSs. Figure 4 clearly shows that the average download delay decreases when the value of S increases with CBCS. However, the delay basically no change with MPP. This is because we place the user in a fixed position nearby the cluster head to simplify, so the distance between user and SBS is fixed with MPP. The average download delay is not changed. The distance between the user and SBS decreases when the value of S increases with CBCS, as a result, the user download delay is reduced.

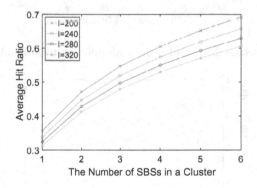

Fig. 2. The average hit ratio with different number of SBSs in a cluster

Figure 5 illustrates the the effect of the Zipf parameter α on the average download delay. We can see from Fig. 5 that when the Zipf exponent increases, the user average download delay decreases. The reason is higher α means that the popularity of the top files in the library will increase, and they are more often required and SBS transmission delay smaller than MBS transmission in general. Furthermore, on account of the SBSs which in a cluster could be allocated orthogonal frequency band ω_S and MPP means that each the bandwidth

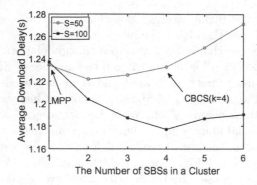

Fig. 3. The average download delay with different k in a cluster

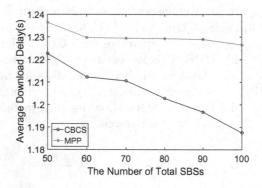

Fig. 4. The average download delay with different total number of SBSs

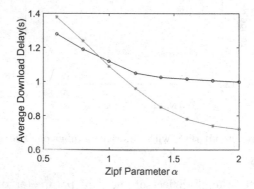

Fig. 5. The average download delay with different Zipf parameter α

of the SBS is ω_S, so MPP less than CBCS SBS transmission delay according to (6). That is the reason for MPP less than CBCS average download delay when $\alpha > 0.9$.

5   Conclusion

In this paper, we focus on the issue of how to cache under limited storage capacity in UDNs. We designed a CBCS algorithm based on SBSs clustering and files fragments. A clustering scheme is proposed based on SBSs's location and load capacity in which every cluster has the same number of SBSs. Then, we split files into fragments and caching them in different SBSs. To find the optimal solution, a traversal algorithm is used to resolve the optimization model. Compared with the MPP algorithm, the simulation results show that the proposed CBCS algorithm can achieve a higher hit ratio under similar average download delays. For future work, in order to improve the performance of the system, variable number of SBSs inside the cluster would be considered, as well as the user movement in UDNs.

Acknowledgements. This paper is sponsored by National Natural Science Foundation of China (Grant 61771070 and 61671088).

References

1. Kamel, M., Hamouda, W., Youssef, A.: Ultra-dense networks: a survey. IEEE Commun. Surv. Tutor. **18**(4), 2522–2545 (2016). Fourthquarter
2. Zhang, J., Zhang, X., Wang, W.: Cache-enabled software defined heterogeneous networks for green and flexible 5g networks. IEEE Access **4**, 3591–3604 (2016)
3. Samarakoon, S., Bennis, M., Saad, W., Latva-aho, M.: Dynamic clustering and on/off strategies for wireless small cell networks. IEEE Trans. Wirel. Commun. **15**(3), 2164–2178 (2016)
4. Chen, Z., Lee, J., Quek, T.Q.S., Kountouris, M.: Cluster-centric cache utilization design in cooperative small cell networks. In: 2016 IEEE International Conference on Communications (ICC), pp. 1–6 (2016)
5. Huang, K., Andrews, J.G.: An analytical framework for multicell cooperation via stochastic geometry and large deviations. IEEE Trans. Inf. Theory **59**(4), 2501–2516 (2013)
6. Al-Habashna, A., Wainer, G., Boudreau, G., Casselman, R.: Distributed cached and segmented video download for video transmission in cellular networks. In: 2016 International Symposium on Performance Evaluation of Computer and Telecommunication Systems (SPECTS), pp. 1–8 (2016)
7. Zhu, K., Zhi, W., Chen, X., Zhang, L.: Socially motivated data caching in ultra-dense small cell networks. IEEE Netw. **31**(4), 42–48 (2017)
8. Hao, P., Yan, X., Li, J., Li, Y.N.R., Wu, H.: Flexible resource allocation in 5g ultra dense network with self-backhaul. In: 2015 IEEE Globecom Workshops (GC Wkshps), pp. 1–6 (2015)
9. Liu, A., Lau, V.K.N.: How much cache is needed to achieve linear capacity scaling in backhaul-limited dense wireless networks? IEEE/ACM Trans. Netw. **25**(1), 179–188 (2017)
10. Breslau, L., Cao, P., Fan, L., Phillips, G., Shenker, S.: Web caching and zipf-like distributions: evidence and implications. In: INFOCOM '99. Eighteenth Annual Joint Conference of the IEEE Computer and Communications Societies. Proceedings, vol. 1, pp. 126–134. IEEE (1999)

11. Gabry, F., Bioglio, V., Land, I.: On energy-efficient edge caching in heterogeneous networks. IEEE J. Sel. Areas Commun. **34**(12), 3288–3298 (2016)
12. Jia, C., Lim, T.J.: Resource partitioning and user association with sleep-mode base stations in heterogeneous cellular networks. IEEE Trans. Wirel. Commun. **14**(7), 3780–3793 (2015)

Image Retrieval Research Based on Significant Regions

Jie Xu[✉], Shuwei Sheng, Yuhao Cai, Yin Bian, and Du Xu

School of Information and Communication Engineering, University of Electronic
Science and Technology of China, Chengdu, China
xuj@uestc.edu.cn

Abstract. Deep Convolution neural networks (CNN) has achieved great success in the field of image recognition. But in the image retrieval task, the global
CNN features ignore local detail description for paying too much attention to
semantic information of images. So the MAP of image retrieval remains to be
improved. Aiming at this problem, this paper proposes a local CNN feature
extraction algorithm based on image understanding, which includes three steps:
significant regions extraction, significant regions description and pool coding.
This method overcomes the semantic gap problem in traditional local characteristic and improves the retrieval effect of global CNN features. Then, we apply
this local CNN feature in the image retrieval task, including the same category
retrieval task by feature fusion strategy and the instance retrieval task by re-
ranking strategy. The experimental results show that this method has achieved
good performance on the Caltech 101 and Caltech 256 classification datasets,
and competitive results on the Oxford 5k and Paris 6k instance retrieval datasets.

Keywords: Significant regions · Image understanding · CNN
Image retrieval

1 Introduction

Content based image retrieval (CBIR) uses the description of image content to search
similar images. Most of the existing methods employ low-level visual features of
image, such as Sift [1], BoW [2], Fisher vector [3] and VLAD [4]. Although CBIR in
the past decade has made a lot of scientific research and set up some research or
commercial image retrieval systems, but most of the image retrieval performance
cannot satisfy the requirement. The main reason is semantic gap problem.

Image descriptors based on the activations within deep convolutional neural networks have emerged as state-of-the-art generic descriptors for visual recognition [5–7].
CNN global characteristics as a kind of high-level semantic representation, used in
other recognition tasks and performed well [8–11]. Razavian [12] studied the characteristics of global CNN feature and used it for different image recognition tasks,
including image retrieval. Yandex [13] achieve a remarkable increase in performance
by fine-tuning CNN model on target dataset and extracting fc6 layer features. Lin [14]
utilized hash code to transform fc6 layer features into a binary sequence, and greatly
improved the retrieval efficiency.

But there are still some problems to be solved. Global CNN feature contains too many high-level semantic information related to the classification, so it tends to ignore the details of images. Recently, some studies began to focus on the image characteristics of granular to improve the global CNN feature. Wang [15] put forward using triplet to increase similarity intra-class and distinction between classes, and proposed a multi-scale network to increase the local detail information in the image. An adaptive region detection method [16] is proposed to eliminate the street snap clothing pictures and store clothing pictures. Clothing attribute dataset is used to mining the fine-grained properties. CKN [17] network was proposed to extract local degeneration characteristics of images. Then Mattis [18] used unsupervised training CKN network to extract local features in image retrieval task.

Combined the advantage of traditional local features, we propose a local CNN features extraction method based on image understanding. This work presents three contributions:

First, we propose a local CNN feature method including three steps: significant regions extraction, significant regions description and coding. Among them, the significant regions are extracted based on image understanding, which can describe the whole attribute of the image and the relation between different entities.

Figure 1 shows eight extraction results of the significant regions. The raw pictures come from four different datasets, which will be used in the retrieval tasks in Sect. 3.

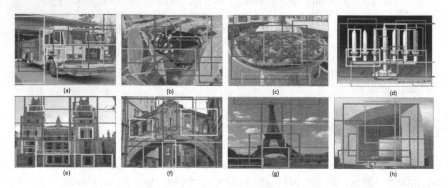

Fig. 1. The results of significant regions. (a) and (b) come from Caltech 101 dataset. (c) and (d) come from Caltech 256 dataset. (e) and (f) come from Oxford 5k dataset. (g) and (h) come from Paris building 6k dataset

Second, we fusion the global CNN feature and local CNN feature and apply it in the same category retrieval task.

Third, we put forward the re-ranking algorithm based on significant regions and employ it to instance retrieval task. The experimental results show that the proposed methods can further improve the accuracy of image retrieval.

The rest of this paper is organized as follows. The details of the proposed local CNN features and image retrieval method are given in Sect. 2. We give the experimental results and analysis in Sect. 3. Finally, we conclude in Sect. 4.

2   Methodology

In this section, we give the details of the proposed local CNN features for image representation and apply it for the image retrieval tasks

2.1   CNN Feature Based on Significant Regions

In traditional image retrieval task, local feature showed greater advantage than global feature, because it can describe more details information and have scale, rotation and brightness invariant. Sift feature is a very common local descriptor. it contains key points detection, key points description and coding three steps to condense the image information into 128 dimensional feature vector. In view of the outstanding characteristics of the sift feature, this paper uses image understanding theories and models to extract the significant regions. Then by significant regions description and coding we generate local CNN feature. Algorithm process as shown in Fig. 2.

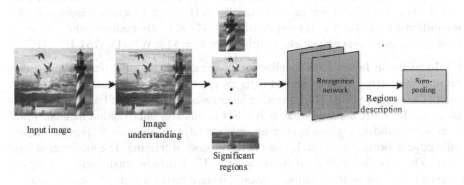

Fig. 2. Local CNN feature extraction process

In Fig. 2, the process of extracting local CNN feature can be divided into four parts

(a) CNN + RPN + LSTM model used to extract the significant regions and the model is trained on image understanding task;
(b) Picking out the highest score of K a significant area;
(c) Describing the K significant regions through identification network and generating feature description;
(d) Coding by sum-pooling.

2.2   Significant Regions Extraction Based on Image Understanding

In order to solve semantic gap problem, this paper attempts extract significant area from the perspective of image understanding

In image caption task, it needs to locate the target area, and also describes the target area in natural language. We use CNN + RPN + LSTM structure to locate the significant regions, then filter these areas and code to generate low dimensional feature vector. The model structure is shown in Fig. 3.

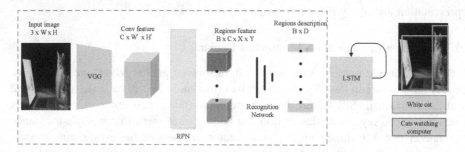

Fig. 3. CNN + RPN + LSTM model

CNN network. We discard the softmax layer and fully connected layers of the original network. Given an input image I of size W × H, the activations (responses) of a convolutional layer form a 3D tensor of W' ×H' × C dimensions, where C is the number of output feature channels, In this paper, C = 512, W' = ⌊W/16⌋, H' = ⌊H/16⌋.

RPN localization layer. RPN localization layer receives the input feature maps, then pinpoints the interest regions and extracts an appropriate length denoting from every region. The structure of RPN localization layer based on the idea of Faster R-CNN. We replace the ROI mechanism in Faster R-CNN to bilateral interpolation method, which can makes candidate regions back propagate the edge information to previous layers. So the edge information can be learned in the process of training. The localization layer accepts a tensor of activations of size C × W' × H'. It then internally selects B regions of interest and returns three output tensors giving information about these regions:

- Region Coordinates: A matrix of shape B × 4 giving bounding box coordinates for each output region.
- Region Scores: A vector of length B giving a confidence score for each output region. Regions with high confidence scores are more likely to correspond to ground-truth regions of interest.
- Region Features: A tensor of shape B × C × X × Y giving features for output regions.

RPN layer is mainly to locate the candidate regions, and filter the regions according to NMS. The rest of the regions are the significant regions in this paper

Recognition network. The recognition network is a fully-connected neural network that processes region features from the localization layer. The features from each region are flattened into a vector and passed through two full-connected layers, each using rectified linear units and regularized using dropout. For each region this produces a description of dimension D = 4096 that compactly encodes its visual appearance. This description of dimension B × 4096 is what we need to code in our task.

LSTM language model. The LSTM model only work in the model training process. We just use it to make the model oriented to image understanding task rather than classification task. We use the Visual Genome dataset [19] to pre-train the model. The aim is to make the model have the ability to locate the significant regions and dig the relation between different entities. The description we need is from recognition network.

2.3 Sum Pooling Coding

In the previous steps, we complete preliminary coding through the recognition network. Then the description of dimension $B \times 4096$ will need to encode into a feature vector in image retrieval task. And the principle for the sum pooling coding is:

First of all, we calculate the sum of feature value and in all significant regions about each dimension

$$F'_k = \sum_{i=1}^{B} C_i \# \tag{1}$$

Then the feature code is the proportion of each dimension

$$F_k = \frac{F'_k}{\sum_{k=1}^{4096} F'_k} \# \tag{2}$$

2.4 Same Category Retrieval

This article takes the algorithm of fusion global CNN features and local CNN in the same category retrieval task. In this part, our local CNN feature based on significant regions aims to improve the global CNN global feature which cannot describe the local details of the image. Algorithm process is shown in Fig. 4.

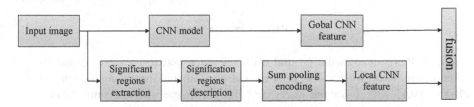

Fig. 4. The process of same category retrieval

In Fig. 4, the upper part employs CNN model as global feature extractor. The second half part is the process of local CNN feature in this paper. Finally, we fuse them together in the same category retrieval task. The dimension of global and local CNN feature vector is 4096.

We use PCA and L2 regularization to fuse them and get the final feature vector. According to the above method, all the pictures can be extracted feature vectors and build features library.

2.5 Instance Retrieval

In instance retrieval, we concentrate on objects in images rather than the class of full image. We employ feature aggregation by cross-dimension weighting proposed by Crow [20] to make initial retrieval and obtain Top-N retrieval results. We propose salient region extraction algorithm to re-rank the Top-N images. The re-ranking algorithm is as follows:

(a) We employ CNN network to extract global query feature q.
(b) We extract salient region in Top-N image and global feature p in every salient region. We re-rank the Top-N based on the similarity between q and p. Images with the highest scores ranking move forward, to further improve the retrieval result.

3 Experiments and Results

3.1 Datasets

We use Caltech 101 and Caltech 256 datasets to verify the results of our method for the same category retrieval task.

Then, we use Oxford Buildings and Paris Buildings to verify the results of our method for the instance retrieval task.

3.2 Experiment in Same Category Retrieval

In this part, we evaluate the ranking of Top-K images with respect to query image q by a precision

$$AP@k = \frac{\sum_{i=1}^{k} Rel(i)}{k} \tag{3}$$

Where Rel(i) denotes the ground truth relevance between a query q and the i-th ranked image. Here, we consider only the category label in measuring the relevance so $Rel(i) \in \{0,1\}$ with 1 for the query and the i-th image with the same label and 0 otherwise.

For each dataset in the experiment, the 5% of the total images are randomly selected as query images. When Top-10 retrieval results are returned, the experimental results are shown in Table 1.

Seen from Table 1, the MAP of traditional algorithms such as BoW is low in the same category retrieval task. The image retrieval algorithm based on deep learning achieves better performance. On Caltech 101, the average retrieval accuracy of global CNN features is 79.14%. The average retrieval accuracy of this algorithm is 80.67%, and the average retrieval accuracy of this algorithm is improved by 1.53%. On the

Table 1. The MAP on Caltech 101 and Caltech 256 datasets using different features

Method	Caltech 101	Caltech 256
BoW	0.223	0.268
Global CNN feature	0.791	0.649
Ours	0.807	0.674
Ours + QE	0.801	0.701

Caltech 256 dataset, the average retrieval accuracy of global CNN features is 64.91%. The average retrieval accuracy of this algorithm is 67.37% and the retrieval accuracy is 2.46%, which proves the accuracy and effectiveness of this algorithm. Finally, we visualize the retrieval results, and choose top 12 most similar results with the query image for display. The result is shown in Fig. 5.

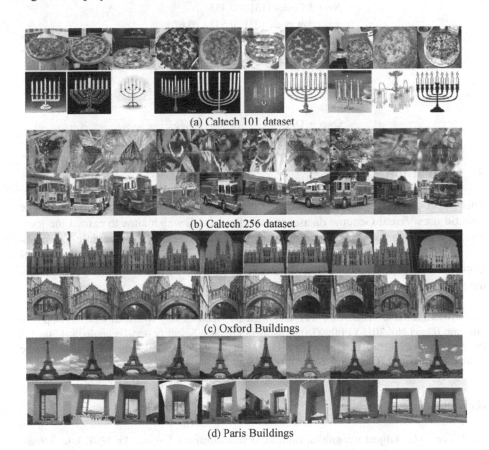

(a) Caltech 101 dataset

(b) Caltech 256 dataset

(c) Oxford Buildings

(d) Paris Buildings

Fig. 5. A same category retrieval example on Caltech 101 and Caltech 256 datasets. An instance retrieval example on Oxford Buildings and Paris Buildings. The left-most image in each row corresponds to the query.

3.3   Experiment in Instance Retrieval

In accordance with of the algorithm Sect. 2.3, this article uses the Oxford and Paris dataset to evaluate the MAP of instance retrieval task. And we compare with some state-of-art algorithms such as R-MAC [21], SPOC [4], the Crow [20]. The experimental results as shown in Table 2. From the table, we can find that the MAP of the algorithm in this paper still be proved competitive compare to these state-of-art algorithms. Finally, we visualize the retrieval results, and choose top 12 most similar results with the query image for display. The result is shown in Fig. 5.

Table 2. The MAP on Oxford 5k and Paris 6k datasets using different features

Method	Oxford	Paris
Tr. Embeding [22]	0.560	—
Neural Codes [13]	0.435	—
Razavian et al. [23]	0.533	0.670
Sum pooling [22]	0.589	—
R-MAC [21]	0.669	0.830
SPoC [4]	0.561	0.729
Crow [20]	0.657	0.7347
Ours	0.682	0.737
Ours + QE	0.703	0.751

4   Conclusion

This paper presents a local CNN feature algorithm based on significant regions. The method uses Visual Genome dataset to train the model, which aims to extract the local information by image understanding. This method overcomes the semantic gap problem in traditional local characteristic and improves the retrieval effect of global CNN features. The experimental results show that this method has achieved good performance both in the same category retrieval task and instance retrieval task.

Acknowledgements. This work was supported by National Key Research and Development Program (Grant No. 2016YFB0800105), Sichuan Province Scientific and Technological Support Project (Grant Nos. 2016GZ0093, 2018GZ0255), the Fundamental Research Funds for the Central Universities (Grant No. ZYGX2015J009).

References

1. Lowe, D.G.: Object recognition from local scale-invariant features. In: IEEE International Conference on Computer Vision (2001)
2. Sivic, J.: A text retrieval approach to object matching in videos. In: Proceedings of IEEE International Conference on Computer Vision (2003)

3. Perronnin, F., Liu, Y., Sánchez, J., Poirier, H.: Large-scale image retrieval with compressed Fisher vectors. In: Proceedings of the IEEE Conference on Computer Vision and Pattern Recognition. Workshops, pp. 3384–3391 (2010)

4. Jegou, H., Perronnin, F., Douze, M., Sánchez, J., Perez, P.: Aggregating local image descriptors into compact codes. IEEE Trans. Pattern Anal. Mach. Intell. **34**(9), 1704–1716 (2012)

5. Krizhevsky, A., Sutskever, I., Hinton, G.E.: Imagenet classification with deep convolutional neural networks. In: Advances in Neural Information Processing Systems (NIPS 2012), pp. 1097–1105 (2012)

6. Simonyan, K., Zisserman, A.: Very deep convolutional networks for large-scale image recognition (2014). arXiv:1409.1556

7. Szegedy, C., Liu, W., Jia, Y., Sermanet, P., Reed, S.: Going deeper with convolutions. In: Proceedings of the IEEE Conference on Computer Vision and Pattern Recognition. Workshops, pp. 1–9 (2015)

8. Zeiler, M.D., Fergus, R.: Visualizing and understanding convolutional networks. In: Proceedings of the European Conference on Computer Vision, pp. 818–833 (2014)

9. Oquab, M., Bottou, L., Laptev, I., Sivic, J.: Learning and transferring mid-level image representations using convolutional neural networks. In: Proceedings of the IEEE Conference on Computer Vision and Pattern Recognition. Workshops, pp. 1717–1724 (2014)

10. Hoang, T., Do, T.T., Tan, D.K.L., Cheung, N.M.: Selective deep convolutional features for image retrieval. In: ACM, pp. 1600–1608 (2017)

11. Xu, J., Shi, C.Z., Qi, C.Z., Wang, C.H., Xiao, B.H.: Unsupervised part-based weighting aggregation of deep convolutional features for image retrieval. In: AAAI2018 (2018)

12. Razavian, A.S., Azizpour, H., Sullivan, J., Carlsson, S.: CNN features off-the-shelf: an astounding baseline for recognition. In: Proceedings of the IEEE Conference on Computer Vision and Pattern Recognition. Workshops, pp. 512–519 (2014)

13. Babenko, A., Slesarev, A., Chigorin, A., Lempitsky, V.: Neural codes for image retrieval. In: Proceedings of the European Conference on Computer Vision, pp. 584–599 (2014)

14. Lin, K., Yang, H.F., Hsiao, J.H., Chen, C.S.: Deep learning of binary hash codes for fast image retrieval. In: Proceedings of the IEEE Conference on Computer Vision and Pattern Recognition. Workshops, pp. 25–37 (2015)

15. Wang, J., Song, Y., Leung, T., Rosenberg, C., Wang, J.: Learning fine-grained image similarity with deep ranking. In: Proceedings of the IEEE Conference on Computer Vision and Pattern Recognition, pp. 1386–1393 (2014)

16. Chen, Q., Huang, J., Feris, R., Brown, LM., Dong, J.: Deep domain adaptation for describing people based on fine-grained clothing attributes. In: Proceedings of the IEEE Conference on Computer Vision and Pattern Recognition, pp. 5315–5324 (2015)

17. Mairal, J., Koniusz, P., Harchaoui, Z., Schmid, C.: Convolutional kernel networks. In: International Conference on Neural Information Processing Systems. MIT Press, pp. 2627–2635 (2014)

18. Paulin, M., Douze, M., Harchaoui, Z., Mairal, J., Perronin, F.: Local convolutional features with unsupervised training for image retrieval. In: IEEE International Conference on Computer Vision, pp. 91–99 (2015)

19. Krishna, R., Zhu, Y., Groth, O., Johnson, J., Hata, K.: Visual genome: connecting language and vision using crowdsourced dense image annotations. Int. J. Comput. Vis. **123**(1), 32–73 (2017)

20. Kalantidis, Y., Mellina, C., Osindero, S.: Cross-dimensional weighting for aggregated deep convolutional features. In: Proceedings of the European Conference on Computer Vision. Workshops, pp. 685–701 (2016)

21. Tolias, G., Sicre, R., Jegou, H.: Particular object retrieval with integral maxpooling of CNN activations. In: Proceedings of the International Conference on Learning Representations, pp. 1–12 (2016)
22. Jegou, H., Zisserman, A.: Triangulation embedding and democratic aggregation for image search. In: Proceedings of the IEEE Conference on Computer Vision and Pattern Recognition, pp. 3310–3317 (2014)
23. Razavian, A.S., Sullivan, J., Maki, A., Carlsson, S.: Visual instance retrieval with deep convolutional networks (2014). arXiv:1412.6574

Partial Systematic Polar Coding

Hongxu Jin[✉] and Rongke Liu

School of Electronic and Information Engineering, Beihang University, Beijing
100191, China
hongxu_jin@buaa.edu.cn

Abstract. Due to having a better performance of bit error rate (BER), systematic polar codes have been potentially applied in digital data transmission. In the systematic polar coding, source bits are transmitted transparently. In this paper, we propose a scheme of novel partial systematic polar coding in which the encoded codeword is only composed of partial source bits with respect to the encoded word of systematic polar codes. To effectively reduce the resource consumption of the systematic encoder/decoder under all-zero frozen bits, the partial systematic polar codes are introduced subsequently. Then the simulation results in terms of core $F = \begin{bmatrix} 1 & 0 \\ 1 & 1 \end{bmatrix}$ are provided to demonstrate the aforementioned analysis with negligible difference of BER performance.

Keywords: Polar codes · Non-systematic polar codes · Systematic polar codes
Partial systematic polar codes

1 Introduction

Polar codes are the first provably capacity-achieving codes for any symmetric binary-input discrete memoryless channel (DMC) [1] with flexible encoding and decoding arrangements. The original manner of polar codes is non-systematic codes. Compared with such codes, the systematic polar codes proposed by Arikan show the better BER performance in the successive cancellation (SC) decoding algorithm [1, 2].

The BER performances of non-systematic/systematic codes are same among the classical linear error-correction codes, such as Bose–Chaudhuri–Hocquenghem (BCH) codes [3] and Low-Density Parity-Check (LDPC) codes [4] etc. As a linear coding strategy, the non-systematic polar codes proposed by Arikan are produced when the information and the frozen bits pass through the generator matrix G [1]. However, systematic polar codes have better BER performance comparable to nonsystematic polar codes, while the two codes have the same frame error rate (FER) under the SC decoding [2]. Systematic polar coding does not simply utilize the SC decoder to recover the source bits like the non-systematic polar codeword. There is an additional preprocessing circuit network after SC decoding process, which is denoted as a de-preprocessing circuit network [2]. As can be seen, compared with the non-systematic polar codes, the improvement of BER performance for systematic polar codes is mainly caused by the de-preprocessing circuit network of polar codes.

© ICST Institute for Computer Sciences, Social Informatics and Telecommunications Engineering 2019
Published by Springer Nature Switzerland AG 2019. All Rights Reserved
X. Liu et al. (Eds.): ChinaCom 2018, LNICST 262, pp. 137–146, 2019.
https://doi.org/10.1007/978-3-030-06161-6_13

Since source bits can appear in the encoded codeword, the bits in information set $\mathcal{A}$, named information bits [1], which are not source bits like non-systematic polar codes but the per-encoded source bits by an additional circuit network as an en-preprocessing process at the encoder input. And after the SC decoding algorithm in the receiver, an additional corresponding circuit network as de-preprocessing process can recover the information bits into the source bits [2]. With all-zero bits in frozen set $\mathcal{A}^c$ named the frozen bits [1], the en-preprocessing process and de-preprocessing process will be employed by G_{AB}^{-1} and G_{AB} which denote a sub-matrix in G^{-1} and G respectively, and will be composed of elements $G_{i,j}^{-1}$ and $G_{i,j}$ with $i \in \mathcal{A}$ and $j \in \mathcal{B}$ respectively [2]. In this letter, the definition of partial systematic polar codes is presented. The difference of the preprocessing process is the employed sub-matrix of $G_{A'B'}^{-1}$ and $G_{A'B'}$ under the condition of all-zero frozen bits, where $\mathcal{A}' \subset \mathcal{A}$. In terms of core $F = \begin{bmatrix} 1 & 0 \\ 1 & 1 \end{bmatrix}$. Next paper, the resource consumption of partial systematic decoder under all-zero frozen bits will be reduced without the BER performance lost.

2 Problem Statement

2.1 Construction of Polar Codes

Polar codes, as a linear block coding scheme, have been proved to achieve the channel capacity at a low encoding and decoding complexity [1]. For polar codes (N, K), pre-encoded word is denoted as $\mathbf{u}$, which is composed of K information-bit word $\mathbf{u}_A$ and $N - K$ frozen-bit word $\mathbf{u}_{A^c}$. Then, the encoded bits can be expressed as: $\mathbf{x} = \mathbf{u}G$, $G = F^{\otimes \log_2 N}$, and code rate is $R = K/N$. Where $F^{\otimes \log_2 N}$ denotes the $\log_2 N$ Kronecker power of F. In the N bit-channels, the bit-channels where decoding result $\hat{u}_i$ equals to pre-encoded bit u_i can be considered as noise-free channels with information set $\mathcal{A}$. Therefore, the rest of the bit-channels are noisy channels with frozen set $\mathcal{A}^c$. Note that $\mathcal{A} + \mathcal{A}^c = \mathcal{N}$ and $\mathcal{N} = [1, 2, \cdots, N]$. SC with the variable format of log-likelihood ratio (LLR) can be expressed as [5]

$$\mathcal{L}_N^{(2i-1)}(y_1^N, \hat{u}_1^{2i-2}) \simeq \mathrm{sign}(\phi)\mathrm{sign}(\varphi)\min(|\phi|, |\varphi|), \tag{1}$$

$$\mathcal{L}_N^{(2i)}(y_1^N, \hat{u}_1^{2i-2}) = (-1)^{\hat{u}_{2i-1}} \phi + \varphi, \tag{2}$$

Where $\phi = L_{N/2}^{(i)}(y_1^{N/2}, \hat{u}_{1,o}^{2i-2} \oplus \hat{u}_{1,e}^{2i-2})$ and $\varphi = L_{N/2}^{(i)}(y_{N/2+1}^N, \hat{u}_{1,e}^{2i-2})$. According to formula (1) and (2), the front decoded bits are used to deduce the sequel bits. Then, the LLR of each decoding bit can be calculated as [6].

$$\mathcal{L}(\hat{u}_i) = ln(\frac{W_N^{(i)}(y_1^N, \hat{u}_1^{i-1}|0)}{W_N^{(i)}(y_1^N, \hat{u}_1^{i-1}|1)}). \tag{3}$$

In the recursive process, the decoding result is decided by

$$\hat{u}_i = \begin{cases} 1 & if\ \mathcal{L}(\hat{u}_i) < 0 \\ 0 & if\ \mathcal{L}(\hat{u}_i) \geq 0 \\ u_i & if\ i \in \mathcal{A}^c \end{cases}.$$

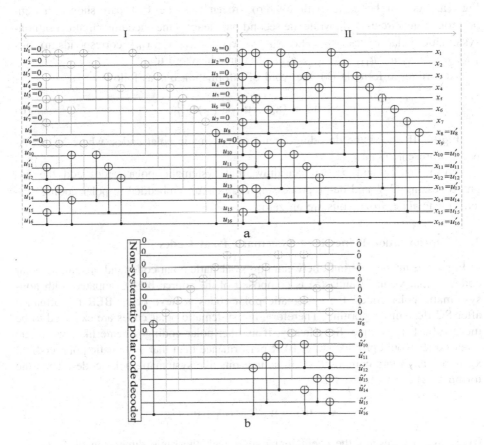

a

b

Fig. 1. The circuit of (16, 8) systematic polar codes, Part I is preprocessing module and Part II is non-systematic polar code encoding module. (a) is decoder and (b) is decoder.

2.2 Systematic Polar Coding Construction

The codeword **u** is composed of information-bit word $\mathbf{u}_A$ and the frozen-bit word $\mathbf{u}_{A^c}$. The encoded codeword **x** can be derived as

$$\mathbf{x} = \mathbf{u}G = \mathbf{u}_A G_A + \mathbf{u}_{A^c} G_{A^c}, \tag{4}$$

where matrix $G_{\mathcal{A}}$ consists of the $\mathcal{A}$ row index vector of matrix G. Matrix $G_{\mathcal{A}^c}$ consists of the $\mathcal{A}^c$ row index vector of matrix G. Then formula (4) can changes as

$$\mathbf{x}_B = \mathbf{u}_{\mathcal{A}} G_{\mathcal{A}B} + \mathbf{u}_{\mathcal{A}^c} G_{\mathcal{A}^c B}, \tag{5}$$

where $\mathcal{B} = \mathcal{A}$, The encoding circuit of $(16, 8)$ systematic polar codes is shown in Fig. 1a. As can be seen, with all-zero frozen bits the first part shows the en-preprocessing circuit $G_{\mathcal{A}B}^{-1}$, while the second part denotes the encoding circuit G of non-systematic polar codes. $\mathbf{x}_B = \{x_8, x_{10}, x_{11}, x_{12}, x_{13}, x_{14}, x_{15}, x_{16}\}$ corresponds to bits $\{u'_8, u'_{10}, u'_{11}, u'_{12}, u'_{13}, u'_{14}, u'_{15}, u'_{16}\}$ of source-bit word $\mathbf{u}'_{\mathcal{A}}$.

According to formula (5), the word of information bits as follow

$$\mathbf{u}_{\mathcal{A}} = (\mathbf{x}_B - \mathbf{u}_{\mathcal{A}^c} G_{\mathcal{A}^c B}) G_{\mathcal{A}B}^{-1}. \tag{6}$$

Figure 1b shows the decoding circuit of systematic polar codes, where XOR network represents $G_{\mathcal{A}B}$.

From Fig. 1a and b, we can draw that systematic polar codes add the en-preprocessing $G_{\mathcal{A}B}^{-1}$ and de-preprocessing $G_{\mathcal{A}B}$ circuit embedded in polar coding system, meanwhile, frozen bits are all zero.

2.3   Optimization Principle of Systematic Polar Codes

In Fig 1, the major difference between non-systematic polar codes and systematic polar codes is that $\mathbf{x}_B$ in formula (5) is composed of the source bits. Compared with non-systematic polar codes, the systematic polar codes achieve better BER performance after SC decoding algorithm. Therefore, the systematic polar codes are expected to be more robust in practice. Specifically, if another polar coding scheme like systematic polar codes would provide better BER performance than that systematic polar coding, $\mathbf{x}_B$ is not any more composed of all source bits like systematic polar codes. Then the formula (4) can be revised as:

$$\mathbf{x} = \mathbf{u}G = \tilde{\mathbf{u}}_{\mathcal{A}} \mathcal{R}^{-1} G_{\mathcal{A}} + \mathbf{u}_{\mathcal{A}^c} G_{\mathcal{A}^c}, \tag{7}$$

where $\mathcal{R}^{-1}$ represents the coefficient matrix with the same dimension of G. Correspondingly, the Part I in Fig. 1 is modified as $\mathcal{R}^{-1}$. Then the formula (6) should be changed as

$$\mathbf{u}_{\mathcal{A}} = (\tilde{\mathbf{x}}_B - \mathbf{u}_{\mathcal{A}^c} G_{\mathcal{A}^c B}) \mathcal{R} G_{\mathcal{A}B}^{-1}. \tag{8}$$

The encoding and decoding of (7) polar coding are based on non-systematic polar coding.

Proposition 1: The decoding performance of another polar coding scheme like systematic polar codes are related to the preprocessing matrix $\mathcal{R}$.

Proof: From Figs. 1 and 2a, the source bits from word $\mathbf{u}'_A$ recovery systems at the receiver can be derived as:

$$\mathbf{u}'_A\mathcal{R} = \tilde{\mathbf{u}}_A \xrightarrow{\text{encoding, decoding and interference}} \hat{\tilde{\mathbf{u}}}_A\mathcal{R}^{-1} = \hat{\mathbf{u}}_A. \tag{9}$$

In the transmitting and receiving systems of (9) and the word $\mathbf{u}'_A\mathcal{R}$ is decoding, we define $\alpha = \mathbf{u}'_A\mathcal{R} = diag(u'_1, u'_2, \ldots, u'_K)$ to be like the transmitting signals and $\beta = \hat{\mathbf{u}}'_A = diag(u'_1, u'_2, \ldots, u'_K)$ to be like the received signals. Hence, in the systems of (9), $\xrightarrow{\text{encoding, decoding and interference}}$ is parameter of systems. Then the systems (9) are transformed into the form

$$\beta = \alpha H + Z. \tag{10}$$

Where Z denotes the system interference. In order to minimize the interference of the decoding process, we minimize the following cost function denotes check bit index:

$$\begin{aligned} J(\hat{H}) &= \left\| \beta - \alpha\hat{H} \right\|^2 \\ &= (\beta - \alpha\hat{H})^H (\beta - \alpha\hat{H}) \\ &= \beta^H\beta - \beta^H\alpha\hat{H} - \hat{H}^H\alpha^H\beta + \hat{H}^H\alpha^H\alpha\hat{H}. \end{aligned} \tag{11}$$

Where $(\cdot)^H$ is operation of matrix transpose. Clearly, the system (9) has the least interference when the cost function (11) takes the minimum value, which can be computed by the partial derivative of $\hat{H}$, namely,

$$\frac{\partial J(\hat{H})}{\partial \hat{H}} = 2\alpha^H\alpha\hat{H} - 2\alpha^H\beta = 0. \tag{12}$$

Then we have $\hat{H} = \alpha^{-1}\beta$. Combining $\hat{H}$ and (14), we can obtain

$$\begin{aligned} \hat{H} &= (diag(u'_1, u'_2, \ldots, u'_K)R)^{-1} diag(\hat{u}'_1, \hat{u}'_2, \ldots, \hat{u}'_K) \\ &= R^{-1}\left\| \hat{\mathbf{u}}'_A \right\|^2. \end{aligned} \tag{13}$$

Therefore, the minimum system interference of (9) is determined by the preprocessing matrix $\mathcal{R}$. Namely, the decoding performance of generalized systematic polar codes are relevant to the preprocessing matrix $\mathcal{R}$.

Proposition 2: In the (7) polar coding, the performance of Arikan's systematic polar coding is optimal while the length becomes more longer.

Proof: For the received codeword $\mathbf{x}$ of (7) polar coding, x_{bi} within word $\tilde{\mathbf{x}}_B$ in (8) has been mistaken by $x_{bi} + \nabla x_{bi}$, where ∇x_{bi} is interference. Then for the zero frozen bits, the formula (6) changes as

$$\hat{\tilde{\mathbf{u}}}_A = (\hat{\tilde{\mathbf{x}}}_B - \mathbf{u}_{A^c} G_{A^c B}) G_{AB}^{-1} \overset{\mathbf{u}_{A^c}=0}{=\!=\!=} \hat{\tilde{\mathbf{x}}}_B G_{AB}^{-1}$$

$$= \begin{bmatrix} x_{b1}g_{11}^{-1} + x_{b1}g_{12}^{-1} + \ldots + (x + \nabla x_{bi})g_{1i}^{-1} + \ldots + x_{bk}g_{1k}^{-1} \\ x_{b1}g_{21}^{-1} + x_{b1}g_{22}^{-1} + \ldots + (x + \nabla x_{bi})g_{2i}^{-1} + \ldots + x_{bk}g_{2k}^{-1} \\ \vdots \\ x_{b1}g_{k1}^{-1} + x_{b1}g_{k2}^{-1} + \ldots + (x + \nabla x_{bi})g_{ki}^{-1} + \ldots + x_{bk}g_{kk}^{-1} \end{bmatrix}, \tag{14}$$

where g_{ii} represents element of $G_{AB}^{-1}, i \in \mathbb{N}$. We define the SC decoded code word as $\hat{\tilde{\mathbf{u}}}_A$. After de-preprocessing, $\hat{\mathbf{u}}'_A$ can be obtained by $\hat{\mathbf{u}}'_A = \hat{\tilde{\mathbf{u}}}_A \mathcal{R}^{-1}$. From (15) in system (9) suppose that the error information $x_{bi} + \nabla x_{bi}$ in $\tilde{\mathbf{x}}_B$ has not been corrected after SC decoding, and then the error bits occur with

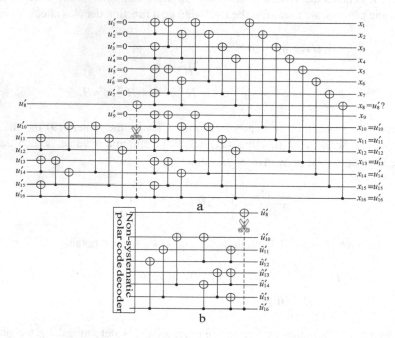

Fig. 2. Circuit of (16, 8) partial systematic polar codes, Part I is preprocessing module and Part II is non-systematic polar code encoding module. In the circuits, the pruned networks mean the reduction of resource consumption. (a) is decoder and (b) is decoder.

$$\hat{\mathbf{u}}_{\mathcal{A}} = (\hat{\mathbf{x}}_{\mathcal{B}} - \mathbf{u}_{\mathcal{A}^c} G_{\mathcal{A}^c \mathcal{B}}) G_{\mathcal{A}\mathcal{B}}^{-1} \overset{\mathbf{u}_{\mathcal{A}^c}=\mathbf{0}}{=\!=\!=} \hat{\mathbf{x}}_{\mathcal{B}} G_{\mathcal{A}\mathcal{B}}^{-1}$$

$$= \begin{bmatrix} x_{b1}g_{11}^{-1} + x_{b1}g_{12}^{-1} + \ldots + (x + \nabla x_{bi})g_{1i}^{-1} + \ldots + x_{bk}g_{1k}^{-1} \\ x_{b1}g_{21}^{-1} + x_{b1}g_{22}^{-1} + \ldots + (x + \nabla x_{bi})g_{2i}^{-1} + \ldots + x_{bk}g_{2k}^{-1} \\ \vdots \\ x_{b1}g_{k1}^{-1} + x_{b1}g_{k2}^{-1} + \ldots + (x + \nabla x_{bi})g_{ki}^{-1} + \ldots + x_{bk}g_{kk}^{-1} \end{bmatrix}, \tag{15}$$

The SC decoding are utilized the channel polarization to transfer the LLR information. Due to (1) and (2) of SC decoding are calculated with odd-even indexes, formula (15) gives the effective error interference in odd-even indexes. Hence, in SC decoder, the error diffusion also occurs in odd-even indexes. Therefore, only when the de-preprocessing matrix satisfies $\mathcal{R} = G_{\mathcal{A}\mathcal{B}}^{-1}$, we can obtain

$$\begin{aligned} \hat{\mathbf{u}}'_{\mathcal{A}} &= [\hat{u}'_1, \hat{u}'_2, \ldots, \hat{u}_i + \Delta, \ldots, \hat{u}_j + \Delta, \ldots, \hat{u}'_K] \\ &\cdot [g_1, g_2, \ldots, g_i = 1, \ldots, g_j, \ldots, \hat{u}_K]^H, \end{aligned} \tag{16}$$

where Δ denotes the error interference and only if there are many error bits, the error will be effectively assembled. Therefore, the long-length systematic polar codes have much error interference. In equation (16), the errors will be counteracted effectively if the number of Δ is abundantly produced by SC decoding. Accordingly, the system achieves the best BER performance when $\mathcal{R}$ equals $G_{\mathcal{A}\mathcal{B}}^{-1}$ based on odd-even indexes of en/de-preprocessing process.

3 Partial Systematic Polar Construction

In section above, systematic polar codes with the long length have been proven to have the best BER performance. However, systematic polar codes with the short length have no enough error to mutually counteract and obtain the best result. Hence, partial systematic polar codes of short length will obtain a better BER than systematic polar codes, such as partial systematic polar codes of (16, 8). In the next section, partial systematic polar coding construction will be represented. The key of partial systematic polar code construction is to cancel partial bits in source word to encode systematically. Figure 2 can well illustrate the process of our partial systematic polar code construction. Firstly, to reduce the circuit resource consumption, the gray figures are deleted in Fig. 1. Secondly, a certain percentage of source bit indexes in $\mathcal{N}$ is selected to cancel systematic polar encoding, then circuit resource consumption is further reduced. Thirdly, those canceled indexes return to non-systematic polar encoding. Finally, in the decoding, those selected source bits are recovered by the non-systematic polar decoder. For instance, Fig. 2, the selected source bit $\hat{u}'_8$ will return to non-systematic polar encoding and $\hat{u}'_8$ does not appear as part of encoded word transparently. Meanwhile, in Fig. 2b, the estimated $\hat{u}'_8$ is recovered as a bit of non-systematic polar codes. Simulation of Fig. 3 demonstrates that the BER performance of Fig. 2 circuit is better than that of the systematic polar coding scheme while pruning unnecessary networks of the circuit $u'_8 \setminus \hat{u}'_8$. Without loss of generality, for word $\mathbf{u}_A$, word $\mathbf{u}_{A-A^c-A'} \subset \mathbf{u}_A$ is selected to cancel systematic polar coding, and then $\mathbf{u}_{A-A^c-A'}$ participates in nonsystematic polar coding. Meanwhile, G_{AB} is updated to $G_{A'B'}$, where $A' \subset A$. For example, in Fig. 2, $\mathbf{u}_{A-A^c-A'} = u'_8$ is supposed. Hence, Eq. (4) can be revised as

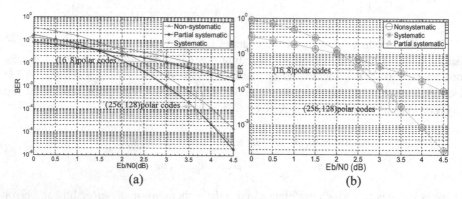

Fig. 3. Error probability of non-systematic polar codes, systematic polar codes, and partial systematic polar codes. (a) is bit error rate and (b) is frame error rate.

$$\mathbf{x}_N = \mathbf{u}_N G_N = \mathbf{u}_{A'} G_{A'} + \mathbf{u}_{N-A'} G_{N-A'}, \tag{17}$$

and the equations of encoded word are derived as

$$\mathbf{x}_{B'} = \mathbf{u}_{A'} G_{A'B'} + \mathbf{u}_{N-A'} G_{(N-A')B'}, \tag{18}$$

$$\mathbf{x}_{N-B'} = \mathbf{u}_{A'} G_{A'(N-B')} + \mathbf{u}_{N-A'} G_{(N-A')B'}, \tag{19}$$

Like Eq. (5), where $B' = A'$, $B' \subset B$ represents the index of appearing $\mathbf{x}_N$ as source bits. According to Eqs. (18) and (19), the equation of unfrozen decoded words with non-systematic decoder is derived as

$$\begin{aligned} \mathbf{u}_A &= \mathbf{u}_{A'} + \mathbf{u}_{N-A^c-A'} \\ &= (\mathbf{x}_{B'} - \mathbf{u}_{N-A'} G_{(N-A')B'}) G_{A'B'}^{-1} + \mathbf{u}_{N-A^c-A'}. \end{aligned} \tag{20}$$

Having obtained a word $\mathbf{u}'_A$ and $\mathbf{u}_{A-A^c-A'}$ in Eq. (20), $\mathbf{u}'_A$ will be calculated to obtain $\mathbf{u}'_{A'}$. Finally, both $\mathbf{u}'_{A'}$ and $\mathbf{u}_{A-A^c-A'}$ are put together as the source-bit word $\mathbf{u}'_A$.

Essentially, the word from the unfrozen set of partial systematic polar codes consists of two parts, one is systematic polar coding part the other is non-systematic polar coding part. These two parts have the respective minimum Hamming distance. If the average minimum Hamming distances among words increase, the error performance will become better [7]. Additionally, the minimum row weight of generator matrix is smaller than the minimum Hamming distance of words [8]. Hence, for systematic polar coding part, the indexes of minimum row weight in G_{AB} are selective. The selected indexes directly participate in the non-systematic polar code encoding. Then, the average minimum Hamming distances of the systematic polar coding part increase and that of the non-systematic polar coding part is no longer equal to zero in the whole coding process. Meanwhile, G_{AB} is updated to $G_{A'B'}$ and its scale becomes smaller.

4   Simulation Results

Firstly, the minimum row weight selecting method of these indexes is verified right by simulations. Comparing with nonsystematic polar coding, Fig. 3 shows that the error probability becomes worse within unfrozen bit indexes of smaller row weight in systematic polar coding. If systematic polar code bits in $A - A^c - A'$ return to non-systematic polar coding scheme, the BER of SC decoding can be reduced and circuits can be simplified along with lower resource consumption. Secondly, the resource consumption further is reduced more than 9.1% in encoder and decoder at 0.5 rate. In simulations, Fig. 4 shows the partial systematic polar codes further reduce the maximum percentage of resource consumption compared with the classical systematic polar coding.

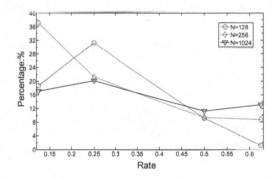

Fig. 4. Reduction of resource consumption for non-systematic polar codes with negligible different BER.

5   Conclusion

In this paper, a partial systematic polar coding is proposed. Due to pruning the coding circuit, the basic circuit architecture of encoder and decoder become concise. Comparing with the systematic coding, the partial systematic polar coding decreases the resource consumption of coding circuit. Meanwhile, the BER is negligible difference between the partial systematic polar codes and systematic polar codes.

References

1. Arikan, E.: Channel polarization: a method for constructing capacity achieving codes for symmetric binary-input memoryless channels. IEEE Trans. Inf. Theory **55**(7), 3051–3073 (2009)
2. Arikan, E.: Systematic polar coding. IEEE Commun. Lett. **15**(8), 860–862 (2011)
3. Lin, S., Costello Jr., D.J.: Error Control Coding. Wiley-Interscience, Upper Saddle River (2004)
4. Ryan, W.E., Lin, S.: Channel Codes Classical and Modern. Wiley-Interscience, Cambridge University, UK (2009)
5. Hashemi, S.A., Balatsoukas-Stimming, A., Giard, P., Thibeault, C., Gross, W.J.: Partitioned successive-cancellation list decoding of polar codes. In: IEEE International Conference on Acoustics, Speech, and Signal Process. (ICASSP) (2016)
6. Zhang, C., Wang, Z., You, X., Yuan, B.: Efficient adaptive list successive cancellation decoder for polar codes. In: IEEE International Asilomar Conference on Signals, Systems, Pacific Grove, CA (2014)
7. Trifonov, P.: Efficient design and decoding of polar codes. IEEE Trans. Commun. **60**(11), 3221–3227 (2012)
8. Shongwe, T., Speidel, U., Swart, T.G., Ferreira, H.G.: The effect of hamming distances on permutation codes for multiuser communication in the power line communications channel. In: Proceedings of the IEEE Africon, Livingstone, Zambia, 13–15 September 2011, pp. 1–5 (2011)

An Adaptive Code Rate Control of Polar Codes in Time-Varying Gaussian Channel

Hongxu Jin[✉] and Bofeng Jiang

School of Electronic and Information Engineering, Beihang University, Beijing 100191, China
hongxu_jin@buaa.edu.cn

Abstract. Under a benchmark of bit error rate (BER) in data transmission, a just perfect trade-off between maximizing code rate (CR) and reliable communication presents a significant coordinated challenge in the time-varying additive white Gaussian noise (T-AWGN) channel. In this paper, based on the guidance of a tight bound as coding parameters of polar code rate R, block length N with the capacity $I(W)$ in channel W of $N \geq \beta/(I(W) - R)^\mu$, a criteria of effectively adjusting the size of the parameter μ will achieve a better trade-off between the CR and the reliability, where β depends only on block error probability. In the circumstance of a round-clock traffic light (RTL) simulation, numerical results show that this scheme has a good preference for the guaranteed reliability for the wireless communication.

Keywords: Polar codes · Code rate · Traffic light (RTL) of wireless communication · Time-varying additive white Gaussian noise

1 Introduction

Polar codes are a major breakthrough in coding theory [1]. In binary-input adaptive white-Gaussian-noise channel (BI-AWGN), successive cancellation (SC) decoding is a vital algorithm can provide an approximate state-of-the-art error probability (EP) for polar codes in a communication system, and its complexity of decoding algorithm is acceptable in all decoding methods. In any error controlled coding strategy, code rate (CR) can adjust the EP. Therefore, many applications have used CR of the codeword to match the reliability of communication. This application is called adaptive code rate control (ACRC) polar codes. ACRC polar coding is also a flexible criteria. Characterized by a CR changes with signal-to-noise ratio (SNR) of T-AWGN channel,[1] the CR of polar codes can be modified solely by only one generator matrix.

---

[1] T-AWGN channel model: $y_i = x_i + n_i(t)$, here x_i is input, y_i is output, and $n_i(t) : N(0, \delta^2(t))$ is a zero mean Gaussian noise and $\delta^2(t)$ is a varying variance with t.

© ICST Institute for Computer Sciences, Social Informatics and Telecommunications Engineering 2019
Published by Springer Nature Switzerland AG 2019. All Rights Reserved
X. Liu et al. (Eds.): ChinaCom 2018, LNICST 262, pp. 147–156, 2019.
https://doi.org/10.1007/978-3-030-06161-6_14

In the normal SC decoding algorithm of polar codes over DMC, rate R, length N and capacity $I(W)$ of channel W are deduced by an inequation (1) [2]:

$$N \geq \beta/(I(W) - R)^{\mu}, \tag{1}$$

here β is a constant that depends only on block error probability p_e, $R < I(W)$ and "=" implies a critical state with unreliability [2–5]. Where in DMC, the capacity $I(W)$ of channel W is $I(W) = \frac{1}{2T}\log(1 + 2R\frac{E_b}{N_0})$, and $\frac{1}{T}$ is symbol rate of speed. In the inequation (1), one of the three parameters can be calculated from the other two among R, N and p_e. Usually, the calculation method of the inequation is deemed as state of equation. And here "=" also means the state of the maximized CR. Therefore, it is significant to calculate CR in the equation state and weaken a critical state of unreliability. However, as methods of adjustment, the conventional $\mu = 3.627$ for binary erasure channel (BEC) [2] and $\mu = 4.001$ for AWGN channel [3] are edge states for their equations without guaranteeing reliability. In addition, packet loss rates on 10^{-3} order of voice link are generally acceptable; while for the data link, bit error rate (BER) of 10^{-6} is regarded as acceptable. In this paper, under constraint of acceptable link and adjusting μ in inequation (1), we fulfill a maximized CR and achieve a better trade-off between the CR and the communication reliability. And in T-AWGN channel, the CR changes will better match the BER of data communication link constraint.

2   Coding and Reliable CR

2.1   Polar Coding

Polar codes, as a linear block coding scheme, have been proved to achieve the channel capacity at a low encoding and decoding complexity [1]. For polar codes, the generator matrix G_2 of size 2×2 can create the two independent copies of a probability W with double channels. Where $G_2 = \begin{bmatrix} 1 & 0 \\ 1 & 1 \end{bmatrix}$. Apply the transform matrix $G_2^{\otimes n}$ (where "$\otimes$" denotes the n-th Kronecker power) to constructed a codeword with length $N = 2^n$. Data will be changed into polar codes as follow: data bits in the information set $\mathcal{A}$ together with known bits in frozen set $\mathcal{A}^c$ compose the un-encoded word $U = u_1^N \triangleq (u_1, u_2, \ldots, u_n)$. The encoded code word is generated as the code word: $UG_2^{\otimes n} = X = x_1^N \triangleq (x_1, x_2, \ldots, x_n)$. After over T-AWGN channel, the received code word $Y = y_1^N \triangleq (y_1, y_2, \ldots, y_n)$ will be decoded by SC decoder. This procedure is a process of bit-channel combining. This encoding procedure corresponds to transforming N copies of channel W into a channel combining: $W_N(Y|U) = \prod_{i=1}^N W(y_i|x_i)$. Then, in the decoding process, its decomposed bit-channels and channel splitting probability $P(\cdot)$ are given by [1]:

$$W_N^{(i)}(y_0^{N-1}, u_0^{i-1}|u_i) \triangleq P(y_0^{N-1}, u_0^{i-1}|u_i) = \sum_{u_{i+1}^{N-1}} \frac{P(y_0^{N-1}|u_0^{i-1})P(u_0^{i-1})}{P(u_i)} = \frac{1}{2^{N-1}}$$

$\sum_{u_{i+1}^{N-1}} P(y_0^{N-1}|u_0^{i-1})$. And SC decoding with the variable format of log-likelihood ratio (LLR) can be derived from [7]:

$$\mathcal{L}_N^{(2i-1)}(y_1^N, \hat{u}_1^{2i-2}) \simeq \text{sign}(\phi)\text{sign}(\varphi)\min(|\phi|, |\varphi|), \tag{2}$$

$$\mathcal{L}_N^{(2i)}(y_1^N, \hat{u}_1^{2i-2}) = (-1)^{\hat{u}_{2i-1}}\phi + \varphi, \tag{3}$$

where $\phi = L_{N/2}^{(i)}(y_1^{N/2}, \hat{u}_{1,o}^{2i-2} \oplus \hat{u}_{1,e}^{2i-2})$ and $\varphi = L_{N/2}^{(i)}(y_{N/2+1}^N, \hat{u}_{1,e}^{2i-2})$. According to formula (1) and (2), the front decoded bits are used to deduce the sequel bits. Then, the LLR of each decoding bit can be decided as [8]. $\mathcal{L}(\hat{u}_i) = ln(\frac{W_N^{(i)}(y_1^N, \hat{u}_1^{i-1}|0)}{W_N^{(i)}(y_1^N, \hat{u}_1^{i-1}|1)})$. In the recursive process, the decoding result is decided by

$$\hat{u}_i = \begin{cases} 1 & if \quad \mathcal{L}(\hat{u}_i) < 0 \\ 0 & if \quad \mathcal{L}(\hat{u}_i) \geq 0 \\ u_i & if \quad i \in \mathcal{A}^c \end{cases}.$$

2.2 An Equation and an Inequation

Consider a binary-input DMC channel W. The channel capacity is $I(W)$ with any CR of $R < I(W)$ and strictly block error probability p_e, a coding scheme is constructed that allows transmission at R with an EP not exceeding p_e. Ideally, given a family of polar coding, the constraint relationship among the three parameters above. There have been two relationships among these three parameters.

The first one is a universal equation for channel coding as a fellow [6]:

$$I_{AWGN} - \sqrt{\frac{V}{N}}Q^{-1}(p_e/RN) + \frac{\log(N)}{2N} = R. \tag{4}$$

where $V \triangleq \frac{P(P+2)}{2(P+1)^2}\log_2^2(e)$, $Q^{-1}(\cdot)$ is the inverse of Q-function. Due to (4) fitting AWGN channel, the construction polar codes utilized by Gaussian (normal) approximation can well match (4). Inequation (1) is the relation of CR and SNR. μ is a defined parameter: Consider the block error probability function $p_e(\cdot)$ of SC decoding algorithm. If there is a function f with a constant $\mu > 0$ as Eq. (4), μ will be defined to satisfy (4) by [3],

$$\lim_{N \to \infty, N^{1/\mu}(I_{AWGN}-R)=\ell} P_N(R, I_{AWGN}) = f(\ell). \tag{5}$$

μ is provided by a heuristic method for computing in BEC by [2]. In [3], the error probability $E(W)$ of channel W of the information set $\mathcal{A}$ satisfies $\sum_{\exists \varepsilon > 0, i \in \mathcal{A}} E(W_N^{(i)}) \leq \delta$.

An inequation $R < I(W) - \beta N^{-\frac{1}{\mu}}$ can revise a size of μ. In [4], because of a 0.5 switching probability of the Arikan's recursions between $Z_{i+1} = Z_i^2$ and $Z_{i+1} = 2Z_i -$

Z_i^2, the eigenvalues of the expanded matrix of the recursions help to calculate μ in the BEC. Where Z is Bhattacharyya parameter and δ is a constant. μ in aforementioned studies can be estimated under the ideal condition. The definition of μ in Eq. (3) is also based on the ideal polarization of polar codes. Hence, the sizes of μ of the inequation (1) are in critical states when the "=" is established. However, Eq. (2) should aid inequation (1) to enhance the reliability of the transmission.

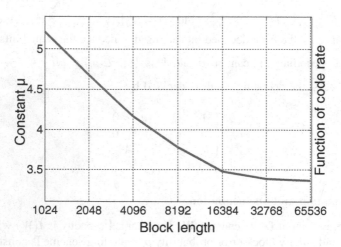

Fig. 1. μ calculation aided by Eq. (4).

3 μ Calculation with Verification

3.1   Polar Coding

Given an encoding and decoding couple, the key of reliable transmission is characterized by CR. Obviously, the more small CR is adjusted that means the more reliability in transmission. However, this trade-off is subtly based on maximal CR and reliability. Define $\mathcal{R}$ as a CR calculated by inequation (1), $\mathcal{R}'$ s CR of Eq. (3). Under the same block error probability p_e constraint, $\mathcal{R} \leq \mathcal{R}'$ of the same condition will be explained by [10] in AWGN channel. However, based on "=" in (1), $\mathcal{R}$ in (1) means the edge of unreliable transmission. Hence, suppose $\mathcal{R}_x$ presents a CR of (1) to escape from the unreliable edge of transmission adjusted by μ and based on "=" calculation, there should be $\mathcal{R}_x \leq \mathcal{R} \leq \mathcal{R}'$ for ACR criteria of polar codes and $\mathcal{R}_x$ is more reliable than others. From the (1) and (2), the deduction is as follows.

Deducing: First, difference of channel capacity:

$$\Delta I \triangleq I_{AWGN} - R \Rightarrow$$

$$\Delta I = (\beta/N)^{1/\mu}, \ \mu > 0, \ \mu_{\mathcal{R}} = \mu, \beta > 0, \tag{6}$$

Similarly:

$$2\Delta I' \triangleq \sqrt{V/N} Q^{-1}(p_e/\mathcal{R}'N), \tag{7}$$

From (6) and (7), we get: $\Delta I' \leq \Delta I$, and then we can get result of $\mathcal{R}_x \leq \mathcal{R} \leq \mathcal{R}'$ as follows:

$$\Delta I' \leq \Delta I$$
$$\Rightarrow \mu_{\mathcal{R}} = \ln(\beta/N)/\ln(\Delta I) \leq \ln(\beta/N)/\ln(\Delta I') = \mu_{\mathcal{R}'}$$
$$\Rightarrow \mathcal{R}_x = I_{AWGN} - (\beta/N)^{1/\mu_{\mathcal{R}'}} \leq \mathcal{R} = I_{AWGN} - (\beta/N)^{1/\mu_{\mathcal{R}'}} \leq \mathcal{R}'$$
$$\Rightarrow \mathcal{R}_x \leq \mathcal{R} \leq \mathcal{R}'$$

Over deducing.

With $\mu_{\mathcal{R}'}$ to govern the CR changes, $\mathcal{R}_x$ means more reliable communication than that of the conventional μ. Hence, $\mathcal{R}_x$ is a suitable CR. Gratified p_e by the transmission over the un-encoded channel, the channel environment will be deemed a noiseless channel on a certain SNR. Hence, if the CR equals to 1, Eq. (4) is still valid, and the size of μ is verified by the known CR with SNR. As polar coding scheme, the strategy is the "provably capacity achieving" code scheme and the CR must show $R \neq 1$ because of the inevitable redundancy.

3.2 μ Calculation

Due to $\Delta I' \leq \Delta I$, the gap to capacity of $\Delta I'$ guarantees to reliably govern changes of ACRC by a size of μ in inequation (1). If $\Delta I'$ substitutes into inequation (1), R equals to 1 and $\beta = \chi p_e$, as a fixed coefficient, and $\chi \in \mathbb{R}$ is aided to verify the point of $R = 1$. μ calculation of polar codes is represented by:

$$\hat{\mu} = \frac{\log(p_e/k)}{\log(\Delta I')}\Bigg|_{k=N, R-1}$$

On $E_b/N_0 = 13\,\mathrm{dB}$ of the AWGN channel, the BER of on-off keying (OOK) modulation, as a modulation scheme, is less than 10^{-6} which is an acceptable communication data link. μ equals to 4.639 by calculation of (4) when block length is 2048. As shown in Fig. 1, the calculated μ also keeps an independent constant character while the block length approaches infinity in AWGN channel. In Fig. 2, the calculated value of μ has shown to govern changes of ACRC as a function $\beta/(I_{AWGN} - R)^{\hat{\mu}} - N$ equals to 0. In [3], Gaussian approximation deduces that μ equals to 4.001. However, from Fig. 2, there is not a large difference between the proposed μ and the $\mu = 4.001$ for the CR changes in AWGN channel. Thus, as block length belongs to finite-block length regime, μ will govern the CR changes with a slight preference for the reliability.

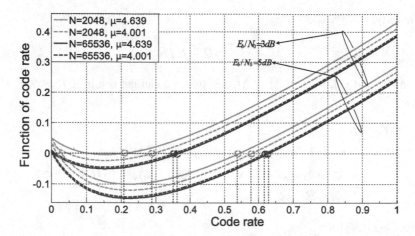

Fig. 2. Changing CR estimation

4 Utilizing RTL Scene to Simulate Polar Codes

More than 70% of total global people will be expected to live in cities by 2050 [9], then, greenhouse gas emissions and energy consumption will increase greatly. As a green communication, the RTL in cities can provide a sufficient light communication service for the waiting, the passing, and other users nearby to alleviate these problems above. Then, a simulation demonstrates an ACRC control of polar codes for RTL wireless transmission in T-AWGN channel environment.

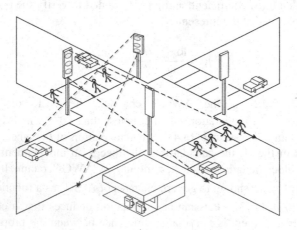

Fig. 3. RTL physical scene.

4.1 RTL Physical Scene

The physical scene of the RTL communication system can be shown in Fig. 3. The desired optical and undesirable optical signals travel through air before reaching RTL receiver. Optical filter is used to minimize the ambient-light noise. In RTL system, error performance variation is occurred in whole daytime. Due to daylight wavelength variation on multi-color channel, the daylight passing the optical filter and the sun can be removed but the color channels of green. Then yellow and red from diffuse reflectance from sun are all distributed itself color-light noise.

4.2 Numerical Results from RTL Simulation

Primarily, we assume that sender and receiver of RTL are ideal. The visible light from the sun is the very significance of photosynthesis, but the visible red light, yellow light and green light in the sunlight are interferences to the visible light transmissions over Gaussian channel model [11]. The light wavelength of green is from 455 nm to 492 nm, yellow is from 577 nm to 597 nm and red is from 622 nm to 780 nm. A fluctuating SNR in one day can be estimated by [10]. Based on the SNR, the frozen set is confirmed with the CR from $N = \beta/(I_{AWGN} - R)^{\hat{\mu}}$ calculation. As a visible light communication in varying channel environment, RTL wireless transmission can be OOK modulation [11]. In the paper of [10], Hu, Wang, and Liu took more than two years to observe and measure the photosynthetic photon flux density (Q_p). In this paper, the literature [10] will be utilized to estimate the varying variance of optical noise in the daytime. In [10], the maximum Q_{p_max} can be obtained in every hour. Hence, the wireless transmission of RTL is considered as a T-AWGN channel model. The users can receive signal power of $S = (E_r/1.09\pi d^2)A_{ph}$ (A fan-shaped angle of RTL wireless emission is about 50°) and the ambient light noise power of $N = \kappa \operatorname{var}(Q_p)\hbar v A_{ph} N_A$, meanwhile the variance can be estimated. Where $\hbar$ is Planck constant, v is green, red or yellow light frequency, N_A is the Avogadro constant and E_r is the rated power of RTL. κ is ratio coefficient (based on sundown or sunrise illumination, the non-direct by sunlight efficiency estimation is $\kappa = Q_{p\_min}/(Q_{p\_max}(hour) - Q_{p\_min})$, all kinds of wavelength from 400 nm to 700 nm, κ is about $10/(1500 - 10) \times 1/(700 - 400)$. The luminous power of red, green or yellow RTL is about limited by $E_r = 10$ W. From 80 m to RTL, if the receiving area of user antenna of visible light is $A_{ph} = 0.8\,\mathrm{cm}^2$, the signal power of $S = 3.98 \times 10^{-7}$ W can be received. Algorithm 1 can calculate the estimated code rate of $\hat{R}ACRC$ by the iterative method.

Algorithm 1: Calculation ACRC

Input: E_b/N_0

Output: $\hat{R}$

1: $SNR(dB)_i, i = 0$;

// initialize the link of OOK SNR and the variable

2: **if** $E_b/N_0 > SNR(dB)_i$ **then**

3: *Give up any channel coding regimes*;

4: **else**

5:    Set initial $R(0) = 0.5, R(1) = 1, \dfrac{E_b}{N_0}\{R(0)\text{ or }R(1)\} = 10\log(\dfrac{1}{2R(0\text{ or }1)(\text{var of nois.})^2})$;

   // initialize the CR and utilize SNR calculation

6:    **While** $(|\hat{R}(i+1) - \hat{R}(i)| > 1/N) \cap (|\dfrac{E_b}{N_0}(i+1) - \dfrac{E_b}{N_0}(i)| > 0.1)$ **do**

7:    $\dfrac{E_b}{N_0}(i) = 10\log(\dfrac{1}{2R(i)(\text{var of nois.})^2})$;

8:    $\hat{R}(i) - \dfrac{1}{T}\log(1 + 2\hat{R}(i)\dfrac{E_b}{N_0}(i)) + (P_e/N)^{1/\hat{\mu}} = 0$; // Calculation $\hat{R}(i)$

   $i = i+1$;

9: **end while**

10: **end if**

In the background of solar irradiance and the data in [10], the diurnal SNR (dB) of green, yellow and red light can be estimated. Under the constraint of BER 10^{-6}, the Fig. 4 shows the simulation of the fixed CR and ACRC of polar codes in the RTL wireless transmission. Based on reliable data communication, if CR is 0.54 of polar

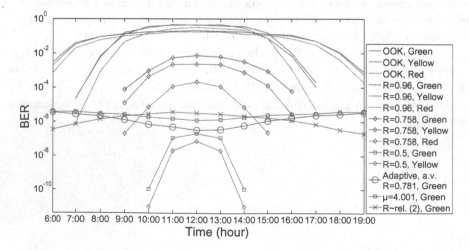

Fig. 4. Seamlessly ACRC by a generator of polar codes in RTL.

codes with block length 2048 in the RTL system, the communication data link works availably for 20 h a day. If CR is fixed 0.758 for polar codes of length 2048, the communication link can't be set up from 8:00 to 17:00 in the sunny days of RTL wireless transmission. However, if the ACRC varies with SNR under the error rate of 10^{-6} constraint, the communication is more perfect with SC decoding algorithm of parameter μ in this paper than that of $\mu = 4.001$. In fact, the polar codes of block length 2048 with fixed 0.758 CR does not match communication link for most of the daytime, but an average polar CR of 0.781 in the adaptive regime can be obtained in an uninterrupted wireless communication connection of a day. Hence, the criteria proposed in this paper can achieve a high average CR in time-varying channel.

5 Conclusion

The trade-off between reliability and CR (or throughput) for frame length is significance. Polar codes have only one generator matrix to cater the variable CR. Nonetheless, in the fluctuating channel environment, the conventional "=" state of single inequation (1) cannot guarantee the reliable communication with any CR from the channel state. However, we find and create a criterion to control variable CR. In T-AWGN channel, μ has been adjusted to satisfy the perfect and reliable mapping for any CR, while preserving the largest CR.

References

1. Arikan, E.: Channel polarization: a method for constructing capacity achieving codes for symmetric binary-input memoryless channels. IEEE Trans. Inf. Theory 55(7), 3051–3073 (2009)
2. Korada, S.B., Montanari, A., Telatar, E., Urbanke, R.: An empirical scaling law for polar codes. In: IEEE International Symposium on Information Theory Proceedings (ISIT), Austin, U.S.A. June 2010
3. Goli, A., Hassani, S.H., Urbanke, R.: Universal bounds on the scaling behavior of polar codes. In: IEEE International Symposium on Information Theory Proceedings (ISIT), Austin, U.S.A. July 2012
4. Hassani, S.H., Alishahi, K., Urbanke, R.: Finite-length scaling for polar codes. IEEE Trans. Inf. Theory 60(10), 8575–5898 (2014)
5. Goldin, D., Burshtein, D.: Improved bounds on the finite length scaling of polar codes. IEEE Trans. Inf. Theory 60(11), 6966–6978 (2014)
6. Polyanskiy, Y., Vincent Poor, H., Verdú, S.: Channel coding rate in the finite blocklength regime. IEEE Trans. Inf. Theory 56(6), 2307–2359 (2010)
7. Hashemi, S.A., Balatsoukas-Stimming, A., Giard, P., Thibeault, C., Gross, W.J.: Partitioned successive-cancellation list decoding of polar codes. In: IEEE International Conference on Acoustics, Speech, and Signal Process (ICASSP) (2016)
8. Zhang, C., Wang, Z., You, X., Yuan, B.: Efficient adaptive list successive cancellation decoder for polar codes. In: IEEE International Asilomar Conference on Signals, Systems, Pacific Grove, CA (2014)

9. Crowther, J., Herzig, C., Feller, G.: The time is right for connected public lighting within smart cities. In: IBSG Cisco System, San Jose, CA, U.S.A. (2012)
10. Hu, B., Wang, Y., Liu, G.: Measurements and estimations of photosystem vertically active radiation in Beijing. Elsevier Atmos. Res. **85**(3–4), 361–371 (2007)
11. Rajagopal, S., Roberts, R., Lim, S.: IEEE 802.15.7 visible light communication: modulation schemes and dimming support. IEEE Commun. Mag. **50**(3), 71–82 (2012)

The Quaternion-Fourier Transform and Applications

Shanshan Li, Jinsong Leng, and Minggang Fei(✉)

School of Mathematical Sciences, University of Electronic Science and Technology of China, Chengdu 611731, People's Republic of China
shanshanli@swun.cn, {lengjs,fei}@uestc.edu.cn

Abstract. It is well-known that the Fourier transforms plays a critical role in image processing and the corresponding applications, such as enhancement, restoration and compression. For filtering of gray scale images, the Fourier transform in $\mathbb{R}^2$ is an important tool which converts the image from spatial domain to frequency domain, then by applying filtering mask filtering is done. To filter color images, a new approach is implemented recently which uses hypercomplex numbers (called Quaternions) to represent color images and uses Quaternion-Fourier transform for filtering. The quaternion Fourier transform has been widely employed in the colour image processing. The use of quaternions allow the analysis of color images as vector fields, rather than as color separated components. In this paper we mainly focus on the theoretical part of the Quaternion Fourier transform: the real Paley-Wiener theorems for the Quaternion-Fourier transform on $\mathbb{R}^2$ for Quaternion-valued Schwartz functions and L^p-functions, which generalizes the recent results of real Paley-Wiener theorems for scalar- and quaternion-valued L^2-functions.

Keywords: Quaternion analysis · Paley-Wiener theorem
Quaternion-Fourier transform

1 Introduction

The original Paley-Wiener theorem [8] describes the Fourier transform of L^2-functions on the real line with support in a symmetric interval as entire functions of exponential type whose restriction to the real line are L^2-functions, which has proved to be a basic tool for transform in various set-ups. Recently, there has been a great interest in the real Paley-Wiener theorem due to Bang in [1] and Tuan in [11], in which the adjective "real" expresses that information about the support of the Fourier transform comes from growth rates associated

The authors were partially supported by NSF of China 11571083.

to the function f on $\mathbb{R}$, rather than on $\mathbb{C}$ as in the classical "complex Paley-Wiener theorem". The Fourier transform of functions with polynomial domain supports, of functions vanishing on some ball, and even in the classical case the result obtained here are also new. The set-up is as follows. For any functions $f \in \mathcal{S}(\mathbb{R}^k)$, there holds

$$\lim_{n \to \infty} \|P^n(iD)f\|_p^{\frac{1}{n}} = \sup_{y \in \mathrm{supp}\hat{f}} |P(y)|$$

and

$$\lim_{n \to \infty} \| \sum_{m=0}^{\infty} \frac{n^m \Delta^m f}{m!} \|_p^{\frac{1}{n}} = exp\Big(- \inf_{y \in \mathrm{supp}\hat{f}} \Big)|y|^2,$$

here $P(y)$ is a non-constant polynomial and $P(iD)$ is the transmutation operator.

In this paper we will consider the real Paley-Wiener theorem for the quaternion Fourier transform (QFT) which is a nontrivial generalization of the real and complex Fourier transform (FT) to quaternion algebra. The four components of QFT separate four cases of symmetry in real signals instead of only two ones in the complex FT. The QFT plays an important role in the representation of signals and transforms a quaternion 2D signal into a quaternion-valued frequency domain signal. There are lots of efforts to devote to many important properties and applications of the QFT (see [2–4, 6, 7, 9, 10]).

Motivated by recent work [5] which derived a real Paley-Wiener theorem to characterize the quaternion-valued L^2-functions whose QFT has compact support, we systematically develop a real Paley-Wiener theorem for QFT on $\mathbb{R}^2$ for quaternion-valued Schwartz functions and L^p-functions, $1 \le p \le \infty$.

The paper is organized as follows. Section 2 is devoted to recalling some definitions and properties for quaternions and their analysis. In Sect. 3, we prove the real Paley-wiener theorems for the QFT.

2   Preliminaries

The quaternion algebra $\mathbb{H}$ and Clifford algebra are extensions of the algebra of complex numbers. The quaternion algebra is given by

$$\mathbb{H} = \{q|q = q_0 + q_1\mathbf{i} + q_2\mathbf{j} + q_3\mathbf{k}, q_0, q_1, q_2, q_3 \in \mathbb{R}\}$$

where the elements $\mathbf{i}, \mathbf{j}, \mathbf{k}$ obey Hamilton's multiplication rules

$$\mathbf{ij} = -\mathbf{ji} = \mathbf{k}, \mathbf{jk} = -\mathbf{kj} = \mathbf{i}, \mathbf{ki} = -\mathbf{ik} = \mathbf{j}, \mathbf{i}^2 = \mathbf{j}^2 = \mathbf{k}^2 = \mathbf{ijk} = -1.$$

The conjugate of a quaternion $q \in \mathbb{H}$ is obtained by changing the sign of the pure quaternion part, i.e., $\bar{q} = q_0 - q_1\mathbf{i} - q_2\mathbf{j} - q_3\mathbf{k}$. This leads to a norm of $q \in \mathbb{H}$, which is defined as

$$|q| = \sqrt{q\bar{q}} = \sqrt{q_0^2 + q_1^2 + q_2^2 + q_3^2}.$$

A quaternion-valued function $f : \mathbb{R}^2 \to \mathbb{H}$ will be written as

$$f(\mathbf{x}) = f_0(\mathbf{x}) + f_1(\mathbf{x})\mathbf{i} + f_2(\mathbf{x})\mathbf{j} + f_3(\mathbf{x})\mathbf{k}, \qquad \mathbf{x} = (x_1, x_2),$$

with real-valued coefficient functions $f_0, f_1, f_2, f_3 : \mathbb{R}^2 \to \mathbb{R}$. We introduce the space $L^p(\mathbb{R}^2)$, $1 \le p \le \infty$, as the left module of all quaternion-valued functions $f : \mathbb{R}^2 \to \mathbb{H}$ satisfying

$$\|f\|_p := \left(\int_{\mathbb{R}^2} |f(\mathbf{x})|^p dx \right)^{1/p} < \infty, \qquad \text{if } 1 \le p < \infty,$$

$$\|f\|_\infty := ess \sup_{\mathbf{x} \in \mathbb{R}^2} |f(\mathbf{x})| < \infty, \qquad \text{if } p = \infty.$$

Definition 1. *The normalized right-sided QFT of a function $f \in L^1(\mathbb{R}^2)$ is defined by*

$$\mathcal{F}_q^r f(\lambda) = \int_{\mathbb{R}^2} f(\mathbf{x}) e^{-ix_1\lambda_1} e^{-jx_2\lambda_2} dx, \qquad \text{for all } \lambda \in \mathbb{R}^2. \tag{1}$$

So the corresponding inversion formula can be given as

$$f(\mathbf{x}) = \frac{1}{(2\pi)^2} \int_{\mathbb{R}^2} \mathcal{F}_q^r f(\lambda) e^{jx_2\lambda_2} e^{ix_1\lambda_1} d\lambda, \qquad \text{for all } \mathbf{x} \in \mathbb{R}^2. \tag{2}$$

Similarly,

Definition 2. *The normalized left-sided QFT of a function $f \in L^1(\mathbb{R}^2)$ is defined through*

$$\mathcal{F}_q^l f(\lambda) = \int_{\mathbb{R}^2} e^{-ix_1\lambda_1} e^{-jx_2\lambda_2} f(\mathbf{x}) dx, \qquad \text{for all } \lambda \in \mathbb{R}^2, \tag{3}$$

and the corresponding inversion formula can be given as

$$f(\mathbf{x}) = \frac{1}{(2\pi)^2} \int_{\mathbb{R}^2} e^{jx_2\lambda_2} e^{ix_1\lambda_1} \mathcal{F}_q^l f(\lambda) d\lambda, \qquad \text{for all } \mathbf{x} \in \mathbb{R}^2. \tag{4}$$

The QFT of a tempered distribution T is defined by

$$\langle \mathcal{F}_q^r T, \phi \rangle = \langle T, \mathcal{F}_q^l \phi \rangle, \qquad \phi \in \mathcal{S}(\mathbb{R}^2), \tag{5}$$

which is compatible with its definition on $L^1(\mathbb{R}^2)$.

In what follows, we recall the following important property of the QFT. For more properties and details, we refer to [5,6].

Proposition 1. (QFT partial derivatives). *If $\frac{\partial^{m_1+m_2}}{\partial x_1^{m_1} \partial x_2^{m_2}} f(\mathbf{x}) \in L^1(\mathbb{R}^2)$, $m_1, m_2 \in \mathbb{N}_0$, then we have*

$$\mathcal{F}_q^r \left\{ \frac{\partial^{m_1+m_2}}{\partial x_1^{m_1} \partial x_2^{m_2}} f(\mathbf{x}) \mathbf{i}^{-m_1} \right\}(\lambda) = \lambda_1^{m_1} \mathcal{F}_q^r f(\lambda) \lambda_2^{m_2} \mathbf{j}^{m_2}, \tag{6}$$

and

$$\mathcal{F}_q^l\{\frac{\partial^{m_1+m_2}}{\partial x_1^{m_1}\partial x_2^{m_2}}\mathbf{j}^{-m_2}f(\mathbf{x})\}(\lambda) = \mathbf{i}^{m_1}\lambda_1^{m_1}\mathcal{F}_q^l f(\lambda)\lambda_2^{m_2}. \tag{7}$$

Proposition 2. (QFT Plancherel). *If $f, g \in L^2(\mathbb{R}^2)$, then there holds*

$$(f, g) = \frac{1}{(2\pi)^2}(\mathcal{F}_q^r f, \mathcal{F}_q^r g). \tag{8}$$

In particular, if $f = g$, we have the following Parseval's Identity:

$$\|f\|_2 = \frac{1}{2\pi}\|\mathcal{F}_q^r f\|_2. \tag{9}$$

3   Real Paley-Wiener Theorems for the Quaternion-Fourier Transform

First, we consider the functions vanishing outside a ball, which is the Paley-Wiener-Type Theorem.

Theorem 1. *Let $P(x) = x_1^{n_1}x_2^{n_2}$ for any fixed nonnegative integers n_1 and n_2. Suppose $P(\partial)^m \in L^p(R^2)$ for all $m \in N_0$ and $1 \le p \le \infty$. Assume further that either $\mathcal{F}_q^r f$ has compact support or that the set $\lambda \in R^2 : |P(\lambda)| \le R$ is compact for all $R \ge 0$. Then in the extended positive real numbers*

$$\lim_{m\to\infty} \|P^m(\partial)f\|_p^{\frac{1}{m}} = \sup_{\lambda\in supp\mathcal{F}_q^r(f)} |P(\lambda)|. \tag{10}$$

Proof. The case for $f \equiv 0$ is trivial, so we assume that $f \not\equiv 0$.

Step 1: If $2 \le p \le \infty$, applying the Hausdorff-Young's inequality with $p^{-1} + q^{-1} = 1$:

$$\begin{aligned}
\|P^m(\partial)f\mathbf{i}^{-mn_1}\|_p &\le C\|P^m(\lambda)\mathcal{F}_q^r(f)\mathbf{j}^{mn_2}\|_q \\
&= C\|P^m(\lambda)\mathcal{F}_q^r(f)\mathbf{j}^{mn_2}\|_{L^q(supp\mathcal{F}_q^r(f))} \\
&= C\|P^m(\lambda)\mathcal{F}_q^r(f)\|_{L^q(supp\mathcal{F}_q^r(f))} \\
&\le C \sup_{\lambda\in supp\mathcal{F}_q^r(f)} |P(\lambda)|^m\|\mathcal{F}_q^r(f)\|_{L^q(supp\mathcal{F}_q^r(f))},
\end{aligned}$$

so we have

$$\begin{aligned}
\lim_{m\to\infty}\sup \|P^m(\partial)f\mathbf{i}^{-mn_1}\|_p^{\frac{1}{m}} &\le \sup_{\lambda\in supp\mathcal{F}_q^r(f)} |P(\lambda)| \lim_{m\to\infty}\sup C^{\frac{1}{m}}\|\mathcal{F}_q^r(f)\|_q^{\frac{1}{m}} \\
&= \sup_{\lambda\in supp\mathcal{F}_q^r(f)} |P(\lambda)|. \tag{11}
\end{aligned}$$

For the case $1 \leq p < 2$, using Hölder's inequality and Plancherel Theorem for the QFT, we get

$$\|f\|_p^p = \int_{R^2} (1+|x|^2)^{-2p}|(1+|x|^2)^2 f(x)|^p dx$$
$$\leq \|(1+|x|^2)^{-2p}\|_{\frac{2}{2-p}} \|(1+|x|^2)^2 f(x)\|_2^p$$
$$\leq C\|(1+|x|^2)^2 f(x)\|_2^p$$
$$= C\|(1-\Delta)^2 \mathcal{F}_q^r(f)\|_2^p, \tag{12}$$

here $\Delta = \frac{\partial^2}{\partial x_1^2} + \frac{\partial^2}{\partial x_2^2}$ denotes the Laplacian.

Substituting f in the above inequality with $P^m(\partial)\mathbf{i}^{-mn_1}$, there holds

$$\|P^m(\partial)\mathbf{i}^{-mn_1}\|_p^p \leq C\|(1-\Delta)^2 P^m(\lambda)\mathcal{F}_q^r(f)\mathbf{j}^{mn_2}\|_2^p.$$

By mathematical induction, we can show that

$$(1-\Delta)^2 (P^m(\lambda)\mathcal{F}_q^r(f)\mathbf{j}^{mn_2}) = P^{m-4}(\omega)\Phi_n(\omega)\mathbf{i}^{mn_2}, \ m > 4,$$

where $\mathrm{supp}\Phi_n \subset \mathrm{supp}\mathcal{F}_q^r(f)$ and $\Phi_n(\omega) \leq Cn^4$.

Hence,

$$\|P^m(\partial)f\mathbf{i}^{-mn_1}\|_p \leq C\|P^{m-4}\Phi_n(\omega)\mathbf{i}^{mn_2}\|_2$$
$$< C \sup_{supp\mathcal{F}_q^r(f)} |P(\omega)|^{m-4}\|\Phi_n(\omega)\mathbf{j}^{mn_2}\|_2$$
$$\leq Cn^4 \sup_{supp\mathcal{F}_q^r(f)} |P(\omega)|^{m-4},$$

which implies

$$\limsup_{m\to\infty} \|P^m(\partial)f\mathbf{i}^{-mn_1}\|_p^{\frac{1}{m}} \leq \sup_{supp\mathcal{F}_q^r(f)} |P(\omega)|. \tag{13}$$

In case $p = \infty$, we have

$$\|f\|_\infty \leq (2\pi)^{-1}\|f\|_1$$
$$= (2\pi)^{-1} \int_{R^2} (1+|x|^2)^{-2}|(1+|x|^2)^2 \mathcal{F}_q^r(f)|dx$$
$$= (2\pi)^{-1}\|(1+|x|^2)^{-2}\|_2\|(1+|x|^2)^2 \mathcal{F}_q^r(f)\|_2$$
$$\leq C\|(1+|x|^2)^2 \mathcal{F}_q^r(f)\|_2.$$

Therefore,

$$\|P^n(\partial)f\mathbf{i}^{-mn_1}\|_\infty \leq C\|(1+|x|^2)^2 P^n(\omega)\mathcal{F}_q^r(f)\mathbf{j}^{mn_2}\|_2$$
$$= C\|(1+|x|^2)^2 \mathcal{F}_q^r(f)P^n(\omega)\mathbf{j}^{mn_2}\|.$$

Consequently,

$$\lim_{m\to\infty}\sup\|P^m(\partial)f\mathbf{i}^{-mn_1}\|_\infty^{\frac{1}{m}} \leq \sup_{\omega\in supp(1+|x|^2)^2\mathcal{F}_q^r(f)}|P(\omega)|$$

$$= \sup_{\omega\in supp\mathcal{F}_q^r(f)}|P(\omega)|. \tag{14}$$

Step 2: Since $f \in S(R^2)$, the function f and its partial derivatives vanish at infinity, therefore, integration by parts gives

$$\int_{R^2}\overline{P^m(\partial)f}P^m(\partial)f(x)dx = \int_{R^2}P^m(\partial)\overline{f(x)}P^m(\partial)f(x)dx$$

$$= -\int_{R^2}\overline{f(x)}P^{2m}(\partial)f(x)dx.$$

Hence, by Hölder inequality, we have

$$\|P^m(\partial)f\|_2^2 \leq \|f\|_q\|P^{2m}(\partial)f\|_p.$$

Replacing f by $P(\partial)f$ in above inequality, we have

$$\|P^{m+1}(\partial)f\|_2^2 \leq \|P(\partial)f\|_q\|P^{2m+1}(\partial)f\|_p.$$

Since $f \in S(R^2)$, we have that $P(iD)f \neq 0$, and consequently,

$$\sup_{\omega\in supp\mathcal{F}_q^r(f)}|P(\omega)| = \lim_{m\to\infty}\|P^{m+1}(\partial)f\|_2^{\frac{1}{m+1}}$$

$$= \lim_{m\to\infty}\|P^{m+1}(\partial)f\|_2^{\frac{2}{2m+1}}$$

$$\leq \lim_{m\to\infty}\|P(\partial)f\|^{\frac{1}{2m+1}}\lim_{m\to\infty}\inf\|P^{2n+1}(\partial)f\|_p^{\frac{1}{2n+1}}$$

$$= \lim_{m\to\infty}\inf\|P^{2m+1}(\partial)f\|_p^{\frac{1}{2m+1}}.$$

For another, applying formula for the proved case $p = 2$, there holds

$$\sup_{\omega\in supp\mathcal{F}_q^r(f)}|P(\omega)| = \lim_{m\to\infty}\|P^m(\partial)f\mathbf{i}^{-mn_1}\|_2^{\frac{1}{m}}$$

$$\leq \lim_{m\to\infty}\|f\|_q^{\frac{1}{2m}}\lim_{m\to\infty}\inf\|P^{2m}(\partial)f\mathbf{i}^{-mn_1}\|_p^{\frac{1}{2m}}$$

$$= \lim_{m\to\infty}\inf\|P^{2m}(\partial)f\mathbf{i}^{-mn_1}\|_p^{\frac{1}{2m}}.$$

In summary, we get

$$\lim_{m\to\infty}\inf\|P^m(\partial)f\|_p^{\frac{1}{m}} \geq \sup_{\omega\in supp\mathcal{F}_q^r(f)}|P(\omega)|. \tag{15}$$

Inequality (15) together with inequalities (11), (13) and (14) give the formula (10). The theorem is proved. $\qquad\square$

Remark 1. Due to the noncommutative property of quaternions, we only consider the special polynomials $P(x) = x_1^{n_1} x_2^{n_2}$. For the general polynomials in $\mathbb{R}^2$, we can only obtain the results in Step 1 in the above theorem.

Second, we consider the functions vanishing on a ball, which is the Boas-Type Theorem.

Theorem 2. *For any function $f \in S(R^2)$, the following equality holds:*

$$\lim_{n \to \infty} \| \sum_{m=0}^{\infty} \frac{n^m \Delta^m f}{m!} \|_p^{\frac{1}{n}} = \exp(- \inf_{y \in supp \mathcal{F}_q^r(f)} |y|^2), \quad 1 \le p \le \infty. \tag{16}$$

Proof. From Proposition 1, we have for any function $f \in S(R^2)$:

$$\mathcal{F}_q^r \Big(\sum_{m=0}^{\infty} \frac{n^m \Delta^m f(x)}{m!} \Big) = \exp(-n|y|^2) \mathcal{F}_q^r(f)(y).$$

Follow the similar proof of the previous theorem, if $2 \le p < \infty$, applying the Hausdorff-Young's inequality with $p^{-1} + q^{-1} = 1$, there holds

$$\| \sum_{m=0}^{\infty} \frac{n^m \Delta^m f}{m!} \|_p \le C \| e^{-n|\lambda|^2} \mathcal{F}_q^r(f) \|_q \le C e^{-n \inf |y|^2} \| \mathcal{F}_q^r(f) \|_q.$$

Therefore,

$$\limsup_{n \to \infty} \| \sum_{m=0}^{\infty} \frac{n^m \Delta^m f}{m!} \|_p^{\frac{1}{n}} \le \exp(- \inf_{y \in supp \mathcal{F}_q^r(f)} |y|^2). \tag{17}$$

For the case $1 \le p < 2$, we first use the inequality (12) to get

$$\| \sum_{m=0}^{\infty} \frac{n^m \Delta^m f}{m!} \|_p \le \| (1 - \Delta)^2 e^{-n|y|^2} \mathcal{F}_q^r(f) \|_2.$$

Second, It's easy to show that

$$(1 - \Delta)^2 \exp(-n|y|^2 \mathcal{F}_q^r(f)) = \exp(-n|y|^2) \Phi_n(y),$$

with $supp \Phi_n \subset supp \mathcal{F}_q^r(f)$ and $\| \Phi_n \|_2 \le C n^4$.
Hence, we can obtain that

$$\limsup_{n \to \infty} \| \sum \frac{n^m \Delta^m f}{m!} \|_p^{\frac{1}{n}} \le \exp(- \inf_{y \in supp \mathcal{F}_q^r(f)} |y|^2). \tag{18}$$

In case $p = \infty$, using the inequality

$$\| f \|_\infty \le C \| (1 + |y|^2)^2 \mathcal{F}_q^r(f) \|_2$$

we get

$$\|\sum_{m=0}^{\infty}\frac{n^m\Delta^m f}{m!}\|_\infty \le C\|\exp(-n|y|^2)\mathcal{F}_q^r(f)(1+|y|^2)^2\|_2.$$

Therefore, we get inequality

$$\lim_{n\to\infty}\sup\|\sum_{m=0}^{\infty}\frac{n^m\Delta^m f}{m!}\|_\infty^{\frac{1}{n}} \le \exp(-\inf_{y\in supp\mathcal{F}_q^r(f)}|y|^2). \tag{19}$$

On the other hand, using the Plancherel theorem for the QFT and Hölder's inequality we have

$$\|\sum_{m=0}^{\infty}\frac{n^m\Delta^m f}{m!}\|_2^2 = \int_{R^2}|\sum_{m=0}^{\infty}\frac{n^m\Delta^m f}{m!}|^2 dx$$

$$= \int_{R^2}e^{-2n|y|^2}|\mathcal{F}_q^r(f)|^2 dy$$

$$= \int_{R^2}\overline{\mathcal{F}_q^r(f)(y)}\exp(-2n|y|^2)\mathcal{F}_q^r(f)(y)dy$$

$$= \int_{R^2}\overline{f(x)}\sum_{m=0}^{\infty}\frac{(2n)^m\Delta^m f(x)}{m!}dx$$

$$\le \|f\|_q\|\sum_{m=0}^{\infty}\frac{(2n)^m\Delta^m f}{m!}\|_p.$$

Similarly,

$$\|\sum_{m=0}^{\infty}\frac{n^m\Delta^m f}{m!}\|_2^2 \le \|\sum_{m=0}^{\infty}\frac{\Delta^m f}{m!}\|_q\|\sum_{m=0}^{\infty}\frac{(2n-1)^m\Delta^m f}{m!}\|_p.$$

In summary, we get

$$\lim_{n\to\infty}\inf\|\sum_{m=0}^{\infty}\frac{n^m\Delta^m f}{m!}\|_p^{\frac{1}{n}} \ge \lim_{n\to\infty}\|\sum_{m=0}^{\infty}\frac{n^m\Delta^m f}{m!}\|_2^{\frac{1}{n}} = \exp(-\inf|\omega|^2). \tag{20}$$

Combining inequalities (17), (18) and (20) we have the final result. □

References

1. Bang, H.H.: A property of infinitely differentiable functions. Proc. Am. Math. Soc. **108**, 73–76 (1990)
2. Bahri, M., Hitzer, E., Hayashi, A., Ashino, R.: An uncertainty principle for quaternion Fourier transform. Comput. Math. Appl. **56**(9), 2398–2410 (2008)
3. Bayro-Corrochano, E., Trujillo, N., Naranjo, M.: Quternion Fourier descriptors for preprocessing and recognition of spoken words using images of spatiotemporal representations. J. Math. Imaging Vis. **28**(2), 179–190 (2007)

4. Bülow, T., Felsberg, M., Sommer, G.: Non-commutative hypercomplex Fourier transform of multidimensional signals. In: Sommer, G. (ed.), Geometric Computing with Clifford Algebras, Springer, Heidelberg, pp. 187–207 (2001)
5. Fu, Y.X., Li, L.Q.: Paley-Wiener and Boas theorems for the quaternion Fourier transform. Adv. Appl. Clifford Algebr. **23**, 837–848 (2013)
6. Hitzer, E.: Quaternion Fourier transform on quaternion fields and generalizations. Adv. Appl. Clifford Algebr. **17**(3), 497–517 (2007)
7. Hitzer, E.: Directional uncertainty principle for quaternion Fourier transform. Adv. Appl. Clifford Algebr. **20**, 271–284 (2010)
8. Paley, R., Wiener, N.: The Fourier transforms in the complex domain, vol. 19, American Mathematical Society Colloquium Publications Service, Providence (1934)
9. Pei, S.C., Ding, J.J., Chang, J.H.: Efficient implementation of quaternion Fourier transform, convolution, and correlation by 2-D complex FFT. IEEE Trans. Signal Process. **49**(11), 2783–2797 (2001)
10. Sangwine, S.J., Ell, T.A.: Hypercomplex Fourier transforms of color images. IEEE Trans. Image Process. **16**(1), 22–35 (2007)
11. Tuan, V.K.: Paley-Wiener-type theorems. Fract. Calc. Appl. Anal. **2**, 135–143 (1999)

An Optimized Algorithm on Multi-view Transform for Gait Recognition

Lingyun Chi, Cheng Dai, Jingren Yan, and Xingang Liu[✉]

School of Information and Communication Engineering, University of Electronic Science and Technology of China, Chengdu, China
Hanksliu@uestc.edu.cn

Abstract. Gait is one of the common used biometric features for human recognition, however, for some view angles, it is difficult to exact distinctive features, which leads to hindrance for gait recognition. Considering the challenge, this paper proposes an optimized multi-view gait recognition algorithm, which creates a Multi-view Transform Model (VTM) by adopting Singular Value Decomposition (SVD) on Gait Energy Image (GEI). To achieve the goal above, we first get the Gait Energy Image (GEI) from the gait silhouette data. After that, SVD is used to build the VTM, which can convert the gait view-angles to 90° to get more distinctive features. Then, considering the image matrix is so large after SVD in practice, Principal Component Analysis (PCA) is used in our experiments, which helps to reduce redundancy. Finally, we measure the Euclidean distance between gallery GEI and transformed GEI for recognition. The experimental result shows that our proposal can significantly increase the richness of multi-view gait features, especially for angles offset to 90°.

Keywords: Gait recognition · Gait energy image · View transform model Principal component analysis

1 Introduction

With the development of information technology, human identification has been widely studied. Recently, human identification bases on biological feature has been the hottest topic in this field, due to the uniqueness of biological characteristics. Biometric based human identification technology refers to identifying the human identity using the body's different inherent physiological characteristics or behavioral characteristics. In our past research, the common used physiological biological features are fingerprint recognition [1], face recognition [2], iris recognition [3]. (1) Fingerprint recognition: fingerprint refers to the ridges on the frontal skin of human fingers. The starting point, the ending point, the joining point and the bifurcation point of the ridge line are feature points, which are different from people to people. The fingerprint recognition is wildly used in many situations, such as unlock the phone and unlock the door. (2) Face recognition, the human face consists of several parts, such as eyes, nose, mouth, and chin. The geometric description of these parts and the structural relationship between them can be used as important features to identify faces. The face recognition is now used in mobile payment and many other fields. (3) Iris recognition. The eye-structure of human consists of sclera, iris, pupil lens, retina and other parts. The iris is an annular portion between the black pupil and the white sclera, which contains many detailed features such as staggered spots, filaments, coronal, stripes and crypts. All these details are unique for human beings, so iris can be used to identify people. Iris is often used in high security requirement situations. In short, all these technologies bring great convenience in our daily life. However, the methods mentioned above are based on the humans' cooperation and the identification task is only suitable for short distance recognition, which may cause masquerade and hidden problems. To overcome these problems, researchers started to focus on the human behavioral characteristics, and the human gait identification is proposed.

After several decades of development, human gait identification has already developed many mature frameworks, in this case, some common problems in gait recognition are solved to some extent such as human with different clothes, human carrying different things, and human in different light conditions. These problems are set in certain condition, one of the limits is they often conduct the research in a certain view angle, however, in practice, the camera is often fixed, and it only get certain gait view angles when the human walks through the captured area in different directions, which makes it difficult to acquire the overall information of human gait, especially for parallel conditions. To solve this problem caused by multi-view, researchers proposed several methods. View Transform Model (VTM), which can realize the mutual transformation between different view angles, makes it possible to obtain more abundant information, hence, the model is widely acknowledged by researchers. Based on VTM, in this paper, we propose an optimized algorithm, which converts all other view angles to 90° thus to obtain more gait features.

The rest of the article is organized as follows. In Sect. 2, we make a summary of the solutions designed for multi-view human gait recognition, and analyze the main challenges of them. In Sect. 3, we introduce the concept involved in View Transform Model (VTM) and demonstrate its application possibility for video-based human gait recognition. In Sect. 4, after demonstrating the experiment result, we discuss the preview of VTM based multi-view gait recognition, which could be utilized under different scenarios. Finally, we conclude the article and discuss the future directions of this research theme in Sect. 5.

2 Technical Background and Related Works

In general, gait recognition has three main steps, (1) gait detection and tracking, (2) gait feature extraction, (3) gait feature matching, after the three main steps, the recognition result can be obtained. As shown in Fig. 1, the captured video sequences containing probe gait are imported into the gait recognition system, and the human silhouettes are extracted from each frames. After that, we can extract the gait features according to the human silhouettes information. Finally, the extracted features are matched with the gallery gait and then recognized. In this case, the key topic of gait recognition is focusing on how to obtain more distinctive features. However, when it comes to multi-view gait recognition, the key topic can be specifically described as matching extracted features from different view angles properly.

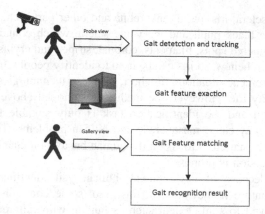

Fig. 1. A typical flow of gait recognition.

After several decades of development, there have been many resolutions proposed in allusion to multi-view gait features extraction, (1) seeking view-invariant gait characteristics; (2) constructing 3D gait model with couples of cameras; (3) establishing view transform model.

For the first method, Liu [4] represented samples from different views as a linear combination of these prototypes in the corresponding views, and extracted the coefficients of feature representation for classification. Kusakunniran [5] extracted a novel view-invariant gait feature based on Procrustes Mean Shape (PMS) and consecutively measure a gait similarity based on Procrustes distance. In general, the common features in different view angles can be extracted in this method, thus, it is possible to realize gait recognition when the view angle span is wide. However, the common feature is often insufficient, which lead that the recognition performance is relatively poor.

For the second method, Kwolek [6] identified a person by motion data obtained by their unmarked 3D motion-tracking algorithm. Wolf [7] presented a deep convolutional neural network using 3D convolutions for multi-view gait recognition capturing spatial-temporal features. Comparably, this kind of methods can achieve higher recognition accuracy, however, constructing 3D model requires a heterogeneous layer learning architecture with additional calibration, as a result, the computational complexity constructing 3D model is significantly higher than other methods.

For the third method, Makihara [8] proposed a method of gait recognition from various view directions using frequency-domain features and a view transformation model. Kusakunniran [9] applied Linear Discriminant Analysis (LDA) for further simplifying computing. Compared with the two methods above, constructing VTM can achieve state-of-the-art accuracy with less cost, in this case, VTM based approaches become the mainstream approach of the multi-view gait recognition. Considering its' advantage, in this paper, our proposal is based on VTM.

Building VTM, however, the recognition effect depends largely on the selection of which angle we transform other angles to. Meanwhile, the computational complexity when building the VTM model is relatively high. Considering the challenges mentioned above, we propose an optimized algorithm, which select a better gallery view angle for exacting more gait features and reduce the computational complexity. In this algorithm, we first get humans' Gait Energy Image (GEI) from videos containing walking sequences. After that, we transform all other view angles to 90° to exact more gait features, and then apply Principal Component Analysis (PCA) to reduce computational complexity. Finally, we measure the Euclidean distance between gallery GEI and transformed GEI for recognition.

3   Related Concept and Proposed Algorithm

3.1   Extracting Gait Energy Image (GEI)

The concept of Gait Energy Image (GEI) was first put forward by Han [10], which presents gait energy by normalizing gait silhouette in periods. As shown in Fig. 2, the higher the brightness, the higher the probability of silhouette appearing in the position.

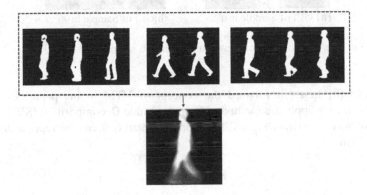

Fig. 2. A sample of gait energy image (GEI)

To get the GEI, as a basis, we need to estimate the number of gait periods contained in each gait sequence. As mentioned in [11], we can determine the gait period by calculating the ratio of height (*H*) to width (*W*) of a person's silhouette. Here, we use *N* to represent the number of periods contained in one gait sequence, and use *T* to represent the number of frames contained in one gait period. Then, we normalize all silhouettes by rescaling them along both horizontal direction and vertical direction to the same Width (*W*) and the same Height (*H*). The GEI can be obtained by:

$$GEI(x, y) = \sum_{n=1}^{N} \sum_{t=1}^{T} S_{n,t}(x, y) / (N + T) \tag{1}$$

Where, $S_{n,t}(x, y)$ represents a particular pixel locating in position (x, y) of t-th ($t = 1, 2, ..., T$) image from n-th ($n = 1, 2, ..., N$) gait period of a gait sequence.

3.2   Constructing an Optimized View Transform Model (VTM)

GEI contains several significant features, including gait silhouette, gait phase, gait frequency etc. However, the richness of GEI features in different view angles is different. As shown in Fig. 3, GEI in $90°$ contains most gait information. Therefore, we propose to transform other different angles to $90°$ for more distinctive features.

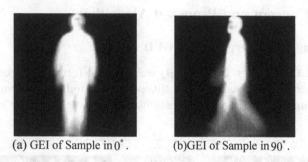

(a) GEI of Sample in $0°$.   (b)GEI of Sample in $90°$.

Fig. 3. Comparison between GEI of one sample in $0°$ and $90°$

Makihara [8] first put forward the concept of VTM in his paper. In order to construct VTM, we apply the method of Singular Value Decomposition (SVD). SVD is an effective way to extract eigenvalue. For any matrix, it can be represented as the following form:

$$A = U \sum V^T \tag{2}$$

If the size of A is $M \times N$, U is a orthogonal matrix of $M \times M$, V is a orthogonal matrix of $N \times N$. Σ is a $M \times N$ matrix, in addition to the diagonal elements are 0, elements on diagonal called singular value.

We create a matrix G_K^M with K rows and M columns, representing gait data containing K angles and M individuals.

$$G_K^M = \begin{bmatrix} g_1^1 & \cdots & g_1^m \\ \vdots & \ddots & \vdots \\ g_k^1 & \cdots & g_k^m \end{bmatrix} = USV^T = \begin{bmatrix} P_1 \\ \vdots \\ P_K \end{bmatrix} \begin{bmatrix} v^1 & \cdots & v^M \end{bmatrix} \tag{3}$$

In formula 3. g_k^m is a column vector of N_g, representing the GEI characteristics of the m-th individual at the k-th angle. The size of U is $KN_g \times M$, the size of V and S are $M \times M.P = \begin{bmatrix} P_1 & \cdots & P_K \end{bmatrix}^T = US$. After the singular value decomposition is completed, we can calculate the g_k^m with this formulation

$$g_k^m = P_k \times v^m \tag{4}$$

Then, suppose G_θ^m representing probe GEI feature with view θ and $\hat{G}_{(\varphi \leftarrow \theta)}^m$ representing the transformed GEI feature with view φ. Firstly, using the probe GEI feature G_θ^m estimate a point on the joint subspace $\hat{G}_{(\leftarrow \theta)}^m$ by

$$\hat{G}_{(\leftarrow \theta)}^m = P(\theta)^+ G_\theta^m \tag{5}$$

$$\text{Where} \quad P(\theta)^+ = \left((P(\theta))^T P(\theta) \right)^{-1} P(\theta)^T \tag{6}$$

Where $\| \cdot \|_2$ denotes the L2 norm

Secondly, the GEI feature $\hat{G}_{(\varphi \leftarrow \theta)}^m$ of transformed view can be generated by projecting the estimated point on the joint subspace.

$$\hat{G}_{(\varphi \leftarrow \theta)}^m = P(\varphi) \hat{G}_{(\leftarrow \theta)}^m \tag{7}$$

3.3 Applying Principal Component Analysis (PCA)

PCA [12] replace the original n features with less number of m for further reducing computational redundancy. New features is a linear combination of the characteristics of the old. These linear combinations maximize the sample variance, and make the new m features mutually related. Mapping from old features to new features captures the inherent variability in the data.

The main processes of PCA are:

Supposing there are N gait samples, and the grayscale value of each sample can be expressed as a column vector x_i with a size of $M \times 1$, the sample set can be expressed as $[x_1, x_2 \cdots x_N]$.

The average vector of the sample set:

$$\bar{X} = \frac{1}{N} \sum_{i=1}^{N} x_i \tag{8}$$

The covariance matrix of the sample set is:

$$\Sigma = \frac{1}{N} \sum_{i=1}^{N} (x_i - \bar{x})(x_i - \bar{x})^T \tag{9}$$

Then, calculating the eigenvectors and eigenvalues of Σ, the eigenvalues of X can be arranged in the following order $\lambda_1 \geq \lambda_2 \geq \lambda_3 \geq \cdots \geq \lambda_N$. We take the eigenvectors corresponding to the important eigenvalues to get a new dataset.

3.4 Contrasting Gait Similarity

In this paper, Euclidean distance is adopted to measure the similarity between gallery GEI (G^i) and transformed GEI (G^i). The Euclidean distance can be obtained by:

$$d(G^i, G^j) = \sum_X \sum_Y |G^i(x,y) - G^j(x,y)| \tag{10}$$

Where, $d(G^i, G^j)$ refers to Euclidean distance between G^i and G^j, (x,y) refers to the location of one specific pixel in GEI. The smaller the Euclidean distance is, the more likely the two GEIs belong to the same person.

4 Experiment

CASIA - B [13] is an extensively used database, which was collected in 2005. It contains a total of 124 objects, each object separately contains 11 view-angles, which take 18° as lad-der, range from 0° to 180°, and meanwhile, there are 6 video sequences for each person in each angle. The sketch map of the database is showed as Fig. 4. In this paper, we construct VTM with 100 objects in one of the video sequences, and use other 24 objects to evaluate the performance of our proposal.

Fig. 4. The sketch map of the database

The overview framework of our proposal is shown in Fig. 5.

Figure 6 illustrates one object transforming GEI view angle from 36° to 90° as example, which reflect the performance of VTM. Table 1 shows the performance of our proposal on the CASIA-B dataset. The first column indicates the probe view angle, and the second column indicates the accuracy of our proposal. Our proposal can significantly increase the gait feature information, in particular, the recognition accuracy is relatively higher when the angles are close to 90°.

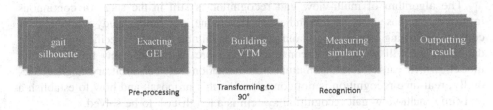

Fig. 5. The structure of the algorithm

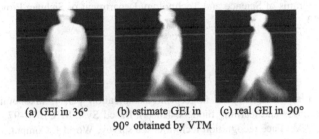

(a) GEI in 36°   (b) estimate GEI in   (c) real GEI in 90°
                          90° obtained by VTM

Fig. 6. Comparison between estimate GEI obtained by VTM with the real GEI in 90°

Table 1. Recognition accuracies with VTM

Probe view angle	Transformed to 90°
0°	81.3%
18°	90.6%
36°	91.5%
54°	95.2%
72°	98.1%
108°	97.6%
126°	93.1%
144°	90.7%
162°	89.4%
180°	80.9%

5 Conclusion and Future Work

The gait identification has gradually stepped into one of the mainstream approaches of biometric identification, considering it's' limitation caused by multi-view in practice, we choose 90° as galley gait view to get more gait features. To transform other view angles to 90°, we first exact GEI from gait silhouettes, then, we construct a view transform model with SVD, and finally, we adopt PCA to further reduce the computational complexity. It can be drawn from Table 1 that our method can significantly improve the multi-view recognition performance.

The algorithm of multi-view gait recognition is still in the stage of continuous improvement, based on the researches we have done, our further study works mainly contains following aspect. Firstly, our proposal is suitable only for specific angles in the database, we will working for constructing a more ubiquitous view transformation model, which can realize the mutual transformation between arbitrary angles. Secondly, real-time recognition is not considered in this proposal, and how to establish a real-time multi-view gait recognition system is a tough task to be solved.

Acknowledgements. This work was supported by the Fundamental Research Funds for the Central Universities on the grant ZYGX2015Z009, and also supported by Applied Basic Research Key Programs of Science and Technology Department of Sichuan Province under the grant 2018JY0023.

References

1. Galar, M., et al.: A survey of fingerprint classification part I: taxonomies on feature extraction methods and learning models. Knowl. Based Syst. **81**(C), 76–97 (2015)
2. Jayakumari, V.V.: Face recognition techniques: a survey. World J. Comput. Appl. Technol. **1**(2), 41–50 (2013)
3. De Marsico, M., Petrosino, A., Ricciardi, S.: Iris recognition through machine learning techniques: a survey. Pattern Recognit. Lett. **82**(2), 106–115 (2016)
4. Liu, N., Lu, J., Tan, Y.-P.: Joint subspace learning for view-invariant gait recognition. IEEE Signal Process. Lett. **18**(7), 431–434 (2011)
5. Kusakunniran, W., et al.: A new view-invariant feature for cross-view gait recognition. IEEE Trans. Inf. Forensics Secur. **8**(10), 1642–1653 (2013)
6. Kwolek, B., Krzeszowski, T., Michalczuk, A., Josinski, H.: 3D gait recognition using spatio-temporal motion descriptor. In: Asian Conference on Intelligent Information and Database Systems, pp. 595–604. Springer, Cham (2014)
7. Wolf, T., Babaee, M., Rigoll, G.: Multi-view gait recognition using 3d convolutional neural networks. In: IEEE International Conference on Image Processing, pp. 4165–4169. IEEE, USA (2016)
8. Makihara, Y., et al.: Gait recognition using a view transformation model in the frequency domain. In: European Conference on Computer Vision, pp. 151–163. Springer, Austria (2006)
9. Kusakunniran, W., et al.: Multiple views gait recognition using view transformation model based on optimized gait energy image. In: IEEE International Conference on Computer Vision Workshops, pp. 1058–1064. IEEE, Japan (2010)
10. Han, J., Bhanu, B.: Individual recognition using gait energy image. IEEE Trans. Pattern Anal. Mach. Intell. **28**(2), 316–322 (2005)
11. Gu, J., Ding, X., Wang, S., et al.: Action and gait recognition from recovered 3-D human joints. IEEE Trans. Syst. Man Cybern. Part B **40**(4), 1021–1033 (2010)
12. Qing-Jiang, W.U.: Gait Recognition Based on PCA and SVM. Comput. Sci. (2006)
13. Yu, S., et al.: View invariant gait recognition using only one uniform model. In: International Conference on Pattern Recognition, pp. 889–894. IEEE, Mexico (2017)

A Novel Real-Time EEG Based Eye State Recognition System

Zijia Zhou, Pan Li[⊠], Jianqi Liu, and Weikuo Dong

University of Electronic Science and Technology of China, Chengdu, China
zhouzijiam@163.com, lipan@ieee.org, liujq@std.uestc.
edu.cn, dongweikuo@gmail.com

Abstract. With the development of brain-computer interface (BCI) technology, fast and accurate analysis of Electroencephalography (EEG) signals becomes possible and has attracted a lot of attention. One of the emerging applications is eye state recognition based on EEG signals. A few schemes like the K* algorithm have been proposed which can achieve high accuracy. Unfortunately, they are generally complex and hence too slow to be used in a real-time BCI framework such as an instance-based learner. In this paper, we develop a novel effective and efficient EEG based eye state recognition system. The proposed system consists of four parts: EEG signal preprocessing, feature extraction, feature selection and classification. First, we use the 'sym8' wavelet to decompose the original EEG signal and select the 5th floor decomposition, which is subsequently de-noised by the heuristic SURE threshold method. Then, we propose a novel feature extraction method by utilizing the information accumulation algorithm based on wavelet transform. By using the CfsSubsetEval evaluator based on the BestFirst search method for feature selection, we identify the optimal features, i.e., optimal scalp electrode positions with high correlations to eye states. Finally, we adopt Random Forest as the classifier. Experiment results show that the accuracy of the overall EEG eye state recognition system can reach 99.8% and the minimum number of training samples can be kept small.

Keywords: Electroencephalogram (EEG) · Eye state identification · Feature extraction · Wavelet transform · Information accumulation algorithm · Random forest

1 Introduction

A brain-computer interface (BCI) [1] is a direct communication system between the human brain and the external world, which supports communication and control between brain and external devices without use of peripheral nerves and muscles. By using BCI, people can directly express ideas or bring them to actions only through their brains. For instance, BCI can enable disabled patients to communicate with the outside world and control external devices. As a new kind of human-computer interaction, BCI has attracted intensive attention in the field of rehabilitation engineering and biomedical engineering in recent years. EEG based eye state recognition is one of the most important research fields of BCI, which has been investigated for many applications,

X. Liu et al. (Eds.): ChinaCom 2018, LNICST 262, pp. 175–183, 2019.
https://doi.org/10.1007/978-3-030-06161-6_17

particularly in human cognitive state classification. For example, EEG based eye state classification has been successfully applied to fatigue driving detection [2], epileptic seizure detection [3], human eye state detection, recognition of infant sleep state [4], classification of bipolar affective disorder [5], human eye blinking detection, etc. These phenomena demonstrate the importance of studying eye state recognition based on EEG.

Previous studies on EEG based eye state recognition can be classified into two categories: improving the accuracy and shortening the computing time. In the first category, one of the most representative works is by Röser and Suendermtann [6], which develops a system to detect a person's eye state based on EEG recordings. The authors test 42 classification algorithms and found that the K* algorithm can get the highest accuracy of 97.3%. As a classic statistical pattern recognition method, the K* algorithm performs classification on a data sample mainly based on the surrounding neighboring samples. However, when the training sample set is large, the computing time of the K* algorithm increases significantly. In order to address this problem, some studies in the second category employ more efficient classification methods to reduce the computing time. For instance, Hamilon, Shahryari, and Rasheed [7] use Boosted Rotational Forest (BRF) to predict eye state with an accuracy of 95.1% and speed of 454.1 instances per second. Reddy and Behera [8] design an online eye state recognition with an accuracy of 94.72% and the classification speed of 192 instances per second.

In this paper, we aim to develop an effective and efficient EEG based eye state recognition system. Different from the above methods which mostly focus on optimizing the classification algorithm, we explore the overall system design consisting of EEG signal preprocessing, feature extraction, feature selection, and classification. Specifically, we first decompose the signal and mitigate noise in it. Then, we conduct feature extraction. We argue that feature extraction of dynamic signals like EEG should consider the information of adjacent time-domain signals rather than only include the information at a certain time instance. Therefore, we propose a novel feature extraction scheme by utilizing the information accumulation 3 algorithm based on wavelet transform. After that, we employ the BestFirst search algorithm to select features, and the Random Forest algorithm to perform classification. The overall EEG eye state recognition system achieves the classification accuracy of 99.8% and the speed of 639.5 instances per second. The rest of this paper is organized as follows. Section II describes the proposed real-time EEG eye state recognition system in detail. Section III demonstrates the experiment results. The last section concludes the paper.

2   System Design

2.1   EEG Signal Proprecessing

EEG signals, which are different from normal electrical signals, are dynamic, random, non-linear bio-electrical signals with high instablity. Traditional de-noising methods including linear filtering and nonlinear filtering, such as wiener filtering [9] and median filtering [10], are inappropriate for EEG signal preprocessing, because the entropy and the non-stationary characteristics of signal transformation cannot be clearly described, and the correlation of signals cannot be easily obtained. Wavelet transform [11] has strong data de-correlation capability, which can make the energy of the signal in the wavelet domain concentrated on a few large wavelet coefficients, and the noise energy distributed in the entire wavelet domain. Moreover, the noise in EEG signal is usually close to white noise. So, this paper uses the wavelet threshold de-noising [12], which can almost completely suppress the white noise, and the characteristics of the original signal are retained well.

Particularly, wavelet threshold de-noising [13] is to employ an appropriate threshold function so that the wavelet coefficients obey certain rules after the wavelet transform to achieve the purpose of de-noising. The selection of the threshold function and the determination of the threshold value are two key problems in the design of wavelet threshold de-noising algorithm, which influence the de-noising result directly. In general, threshold functions can be divided into two categories: hard threshold and soft threshold. Currently, fixed threshold [14], Stein unbiased likelihood estimation threshold, heuristic threshold [15] and minimum maximum criterion threshold are the four most frequently used selection rules. Because Stein unbiased likelihood estimation threshold and min-max criterion threshold often result in incomplete de-noising, in this paper we adopt the heuristics threshold.

2.2 Feature Extraction

In feature extraction, fast Fourier transform (FFT), autoregressive (AR) models, wavelet transform (WT) and short-time Fourier transform (STFT) are widely used to extract features of EEG signals. But transient features cannot be captured by AR models or FFT models. Both SFT transform and wavelet transform are time-frequency analysis methods, and have a unified time window to simultaneously locate different frequency ranges and time intervals. Studies have shown that the combination of time domain information and frequency information can improve the classification performance of the EEG recognition system, and that for non-stationary transient signals such as EEG, WT is more effective than SFT. Therefore, we develop a wavelet packet decomposition (WPD) based approach to extract features of EEG signals. The coefficients of WPD and the wavelet packet energy of special sub-bands are taken as the original features.

Table 1. The correspondence between components of the wavelet and frequencies of the EEG signals

Wavelet component	EEG signal frequency
A5	Delta waves
D5	Theta waves
D4	Alpha waves
D3	Beta waves

Note that in the proposed scheme determining suitable wavelet and the number of decomposition levels is critical. In particular, different types of wavelets are usually used in testing to find the wavelet with the highest efficiency for a particular application. The smoothing characteristic of the db4 wavelet is more suitable for detecting changes of EEG signals. Thus, we employ this scheme to compute the wavelet coefficients in this paper. Moreover, the number of decomposition levels is usually chosen based on the main frequency components of the signal. According to previous studies, the number of decomposition levels is set to 5 because EEG signals do not have any useful information above 30 Hz. Then, the EEG signals were decomposed into details D3-D5 and one final approximation, A5. Table 1 shows the correspondence between components of the wavelet and frequencies of the EEG Signals.

Furthermore, after analyzing the EEG signal changes corresponding to more than 50 eye state changes, we find that changes in EEG usually happen before eye movement as shown in Fig. 1. The reason for this phenomenon may be that there is a process of brain consciousness formation before people perform physiological activities. This process is related to Event-related potentials (ERP [16]), a special kind of brain evoked potentials. Evoked Potentials (EPs [17]), also known as Evoked Responses, refer to the specific stimulation of the nervous system (from the receptor to the cerebral cortex) or the processing of information about the stimulus (positive or negative). EPs are bioelectrical responses that are detectable in a system and at a corresponding portion of the brain with a relatively fixed time interval (lock-time relationship) and a specific phase. Experimental psychologists and neuroscientists have discovered many different stimuli that elicit reliable ERPs from participants. The timing of these responses can provide a measure of the timing of the brain's communication or timing of information processing. Therefore, we attempt to improve our algorithm by exploring the occurrence of brain event-related potential (ERP) in the case of human eye movement and finding out the response time of the brain to eye movement consciousness through experiments.

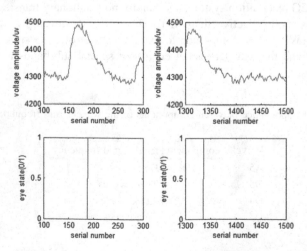

Fig. 1. The changes of signal potential when the change of eye state change

Specifically, according to the Parseval theorem, we can obtain the energy of each component after wavelet transform. Let f_x represent the energy of the x component. Then the feature vector f can be described as: $f = [f_{D3}, f_{D4}, f_{D5}, f_{A5}]^T$. As mentioned above, it has been shown that the voltage amplitude of the EEG signals starts to rise or descend before the change of eye state. Thus, the vector f only includes the energy at a certain time, but neglects the useful information of adjacent time or previous time of the signal. So, we employ the information accumulation algorithm to extend the feature vector f to f' which can better represent the characteristics of the EEG signal as in (1) and (2).

$$f_i' = \sum_T f_i \ (for \ i = 1, 2, \cdots, n) \tag{1}$$

$$f' = [f_1', f_2', \cdots, f_n']^T \tag{2}$$

Here, T represents a time window whose optimal value can be determined by experiments, and n represents the number of data samples.

2.3 Feature Selection

In machine learning, if the number of features is too many, there may exist irrelevant features and may be interdependency among features. So, it is necessary to select features before classification. This paper uses CfsSubsetEval evaluator based on the BestFirst search method [18] derived in Weka toolkit [19] to select features.

2.4 Classification

The Random Forest classification (RFC) [20] is a classification model that is composed of many decision tree classification models. Specifically, given one variable X, in each decision tree classification model, the optimal classification result depending on one vote. In contrast RFC works as follows. First, it uses the bootstrap sample method to extract k samples from the original training set. Second, k decision tree models are established from k samples, and k classification results are obtained. Finally, the classification result is obtained by following the plurality rule, i.e.,

$$H(x) = \arg\max_Y \sum_{i=1}^k I(h_i(x) = Y) \tag{3}$$

In (3), $H(x)$ is the final classification result, $h_i(x)$ is the classification result of a single decision tree, Y is target classification, and $I(\bullet)$ is the indicator function.

3   Experiment Results

3.1   Dataset

This paper uses EEG Eye State Data Set [21] from the UCI database. All EEG signals were recorded by Emotiv EEG Neuroheadset [22]. Each sample consists of 14 values from 14 electrode positions, and a label indicating the eye state ('1' indicates the eye-closed state and '0' the eye-open state). The duration of the EEG recording was 117 s.

3.2   Results

3.2.1   EEG Signal Proprecessing Results

In the SIGNAL toolbox of the MATLAB2010 platform, the signal is decomposed by the 'sym8' wavelet. On the 5th floor of the decomposition, the heuristic SURE threshold is used to de-noise the signal. The Fig. 2 shows the difference between the signal before de-noising and after de-noising in channel AF3. From the Fig. 2, it can be seen that the shaking of the waveform after de-noising is reduced, and a larg- e proportion of noises have been removed. Consequently, the wavelet threshold de-noising is a useful method to the EEG signal.

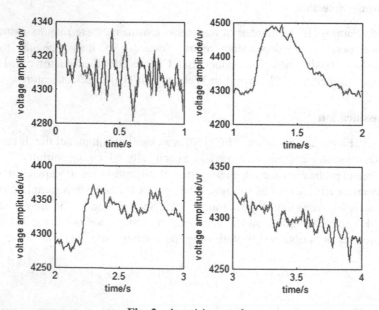

Fig. 2. de-noising results

3.2.2   Classification Results

In the experiments, all the patterns were partitioned for training and testing with the division of 66% and 34% (Röser & Suendermann's work use 10-fold cross validation).

In the step of feature extraction, there are 8 methods shown in Table 2 which lead to 8 results of feature selection and classification shown in Table 3. The classification speed v is defined as:

$$v = \frac{n}{t}(instance/s) \tag{4}$$

In (4), n is the number of instance, t is the cpu execution time. The cost time didn't include the obtainment and transmission time in hardware.

Table 3 shows that the best approach is the No.5 which has an accuracy rate of 99.8% and speed of 639.5 instances/s if taking all aspects into consideration. The best selection of the parameter T is the time period including 49 points before the current point and 50 points after it. The information in this time period can reflect eye state most effectively. The margin curve of this method is shown in Fig. 6. From this figure, we can find that, when the number of samples is larger than 1563, the classification result tends to be stable and the calculation cost is low. When the number of samples is less than 210, the classification accuracy is low and the calculation cost is high. When the number of samples is larger than 210, the accuracy increases rapidly and the calculation cost is reduced greatly. So, the minimum number of training samples is 1563. Compared not using feature extraction with only using wavelet transform when select RF as classifier, the accuracy rate increases 7.9% and 1.9% respectively. When using K* classification algorithm, the accuracy rate increases 3.7% and the speed increases to ten times. Moreover, by feature selection, it is proved that AF3(A5), F7 (A5), T7(A5), O1(A5) and FC6(D5) are 5 scalp electrode positions with high correlation to eye state. So, the recognition of eye state based on the EEG signal only needs the information of 5 channels and frequency components of delta waves and alpha waves.

Table 2. 8 feature extraction methods

Number	Method of feature extraction
1	With no feature extraction
2	WT
3	WT and IAA($T = [t_{i-49}, t_i]$)
4	WT and IAA($T = [t_{i+1}, t_{i+50}]$)
5	WT and IAA($T = [t_{i-49}, t_{i+50}]$)
6	WT and IAA($T = [t_{i-99}, t_i]$)
7	WT and IAA($T = [t_{i+1}, t_{i+100}]$)
8	WT and IAA($T = [t_{i-99}, t_{i+100}]$)

Table 3. 8 results of feature selection and classification

No.	Selected features	Speed	Accuracy	Minimum number of training samples
1	ALL	2358.4instances/s	91.9%	5093
2	AF3(A5),F7(A5),T7(A5),O1(A5),FC6(A5)	2573.2instances/s	97.9%	3552
3	AF3(A5),F7(A5),T7(A5),O1(A5)	548.2istances/s	98.6%	2737
4	AF3(A5),F7(A5),T7(A5),O1(A5)	641.1instance/s	99.3%	2189
5	AF3(A5),F7(A5),T7(A5),O1(A5),FC6(D4)	639.5instances/s	99.8%	1563
6	AF3(A5),F7(A5),T7(A5),O1(A5),FC6(D4)	616.3instances/s	99.8%	1600
7	AF3(A5),F7(A5),T7(A5),O1(A5)	654.0instances/s	99.6%	1345
8	F7(A5),T7(A5), O1(A5)	654.5instances/s	99.6%	1047

4 Conclusions

This paper develops a novel efficient EEG eye state recognition system. It has a significantly faster classification speed and higher accuracy compared with the K* algorithm. Compared with the K*algorithm, the optimal performance in this study reaches the accuracy of 99.8% and the classification speed of at-least 639.5 samples per second, making it appropriate to real-time BCI systems. We hope that this study will help more scientists and engineers understand brain activities and develop BCI systems for improving human lives.

References

1. Kim, Y.-J., Kwak, N.-S., Lee, S.-W.: Classification of motor imagery for Ear-EEG based brain-computer interface. In: 2018 6th International Conference on Brain-Computer Interface (BCI). IEEE (2018)
2. Hailin, W., Hanhui, L., Zhumei, S.: Driving detection system design based on driving behavior. In: 2010 International Conference on Optoelectronics and Image Processing. IEEE (2010)
3. Sinha, A.K., Loparo, K.A., Richoux, W.J.: A new system theoretic classifier for detection and prediction of epileptic seizures. In: The 26th Annual International Conference of the IEEE Engineering in Medicine and Biology Society. IEEE (2005)
4. Fraiwan, L., Lweesy, K.: Neonatal sleep state identification using deep learning autoencoders. In: 2017 IEEE 13th International Colloquium on Signal Processing and its Applications (CSPA). IEEE (2017)
5. Leow, A., et al.: Measuring inter-hemispheric integration in bipolar affective disorder using brain network analyses and HARDIA. In: 2012 9th IEEE International Symposium on Biomedical Imaging (ISBI) In 4th International Conference. IEEE, 2012

6. RoSler, O., Suendermann, D.: A first step towards eye state prediction using EEG. In: AIHLS 2013: The International Conference on Applied Informatics for Health and Life Sciences. IEEE (2013)
7. Hamilton, C.R., Shahryari, S., RasheedEye, K.M.: State prediction from EEG data using boosted rotational forests, In: IEEE International Conference on Machine Learning. IEEE (2016)
8. Reddy, T.K., Behera, L.: Online eye state recognition from EEG data using deep architectures.: 2016 IEEE International Conference on Systems, Man, and Cybernetics (SMC 2016). IEEE (2016)
9. Girault, B., Gonc,alves, P., Fleury, E., Mor, A.S.: Semisupervised learning for graph to signal mapping: A graph signal wiener filter interpretation. In: Proceedings of IEEE International Conference Acoustic, Speech and Signal Process. (ICASSP), pp.1115–1119. IEEE (2014)
10. Huang, T.S. (ed.): Two-Dimensional Digital Signal Processing II: Transforms and Median Filters. Springer-Verlag, Berlin (1981)
11. Grossmann, A.: Wavelet transforms and edge detection to be published in Stochastic Processes in Physics and Engineering, Ph.Blanchard, L. Streit, and M. Hasewinkel, Eds
12. Jin, Z., Jia-lunl, L., Xiao-ling, L., wei-quan, W.: ECG signals denoising method based on improved wavelet threshold algorithm. In: Advanced Information Management, Communicates, Electronic and Automation Control Conference (IMCEC). IEEE (2016)
13. Saxena, S., Khanduga, H.S., Mantri, S.: An efficient denoising method based on SURE-LET and Wavelet Transform. In: International Conference on Electrical, Electronics, and Optimization Techniques (ICEEOT). IEEE (2016)
14. Movahednasab, M., Soleimanifar, M., Gohari, A.: Adaptive transmission rate with a fixed threshold decoder for diffusion-based molecular communication. IEEE Trans. Commun. **64**, 236–248 (2016)
15. Wang, Y., He, S., Jiang, Z.: Weak GNSS signal acquisition based on wavelet de-noising through lifting scheme and heuristic threshold optimization.In: 2014 International Symposium on Wireless Personal Multimedia Communications (WPMC). IEEE (2015)
16. Zhifeng Lin, Zhihua Huang". Research on event-related potentials in motor imagery BCI": 10th International Congress on Image and Signal Processing, p. 2017. BioMedical Engineering and Informatics (CISP-BMEI), IEEE (2017)
17. Kidmose, P., Looncy, D., Ungstrup, M., Rank, M.L., MandicA, D.P.: Study of evoked potentials from ear-EEG. IEEE Trans. Biomed. Eng. **60**, 2824–2830 (2013)
18. Saad, S., Ishtiyaque, M., Malik, H.: Selection of most relevant input parameters using WEKA for artificial neural network based concrete compressive strength prediction model. In: Power India International Conference (PIICON). IEEE (2016)
19. Kumar, N., Khatri, S.: Implementing WEKA for medical data classification and early disease prediction. In: 3rd International Conference on Computational Intelligence and Communication Technology (CICT). IEEE (2017)
20. Kooistra, I., Kuilder, E.T., Mücher, C.A.: Object-based random forest classification for mapping floodplain vegetation structure from nation-wide CIR AND LiDAR datasets. In: Hyperspectral Image and Signal Processing: Evolution in Remote Sensing (WHISPERS). IEEE (2017)
21. Frank, A., Asuncion, A.: UCI Machine Learning Repository. (2010) http://archive.ics.uci.edu/ml/
22. Diego, S., Toscano, B.S., Silva, A.: On the use of the Emotiv EPOC neuroheadset as a low cost alternative for EEG signal acquisition. In: 2016 IEEE Colombian Conference on Communications and Computing (COLCOM). IEEE (2016)

Tracking Performance of Improved Convex Combination Adaptive Filter Based on Maximum Correntropy Criterion

Wenjing Wu, Zhonghua Liang$^{(\boxtimes)}$, Qianwen Luo, and Wei Li

School of Information Engineering, Chang'an University, Xi'an 710064, P. R. China
lzhxjd@hotmail.com

Abstract. A convex combination adaptive filter based on maximum correntropy criterion (CMCC) was widely used to solve the contradiction between the step size and the misadjustment in impulsive interference. However, one of the major drawbacks of the CMCC is its poor tracking ability. In order to solve this problem, this paper proposes an improved convex combination based on the maximum correntropy criterion (ICMCC), and investigates its estimation performance for system identification in the presence of non-Gaussian noise. The proposed ICMCC algorithm implements the combination of arbitrary number of maximum correntropy criterion (MCC) based adaptive filters with different adaption steps. Each MCC filter in the ICMCC is capable of tracking a specific change speed, such that the combined filter can track a variety of the change speed of weight vectors. In terms of normalized mean square deviation (NMSD) and tracking speed, the proposed algorithm shows good performance in the system identification for four non-Gaussian noise scenarios.

Keywords: Convex combination · Maximum correntropy criterion (MCC); Non-Gaussian noise; Normalized mean square deviation (NMSD); System identification

1 Introduction

Due to the low computational complexity and ease of implementation of the least mean square Algorithm (LMS), it is widely used in signal processing, system identification, acoustic echo cancellation, blind equalization, and so on [1]. However, the output of the LMS filter is not only sensitive to the amount of scaling of the input [2,3], but also degrades under non-Gaussian noise [4–6].

This work was supported in part by the National Natural Science Foundation of China under Grant No. 61271262, and in part by the Special Fund for Basic Scientific Research of Central Colleges, Chang'an University (310824171004).

© ICST Institute for Computer Sciences, Social Informatics and Telecommunications Engineering 2019
Published by Springer Nature Switzerland AG 2019. All Rights Reserved
X. Liu et al. (Eds.): ChinaCom 2018, LNICST 262, pp. 184–193, 2019.
https://doi.org/10.1007/978-3-030-06161-6_18

Therefore, the least mean fourth (LMF) algorithm [7,8], least mean p-norm algorithm (LMP) [9], and recursive least p-norm algorithm (RLP) [10] based on the gradient algorithm are proposed to improve the performance degradation under non-Gaussian noise. Recently, a more robust adaptive algorithm from the information theoretic (IT) has been proposed by Principle and et al, where the algorithm includes entropy [11], mutual information [12], and correntropy [13]. Due to its simplicity and robustness to non-Gaussian environments, the maximum correntropy criterion (MCC) [14] and the minimum error entropy (MEE) [15] have been paid attention in recent years. Although the performance of MCC and MEE is similar, the computational complexity of MEE is relatively high compared to MCC.

Recently, MCC has been used as an adaptive criterion for non-gaussian signal processing in [4]. At the same time, the tracking analysis and steady-state mean square error analysis of MCC were proposed in [16,17]. The steady-state error of the MCC algorithm depends mainly on the step size, and its convergence speed is mainly based on step length and kernel width. When the step length is fixed, the contradiction between convergence speed and steady-state mean square error can be overcome by changing the kernel width. In [18], Weihua Wang et al. proposed a switch width based on maximum correntropy. In [19], Yicong He et al. proposed a new adaptive algorithm based on generalized correntropy, using generalized Gaussian density instead of traditional width. When the width is certain, the step length is inversely proportional to the misalignment. In [20], Ren Wang et al. proposed a variable step maximum correntropy adaptive filter. In [21], the convex combination is introduced into an MCC-based adaptive filter, so that the combination filter simultaneously gets the fast convergence speed of the filter with large step size as well as the low misadjustment of the filter with small step size. However, in the convex combination of maximum correntropy criterion (CMCC) filter, the tracking and convergence performance of the combined filter are reduced. Therefore, an improved convex combination filter based on maximum correntropy criterion (ICMCC) is presented in this paper. Compared with the CMCC, ICMCC not only has fast convergence speed and low misalignment, but also can track the optimal value fast in any weight coefficient changes.

The rest of the paper is organized as follows. we review briefly MCC-based algorithms in Sect. 2. Then, ICMCC, adding weight transfer algorithm to enhance convergence rate, is proposed in Sect. 3. The simulation results are given in Sect. 4, and the conclusion is presented in Sect. 5.

2   MCC-Based Algorithms

Correntropy is a measure of local similarity between two random variables, and can also be used as a cost function in adaptive filtering [22]. Considered X and Y two random variables, the correntropy is [23]

$$v(X,Y) = E[k(X,Y)] = \int k(X,Y)\,dF_{XY}F(x,y), \tag{1}$$

where $\kappa(\cdot,\cdot)$ is a shift-invariant Mercer kernel, and F_{XY} denotes the joint distribution function of (X,Y). The most widely used kernel in correntropy is the Gaussian kernel

$$k(X,Y) = G_\sigma(e) = \frac{1}{\sqrt{2\pi}\sigma} exp(\frac{-e^2}{2\sigma^2}), \tag{2}$$

where σ is the kernel width, and $e = x - y$. The MCC algorithm finds the optimal value by maximizing the correntropy.

According to the stochastic gradient principle of adaptive algorithm, the updating equation of weight coefficient based on maximum correntropy is [21]

$$w[k] = w[k-1] + \lambda \exp(\frac{-e[k]^2}{2\sigma^2})e[k]X[k], \tag{3}$$

where λ is the step size and $X[k]$ is the input at the moment k.

3   Improvement of MCC-Based Adaptive Filter

CMCC adaptive filtering algorithm is the latest development based on the maximum correntropy criterion in the contradiction between convergence speed and misalignment. This method combines two adaptive filters by convex combination, thus obtaining the fast convergence speed of large step and the low offset of small step length. But this method has a major challenge, that is, the tracking ability of the combinational filter is reduced. In this part, we first introduce the CMCC algorithm, and then extend the method to arbitrary number adaptive filters through the maximization of the correntropy of the combined filter to improve the tracking ability and convergence speed of the combined filter.

3.1   CMCC Algorithm

The implementation of the CMCC first requires two filters with different step sizes, and then the two filters update their own weights according to their own criteria and errors. However, the update criterion of the mixing coefficient is to maximize the correntropy of combined filter. The combination weight of CMCC can be expressed as [25]

$$w[k] = v[k]w_1[k] + (1 - v[k])w_2[k], \tag{4}$$

where the mixing coefficient $v[k]$ can be denoted as $v[k] = sgma[k] = 1/(1 + e^{-\alpha[k]})$. $w_1[k]$ and $w_2[k]$ are the weights of large step and small step respectively, and they are expressed as $w_i[k] = w_i[k-1] + \lambda_i \exp(\frac{-e_i[k]^2}{2\sigma^2})e_i[k]X[k], i = 1,2$. And $e_i[k] = d[k] - y_i[k], i = 1,2$ is the error incurred by the component adaptive filter. Similarly, the combined filter output can be obtained and expressed as

$$y[k] = v[k]y_1[k] + (1 - v[k])y_2[k], \tag{5}$$

where $y_1[k] = X^T w_1[k]$ and $y_2[k] = X^T w_2[k]$ represent the output of a large step filter and a small step filter, respectively. The parameter $\alpha[k]$ is used to indirectly adjust the mixing coefficient, and updated by gradient algorithm that maximizes the correntropy of the combined filter, that is [21]:

$$\alpha[k+1] = \alpha[k] + \mu_\alpha \sigma^2 \frac{\partial \exp(\frac{-e^2[k]}{2\sigma^2})}{\partial \alpha[k]}$$

$$= \alpha[k] + \mu_\alpha v[k](1 - v[k])(y_1[k] - y_2[k]) \exp(\frac{-e^2[k]}{2\sigma^2}) e[k],$$

$$(6)$$

where μ_α represents the step size of the parameter $\alpha[k]$. In order to ensure that the adaptive speed of the combined filter is faster than that of the large step filter, μ_α must be larger than μ_1. In order to prevent the $v[k]$ from approaching 0 or 1, the range of $\alpha[k]$ is limited to $[-4, 4]$ [25].

Following [20], when the fast filter is significantly better than the slow filter, we can accelerate the convergence performance of the algorithm by the following formula. The modified small step filter can be expressed as [21]

$$w_2[k] = \beta w_2[k-1] + \lambda_2 \exp(\frac{-e_2^2[k]}{2\sigma^2}) e_2[k]x[k] + (1 - \beta)w_1[k], \qquad (7)$$

where β is the transfer coefficient.

3.2 ICMCC Algorithm

In order to improve the disadvantage of poor tracking performance of CMCC, this paper makes a convex combination of arbitrary number filters with different steps and obtains the ICMCC algorithm. When the number of L adaptive filters based on MCC is combined, the weight of the combined filter is obtained as follows [25]:

$$w_{eq}[k] = \sum_{i=1}^{L} v_i[k]w_i[k], \qquad (8)$$

where $w_i[k]$ represents the weight of the component filter whose the step size is denoted by μ_i $(\mu_1 > \mu_2 > \cdots \mu_L)$. Each filter updates its weight based on its own error and can be expressed as $w_i[k] = w_i[k-1] + \lambda_i \exp(\frac{-e_i[k]^2}{2\sigma^2})e_i[k]X[k], i = 1, 2 \cdots L$. $v_i[k]$ is the mixing coefficient and satisfies $\sum_{i=1}^{L} v_i[k] = 1$.

In the case of combining arbitrary number of filters, the L auxiliary parameters $\alpha_i[k]$ were updated with stochastic positive gradient method to adjust the L mixing parameters. We use softmax activation function to define the relationship between $v_i[k]$ and $\alpha_i[k]$ to ensure the stability of $v_i[k]$. $v_i[k]$ can be expressed as [25]

$$v_i[k] = \frac{\exp(\alpha_i[k])}{\sum_{j=1}^{L} \exp(\alpha_j[k])}, \qquad (9)$$

By multiplying both sides of formula (8) by $X^T[k]$, it is concluded that the output of ICMCC filter is a convex combination of all filter outputs.

$$y_{eq}[k] = \sum_{i=1}^{L} v_i[k]y_i[k], \tag{10}$$

where $y_i[k] = X^T w_i[k], i = 1, 2 \cdots L$ is the component filter output.

As for the CMCC filter, the parameter $\alpha_i[k]$ is updated using MCC rules to maximize the overall correntropy

$$\alpha_i[k+1] = \alpha_i[k] + \mu_\alpha v_i[k](y_i[k] - y_{eq}[k]) \exp(\frac{-e^2[k]}{2\sigma^2})e[k]. \tag{11}$$

In (11), We must qualify μ_α larger than the step size of any component filter. To prevent the ICMCC algorithm from stopping, we usually limit $|v_i[k]| < 0.95$. Since $v_i[k]$ is regulated by $\alpha_i[k]$, the range of $\alpha_i[k]$ is $|\alpha_i[k]| \leq 0.5 \ln(19(L-1))$.

Algorithm 1 ICMCC ALGORITHM

Initialization:
1. Parametres: μ_α, β, σ, L, γ, μ_i, $i = 1, \cdots L$
2. Initialize $\alpha_i[0] = 0$, $v_i[0] = 1/L$, $w_i[0] = 0$, $i = 1, \cdots L$
Update:
for $k = 0, 1, 2, \cdots$.
$y_i[k] = w_i^T[k]X[k]$, $i = 1, \cdots L$
$e_i[k] = d[k] - y_i[k]$, $i = 1, \cdots L$
$y_i[k] = \Sigma_{i=1}^T w_i^T[k]X[k]$, $i = 1, \cdots L$
$e[k] = d[k] - y_{eq}[k]$
$\alpha_i[k+1] = \alpha_i[k] + \mu_\alpha v_i[k](y_i[k] - y_{eq}[k]) \exp(\frac{-e^2[k]}{2\sigma^2})e[k]$
$v_i[k] = \frac{\exp(\alpha_i[k])}{\sum_{j=1}^{L} \exp(\alpha_j[k])}$
if $\gamma_i \geq cor(e^2[k])/cor(e_i^2[k])$
$w_i[k] = \beta w_i[k-1] + \mu_i \exp(\frac{-e_i^2[k]}{2\sigma^2})e_i[k]X[k] + (1-\beta)w_{eq}[k]$
else
$w_i[k] = w_i[k-1] + \lambda_i \exp(\frac{-e_i[k]^2}{2\sigma^2})e_i[k]X[k]$
end
End

In order to increase the tracking performance of the combined filter, we can determine whether to accelerate by calculating the ratio of the estimated correntropy of each component filter to the estimated correntropy of the combined filter. The estimated correntropy of the component filter and the combined filter are

$$cor(e_i[k]) = 0.98cor(e_i[k-1]) + 0.02\exp(\frac{-e_i^2[k]}{2\sigma^2}), \tag{12}$$

$$cor(e[k]) = 0.98cor(e[k-1]) + 0.02\exp(\frac{-e^2[k]}{2\sigma^2}). \tag{13}$$

When $\gamma \geq cor(e^2[k])/cor(e_i^2[k]), \gamma > 1$, we transfer a certain proportion of the combined filter to component filters that are worse than the combined filter. The modified adaption rule for $w_i[k]$ becomes

$$w_i[k] = \beta w_i[k-1] + \mu_i \exp(\frac{-e_i^2[k]}{2\sigma^2})e_i[k]x[k] + (1-\beta)w_{eq}[k], \qquad (14)$$

where β is the transfer coefficient. The condition for using Eq. (14) is that the large step size filter is significantly better than the small step size. Through a large number of experiments, γ and β were selected to be 2 and 0.8 respectively, to achieve the best transfer effect. The closer the choice of β is to 1, the more likely it is that the convex combination filter does not have a transfer coefficient. The pseudocodes of the proposed ICMCC are presented in Algorithm 1.

4   Simulation Results in System Identification Scenarios

In this section, we simulate the non-stationary system identification under non-Gaussian noise to verify the tracking performance of CMCC and ICMCC, and quantify each algorithm by normalized mean square deviation (NMSD) calculation which is expressed as $NMSD = 10\log_{10}(\| w_i - w_0 \|_2)/(\| w_0 \|_2)$.

The length of the unknown system is 10, which is the same length as the adaptive filter, and the input signal is a Gaussian signal with zero-mean and unit power. At the output of the plant we add measurement noise $N[k]$, we give four different distributions for measurement noise including (1) uniform noise, where the uniform noise is distributed over $[-1, 1]$; (2) Laplace noise, where the probability density function of Laplace noise is $p(n) = 1/\pi(1 + n^2)$; (3) binary noise, where the binary noise is either -1 or 1 (each with probability 0.5); (4) mixed Gaussian noise, where the mixed Gaussian noise is $N[k] = (1 - \theta)N(\zeta_1, \delta_1^2) + (\theta)N(\zeta_2, \delta_2^2)$ (ζ_i and δ_i^2 represent mean and variance, respectively). In this paper, the parameter is set to (0, 0, 0.001, 10, 0.1).

For the sake of simplicity, $L = 4$ MCC filters with step sizes $\mu_1 = 0.1$, $\mu_2 = 0.03$, $\mu_3 = 0.01$ and $\mu_4 = 0.002$ are considered as component filters. Simultaneously, the two steps in the CMCC algorithm are selected as $\mu_1 = 0.1$ and $\mu_4 = 0.002$. The step size μ_α of the parameter $\alpha[k]$ is fixed at 30. The initial value of unknown system is random values between -1 and 1, and the random-walk model introduces different rate of change of weight vectors. The random-walk model can express as:

$$w_0[k+1] = w_0[k] + q[k], \qquad (15)$$

where $q[k]$ is an i.i.d. vector, with autocorrelation matrix $Q = E\{q[k]q^T[k]\}$. $Tr(Q)$ is a measure of the speed of the weight vector. In this paper, we consider that q[k] is an independent Gaussian distribution.

In order to better embody the tracking performance of ICMCC, we choose $Tr(Q_1) = 10^{-6}$ and $Tr(Q_2) = 10^{-7}$ two different speed of change for weighting coefficients in $5000 < k < 10000$ and $15000 < k < 20000$ respectively. From

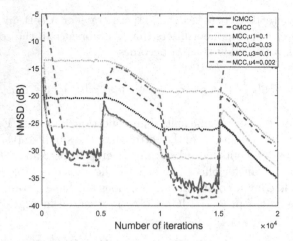

Fig. 1. Comparison of the convergence curve of a ICMCC filter and a CMCC filter with uniform noise.

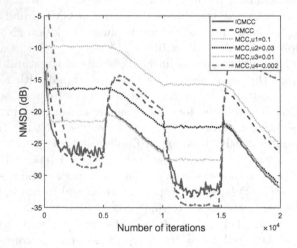

Fig. 2. Comparison of the convergence curve of a ICMCC filter and a CMCC filter with Laplace noise.

Figs. 1, 2, 3 and 4, when $k < 5000$, the adaptive filter of μ_1 has the fastest convergence speed but the highest amount of offset; μ_4's adaptive filter has the lowest amount of offset, but the slowest convergence rate. Although the CMCC adaptive filter has a fast convergence rate and a slow offset amount, the tracking performance is not as good as that of the ICMCC. When $5000 < k < 10000$ and $15000 < k < 20000$, since the two weight coefficients of $Tr(Q_1)$ and $Tr(Q_2)$ are added to change the speed, the optimal value of the weight coefficient changes. Compared to the CMCC, the ICMCC not only quickly tracks the optimal value,

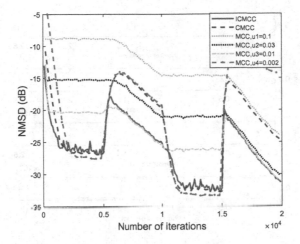

Fig. 3. Comparison of the convergence curve of a ICMCC filter and a CMCC filter with binary noise.

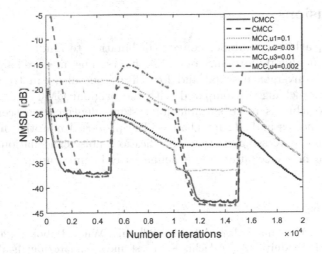

Fig. 4. Comparison of the convergence curve of a ICMCC filter and a CMCC filter with mixed Gaussian noise.

but also maintains a lower amount of misalignment. In summary, ICMCC gathers fast convergence speed, low misalignment and good tracking ability.

Simultaneously, we can study the tracking ability of ICMCC from four variations of the mixing coefficients. As shown in Fig. 5, the change process of the ICMCC four mixed parameters is indicated. We can see that in $5000 < k < 10000$ and $15000 < k < 20000$, ICMCC will use the hybrid coefficient adaptive filter to select the optimal performance as the primary role. Therefore, ICMCC shows the same performance as the optimal partial filter at any given moment. Therefore, ICMCC has good tracking performance and tracks a variety of changes.

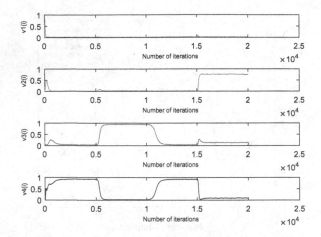

Fig. 5. Evolution of the mixing coefficients $v_1(i), v_2(i), v_3(i)$ and $v_4(i)$

5   Conclusions

In this paper, arbitrary number convex combination technique is employed to improve the tracking performance of CMCC filters. The improved algorithm not only has fast convergence speed and low offset, but also can track a variety of weight vector changes. Compared with the original CMCC algorithm, the improved algorithm is more suitable for system identification scenarios with non-Gaussian noises and abrupt change. The proposed adaptive filter can be applied to signal processing, system identification, noise cancellation, automatic equalization, echo cancellation and antenna array beamforming.

References

1. Sayed, A.H.: Fundamentals of Adaptive Filtering. Wiley, Hoboken (2003)
2. Gui, G., Mehbodniya, A., Adachi, F.: Least mean square/fourth algorithm for adaptive sparse channel estimation. In: 2013 IEEE 24th International Symposium on Personal Indoor and Mobile Radio Communications (PIMRC), pp. 296–300 (2013)
3. Li, Y., Wang, Y., Jiang, T.: Sparse-aware set-membership NLMS algorithms and their application for sparse channel estimation and echo cancelation. AEU-Int. J. Electron. Commun. **70**, 895–902 (2016)
4. Chen, B., Xing, L., Zhao, H.: Generalized correntropy for robust adaptive filtering. IEEE Trans. Signal Process. **64**, 3376–3387 (2016)
5. Li, Y., Wang, Y., Yang, R., Albu, F.: A soft parameter function penalized normalized maximum correntropy criterion algorithm for sparse system identification. Entropy **19**, 1–16 (2017)
6. Li, Y., Jin, Z.: Sparse normalized maximum correntropy criterion algorithm with L1-norm penalties for channel estimation. In: 2017 Progress in Electromagnetics Research Symposium, pp. 1677–1682 (2017)

7. Walach, E., Widrow, B.: The least mean fourth (LMF) adaptive algorithm and its family. IEEE Trans. Inf. Theory **30**, 275–283 (1984)
8. Hajiabadi, M., Zamiri-Jafarian, H.: Distributed adaptive LMF algorithm for sparse parameter estimation in Gaussian mixture noise. In: 7th International Symposium on Telecommunications, pp. 1046–1049. Tehran (2014)
9. Lu, L., Zhao, H., Wang, W.: Performance analysis of the robust diffusion normalized least mean P-power algorithm. IEEE Trans. Circuits Syst. II: Express Briefs 1–1 (2018)
10. Lu, L., Zhao, H., Chen, B.: Improved variable forgetting factor recursive algorithm based on the logarithmic cost for volterra system identification. IEEE Trans. Circuits Syst. II: Express Briefs **63**, 588–592 (2016)
11. Principe, J.C.: Information Theoretic Learning: Renyis Entropy and Kernel Perspectives. Springer, Berlin (2010)
12. Erdogmus, D., Principe, J.C.: Convergence properties and data efficiency of the minimum error entropy criterion in adaline training. IEEE Trans. Signal Process. **51**, 1966–1978 (2003)
13. Chen, B., et al.: Stochastic gradient identification of wiener system with maximum mutual information criterion. IET Signal Process **5**, 589–597 (2011)
14. Liu, Y., Fan, Y., Zhou, L.C.: Ensemble correntropy based mooney viscosity prediction model for an industrial rubber mixing process. Chem. Eng. Technol. **39**, 1804–1812 (2016)
15. Han, S., Jeong, K.H., Principe, J.: Robust adaptive minimum entropy beam former in impulsive noise. In: IEEE Workshop on Machine Learning for Signal Processing (2007)
16. Chen, B., Xing, L., Liang, J.: Steady-state mean-square error analysis for adaptive filtering under the maximum correntropy criterion. IEEE Signal Process. Lett. **21**, 880–884 (2014)
17. Khalili, A., Rastegarnia, A.: Steady-state tracking analysis of adaptive filter with maximum correntropy criterion. Circuits Syst. Signal Process (2016)
18. Wang, W., Zhao, J., Qu, H., Chen, B.: A switch kernel width method of correntropy for channel estimation. In: International Joint Conference on Neural Networks, pp. 1–7 (2015)
19. He, Y., Wang, F.: Kernel adaptive filtering under generalized maximum correntropy criterion. In: International Joint Conference on Neural Networks, pp. 1738–1745 (2016)
20. Wang, R., Chen, B.: A variable step-size adaptive algorithm under maximum correntropy criterion. In: IEEE International Conference on Acoustic, Speech and Signal Processing (ICASSP), pp. 1–5 (2015)
21. Shi, L.: Convex combination of adaptive filters under the maximum correntropy criterion in impulsive interference. IEEE Signal Process. Lett. **21**, 1385–1388 (2014)
22. Liu, W.: Correntropy: properties and applications in non-Gaussian signal processing. IEEE Trans. Signal Process. **55**, 5286–5298 (2007)
23. Liu, C., Qi, Y., Ding, W.: The data-reusing MCC-based algorithm and its performance analysis. Chin. J. Electron. **25**, 719–725 (2016)
24. Arenas-Garcia, J., Figueiras-Vidal, A.R., Sayed A.H.: Improved adaptive filtering schemes via adaptive combination. In: 2009 Conference Record of the Forty-Third Asilomar Conference on Signals, Systems and Computers, pp. 1–4 (2009)
25. Arenas-Garca, J., Gmez-Verdejo, V., Figueiras-Vidal, A.R.: Combinations of adaptive filters: performance and convergence properties. IEEE Signal Process. Mag. **33**, 120C–140 (2016)

Module Selection Algorithm Based on WSS/SSS-Hybrid AoD Node in Dynamic Elastic Optical Networks

Ziqin Li[1], Xiaosong Yu[1(✉)], Shimulin Xie[2], Yan Wang[1],
Yuhui Wang[1], Yongli Zhao[1], and Jie Zhang[1]

[1] State Key Laboratory of Information Photonics and Optical Communication,
Beijing University of Posts and Telecommunications (BUPT), Beijing 100876,
China
{xiaosongyu, zhaoyongli}@bupt.edu.cn
[2] State Grid Info-Telecom Great Power Science and Technology Co., LTD,
Fujian 350003, China

Abstract. Driven by the emerging applications based on Internet, optical backbone networks need to improve their transmission capabilities while ensuring high reliability, flexibility, and scalability. Elastic optical networks and space-division multiplexing optical networks are seen as the potential solutions. In order to implement these technologies, innovative nodes are required to provide flexibility, reliability, and scalability for the optical networks. Architecture on Demand (AoD) node is a new type of elastic optical node structure proposed in the recent years and can dynamically provide a customizable structure according to the exchange and processing requirements of the network traffic. Spectrum Selector Switches (SSS) is one of the key modules but has not been widely used because of its excessive cost. To solve the problem of how to select the Wavelength Selective Switch (WSS)/SSS coexistence in the current network, we propose a pre-built algorithm for the modules in the AoD nodes. Simulation results show that the proposed algorithm performs better than the benchmarks in different network scenarios and provides a solution to the gradual upgrade of AoD nodes.

Keywords: AoD node · WSS/SSS selection · Elastic optical networks

1 Introduction

Driven by the growing of Internet traffic as well as emerging applications such as cloud computing and big data, fiber optic backbone networks need to improve its transmission capacity, reliability, flexibility, scalability and cost-efficiency. In this case, Elastic Optical Network (EON) and Spatial Division Multiplexing (SDM) optical networks have been studied extensively and became promising solutions. The appearance of optical fibers based on SDM technology and the increase in the number of basic optical fibers, which are aimed at expanding the transmission capacity, are expected to magnify and complicate the elastic optical nodes in the future optical fiber backbone networks. However, the work on flexible optical node architectures based on these

X. Liu et al. (Eds.): ChinaCom 2018, LNICST 262, pp. 194–203, 2019.
https://doi.org/10.1007/978-3-030-06161-6_19

flexible technologies is limited. This optical node that are used in SDM based EON need to provide flexibility, reliability and scalability for the entire network. Therefore, elastic optical node architectures are an interesting topic for current optical fiber backbone networks.

Among the different solutions proposed to the elastic node architectures, broadcast-and-select and spectrum-routing are common elastic optical node architectures that have sufficient flexibility to implement a completely flexible optical network [1, 2]. In these architectures, the Spectrum Selective Switch (SSS) is not only the main building module, but also the dominant module in terms of cost and power consumption. Due to the limited port number of SSSs, the cost and power consumption of these elastic optical nodes will increase significantly with the increase of the number of SSSs in large-scale networks. Another hierarchical optical switch node architecture using a small Optical X-Connect (OXC) as a subsystem module has been proposed to suppress the increasing of building modules [3], but this architecture has a negative impact on the transmission success rate. Therefore, it is necessary to consider cost, power reduction and successful transmission rate in the design of the optical node architecture.

The solution based on Architecture on Demand (AoD) [4] reduces not only the number of building modules implemented, but also exhibits remarkable flexibility [5], reliability [6] and scalability [4, 7] compared to the existing alternatives (such as MG-OXC [8], BV-ROADM [9]). However, due to the fact that key enabling devices for AoD nodes, such as SSSs that support multi-granularity service switching, are now expensive, it is not feasible to use SSS extensively across the entire network. In this paper, we propose a selection algorithm of WSS/SSS in AoD nodes to address the problem of how to choose WSS/SSS coexistence of the current network.

2 Optical Node Architecture

2.1 Traditional Optical Node Architecture

In EON or SDM, the elastic optical node will process the transmitted signal with better flexibility and finer granularity than the traditional optical network. SSS is the most important building module in EON and SDM elastic optical nodes, as Fig. 1. It is also called flexible Wavelength Selective Switch (WSS). The SSS can filter the input signals with an arbitrary width spectrum and switch them to another arbitrary port without copying the signals. The choice of SSS enables the flexible networking of EON and SDM.

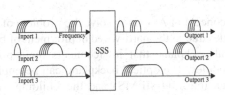

Fig. 1. The concept of SSS.

In the broadcast-and-select architecture, as shown on Fig. 2(a), the input signal is first copied in the splitter and then broadcasted to all output ports. On each output port, use the SSS to select the appropriate spectrum and sending the multiplexed signal through the output port. If there are many ports in a node, the replication of these splitters will seriously degrade the transmitted signal [10]. In the spectrum-routing architecture, which is shown in Fig. 2(b), the input signal is first de-multiplexed by the SSS at the input ports and not copied. These de-multiplexed signals, then, are routed to different output ports and finally multiplexed at the output ports through the SSS. Unlike broadcast-and-select architectures, the spectrum-routing architecture does not produce splitter signal degradation, but due to the number of equipped SSSs doubles at the input, the cost is much higher than broadcast-and-select architectures.

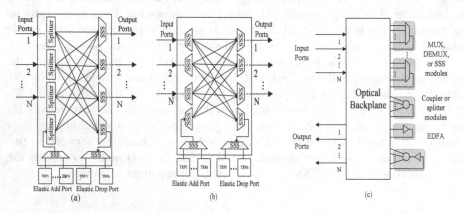

Fig. 2. (a) Broadcast-and-select architecture. (b) Spectrum-routing architecture. (c) Architecture of AoD nodes

In both architectures, the cost of elastic optical nodes will be an important issue. If the number of input/output ports of an elastic optical node is large, it is necessary to match SSSs having the same number of ports in these traditional node architectures. However, the number of SSS ports has certain limitations. In order to meet the requirements of the node architectures, the number of SSSs may increase explosively, and thus the cost of elastic optical nodes also increases dramatically.

2.2   AoD Node Architecture

The introduction to the concept of AoD is to solve the problem of insufficient flexibility in traditional optical nodes due to the hard-wired deployment of building modules. The AoD includes an optical backplane that interconnects input ports, output ports and architecture-building modules. The optical backplane can be implemented with a large port-count optical switch (e.g. 3D-MEMS) and the building modules can be either single devices for optical processing (e.g., MUX, DEMUX, WSS, SSS, amplifier, etc.) or subsystems that are composed of several devices, as shown in Fig. 2(c). AoD nodes can dynamically reconfigure the overall architecture based on the network's signals switching or processing requirements. Because these building modules are dynamically configured, they can provide additional functionalities for the nodes. Therefore, AoD nodes have greater flexibility and scalability than traditional static optical node architectures.

Reference [5] defines the flexibility of the node architecture according to the entropy of the system and compares the traditional static node architecture of AOD nodes and EON. In Reference [11], multi-granular transmission of space/frequency/ time domain has been demonstrated by AoD nodes and Multi-Core Fiber (MCF). The experiment proves that the AoD node in EON based on MCF has great flexibility. The advantages of AoD node flexibility have been proved theoretically and experimentally.

Moreover, the flexibility of AoD nodes can be used to reduce the power consumption of optical nodes. Because the AoD node is dynamically constructed based on the switching request, it only needs to use a minimum amount of required building modules. The traditional optical node structure always needs to fix the maximum number of hard-wired modules without considering the request. The use of AoD nodes can reduce the number of building modules and power consumption. References [7, 12] analyze the power consumption of AoD nodes. Reference [7] analyzes the benchmark for power reduction based on the granularity of the switching request. Reference [12] shows that in the dynamic scenario, using the ILP model to find the optimal solution for AoD construction is extremely complicated, thus AoD nodes can use heuristic algorithms to build AoD nodes and eventually reduce the total power consumption of the network by more than 25%.

Although the AoD node can provide many advantages, it still has some challenges. The high flexibility and scalability of AoD nodes have a directly relationship with the number of core building modules SSSs. The number of SSSs will directly affect the cost and energy consumption of AoD nodes. Due to the high cost of SSSs, WSS is still widely used in the current network to provide network flexibility. In the following section, we will study how to choose WSS/SSS reasonably in the WSS/SSS coexistence network and propose a Pro-built algorithm to solve this problem.

3   Pre-built Algorithm for AoD Node

3.1   Module Selection and AoD Construction

Our proposed Pre-built Algorithm is based on a given set of feasible requests for AoD module construction. The algorithm flow chart is shown in Fig. 3. Pre-built has five steps: one step to calculate shortest paths, three steps for switching function to switch from coarser granularity to finer granularity (that is, fiber switching, super channel, single wavelength and sub-wavelength level), the last step of AoD Module building.

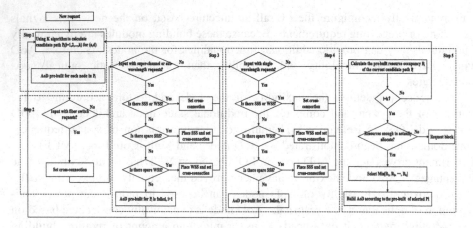

Fig. 3. The flow chart of Pre-built Algorithm.

Step 1: Calculate K alternative paths using KSP algorithm based on the source-destination node pair in a request and perform AoD pre-construction on them.

Step 2: Check if the request is a fiber switch request. If so, check the destination of all signals from each input and set the cross-connect directly if they all are the same output.

Step3: Check if the request is a super-channel or sub-wavelength request. If so, Check if there is SSS or WSS at the input port. If so, preferentially use the existing module; if not, give priority to placing SSS (due to the SSS arbitrary bandwidth switching capability), and then consider placing WSS. If SSS or WSS cannot be provided for the request at the node, the AoD pre-build on the current alternate path fails and it will return to re-build the next path.

Step 4: Check whether the request is a single-wavelength request. Similar to the third step, pre-existing modules at the port are used. The difference is that if there is no existing module, it is preferable to place the WSS, and secondly to place the SSS.

Step 5: Calculate the resource metrics R_i of the pre-built successful path. At the same time, it should satisfy all candidate paths traversed. In this case, the indexes of alternative paths are compared, and the suitable path is selected as the final AoD construction. "Ri" satisfied the following:

$$Ri = \frac{Sall + S'_{all}}{Wall} \tag{1}$$

where Ri represents the spectrum slots usage rate of the candidate path P_i, $Wall$ is the sum of the spectrum slots in each link of the P_i, $Sall$ is the sum of the spectrum slots already been occupied in each link of the P_i, S'_{all} is the sum of the spectrum slots will be occupied when pre-built in each link of the P_i.

3.2 Illustration for Pre-built Algorithm

To better to illustrate the Pre-built algorithm, we can look at the example in Fig. 4. Different businesses are shown in the table. These services include different granularity services such as fiber switching services, sub-wavelength services, single-wavelength services, and super-channel services. Taking Request 1 as an example, the service source/destination node pair is (1, 4). In this case, the K algorithm (K = 3) is first used to select a candidate path, as shown in Step 1. After that, we proceed with Step 2, which will pre-build AoD modules for this candidate path. Since this is a 100 G super-channel service, we will configure the SSS for it, and each node in the candidate path node set (1, 2, 3, 4) of the candidate path 1 will configure SSS for it. Of course, if this is a single-wavelength service (like Request 2) we will configure it as WSS. If there is already a module at the port, consider reusing the module (Request 3). In this way, the three candidate paths are pre-build in sequence, and perform the calculation in step 3, which the resources on the three alternative paths after the construction according to the pre-built algorithm is calculated. By comparison, the more available resources are in the candidate path 1, so the construction method of candidate path 1 will be used to configure the functional modules of the final AoD node. If there is no module available for use, this service will be blocked (Request 5).

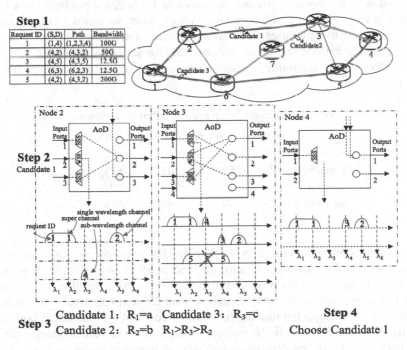

Fig. 4. Illustration for algorithm.

4   Simulations and Results

We evaluated the proposed algorithm by software simulation. The simulation uses an US network topology, which has 28 nodes and 45 links. Each bidirectional link is configured with 400 spectrum slots and each spectrum slot is 12.5 GHz. 100000 services requests following Poisson distribution are given, and their bandwidth requirements are randomly generated from 40 to 400 Gbps. For each service type, the number of spectrum slots occupation is summarized in the Table 1. Requests are processed one by one and the K is KSP is set as 3, and first-fit spectrum assignment are used for each link.

Table 1. Required spectrum for different demands

Channel	Flexible grid	Slots
40 Gbps	25 GHz	2
100 Gbps	37.5 GHz	3
200 Gbps	75 GHz	6
400 Gbps	125 GHz	10

According to the demand proportions shown in Table 2, the pre-built algorithm is evaluated in four different traffic scenarios. Among them, scenarios 1 and 2 focus on small service, scenario 3 focuses on balancing services, and scenario 4 focuses on large service for testing performance of algorithms. The four scenarios follow uniform traffic models, which means that traffic is evenly distributed among all nodes in the network.

Table 2. Demands proportion in different scenarios

	40 Gbps	100 Gbps	200 Gbps	400 Gbps
Scenario 1	50%	30%	15%	5%
Scenario 2	10%	50%	30%	10%
Scenario 3	25%	25%	25%	25%
Scenario 4	0%	40%	40%	20%

Simulation results are shown in the Fig. 5. The proposed pre-built algorithm is compared with the First-Fit algorithm for the WSS or SSS modules in Fig. 5(a). In four different traffic scenarios, the proposed algorithm performs better than FF algorithm, and the two algorithms are better in scenarios 1 and 2. From Fig. 5(b), it can be further found that the traffic blocking ratio in the network increases with the traffic load, which is mainly due to the increase in the number of large-bandwidth super-channel services in the network. The increase in the occupancy of spectrum resources has led to an increase in the blocking ratio.

Figure 6(a), (b) show the effect of k-values on the effects of the proposed algorithm and the FF algorithm in the K-shortest routing algorithm respectively in different scenarios. From the result point of view, the pre-built algorithm works best when k = 3.

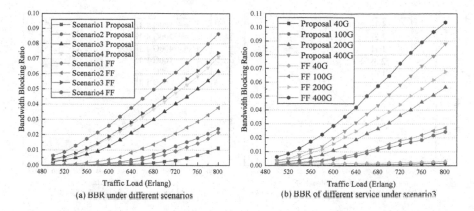

(a) BBR under different scenarios   (b) BBR of different service under scenario3

Fig. 5. Simulation results with two algorithms under different scenarios.

This is because the algorithm compares the resource conditions of the candidate paths in advance and selects the optimal situation to construct the AoD module. At the same time, it can be inferred that when k increases, the algorithm performs better.

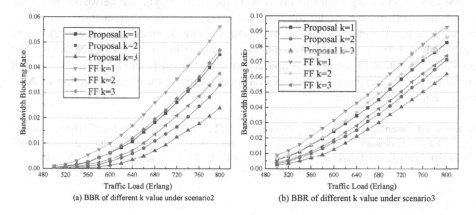

(a) BBR of different k value under scenario2   (b) BBR of different k value under scenario3

Fig. 6. BBR of different k values under different scenarios.

Figure 7(a), (b) show the impact of the number of WSS and SSS on the blocking ratios in the AoD node under the fourth traffic scenario. From Fig. 7(a), it is easy to find that the blocking ratio of the network decreases with the increase of the number of SSSs in the nodes. This is because the SSS can provide more flexibility and allow more large-bandwidth super-channel services to be configured successfully. Figure 7(b) shows the blocking ratio of different services under different module proportions. This further reflects the direct impact on the number of SSSs on super-channel services. At the same time, it also indicates that in the WSS/SSS coexistence networks, we can upgrade the AoD node in a gradual upgrade manner under the premise of ensuring certain flexibility. This also provides a solution to the construction and upgrade of AoD nodes.

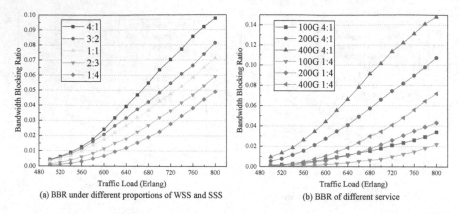

(a) BBR under different proportions of WSS and SSS   (b) BBR of different service

Fig. 7. BBR under different proportions of WSS and SSS.

5   Conclusion

This paper studied how to select the building modules in the current WSS/SSS coexistence optical network. Based on AoD nodes architecture, we proposed a WSS/SSS selection algorithm to build AoD modules in dynamic networks and compared their performance under different traffic scenarios. In different traffic scenarios, we can see that the proposed algorithm performs better in small service scenarios, and the number of WSS and SSS plays an important role in large service scenarios. This provides a solution to the gradual upgrade of AoD nodes. In the future work, we will continue to try to explore the factors that more influence on the construction of AoD nodes under such scenario.

Acknowledgements. This work is supported by NSFC project (61601052, 61571058), Fund of State Key Laboratory of Information Photonics and Optical Communications, BUPT (IPOC2017ZT10), the Fundamental Research Funds for the Central Universities (2018RC24).

References

1. Kozicki, B., Takara, H., Tsukishima, Y., Yoshimatsu, T., Yonenaga, K., Jinno, M.: Experimental demonstration of spectrum-sliced elastic optical path network (SLICE). Opt. Express **18**(21), 22105–22118 (2010)
2. Collings, B.: The next generation of ROADM devices for evolving network applications. European Conference Optical Communication (ECOC) Exhibition, Geneva (2011)
3. Ishida, H., Hasegawa, H., Sato, K.: Hardware scale and performance evaluation of a compact subsystem modular optical cross connect that adopts tailored add/drop architecture. J. Opt. Commun. Netw. **7**(6), 586–596 (2015)
4. Garrich, M., Amaya, N., Zervas, G., Giaccone, P., Simeonidou, D.: Architecture on demand: synthesis and scalability. 16th International Conference on Optical Network Design and Modeling (ONDM), pp. 1–6. London (2012)

5. Amaya, N., Zervas, G., Simeonidou, D.: Introducing node architecture flexibility for elastic optical networks. J. Opt. Commun. Netw. 5(6), 593–608 (2013)
6. Dzanko, M., Furdek, M., Gonzalez, N., Zervas, G., Mikac, B., Simeonidou, D.: Self-healing optical networks with architecture on demand nodes. In: 39th European Conference and Exhibition on Optical Communication (ECOC), pp. 1–3. London (2013)
7. Garrich, M., Amaya, N., Zervas, G., Oliveira, J., Giaccone, P., Bianco, A., Simeonidou, D.: Architecture on demand design for high-capacity optical SDM/TDM/FDM switching. IEEE/OSA J. Opt. Commun. Netw. 7(1), 21–35 (2015)
8. Wang, Y., Cao, X.: Multi-granular optical switching: a classified overview for the past and future. IEEE Commun. Surv. Tutor. 14(3), 698–713 (2012)
9. Jinno, M., Takara, H., Kozicki, B., Tsukishima, Y., Sone, Y., Matsuoka, S.: Spectrum-efficient and scalable elastic optical path network: architecture, benefits, and enabling technologies. IEEE Commun. Mag. 47(11), 66–73 (2009)
10. Morea, A., Renaudier, J., Zami, T., Ghazisaeidi, A., Bertran-Pardo, O.: Throughput comparison between 50-GHz and 37.5-GHz grid transparent networks. J. Opt. Commun. Netw. 7(2), A293–A300 (2015)
11. Amaya, N., Irfan, M., Zervas, G., Nejabati, R., Simeonidou, D., Sakaguchi, J., Klaus, W., Puttnam, B.J., Miyazawa, T., Awaji, Y., Wada, N., Henning, I.: Fully-elastic multi-granular network with space/frequency/time switching using multi-core fibres and programmable optical nodes. Opt. Express 21(7), 8865–8872 (2013)
12. Muhammad, A., Zervas, G., Amaya, N., Simeonidou, D., Forchheimer, R.: Introducing flexible and synthetic optical networking: Planning and operation based on network function programmable ROADMs. J. Opt. Commun. Netw. 6(7), 635–648 (2014)

Joint Optimization of Energy Efficiency and Interference for Green WLANs

Zhenzhen Han[1], Chuan Xu[1], Guofeng Zhao[1], Rongtong An[1], Xinheng Wang[1], and Jihua Zhou[2(✉)]

[1] School of Communication and Information Engineering,
Chongqing University of Posts and Telecommunications, Chongqing, China
Hanzhenzhen512@gmail.com, {xuchuan,zhaogf}@cqupt.edu.cn,
artavrillavigne@gmail.com, xhwangcqupt@gmail.com
[2] Chongqing Jinmei Communication Co., Ltd., Chongqing, China
jhzhou@ict.ac.cn

Abstract. In the past years, the issues of energy efficiency and inter-ference are becoming increasingly serious in wireless local area network (WLAN) since lots of access points (AP) are deployed densely to pro-vide high-speed users access. However, current works focus on solving the two issues separately and the influence of each other is rarely consid-ered. To address these problems, we propose a joint optimization scheme of energy efficiency and interference to reduce energy consumption and interference together without sacrificing users' traffic demands. Firstly, based on energy consumption measurement of AP and network inter-ference analysis, we establish energy efficiency and interference models respectively. Then, the weighting method is introduced to build the joint optimization to quantify the effects of user-AP association, AP switch, AP transmit power and AP channel on energy consumption and inter-ference. Lastly, we formulate the joint optimization as an Mixed Integer Non-Linear Programming (MINLP) problem. Since the MINLP prob-lem is NP-hard, we proposed an Joint Optimization of Energy Efficiency and Interference (JOEI) algorithm based on greedy method to simplify its computational complexity. The evaluation results show that the pro-posed algorithm can effectively reduce the network energy consumption while improve the capacity of WLANs.

Keywords: Energy efficiency · Interference · Joint optimization · Green WLAN

1 Introduction

As a high-speed mobile network access scheme, wireless local area network (WLAN) has been densely deployed in the enterprises, schools and other public

Supported by the National Natural Science Foundation of China (Grant No.61701058), the Innovation Funds of Graduate PhD (Grant No.BYJS2017002), and Chongqing Graduate Research and Innovation Project(Grant No.CYB18167).

X. Liu et al. (Eds.): ChinaCom 2018, LNICST 262, pp. 204–213, 2019.
https://doi.org/10.1007/978-3-030-06161-6_20

areas to meet the traffic demand at peak hours [1]. However, the peak period rarely happens, and the utilization of access points (AP) during the off-peak period are reduced to low or idle, which leads to a serious energy waste [2]. Moreover, the channel overlapping in density deployment scene will cause heavy interference and degradation of user's quality of experience (QoE) [3].

Recently,the issue of energy saving and interference have been studied independently from different aspects [4]. To reduce energy wastage, researchers try to adopt switching strategies (also called 'sleep-awake') to turn off/on the low-utilization or idle APs for adapting the active capacity [1,5,6]. A green clustering algorithm was introduced to initiate a cycle of estimating user demand and performance to power on or off APs, then adjust the transmit power [1]. In reference [5], based on a centralized control framework, the actual network conditions in terms of both user density and traffic patterns are monitored and used to tune the energy consumption through a flexible energy-saving decision algorithm. Similarly, a context-aware power management framework and adaptive algorithms were proposed to dynamically configure different network elements according to user needs [6]. Actually, those existing energy-saving studies mainly focus on switching off unnecessary APs to reduce the energy consumption, but the interference between adjacent APs has not yet been considered specifically. However, according to the research [7], interference not only affects the stability of wireless network, but also increases the energy consumption of wireless systems. Therefore, to achieve more effective energy-saving and sacrifice little on users' QoE, the interference and energy consumption should be taken into account together.In work [8], the authors indicated that the interference results from the aggressive spectral reuse and high power transmission severely limits the system performance, then they use a non-cooperative game to optimize energy-efficient power for interference-limited wireless communication. However, in those works, the relationship between interference and energy consumption has not yet been quantitative analyzed accurately, which is the basis of resource scheduling and optimization in wireless system [9].

In this paper, we aim to build a quantitative optimization model which could reduce network energy consumption and interference with guaranteeing users' QoE in WLAN system. Firstly, we set experiments to quantify the influence of transmit power and throughput to energy efficiency, and analyze the relationship between interference and transmit power, establish optimization models respectively. Secondly, we introduce the weighting method to establish a joint optimization objective function, optimized energy consumption and capacity. Lastly, we put forward JOEI algorithm and verify the validity compared with three popular algorithms.

2   System Model

2.1   Network model

As shown in Fig. 1, we consider a centralized WLAN system, the reasonable and effective optimization scheme can be achieved based on the information from all wireless access points collected by controller, which can be used to operate the state of user association matrix β,AP switch on/off α,AP transmit power p and AP channel f to reduce interference and save energy.

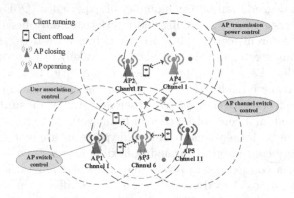

Fig. 1. The simplify scenario for energy consumption and interference joint optimization of WLAN system.

2.2 Energy Consumption Model

In order to determine the relationship among AP load, transmit power and energy consumption, we conduct an experiment with real NETGEAR WNDR 3800 WiFi devices which were deployed in common application scenario working at 2.4 GHz with 802.11 n mode, 20 MHz HT mode and random channel. As the real AP load mainly from the downlink service, we gradually change AP's downlink data transmission rate (the data from the wired side to the wireless side) and transmit power, and record AP operating power in each scene by the power tester (TECMAN-TM6). Finally, we quantitative analysis the relationship of AP load-AP transmit power-AP energy consumption according to the measurement

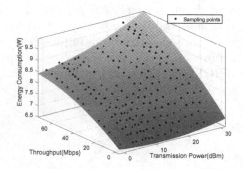

Fig. 2. Energy efficiency of AP versus throughput and transmission power.

result. The energy efficiency relation model of AP is obtained by fitting the least squares polynomial of all discrete relation data as Fig. 2.

And the relation function as follows:

$$F(L,p) = a_1 + a_2 * p + a_3 * L + a_4 * p^2 + a_5 * p * L + a_6 * L^2 \qquad (1)$$

where L represents the throughput of AP, which equals to the sum traffic requirement of all associated users with this AP, p represents the actual AP transmit power, $F(L,p)$ represents AP power consumption, and a_1-a_6 represent the fitted polynomial coefficient.

2.3 Interference Model

The interference accumulation effect between APs is described by calculating the SINR of user links [10,11]. This physical interference model considers all the links in wireless system to interfere users, which can be used to determine whether the user node meet its demands.

Assuming that $user_j$ is associated with AP_i, the SINR of the link that $user_j$ received is expressed as:

$$SINR_j^i = \frac{p_i g_{ij}}{N_0 + I_j^{-i}} \qquad (2)$$

where N_0 represents the system thermal noise power, p_i represents the transmit power of AP_i, and g_{ij} represents the path attenuation factor from AP_i to $user_j$,

$$g_{ij} = d_{ij}^{-r} \qquad (3)$$

where d_{ij} represents the Euclidean distance between AP_i and $user_j$. r represents the attenuation coefficient, and the usual value is 2–4 [12].

I_j^{-i} represents the sum interference that $user_j$ receives from surrounding APs except for the currently associated AP_i. Furthermore, I_j^x indicates the same channel interference received by $user_j$ from the AP_x which adjacent to AP_i.

$$I_j^x = e_{ix} \Delta(f_i, f_x) p_x g_{xj} \qquad (4)$$

e_{ix} represents the adjacency relationship between AP_i and AP_x. When the AP_i and AP_x have the overlapping coverage area, they are considered to be adjacent, and the value of e_{ix} is 1, otherwise 0.

$\Delta(f_i, f_x)$ represents the channel relationship between AP_i and AP_x. if f_i is equal to f_x, which means AP_x is in the same channel with AP_i, the value of $\Delta(f_i, f_x)$ is 1, otherwise 0.

It can be seen that the total interference received by $user_j$ is expressed as:

$$I_j^{-i} = \sum_{x=1}^{n} I_j^x = \sum_{x=1}^{n} \{e_{ix} \Delta(f_i, f_x) p_x g_{xj}\} \qquad (5)$$

At the same time, when $user_j$ associated with AP_i, the channel capacity can be obtained according to Shannon formula.

3 The Joint Optimization of Energy Consumption and Interference

During the off-peak period, there are lots of low-utilization or idle APs,which will lead to heavy energy waste. Meanwhile, these idle APs will cause much interference to users associated with adjacent APs. Therefore, the energy consumption and interference can be reduced simultaneously through switch off idle APs. To reduce energy consumption and improve performance of system, we try to make more APs idle through users' reassociation without sacrificing users' traffic demands.

Furthermore, according to the model analysis in Sect. 2 the energy consumption and interference can be reduced simultaneously by adjusting the AP power, then improve the system capacity. Therefore, in order to reduce energy consumption with ensuring network performance, the energy consumption and the interference should be considered together. Because of the adjustment of user-AP association, AP switch, AP transmit power and AP channel have the similar influence on energy consumption and interference, these two optimization models can be combined to achieve joint optimization.

Assuming that there are n APs and m users in the WLAN system, $N = \{1, 2, \cdots, n\}$, $M = \{1, 2, \cdots, m\}$, where $AP_i \in N$, $User_j \in M$. Meanwhile, user's traffic demand is defined as vector l, where l_j represents the traffic demand of $user_j$. In order to make the energy consumption - interference joint optimization model more reasonable, we introduce a weighting factor θ, and define the joint optimization objective function as follows:

$$\Im_1 : \min_{\alpha, \beta, p, f} \sum_{i=1}^{n} \left\{ F\left(\sum_{j=1}^{m} \beta_{ij} l_j, p_i \right) \alpha_i \right\}$$

$$+ \theta \sum_{i=1}^{n} \left\{ \alpha_i \sum_{j=1}^{m} \left(\beta_{ij} \sum_{x=1}^{n} e_{ix} \Delta\left(f_i, f_x \right) p_x g_{xj} \right) \right\} \quad (6)$$

$$s.t. \ \alpha_i, \beta_{ij} \in \{0, 1\},$$

$$p_i \in \{0, 1, 2, \cdots, 30\}, f_i \in \{1, 6, 11\} \quad (7)$$

$$Cap_j > l_j \quad (8)$$

$$\sum_{j=1}^{m} \beta_{ij} l_j < L_{\max} \quad (9)$$

$$\beta_{ij} \leq c_{ij}, c_{ij} = \begin{cases} 1 & d_{ij} < R_i \\ 0 & otherwise \end{cases} \quad (10)$$

$$\sum_{i=1}^{n} \beta_{ij} = 1 \quad (11)$$

$$\sum_{i=1}^{n} \alpha_i \beta_{ij} = 1 \quad (12)$$

The first part of objective (6) shows that the AP energy consumption model F is used to optimize the overall energy consumption of the system by adjusting user-AP association β, AP switch α and AP transmit power p. The second part of objective (6) denotes the sum of the interference received by all users is used as an indicator to assess the interference of the system.

The optimal convergence state of the joint optimization model can be dynamically controlled by setting the weight factor θ, so that WLAN service providers can change the size of θ in the algorithm according to the specific network situations and different optimization requirements, and obtain the more scientific and effective optimal solution.

The objective (6) measures the sum of energy consumption and interference. Solving problem $\Im_1$ means that the corresponding algorithm should return the optimal active AP vector α^*, the user-AP association matrix β^*, the AP transmit power vector p^* and the AP channel selection vector f^*. The constraint (7) presents the feasible domain of α_i, β_{ij}, p_i and f_i. The constraint (11) and (12) ensure that user j only can associate to one AP, and the constraint (10) ensures that only the user that within the coverage of the AP can connect to it. The constraint (9) shows that the total load of every AP is within its transmission capacity L_{max}. And the constraint (8) ensure that channel capacity between user and AP must satisfy the traffic demand of users.

From the convex function definition, we can obtain that The objective function $\Im_1$ is convex, and area bounded by the constraint functions (7)–(12) is convex.we confirm that our joint function exist an optimal solution under all of constraints. However, as we have integrated four unknown variables (i.e. user-AP association, AP switch, AP channel, AP transmit power) into the one objective function, which make it become an MINLP problem. As far as we know, solving the optimal solution of an MINLP problem is a difficult work which requires a lot of computation and time complexity. Hence, in order to solve the NP-hard problem, we propose an efficient algorithm which is based on the idea of greedy algorithm in the next section.

4   The JOEI Algorithm Based on Greedy Method

We propose an JOEI algorithm to solve the Joint Optimization of Energy consumption - Interference inspired by the idea of greedy algorithm.In energy-saving aspect, the transmit power has smaller influence than AP switch [13]. Meanwhile in interference aspect, channel selection can decrease interference more effective compared with transmit power [7]. So, in the JOEI algorithm we firstly consider transmit power as a default value to compute other three variables to reduce computational complexity. Furthermore, the optimal transmit power is obtained by solving the objective function with other three variables that have been obtained. The details of JOEI algorithm based on greedy method is given in Algorithm 1.

Algorithm 1. The JOEI algorithm

Ensure: $\alpha, \beta \in \{0, 1\}, f \in \{1, 6, 11\}, p \in \{0, 1, 2, \cdots, 30\}$.

1: **while** (1) **do**
2:    Set c, l, g and e according to the network status;
3:    Compute the number of adjacent APs for each AP and store them in array ap_adjace by ascending order;
4:    **for** $i = 1$ to n by ap_adjace order **do**
5:       Select a channel from $\{1, 6, 11\}$ as f_i which minimize the objective function $\Im_1$;
6:    **end for**
7:    Compute the interference value of all users under each AP_i by
$$\sum_{j=1}^{m} \left(\beta_{ij} \sum_{x=1}^{n} e_{ix} \Delta\left(f_i, f_x\right) p_x g_{xj} \right);$$
8:    Sort interference values of all APs in descending order and store them in array ap_interf;
9:    **for** $i = 1$ to n by ap_interf order **do**
10:      **if** all users in AP_i can be offloaded to adjacent APs **then**
11:         $k \leftarrow$ the number of users associated with APi;
12:         Compute the number of users' associative APs and store them in array $user\_ass$ by ascending order;
13:         **for** $j = 1$ to k by $user\_ass$ order **do**
14:            Offload $user_j$ to an associative AP which minimize the objective function $\Im_1$ based on change of β;
15:         **end for**
16:         Turn off AP_i ,update α, β and return to step 2;
17:      **end if**
18:   **end for**
19:   **if** all APs can't be switched off **then**
20:      break;
21:   **end if**
22: **end while**
23: Introduce the optimal solution of α^*, β^*, f^* to objective function $\Im_1$, compute the set of transmit power p^* as an integer programming with single-variable.

---

5   Performance Evaluation

In this section, we conduct experiments to validate the efficiency of JOEI algorithm compared to two classical energy-saving schemes, green-clustering algorithm [1] and cooperative energy-efficient method [14], which don't consider the influence to interference in energy-saving process. Moreover, we conduct simulations to compare the performance with another energy - interference joint study algorithm [7]. Because we take into account the more sophisticated constraints to ensure users' demands and more measures to reduce interference, our algorithm shows better performance in interference aspect, although there is a small gap in energy saving.

It is noteworthy that we solve the integer programming problem for p by the SCIP [15] optimization toolbox which uses a branch-cut-and-price method.

Meanwhile, in the experiments, we set the weight factor θ be 100 which let the part of interference in the same order of magnitude with energy consumption. More parameters are shown in the Table 1.

Table 1. Experiment parameter

Parameter	Value	Commons
r	2	Attenuation coefficient
p	[1 mW/0 dBm, 1W/30 dBm]	Limit of AP transmit power
L_{max}	70 Mbps	Limit of AP load
R	40 m	Coverage radius of AP
f_c	2.4 GHz	Carrier spectrum
N_0	10^{-13} W/-100 dBm	Thermal noise power
B	20 MHz	Bandwidth of channel

In order to get the performance of greedy algorithm, we design a set of network typologies, where 100 APs are regularly deployed and a number of users are randomly placed in a 300 m-by-300 m area. The effective coverage radius of AP is set to 40 m, the throughput demand of each user is set to a range from 2 Mbps to 4 Mbps. To simulate a real and comprehensive network condition, we change the number of users increases from 50 to 800 in steps of 50. Before the experiments, AP transmission power is initialized to 30 dBm and AP channel is set to a random value among 1, 6 and 11.

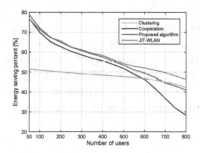
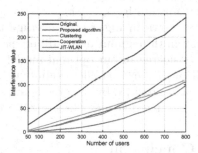

Fig. 3. Energy saving percent compared with three classical algorithm.

Fig. 4. Interference compared with three classical algorithm.

As shown in Fig. 3, the proposed algorithm has a better performance of energy efficiency in all types of scenes. When there are only 50 users with throughput demand, proposed algorithm obtain a high energy-saving rate close to cooperative method, which is almost 30 percent higher than clustering algorithm. As the number of users increases, the rate of energy saving reduce gradually. When the number of users increases to 800, the energy-saving rate of both proposed algorithm and clustering algorithm are decreased to 45%, but that of cooperative method is down to the percent of 30.

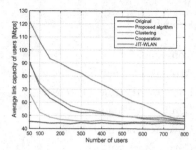

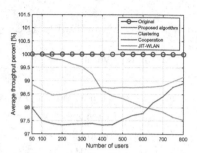

Fig. 5. Average link capacity of users compared with three classical algorithm.

Fig. 6. Average throughput percent compared with three classical algorithm.

As shown in Fig. 4, it is the interference comparison of proposed algorithm, clustering algorithm and cooperative algorithm. The original interference represents the system's interference in the initial scenario before running algorithms. With the increase in the number of users, the initial interference is increasing. Although the two classic algorithm don't consider interference in energy-saving process, they all be helpful in reducing interference as idle APs are switch off. Obviously, Our algorithm has a better performance in all scenarios. As the channel capacity mainly affected by interference, our algorithm have the largest average link capacity of users than the original situation and other two energy-saving algorithms in Fig. 5.

As shown in Fig. 6, the average throughput percent represents the change of system throughput under different algorithms. As user throughput protection are considered in our optimal model, the average user throughput provided by JOEI algorithm is stable around the original throughput, which is 2.6% higher than that provided by the green-clustering and cooperative algorithm at most. It demonstrates that our scheme will not cause any influence on user throughput.

6   Conclusion

In this paper, we have addressed the issue of energy consumption and interference in dense WLAN. Based on real test trace, we determine the mathematical model of throughput, transmit power with energy consumption. Meanwhile, we define the interference model by analyzing the sources of interference to users in the network. We further design a joint optimization model of energy consumption and interference, which adjust user-AP association, AP switch, AP transmit power and AP channel, to reduce energy consumption as well as interference without sacrificing users' QoE. However, the solution to joint optimal is formulated as an NP-hard problem. To simplify computational complexity, an JOEI algorithm is proposed based on greedy method. The comparison experiments show that the proposed algorithm has good performance in reducing interference and ensuring user's demand with energy saving. We believe that our research will promote the development of green WLANs.

References

1. Jardosh, A.P., Papagiannaki, K., Belding, E.M., Almeroth, K.C., Iannaccone, G., Vinnakota, B.: Green WLANs. on-demand WLAN infrastructures. Mob. Netw. Appl. **14**(6), 798 (2009)
2. Jardosh, A.P., Iannaccone, G., Papagiannaki, K., Vinnakota, B.: Towards an energy-star WLAN infrastructure. In: Eighth IEEE Workshop on Mobile Computing Systems and Applications. HotMobile 2007, pp. 85–90 (2007)
3. Su, Y., Wang, Y., Zhang, Y., Liu, Y., Yuan, J.: Partially Overlapped Channel interference measurement implementation and analysis. In: Computer Communications Workshops, pp. 760–765 (2016)
4. Budzisz, L., Ganji, F., Rizzo, G., Marsan, M.A.: Dynamic resource provisioning for energy efficiency in wireless access networks: a survey and an outlook. Commun. Surv. Tutor. IEEE **16**(4), 2259–2285 (2014)
5. Debele, F.G., Meo, M., Renga, D., Ricca, M., Zhang, Y.: Designing resource-on-demand strategies for dense WLANs. IEEE J. Sel. Areas Commun. **33**(12), 24594–2509 (2015)
6. Bayer, N., Gomez, K., Sengul, C., Hugo, D.V., Gndr, S., Uzun, A.: Load-adaptive networking for energy-efficient wireless access. Comput. Commun. **72**(C), 107–115 (2015)
7. Lee, K., Kim, Y., Kim, S., Shin, J., Shin, S., Chong, S.: Just-in-time WLANs: On-demand interference-managed WLAN infrastructures. In: IEEE INFOCOM 2016 - the IEEE International Conference on Computer Communications, pp. 1–9 (2016)
8. Zeng, Y., Pathak, P.H., Mohapatra, P.: Just-in-time WLANs: A first look at 802.11ac in action: Energy efficiency and interference characterization. In: NETWORKING Conference, pp. 1–9 (2014)
9. Miao, G., Himayat, N., Li, G., Talwar, S.: Distributed interference-aware energy-efficient power optimization. IEEE Trans. Wirel. Commun. **10**(4), 1323–1333 (2011)
10. Zhou, Y., Li, X., Liu, M., Mao, X., Tang, S., Li, Z.: Throughput optimizing localized link scheduling for multihop wireless networks under physical interference model. IEEE Trans. Parallel Distrib. Syst. **25**(10), 2708–2720 (2013)
11. Gupta, P., Kumar, P.R.: The capacity of wireless networks. IEEE Trans. Inf. Theory **46**(2), 388–404 (2000)
12. Zhou, K., Jia, X., Xie, L., Chang, Y.: Channel assignment for WLAN by considering overlapping channels in SINR interference model. In: International Conference on Computing, NETWORKING and Communications, pp. 1005–1009 (2012)
13. Xu, C., Jin, W., Zhao, G., Tianfield, H., Yu, S., Qu, Y.: A novel multipath-transmission supported software defined wireless network architecture. IEEE Access **5**(99), 2111–2125 (2017)
14. Rossi, C., Casetti, C., Chiasserini, C.F., Borgiattino, C.: Cooperative energy-efficient management of federated WiFi networks. IEEE Trans. Mob. Comput. **14**(11), 2201–2215 (2015)
15. Vigerske, S., Gleixner, A.: SCIP: global optimization of mixed-integer nonlinear programs in a branch-and-cut framework. Optim. Methods Softw. **33**, 1–31 (2016)

User Assisted Dynamic RAN Notification Area Configuration Scheme for 5G Inactive UEs

Chunyan Cao[✉], Xiaoge Huang, Xiayan Fan, and Qianbin Chen

School of Communication and Information Engineering, Chongqing University of
Posts and Telecommunications, Chongqing, China
{Caocyan,Fanxiayan}@yeah.net
{Huangxg,Chenqb}@cqupt.edu.cn

Abstract. The new radio resource control inactive state has become
the main status of user equipments (UEs) in 5G networks, because of
its low power consumption and energy saving features. To deal with the
massive signaling overhead in 5G networks, in this paper, we introduce a
UE assisted dynamic RAN notification area (RNA) configuration scheme
to effectively reducing the paging and the RNA update overhead of inactive UEs. Especially, UEs are divided into two categories, namely, the
speed-priority type and the rate-priority type based on their communication rate, mobility, as well as the location. Accordingly, we further
extensively investigate the dynamic RNA configuration update process in
both the theoretical and the practical manner. The performance of proposed schemes is evaluated via simulations and the results demonstrate
the effectiveness and the efficiency in achieving the design goals, which
could achieve a considerable performance improvement with respect to
schemes in literatures.

Keywords: Small cell · 5G · Inactive state · RAN notification area

1 Introduction

In order to satisfy the low-latency, high-reliability and low-power transmission
performance requirements to support various scenarios in 5G networks, the ultra
dense small cell network is considered as a feasible solution [1]. Recently a new
state called RRC Inactive is designed as a main state for inactive UEs in 5G
networks. The main characteristics of the inactive state is that the interface
between the Radio Access Network (RAN) and Core Network (CN) is kept [2].

This work is supported by the National Natural Science Foundation of China (NSFC)
(61401053), the 863 project No. 2014AA01A701, and Changjiang Scholars and Innovative Research Team in University (IRT1299), Special Fund of Chongqing Key Laboratory (CSTC)

X. Liu et al. (Eds.): ChinaCom 2018, LNICST 262, pp. 214–224, 2019.
https://doi.org/10.1007/978-3-030-06161-6_21

However, the small coverage of cells increases the total number of paging and location update signaling messages due to the movement of Inactive User Equipments (UEs) [3]. Paging is a system access functionality that is triggered by the network to locate a UE when there is downlink packet for the UE.

In Long Term Evolution (LTE) networks, the location tracking of UEs is based on the tracking area (TA) and the tracking area list (TAL). Several works have been done for reducing the overhead due to the mobility of UEs during the last years [4–7]. In [4], an experience-based scheme was used to predict the movement trend of UEs, and the TAL is allocated to UEs according to movement type. An automatic scheme which could obtain the optimal TA update scheme and minimize network signaling was presented in [5], the scheme detected periods of similar trends of UEs that could share the same TA update scheme. In [6], the authors proposed an effective TAL design to reduce the overall signaling overhead while taking into account the TAL overlapping. In this case, a cell could be included in more than one TAL. When UEs moved from cell i to cell j, the location update is performed if cell j is in none of those TALs of UEs. The authors designed an effective tracking area list management scheme to optimize TAs in the form of TALs for each UE, which could minimize the number of the paging and tracking area updates signaling message [7].

The 3rd generation partnership project (3GPP) standard proposed three different granularity RNA configuration schemes for inactive UEs, that is, the cell list, the radio access network (RAN) area identification (ID) list and tracking area identity (TAI) list configuration scheme, respectively [8]. However, the small configuration granularity of the cell list-based RNA configuration scheme is only applicable to UEs with high paging rates and low-speed. The larger granularity of the RNA configuration scheme is based on the RAN area ID list/TAI list, which is only applicable to UEs with a high-speed and low paging rate.

In this paper, a UE assisted information-based dynamic RNA configuration scheme is proposed to minimize the total number of paging and RNA update signaling in 5G small cell network. The main contributions can be summarized as: First, UEs are divided into different categories based on their own state informations. In particular, UEs could recommend an appropriate RNA and report it to the anchor gNB as the RNA configuration assisted information. Finally, the anchor gNB will determine whether to reconfigure the recommended RNA based on assisted information from UEs. Second, we aim to minimize the total number of messages in paging and location tracking of UEs in Inactive mode. UEs could report the recommended RNA base on the state informations of UEs, thereby reducing the total overhead in the scheme we proposed. In addition, the RNA for a UE is dynamic update according to UEs activities. such as the current communication frequency and mobility, to further improve the performance of the network.

The rest of this paper is structured as follows: the system model is presented in Sect. 2. The detail procedure of the dynamic RNA configuration scheme is presented in Sect. 3. Section IV introduces the problem formulation of the proposed RNA configuration scheme. The simulation results are presented in Sect. 5. Finally, the paper is concluded in Sect. 4.

2 System Model

In this paper, we consider a 5G small cell network consists of the RAN and next generation core network (NGC), where RAN includes a set of small cell gNBs (i.e., base station of the 5G network), and UEs are randomly located in the coverage of gNBs. NGC includes the access and mobility management function (AMF), the user plane gateway (UP-GW) and the control plane gateway (CP-GW) as shown in Fig. 1. Each gNB covers a different number of cells and inter-gNBs are connected via the X_n interface. The anchor gNB maintains the connection of the user plane and the control plane with the NGC, which is in charge of the paging and location tracking of UEs. In order to avoid a large number of paging and RNA updates signaling message between gNBs, we assume there is a X_n interface between gNBs and the anchor gNB where the UE is located. Alternatively, when a UE moves to a gNB that does not have interface with the anchor gNB, the anchor gNB will be replaced by the gNB where the UE is currently located. In this case, the intra-gNB mobility is hidden from the anchor gNB enabling the tracking of the cell level location of UEs to minimize the signaling overhead.

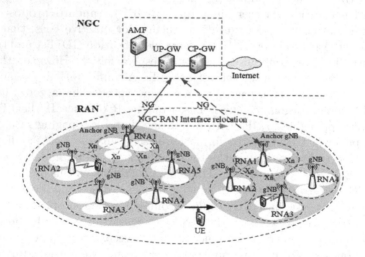

Fig. 1. System model

Generally, the initial configuration of the RNA is depended on the mobility of UEs as well as on their geographical distribution. gNBs will provide UEs with the initial configuration of the RNA when UEs firstly connect to the network. The initial RNA configuration varies from scene to scene and one RNA can contain multiple gNBs regions. We denote $R = \{1, 2...r\}$ as the set of RNAs, and $U = \{1, 2...u\}$ as the set of UEs. Let Υ denote the set of all possible RNA lists in the network, Υ_i denote the set of all possible RNA lists that can be assigned to UEs in RNA i. $m_{i\ell}$ represents the probability to assign RNA list in RNA i to UEs, h_{ij} represents the probability that UEs move from RNA i to RAN j. When UEs move to a new RNA does not belong to it is own RNA list, there is a RNA update procedure for UEs and a signaling message to the anchor gNB. Once the anchor gNB received the RNA update message, it would calculate and forward the set of possible RNA lists to UEs to generate new RNA. Symbols used in this paper are listed in Table 1.

Table 1. Notations of Symbols

Symbol	Description
R	The set of RNAs in the network
U	The set of UEs in the network
Υ	The set of all possible RNA lists
Υ_i	The set of RNA lists can be assigned to UEs in RNA i
$m_{i\ell}$	The probability to assign RNA list in RNA i to UEs
h_{ij}	The probability UEs move to RAN j from RNA i
τ	Signaling overhead of one paging process
λ_0	The necessary overhead generated by cells UE camped
λ_1	The unnecessary overhead generated by cells except UE camped
n_u^k	The cells of the RNA of UE other than the cell UE camped
α_u	The paging probability of UE i during period k
ρ	Signaling overhead of one RNA updating process
β_u	The RNA update probability of UE i during period k
t_{ui}	The duration time of UE u in RNA i

3   Dynamic RNA Configuration Scheme

3.1   Dynamic RNA Configuration

According to the dynamic changes of UEs, the initial RNA configuration needs to be reconfigured timely. The user assisted dynamic RAN notification area configuration scheme consists of two steps: the decision making step and the RNA adjustment step, as shown in Fig. 2. Firstly, in the decision making step, UEs will periodically collect their own state informations for time interval T, which include the paging rate, mobility rate and the time duration of UEs in the visited RNA etc. Based on these information UE could determine the categories of UEs, namely, rate-priority or speed-priority. Formally, the category of UE u is depended on the ratio of paging overhead and RNA update overhead in the network within the kth period T (T_k), which is given as follows:

$$O_u^k = \frac{\tau(\lambda_0 + \lambda_1 n_u^k)\alpha_u}{\tau(\lambda_0 + \lambda_1 n_u^k)\alpha_u + \rho\sigma_u\beta_u} \tag{1}$$

where $0 < O_u^k < 1$, $\sigma_u = T_k/t_{ui}$. $\tau(\lambda_0 + \lambda_1 n_u^k)T_k\alpha_u$ denotes the paging overhead of UE u during T_k. $\rho T_k/t_{ui}\beta_u$ denote the RNA update overhead incurred by UE u during T_k. T_k/t_{ui} denotes the RNA update frequency of UE u in the T_k. UE with larger O_u^k causes larger number of the paging message than the RNA update message, which is considered as a rate-priority type UE. Consequently, the smaller the value of O_u^k the more RNA update message than the paging message, which is considered as a speed-priority type UE.

Secondly, in the dynamic adjustment step, according to the category of UEs determined by the previous step, the set of possible RNA lists can be assigned to UEs to generate a new RNA. In this step, rate-priority type UEs would select a RNA list with small number of RNAs to reduce the paging overhead. Similarly, speed-priority type UEs would select a RNA list with large number of RNAs to mitigate the impact of the RNA update on the network performance. Then, UEs report the assisted information to the anchor gNB to update the RNA configuration.

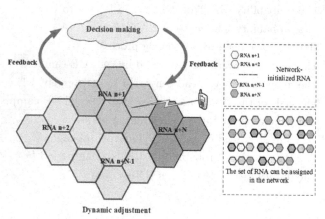

Fig. 2. Dynamic RNA configuration scheme

3.2   RNA Lists Update Process

We give an example to show the update process of the RNA of UEs. Assume that there are three RNAs in the network, named $R1$, $R2$ and $R3$, UEs can freely move in any RNAs. Notice that, when UEs are in one of RNAs, there would form a set of RNA lists that can be assigned to UEs with the neighboring RNAs. For example, if UEs camp in $R1$, all the RNA lists that can be assigned to UEs are denoted as Υ_1, $\Upsilon_1=\{\{R1\},\{R1,R2\},\{R1,R3\},\{R1,R2,R3\}\}$. Additionally, all the RNA lists in the network are denoted as $\Upsilon=\bigcup_{i\in R}\Upsilon_i$, $\Upsilon=\{\{R1\},\{R2\},\{R3\},\{R1,R2\},\{R1,R3\},\{R2,R3\},\{R1,R2,R3\}\}$.

Assume that each RNA list is assigned to UEs with a specific probability, denotes as $P_i(L)$. All RNA lists are in descending order by the number of included RNAs. Furthermore, the category of UEs should be taken into consideration during RNA update process. Speed-priority type UEs should be assigned to a large

RNA list as the notification area to reduce RNA update signaling. Otherwise, rate-priority type UEs should be assigned to a small RNA list as the notification area to reduce paging signaling. Thus, speed-priority type UEs has high probability to be assigned to $\{R1, R2, R3\}$, while $\{R1\}$ is probably be assigned to rate-priority type UEs.

We define a poisson cumulative distribution function $F(r_u, t_{ui})$, where $r_u = 1/(O_u^k) \geq 1$. Especially, the cumulative distribution probability of $\{R1, R2, R3\}$ is the smallest, while $\{R1\}$ is the biggest. UEs will select the appropriate RNA list when the following condition are satisfied:

$$\sum_{m=1}^{L-1} P_i(m) < F(r_u, t_{ui}) < \sum_{m=1}^{L} P_i(m) \tag{2}$$

From above, larger O_u^k results in smaller r_u and indicates that UE u, which called rate-priority type. It is better to assign a RNA list with small number of RNAs for this UE u. In contrast, smaller O_u^k results in bigger r_u and indicates that UE u, which called speed-priority type. It is better to assign a RNA list with a number of RNAs for this UE u. In addition, the small t_{ui} indicates the high speed of UEs, which will be assigned to a large RNA list. As shown in Fig. 3, when r_u is small, UEs have a higher probability to select a smaller RNA list. Otherwise, when r_u is large, UEs have a higher probability to select a larger RNA list. In addition, the time duration t_{ui} indicates the speed of UEs, the speed-priority type UEs will select a large RNA list.

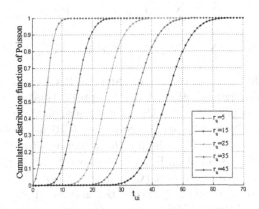

Fig. 3. t_{ui} versus the cumulative distribution function of poisson

4   Problem Formulation

In this paper, we consider a UE assisted information-based dynamic RNA configuration scheme to minimize the total number of paging and RNA update signaling message in 5G small cell network. The multi-objectives optimization technique is used to minimize both the paging and the RNA update overhead, find the optimal distributions of the RNA, and find a fair tradeoff of these two overhead. The RNA update overhead which generates in RNA i and j (not in the same RNA list) can be expressed as follows:

$$f_{RNAU(i,j)} = \rho \sum_{\ell \in \gamma_i \land \ell \notin \gamma_j} h_{ij} m_{i\ell} + \rho \sum_{\ell \in \gamma_j \land \ell \notin \gamma_i} h_{ji} m_{j\ell} \tag{3}$$

where ℓ is the current RNA list of UEs. The overhead of paging generated in RNA i can be expressed as:

$$f_{paging(i)} = \tau m_{i\ell} \sum_{u \in U} \alpha_u t_{ui} \tag{4}$$

where $\sum_{u \in U} \alpha_u t_{ui}$ is a constant that represents the paging overhead in RNA i. When there is data transmission requirement, the anchor gNB will send paging message to gNBs in the RNA of UE u. Based on the above discussion, the optimization problem can formulate as:

$$\min \quad \omega \sum_{i \in R} \sum_{j \in R \land i \neq j} f_{RNAU(i,j)} + (1 - \omega) \sum_{\ell \in \Upsilon} \sum_{i \in \ell} f_{paging(i)}$$

$$\text{s.t.} \quad (C1): \quad i, j \in R, 0 \leq h_{ij} \leq 1$$

$$(C2): \quad i \in R, \ell \in \Upsilon, 0 \leq m_{i\ell} \leq 1$$

$$(C3): \quad \forall i \in R, \sum_{\ell \in \gamma} m_{i\ell} = 1$$

$$(C4): \quad \sum_{i \in R} \sum_{j \in R \land i \neq j} f_{RNAU(i,j)} \leq RNAU_{\max}$$

$$(C5): \quad \sum_{i \in R} f_{paging(i)} \leq PAGING_{\max} \tag{5}$$

where ω is the weight factor of the paging and the RNA update overhead in the network, which could dynamically adjust the proportion of these two factors. Specifically, the speed-priority type UEs will be assigned with a larger ω than the rate-priority type UEs. The set of constraint (C3) assures that the sum of the proportional usage of the RAN lists in RNA i equals to one. The constraint (C4) guarantees that the sum of all RNA update overhead should not exceed the maximum $RANU_{\max}$, the constraint (C5) indicates that the sum of all paging overhead in the network should not exceed the maximum $PAGING_{\max}$.

In this paper, the discrete Markov chain is used to analyze the RNA update and the paging overhead in the network, which can be obtained by the solution of the balance equations of the discrete Markov chain. The detail process refers to [7].

5   Simulation results

In this section, we verify the performance of the proposed user assisted dynamic RAN configuration scheme (UD-RNA) algorithm by simulations. We focus on the impact of RNA update and paging overhead on the UD-RNA scheme. In order to qualify the proposed algorithm, the following two algorithms are used for comparison: the cell list configuration scheme (C-RNA) and the RAN area ID list configuration scheme (R-RNA). C-RNA scheme could optimize the paging overhead while R-RNA could optimize the RNA update overhead. The RNA update rate of UEs is represented by ν, $\frac{1}{\nu}$ refers to the average duration time of UEs in each RNA. The paging rate of UEs is represented by p, which is the average arrival packets of UEs. The following three scenarios is considered in the simulation: scenario one, $1 \leq p \leq 10$, $\nu = 5$ and $\omega = 5$; scenario two $1 \leq \nu \leq 10$, $p = 0.5$ and $\omega = 0.5$; scenario three, we vary the $0.1 \leq \omega \leq 1$.

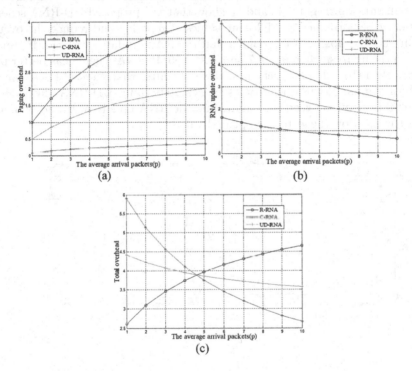

Fig. 4. Performance comparison of three schemes as a function of average arrival packets p

Figures 4 and 5 show the paging overhead, the RNA update overhead and the total overhead of C-RNA, R-RNA and UD-RNA algorithm versus the average arrival packets p and the average time duration of UEs in each RNA, respectively. Figure 4a,b show that the average arrival packets p of UEs have a negative impact on paging but a positive impact on RNA update overhead, this is due to the rate-priority type UEs generate more paging overhead than the RNA update overhead. Thus, the larger the p value is, the higher the paging overhead becomes. As depicted in Fig. 5a,b, the increase of the average duration time UE also has a negative impact on paging and a positive impact on RNA update overhead. The longer a UE resides in a RNA, the lower its mobility and the less RNA update overhead is generated.

Figures 4a and 5a illustrate that C-RNA has a better performance than UD-RNA and R-RNA in terms of paging overhead regardless of the vary of p and $\frac{1}{\nu}$. It proves that the C-RNA scheme can effectively reduce the paging overhead because of the small granularity of RNA configuration. Figures 4b and 5b demonstrate that R-RNA has a better performance than UD-RNA and C-RNA in terms of the paging overhead regardless of the vary of p and $\frac{1}{\nu}$. R-RNA scheme can effectively reduce the RNA update overhead because of the big granularity of RNA configuration. Figures 4c and 5c show that the proposed UD-RNA scheme has a better performance for different values of p and $\frac{1}{\nu}$ and reduce the overall signaling message for both the paging and the RAN update overhead. It can be seen from the figures, when $p < 4.7$ and $\frac{1}{\nu} < 0.55$, R-RNA scheme has better performance than the other two schemes. However, after that point C-RNA scheme is the best compared with other two schemes. Especially, UD-RNA always finds an optimal tradeoff between the paging and the RNA update overhead by maintaining the total overhead to the optimal value.

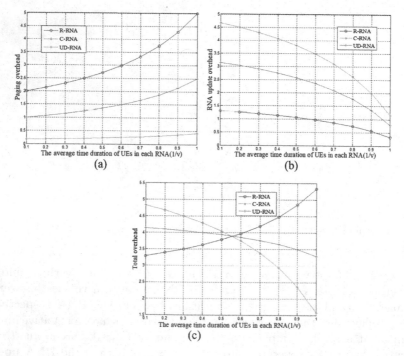

Fig. 5. Performance comparison of three schemes as a function of average duration time ν

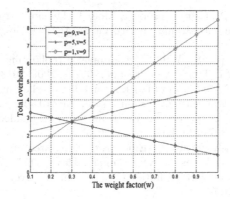

Fig. 6. The impact of the weight factor on the total overhead

Figure 6 illustrates the change of the total overhead in the network versus the weight factor. The following three scenarios are used for comparison: (1) $p = 9$ and $v = 1$, when the paging overhead is larger than the RNA update overhead; (2) $p = 5$ and $v = 5$, when the paging overhead equals to the RNA update overhead; (3) $p = 1$ and $v = 9$, when the paging overhead is smaller than the RNA update overhead. With the increase of ω, the proportion of the RNA update overhead increases, while the proportion of the paging overhead decrease. For scenario 1, the larger the ω is, the small the paging overhead achieves, thus the total overhead is declined. For scenario 2, with the increase of ω, the RNA update overhead is increasing and the paging overhead is decreasing, result in slowly increasing of the total overhead. For scenario 3, the larger the ω is, the bigger the paging overhead becomes, result in increasing of the total overhead.

6   Conclusions

In this paper, a UE assisted based dynamic RNA configuration scheme was proposed to reduce the total overhead of the paging and the RNA update signaling. Simulation results illustrated the effectiveness of the proposed scheme.

References

1. Osseiran, A., Boccardi, F., Braun, V.: Scenarios for 5G mobile and wireless communications: the vision of the METIS project. IEEE Commun. Mag. **52**(5), 26–35 (2014)
2. Da Silva, I.L., Mildh, G., Saily, M., Hailu, S.: A novel state model for 5G radio access networks. In: 2016 IEEE International Conference on Communications Workshops (ICC), PP. 632–637. Kuala Lumpur (2016)
3. Hailu, S., Saily, M.: Hybrid paging and location tracking scheme for inactive 5G UEs. In: 2017 European Conference on Networks and Communications (EuCNC), pp. 1–6. Oulu (2017)

4. Razavi, S.M., Yuan, D., Gunnarsson, F., Moe, J.: Dynamic tracking area list configuration and performance evaluation in LTE. In: IEEE Globecom Workshops, pp. 49–53. Miami (2010)
5. Toril, M., Luna-Ramírez, S., Wille, V.: Automatic replanning of tracking areas in cellular networks. IEEE Trans. Veh. Technol. **62**(5), 2005–2013 (2013)
6. Widaa, L.O., Sharif, S.M.: Effect of tracking area list overlapping in reducing overall signaling overhead in long term evolution system. In: 2015 International Conference on Computing, Control, Networking, Electronics and Embedded Systems Engineering (ICCNEEE), pp. 397–400. Khartoum (2015)
7. Bagaa, M., Taleb, T., Ksentini, A.: Efficient tracking area management framework for 5G Networks. IEEE Trans. Wirel. Commun. **15**(6), 4117–4131 (2016)
8. RAN3, LS on definition of RAN notification area in inactive state. 3GPP R2 WG, Tdoc R2–1710036 (2017)

Transmission Capacity Analysis of Distributed Scheduling in LTE-V2V Mode 4 Communication

Jie Lv[1($\boxtimes$)], Xinxin He[1], Jianfeng Li[1], Huan Wang[2], and Tao Luo[1]

[1] Beijing Key Laboratory of Network System Architecture and Convergence, Beijing University of Posts and Telecommunications, Beijing 100876, China
lvj@bupt.edu.cn

[2] DOCOMO Beijing Communication Laboratories Co., Ltd, Beijing 100080, China

Abstract. LTE-V2X sidelink/PC5 communication aimed at supporting device-to-device (D2D) communications in vehicular scenario has been developed as an appropriate technology by 3GPP. Particularly, mode 4 operating without cellular coverage permits vehicles autonomously to select resources and has the potential to achieve an efficient and reliable transmission for vehicle safety applications. However, there is very little research conducted on theoretical understanding of the characteristics and performance of mode 4. In this work, we propose a tractable mathematical analysis to evaluate the performance of LTE-V2V in mode 4. Specifically, we assume that vehicles driving on 1-D abstract lane follow a Poisson Point Process (PPP). By means of probability model, we analyze the event that vehicles randomly select the same resource inducing collision, and investigate the failure probability of transmission. Also, the distance between adjacent vehicles is log-normally distributed and the transmission outage probability under a fixed threshold is given. Furthermore, we derive the expression of transmission capacity. To this end, numerical results verify that the transmission capacity of mode 4 can be improved to a certain extent with the increasing of density of vehicles.

Keywords: LTE-V2V mode 4 communication · Distributed scheduling Collision · Interference · Transmission capacity

1 Introduction

In recent years, the Intelligent Transportation System (ITS) considered as an effective method can be push forward by the revolution of automotive industry and vehicle wireless communication. Exchanging information acquired by on-board sensors with neighbors including location, direction and speed, plays a crucial role in vehicle safety application. Meanwhile, in order to ensure the reliability of vehicular communication, different standardizations have devoted efforts into normalizing vehicle-to-everything (V2X) wireless technologies. To date, there are two main standards, i.e. IEEE 802.11p

X. Liu et al. (Eds.): ChinaCom 2018, LNICST 262, pp. 225–234, 2019.
https://doi.org/10.1007/978-3-030-06161-6_22

[1] and 3GPP's LTE-V2X [2]. At the end of 2016, 3GPP developed LTE sidelink/PC5 communication as a potential technology for vehicular applications in the first version of Release14 [2].

In Release 14, the standard concludes two modes for sidelink/PC5 communication, in which vehicles directly communicate with each other. In mode 3, the management or scheduling of resources is controlled by the base station (i.e. eNB), but vehicles autonomously select and utilize resources in a distributed manner in mode 4. Currently, some referable results of reliability of LTE-V2V have been acquired by simulations [3–6] and rarely theoretical studies. Especially, mode 4 has been considered the baseline mode and represents an alternative to IEEE 802.11p. The performance of mode 4 for V2 V communications has increasingly being discussed in last few years. So far, most of these studies focus on simulation analysis of reliability that is considered as the important performance indicator measured by Packet Delivery Ratio, such as [3, 4]. Additionally, the presented model in [7] has characterized an upper performance of vehicular D2D communication relying on LTE-D2D (Release 12) as function of the application traffic pattern, without the effect of interference. In [6], the authors propose 'spatial capacity' defined as the total amount of bits that can be successfully delivered in 1 km and 1 s to assess the performance of LTE-V2 V mode 4 by using the vehicle traffic simulator. As far as we know, the study of capacity on mode 4 is still a blank field.

Since Gupta and Kumar's remarkable study on the capacity of ad hoc network [8], S. Weber and J. G. Andrews proposed the notion that the spatial density of successful transmission is defined as transmission capacity (TC) [9]. After that, some researchers have conducted a series of studies on transmission capacity for Vehicle Ad Hoc Network (VANET) [10–12]. These related studies consider 1-D linear road and discuss the concurrent transmitters and interference on distributed access scheme. For example, they are mostly based on Carrier Sense Multiple Access/Collision Avoidance (CSMA/CA) mechanism that vehicles successfully access the channel after going through the sensing phase and the finite back-off process. Different from this typical distributed scheme, vehicles in mode 4 use the time-frequency resources determined by sensing results and randomly select some resources to transmit immediately. Vehicles may conflict with others when they choose the same resource, leading to the failure of transmission. Thus, the existing models of capacity for distributed scheme in VANET cannot be directly applied to mode 4.

To better understand characterizes of decentralized scheme in mode 4, we undertake to develop a novel transmission capacity analysis model of mode 4 by taking its resource collision avoidance and vehicle scenarios into consideration. In this paper, we firstly analyze the collision probability due to the resource competition between transmitters, further obtain the density of concurrent transmitters. Then, the transmission outage probability is derived, considering interference to receivers under a simplified channel. Our ultimate goal is to extend a theoretical study on the capacity for LTE-V2V mode 4.

The rest of the paper is organized as follows. System model is built in Sect. 2. In Sect. 3, theoretical analysis of collision probability, interference analysis, and transmission capacity is given. Numerical results are presented in Sect. 4. Finally, in Sect. 5, the conclusion is obtained.

2   System Model

2.1   Network Model

As represented in Fig. 1(a), a multi-lane highway scenario constituted from N parallel traffic lanes separated by a fixed distance. Compared with the transmission range of vehicles, the road width is much smaller. Thus the 2-D multiple lanes are always approximated by a 1-D single lane in previous works, such as [10–12].

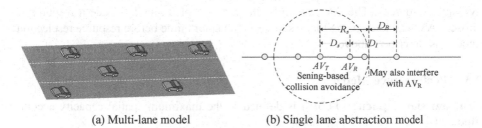

| (a) Multi-lane model | (b) Single lane abstraction model |

Fig. 1. Highway scenario

In order to simplify the analytical model, we also consider a 1-D straight lane that is shown in Fig. 1(b). From the independent theorem of PPP [13], the set of vehicles Π_λ within the abstract 1-D line follows a one-dimensional PPP of density λ vehicles/km.

Additionally, for vehicle mobility, we utilize the well-known Car-following model [14], which models the mean distance between any two adjacent vehicles as a log-normal distribution with the main parameters μ and σ.

$$D_X(x) = \frac{1}{\sqrt{2\pi}\sigma x}\exp\left(\frac{-(\ln x - \mu)^2}{2\sigma^2}\right).$$ (1)

Each vehicle, hereafter Autonomous Vehicle (AV), broadcasts beacons periodically for V2V safety application by a D2D type of radio link (sidelink).

Channel model: All AVs are assumed to transmit with a fixed power P_t and have the same transmission rate R. The signal power decays with the distance according to the large-scale fading model with a path-loss exponent $\alpha > 1$. Besides path loss attenuation, the signal power experiences with the small-scale Rayleigh fading which follows an exponential distribution with mean equal to 1. Consequently, the power received is $P_t H d^{-\alpha}$, where H represents the channel fading, d represents the transmission distance that is defined the Euclidean distance.

2.2   Distributed Scheduling Scheme in Mode 4

In LTE-V2V mode 4 communication, AVs transmit to each other in direct mode, with resource allocation performed by distributed scheduling. According to the framework in R14 specification, radio resources are selected from a selection window, which is the beacon generation period (here equal to $t_p = 100$ ms). We assume that all the resources in the selection window are available, which is defined as $L = \{1, 2, \ldots, M\}$. Each AV continuously senses the radio channel to learn about the periodic transmission status of the neighboring AVs. When a packet needs to be transmitted, the last 1000 ms history, referred as the sensing window, is used to make a process of resource exclusion and selection. In detail, the AV decodes the SCI information and measures the energy level of related signals. We define that the maximum distance at which the neighbors are able to decode sidelink control information (SCI) messages is R_s, i.e. sensing distance. Thus, the final candidate resource pool $L_1 = \{1, 2, \ldots, m\}$ is composed of the remaining available resources. The AV randomly chooses one of the in L_1. Once a resource is chosen, AV may keep using that resource for a random time before resource reselection occurs as described above.

2.3   Performance Metric

Transmission capacity (TC): It is defined as the maximum spatial capacity accommodated in the network [9], that is

$$C_T = \lambda P_s (1 - \tau) R, \tag{2}$$

Where λ describes the density of the potential transmitters in the network. P_s represents the probability of transmission. τ is the outage probability of transmission when the signal-to-interference ratio (SIR) is smaller than the threshold β. R represents the transmission rate and equals 1 bps. In this case, we would assess the normalized TC in the final results.

3   Theoretical Analysis

In the context of V2 V mode 4 operation, given these assumptions in Sect. 2.2, the number of vehicles that can be allocated initially in the selection window is equal to $M = \lfloor t_p / t_{sfr} / n_{sfr\_bcn} \rfloor$, where $nsfr\_bcn$ equals to 1 that means one sub-frame needed to transmit each beacon. Through the resource selection procedure, the size of resource pool available L_1 is denoted as

$$m = E \left[\sum_{i=1}^{M} X_i \right], \tag{3}$$

where X_i is a random variable that determines whether a resource i is selected into the resource L_1, as

$$X_i = \begin{cases} 1, & i \ is \ in \ L_1 \\ 0, & i \ is \ not \ in \ L_1 \end{cases}. \tag{4}$$

Note that for the tagged AV, each resource is selected with different probability. Nevertheless, this issue will not be discussed in this paper that may be investigated in our future work.

3.1 Collision Probability

In this subsection, we present a model to capture the collision probability. The approach of resource reservation in LTE-V2V mode 4 does not completely resolve collisions when trying to reserve an idle slot.

Lemma 1. The target probability of failure probability due to collisions is

$$P_{failure} = 1 - e^{-\bar{n}} \cdot \sum_{k=0}^{m} \frac{\bar{n}^k}{m^k} \cdot C_m^k. \tag{5}$$

Proof. AVs use sensing to determine transmission opportunity, i.e. a suitable resource for transmission. The resource occupancy and energy detection level are two important conditions to conduct resource exclusion. We convert these into spatial constraints and infer the density of the competitive AVs that cause resource collision with the tagged AV.

We arbitrarily choose an AV that constructs a resource utilization map based on the occupancy of resources for each interfering AV indicated by decoding SCI. Based on the sensing-based protocol, the subset of competing AVs can be expressed as

$$\Pi_c = \{AV_1, AV_2, \ldots | AV_j \in \Pi_\lambda, a_j = 0\}. \tag{6}$$

where $a_j = 0$ represents AV_i hasn't occupied resources. Note that the density is $Q \cdot \lambda$, Q represents the probability of competition that AVs need to select resources at the same slot.

Obtaining through the independent thinning from Π_c, the average number of vehicles competing in t_p within the sensing range R_s is $\bar{n} = R_s \cdot Q \cdot \lambda$. A segment l_r of sensing range R_s have k competing vehicles,

$$P_{l_r}(k) = \frac{\bar{n}^k \cdot e^{-\bar{n}}}{k!}. \tag{7}$$

Then, we calculate the collision probability $P_c(k)$ conditioned to having a number of AVs equal to k. Provided that AVs exceed m, AVs have not sufficient resources inducing collision. The number of AVs is smaller than available resources, collisions happen when more than two vehicles randomly select the same reserved slot in the same one-hop set. The detail expression can be given as

$$P_c(k) = \begin{cases} 1 - \frac{m!}{(m-k)!m^k}, & k \leq m \\ 1, & k > m \end{cases}. \tag{8}$$

According to the Total Probability Theorem, the collisions under the given number of AVs from zero to infinity make up a complete event. Thus, the failure probability is derived as follows

$$P_{failure}\underset{=}{(a)}\sum_{k=0}^{\infty}P_{l_r}(k)\cdot P_c(k)\underset{=}{(b)}\sum_{k=0}^{\infty}P_{l_r}(k)\cdot(1-P_{nc}(k))$$

$$\underset{=}{(c)}1-\sum_{k=0}^{m}P_{l_r}(k)\cdot P_{nc}(k)\underset{=}{(d)}1-e^{-\bar{n}}\cdot\sum_{k=0}^{m}\frac{\bar{n}^k}{m^k}\cdot C_m^k, \tag{9}$$

where $P_{nc}(k)$ in (b) and (c) is defined as the condition probability that the k AVs do not collide; (a) comes after applying the Total Probability Theorem; (c) after invoking the property of probability that the sum is 1; and (d) after substituting Eqs. (7) and (8).

3.2 Interference Analysis

In subsection 3.1, we have considered the transmission collision. When the tagged AV successfully selects some suitable resources, other concurrent transmitters out of the sensing range R_s still may cause interference to the receivers in the transmission range D_s. We define the sensing range R_s as exclusive range, hereafter ER.

To simplify the calculation, we consider the closest AV outside of ER as an active node transmit packet using the same resource that is illustrated as Fig. 1(b), D_s represents the distance between transmitter AV_T and receiver AV_R. The notation D_B is used to represent the distance between the right boundary of AV_T's ER and AV_R's interferer AV_I. D_I is the distance between AV_T and AV_R.

Lemma 2. Under Rayleigh fading, we can approximate the Laplace transform of the interference I from AV_I as

$$L_I(s)==\int_0^\infty f_{D_B}(d-R_s+D_s)\frac{1}{sx^{-\alpha}+1}dx, \tag{10}$$

where $f_{D_B}(\cdot)$ represents the Probability Distribution Function (PDF) of ER's boundary location.

Proof Combining channel model and node distribution in subsection 2.1, the interference can be derived as follows

$$L_I(s)=E[\exp(-sI)]=E\big[\exp(-sHD_I^{-\alpha})\big]$$

$$=\int_0^\infty f_{D_I}(x)E_H[\exp(-sHx^{-\alpha})]dx$$

$$\underset{=}{(e)}\int_0^\infty f_{D_I}(x)\frac{1}{sx^{-\alpha}+1}dx\underset{=}{(f)}\int_0^\infty f_{D_B}(d-R_s+D_s)\frac{1}{sx^{-\alpha}+1}dx, \tag{11}$$

where (e) follows by $f_H(x)=\exp(-x)$, and (f) comes after substituting the geometrical relationship (12) into (e).

$$D_I = R_S - D_S + D_B. \tag{12}$$

Since the boundary of ER can locate anywhere between the two neighboring nodes uniformly, the random variable D_B can be treated as $D_B = Y \cdot U$, where Y is a random variable following the log-normal distribution mentioned in Eq. (1), and U is a random variable following a uniform distribution within [0,1]. With omitting the specific derivation process, the PDF of D_B is

$$
\begin{aligned}
f_{D_B}(z) = \frac{dF_{D_B}(z)}{dz} = \frac{d\,\Pr\{Y \cdot U \le z\}}{dz} &= \exp\left(\frac{\sigma^2}{2} - \mu\right) \cdot \Phi\left(\frac{\mu - \sigma^2 - \ln z}{\sigma}\right) \\
&- \frac{1}{\sqrt{2\pi}\sigma} \exp\left(\frac{\sigma^2}{2} - \mu - \frac{(\mu - \sigma - \ln z)^2}{2\sigma^2}\right) + \frac{1}{\sqrt{2\pi}\sigma z} \exp\left(-\frac{(\ln z - \mu)^2}{2\sigma^2}\right)
\end{aligned} \tag{13}
$$

3.3   Transmission Capacity

Based on the previous analysis, the outage probability τ is calculated as

$$
\begin{aligned}
\tau = \Pr(SIR \le \beta) = \Pr\left(\frac{P_t H d_s^{-\alpha}}{P_t I} \le \beta\right) &= \Pr\left(H \le \beta D_s^\alpha I\right) \\
= 1 - \mathrm{E}\left[\exp\left(-\beta D_s^\alpha I\right)\right] &= 1 - L_I(\beta D_s^\alpha).
\end{aligned} \tag{14}
$$

The success probability is defined as the probability that the tagged transmitter has the access right and the data is received by the tagged receiver without outage. Finally, transmission capacity that is derived as

$$
\begin{aligned}
C_T = \lambda P_s (1 - \tau) R = \lambda (1 - P_{failure})(1 - \tau) R \\
= \lambda \cdot e^{-\bar{n}} \cdot \sum_{k=0}^{m} \frac{\bar{n}^k}{m^k} \cdot C_m^k \cdot L_I(\beta D_s^\alpha) \cdot R,
\end{aligned} \tag{15}
$$

where R equals to 1 bps that is defined as a normalized value.

4   Numerical Results

We have conducted a series of experiments with MATLAB to verify our previous analytical models. Main parameters in LTE-V2V standard are presented in Table 1.

Figure 2 represents the failure probability caused by collisions varying the density of AVs, for two various sensing range R_s = 300, 400 m. A 6 lane highway with a density of 0 to over 30 vehicles/km/lane is assumed. As expected, the curves are monotonically increasing as the density increases. The failure probability is increased due to the increment of AVs competing for the limited resources. Figure 3 compares the failure probability versus the density of AVs when Q = 0.1, 0.3, 0.5, respectively, and R_s = 300 m. By the analytical results illustrated by Figs. 2 and 3, it is seen that the collision is more likely to occur when the density of concurrent AVs increases.

Table 1. Main parameters

Parameter	Value	Parameter	Value
AV density (λ)	Variable input	Beacon period (t_p)	100 ms
Distance between source and destination (D_s)	Variable input	Beacon packet size	190 bytes
Sensing range (R_s)	Variable input	Equivalent transmission power (P_t)	23 dBm
Bandwidth (W)	10 MHz	Loss exponent (α)	2.75

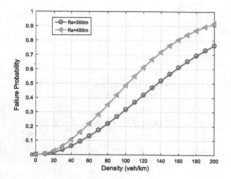

Fig. 2. Failure probability vs. density

Fig. 3. Failure probability vs. density

Figure 4 represents the outage probability varying with the density of AVs, when $D_s = 100$ m and $R_s = 300, 400$ m are assumed. Generally, when the sensing range or ER increases, the probability for the receiver AV_R to have a high SIR is reduced due to the interferer is closer. However, the increase of outage probability is not obvious as the density increases. In Fig. 5, the outage probability versus the distance between the transmitter and the receiver D_s are presented, when $\lambda = 10$ veh/km, $R_s = 100, 300, 500$ m are assumed. As observed, the farther away the receiver is from the transmitter

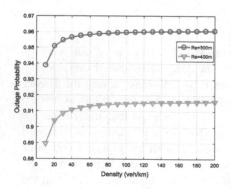

Fig. 4. Outage probability vs. density

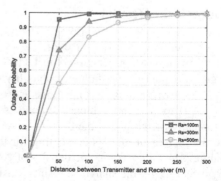

Fig. 5. Outage probability vs. D_s

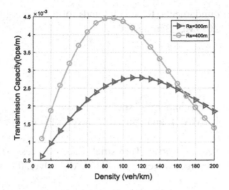

Fig. 6. Transmission capacity vs. density.

AV_T, the lower SIR is. Therefore, under the same condition, the near-destination AVs have better packet reception performance.

Finally, Fig. 6 depicts the transmission capacity over the density with given $D_s = 100$ m and $R_s = 300, 400$ m. It can be seen that two curves have the same two tendencies. As the density of AVs increases, the transmission capacity is improved to some extent (e.g., up to 90 and 110 veh/km for $R_s = 300, 400$ m, respectively), and then is degraded. The reason for the increasing part is that the transmission capacity increases with the number of transmitters; while the decreasing part is that the collision probability and the outage probability all increase that have been illustrated in Figs. 2 and 4, respectively. The peak point for the transmission capacity curve is also moved to the high density of AVs with the decreasing of R_s. In short, the transmission capacity depends on the interaction between AVs' density and sensing capability.

5 Conclusion

In this paper, we analyze the transmission capacity for LTE-V2V sidelink distributed communication (mode 4). When the tagged AV_T acquire available resources to deliver the packet before the deadline, resource collisions would result to fail in delivering. Simultaneously, the transmitting AVs can lead to interference for the tagged AV_R. Based on these analyses, some simple expressions have been obtained for collision probability, outage probability and transmission capacity. Integrating these expressions and simple numerical simulation, we can quickly evaluate the performance of the LTE-V2V mode 4. As a result, vehicle density, road model and resource allocation scheme have a significant impact on the transmission capacity of LTE-V2V mode 4.

On the basis of this paper, the simulation for validating the proposed model should be conducted. In addition, the performance of centralized scheduling (mode 3) in LTE-V2X also needs to be assessed. We will target these topics for further works.

Acknowledgments. This work is supported in part by the National Natural Science Foundation of China under Grant No. 61571065, the China Postdoctoral Science Foundation

No. 2017M620695, and the Fundamental Research Funds for the Beijing University of Posts and Telecommunications and DOCOMO Beijing Labs.

References

1. Standard for information technology—specific requirements—part 11: Wireless LAN medium access control (MAC) and physical layer (PHY) specifications. IEEE Std. No. 802.11-2016 (2016)
2. GPP TS 36.300: Evolved Universal Terrestrial Radio Access (EUTRA) and Evolved Universal Terrestrial Radio Access Network (EUTRAN); Overall description; Stage 2. Rel-14 V14.0.0 (2016)
3. Yang, J., Pelletier, J., Champagne, B.: Enhanced autonomous resource selection for LTE-based V2V communication. In: 2016 IEEE Vehicular Networking Conference (VNC), pp. 1–6. Columbus (2016)
4. Molina-Masegosa, R., et al.: System level evaluation of LTE-V2V mode 4 communications and its distributed scheduling. In: 2017 IEEE VTC 2017-Spring (2017)
5. Molina-Masegosa, R., Gozalvez, J.: LTE-V for sidelink 5G V2X vehicular communications: a new 5G technology for short-range vehicle-to-everything communications. IEEE Veh. Technol. Mag. 12(4), 30–39 (2017)
6. Park, Y., Weon, S., Hwang, I., Lee, H., Kim, J., Hong, D.: Spatial capacity of LTE-based V2V communication. In: Electronics Information and Communication (ICEIC) 2018 International Conference, pp. 1–4 (2018)
7. Piro, G., Orsino, A., Campolo, C., Araniti, G., Boggia, G., Molinaro, A.: D2D in LTE vehicular networking: system model and upper bound performance. In: 2015 7th International Congress on Ultra Modern Telecommunications and Control Systems and Workshops (ICUMT), pp. 281–286. Brno (2015)
8. Gupta, P., Kumar, P.R.: The capacity of wireless networks. IEEE Trans. Inf. Theory 46(2), 388–404 (2000)
9. Weber, S.P., Yang, X., Andrews, J.G., et al.: Transmission capacity of wireless ad hoc networks with outage constraints. IEEE Trans. Inf. Theory 51(12), 4091–4102 (2006)
10. Ni, M., Pan, J., Cai, L., Yu, J., Wu, H., Zhong, Z.: Interference-based capacity analysis for vehicular ad hoc networks. IEEE Commun. Lett. 19(4), 621–624 (2015)
11. He, X., Zhang, H., Shi, W., Luo, T., Beaulieu, N.C.: Transmission capacity analysis for linear VANET under physical model. China Commun. 14(3), 97–107 (2017)
12. Hassan, M.I., Vu, H.L., Sakurai, T.: Performance analysis of the IEEE 802.11 MAC protocol for DSRC safety applications. IEEE Trans. Veh. Technol. 60(8), 3882–3896 (2011)
13. Stochastic Geometry for Wireless Networks, ser. Stochastic Geometry for Wireless Networks. Cambridge University Press, Cambridge (2013)
14. Brackstone M., McDonald, M.: Car-following: a historical review. Transp. Res. Part F, Traffic Psychol. Behav. 2(4), 181–196 (1999)

A Time-slot Based Coordination Mechanism Between WiFi and IEEE 802.15.4

Xiao Wang$^{(\boxtimes)}$ and Kun Yang

School of Computer Science and Electronic Engineering, University Of Essex,
Colchester CO4 3SQ, United Kingdom
{xwangai,kunyang}@essex.ac.uk

Abstract. Both WiFi and IEEE 802.15.4 are wide-spread wireless communication technologies utilized particularly in indoor environments such as home, offices and buildings. Since these wireless networks are normally operating in the license-free Industrial Scientific Medical (ISM) frequency band and share the same wireless medium, where no coordination mechanism is available to guarantee communications, unavoidably it leads to interference among them. In order to address this problem, this paper proposes a time-slot based coordination mechanism between WiFi and IEEE 802.15.4, which is achieved by introducing Access Suppression Notification (ASN) frame into IEEE 802.15.4. The static scheduling algorithm is designed and the experiments show that proposed coordination mechanism demonstrates an overall improvement in both IEEE 802.15.4 packet loss ratio and packet transmission rate.

Keywords: WiFi · IEEE 802.15.4 · Cross interference
Coordination mechanism · Time-slot

1 Introduction

Wireless networks are making life easier, smarter and more convenient. With the development of Internet of Thing (IoT), various network technologies are introduced to meet different performance requirements in term of data throughput, communication distance and power consumption. One popular wireless technology is IEEE 802.15.4 which is a Low-Rate WPAN (LR-WPAN) standard in physical and link layers with features of low-power consumption, flexible topology, high receiver sensitivity and long commutation distance [6]. Base on IEEE 802.15.4, many wireless protocol network stacks are developed, such as ZigBee [3], 6LoWPAN [1], Rime [7], Thread [2] and so on. WiFi (IEEE 802.11) is a widely adopted WLAN technology that is a simple and universal way to connect wireless devices, e.g. smart phones, laptops, TVs and digital camera to the Internet. Thus, it is necessary to have both IEEE 802.15.4 and WiFi available in IoT. However, most of those wireless networks operate in the license-free Industrial

© ICST Institute for Computer Sciences, Social Informatics and Telecommunications Engineering 2019
Published by Springer Nature Switzerland AG 2019. All Rights Reserved
X. Liu et al. (Eds.): ChinaCom 2018, LNICST 262, pp. 235–244, 2019.
https://doi.org/10.1007/978-3-030-06161-6_23

Scientific Medical (ISM) frequency band and share the same wireless medium, where no coordination mechanism is available to guarantee communications, unavoidably leading to interference among them. Many works are investigated to address the interference by analysing Packet Error Rate (PER) [8,10,15,20]. IEEE 802.15.4 has negligible impacts on WiFi [12], while IEEE 802.15.4 is excessively interrupted by WiFi network due to the much higher transmission power of WiFi [9,14,16]. In some worst-case scenarios, WiFi devices possibly jams IEEE 802.15.4 communications [9].

There are extensive studies about the interference mitigation among the wireless technologies in 2.4 GHz band. For example, [19] suggests that a wireless node can detect the interference and retreat from the interference by dynamic channel switching and dynamic power management. Similarly, [13] claims that dynamically adjusting radio transmission power helps maximize spectrum utilization and avoid interference. However, in some situations where these two radio modules are designed into one box and near each other, IEEE 802.15.4 channel switching and WiFi radio power adjusting are not an effective way to mitigate the interference [18]. Thus, a CTS (Clear To Send) blocking way [9] is proposed to protect the IEEE 802.15.4 communication, in which a WiFi CTS frame is sent to block WiFi traffic before transmitting any IEEE 802.15.4 data. Reference [11] optimizes the idea of [9] and resolved possible hidden nodes problem by introducing a helper AP. Unlike the method in [9], the node sends RTS (Request To Send) instead of CTS. However, it is not realistic to put an extra WiFi radio in to a tiny IEEE 802.15.4 sensor. Our method is to introduce time slot concept to coordinate their transmissions. Namely, they are transmitted at different time slots.

The major contributions of the paper are as follows. Firstly the paper proposes a time-slot based method to separate these two types of traffic. Secondly a new control frame called Access Suppression Notification (ASN) is introduced into IEEE 802.11.5. finally a unified coordination control architecture is designed and practically implemented to fulfil the above method and control.

The following paper is organized as follows. Section 2 introduces the newly proposed ASN frame, based on which Sect. 3 presents the overall time-slot frame structure and coordination control architecture. A static time-slot scheduling algorithm is also described in this section. Section 4 gives an experimental evaluation of the overall proposal with respect to the IEEE 802.15.4 performance under WiFi traffic from two aspects: Packet Loss Ratio (PLR) and Packet Transmission Rate (PTR). The paper is concluded in Sect. 5.

2   ASN Command Frame

The Access Suppression Notification (ASN) frame is designed in IEEE 802.15.4 to be sent by coordinator to notify its child nodes that following specific duration is suppressed to access medium. Such a duration is called Suppressed Medium Access Duration (SMAD). In order to be compatible with the nodes that do not support ASN MAC command, IEEE 802.15.4 communication is not thoroughly blocked when in SMAD, but access medium in SMAD may experience reduced channel quality and high PLR, because of the possible cross interference from WiFi or other wireless technology. In other word, the transmission in SMAD is more like to be interrupted. The ASN frame from the coordinator allows child nodes to pick advised, safe and guaranteed time slots when sending data.

According to [4], four frame types are defined in IEEE 802.15.4, including Beacon frame, Data frame, ACK frame and MAC command frame. There are nine command frame types, with command frame identifier from 0x01 to 0x09. The ASN frame is added as a command frame that uses a reserved command frame identifier 0x0a.

Table 1. ASN Command frame format

Bytes: 7/9	1	1	Variable
MHR field	Command frame identifier (0x0a)	Suppressed medium access duration	Schedule list

This ASN MAC command format is illustrated in Table 1:

- The Frame Type subfield of the Frame Control field shall be set to three, indicating MAC command frame.
- The Security Enable field, Frame Pending field and Acknowledgement request of the Frame Control field shall be set to zero.
- The PAN ID Compression subfield of the Frame Control field shall be set to one. In accordance with this value of the PAN ID compression subfield, the Source PAN Identifier field shall be omitted. If the Destination PAN Identifier is not a broadcast PAN identifier (i.e., 0xffff), the ASN is just limited in one PAN. Otherwise, all the nearby PANs are affected.
- The Source Address field should contain the short address of the coordinator.
- The Supressed Medium Access Duration is an unsigned 8-bit integer, covering 1 255 ms. 0 is reserved. The time beyond SMAD is called Normal Medium Access Duration (NMAD).
- Schedule list is a serial of unsigned 8-bit integers, covering 1 255 ms for each byte. 0 is reserved. This list indicates several possible moments that next ASN frame will occur. This helps to implement Energy efficient algorithms on child nodes with ASN MAC command support. The Schedule list not always exists. An ASN frame without schedule list means the next ASN frame may be sent anytime. The child nodes, who enabled ASN support, must listen this MAC command in the entire NMAD.

3   Proposed Time-slot Based Coordination Mechanism

3.1   Time-slot Design Based on WiFi CTS and IEEE 802.15.4

CTS frame is used to block nearby WiFi nodes from sending any data, except for the one that is chosen and allowed to send without worry about interference from other nodes. In this paper, CTS frame is forced to send and block nearby WiFi communications. With the introduction of ASN frame, if these two frames is sent in a coordinated way, they can be used to synchronize WiFi and IEEE 802.15.4 traffic and allow these two wireless traffic to access medium in separated time slots without interference.

Figure 1 depicts how to realize time slots by using CTS and ASN command frames. Basically, IEEE 802.15.4 traffic is depressed when in WiFi slot by Sending an ASN frame. WiFi CTS should be sent just before the finish of the SMAD, so WiFi transaction is blocked and IEEE 802.15.4 starts working without worry about cross interference from WiFi. Before WiFi activates again, another ASN frame shall be sent to curb IEEE 802.15.4 communication. Thus the WiFi and IEEE 802.15.4 are separated in different time slot. The time slots with only WiFi traffic is WiFi slots and the slots with only IEEE 802.15.4 traffic is called 802.15.4 slots. By dynamic and cautious adjustment of both kinds of time slots, time-slot based resource scheduling algorithms can be implemented.

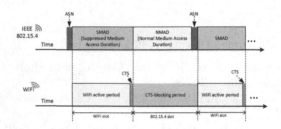

Fig. 1. Time-slot based coordination by using WiFi CTS and IEEE802.15.4 ASN

3.2   Overall Coordination Control Architecture

Figure 2a describes the overall architecture of the gateway and IEEE 802.15.4 end node. The gateway, integrating WiFi radio module and IEEE 802.15.4 radio module, runs a Coordination Controller (CC). By utilizing CTS frame and proposed ASN frame, CC is able to schedule WiFi and IEEE 802.15.4 traffic into separated time slots.

The detail design of the architecture is in Fig. 2b. Generally, CTS is directly controlled by hardware. Thus OS does not provide any interface to send CTS. In order to send arbitrary CTS frame, OS kernel modifications are necessary. The implementation is based on AR9331 [5] SoC with OpenWRT OS. AR9331 integrates a 802.11 b/g/n radio module. As shown in Fig. 2b, MAC80211 and ath9k driver for the WiFi module are modified to enable WiFi CTS frame injection. The proposed IEEE 802.15.4 ASN frame is implemented in Contiki-OS, thanks to its flexible radio control, rich features, support of simulation. We design new Radio Duty-Cycle (RDC) layer to enable ASN support. the RDC layer in IEEE 802.15.4 coordinator module provides control API by serial port, allowing CC to send signal to it. Once the RDC layer receives the signal from serial port, it sends out proper ASN fame and blocks its upper MAC layer data transmission for a specific period. The transmitted ASN frame will be received by IEEE 802.15.4 end node and the RDC layer in the end node will also block its upper MAC layer data transmission. Thus the ASN is implemented and able to suppress IEEE 802.15.4 traffic. Specifically, the ASN command frame is implemented on CC2538 [17] chip. The CC2538 chip is connected to the AR9331 by serial port. Test applications are also designed to evaluation the coordination mechanism performance.

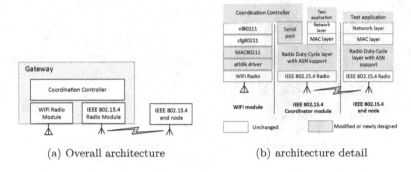

(a) Overall architecture   (b) architecture detail

Fig. 2. Coordination control architecture

3.3 Static Time-slot Scheduling

The key is to decide the length of WiFi slot and that of 802.15.4 respectively. Since this paper focuses on the description of the operational procedure of the coordination scheme it adopts a static method to decide the lengths. Namely, they are fixed to 32 ms, as depicted in Fig. 3.

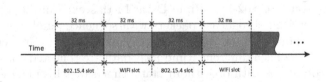

Fig. 3. Static Time-Slot Scheduling

4   Experimental Evaluation

4.1   Experiment Setup

Figure 4a demonstrates a dedicated gateway implementation dedicated designed for static time-slot scheduling algorithm. Specifically, the gateway is built with

three components: (A) CC2538 radio module running specially designed IEEE
802.15.4 MAC with ASN command frame support, (B) dedicated WiFi module
with modified kernel and able to send CTS control frames, (C) normal WiFi
module running WiFi Access Point (AP) and generating WiFi traffic. Of these
three modules, Module B and C are able to integrated into one WiFi module. The
reason why these two modules are separated is to get more accurate experiment
results by minimizing impact of limited CPU computing power.

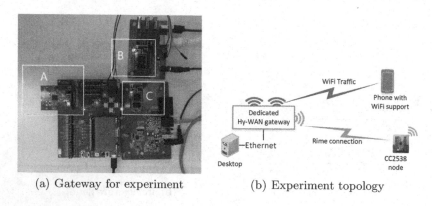

(a) Gateway for experiment   (b) Experiment topology

Fig. 4. Test bed

The experiment adopts Rime as an upper network layer on IEEE 802.15.4. As
shown in Fig. 4b, the experiment platform consists of a desktop PC, a gateway,
a phone and a CC2538 node. The desktop is used to access the gateway and
also retrieve statics from the gateway. The WiFi traffic is generated by iperf3
which is installed in both the gateway and the phone. IEEE 802.15.4 traffic is
generated and recorded by customized program running on CC2538 module. We
designed two programs to evaluate Rime PLR and PTR. They are:

- Packet Loss Ratio of 802.15.4 Rime Data: In this experiment, the CC2538
  node sends out Rime broadcast packets with sequence number every 15 ms,
  and the packet number is set to 1000 for each run. Three runs are performed at
  each WiFi speed. Then different Rime PLR data are collected and calculated
  under two conditions with or without static time-slot scheduling algorithm.
- Unicast Packet Transmission Rate of 802.15.4 Rime Data: In this experiment,
  the CC2538 node sends out Rime reliable unicast packet to the gateway.
  The unicast packets are transmitted under the guarantee of an up-four-times
  retransmission mechanism. The transmission is successful if the packet is
  transmitted within four retransmissions; otherwise, the transmission is timeout. The successful transmissions are counted with a time period of 100 seconds in each run. Similarly, three runs are performed at each WiFi speed.
  Then Rime PTR with static time-slot scheduling algorithm is compared with
  the PTR without the algorithm.

4.2   Packet Loss Ratio of 802.15.4 Rime Data

As shown in Fig. 5a, this experiment introduces desired WiFi speed and real WiFi speed concepts. The desired WiFi speed means the WiFi traffic that the iperf3 program intends to generate. However, the real generated wifi traffic may be different, because the channel capacity is limited and there are possibly multiple wireless technologies sharing the same medium and they may compete with each other. The experiment is conducted in WiFi channel 6 and ZigBee channel 17, and these two channels overlap in spectrum and are expected to interfere each other. Desired WiFi speed is set from 0 to 40 Mbits/s with a speed step of 2 Mbits/s. At each WiFi speed, statistics are collected and Rime PLR is calculated. Meanwhile, the real WiFi speed is also recorded. Similarly, we test and collect data when running static time slot based resource scheduling algorithm. Thus, Fig. 5a is acquired.

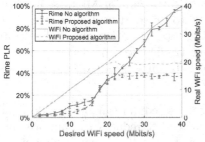
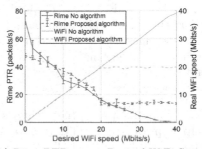

(a) Rime PLR versus Desired WiFi Speed   (b) Rime PTR versus Desired WiFi Speed

Fig. 5. Experimental results

The figure depicts that overall Rime PLR with proposed algorithm is lower than that without any algorithm. More detailed analysis can be done by dividing the graph into three parts according to various desired WiFi speed ranges:

- Effective phase (0–16 Mbits/s): WiFi traffic is not crowded and the real WiFi speed in both situations is basically equal with the desired WiFi speed. Average of Rime PLR without scheduling is 7.41%, while Average of Rime PLR with scheduling is 4.39%. The proposed algorithm is obviously effective by reduce Rime PLR from 7.41% to 4.39% without affecting WiFi traffic.
- Transition phase (18–22 Mbits/s): WiFi traffic is medium. Both PLR lines raise rapidly over this range. There is no obvious difference between two PLR lines. At the end of this range, scheduled WiFi time slot are almost fully used.
- Stable phase (24–40 Mbits/s): WiFi traffic becomes more crowded. The PLR without proposed algorithm keeps growing rapidly, while the PLR with the scheduling algorithm keeps at around 37%. Proposed algorithm shows improved performance of Rime PLR. When taking real WiFi speed into consider, the low PLR for proposed algorithm is at the cost of WiFi speed reduction. The scheduling algorithm periodically sends out CTS frames, which

blocks WiFi traffic periodically and creates time slots for Rime to send out broadcast. With scheduling algorithm, WiFi has used up all allocated time slot, thus real WiFi limited at around 19 Mbits/s. Such a mechanism works like applying protection to Rime traffic when there are two much WiFi traffic.

Although the figure shows that static time-slot based resource scheduling algorithm has an improved performance in term of Rime PLR, there still are some limitations and flaws on algorithm implementation. In an ideal implementation of the algorithm, a very low PLR, such as 2%, over all the WiFi speed ranged is expected, because the ideal scheduling will effectively separate Rime traffic from WiFis and they will never interfere each other. The experiment shows that high WiFi traffic affects the scheduling and results in an increase of PLR.

4.3   Unicast Packet Transmission Rate of 802.15.4 Rime Data

Different from WiFi of which speed is measured in unit of Mbits/s, however, in IEEE 802.15.4, application message tends to be short and only care about whether the message is successful transmitted or not, rather than the bitrate. Thus it is more reasonable to measure Rime traffic speed by successful transmitted packets per second (packets/s). Figure 5b can be study over four desired WiFi speed ranges:

- ASN suppress phase (0–8 Mbits/s): Static scheduling algorithm does not help improve Rime PTR performance, instead it reduces the performance. This is caused by periodical ASN frames that static scheduling algorithm sends to suppress Rime traffic. When no scheduling algorithm is introduced, Rime PTR reach a fast speed at 73.32 packets/s while desired WiFi speed is zero, but PTR drop dramatically once any WiFi traffic is introduced. With the scheduling algorithm, PTR is limited at 48.08 packets/s, and this value slightly drop when there is a little WiFi traffic. These two lines meet and merge quickly when WiFi traffic at 6 Mbits/s.
- Effective phase (10–16 Mbits/s): Advantages of the static scheduling algorithm starts showing up in this range. The scheduling algorithm separates WiFi and Rime traffic in different time slots. This improves PRT by avoiding potential cross interference.
- Transition Phase (18–20 Mbits/s): with proposed algorithm, the WiFi slot usage starts reaching its limit and PTR decreases rapidly, and very soon it is stable at 13.9 packets/s.
- Stable phase (22–40 Mbits/s): Under the static scheduling, WiFi has fully used its time slot and reached an stable state that real WiFi speed is about 19 Mbits/s and PTR is about 13.9 packets/s. If without scheduling, the PTR keeps dropping until almost 0 packets/s with the increase of the WiFi traffic. The proposed algorithm improves Rime PTR at the cost of reduction of real WiFi speed. The scheduling protects Rime traffic and avoids that WiFi takes up all the medium. If no scheduling protection, WiFi traffic blocks Rime transmission and causes Rime network failure.

In summary, the scheduling algorithm implementation demonstrates an overall improvement of Rime unicast PTR. However, the implementation still needs to be improved. An ideal implementation should give an stable Rime PTR that should be half of maximum PTR. In this experiment, the maximum PTR is 73.32 packets/s and ideal implementation should give a stable PRT at about 36.66 packets/s, but the PTR drops to around 14.9 packets/s when WiFi traffic uses up all its scheduling time slot. In result, the algorithm works as expected in low WiFi traffic situation, but it is not efficient enough when there is too many WiFi traffic.

5   Conclusion

This paper introduces ASN fame into IEEE 802.15.4 and proposes a time-slot based coordination mechanism between WiFi and IEEE 802.15.4. ASN frame is implemented on Contiki-OS by designing new RDC sublayer. Rime is chosen as a network layer on top of IEEE 802.15.4. The mechanism is evaluated in term of Rime broadcast PLR and Rime unicast PTR over various WiFi transmission speed. According to the experiment, proposed coordination mechanism demonstrates an overall improvement in both IEEE 802.15.4 Rime PLR and Rime PTR. Especially, the mechanism shows an distinct improvement in effective phase. However, current implementation still has limitation, because there is no obvious improvement in transition phase. In stable phase, the Rime performance is also improved but at cost of reducing WiFi speed. Currently implementation of coordination mechanism is in an static manner. In the future, the additive scheduling algorithms will be studied.

References

1. Contiki: The open source os for the internet of things (2018). Accessed 20 March 2018
2. Thread group (2018). Accessed 20 March 2018
3. The zigbee alliance (2018). Accessed 25 March 2018
4. Part: 15.4: Wireless medium access control (mac) and physical layer (phy) specifications for low-rate wireless personal area networks (wpans). IEEE Std. 802.15.4TM-2006 (2006)
5. Atheros Communications, Inc.: AR9331 Highly-Integrated and Cost Effective IEEE 802.11n 1x1 2.4 GHz SoC for AP and Router Platforms, 12 (2010)
6. De Nardis, L., Di Benedetto, M.G.: Overview of the IEEE 802.15. 4/4a standards for low data rate wireless personal data networks. In: 4th Workshop on Positioning, Navigation and Communication, 2007. WPNC'07, pp. 285–289. IEEE (2007)
7. Dunkels, A.: Rime-a lightweight layered communication stack for sensor networks. In Proceedings of the European Conference on Wireless Sensor Networks (EWSN), Poster/Demo session, Delft, The Netherlands (2007)
8. Golmie, N., Cypher, D., Rébala, O.: Performance analysis of low rate wireless technologies for medical applications. Comput. Commun. 28(10), 1266–1275 (2005)

9. Hou, J., Chang, B., Cho, D.K., Gerla, M.: Minimizing 802.11 interference on zigbee medical sensors. In Proceedings of the Fourth International Conference on Body Area Networks, p. 5. ICST (Institute for Computer Sciences, Social-Informatics and Telecommunications Engineering) (2009)
10. Howitt, I., Gutierrez, J.A.: IEEE 802.15. 4 low rate-wireless personal area network coexistence issues. In Wireless Communications and Networking, 2003. WCNC 2003. 2003 IEEE 3, pp. 1481–1486. IEEE (2003)
11. Ishida, S., Tagashira, S., Fukuda, A.: Ap-assisted cts-blocking for wifi-zigbee coexistence. In: 2015 Third International Symposium on Computing and Networking (CANDAR), pp. 110–114. IEEE (2015)
12. Myoung, K.-J., Shin S.-Y., Park, H.-S., Kwon, W.-H.: IEEE 802.11 b performance analysis in the presence of ieee 802.15. 4 interference. IEICE Trans. Commun. 90(1), 176–179 (2007)
13. Phunchongharn, P., Hossain, E., Camorlinga, S.: Electromagnetic interference-aware transmission scheduling and power control for dynamic wireless access in hospital environments. IEEE Trans. Inf. Technol. Biomed. 15(6), 890–899 (2011)
14. Rihan, M., El-Khamy, M., El-Sharkawy M.: On zigbee coexistence in the ism band: Measurements and simulations. In: 2012 International Conference on Wireless Communications in Unusual and Confined Areas (ICWCUCA), pp. 1–6. IEEE (2012)
15. Shin, S.Y., Park, H.S., Kwon, W.H.: Packet error rate analysis of IEEE 802.15. 4 under saturated IEEE 802.11 b network interference. IEICE Trans. Commun. 90(10), 2961–2963 (2007)
16. Sikora, A., Groza, V.F.: Coexistence of ieee802. 15.4 with other systems in the 2.4 ghz-ism-band. In: Proceedings of the IEEE, Instrumentation and Measurement Technology Conference, 2005. IMTC 2005, vol. 3, pp. 1786–1791. IEEE (2005)
17. Texas Instruments Inc.: CC2538 Powerful Wireless Microcontroller System-On-Chip for 2.4-GHz IEEE 802.15.4, 6LoWPAN and ZigBee Application, 4 (2015)
18. Wang, X., Yang, K.: A real-life experimental investigation of cross interference between wifi and zigbee in indoor environment. In: 2017 IEEE International Conference on Internet of Things (iThings) and IEEE Green Computing and Communications (GreenCom) and IEEE Cyber, Physical and Social Computing (CPSCom) and IEEE Smart Data (SmartData), pp. 598–603. IEEE (2017)
19. Wenyuan, X., Ma, K., Trappe, W., Zhang, Y.: Jamming sensor networks: attack and defense strategies. IEEE Netw. 20(3), 41–47 (2006)
20. Yoon, D.G., Shin, S.Y., Kwon, W.H., Park, H.S.: Packet error rate analysis of ieee 802.11 b under ieee 802.15. 4 interference. In: Vehicular Technology Conference, 2006. VTC 2006-Spring. IEEE 63rd, vol. 3, pp. 1186–1190. IEEE (2006)

Modeling a Datacenter State Through a Novel Weight Corrected AHP Algorithm

Weiliang Tan, Yuqing Lan[(⊠)], and Daliang Fang

School of Computer Science and Technology, Beihang University, Beijing
100191, China
{tanweiliang, lanyuqing, fdll5}@buaa.edu.cn

Abstract. Analytic Hierarchy Process (AHP) is an effective algorithm for determining the weight of each module of a model. It is generally used in the process of multi-indicator decision making. But, when using AHP for evaluation, it is inevitable to introduce the evaluator's subjectivity. In this paper, an algorithm based on Bayes' formula is proposed for correcting the weights determined by the analytic hierarchy process. This algorithm can reduce the subjectivity of the evaluator introduced during the evaluation process. At the same time, the common operational indicators of a data center are summarized and classified. I chose some relatively important indicators and established an evaluation model for the operational status of the data center. The weight of the modules of the established model is corrected using this improved algorithm.

Keywords: Analytic hierarchy process · Datacenter indicators Cloud datacenter evaluation model · Bayes' formula

1   Introduction

1.1   A Subsection Sample

The concept of cloud computing [7] was first proposed by Google CEO Eric Schmidt at the 2006 Search Engine Conference. Cloud computing typically provides computing, networking, storage, and other resources to users on a rental basis. The carrier of cloud computing is a wide variety of cloud datacenters. After decades of development, the traditional computing center has gradually transformed from the traditional "computer room" mode of storing mainframes to the modern large-scale cloud computing center with highly standardized modularization. [8]. With the development of virtualization [9, 10] and other technologies, the datacenter is the key to ensuring that users can use cloud computing anytime, anywhere, like using water, electricity, gas and other resources as a public service infrastructure. Therefore, how to ensure the safe and stable of the data center has become a topic worthy of further study.

At present, the monitoring method of the operation state of the data center generally sets the threshold value for each key indicator, and if the threshold value is exceeded or continuously exceeds the target value within a time window, the operation of alarm will be taken [1]. Haryadi S. Gunawi, Agung Laksono and others classified failures into upgrade, network failures, bugs, misconfigurations, traffic load, cross-service

X. Liu et al. (Eds.): ChinaCom 2018, LNICST 262, pp. 245–251, 2019.
https://doi.org/10.1007/978-3-030-06161-6_24

dependencies, power outages, security attacks, human errors, storage failures, natural factors, etc. According to the data published by the data center, the frequency and impact of the above faults were statistically analyzed [2]. Dong Seong Kim1, Fumio Machida develops an availability model of a virtualized system using continuous time Markov chains [4].

This paper first summarizes the key operational data indicators of the datacenter. I use principal component analysis [5] to select key impact factors in the monitoring indicators commonly used in datacenters and constructed an evaluation model of the operational status of the datacenter using the Analytic Hierarchy Process [3]. Aiming at the subjective problem of the analytic hierarchy process, I proposed an improved AHP scheme based Bayes' theorem.

2   Cloud Datacenter Operation and Maintenance Indicators

The monitoring indicators of the entire cloud data center are very complex. After a long period of investigation and research, the daily monitoring data of the data center are divided into two categories, which are operation indicators and operation and maintenance indicators. Operational indicators refer to the daily operations of the data center, mainly including tax, site cost, human cost, power cost, water cost and other indicators. The operation indicators are not detailed in this paper. The daily operation and maintenance monitoring of the data center mainly includes eight indicators: moving ring indicators, physical machine indicators, virtual machine indicators, middleware, security, log system, operation and application monitoring, operation and maintenance human management. The indicators and their sub-indicators are listed in Table 1:

Table 1. Operation and maintenance indicators and their sub-indicators

Types	Indicators
Power and environmental indicators	UPS, Distribution Cabinet, Precision air conditioner, Fresh air equipment, Air quality indicators
Physical machine indicators	CPU temperature, CPU usage, Power status, Network bandwidth capacity, Memory usage
Virtual machine indicators	vCPU usage, Network usage, Network throughput, Network delay, Disk operation
Middleware	SQL database, NoSQL database, Message queue, Docker
Log system	Hardware log, Firewall log, Operation log
Safety	Network firewall, Intrusion detection, Gateway monitoring, VPN, Vulnerability scanning

However, building the datacenter operational state evaluation model directly with the above indicators will be to complex. At the same time, the weight of the model components cannot be well determined. Moreover, the calculation work of AHP will increases dramatically with the increase of criteria layer targets. Furthermore, there are also problems of consistency and objectivity when adopting the analytic hierarchy process. In this paper, the principal component analysis method is used to solve the problem of too many targets in the criterion layer, and the consistency evaluation method is used to remove the evaluation matrix when it is inconsistent. The weights obtained by the evaluation are corrected using a weight correction algorithm based on the Bayes' formula.

3   Modeling Process and Method

3.1   Objective Measure

Evaluation with AHP will inevitably disturbed by the subjectivity of the evaluator. Subjectivity can significantly affect the accuracy of the assessment result. Therefore, the key step in the evaluation using AHP is to conduct the consistency test. The process of the consistency check is to calculate the maximum eigenvalue of the evaluation matrix and its corresponding eigenvector. Use the formula.

$$C.I. = (\lambda\_max - n)/(n - 1) \tag{1}$$

$$C.R. = (C.I.)/(R.I.) \tag{2}$$

to calculate the value of C.I. and R.I. Where n is the number of items evaluated. C.I. is the consistency indicator, R.I. is the average random consistency indicator, and C.R. is the ratio consistency [6]. If the value of the ratio consistency is less than 0.1, it is generally considered that the weight vector constructed by the AHP method is reasonable. The consistency test can only indicate that the evaluation is more reasonable for the distribution of each weight, but this does not mean that the evaluation is objective. So how should we define objectivity? In the process of evaluating multiple evaluation indicators of a system, we can assume that the evaluator with complete domain knowledge evaluates each indicator reasonably. In the case of the existence of the objective weight of the indicator i, the subjective weight obtained by the different evaluators (suppose the number of evaluators is k) for the evaluation of the indicator i should obey the normal distribution of the objective weight value. At this point, the process of using AHP to score the indicators can be abstracted into selecting one subjective and most important indicator from the k indicators. We assume that in an evaluation process, the evaluators' subjective degree is the same for each evaluation indicator, that is, the standard deviation of the normal distribution of each indicator is σ.

According to the nature of the normal distribution, the probability that the index i is more important than the index j in an evaluation process is [11]:

$$\left(\int_{-\infty}^{\frac{(w_{iob} + w_{job})}{2}} \frac{1}{\sigma\sqrt{2\pi}} e^{-\frac{(x - w_{iob})^2}{2\sigma^2}} dx \right)^2 \tag{3}$$

The derivation process is omitted here. It is easy to know that (3) is a function related only to w_{iob} and w_{job}. In the system of promotion to multiple indicators, the probability i is evaluated as the most important probability is $P(w_{iob}, w_{mob})$. Where

w_{mob} is the weight of the objective most important indicator among all the indicators involved in the evaluation. Let event T_i be the indicator A is chosen as the most important indicator in an evaluation process. Then, $P(T_i) = P(w_{iob}, w_{mob})$ is indicator i's weight w_i. That is $w_i = P(T_i)$.

3.2   Weight Correction Algorithm Based on Bayes Formula

Defining that event O is evaluating objectively in an evaluation process, then event $\overline{O}$ is evaluating not objectively in an evaluation process. As can be seen from the above discussion, $P(T_i) = w_i$. Where $P(T_i)$ is a unconditional probability. This means that this evaluation may be objective or non-objective. Then we can calculate the prior probability from the posterior probability. According to Bayes' formula:

$$P(T_i|O) = \frac{P(T_i)P(O|T_i)}{P(O)} \tag{4}$$

$P(O|T_i)$ refers to the objective probability under the premise of obtaining the weight of indicator i. We assume that there is a linear correlation between the consistency indicator C.I. and the objective probability of an assessment. So, $P(O|T_i) = \theta * C.I. + b$. Correlation coefficient θ is less than 0. $P(O)$ in the above formula refers to the degree of overall objectivity of an evaluating process, and here refers to the mean probability of objectiveness of all participating evaluators. According to this, the following algorithm can be proposed. Calculate the ratio consistency C.R. for each evaluation matrix and judge whether the value is less than 0.1, and select the evaluation matrix that conforms to the consistency to perform the weight correction. Calculate the weight of each indicator by AHP, and record it as W $(w_1, w_2, w_2 \ldots w_n)$. For each w_i in W use the formula (5) to correct its weight.

$$P(T_i|O) = \frac{w_i * (\theta * C.I. + b)}{P(O)} \tag{5}$$

4   Analysis and Verification

I invite cloud computing service providers, cloud computing experts, and research scholars to evaluate the importance of indicators in A by means of questionnaires. In the end, I collected a total of 23 valid evaluation samples. I use principal component analysis to select the three factors with the highest cumulative contribution rate, namely, power and environmental indicators, physical machine indicators, and virtual machine indicators for algorithm verification. I will record the three indicators taken as a_1, a_2, a_3, and use the 1–9 scale method to evaluate the three indicators. The detailed meaning of the scale method is listed in Table 2 [5].

Table 2. Importance degree

Difference in importance level	Description of importance degree
1	Equally important
3	Slightly important
5	Obviously important
7	Quite important
9	Extremely important

The evaluation matrix that can pass the consistency test is used as an effective evaluation sample. The weight result calculated by the AHP is shown in Fig. 1. Where the ordinate is the weight of each indicator and the abscissa is the ith assessment:

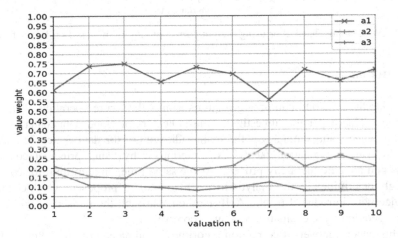

Fig. 1. Raw weight calculated using AHP

The weights obtained after the correction are shown in Fig. 2 (Here, the value of the correlation coefficient θ is -1, and the value of b is taken as 2):

The variance of the raw data of the three indicators a_1, a_2, a_3 is [0.0033899, 0.00238617, 0.00085738]. The variance of the three data processed by the algorithm is reduced to [0.00289193, 0.0028247, 0.00083877], which is reduced by 17.22%, 15.52%, and 2.22%, respectively. The volatility of this three indicators have been reduced to some extent after being revised. According to the discussion above, the less volatility means subjective. This shows that this algorithm effectively reduces the subjectivity in the evaluation of the analytic hierarchy process. Finally, the weights of the three indicators a_1, a_2, a_3 are [0.68200825 0.21602911 0.10196264]. Using these weights can easily constructs a data center model.

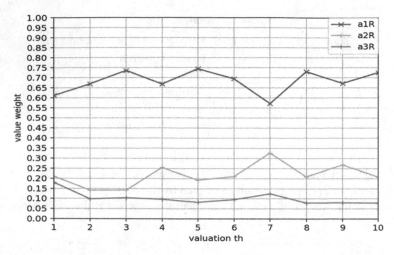

Fig. 2. Weight calculated after using weight correction algorithm based on Bayes formula

5 Conclusion

The availability of cloud computing requires a smooth, secure operation of the cloud data center. This paper first investigates the indicators of cloud data center operation and maintenance, and summarizes the main indicators of the data center. The weights of the individual evaluation factors are determined by using principal component analysis and analytic hierarchy process. At the same time, an algorithm for correcting the weight determined by the analytic hierarchy process is proposed, which can reduce the subjectivity introduced by the AHP method. Through the revised weights, a datacenter operation state evaluation model is constructed.

In this paper, the weight correction algorithm is validated only in the case of three evaluation factors, and no further verification is entered in the case of more indicators and greater volatility of weights. In addition, the values of the parameters θ, b can be determined using a gradient descent algorithm. These related work can continue in the future.

References

1. LNCS Homepage. http://www.springer.com/lncs. Accessed 21 Nov 2016
2. Meng, S., Liu, L., Wang, T.: State monitoring in cloud datacenters[J]. IEEE Trans. Knowl. Data Eng. **23**(9), 1328–1344 (2011)
3. Gunawi, H.S., Hao, M., Suminto, R.O., et al.: Why does the cloud stop computing?: Lessons from hundreds of service outages[C]. In: Proceedings of the Seventh ACM Symposium on Cloud Computing, pp. 1–16. ACM (2016)
4. Saaty, T.L.: Analytic hierarchy process[M]. Encyclopedia of operations research and management science, pp. 52–64. Springer, Boston, MA (2013)

5. Kim, D.S., Machida, F., Trivedi, K.S.: Availability modeling and analysis of a virtualized system[C]. In: 15th IEEE Pacific Rim International Symposium on Dependable Computing, 2009. PRDC 2009, pp. 365–371. IEEE (2009)
6. Jolliffe, I.: Principal component analysis[M]. International encyclopedia of statistical science, pp. 1094–1096. Springer, Berlin, Heidelberg (2011)
7. Saaty, T.L.: How to make a decision: the analytic hierarchy process[J]. Eur. J. Oper. Res. **48** (1), 9–26 (1990)
8. Armbrust, M., Fox, A., Griffith, R., et al.: A view of cloud computing[J]. Commun. ACM **53** (4), 50–58 (2010)
9. Fox, A., Griffith, R., Joseph, A., et al.: Above the clouds: a berkeley view of cloud computing[J]. Dept. Electr. Eng. Comput. Sci.**28**(13) (2009). University of California, Berkeley, Rep. UCB/EECS
10. Rimal, B.P., Choi, E., Lumb, I.: A taxonomy and survey of cloud computing systems[C]. In: Fifth International Joint Conference on INC, IMS and IDC, 2009. NCM2009, pp. 44–51. IEEE (2009)
11. Barroso, L.A., Clidaras, J., Hölzle, U.: The datacenter as a computer: an introduction to the design of warehouse-scale machines[J]. Synth. Lect. Comput. Arch. **8**(3), 1–154 (2013)
12. Anderson, T.W., Anderson, T.W., Anderson, T.W., et al.: An introduction to multivariate statistical analysis[M]. Wiley, New York (1958)
13. Wang, L., Khan, S.U.: Review of performance metrics for green data centers: a taxonomy study[J]. J. Supercomput. **63**(3), 639–656 (2013)

Research on Semantic Role Labeling Method

Bo Jiang[(✉)] and Yuqing Lan

School of Computer Science and Technology, Beihang University, Beijing
100191, China{jiangbo1,lanyuqing}@buaa.edu.cn

Abstract. Semantic role labeling task is a way of shallow semantic analysis. Its research results are of great significance for promoting Machine Translation [1], Question Answering [2], Human Robot Interaction [3] and other application systems. The goal of semantic role labeling is to recover the predicate-argument structure of a sentence, based on the sentences entered and the predicates specified in the sentence. Then mark the relationship between the predicate and the argument, such as time, place, the agent, the victim, and so on. This paper introduces the main research directions of semantic role labeling and the research status at home and abroad in recent years. And summarized a large number of research results based on statistical machine learning and deep neural networks. The main purpose is to analyze the method of semantic role labeling and its current status. Summarize the development trend of the future semantic role labeling.

Keywords: Semantic role labeling · Semantic analysis · Deep neural networks

1 Introduction

Natural language processing is an interdisciplinary subject that integrates multiple disciplines such as computer science and linguistics. It is one of the core research directions of artificial intelligence. It has a history of more than 70 years since the last century. The research on natural language understanding has made great breakthroughs through the efforts of many generations. Natural language understanding is broadly divided into three levels, lexical analysis, syntactic parsing and semantic analysis. Chinese lexical analysis mainly include Chinese word segmentation, named entity recognition and word sense disambiguation. The main task of syntactic parsing is to identify the syntactic structure of a sentence. And the semantic analysis task is to let the machine understand the meaning of natural language, which is divided into shallow semantic analysis and deep semantic analysis. The goal of deep semantic analysis is to understand the content of the entire sentence. Convert the sentence into a formal representation is the general approach. Its main methods are semantic analysis based on knowledge, supervised semantic analysis and semi-supervised or unsupervised semantic analysis. However, deep semantic analysis still faces many difficulties. For example, the mapping between ordinary text and relational predicates is more difficult to solve, and the results of semantic analysis for the open field are not satisfactory. Shallow semantic analysis proposes a simple solution, its goal is not to completely analyze the meaning of the whole sentence. It only specifies which components of the

X. Liu et al. (Eds.): ChinaCom 2018, LNICST 262, pp. 252–258, 2019.
https://doi.org/10.1007/978-3-030-06161-6_25

words or phrase are in the sentence, and then identifies the component of the words in the sentence. Compared with deep semantic analysis, it can perform semantic analysis more quickly and obtain higher correct rate results. Semantic role annotation (SRL) is a shallow semantic analysis technique. It can achieve higher efficiency and correct rate in semantic analysis if compared with deep semantic analysis. Semantic role labeling (SRL) is a shallow semantic analysis technique.

Firstly, this paper introduces the research content and importance of the semantic role labeling task. Then investigate the main research directions in recent years and the research status at home and abroad. Secondly, this paper analyzes the development of semantic role labeling tasks, and studies a lot of research results based on statistical machine learning and deep neural network. It is intended to analyze the processing methods of semantic role labeling tasks and their current developments. Finally, the paper summarizes the development trend of future semantic role labeling tasks.

2 Semantic Role Labeling

Semantic role labeling is an intermediate process of natural language processing. Its results can not be used to express the meaning of natural language, but it can help many application systems such as Machine Translation, Question Answering, Human Robot Interaction. The high accuracy rate of labeling results can improve the precision and recall rate of these systems, which is of great significance for performance improvement. Therefore, semantic role labeling has very high research value. The goal of semantic role labeling is to recover the predicate-argument structure of a sentence, based on the sentences entered and the predicates specified in the sentence. Then mark the relationship between the predicate and the argument, such as time, place, the agent, the victim, and so on.

The development of semantic role labeling, like most natural language processing tasks, has evolved from rule-based to statistical-based to deep-based learning. The early rule-based semantic role labeling requires people with certain linguistic knowledge or computer science to manually formulate language rules. The advantage of this method is that the rules are more flexible and easy to understand, but it is difficult to cover all language rules. High cost and inefficient processing results do not meet demand. After that, the statistical natural language processing becomes mainstream. Many statistical machine learning models can deal with the semantic role labeling problem, the sequence labeling problem is regarded as the classification problem. But it also has certain limitations in model optimization and performance. With the development of deep learning, which has swept the various fields of artificial intelligence, it also brought a new development direction for semantic roles labeling. Researchers used convolutional neural networks and LSTM neural network frameworks to deal with sequence labeling problems. In recent years, some research results have been achieved, and more and more people are beginning to shift their research direction to deep learning.

2.1 Semantic Role Labeling Based on Statistics

With the developing of statistics-based natural language processing technology, researchers built a large number of corpus base and corpus linguistics. It enables the rapid development of statistical natural language processing and greatly improves the correct rate of model processing results.

According to the three main methods of syntactic parsing: phrase structure parsing, shallow parsing, and dependency parsing. Semantic role labeling can be divided into semantic role labeling method based on phrase structure tree, semantic role labeling method based on shallow parsing result, and semantic role labeling method based on dependency parsing result. However, the basic processes of these methods are similar. After inputting the syntax analysis tree into the semantic role labeling system, the following four processes are performed: candidate arguments pruning, argument identification, argument labeling, and post-processing. The candidate argument pruning is to select a word or phrase that may be an argument from the leaf nodes of the input grammatical structure tree. Then delete the words that cannot be arguments and get the final set of arguments. Argument recognition is to judge the candidate argument and judge whether it is argument or not. It is a two-class problem. Argument labeling is to construct some features from the tree, and obtain a semantic role description for each argument through the multi-classification problem. Finally, get the result of the semantic role labeling through the post-processing process. The process is shown in the following and Fig. 1:

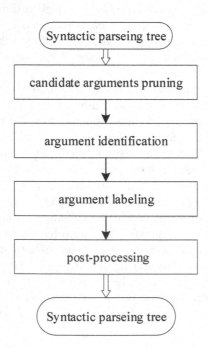

Fig. 1. Semantic role labeling process based on statistical machine learning

The commonly used machine learning models are Support Vector Machine (SVM), Maximum Entropy Model (ME), etc. It was first proposed by Gildea and Jurafsky (2002) [4] to use statistical machine learning models to deal with semantic role labeling tasks. They use supervised training methods to train models and training data is manually labeled. The model can achieve 65% precision and 61% recall rate when processing argument segmentation and semantic role labeling tasks. In the pre-segmented component statement, the semantic role tagging task can achieve an pre-cision rate of 82%. Afterwards, a large number of researchers began there works on machine learning to handle semantic role labeling tasks. There are also scholars in China who have done research in this area. Liu Ting of Harbin Institute of Technology proposed a semantic role labeling system based on the maximum entropy classifier [5]. The system uses the maximum entropy classifier to identify and classify the semantic roles of predicates in sentences and obtains 75.49% and 75.60% of F1 values on the development set and test set respectively. He pointed out that various machine learning methods are mature in the feature engineering of Chinese semantic role labeling [6], and it is difficult to improve the performance of semantic role labeling by improving the machine learning model. Research rich features are more important for semantic role labeling. Li and Qian [7] of Suzhou University mainly studied the semantic role labeling of noun predicate. They further proposed the relevant feature set of noun predicate and obtained better STL processing performance.

However, the statistical role-based semantic role labeling method is highly dependent on the result of syntactic parsing. Its input is a parsing tree but syntactic parsing is a very difficult task in natural language processing. Its results generally do not have high accuracy. Further, semantic role labeling has been broken into several sub-problems. Even if each process has high accuracy, the final result may still be lower accuracy. Moreover, different machine learning models and optimization algorithms need to be selected for different processing tasks, which makes the labeling task complicated. Finally, the statistical-based semantic role labeling method has reached the technical bottleneck, and it is difficult to improve the performance through model optimization.

2.2 Semantic Role Labeling Based on Deep Learning

Semantic role labeling task based on deep learning is to use complex deep neural network to deal with semantic labeling problem. It uses deep learning methods to solve some problems, including traditional semantic role annotation dependent syntax analysis and needs to decompose the annotation task into multiple subtasks. It use an end-to-end manner to performs semantic role labeling in one step. This process takes an integrated approach and no longer relies on syntactic parsing, reducing the need for syntactic parsing and the risk of error accumulation.

Collobert [8] first proposed the application of deep learning to semantic role labeling tasks. They used a general convolutional neural network to treat semantic role labeling as a sequence labeling problem. The system uses the CNN network to obtain the context representation of the current word. The information used includes the current word, the distance between the current word and the predicate, the distance between the context word and the current word. The model uses a large amount of unlabeled data for training. However, if you want to make the model have higher accuracy, you still need to use the results of syntactic parsing. And the performance of the system is lower than the traditional statistical machine learning based method.

Zhou and Xu [9] used the deep Bi-LSTM CNN as an end-to-end system. Only the original text information was used as input information and without using any grammatical knowledge in their research. The model implicitly captures the syntactic structure of the sentence. This strategy is superior to Collobert's practice and has achieved very good test results on the public test set, reaching an F1 value of 81.27%. However, this model has the problem of vanishing gradient and exploding gradient when the number of layers is high. Therefore, the model network layer can only be controlled within a range, which limits the improvement of network performance. The LSTM memory block is shown in Fig. 2:

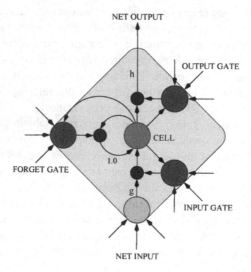

Fig. 2. LSTM memory block with a single cell [10].

He and Lee [11] used the deep highway bidirectional LSTMs with constrained decoding to significantly improve the performance of models. They also analyzed the advantages and disadvantages of the model in detail, and the 8-layer depth model finally achieved an F1 value of 83.4% on the test set. Their results show that the deep LSTM model has excellent performance in recovering long-distance semantic dependencies. But there are still obvious errors in this model. So there is still much room for improvement in semantic analysis techniques.

Wang and Liu [12] equipped the deep LSTM neural network with a novel "straight ladder unit" (EU), which can linearly connect the input and output of the unit. So the information can be transmitted between different layers. This design solved the problem that the number of model layers cannot be too high. The model does not require any additional feature input. It has achieved an 81.53% F1 value on the public test set with improved performance. Table 1 compares the performance of the deep neural network model and the statistical machine learning model studied in recent years.

Table 1. Performance comparison of semantic role labeling models.

	F/%		
	Development	WSJ test	Brown test
Toutanova	78.6	80.3	68.8
Koomen	77.35	79.44	67.75
Pradhan	78.34	78.63	68.44
Collobert(w/parser)	75.42	–	–
Collobert(w/o parser)	72.29	–	–
Zhou	79.6	82.8	69.4
He	82.7	84.6	73.6
Wang	81.26	82.95	72.61

3 Conclusion and Future Work

Semantic role labeling task is a way of shallow semantic analysis. Its research results are of great significance for promoting Machine Translation, Question Answering, Human Robot Interaction and other application systems. This paper introduces the semantic role labeling task and its main research directions at home and abroad in recent years. Then it summarizes a lot of research results based on statistical machine learning and deep neural network. It can be said that the research on semantic role labeling task has achieved very fruitful results, but there is still room for improvement.

Applying deep neural networks to natural language processing tasks is a research hotspot in currently because of its end-to-end characteristics and its reliance on syntactic analysis. The research work to be broken has several aspects:

(1) Construction of a deep neural network model. Among the deep learning models, it is found that the bidirectional LSTMs neural network has better performance in dealing with semantic role labeling. Its modeling idea conforms to the language generation process, and has great advantages in the processing of sequence labeling problems. However, there are many parameters in the model, and it is necessary to find a more suitable optimization method to build the model, so that the model has higher performance and better generalization ability.
(2) The construction of the corpus base. Corpus base and corpus linguistics are indispensable tools for natural language processing tasks. Its richness also plays a key role in semantic role labeling. The word vector representation trained by large-scale corpus can better help model processing label task.
(3) Chinese semantic role labeling. At present, the semantic role labeling system for training and testing is researched on English corpus, but the model for Chinese corpus is very few. Therefore, how to use the model to efficiently deal with Chinese semantic role labeling is also an important research direction.

References

1. Knight, K., Luk, S.K.: Building a large-scale knowledge base for machine translation. Comput. Sci. 773–778 (1994)
2. Shen, D., Lapata, M.: Using semantic roles to improve question answering. EMNLP-CoNLL 2007. In: Proceedings of the 2007 Joint Conference on Empirical Methods in Natural Language Processing and Computational Natural Language Learning, pp. 12–21, June 28–30, 2007, Prague, Czech Republic. DBLP (2007)
3. Bastianelli, E., Castellucci, G., Croce, D., et al.: Textual inference and meaning representation in human robot interaction. Newdesign.aclweb.org (2013)
4. Gildea, D., Jurafsky, D.: Automatic labeling of semantic roles. Comput. Linguist. **28**(28), 245–288 (2002)
5. Liu, T., Che, W., Li, S., et al.: Semantic role labeling system using maximum entropy classifier. In: Conference on Computational Natural Language Learning. Association for Computational Linguistics, pp. 189–192 (2005)
6. Liu, H., Che, W., Liu, T.: Feature engineering for Chinese semantic role labeling. J. Chin. Inf. Process. **21**(1), 79–84 (2007)
7. Li, J., Zhou, G., Zhu, M., et al.: Semantic role labeling of nominalized predicates in Chinese. J. Softw. **22**(8), 1725–1737 (2011)
8. Collobert, R., Weston, J., Karlen, M., et al.: Natural language processing (almost) from scratch. J. Mach. Learn. Res. **12**(1), 2493–2537 (2011)
9. Zhou, J., Xu, W.: End-to-end learning of semantic role labeling using recurrent neural networks. In: Proceedings of the 53rd Annual Meeting of the Association for Computational Linguistics and the 7th International Joint Conference on Natural Language Processing (vol. 1: Long Papers), vol. 1, pp. 1127–1137 (2015)
10. Graves, A., Liwicki, M., Fernã n S, et al.: A novel connectionist system for unconstrained handwriting recognition. IEEE Trans. Pattern Anal. Mach. Intell. **31**(5), 855–868 (2009)
11. He, L., Lee, K., Lewis, M., et al.: Deep semantic role labeling: what works and what's next. In: Meeting of the Association for Computational Linguistics, pp. 473–483 (2017)
12. Wang, M., Liu, Q.: Semantic role labeling using deep neural networks. J. Chin. Inf. Process. **32**(2) (2018)

Network Load Minimization-Based Virtual Network Embedding Algorithm for Software-Defined Networking

Desheng Xie$^{(\boxtimes)}$, Rong Chai, Mengqi Mao, Qianbin Chen, and Chun Jin

Key Lab of Mobile Communication Technology,
Chongqing University of Posts and Telecommunications, Chongqing 400065, China
{1050386890,1747495352}@qq.com, {chairong,chenqb,jinchun}@cqupt.edu.cn

Abstract. In a network virtualization-enabled software-defined networking (SDN), the problem of virtual network embedding (VNE) is a major concern. Although a number of VNE algorithms have been proposed, they fail to consider the efficient utilization of substrate resources or the network load extensively, thus resulting in less efficient utilization of substrate resources or higher blocking ratio of the virtual networks. In this paper, we study the problem of mapping a number of virtual networks in SDN and formulate the VNE problem as a network load minimization problem. Since the formulated optimization problem is NP-hard and it cannot be solved conveniently, we propose a two-stage VNE algorithm consisting of node mapping stage and link mapping stage. Numerical results demonstrate that the effectiveness of our proposed algorithm.

Keywords: Software-defined networking · Network virtualization · Virtual network embedding · Network load

1 Introduction

The emerging demands for mobile Internet applications and multimedia contents bring challenges and difficulties to traditional Internet architecture and technologies. To tackle this problem, software-defined networking (SDN) is proposed as a new networking paradigm which decouples network control from data forwarding functions to enable the flexible network service offering [1]. As another key enabler of the future Internet, network virtualization (NV) is capable of shielding the complexity of the underlying network deployment, and facilitating the flexible management of resources by creating and operating multiple logically-isolated virtual networks on top of substrate network [2]. To achieve NV, the problem of virtual network embedding (VNE) should be considered, which is mapping virtual networks onto a shared substrate network [3].

© ICST Institute for Computer Sciences, Social Informatics and Telecommunications Engineering 2019
Published by Springer Nature Switzerland AG 2019. All Rights Reserved
X. Liu et al. (Eds.): ChinaCom 2018, LNICST 262, pp. 259–270, 2019.
https://doi.org/10.1007/978-3-030-06161-6_26

The problem of VNE has been studied in some recent works. References [4–6] considering the energy consumption issue in designing the VNE algorithms. The authors in [4] proposed an energy-cost model for conducting virtual network embedding in SDN and formulated the energy-aware virtual network embedding problem as an integer linear programming. To stress the problem of non-optimality in energy consumption and the utilization of the substrate network, the authors in [5] proposed an integer linear programming based green mapping algorithm which facilitated the migration of virtual routers and links to achieve the minimum energy consumption. In [6], the authors formulated an energy efficient virtual node embedding model and proposed a minimal element method-based heuristic algorithm to solve the formulated optimization problem and obtain the near-optimal virtual network embedding strategy.

To solve the location-constrained virtual network embedding (LC-VNE) problem efficiently, the authors in [7] reduced LC-VNE to the minimum-cost maximum clique problem and proposed a greedy LC-VNE algorithm and load-balance enhanced LC-VNE algorithm to achieve the integrated node and link mapping strategy with significantly reduced time complexity. To achieve the tradeoff between the profits of physical infrastructure providers and the waiting time for virtual network requests, the authors in [8] proposed an auction-based joint node and link embedding algorithm by using column generation approach.

To support multicast services over the virtual networks, reliability should be a major concern. The authors in [9] formulated a mixed integer linear programming model to determine the upper bound on the reliability of the network with the max-min fairness and proposed a reliability-aware genetic algorithm to achieve reliable multicast VN mapping under low computational complexity. Leveraging the consensus-based resource allocation schemes, the authors in [10] proposed a general distributed auction mechanism for the virtual network embedding problem and demonstrated that the obtained solutions guarantee a worst-case efficiency relative to the optimal virtual network embedding. To solve the cost-effective VNE problem in SDN, the authors in [11] formulated an integer linear programming model and designed a time-efficient heuristic algorithm to achieve the minimum resource consumption.

Though the aforementioned research works consider various VNE algorithms and aim to design the optimal embedding strategies which minimizes the energy consumption or enhances the embedding reliability, they fail to consider the network load or the utilization of the substrate resource extensively, which may result in less efficient utilization of the substrate resources or the high blocking rate of the virtual networks.

In this paper, we study the problem of mapping a number of virtual networks in SDN. Stressing the importance of efficient utilization of substrate resources, we define the network load of the substrate network, and formulate the VNE problem as a network load minimization problem. Since the formulated optimization problem is NP-hard and it cannot be solved conveniently, we propose a two-stage VNE algorithm consisting of node mapping stage and link mapping stage. During the node mapping stage, we examine and rank the performance metrics of substrate nodes and virtual nodes by using recursive method, and then perform greedy strategy-based node mapping. And finally perform link mapping procedure by applying the K-shortest-path algorithm.

2   Network Model

2.1   Substrate Network

In this work, we model the substrate network as a weighted undirected graph which is denoted by $G^{\mathrm{s}} = (N^{\mathrm{s}}, E^{\mathrm{s}}, A^{\mathrm{s}}_N, A^{\mathrm{s}}_E)$, where $N^{\mathrm{s}} = \{\mathrm{n}^{\mathrm{s}}_i, 1 \leq i \leq M\}$ denotes the set of substrate nodes, $\mathrm{n}^{\mathrm{s}}_i$ denotes the ith substrate node, M is the total number of the substrate nodes, $E^{\mathrm{s}} = \{\mathrm{e}^{\mathrm{s}}_{i,j}, 1 \leq i, j \leq M, i \neq j\}$ denotes the set of substrate links, and $\mathrm{e}^{\mathrm{s}}_{i,j}$ denotes the substrate link.

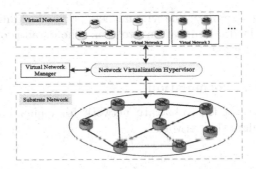

Fig. 1. System model of SDN

$A^{\mathrm{s}}_N = \{\{C(\mathrm{n}^{\mathrm{s}}_i), S(\mathrm{n}^{\mathrm{s}}_i), T(\mathrm{n}^{\mathrm{s}}_i)\}, \mathrm{n}^{\mathrm{s}}_i \in N^{\mathrm{s}}\}$ denotes the attribute set of the substrate nodes, where $C(\mathrm{n}^{\mathrm{s}}_i)$, $S(\mathrm{n}^{\mathrm{s}}_i)$ and $T(\mathrm{n}^{\mathrm{s}}_i)$ denote the CPU capacity, the storage capacity and the ternary content addressable memory (TCAM) capacity of $\mathrm{n}^{\mathrm{s}}_i$, respectively. $A^{\mathrm{s}}_E = \{\{B(\mathrm{e}^{\mathrm{s}}_{i,j}), D(\mathrm{e}^{\mathrm{s}}_{i,j})\}, \mathrm{e}^{\mathrm{s}}_{i,j} \in E^{\mathrm{s}}\}$ denotes the attribute set of substrate links, where $B(\mathrm{e}^{\mathrm{s}}_{i,j})$ and $D(\mathrm{e}^{\mathrm{s}}_{i,j})$ denote the bandwidth capacity and the propagation delay of $\mathrm{e}^{\mathrm{s}}_{i,j}$, respectively.

2.2   Virtual Network

Denoting the number of virtual networks as K_0, we also characterize each virtual network as a weighted undirected graph. For the kth virtual network, we define the weighted undirected graph $G^{\mathrm{v}}_k = (N^{\mathrm{v}}_k, E^{\mathrm{v}}_k, A^{\mathrm{v}}_{k,N}, A^{\mathrm{v}}_{k,E})$, where $N^{\mathrm{v}}_k = \{\mathrm{n}^{\mathrm{v}}_{k,u}, 1 \leq u \leq M_k\}$ and $E^{\mathrm{v}}_k = \{\mathrm{e}^{\mathrm{v}}_{k,u,r}, 1 \leq u, r \leq M_k, u \neq r\}$ denote the set of virtual nodes and virtual links in the kth virtual network, respectively, $\mathrm{n}^{\mathrm{v}}_{k,u}$ denotes the uth virtual node of the kth virtual network, M_k denotes the total number of the virtual nodes of the kth virtual network, $\mathrm{e}^{\mathrm{v}}_{k,u,r}$ denotes the virtual link.

$A_{k,N}^{\mathrm{v}} = \{\{C(\mathrm{n}_{k,u}^{\mathrm{v}}), S(\mathrm{n}_{k,u}^{\mathrm{v}}), T(\mathrm{n}_{k,u}^{\mathrm{v}})\}, \mathrm{n}_{k,u}^{\mathrm{v}} \in N_k^{\mathrm{v}}\}$ denotes the attribute set of the virtual nodes of the kth virtual network, $C(\mathrm{n}_{k,u}^{\mathrm{v}})$, $S(\mathrm{n}_{k,u}^{\mathrm{v}})$ and $T(\mathrm{n}_{k,u}^{\mathrm{v}})$ denote the required CPU capacity, storage capacity and TCAM capacity of $\mathrm{n}_{k,u}^{\mathrm{v}}$, $A_{k,E}^{\mathrm{v}} = \{\{B(\mathrm{e}_{k,u,r}^{\mathrm{v}}), D(\mathrm{e}_{k,u,r}^{\mathrm{v}})\}, \mathrm{e}_{k,u,r}^{\mathrm{v}} \in E_k^{\mathrm{v}}\}$ denotes the attribute set of the virtual links of the kth virtual network, $B(\mathrm{e}_{k,u,r}^{\mathrm{v}})$ and $D(\mathrm{e}_{k,u,r}^{\mathrm{v}})$ denote the required bandwidth capacity and the tolerable propagation delay of $\mathrm{e}_{k,u,r}^{\mathrm{v}}$, respectively.

In this paper, we introduce a functional model called network virtualization hypervisor (NVH) to conduct virtual network embedding. We assume that the NVH is capable of collecting network state information. Upon receiving the virtual networks from the tenants, the NVH conducts the proposed virtual network embedding, and sends the strategy to the switches in the network. Figure 1 shows the system model considered in this paper.

3   Network Load Optimization Problem Formulation

In this section, we formulate the virtual network embedding problem of the SDN as a network load minimization problem. The detail problem formulation is described in this section.

3.1   Network Load Formulation

Considering the load of the substrate network for embedding all the virtual networks, we formulate the network load Ψ as

$$\Psi = \sum_{k=1}^{K_0} \Psi_k \tag{1}$$

where Ψ_k denotes the load of the substrate network caused by mapping the kth virtual network, and can be expressed as

$$\Psi_k = \Psi_k^{\mathrm{N}} + \Psi_k^{\mathrm{E}} \tag{2}$$

where Ψ_k^{N} and Ψ_k^{E} denote the load of the substrate nodes and substrate links, respectively, caused by mapping the kth virtual network.

Ψ_k^{N} in (2) can be expressed as

$$\Psi_k^{\mathrm{N}} = \sum_{\mathrm{n}_{k,u}^{\mathrm{v}} \in N_k^{\mathrm{v}}} \sum_{\mathrm{n}_i^{\mathrm{s}} \in N^{\mathrm{s}}} \alpha_{k,u,i} \frac{C(\mathrm{n}_{k,u}^{\mathrm{v}}) + S(\mathrm{n}_{k,u}^{\mathrm{v}}) + T(\mathrm{n}_{k,u}^{\mathrm{v}})}{C_k(\mathrm{n}_i^{\mathrm{s}}) + S_k(\mathrm{n}_i^{\mathrm{s}}) + T_k(\mathrm{n}_i^{\mathrm{s}})} \tag{3}$$

where $\alpha_{k,u,i}$ denotes the binary node embedding variable that equals to 1 when $\mathrm{n}_{k,u}^{\mathrm{v}}$ is mapped to $\mathrm{n}_i^{\mathrm{s}}$, and 0 otherwise, $C_k(\mathrm{n}_i^{\mathrm{s}})$, $S_k(\mathrm{n}_i^{\mathrm{s}})$ and $T_k(\mathrm{n}_i^{\mathrm{s}})$ denote respectively the residual CPU capacity, storage capacity and TCAM capacity of $\mathrm{n}_i^{\mathrm{s}}$ after conducting the virtual node mapping for the first $k-1$ virtual networks. In this paper, we assume that the K_0 virtual networks will be mapped successively from the first one to the K_0th one, hence, we obtain

$$C_k(\mathrm{n}_i^{\mathrm{s}}) = C(\mathrm{n}_i^{\mathrm{s}}) - \sum_{l=1}^{k-1} \sum_{\mathrm{n}_{l,u}^{\mathrm{v}} \in N_l^{\mathrm{v}}} \alpha_{l,u,i} C(\mathrm{n}_{l,u}^{\mathrm{v}}), \tag{4}$$

$$S_k(n_i^s) = S(n_i^s) - \sum_{l=1}^{k-1} \sum_{n_{l,u}^v \in N_l^v} \alpha_{l,u,i} S(n_{l,u}^v). \tag{5}$$

$$T_k(n_i^s) = T(n_i^s) - \sum_{l=1}^{k-1} \sum_{n_{l,u}^v \in N_l^v} \alpha_{l,u,i} T(n_{l,u}^v) - \sum_{l=1}^{k-1} \sum_{e_{l,u,r}^v \in E_l^v} \beta_{l,u,r,i} T(e_{l,u,r}^v) \tag{6}$$

where $\beta_{l,u,r,i}$ denotes the binary embedding variable of intermediate node that equals to 1 when n_i^s is an intermediate node of the substrate path that $e_{l,u,r}^v$ is embedded onto, and 0 otherwise.

Similarly, Ψ_k^E in (2) can be expressed as

$$\Psi_k^E = \sum_{e_{k,u,r}^v \in E_k^v} \sum_{e_{i,j}^s \in E^s} \gamma_{k,u,r,i,j} \frac{B(e_{k,u,r}^v)}{B_k(e_{i,j}^s)} \tag{7}$$

where $\gamma_{k,u,r,i,j}$ denotes the binary link embedding variable that equals to 1 when $e_{k,u,r}^v$ is mapped to $e_{i,j}^s$, and 0 otherwise, $B_k(e_{i,j}^s)$ represents the residual bandwidth capacity of $e_{i,j}^s$ after conducting the virtual link mapping for the first $k-1$ virtual networks, which can be expressed as

$$B_k(e_{i,j}^s) = B(e_{i,j}^s) - \sum_{l=1}^{k-1} \sum_{e_{l,u,r}^v \in E_l^v} \gamma_{l,u,r,i,j} B(e_{l,u,r}^v). \tag{8}$$

3.2 Optimization Constraints

The optimization problem of virtual network embedding should subject to a number of constraints, as discussed in this subsection in detail.

Node Mapping Constraints In this paper, we assume that one virtual node of a particular virtual network can only be mapped onto one substrate node, i.e.,

$$C1: \sum_{n_i^s \in N^s} \alpha_{k,u,i} = 1. \tag{9}$$

Similarly, we assume that a substrate node can only map to at most one virtual node from a particular virtual network, hence, we obtain

$$C2: \sum_{n_{k,u}^v \in N_k^v} \alpha_{k,u,i} \leq 1. \tag{10}$$

Resource Constraints To map virtual nodes onto substrate nodes, the resource constraints on the substrate nodes, including the total CPU usage, the storage usage and the TCAM usage, should be satisfied, i.e.,

$$C3: \sum_{k=1}^{K_0} \sum_{n_{k,u}^v \in N_k^v} \alpha_{k,u,i} C(n_{k,u}^v) \leq C(n_i^s), \tag{11}$$

$$C4 : \sum_{k=1}^{K_0} \sum_{n_{k,u}^v \in N_k^v} \alpha_{k,u,i} S(n_{k,u}^v) \leq S(n_i^s), \tag{12}$$

$$C5 : \sum_{k=1}^{K_0} \left(\sum_{n_{k,u}^v \in N_k^v} \alpha_{k,u,i} T(n_{k,u}^v) + \sum_{e_{k,u,r}^v \in E_k^v} \beta_{k,u,r,i} T(e_{k,u,r}^v) \right) \leq T(n_i^s). \tag{13}$$

And the bandwidth constraint on the substrate links should also be satisfied, i.e.,

$$C6 : \sum_{k=1}^{K_0} \sum_{e_{k,u,r}^v \in E_k^v} \gamma_{k,u,r,i,j} B(e_{k,u,r}^v) \leq B(e_{i,j}^s). \tag{14}$$

Flow Conservation Constraint While mapping a virtual link to a substrate link, the flow conservation constraints should be met, which can be expressed as

$$C7 : \sum_{n_j^s \in N(n_i^s)} \gamma_{k,u,r,i,j} - \sum_{n_j^s \in N(n_i^s)} \gamma_{k,u,r,j,i} = \alpha_{k,u,i} - \alpha_{k,r,i}. \tag{15}$$

Intermediate Node Constraint To ensure that the intermediate nodes of the substrate path that carries $e_{k,u,r}^v$ are identified correctly, we obtain

$$C8 : \sum_{n_j^s \in N(n_i^s)} \gamma_{k,u,r,i,j} - \beta_{k,u,r,i} = \alpha_{k,u,i}. \tag{16}$$

Tolerable Propagation Delay Constraint The total propagation delay of the substrate path that carries $e_{k,u,r}^v$ should not exceed the tolerable propagation delay of $e_{k,u,r}^v$, i.e.,

$$C9 : \sum_{e_{i,j}^s \in E^s} \gamma_{k,u,r,i,j} D(e_{i,j}^s) \leq D(e_{k,u,r}^v). \tag{17}$$

3.3 Optimization Problem Modeling

Aiming to minimize the total network load subject to the constraints, we formulate the optimization problem as

$$\min_{\alpha_{k,u,i}, \beta_{k,u,r,i}, \gamma_{k,u,r,i,j}} \Psi \tag{18}$$

$$\text{s.t.} \quad C1 - C9.$$

4 Solution to the Optimization Problem

The optimization problem formulated in (18) is NP-hard, the optimal solution of which is very difficult to obtain in general. In this paper, to tackle the resource sharing problem among the virtual networks, we propose a sequential VNE method by ordering the virtual networks according to their resource requirement. Then for individual virtual network, we propose a two-stage VNE algorithm consisting of node mapping stage and link mapping stage.

4.1 Ordering the Virtual Networks

In this subsection, we examine the resource requirement of the virtual networks by summing up the required resources of all the virtual nodes and links contained in the virtual network. For the kth virtual network, the amount of the required resources can be calculated as

$$R(G_k^{\mathrm{v}}) = \sum_{\mathrm{n}_{k,u}^{\mathrm{v}} \in N_k^{\mathrm{v}}} (C(\mathrm{n}_{k,u}^{\mathrm{v}}) + S(\mathrm{n}_{k,u}^{\mathrm{v}}) + T(\mathrm{n}_{k,u}^{\mathrm{v}})) + \sum_{e_{k,u,r}^{\mathrm{v}} \in E_k^{\mathrm{v}}} B(\mathrm{e}_{k,u,r}^{\mathrm{v}}). \quad (19)$$

Based on $R(G_k^{\mathrm{v}})$, $1 \leq k \leq K_0$, we can then rank the virtual networks according to the amount of the required resources. Aim to utilize the substrate resources in an efficient manner, we first conduct the VNE for the virtual network with the largest resource requirement, and then conduct the VNE for the virtual network with the second largest resource requirement. This process repeats until all the virtual networks have been embedded or there is no enough substrate resources being available.

4.2 Node Mapping Stage

For certain virtual network, we first conduct node mapping. In this paper, we design a recursive method-based performance metrics evaluation algorithm for both the substrate nodes and virtual nodes, then rank the substrate nodes and virtual nodes according to the obtained performance metrics and conduct node mapping according to the rank list based on the greedy strategy.

Recursive Method-Based Performance Metrics Evaluation Algorithm
To examine the performance metric of one substrate node, we consider the amount of the resources that the node can offer and define the performance metric of $\mathrm{n}_i^{\mathrm{s}}$, denoted by $V(\mathrm{n}_i^{\mathrm{s}})$ as

$$V(\mathrm{n}_i^{\mathrm{s}}) = (1 - \delta)\bar{R}(\mathrm{n}_i^{\mathrm{s}}) + \delta \sum_{\mathrm{n}_j^{\mathrm{s}} \in N(\mathrm{n}_i^{\mathrm{s}})} \frac{\bar{R}(\mathrm{n}_j^{\mathrm{s}})}{\sum_{\mathrm{n}_h^{\mathrm{s}} \in N(\mathrm{n}_i^{\mathrm{s}})} \bar{R}(\mathrm{n}_h^{\mathrm{s}})} V(\mathrm{n}_j^{\mathrm{s}}) \quad (20)$$

where $\delta \in [0,1)$ is the relative weight of the neighbor performance metric, $N(n_i^s)$ denotes the set of neighbors of n_i^s, and $\bar{R}(n_i^s)$ is the amount of the normalized resources on n_i^s, which can be defined as

$$\bar{R}(n_i^s) = \frac{R(n_i^s)}{\sum\limits_{n_j^s \in N^s} R(n_j^s)} \tag{21}$$

where $R(n_i^s)$ denotes the amount of resources on n_i^s. To evaluate the amount of resources on n_i^s, we jointly consider node degree, the characteristics of link resources and node resources, and formulate $R(n_i^s)$ as

$$R(n_i^s) = d(n_i^s)l(n_i^s)r(n_i^s) \tag{22}$$

where $d(n_i^s)$ denotes the degree of n_i^s, which can be expressed as

$$d(n_i^s) = |N(n_i^s)| \tag{23}$$

where $|x|$ denotes the number of elements in set x.

$l(n_i^s)$ in (22) denotes the characteristics of the link resource of n_i^s, which is defined as

$$l(n_i^s) = \sum_{n_j^s \in N(n_i^s)} B(e_{i,j}^s)(\frac{D^{\max} - D(e_{i,j}^s)}{D^{\max} - D^{\min}}) \tag{24}$$

where $D^{\max}$ and $D^{\min}$ denote respectively the maximum and the minimum propagation delay of the links in the substrate network, which can be expressed as

$$D^{\max} = \max\left\{D(e_{i,j}^s), 1 \leq i,j \leq M, i \neq j\right\}, \tag{25}$$

$$D^{\min} = \min\left\{D(e_{i,j}^s), 1 \leq i,j \leq M, i \neq j\right\}. \tag{26}$$

$r(n_i^s)$ in (22) denotes the characteristics of the node resource of n_i^s, which can be calculated as

$$r(n_i^s) = C(n_i^s) + S(n_i^s) + T(n_i^s). \tag{27}$$

Matrix Form-Based Performance Metrics Calculating Expressing in matrix form, we can rewrite the performance metrics of the substrate nodes as

$$\mathbf{V} = (1-\delta)\mathbf{R} + \delta\mathbf{Q}\mathbf{V} \tag{28}$$

where $\mathbf{V} = (\bar{V}(n_1^s), ..., \bar{V}(n_i^s), ..., \bar{V}(n_M^s))^T$, $\mathbf{R} = (\bar{R}(n_i^s), ..., \bar{R}(n_i^s), ..., \bar{R}(n_M^s))^T$, $\mathbf{Q} = [Q(n_i^s, n_j^s)]$ is an $M \times M$ matrix, and $Q(n_i^s, n_j^s)$ is defined as

$$Q(n_i^s, n_j^s) = \begin{cases} \dfrac{\bar{R}(n_j^s)}{\sum\limits_{n_h^s \in N(n_i^s)} \bar{R}(n_h^s)} & n_j^s \in N(n_i^s) \\ 0 & n_j^s \notin N(n_i^s) \end{cases}. \tag{29}$$

Since the calculation of performance metrics vector could be time-consuming as the size of substrate network becomes large, in this subsection, we develop a simple iterative calculation strategy as: $\mathbf{V}^{(t+1)} = (1-\delta)\mathbf{R} + \delta\mathbf{Q}\mathbf{V}^{(t)}$, where $\mathbf{V}^{(t)}$ is the performance metrics vector at the tth iteration. For initialization, we set $\mathbf{V}^{(0)}$ as: $\mathbf{V}^{(0)} = \mathbf{R}$.

Greedy Strategy-Based Node Mapping Method we can also calculate the performance metrics of virtual nodes by conducting similar procedure. Based on the performance metrics, we rank the substrate nodes and the virtual nodes in non-increasing order and conduct greedy strategy-based node mapping.

Algorithm 1. Recursive method-based performance metrics evaluation algorithm for substrate nodes

---

Input: Network topology $G^{\mathrm{s}} = (N^{\mathrm{s}}, E^{\mathrm{s}}, A_N^{\mathrm{s}}, A_E^{\mathrm{s}})$, the maximum number of iterations $t_{\max}$, and the maximum tolerance ϵ;
Output: Performance metrics vector $\mathbf{V}$;
1: Calculate $\mathbf{R}$ and $\mathbf{Q}$ according to (21) and (29), respectively;
2: Initialization: $\mathbf{V}^{(0)} = \mathbf{R}$, $t = 0$, $\mu = \infty$;
3: **while** $t < t_{\max}$ and $\mu > \epsilon$ **do**
4:     $\mathbf{V}^{(t+1)} = (1 - \delta)\mathbf{R} + \delta \mathbf{Q}\mathbf{V}^{(t)}$;
5:     $\mu = \left\| \mathbf{V}^{(t+1)} - \mathbf{V}^{(t)} \right\|$;
6:     $t = t + 1$;
7: **end while**
8: $\mathbf{V} = \mathbf{V}^{t+1}$.

4.3   Shortest-Path Algorithm-Based Link Mapping Stage

With embedding all virtual nodes of the certain virtual network, the virtual links of the virtual network demand to be embedded. For the link mapping stage of the certain virtual network, we order the virtual links in non-increasing order according to the bandwidth requirement, i.e., $B(\mathrm{e}_{k,u,r}^{\mathrm{v}})$, and then conduct link mapping from the one of the largest bandwidth requirement to that with the least bandwidth requirement.

To map one certain virtual link, we jointly consider substrate link bandwidth consumption, which can be characterized by the hops of the substrate paths, and the substrate link bandwidth capacity.

In this paper, we propose to use the K-shortest-path algorithm to determine the shortest paths between the two corresponding substrate nodes of one virtual link. For convenience, we characterize the substrate network as a weighted graph $G^{\mathrm{s}} = (N^{\mathrm{s}}, E^{\mathrm{s}}, W^{\mathrm{s}})$, where $W^{\mathrm{s}} = \{\mathrm{w}_{i,j}^{\mathrm{s}}, 1 \leq i, j \leq M, i \neq j\}$ denotes the weight set of the substrate links, and $\mathrm{w}_{i,j}^{\mathrm{s}}$ denotes the weight of $\mathrm{e}_{i,j}^{\mathrm{s}}$, which is set to be 1. By applying the K-shortest-path algorithm in G^{s}, the K candidate shortest paths can be obtained. Let $P_{k,u,r,f}$ denote the fth candidate substrate path of $\mathrm{e}_{k,u,r}^{\mathrm{v}}$, $1 \leq f \leq K$, we then select the substrate path with the largest minimum link bandwidth capacity from $\{P_{k,u,r,f}, 1 \leq f \leq K\}$, i.e.,

$$f^* = \arg\max \left\{ \min_{f \in \{1,\dots,K\}} \left\{ B(\mathrm{e}_{i,j}^{\mathrm{s}}), \mathrm{e}_{i,j}^{\mathrm{s}} \in P_{k,u,r,f} \right\} \right\}. \tag{30}$$

5   Performance Evaluation

In this section, we perform numerical simulations to examine the performance of the proposed algorithm. In the simulation, the substrate nodes are uniformly distributed in a square region with the size being Lkm $\times$ Lkm, and any two substrate nodes are connected with the probability of $P = a\exp(\frac{-d^5}{bL})$, where a and b are network characteristic parameters, d is the Euclidean distance between two nodes and L denotes the length of the region, which equals 100. The number of the substrate nodes is chosen as 30 and 40, respectively. The CPU, storage, TCAM capacity of the substrate nodes and the bandwidth capacity of the substrate links are real numbers which are uniformly chosen between 40 and 50. The propagation delay of each substrate link is proportional to the geographical distance between substrate nodes.

For each virtual network, the virtual nodes are uniformly distributed in a square region with the size being Lkm $\times$ Lkm. Similar to the substrate nodes, any two virtual nodes are connected with the probability of $P = a\exp(\frac{-d^5}{bL})$ and the number of virtual nodes in each virtual network is randomly chosen between 4 and 8. The CPU, storage, TCAM demand of the virtual nodes and the bandwidth demand of the virtual links are real numbers uniformly chosen between 5 and 10. Simulation results are averaged over 1000 independent processes involving different simulation parameters.

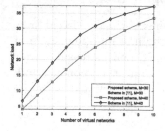

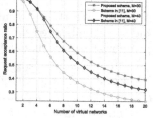

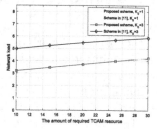

Fig. 2. Network load versus the number of virtual networks

Fig. 3. Request acceptance ratio versus the number of virtual networks

Fig. 4. Network load versus required TCAM resource

Figure 2 shows the network load versus the number of virtual networks. For comparison, we examine the performance of our proposed algorithm and the algorithm proposed in [11]. It can be observed from the figure that the network load increases with the increase of the number of virtual networks for both schemes. This is because the increase in the number of virtual networks results in the increased resource consumption in the substrate network, thus causing the increase of the network load in turn. It can also be seen from the figure that as the number of the substrate nodes increases, the network load decreases. This is because the available substrate resources increase with the increase of the number of substrate nodes, thus offering better network load performance. Comparing the results obtained from our proposed scheme and the scheme proposed in [11], we can see that our proposed scheme offers better network load performance.

In Fig. 3, we plot the request acceptance ratio versus the number of virtual networks. For comparison, we examine the performance of our proposed algorithm and the algorithm proposed in [11]. It can be seen from the figure that the request acceptance ratio decreases with the increase of virtual networks for both schemes. This is because the consumption of substrate network resources increases with the increase of virtual networks. Comparing the results obtained from our proposed scheme and the scheme proposed in [11], we can see that our proposed scheme can accommodate more virtual networks and obtain higher request acceptance ratio.

In Fig. 4, we consider different number of virtual networks and examine the network load versus the amount of required TCAM resources. It can be seen from the figure that the network load obtained from both schemes increases with the increase of the amount of required TCAM resources and our proposed algorithm can achieve lower network load than the algorithm proposed in [11].

6   Conclusions

In this paper, we study the problem of mapping a number of virtual networks in SDN. To stress the importance of efficient utilization of substrate resources, we formulate the VNE problem as a network load minimization problem. To solve the formulated problem, a two-stage VNE algorithm is then proposed consisting of node mapping stage and link mapping stage. Numerical results demonstrate the effectiveness of our proposed algorithm.

References

1. Chowdhury, N.M.K., Boutaba, R.: Network virtualization: state of the art and research challenges. IEEE Commun. Mag. **47**(7), 20–26 (2009)
2. Mijumbi, R., Serrat, J., Gorricho, J.L., Bouten, N., De Turck, F., Boutaba, R.: Network function virtualization: state-of-the-art and research challenges. IEEE Commun. Surv. Tutor. **18**(1), 236–262 (2016)
3. Fischer, A., de Meer, H.: Generating virtual network embedding problems with guaranteed solutions. IEEE Trans. Netw. Serv. Manag. **13**(3), 504–517 (2016)
4. Su, S., Zhang, Z., Liu, A.X., Cheng, X., Wang, Y., Zhao, X.: Energy-aware virtual network embedding. IEEE Trans. Netw. **22**(5), 1607–1620 (2014)
5. Rodriguez, E., Alkmim, G.P., Fonseca, N., Batista, D.: Energy-aware mapping and live migration of virtual networks. IEEE Syst. J. **11**(2), 637–648 (2017)
6. Chen, X., Li, C., Jiang, Y.: Optimization model and algorithm for energy efficient virtual node embedding. IEEE Commun. Lett. **7**(9), 1327–1330 (2015)
7. Gong, L., Jiang, H., Wang, Y., Zhu, Z.: Novel location-constrained virtual network embedding LC-VNE algorithms towards integrated node and link mapping. IEEE/ACM Trans. Netw. **24**(6), 3648–3661 (2016)

8. Jarray, A., Karmouch, A.: Decomposition approaches for virtual network embedding with one-shot node and link mapping. IEEE/ACM Trans. Netw. **23**(3), 1012–1025 (2015)
9. Gao, X., Ye, Z., Fan, J., Zhong, W., Zhao, Y.: Virtual network mapping for multicast services with max-min fairnessof reliability. IEEE/OSA J. Opt. Commun. Netw. **7**(9), 942–951 (2015)
10. Esposito, F., Paola, D.D., Matta, I.: On distributed virtual network embedding with guarantees. IEEE/ACM Trans. Netw. **24**(1), 569–582 (2016)
11. Huang, H., Li, S., Han, K., Sun, Q., Hu, D., Zhu, Z.: Embedding virtual software-defined networks over distributed hypervisors for vDC formulation. In: IEEE ICC, Paris, pp. 1–6 (2017)

Joint User Association and Content Placement for D2D-Enabled Heterogeneous Cellular Networks

Yingying Li[✉], Rong Chai, Qianbin Chen, and Chun Jin

Key Lab of Mobile Communication Technology,
Chongqing University of Posts and Telecommunications, Chongqing 400065, China
17783195324@163.com, {chairong,chenqb,jinchun}@cqupt.edu.cn

Abstract. The explosive increase of the multimedia traffic poses challenges on mobile communication systems. To stress this problem, caching technology can be exploited to reduce backhaul transmissions latency and improve content fetching efficiency. In this paper, we study the user association and content placement problem of device-to-device-enabled (D2D-enabled) heterogeneous cellular networks (HCNs). To stress the importance of the service delay of all the users, we formulate the joint user association and content placement problem as an integer-nonlinear programming problem. As the formulated NP-hardness of the problem, we apply the McCormick envelopes and the Lagrangian partial relaxation method to decompose the optimization problem into three subproblems and solved it by using Hungarian method and unidimensional knapsack algorithm. Simulation results validate the effectiveness of the proposed algorithm.

Keywords: Heterogeneous cellular networks · User association · Content placement · D2D communication

1 Introduction

The explosive increase of diversified high-speed traffics poses challenges on the transmission performance of the radio access networks (RANs) and backhaul links of the mobile networks [1, 2]. To stress this problem, heterogeneous cellular networks (HCNs) with caching functionality are expected to offer users more high-quality communication links and locally fetch request content by interacting with the small cells and thus significantly reducing redundant content downloads through the backhaul links.

To further improve user QoS, device-to-device (D2D) communication technology which allows UEs communicate directly without the data forwarding by the BSs is proposed [3]. if D2D communication technology is considered in HCNs, transmission performance will be further improved. In D2D-enabled

© ICST Institute for Computer Sciences, Social Informatics and Telecommunications Engineering 2019
Published by Springer Nature Switzerland AG 2019. All Rights Reserved
X. Liu et al. (Eds.): ChinaCom 2018, LNICST 262, pp. 271–282, 2019.
https://doi.org/10.1007/978-3-030-06161-6_27

HCNs, UEs may associate with the macro BS (MBS), small cell BS (SBS) or D2D peer for information interaction. It is apparent that different user association or mode selection strategies may result in various network transmission performance based on different channel characteristics and network resources.

Some recent research works study the content placement strategy for cellular networks [4–6]. The authors in [4] consider the content placement in a femtocell network and design caching strategy which minimizes the average downloading delay of all the UEs. In [5], the optimal content placement problem in a femtocell network is formulated as an average bit error rate (BER) minimization problem and use greedy algorithm to solve it. In [6], the authors design a collaborative multi-tier caching framework and propose a joint content placement and routing scheme to maximize traffic offloading.

Content placement in D2D-enabled networks is studied as well. The authors in [7] examine the average caching failure rate in D2D communication network and propose a dual-solution search algorithm to solve content placement problem. In [8], the authors formulate the access selection and spectrum allocation problem as a utility function maximization problem and propose an efficient algorithm to obtain the optimal strategy. User association and content placement problem are jointly considered for HCNs [9–11]. In [9], the author formulate the joint content caching, routing and channel assignment problem as a throughput maximization problem and propose the column generation method. In [10], the authors examine the tradeoff between load balancing and backhaul traffic reduction and solving the problem iteratively. In [11], the authors consider the various condition of the backhaul links and propose a near-optimal distributed algorithm solving it.

However, few previous works consider the D2D-enabled HCNs, Meanwhile, most of these works focus on the performance optimization of throughput or network utility but fail to stress the importance of service delay, which should become a major concern especially for delay-sensitive services.

In this paper, we study joint user association and content placement problem of D2D-enabled HCNs. Jointly considering the constraints on wireless resources, storage capacity as well as user QoS requirements, we formulate the joint user association and cache content placement problem as a service delay minimization problem and propose an efficient algorithm by applying the McCormick envelopes and the Lagrangian partial relaxation method to obtain the solution.

This paper is organized as follows. The system model is described in Sect. 2. The proposed optimization problem is formulated in Sect. 3. In Sect. 4, the solution to the formulated optimization problem is presented. Simulation results are discussed in Sect. 5. Finally, the conclusions are drawn in Sect. 6.

2   System Model

In this paper, we consider the downlink transmission in a D2D-enabled HCN consisting of a MBS, N SBSs and a number of UEs. We assume that the MBS connects to IP network through a wired backhaul link and the SBSs access core network by associating to the MBS as shown in Fig. 1.

By applying local caching scheme, we assume SBSs and some users has certain caching functionality. For simplicity, we refer cache-enabled users as serving users (SUs). Therefore, the RUs can access contents via three user association modes, i.e., MBS association, SBS association and D2D transmission mode.

In this paper, we denote the set of the SBSs as SBS= $\{SBS_1, \ldots, SBS_N\}$, where SBS_n represents the nth SBS, $1 \leq n \leq N$. Denote the storage capacity of SBS_n as S_n^b. Let RU = $\{RU_1, \ldots, RU_M\}$ denote the set of the RUs, where RU_m denotes the mth RU, $1 \leq m \leq M$, M is the number of RUs. Let SU= $\{SU_1, \ldots, SU_K\}$ denote the set of SUs where SU_k denotes the kth SU, $1 \leq k \leq K$, K is the number of the SUs. We assume RU_m can make random requests from the content library and let $F = \{F_1, \ldots F_L\}$ denotes the set of content files, where F_l represents the lth content file, $1 \leq l \leq L$, L is the number of content files.

Let binary variable $a_{m,l} \in \{0,1\}$ describe the content request variable of RU_m, i.e., $a_{m,l} = 1$, if RU_m requests content F_l; otherwise, $a_{m,l} = 0$. We assume each user can request at most one content and SU_k can serve at most one RU during a time period. Denote $y_{k,l}^d$ as the caching variable of SU_k, i.e., $y_{k,l}^d = 1$, if content F_l is stored in SU_k; otherwise, $y_{k,l}^d = 0$.

To avoid transmission interference, we assume the bandwidth of the MBS and SBSs is divided into a number of subchannels with equal bandwidth and each RU can only be allocated one subchannel. Let W^d denotes the available bandwidth of each D2D communication link, and $W_0^{\max}$ and $W_n^{\max}$ denote the maximum available bandwidth of the MBS and SBS_n and W_0 and W_n denote the subchannel bandwidth of them. The maximum number of users associated to MBS and SBS_n can be calculated respectively as $A_0 = \lfloor W_0^{\max}/W_0 \rfloor$ and $A_n = \lfloor W_n^{\max}/W_n \rfloor$.

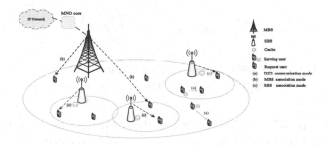

Fig. 1. System model

3 Optimization Problem Formulation

In this section, we formulate joint user association and content placement problem as a service delay minimization problem.

3.1   Objective Function

We express the service delay of the RUs as

$$D = D^{\mathrm{d}} + D^{\mathrm{m}} + D^{\mathrm{s}} \tag{1}$$

where D^{d}, D^{m} and D^{s} denote the service delay of the RUs when acquiring required contents through D2D transmission mode, MBS association mode and SBS association mode. The expressions of D^{d}, D^{m} and D^{s} will be described in following subsections.

Service Delay of RUs in D2D Communication Mode   The service delay of the RUs in D2D communication mode, denoted by D^{d}, can be calculated as

$$D^{\mathrm{d}} = \sum_{m=1}^{M} \sum_{k=1}^{K} \sum_{l=1}^{L} a_{m,l} x_{m,k}^{\mathrm{d}} D_{m,k,l}^{\mathrm{d}} \tag{2}$$

where $x_{m,k}^{\mathrm{d}}$ denotes the binary association variable for D2D transmission mode, i.e., $x_{m,k}^{\mathrm{d}} = 1$, if RU_m is associated with SU_k; otherwise, $x_{m,k}^{\mathrm{d}} = 0$, $D_{m,k,l}^{\mathrm{d}}$ denotes the service delay of RU_m acquiring F_l through associating with SU_k. We express $D_{m,k,l}^{\mathrm{d}}$ as

$$D_{m,k,l}^{\mathrm{d}} = y_{k,l}^{\mathrm{d}} \frac{S_l}{R_{m,k}^{\mathrm{d}}} \tag{3}$$

where $y_{k,l}^{\mathrm{d}}$ denotes the binary content caching index of the SUs, i.e., $y_{k,l}^{\mathrm{d}} = 1$, if F_l is cached in SU_k; otherwise, $y_{k,l}^{\mathrm{d}} = 0$, S_l denotes the size of F_l and $R_{m,k}^{\mathrm{d}}$ denotes the data rate of the transmission link between RU_m and SU_k, which is given by

$$R_{m,k}^{\mathrm{d}} = W^{\mathrm{d}} \log_2 \left(1 + \frac{P_k^{\mathrm{d}} g_{m,k}^{\mathrm{d}}}{\sigma^2} \right) \tag{4}$$

where P_k^{d} is the transmission power of SU_k, $g_{m,k}^{\mathrm{d}}$ is the channel gain between SU_k and RU_m, and σ^2 is the power of Guassian white noise.

Service Delay of RUs in MBS Association Mode   The service delay of the RUs in MBS association mode, denoted by D^{m}, can be calculated as

$$D^{\mathrm{m}} = \sum_{m=1}^{M} \sum_{l=1}^{L} a_{m,l} x_m D_{m,l} \tag{5}$$

where x_m denotes the binary association variable for MBS association mode, i.e., $x_m = 1$, if RU_m is associated with the MBS; otherwise, $x_m = 0$, $D_{m,l}$ denotes the service delay of RU_m acquiring F_l by associating with the MBS. We express $D_{m,l}$ as

$$D_{m,l} = D_{m,l}^{\mathrm{t}} + D_{m,l}^{\mathrm{B}} + D_{m,l}^{\mathrm{w}} \tag{6}$$

where $D_{m,l}^{\mathrm{t}}$ denotes the transmission delay of RU_m acquiring F_l by associating with the MBS, $D_{m,l}^{\mathrm{w}}$ denotes the queuing delay at the MBS when RU_m acquiring F_l and $D_{m,l}^{\mathrm{B}}$ denotes the backhaul delay of the MBS. In this paper, we model the backhaul delay as an exponentially distributed random variable with a given mean value [11].

$D_{m,l}^{\mathrm{t}}$ in (6) can be expressed as

$$D_{m,l}^{\mathrm{t}} = \frac{S_l}{R_m} \tag{7}$$

where R_m denotes the data rate of the transmission link of between the MBS and RU_m, which can be expressed as

$$R_m = W_0 \log_2 \left(1 + \frac{P_m g_m}{\sigma^2} \right) \tag{8}$$

where P_m is the transmission power of the MBS when sending the content to RU_m and g_m is the channel gain between the MBS and RU_m.

$D_{m,l}^{\mathrm{w}}$ in (6) can be calculated as

$$D_{m,l}^{\mathrm{w}} = \frac{1}{\mu - \lambda} \tag{9}$$

where μ and λ are the service rate and arrival rate of the MBS, respectively.

Service Delay of RUs in SBS Association Mode where $x_{m,n}^{\mathrm{s}}$ denotes the binary association variable for SBS association mode, i.e., $x_{m,n}^{\mathrm{s}} = 1$ if RU_m is associated with SBS_n; otherwise, $x_{m,n}^{\mathrm{s}} = 0$, $D_{m,n,l}^{\mathrm{s}}$ denotes the service delay of RU_m when acquiring F_l through associating with SBS_n and can be computed as

$$D_{m,n,l}^{\mathrm{s}} = D_{m,n,l}^{\mathrm{s,t}} + \left(1 - y_{n,l}^{\mathrm{s}} \right) \left(D_{n,l}^{\mathrm{s,t}} + D_{m,l}^{\mathrm{B}} + D_{m,l}^{\mathrm{w}} \right) \tag{10}$$

where $D_{m,n,l}^{\mathrm{s,t}}$ denotes the transmission delay of RU_m acquiring F_l by associating with SBS_n, $D_{n,l}^{\mathrm{s,t}}$ denotes the transmission delay between MBS_n and the SBS acquiring F_l, $y_{n,l}^{\mathrm{b}}$ denotes the binary content placement variable of the SBSs, i.e., $y_{n,l}^{\mathrm{b}} = 1$ if F_l is placed at SBS_n; otherwise, $y_{n,l}^{\mathrm{b}} = 0$.

$D_{m,n,l}^{\mathrm{s,t}}$ in (10) can be computed as

$$D_{m,n,l}^{\mathrm{s,t}} = \frac{S_l}{R_{m,n}^{\mathrm{s}}} \tag{11}$$

where $R_{m,n}^{\mathrm{s}}$ denotes the data rate of the transmission link between RU_m and SBS_n, which can be expressed as

$$R_{m,n}^{\mathrm{s}} = W_n \log_2 \left(1 + \frac{P_n^{\mathrm{s}} g_{m,n}^{\mathrm{s}}}{\sigma^2} \right) \tag{12}$$

where P_n^{s} denotes the transmission power of SBS$_n$ and $g_{m,n}^{\mathrm{s}}$ is the channel gain between RU$_m$ and SBS$_n$.

$D_{n,l}^{\mathrm{s,t}}$ in (10) can be expressed as

$$D_{n,l}^{\mathrm{s,t}} = \frac{S_l}{R_n^{\mathrm{s}}} \tag{13}$$

where R_n^{s} denotes the data rate of the transmission link between SBS$_n$ and the MBS, which can be computed as

$$R_n^{\mathrm{s}} = W_0 \log_2 \left(1 + \frac{P_n^{\mathrm{s}} g_n^{\mathrm{s}}}{\sigma^2} \right) \tag{14}$$

where P_n^{s} is the transmission power of the MBS when transmitting to SBS$_n$ and g_n^{s} is the channel gain between the MBS and SBS$_n$.

3.2   Optimization Constraints

To design the optimal joint user association and content placement policy which minimizes the service delay of all the RUs, a number of optimization constraints have to be considered.

User Association Constraints In this paper, we assume that each RU can acquire the required content by means of at most one association mode, i.e.,

$$\mathrm{C1}: \sum_{k=1}^{K} x_{m,k}^{\mathrm{d}} + \sum_{n=1}^{N} x_{m,n}^{\mathrm{s}} + x_m \leq 1. \tag{15}$$

As for D2D communication, we assume each SU can only offer service for at most one RU provided that the SU has cached the required content of the RU. Hence, we can express the constraints as

$$\mathrm{C2}: \sum_{m=1}^{M} x_{m,k}^{\mathrm{d}} \leq 1, \tag{16}$$

$$\mathrm{C3}: a_{m,l}\, x_{m,k}^{\mathrm{d}} \leq y_{k,l}^{\mathrm{d}}. \tag{17}$$

Accounting for the bandwidth capacity constraints of the MBS and the SBSs, the number of RUs associating with each SBS or the MBS should not exceed the maximum number of the subchannels of the corresponding BS, which can be formulated as

$$\mathrm{C4}: \sum_{m=1}^{M} x_{m,n}^{\mathrm{s}} \leq A_n, \tag{18}$$

$$\mathrm{C5}: \sum_{m=1}^{M} x_m \leq A_0. \tag{19}$$

Data Rate Constraints We assume that the RUs with certain content demand may have different minimum data rate requirements, thus the data rate constraint of RU_m can be expressed as

$$C6 : \sum_{k=1}^{K} x_{m,k}^{d} R_{m,k}^{d} + \sum_{n=1}^{N} x_{m,n}^{s} R_{m,n}^{s} + x_m R_m \geq R_m^{min} \tag{20}$$

where R_m^{min} denotes the minimum data rate requirement of RU_m.

Caching Storage Constraints of the SBSs Considering the limited and various cache capacity of the SBSs, the number of contents placed in the cache of the SBSs should be limited to the maximum cache storage constraint, which can be expressed as

$$C7 : \sum_{l=1}^{F} y_{n,l}^{s} S_l \leq S_n^{s} \tag{21}$$

where S_n^{s} denotes the maximum cache capacity of SBS_n.

User Content Request Constraints In this paper, we assume that each RU can only access one content, i.e., $\sum_{l=1}^{L} a_{m,l} \leq 1$. It is apparent that user association and content placement should subject to user requirement on certain content. More specifically, in the case that one RU does not pose requirement on one particular content, no corresponding user association and content placement strategy should be designed, i.e., if there have no request for F_l, we set $x_{m,k}^{d}$, $x_{m,n}^{s}$, x_m and $y_{n,l}^{s} = 0$, otherwise, $x_{m,k}^{d}$, $x_{m,n}^{s}$, x_m and $y_{n,l}^{s} = 0$ or 1. We can rewrite the constraints of user content request as

$$C8 : (1 - \sum_{l=1}^{L} a_{m,l}) x_{m,k}^{d} \leq 0, \tag{22}$$

$$C9 : (1 - \sum_{l=1}^{L} a_{m,l}) x_{m,n}^{s} \leq 0, \tag{23}$$

$$C10 : (1 - \sum_{l=1}^{L} a_{m,l}) x_m \leq 0, \tag{24}$$

$$C11 : (1 - \sum_{l=1}^{L} a_{m,l}) y_{n,l}^{s} \leq 0. \tag{25}$$

3.3 Optimization Problem

Jointly considering the optimization objective and the constraints, we can formulate the optimization problem as follows.

$$\min_{x^d_{m,k}, x^s_{m,n}, x_m, y^s_{n,l}} D \tag{26}$$

$$\text{s.t.} \qquad \text{C1} - \text{C11}.$$

4 Solution of the Optimization Problem

The problem in (26) is an integer-nonlinear programming problem, it is difficult to solve it directly. To solve the problem, in this section, we apply McCormick envelopes to remove the coupling among optimization variables in (26) and equivalently transform the optimization problem into three subproblems by applying Lagrangian partial relaxation, then we solve the subproblems by using Hungarian method and unidimensional knapsack algorithm respectively.

4.1 Reformulation of the Optimization Problem

To decouple the user association variables $x^s_{m,n}$ and the content placement variables $y^s_{n,l}$ in the objective function in (26), we introduce a new variable, $z^s_{m,n,l} = x^s_{m,n} y^s_{n,l}$ and rewrite the optimization problem by using McCormick envelopes [11]. For convenience, we set

$$\mathbf{X} = \{x^d_{m,k}, x^s_{m,n}, x_m | \text{RU}_m \in \text{RU}, \text{SBS}_n \in \text{SBS}, \text{SU}_k \in \text{S}_U\}, \tag{27}$$

$$\mathbf{Y} = \{y^s_{n,l} | \text{SBS}_n \in \text{SBS}, F_l \in F\}, \tag{28}$$

$$\mathbf{Z} = \{z^s_{m,n,l} | \text{RU}_m \in \text{RU}, \text{SBS}_n \in \text{SBS}, F_l \in F\}. \tag{29}$$

The original optimization problem can be rewritten as

$$\min_{\mathbf{X},\mathbf{Y},\mathbf{Z}} \sum_{m=1}^{M} \sum_{k=1}^{K} \sum_{l=1}^{L} a_{m,l} x^d_{m,k} y^d_{k,l} D^d_{m,k,l} + \sum_{m=1}^{M} \sum_{l=1}^{L} a_{m,l} x_m \{D^t_{m,l} + D^B_{m,l} + D^w_{m,l}\}$$

$$\sum_{m=1}^{M} \sum_{n=1}^{N} \sum_{l=1}^{L} \{a_{m,l} x^s_{m,n} (D^{s,t}_{m,n,l} + D^{s,t}_{n,l} + D^B_{m,l} + D^w_{m,l}) -$$

$$a_{m,l} z^s_{m,n,l} (D^{s,t}_{n,l} + D^B_{m,l} + D^w_{m,l})\} \tag{30}$$

$$\text{s.t.} \quad \text{C1} - \text{C11 in (26)},$$

$$\text{C12}: z^s_{m,n,l} \geq 0,$$

$$\text{C13}: z^s_{m,n,l} \geq x^s_{m,n} + y^s_{n,l} - 1,$$

$$\text{C14}: z^s_{m,n,l} \leq y^s_{n,l},$$

$$\text{C15}: z^s_{m,n,l} \leq x^s_{m,n}.$$

4.2 Lagrangian Partial Relaxation and Dual Problem Formulation

Obviously the optimization problem in (30) is convex which can be solved using traditional optimization tools. In this subsection, we apply the method of Lagrange partial relaxation [12] to incorporate C13–C15 into the function (30), which can be calculated as

$$
\max_{\varphi,\nu,\eta} \min_{\mathbf{X},\mathbf{Y},\mathbf{Z}} L\left(\varphi_{m,n,l}, \nu_{m,n,l}, \eta_{m,n,l}, x^{\mathrm{d}}_{m,k}, x^{\mathrm{s}}_{m,n}, x_m, y^{\mathrm{s}}_{n,l}, z^{\mathrm{s}}_{m,n,l}\right).
$$

$$
= \sum_{m=1}^{M}\sum_{n=1}^{N}\sum_{l=1}^{L}\{a_{m,l}x^{\mathrm{s}}_{m,n}(D^{\mathrm{s,t}}_{m,n,l} + D^{\mathrm{s,t}}_{n,l} + D^{\mathrm{B}}_{m,l} + D^{\mathrm{w}}_{m,l})-
$$

$$
a_{m,l}z^{\mathrm{s}}_{m,n,l}(D^{\mathrm{s,t}}_{n,l} + D^{\mathrm{B}}_{m,l} + D^{\mathrm{w}}_{m,l})+
$$

$$
\sum_{m=1}^{M}\sum_{k=1}^{K}\sum_{l=1}^{L} a_{m,l}x^{\mathrm{d}}_{m,k}y^{\mathrm{d}}_{k,l}D^{\mathrm{d}}_{m,k,l} + \sum_{m=1}^{M}\sum_{l=1}^{L} a_{m,l}x_m\{D^{\mathrm{t}}_{m,l}+D^{\mathrm{B}}_{m,l} + D^{\mathrm{w}}_{m,l}\}+
$$

$$
\sum_{m=1}^{M}\sum_{n=1}^{N}\sum_{l=1}^{F}\{a_{m,l}\varphi_{m,n,l}(x^{\mathrm{s}}_{m,n} + y^{\mathrm{s}}_{n,l} - 1 - z^{\mathrm{s}}_{m,n,l}) + a_{m,l}\nu_{m,n,l}(z^{\mathrm{s}}_{m,n,l} - y^{\mathrm{s}}_{n,l})+
$$

$$
a_{m,l}\eta_{m,n,l}(z^{\mathrm{s}}_{m,n,l} - x^{\mathrm{s}}_{m,n})\}
$$

$$(31)$$

where $\varphi > 0$, $\nu > 0$ and $\eta > 0$ are the corresponding lagrange multipliers for C13, C14 and C15.

4.3 Dual Decomposition and Solution

By examining the optimization problem formulated in (31), it can be validated that both the objective problem and the constraints are separable in terms of $x^{\mathrm{d}}_{m,k}$, $x^{\mathrm{s}}_{m,n}$, x_m, $y^{\mathrm{s}}_{n,l}$ and $z^{\mathrm{s}}_{m,n,l}$. We can further decompose the problem into three subproblems, that is

$$
P_1 : \min_{\mathbf{X}} \sum_{m=1}^{M}\sum_{k=1}^{K}\sum_{l=1}^{L} a_{m,l}x^{\mathrm{d}}_{m,k}y^{\mathrm{d}}_{k,l}D^{\mathrm{d}}_{m,n,l} + \sum_{m=1}^{M}\sum_{l=1}^{L} a_{m,l}x_m\{D^{\mathrm{t}}_{m,l} + D^{\mathrm{B}}_{m,l}+D^{\mathrm{w}}_{m,l}\}
$$

$$
\sum_{m=1}^{M}\sum_{n=1}^{N}\sum_{l=1}^{L} a_{m,l}x^{\mathrm{s}}_{m,n}\left(D^{\mathrm{s,t}}_{m,n,l}\ D^{\mathrm{s,t}}_{n,l} + D^{\mathrm{B}}_{m,l} + D^{\mathrm{w}}_{m,l} + \varphi_{m,n,l} - \eta_{m,n,l}\right)
$$

$$(32)$$

s.t. C1 − C6, C8 − C10.

$$
P_2 : \max_{\mathbf{Y}} \sum_{m=1}^{M}\sum_{n=1}^{N}\sum_{l=1}^{L} a_{m,l}y^{\mathrm{s}}_{n,l}\left(\nu_{m,n,l} - \varphi_{m,n,l}\right)
$$

$$(33)$$

s.t. C7, C11.

$$P_3 : \min_{\mathbf{Z}} \sum_{m=1}^{M} \sum_{n=1}^{N} \sum_{l=1}^{L} a_{m,l} z_{m,n,l}^{s}(v_{m,n,l} + \eta_{m,n,l} - \varphi_{m,n,l} - D_{n,l}^{s,t} - D_{m,l}^{B} - D_{m,l}^{w})$$

$$(34)$$

s.t. C12.

The subproblem P_1 involves only the association variables $x_{m,k}^{d}$, $x_{m,n}^{s}$, x_m, which can be transformed into a balanced assignment problem and solved by using Hungarian method [13]. The subproblem P_2 involves the content placement variables $y_{n,l}^{s}$, which can be decomposed into $|N|$ unidimensional knapsack problems, one for each SBS_n, which can be solved independently. Similar to subproblem P1, P3 can also be solved by using Hungarian method.

4.4   Updating Lagrange Multipliers

Lagrange multipliers can be updating based on a subgradient method for finding the locally optimal solution of above three subproblems via the form of an iterative operation, i.e.,

$$\varphi_{m,n,l}(t+1) = [\varphi_{m,n,l}(t) + \alpha(t)d(\varphi_{m,n,l}(t))]^{+} \quad (35)$$

$$v_{m,n,l}(t+1) = [v_{m,n,l}(t) + \alpha(t)d(v_{m,n,l}(t))]^{+} \quad (36)$$

$$\eta_{m,n,l}(t+1) = [\eta_{m,n,l}(t) + \alpha(t)d(\eta_{m,n,l}(t))]^{+} \quad (37)$$

where $[z]^{+} = \max\{0, z\}$ and $\alpha(t) = \varepsilon \frac{u_b - l_b}{||g(t)||^2}$ [11] is the step-size in tth iteration. ε is the positive control parameter. u_b and l_b are the upper bound and lower bound respectively. Conducting the above process iteratively, the algorithm will achieve convergence and then obtain the globally near-optimal user association and content placement strategy.

5   Simulation Results

In this section, we examine the performance of the proposed algorithm and compare the algorithm with other two algorithms via simulation. In the simulation, we consider a D2D-enabled HCN consisting of one MBS, two SBSs and a number of UEs. We consider 10 SUs and other parameters are summarized in Table 1. We initially set ε to 2.0 and update it to $\varepsilon = \varepsilon/2$ if there is no variation in the upper bound for about 50 successive iterations.

Figure 2 shows the upper bound and lower bound versus the number of iterations obtained from our proposed algorithm. The number of RUs is chosen as 40 and we set subchannel bandwidth is 1 MHz. We assume each RU makes random requests from the content library and the SUs randomly pre-caching some contents. It can be observed that the upper bound and lower bound nearly simultaneously converges within less than 140 iterations.

Table 1. Simulation parameters

Parameters	Value
Transmission power of SU_k (P_k^d), $SBS_n(P_n^s)$, MBS (P_n)	0.25 W, 2 W, 40 W
Storage capacity (SBS_n^s)	10
The number of the contents (L)	50
Content size (S_l)	[1, 10] Mbits
Noise power (σ^2)	−174 dBm/HZ
Maximum numbers of associated users of the MBS (A_0), SBS_n (A_n)	40, 10
Minimum data rate requirements of users (R_m^{min})	100 Kbps
Backhaul delay $(D_{m,l}^B)$	$\{1, 2\}$
Channel path loss model	128.1 + 27log(d) dB, d denotes the distance

Figure 3 shows the service delay versus different number of RUs. we compare the performance of proposed scheme with two algorithms: Scheme 1 proposed in [8] and Scheme proposed in [10]. From the figure, we can see that the service delay increases with the increase of backhaul delay and our proposed scheme outperforms them and the performance gap is lager with the increasing number of the RUs. This is because Scheme 1 only considers cache-enabled BSs, no D2D transmission is allowed. In Scheme 2, the popular contents are pre-caching into BSs and UEs and request UEs associate to BSs or UEs providing its maximum received power. However, separately designing content placement without jointly consider user association will not efficient improve system performance.

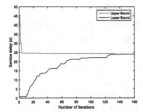 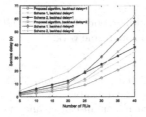 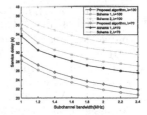

Fig. 2. Service delay vs number of iterations.

Fig. 3. Service delay vs the number of RUs.

Fig. 4. Service delay vs the subchannel bandwidth

Figure 4 shows the service delay versus the subchannel bandwidth of the MBS. We set backhaul delay as 1, the service rate of the MBS, i.e., μ=110 and examine the service delay performance corresponding to different packet arrival rate. From the figure we can see that the service delay increases with the increase of the arrival rate and our proposed algorithm achieves better performance than other two schemes. This is because our proposed algorithm can make a suitable adjustment to deal with different bandwidth status while the other two schemes fail to consider it, thus result in relatively lower performance.

6   Conclusions

In this paper, we investigated the joint user association and content placement strategy in D2D-enabled HCN and formulate joint user association and content caching problem as the service delay minimization problem of all the RUs. By applying McCormick envelopes and Lagrangian partial relaxation, we decoupling the original problem into three subprolbems and then solve it by using Hungarian method and unidimensional knapsack algorithm. Numerical results demonstrated the proposed algorithm outperforms previously proposed schemes.

References

1. Wang, X., Li, X., Leung, V.C.M., Nasiopoulos, P.: A framework of cooperative cell caching for the future mobile networks. In: Hawaii International Conference on System Sciences, Kauai, HI, pp. 5404–5413 (2015)
2. Li, X., Wang, X., Li, K., Leung, V.C.M.: Collaborative hierarchical caching for traffic offloading in heterogeneous networks. In: IEEE ICC, pp. 1–6(2017)
3. Tan, Z., Li, X., Yu, F.R., Chen, L., Ji, H., Leung, V.C.M.: Joint access selection and resource allocation in cache-enabled HCNs with D2D communications. In: IEEE Wireless Communications and Networking Conference (WCNC), pp. 1–6 (2017)
4. Shanmugam, K., Golrezaei, N., Dimakis, A., Molisch, A.F., Gaire, G.: FemtoCaching: wireless content delivery through distributed caching helpers. IEEE Trans. Inform. Theory 59(12), 8402–8413 (2013)
5. Song, J., Song, H., Choi, W.: Optimal content placement for wireless Femto-caching network. IEEE Trans. Wirel. Commun. 16(7), 4433–4444 (2017)
6. Li, X., Wang, X., Li, K., Han, Z., Leung, V.C.M.: Collaborative multi-tier caching in heterogeneous networks: modeling, analysis, and design. IEEE Trans. Wirel. Commun. 16(10), 6926–6939 (2017)
7. Kang, H.J., Kang, C.G.: Mobile device-to-device (D2D) content delivery networking: a design and optimization framework. J. Commun. Netw. 16(5), 568–577 (2014)
8. Dai, B., Yu, W.: Joint user association and content placement for cache-enabled wireless access networks. In: IEEE ICASSP, pp. 3521–3525 (2016)
9. Khreishah, A., Chakareski, J., Gharaibeh, A.: Joint caching, routing, and channel assignment for collaborative small-cell cellular networks. IEEE J. Sel. Areas Commun. 34(8), 2275–2284 (2016)
10. Yang, C., Yao, Y., Chen, Z., Xia, B.: Analysis on cache-enabled wireless heterogeneous networks. In IEEE Trans. Wirel. Commun. 15(1), 131–145 (2016)
11. Wang, Y., Tao, X., Zhang, X., Mao, G.: Joint caching placement and user association for minimizing user download delay. In: IEEE Access, vol. 4, pp. 8625–8633 (2016)
12. Bertsekas, D., Nedic, A., Ozdaglar, A.: Convex analysis and optimization. In: Athena. Scientific Press (2003)
13. Kuhn, H.W.: The Hungarian method for the assignment problem. Nav. Res. Logist. 52(1), 7–21 (2005)

Hybrid Caching Transmission Scheme for Delay-sensitive Service in Vehicular Networks

Rui Shi[✉], Xi Li, Hong Ji, and Heli Zhang

Key Laboratory of Universal Wireless Communications, Ministry of Education,
Beijing University of Posts and Telecommunications,
Beijing, People's Republic of China
{sr_bj,lixi,jihong,zhangheli}@bupt.edu.cn

Abstract. With the inspiring development of vehicular networks, caching popular contents in the network edge nodes could greatly enhance the quality of user experience. However, the highly dynamic movements of vehicles make it difficult to maintain stable wireless transmission links between vehicle-to-vehicle pairs or vehicle-to-road side units (RSUs) pairs, and then resulting in unbearable transmission delays or even transmission interruptions. In this paper, we proposed a predictive and hybrid caching transmission scheme for delay-sensitive services in vehicular networks. In order to select the most proper node for transmitting desired content from the nearby RSUs or vehicles, we evaluate the candidate nodes from the prediction on effective communication range and connection time based on the relative velocities and SINR threshold. Then the end-to-end delay for respective nodes is compared which includes two parts: waiting period and transmitting period. Waiting period is predicted based on the relative distance and relative velocities between two nodes at the starting position. Transmitting period is calculated from the transmission rate and effective communication range. The candidate node with the lowest delay is selected to transmit the desired content to the destination vehicle. Simulation results show that the proposed scheme could significantly reduce time delays in data transmission, especially when the requesting vehicle is far from the nearest RSU.

Keywords: Vehicular networks · Cached data distribution · Transmission delay

1 Introduction

The development of intelligent transportation systems (ITSs) has attracted increasing attention to the communication in vehicular networks. As more and more data needs to be exchanged within moving vehicles in various scenarios, the throughput and delay requirements have also been enhanced for users quality of

X. Liu et al. (Eds.): ChinaCom 2018, LNICST 262, pp. 283–293, 2019.
https://doi.org/10.1007/978-3-030-06161-6_28

services (QoSs). Then caching technology is introduced into vehicular networks to provide higher local throughput and lower end-to-end transmission delay. It has the potential to reduce vehicle data access times, improve the utilization of vehicular storage space, and reduce network bandwidth consumption [2,3]. There are still several open issues in the cached-based vehicular networks. How to provide efficient transmission for the delay-sensitive services for moving vehicles is an interesting and important problem that has attracted many researchers' attention.

Most of the existing research focus on pre-caching, data replacement, multi-hop relay and throughput maximization. But there is a practical constraint that has been ignored, which is the short connection time between the source and destination due to the mobility of vehicles. The connection time may be shorter than the time required to transmit the content, thus resulting in connection break. Even if there is no connection break, the time delay may be unbearable. In V2I communication, although a vehicle is within the communication range of a road side unit (RSU), the distance could be too long to establish a stable link. Thus the transmission rate or delay may be degraded or even cause user frustration and network congestion. In this paper, we investigate the problem of how to reduce data transmission delays in hybrid V2V and V2I scenarios by mobility prediction. We focus on the following issue: during V2I communication, a large distance between the vehicle and the nearest RSU could result in long delay in data transmission. In addition, if the connection time is not enough for content transmitting, there may be connection break.

Valuable research has already been conducted on the direction of where and how to efficiently cache content. Most of these have focused on where to cache, how to cache and cache replacement. For example, Ding et al. [4] proposed three algorithms to allocate files to RSUs to minimize the average transmission time delay. Ma et al. [5] proposed a caching placement policy which jointly considered caching at the vehicular and RSUs. The caching placement problem is modeled as an optimization problem to minimize the average latency while satisfying the QoE requirements of vehicles. Wei et al. [6] proposed a layered cooperative cache management to select neighbor nodes within broadcast range to cache proper contents to reduce retrieval time and prevent stalls of the video playback. Alotaibi et al. [7] proposed the Area Defer Transmission dissemination algorithm to enable each vehicle to independently decide whether to transmit considering heterogeneous transmission ranges and the amount of area that would be covered by potential new transmission. Deng et al. [8] proposed a Prior-Response-Incentive-Mechanism to stimulate vehicles to take part in cooperative downloading in VANETs-LTE heterogeneous networks.

In this paper, we propose a novel scheme to reduce transmission delays in hybrid V2I and V2V communications by evaluating the ability of the candidate nodes for maintaining a stable wireless link, based on the relative velocities and SINR threshold. Then we predict the waiting period and transmitting period of each nodes, using the relative distance and relative velocities between two nodes at the starting position, and the transmission rate and effective communication range. The waiting period together with the transmitting period composes the transmission delays. Finally we select the node which has the lowest delays to transmit the content. Simulation results show that the proposed hybrid caching transmission scheme achieves good performance.

The rest of this paper is organized as follows, In Sect. 2, we provide the system model we use. In Sect. 3, we present the proposed scheme. In Sect. 4, The Simulation scenarios and results are presented. The conclusions are stated in Sect. 5.

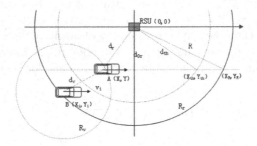

Fig. 1. Relative positions and distance between nodes

2   System Model

2.1   Vehicle Mobility Model

The communication scenario we consider in this paper is shown in Fig. 1. We focused on a one-way street which has road side units (RSUs) deployed in an urban setting. The traffic is assumed to flow freely, and the vehicle arrival process follows a Poisson distribution [9]. We further assumed no congestion, that arrival processes are independent and that the vehicle speed follows uniform distribution [10].

For V2I communication scenario let $(0,0)$ be the location of the RSU, let (x, y) be the location of vehicle A which sends a request for specific content, and let (x_i, y_i) be the location of vehicle B which has the requested content cached and is within communication range of A. Where the communication range of the RSU is R_r then the point (x_r, y_r),where vehicle A would lose connection with RSU can be calculated as:

$$(x_r, y_r) = (\sqrt{R^2 - y^2}, y) \tag{1}$$

For V2V communication scenario, where the communication range of a vehicle is R_v, consider the relative motion between A and B, then let the relative velocity of B to A be $v_B = v_i - v$, and relative location be $(x_B, y_B) = (x_i - x, y_i - y)$. Here we only consider the horizontal distance of A and B, because the two vehicles are assumed to be moving on the same road. The minimum distance of two vehicle is set at 1 meter.

Assuming that the velocity of a vehicle stays the same during the data distribution process, then the horizontal location of a vehicle also follows uniform distribution. The probability distribution is given by:

$$y = \begin{cases} 1/(x_{t+\Delta t} - x_t), & \Delta t \geq 0 \\ 0, & \Delta t < 0 \end{cases} \tag{2}$$

2.2 Communication Channel Model

We considered both V2V and V2I communications over Rayleigh fading channel. Let $\Upsilon_{u,v}$ denote the SINR at node v when node u transmits data, let N_0, B, P_t, $I_{u,v}$ $d_{u,v}$, α denote the power spectral density of additive Gaussian white noise, the channel bandwidth, the transmit power of node u, the inter-cell interference, the distance between node u and node v, and the path loss exponent. Then the SINR is given by:

$$\Upsilon_{u,v} = \frac{P_t \cdot d_{u,v}^{-\alpha}}{N_0 \cdot B + I_{u,v}} \tag{3}$$

3 Proposed Scheme

3.1 Overview of the Scheme

Data transmission rates are inversely proportional to the distance between RSUs and vehicles, meaning that vehicles far from the RSU may suffer longer delays when transmitting content. As the relative velocity between two vehicles moving in the same direction on the same road is relatively low, any two such vehicles are likely to stay close to each other over a short period of time. The data transmission rate in this case is therefore relatively high, and the vehicles are able to establish a connection which is able to transmit data with a short time delay. In the proposed scheme, when vehicle A requests content, it could get the content from either the RSU or from another nearby vehicle B, which has the requested content cached. We first determine if a candidate node is capable of successfully transmitting the content based on the relative velocities and SINR threshold. Then predict the time delays using the relative distance and relative velocities between two nodes at the starting position and the transmission rate and effective communication range. Finally, we compared the time delays in each case, allowing for a selection of the node that has lower delay as the transmission node.

3.2 Effective Connection Time

To ensure successful transmission, a receiver would only begin to receive data when the SINR is greater than a specific threshold Ψ [10]. Using Eq. 3, the maximum distance between two nodes for maintaining a link is given by:

$$d_{th} = \sqrt[\alpha]{\frac{P_t}{\Psi \cdot N_0 \cdot B + I_{u,v}}} \tag{4}$$

Then the range of horizontal location for A to successfully receive content from RSU is $[-x_{th}, x_{th}]$, where $x_{th} = \sqrt{d_{th}^2 - y^2}$. And the range of relative distance between A and B to successfully transfer content is $[-d_{th}, d_{th}]$.

For V2I communication, the maximum effective connection time T_r depends on the location (x, y) of A when it sends the request for content. Let this location also be the observation start point. We distinguished between communication range (where communication between vehicle and RSU is possible, but where the SINR may be not big enough to ensure a stable link) and effective communication range (where communication between vehicle A and RSU is possible over a stable connection). Therefore, when the location of A is within communication range $[-x_r, x_r]$ of the RSU but out of the effective communication range $[-x_{th}, x_{th}]$, as soon as A moves into $[-x_{th}, x_{th}]$, it is able to form an effective connection with the RSU. When A is within the range $[-x_{th}, x_{th}]$, the connection is established immediately and will be maintained to the point x_{th}. Then T_r is given by:

$$
T_r = \begin{cases} \frac{x_{th} - x}{v}, & -x_{th} \leq x \leq x_{th} \\ \frac{2x_{th}}{v}, & -x_{th} \leq -x_r \leq -x_{th} \\ 0, & otherwise \end{cases} \tag{5}
$$

For V2V communication, the maximum effective connection time T_v depends on both the relative distance d_{0v} and velocity of B to A. The relative velocity of B to A is positive when B is moving faster than A, hence the relative distance threshold should also be positive. Similarly, when B moves slower than A, the relative distance threshold should be negative. Therefore T_v is given by:

$$
T_v = \begin{cases} \frac{-d_{th} - d_0 v}{v_B - v_A}, & v_B - v_A < 0 \\ \frac{d_{th} - d_0 v}{v_B - v_A}, & v_B - v_A \geq 0 \end{cases} \tag{6}
$$

Let R_b be the minimum data transmission rate, and assume it equals channel bandwidth [11]. Let q be the size of requested content. Then the maximum time delay T_m that A requires in order to receive the content is given by:

$$
T_m = \frac{q}{R_b} \tag{7}
$$

3.3   Transmission Delays Prediction

In order to calculate the data transmission rate, we first need to know the transmission location range according to the starting location (x, y) of A, velocity v and the maximum time delay R_b. In V2I scenario, the transmission location range is given by:

$$
[x, x_w] = [x, x + v \cdot T_m] \tag{8}
$$

In the V2V scenario, using the relative starting distance d_{0v} between B and A, the relative velocity of B to A, and the maximum time delay T_m, the relative transmission distance range is given by:

$$
[d_{0v}, d_w] = [d_{0v}, d_{0v} + (v_B - v) \cdot T_m] \tag{9}
$$

According to Shannon's capacity formula, the maximum transmission rate equals the instantaneous capacity of the channel. Let $R_{u,v}$ be the transmission rate from node u to node v:

$$R_{u,v} = B \cdot \log_2 (1 + \Upsilon_{u,v}) \tag{10}$$

Where B is the bandwidth of the channel and $\Upsilon_{u,v}$ is the SINR. By combining Eqs. 3 and 10, the transmission rate is denoted by:

$$R_{u,v} = B \cdot \log_2 \left(1 + \frac{P_t \cdot d_{u,v}^{-\alpha}}{N_0 \cdot B + I_{u,v}}\right) \tag{11}$$

Let R_{avg} denote the average transmission rate, we have:

$$R_{avg} = E[R_{u,v}] =$$
$$\int_{d_0}^{d} f(d) \cdot B \cdot \log_2 \left(1 + \frac{P_t \cdot d^{-\alpha}}{N_0 \cdot B + I_{u,v}}\right) dd \tag{12}$$

Where $f(d)$ is the probability distribution of the distance between two nodes. According to Eqs. 8 and 9, the average transmission rate from RSU to A is:

$$R_{avg} = \begin{cases} \dfrac{\int_0^{|x|} B \cdot \log_2 \left(1 + \frac{P_t \cdot (\sqrt{y^2 + d^2})^{-\alpha}}{N_0 \cdot B + I_{u,v}}\right) dd}{x_w - x} + \\ \dfrac{\int_0^{|x_w|} B \cdot \log_2 \left(1 + \frac{P_t \cdot (\sqrt{y^2 + d^2})^{-\alpha}}{N_0 \cdot B + I_{u,v}}\right) dd}{x_w - x} \\ \quad , \quad x \cdot x_w < 0 \\ \dfrac{\int_{min(|x|,|x_w|)}^{max(|x|,|x_w|)} B \cdot \log_2 \left(1 + \frac{P_t \cdot (\sqrt{y^2 + d^2})^{-\alpha}}{N_0 \cdot B + I_{u,v}}\right) dd}{x_w - x} \\ \quad , \quad x \cdot x_w \geq 0 \end{cases} \tag{13}$$

And the average transmission rate from B to A is given by:

$$R_{avg} = \begin{cases} \dfrac{\int_1^{|d_w|} B \cdot \log_2 \left(1 + \frac{P_t \cdot d^{-\alpha}}{N_0 \cdot B + I_{u,v}}\right) dd}{|d_w - d_{0v}|} + \\ \dfrac{\int_1^{|d_{0v}|} B \cdot \log_2 \left(1 + \frac{P_t \cdot d^{-\alpha}}{N_0 \cdot B + I_{u,v}}\right) dd}{|d_w - d_{0v}|} \\ \quad , \quad d_w \cdot d_{0v} < 0 \\ \dfrac{\int_{min(|d_{0v}|,|d_w|)}^{max(|d_{0v}|,|d_w|)} B \cdot \log_2 \left(1 + \frac{P_t \cdot d^{-\alpha}}{N_0 \cdot B + I_{u,v}}\right) dd}{|d_w - d_{0v}|} \\ \quad , \quad d_w \cdot d_{0v} \geq 0 \end{cases} \tag{14}$$

Let Q be the size of requested content, and let D_r, D_v be the average time delay incurred when A receives content from RSU and from B. When the starting location of A is out of the effective communication range $[-x_{th}, x_{th}]$, A must wait until it gets into this range. This waiting time is denoted as:

$$T_w = \frac{-x_{th} - x}{v} \tag{15}$$

Then D_r, D_v are given by:

$$D_r = \begin{cases} \frac{Q}{R_{avg}}, & -x_{th} \leq x \leq x_{th} \\ T_w + \frac{Q}{R_{avg}}, & -x_{th} \leq -x_r \leq -x_{th} \end{cases} \tag{16}$$

$$D_v = \frac{Q}{R_{avg}} \tag{17}$$

3.4 Node Selection Algorithm

The algorithm to select the transmission node with the lowest time delay follows the following steps:

Step1: give both RSU and vehicle B flags which start as zero. Calculate the distance threshold between A and RSU and between A and B. Get the achievable effective connection time by comparing the distance threshold to the maximum time delay needed to receive content q. If achievable effective connection time between A and node U is larger than maximum time delay, the value of U's flag becomes 1.

Step 2: node U is capable of content transmission only if node U's flag is 1. If the flag of a node is zero, set the time delay as infinite. Calculate the transmission rates from RSU and B according to the predicted distances, then calculate the time delays for both the RSU and B according to the predicted location of A.

Step 3: compare the transmission time delays for transmission from RSU and B, and choose the node which has shortest delay to be the transmitter for content q.

4   Simulation Results and Discussions

In this section, we present the simulation results of the proposed scheme. The simulation setup and detailed experimental results are as follows:

4.1 Simulation Setup

We conducted the simulation using MATLAB. The simulation parameters are shown in Table 1. In order to analyse the performance of our scheme, we set the location range of vehicle A as $[-x_r, x_r]$, and the velocity of each vehicle randomly between 10m/s and 20m/s. Our model assumed traffic was going in one direction, as vehicles moving at two opposite direction would have lower connection times. We simulated connection time, average transmission rate, average transmission delay and reduction in transmission delay to examine the effectiveness of our scheme and how it performs under different levels of transmission power.

4.2   Simulation Results

Average Transmission Delay Figure 2 shows the average delay incurred when vehicle A receives content q from: (1) only the RSU; (2) only a vehicle nearby. The time delay incurred when content is received from the RSU is related to the location of A. When A is far from RSU (outside of the effective communication range), the time delay is relatively high. As A moves closer to RSU, the time delay decreases. By contrast, the average time delay incurred when A receives content from another vehicle is stable over the observation area. Our results show that after adopting our scheme, when A is far from the RSU it receives content from a nearby vehicle, and when it is close to the RSU it receives content from the RSU. This ensures that the time delay is minimized throughout the observation area.

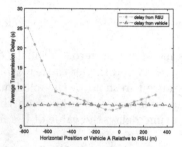

Fig. 2. Delay of a getting content from RSU and from vehicles on different position

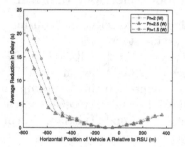

Fig. 3. Relationship between vehicle position, average delay reduction and RSU transmission power

Average Reduction in Transmission Delay Figure 3 shows the average reduction in transmission delay. When A is outside of effective communication range of the RSU, there are relatively long delays in receiving content from the RSU. As A moves closer to the RSU and into effective communication range, the average transmission delay falls due to increases in the transmission rate between A and RSU. This indicates that the greatest reductions in transmission delay from adoption of our scheme are likely to occur when vehicle A is outside the effective communication range. These results also indicate that when the RSU is using lower levels of transmission power, adoption of our scheme could result in greater reductions in transmission delay.

Average Transmission Rate Figure 4 shows the average transmission rate in both V2V communication and V2I communication. In V2V communication, the random nature of inter-vehicle distance implies that the average transmission rate is mainly a function of the transmission power of the vehicles involved. In our experiment, as vehicle A and vehicle B are assumed to have the same

Table 1. Simulation parameters

Parameter	Value
RSU coverage range R_r (m)	800
Communication range of vehicle R_v (m)	250
RSU transmission power P_r (W)	2
Vehicle transmission power P_v (mw)	100
Vehicle speed v (m/s)	[10, 20]
Channel bandwidth B_w (MHz)	10
Path loss exponent α	3
Packet length Q (mB)	100
SNR threshold Ψ	400
Inter-Cell Interference (dBm)	−75
Addictive Gaussian Noise (dBm)	−105

level of transmission power, the transmission rate fluctuates around $1.5 * 10^8 b/s$. In V2I communication, the fixed transmission power of the RSU implies that the average transmission rate depends on the distance between A and RSU. As shown in Fig. 4, the transmission rate is highest when A is at the nearest point to the RSU, and as A gets farther away from RSU the transmission rate falls.

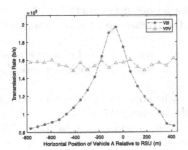

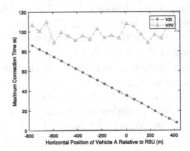

Fig. 4. Transmission rate in V2V and V2I communications on different position

Fig. 5. Maximum effective connection time in V2V and V2I communications on different position

Maximum Connection Time Figure 5 shows the average transmission rate in both V2V communication and V2I communication. In V2V communication, the relative velocity between vehicles on the same road over a short period of time can be assumed not to change substantially, with the result that the distance between the two vehicles also remains relatively stable. The maximum connection time therefore fluctuates around 90s. In V2I communication, the maximum connection time between RSU and A depends on two factors: first, it depends on the distance from A to the point x_{th} where A loses contact with RSU. Second, it depends on the velocity of A. As the velocity of A is assumed to remain the same over the observation period, as A gets closer to x_{th} the maximum connection time falls.

5 Conclusion

In this paper, we proposed a predictive and hybrid caching transmission scheme to reduce time delays in transmitting content within vehicular networks. The scheme first decides the reliability of the connections between the requesting vehicle and an RSU and between the requesting vehicle and any vehicle nearby which has the requested content cached. Then compares the predicted time delay needed to receive the content from the RSU and the nearby vehicle. Finally, this information is used to choose the node which has the shortest time delay to be the transmitter. Simulation results show that our scheme could substantially reduce data transmission time delays.

Acknowledgement. This paper is sponsored by the National Science and Technology Major Project of China (Grant No.2017ZX03001014).

References

1. Yousefi, S., Mousavi, M.S., Fathy, M.: Vehicular ad hoc networks (VANETs): challenges and perspectives[C]. In: International Conference on ITS Telecommunications Proceedings. IEEE (2006)
2. Song, H.B., Xiao, X.Q., Ming, X.U. et al.: Data caching algorithm in metropolitan vehicle network: data caching algorithm in metropolitan vehicle network[J]. J. Comput. Appl. (2010)
3. Zhao, W., Qin, Y., Gao, D, et al.: An efficient cache strategy in information centric networking vehicle-to-vehicle scenario[J]. IEEE Access (2017)
4. Ding, R., Wang, T., Song, L., et al.: Roadside-unit caching in vehicular ad hoc networks for efficient popular content delivery[C]. In: Wireless Communications and Networking Conference. IEEE (2015)
5. Ma, J., Wang, J., Liu, G., Fan, P.: Low latency caching placement policy for cloud-based VANET with both vehicle caches and RSU caches[C]. 2017 IEEE Globecom Workshops (GC Wkshps). Singapore (2017)
6. Wei, Y., Xu, C., Wang, M., Guan, J.: Cache management for adaptive scalable video streaming in vehicular content-centric network. In: 2016 International Conference on Networking and Network Applications (NaNA). Hakodate (2016)
7. Alotaibi, M.M., Mouftah, H.T.: Data dissemination for heterogeneous transmission ranges in VANets. In: 2015 IEEE 40th Local Computer Networks Conference Workshops (LCN Workshops). Clearwater Beach (2015)
8. Deng, G., Li, F., Wang, L.: Cooperative downloading in VANETs-LTE heterogeneous network based on named data[C]. In: Computer Communications Workshops. IEEE (2016)

9. Neelakantan, P.C., Babu, A.V.: Selection of minimum transmit power for network connectivity in vehicular ad hoc networks[C]. In: Fourth International Conference on Communication Systems and Networks. IEEE (2012)
10. Shelly, S., Babu, A.V.: A probabilistic model for link duration in vehicular ad hoc networks under rayleigh fading channel conditions[C]. In: Fifth International Conference on Advances in Computing and Communications. IEEE (2016)
11. Su, Z., Ren, P., Chen, Y.: Consistency control to manage dynamic contents over vehicular communication networks[C]. In: Global Telecommunications Conference. IEEE (2011)

Predictive Time Division Transmission Algorithm for Segmented Caching in Vehicular Networks

Rui Shi[✉], Xi Li, Hong Ji, and Heli Zhang

Key Laboratory of Universal Wireless Communications, Ministry of Education,
Beijing University of Posts and Telecommunications, Beijing, People's Republic of
China
{sr_bj,lixi,jihong,zhangheli}@bupt.edu.cn

Abstract. With the increasing number of different types of applications
for road safety and entertainment, it demands more flexible solutions for
caching and transmitting large files in vehicular networks. In order to
decrease the transmission delay and raise the hit ratio of cached files,
there is already a lot of research on caching technology, including segmented caching technology. But the problem of long transmission delay
and low successful transmission ratio caused by the high dynamic of
vehicles still needs to be solved. In this paper, we proposed an algorithm
named Predictive Time Division Transmission (PTDT) to reduce transmission delay and raise the ratio of successful transmission for segmented
cached file in vehicular networks. Our algorithm predicts the link duration between requesting vehicle and neighboring vehicles according to the
relative inter-vehicle distances and velocities. By predicting the transmit
rate of each vehicle on different time point, we divide the link duration
into slices for subsequent transmitter selections. And finally we compare
those time points and select the vehicles that make the transmitting
delay the lowest. In the mean time, we arrange the transmitting order of
those vehicles to guarantee the success of full file transmission process.
The simulation results show that after applying our algorithm, transmission delay has reduced and successful transmission rate has increased
substantially.

Keywords: Vehicular networks · Segment caching · Transmission
delay · Successful transmission ratio

1 Introduction

With the development of vehicular networks and the increasing demands for
many applications that require data of big size, caching technologies and transmission schemes are attracting more and more attentions. There are two types

© ICST Institute for Computer Sciences, Social Informatics and Telecommunications Engineering 2019
Published by Springer Nature Switzerland AG 2019. All Rights Reserved
X. Liu et al. (Eds.): ChinaCom 2018, LNICST 262, pp. 294–304, 2019.
https://doi.org/10.1007/978-3-030-06161-6_29

of communication systems in vehicular networks, which are Vehicle to Vehicle (V2V) and Vehicle to Infrastructure (V2I). In V2I communication, the limited storage of road side units (RSUs) and the great expenses of building them impel the emerge of distributed caching technology. In order to achieve better performance, there have been a lot of research on V2V caching technology [3–5]. To raise the hit ratio of cached files, and make the best of the storage at the same time, a method named segmented caching appeared. By applying this technology, a file is divided into pieces, the pieces of a file are distributed over multiple vehicles to be cached and forwarded to the requesting vehicle [6]. However, for the transmission of segmented file, due to the highly dynamic of vehicles, long transmission delay caused by big relative velocity, and high probability of transmission failure caused by connection break are still two problems that need to be solved.

There are many researches on how to maximize throughput and minimize transmission delay in vehicular networks. Basing on [7], an architecture which improves the throughput and resolve the problem of increasing traffic by using clustering technology in D2D links, [8,9] proposed an algorithm named CSVD which divides files into pieces and file pieces are cached at multiple SMs, selected UEs in each cluster are used for caching to reduce inter-cluster interference. And to maximize total throughput, [10] proposed a cooperative downloading strategy which utilizes both V2I and V2V communication. To minimize the transmission delay, [11] proposed a scheme which automatically choose the most appropriate mobility information when deciding next data-relays. But some thorough research on the above mentioned problems is still required.

In this paper, we focus on reducing transmission delay and raising successful transmission ratio for segmented cached files in vehicular networks. Since a requested file is divided into pieces and cached in different vehicles, the vehicles which are to be chosen as transmitters and the order of them to start transmitting matter a lot. Due to the highly dynamic of vehicles, connection break between vehicles happens all the time, if a vehicle starts transmitting file without analysing the ability of success, there is a big chance that the connection would break during the process of transmitting. And even though the transmitting process succeed, without analysing the delay each vehicle needs, the delay would be too long. And a big part of the transmission delay comes from the process of collecting data from all the vehicles in the communication range of the control center repeatedly.

To solve the two problems mentioned above, we proposed an algorithm named Predictive Time Division Transmission (PTDT). Our algorithm predicts the link duration between requesting vehicle and other vehicles, by using the relative inter-vehicle distances and velocities. Then analyse the probability of successful transmission to avoid transmission failure. By predicting the transmit rate of each vehicle on different time point, we divide the link duration into slices for subsequent transmitter selections. And we select vehicles with lowest transmission delay, at the same time arrange the transmitting order of those vehicles to ensure successful transmission.

The rest of this paper is organized as follows, In Sect. 2, we provide the system model we use. In Sect. 3, we present the proposed algorithm. In Sect. 4, The Simulation scenarios and results are present. The conclusions are stated in Sect. 5.

2   System Model

The communication scenario is considered as in a two-way street which is within the communication range of a base-station in an urban setting, and we assume there is no congestion, the velocity of each vehicle is not affected by other vehicles.

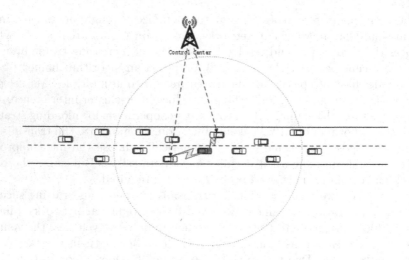

Fig. 1. Communication scenario

Consider the communication channel between vehicles as small-scale fading channel, The signal to noise ratio at the receiver node is denoted by $\frac{Z^2\beta \cdot P_t}{r^\alpha \cdot P_{noise}}$, where r is the distance between transmitter and receiver, Z is the fading coefficient, β is a constant associated with path loss model, P_t is the transmit power, α is the path loss exponent, and P_{noise} is the total additive noise power [1]. Here $\beta = \frac{G_T G_R \lambda^2}{2\pi d_0{}^2}$, where G_T and G_R are the gain of transmitter and receiver antennas, we assume all the antennas of vehicles are omni directional, so that $G_T = G_R = 1$. The total addictive noise power is P_{noise}, given by $P_{noise} = Fk_B T_0 r_b$, where F is the receiver noise figure, $k_B = 1.38 \cdot 10^{23} J/K$ is the Boltzmann constant, T_0 is the room temperature, $T_0 = 300K$ and r_b is the data transmission rate [2]. Assuming that $E[Z^2] = 1$, the average SNR can be written as

$$\Upsilon = \frac{\beta \cdot P_t}{r^\alpha \cdot P_{noise}} \tag{1}$$

As for communication range, we assume a transmitting process would success only if the signal to noise ratio (SNR) at the receiver node is above a specific threshold Υ during the process. Thus the communication range can be given by

$$r = \sqrt[\alpha]{\frac{\Upsilon \cdot P_{noise}}{\beta \cdot P_t}} \tag{2}$$

And for transmit rate, according to Shannon's equation, the maximum transmit rate can be denoted as the channel capacity, which is $R_{max} = B \cdot \log_2 (1 + \Upsilon)$, where B is the bandwidth of the communication channel. Using Eq. 1 the transmit rate can be given by

$$R = B \cdot \log_2 (1 + \frac{\beta \cdot P_t}{r^\alpha \cdot P_{noise}}) \tag{3}$$

We assume the files are segmented into many pieces, and each vehicle can cache any pieces of any files if it has enough space. And a vehicle can communicate with any vehicles in its communication range.

3   Proposed Algorithm

Our algorithm first group vehicles by the file pieces, and predict the link duration between requesting vehicle and other vehicles using the relative velocity and SNR threshold. Then we divided the link durations into different slices according to the inter-vehicle distance (to get the transmission rate at that distance) and the size of file pieces, each slice represent the delay of transmitting the file piece for one time. The link duration which has been divided into slices called time scale. Then according to the time scale, we select one vehicle in each group as the transmitter of each file piece. The main goals of our algorithm are reducing transmission delay of segmented file by selecting the most suitable vehicles as transmitter and increasing the successful transmission ratio by arranging the order of each transmitter we selected.

3.1   Grouping

When a requesting vehicle sends request to the control center, the control center first collects parameters of all the vehicles in its communication range R, including position, velocity, transmit power, the index of file piece it has. First our algorithm will eliminate those vehicles which don't have any piece of the requested file cached, then analyse the ability of vehicles for successfully transmitting file piece at least once. The ability is determined by comparing the maximum time needed for transmitting the piece and the maximum link duration, the maximum time needed for transmitting is calculated by suing the minimum transmitting rate which we assume to be $6 * 10^6 b/s$, the maximum link duration is calculated by using SNR threshold Υ and Eq. 2. Vehicles that cached the same file piece and are able to successfully transmit file piece will be grouped together. Every group has five parameters, which are index of vehicles that have the same file piece, the relative velocity of its members to the requesting vehicle, the distance of its members to the requesting vehicle at the starting point, the maximum transmitting round and max waiting time.

3.2   Transmitting Round Calculation

Every vehicle in the group has a link duration T according to the starting distance to the requesting vehicle and the SNR threshold, and a required transmitting time T_t at any distance to the requesting vehicle. By using T_t we can divide T into many slices, the process is as follows:

At any time point t_0 in the time scale of any vehicle we have the distance d_0 between this vehicle and the requesting vehicle, so that using Eq. 1 we get the required transmitting time Tt_0, Tt_0 means if this vehicle starts transmitting file at time point t_0, how long it will take for this vehicle to finish transmitting the file piece. Then move to the next time point which is $t_1 = t_0 + Tt_0$, using relative velocity of this vehicle to the requesting vehicle we get the inter-vehicle distance at t_1, then start the above process again till the time point moves to the end of link duration. Thus we'll have many points which represent the time point where vehicle could finish transmitting the file piece if it start transmitting at the time point ahead of this time point. We assume a vehicle would only start transmitting file at any of the points we calculated, this link duration with time points is called time scale, every vehicle has a time scale for the process of our algorithm. The point of dividing T into slices is that we want to choose the vehicle which has the lowest delay in every group to transmit file piece without calculating every time. Because in each round, it's not clear which vehicle will be chosen to transmit file piece, it is not clear what time it will finish transmitting, but the transmit rate of each vehicle is related to the distance to the requesting vehicle, which means if we don't divide T into slices and specify the transmit time point, every time a vehicle is chosen, we'll have to calculate the transmit delay of other vehicles according to the time that the chosen vehicle finishes transmitting.

3.3   Maximum Waiting Time

After grouping and calculating the time points of all vehicles, we now have the maximum waiting time. The maximum waiting time is determined by the vehicle which has the longest link duration in a group, and the maximum waiting time of this group is the last time point of that vehicle. The meaning of this is when the process is about to reach the maximum waiting time, if none of the vehicles in this group has ever been chosen to transmit file piece, we have to pick one vehicle in this group to transmit file piece otherwise we'll miss the chance of transmitting this file piece.

3.4   Algorithm Flow

Suppose the requested file is divided into M pieces, so there are M groups of vehicles, vehicles in the same group has the same file piece cached. The total transmitting delay is denoted as t and the transmitting round is denoted as R, at the beginning $t = 0$, $R = 1$, all the vehicles are in the left vehicle queue, assuming vehicle p is the one that has the lowest transmit delay in round 1, then vehicle p is chosen to be the first vehicle to transmit file piece. Then we move the transmit delay t to the finishing time point of vehicle p. Then those vehicles in the same group would be eliminated from the left vehicle queue. After first round, time delay t moves to t_{p1}, and the next choosing round starts. Now our algorithm needs to pick the vehicle which has the lowest transmitting delay in the second round, which should be the vehicle whose second time point behind t_{p1} is the earliest if t_{p1} is not about to reach any of the maximum waiting time of any group. The number of picking round is supposed to be the same as the number of file pieces, which means every piece of the file should only be transmit by one vehicle. Once there is no vehicle left in the left vehicle queue, we need to check if the number of the vehicles we picked is the same as the number of groups, and the vehicles we picked are in deferent groups. If not, we consider this process a failure, and set the time delay of using our algorithm as infinite, but as the experiment result shows this situation is rare to arise. The algorithm flow is shown in Algorithm 1.

Algorithm 1 Predictive Time Divided Transmission

Input: M: the number of file pieces; t: total transmission delay; $timeScale$: timescale of all the vehicles; $R = 1$: current picking round; $maxWait$: the maximum waiting time of all groups;

Output: t

1: **while** $(R \leq M) and (t \neq Inf)$ **do**
2:      **if** the row number of $timeScale$ is zero **then**
3:          $t = Inf$ **return**
4:      **end if**
5:      **if** $R = 1$ **then**
6:          find the minimum time point t_1 behind 0 in all $timescale[i]$.
7:          $t \leftarrow t_1$.
8:      **else**
9:          **if** $maxWait[i]$ close to t **then**
10:             find the minimum second time point t_R behind t in $timescale$ that belong to those vehicles in group i.
11:         **else**
12:             find the minimum second time point t_R behind t in all $timescale[i]$.
13:         **end if**
14:         $t \leftarrow t_R$.
15:     **end if**
16:     $R = R + 1$.
17:     eliminate rows which belong to vehicles in the same group with the one just picked from $timescale$.
18: **end while**

---

4 Simulation Results and Discussions

In this section, we conducted simulations to see the performance of our algorithm, we compared the delays to transmit files of different sizes and the successful transmit ratios after applying our algorithm.

4.1 Simulation Setup

We implemented the algorithm in MATLAB. The parameters are shown in Table 1. For the control center, the communication range is set as 2 km, which means it collects data in this area every time it choose a vehicles as transmitter for a piece of file. And the size of file piece is set as 5 Mb, we assume every piece of file has the same size for convenient application. And to make sure that the numbers of vehicles which have the same piece of file would not influence the simulation results, in different situation(applied our algorithm and without algorithm), the numbers of vehicles with the same file pieces in the communication range of requesting vehicle are the same. The range of velocity is set as [−20 m/s, 20 m/s].

4.2 Simulation Results

Average Reduction in Transmission Delay The average transmission delays and average reduction in transmission delays of transmitting a file which is segmented into four pieces are shown in Figs. 2, 3, 4. As shown in the figures, after applying our algorithm the average transmission delay has dropped in all situations. The main reason is that when using our algorithm the control center only needs to collect data from all the vehicles in its communication range once, but without our algorithm, if a vehicle requests a specific file, every time before transmitting a piece of the file, the control center needs to collect data from other vehicles and choose one to be the transmitter. If the density of vehicles in its communication range is high, the time of collecting data would make a big difference. Our algorithm avoids this repeating collections by predicting the relative distance between vehicles that have the file pieces cached and the requesting vehicle at the beginning.

Performance Under Different Path Loss Exponent Figure 5 shows the average transmission delays under different path loss exponent. As shown in the figure, with the increasing of path loss exponent, the average transmission delays under our algorithm is getting lower. When the path loss exponent increases, the communication range of vehicles under the same SNR threshold gets smaller, which means the link durations between vehicles will get shorter. And in order to ensure that the number of vehicles which cached the file pieces we need is the same in two different situation, when communication range gets smaller, the density of vehicles gets higher, which makes the collecting time of control center longer (transmitting delays bigger).

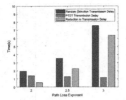
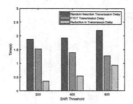
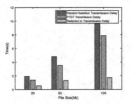

Fig. 2. Reduction in transmission delays under different path loss exponent

Fig. 3. Reduction in transmission delays under different SNR threshold

Fig. 4. Reduction in transmission delays under different file size

Figure 6 shows the average successful transmission ratios under different path loss exponent. As shown in the figure, after applying our algorithm, the average successful transmission ratio has increased to around 99%. That is because before we select the vehicles with different file pieces as transmitters, we first analyse the ability of those vehicles to finish transmitting file pieces to ensure successful transmission. With the increasing of path loss exponent, the communication range of vehicles under the same SNR threshold gets smaller, so that the probability of the connection break gets higher.

Table 1. Simulation parameters

Parameter	Value
Control center coverage range (m)	2000
Vehicle transmission power P_t (mw)	100
Vehicle speed v (m/s)	[−20, 20]
Channel bandwidth B_w (MHz)	10
File piece length Q (mB)	5

Performance Under Different File Size Figure 7 shows the average transmission delays of transmitting files with different sizes. As shown in the picture, with the increase of file size, the average transmission delays increase no matter with or without our algorithm. Reason is obvious, with the file size gets bigger, the time for transmitting it increases. But still the average transmission delays after applying our algorithm is lower than not using algorithm.

Figure 8 shows the average successful transmission ratios under different file sizes. As shown in the figure, after using our algorithm, the average hit ratio stays flat and nearly reach 100%. The reason is when the file size increases, the number of pieces increases too, which means the probability of select a vehicle

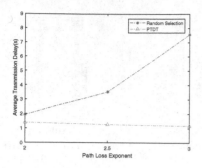

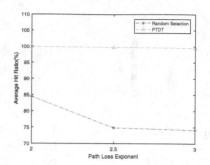

Fig. 5. Transmission delays under different path loss exponent

Fig. 6. Successful transmission ratios under different path loss exponent

whose link duration is not long enough to finish transmitting the piece of file gets higher, but this situation is considered fully in our algorithm.

Performance Under Different SNR Threshold Figure 9 shows the average transmission delays under different SNR threshold. As shown in the picture, with the increase of SNR threshold, the average transmission delays is getting lower after applying our algorithm, but is getting higher without the algorithm. The reason is when SNR threshold gets higher, the communication range of vehicles gets smaller, to ensure the numbers of vehicles with different file pieces stay the same, the density of vehicles gets higher, so the collection time of the control center gets higher.

Figure 10 shows the average successful transmission ratios under different SNR threshold. As shown in the figure, after applying our algorithm, the average successful transmission ratio stays close to 100%. Even though the SNR threshold influences the communication range of vehicles, as mentioned before, our algorithm has considered the link duration of any vehicles it chooses, so the average successful transmission ratio would not be influenced.

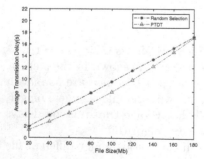

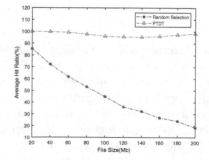

Fig. 7. Transmission delays under different SNR threshold

Fig. 8. Successful transmission ratios under different SNR threshold

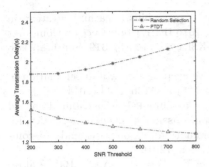

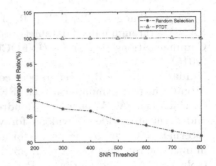

Fig. 9. Transmission delays under different file size

Fig. 10. Successful transmission ratios under different file size

5   Conclusion

In this paper, we proposed an algorithm named Predictive Time Divided Transmission (PTDT) for reducing the transmission delay and increasing the successful transmission ratio of segmented cached file in distributed vehicular networks. We predict the link duration of vehicles and divide it into time slices, the size of time slices are determined by the predicted inter-vehicle distance at the starting point of the time slice. Using those time slices of each vehicle, we analyse the ability of those vehicles for transmitting file pieces successfully, then select vehicles and arrange the transmit order of them, to ensure successful transmission and make the total delay the lowest. Simulation results show that our algorithm strongly improves the performance of transmission for segmented cached file in vehicular networks.

Acknowledgement. This paper is sponsored by the National Science and Technology Major Project of China (Grant No.2017ZX03001014).

References

1. Neelakantan, P.C., Babu, A.V.: Selection of minimum transmit power for network connectivity in vehicular ad hoc networks. In: 2012 Fourth International Conference on Communication Systems and Networks (COMSNETS 2012), pp. 1–6. Bangalore (2012)
2. Shelly, S., Babu, A.V.: A probabilistic model for link duration in vehicular ad hoc networks under rayleigh fading channel conditions. In: 2015 Fifth International Conference on Advances in Computing and Communications (ICACC), pp. 177–182. Kochi (2015)
3. Ota, K., Dong, M., Chang, S., Zhu, H.: MMCD: cooperative downloading for highway VANETs. IEEE Trans. Emerg. Top. Comput. **3**(1), 34–43 (2015)

4. Deng, G., Wang, L., Li, F., Li, R.: Distributed probabilistic caching strategy in VANETs through named data networking. In: 2016 IEEE Conference on Computer Communications Workshops (INFOCOM WKSHPS), pp. 314–319. San Francisco (2016)
5. Kumar, N., Lee, J.H.: Peer-to-peer cooperative caching for data dissemination in urban vehicular communications. IEEE Syst. J. 8(4), 1136–1144 (2014)
6. Al-Habashna, A., Wainer, G., Boudreau, G., Casselman, R.: Distributed cached and segmented video download for video transmission in cellular networks. In: International Symposium on Performance Evaluation of Computer and Telecommunication Systems (SPECTS), pp. 1–8. Montreal (2016)
7. Golrezaei, N., Mansourifard, P., Molisch, A.F., Dimakis, A.G.: Base-station assisted device-to-device communications for high throughput wireless video networks. IEEE Trans. Wirel. Commun. 13(7), 3665–3676 (2014)
8. Al-Habashna, A., Wainer, G., Boudreau, G., Casselman, R.: Improving wireless video transmission in cellular networks using D2D communication. Canada. Provisional patent P47111. May 2015
9. Al-Habashna, A., Wainer, G., Boudreau, G., Casselman, R.: Cached and segmented video download for wireless video transmission. In: Proceedings of the ANSS, pp. 1–8 (2016)
10. Yang, S., Yeo, C.K., Lee, B.S.: MaxCD: Efcient multi-flow scheduling and cooperative downloading for improved highway drive-thru Internet systems. Comput. Netw. 57(8), 1805–1820 (2013)
11. Zhu, H., Dong, M., Chang, S., Zhu, Y., Li, M., Sherman Shen, X.: ZOOM: scaling the mobility for fast opportunistic forwarding in vehicular networks. In: 2013 Proceedings IEEE INFOCOM, pp. 2832–2840. Turin (2013)

Secrecy Sum Rate Optimization in MIMO NOMA OSTBC Systems with Imperfect Eavesdropper CSI

Jianfei Yan[✉], Zhishan Deng, and Qinbo Chen

School of Electronics and Information Technology, Sun Yat-sen University,
Guangzhou 510006, Guangdong, China
{yanjf5, dengzhsh}@mail2.sysu.edu.cn, dicbldicbl@gmail.com

Abstract. In this research, we investigate the secrecy sum rate optimization problem for a multiple-input multiple-output (MIMO) non-orthogonal multiple access (NOMA) system with orthogonal space-time block codes (OSTBC). We construct a model where the transmitter and the relay send information by employing OSTBC, while both the source and the relay have imperfect channel state information (CSI) of the eavesdropper. The precoders and the power allocation scheme are jointly designed to maximize the achievable secrecy sum rate subject to the power constraints and the minimum transmission rate requirements of the weak user. To solve this non-convex problem, we propose the constrained concave convex procedure (CCCP)-based iterative algorithm and the alternative optimization (AO) method, where the closed-form expression for power allocation is derived. The simulation results demonstrate the superiority of our proposed scheme.

Keywords: Multiple-input Multiple-output (MIMO) · Relay
Non-orthogonal multiple access (NOMA) · Orthogonal space-time
block codes (OSTBC) · Secrecy sum rate · Imperfect CSI

1 Introduction

With the popularization of smart terminals and the rapid development of 5G wireless communication technology, people are looking for a method to take full use of spectrum resources. Nowadays, non-orthogonal multiple access (NOMA) scheme can not only meet the requirements of wireless communication for spectrum utilization, but also achieve better throughput gain. In [1], the NOMA technique is applied in the multiple-input multiple-output (MIMO) systems.

With the aid of linear processing at the receiver, the orthogonal space-time block codes (OSTBC) technique for MIMO NOMA systems can achieve full diversity and full rate using [2]. Meanwhile, since the existence of potential

© ICST Institute for Computer Sciences, Social Informatics and Telecommunications Engineering 2019
Published by Springer Nature Switzerland AG 2019. All Rights Reserved
X. Liu et al. (Eds.): ChinaCom 2018, LNICST 262, pp. 305–314, 2019.
https://doi.org/10.1007/978-3-030-06161-6_30

eavesdroppers in the current communication system, wireless information is susceptible to be eavesdropped. Thus, effective measures should be taken to deal with the threat of eavesdropping events in MIMO NOMA systems [4]. However, to the best of our knowledge, the secure precoding scheme in the MIMO NOMA robust systems with OSTBC has not been studied yet.

In this paper, a MIMO NOMA robust system is investigated to ensure secure information transmission between the transmitter and relay, which employ OSTBC. The precoder and power allocation scheme are jointly designed to maximize achievable secrecy sum rate. Since the optimization problem is non-convex, we propose the constrained concave convex procedure (CCCP)-based iterative algorithm and the AO method, where the closed-form expression for power allocation is derived. The remainder of the paper is organized as follows. In Sect. 2, we describe the system model and problem formulation. Section 3 presents the proposed algorithm to solve the precoder and power allocation problem. Simulation results are provided in Sects. 4 and 5 concludes this paper.

2 System Model and Problem Formulation

As shown in Fig. 1, we consider a MIMO NOMA system consisting of one transmitter, one relay, two legal users U_1, U_2, and a potential eavesdropper U_3, equipped with M_s, M_r, M_1, M_2, M_3 antennas, respectively. The channel coefficient from the transmitter to relay is denoted as H_r, and the ones from the relay to U_1, U_2 and U_3 are H_1, H_2 and H_3, respectively. In this system, the transmitter tends to send information to U_1 and U_2 safely.

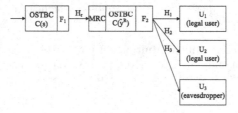

Fig. 1. Precoding of OSTBC for MIMO relay system.

We assume that the transmitter knows the perfect CSI of the relay, U_1 and U_2 while the CSI of U_3 is imperfect. Thus, the channel response H_3 is defined as

$$H_3 = \hat{H}_3 + \triangle H_3 \tag{1}$$

where $\hat{H}_3$ denotes the estimation of the channel from the transmitter to the eavesdropper, $\triangle H_3$ refers to the channel uncertainty of H_3, which is bounded by the elliptical regions, i.e., $\mathcal{H}_3$

$$\mathcal{H}_3 = \{\triangle H_3 | Tr(\triangle H_3^\dagger Q_3 \triangle H_3) \leq 1\} \tag{2}$$

where Q_3 is assumed to be known and determine the qualities of CSI.

The S-R-D signal transmission procedure takes place in two steps. In the first phase, the received signal at the relay is expressed as $Y_r = H_r F_1 C(s) + N_r$, where $F_1 \in \mathbb{C}^{M_s \times M_s}$ denotes the transmitter's precoder; N_r refers to the Gaussian noise with $N_r \sim \mathcal{CN}(0, \sigma_r^2 I)$; $C(s) \in \mathbb{C}^{M_s \times T}$ $(C(s)C^H(s) = \tilde{a}P_s I_{M_s})$ is the OSTBC matrix with a code specific constant $\tilde{a}$.

We assume that the U_1's information corresponds to the symbols $x_{1,1}$ and $x_{1,2}$, the U_2's information corresponds to the symbols $x_{2,1}$ and $x_{2,2}$. The encoding and transmission sequence at the transmitter for the MIMO NOMA system with OSTBC is shown in Table 1 [3], where $T = 2$, $M_s = 2$.

Table 1. Transmitter schematic of OSTBC

Antenna 1	Antenna 2
$t_1 : s_1 = \sqrt{\phi_1 P_s} x_{1,1} + \sqrt{\phi_2 P_s} x_{2,1},$	$s_2 = \sqrt{\phi_1 P_s} x_{1,2} + \sqrt{\phi_2 P_s} x_{2,2}$
$t_2 : -s_2^* = -\sqrt{\phi_1 P_s} x_{1,2}^* - \sqrt{\phi_2 P_s} x_{2,2}^*,$	$s_1^* = \sqrt{\phi_1 P_s} x_{1,1}^* + \sqrt{\phi_2 P_s} x_{2,1}^*$

As presented in Table 1, s_1 and s_2 are simultaneously transmitted to the relay in the time slot t_1. In the next time slot t_2, s_1^* and $-s_2^*$ are transmitted. ϕ_1 and ϕ_2 are the power allocation factors of U_1, U_2, which satisfy $\phi_1 + \phi_2 = 1$. P_s is the transmission power of the transmitter and $E\{|x_{j,i}|^2\} = 1$, $i, j \in \{1, 2\}$.

Since the relay processes Y_r with maximum ratio combiner (MRC), thus, the received signal at the relay can be written as [5]

$$y_k^R = \|H_r F_1\| z_k + n_{r,k}, i \in \{1, ..., K\} \tag{3}$$

where $z_k = \sqrt{\frac{\phi_1 P_s}{K}} x_{1,k} + \sqrt{\frac{\phi_2 P_s}{K}} x_{2,k}$, $n_{r,k} \sim \mathcal{CN}(0, \sigma_r^2)$ refers to the AWGN at the relay for the kth S-R SISO channel and follows complex Gaussian distribution with zero-mean and variance σ_1^2. The relay normalizes $\{y_k^R\}_{k=1}^K$ yielding

$$\tilde{y}_k^R = \frac{y_k^R}{\sqrt{E\{|y_k^R|^2\}}} = \frac{\|H_r F_1\| z_k + n_{r,k}}{\sqrt{\|H_r F_1\|^2 + \sigma_r^2}} \tag{4}$$

In the second phase, the destination receive signal $Y_d = H_j F_2 C(\tilde{y}^R) + N_j^d$, where $C(\tilde{y}^R) \in \mathbb{C}^{M_r \times T}$ is the OSTBC formed from $\tilde{y}^R = [\tilde{y}_1^R, ..., \tilde{y}_K^R]$; $F_2 \in \mathbb{C}^{M_r \times M_r}$ is the relay precoder; H_j, $j \in \{1, 2, 3\}$ is the R-D MIMO channel and N_j^d is the matrix of AWGN samples at the destination with zero mean and variance $\tilde{\sigma}_j^2$.

As in the case of kth S-R SISO channel, the signal received by the destination in the kth R-D SISO channel is given by

$$y_k^D = \|H_j F_2\| \tilde{y}_k^R + n_{2,k}, \forall k \tag{5}$$

where $n_{j,k} \sim \mathcal{CN}(0, \sigma_j^2)$ is the AWGN at the relay for the kth S-R SISO channel. Using (4), (5) can be expressed as

$$y_k^D = \frac{\|H_j F_2\|\|H_r F_1\|z_k + \|H_j F_2\|n_{r,k}}{\sqrt{\|H_r F_1\|^2 + \sigma_1^2}} + n_{j,k} \tag{6}$$

Therefore, the SNR at the destination is given as

$$y_k^D = \frac{\|H_j F_2\|\|H_r F_1\|z_k + \|H_j F_2\|n_{r,k}}{\sqrt{\|H_r F_1\|^2 + \sigma_1^2}} + n_{j,k} \tag{7}$$

We assume that $\|H_1 F_2\|^2 \geqslant \|H_2 F_2\|^2 \geqslant \|H_3 F_2\|^2$, user U_1 is the strong user and user U_2 is the weak user. At the strong user U_1, the signal-to-noise ratio (SNR) to decode signals of the weak user U_2 is

$$\gamma_1^2 = \frac{\frac{\phi_2 P_s}{K}\|H_1 F_2\|^2\|H_r F_1\|^2}{\frac{\phi_1 P_s}{K}\|H_r F_1\|^2\|H_1 F_2\|^2 + \|H_r F_1\|^2\sigma_1^2 + \|H_1 F_2\|^2\sigma_2^2 + \sigma_1^2\sigma_2^2} \tag{8}$$

$$= \frac{\frac{\phi_2 P_s}{K}\gamma_1\gamma}{\frac{\phi_1 P_s}{K}\gamma_1\gamma + \gamma_1 + \gamma + 1}. \tag{9}$$

At the weak user U_2, the SNR to detect its own signals is

$$\gamma_2^2 = \frac{\frac{\phi_2 P_s}{K}\gamma_2\gamma}{\frac{\phi_1 P_s}{K}\gamma_2\gamma + \gamma_2 + \gamma + 1}. \tag{10}$$

The minimum SNR requirement of the weak user U_2 is denoted as γ_0, due to $\gamma_1^2 \geq \gamma_2^2$, the strong user U_1 can remove signals of U_2. After successive interference cancelation (SIC), the SNR of U_1 to detect its own signals is given by

$$\gamma_1^1 = \frac{\frac{\phi_1 P_s}{K}\gamma_1\gamma}{\gamma_1 + \gamma + 1} \tag{11}$$

Then at the eavesdropper U_3, the SNR to detect the signals of U_1 and U_2 is

$$\gamma_3^{1,2} = \frac{\frac{P_s}{K}\gamma_3\gamma}{\gamma_3 + \gamma + 1}. \tag{12}$$

where $\gamma = \frac{\|H_r F_1\|^2}{\sigma_r^2}$, $\gamma_j = \frac{\|H_i F_2\|^2}{\sigma_j^2}, j \in \{1, 2, 3\}$. The transmit powers of the source and relay per OSTBC block can be given by $P_s = \tilde{a} P_s Tr(F_1 F_1^H)$ and $P_r = \tilde{a} P_s Tr(F_2 F_2^H)$, where $P_s = P_r$.

Based on (11), (2) and (12), the achievable secrecy sum rate can be calculated as

$$R_s = \log_2(1 + \gamma_1^1) + \log_2(1 + \gamma_2^2)$$
$$- \log_2(1 + \gamma_3^{1,2}) \tag{13}$$

3   Precoder and Power Optimization

The secrecy sum rate optimization problem can be formulated as

$$\max_{\{W_1, W_2, \gamma_1, \gamma_2, \phi_1, \phi_2\}} \log_2(1 + \gamma_1^1) + \log_2(1 + \gamma_2^2)$$

$$- \log_2(1 + \gamma_3^{1,2}) \tag{14a}$$

$$\text{s.t.} \quad \gamma_2^2 \geq \gamma_0, \tag{14b}$$

$$\phi_1 + \phi_2 = 1, \tag{14c}$$

$$0 \leq \phi_1 \leq 1, \tag{14d}$$

$$0 \leq \phi_2 \leq 1, \tag{14e}$$

$$\frac{\|H_r F_1\|^2}{\sigma_r^2} = \gamma, \tag{14f}$$

$$\frac{\|H_1 F_2\|^2}{\sigma_1^2} \geq \gamma_1, \tag{14g}$$

$$\frac{\|H_2 F_2\|^2}{\sigma_2^2} \geq \gamma_2, \tag{14h}$$

$$Tr(F_1 F_1^H) \leq \frac{1}{\tilde{a}}, Tr(F_2 F_2^H) \leq \frac{1}{\tilde{a}}, \tag{14i}$$

$$Tr(H_1^H H_1 W_2) \geqslant Tr(H_2^H H_2 W_2), \tag{14j}$$

$$Tr(H_2^H H_2 W_2) \geqslant Tr(H_3^H H_3 W_2) \tag{14k}$$

$$\frac{\|H_3 F_2\|^2}{\sigma_3^2} \leq \gamma_3 \tag{14l}$$

Next, we employ the S-procedure to convert the constraints (14l) into linear matrix inequalities (LMIs).

$$\begin{cases} \forall \triangle H_3 : Tr(\triangle H_3^\dagger Q_3 \triangle H_3) \leq \theta_3 \\ Tr(H_3^\dagger W_2 H_3) - \gamma_3 \sigma_3^2 \leq 0 \end{cases} \tag{15}$$

Then we rewrite it as follows:

$$\begin{cases} \forall \triangle h_3 : \triangle h_3^\dagger (I \otimes Q_3) \triangle h_3 - \theta_3 \leq 0 \\ \triangle h_3^\dagger (I \otimes Q_3) \triangle h_3 + 2Re\{((I \otimes Q_3) h_3^\dagger) \triangle h_3^\dagger\} + h_3^\dagger (I \otimes Q_3) h_3 - \gamma_3 \sigma_3^2 \leq 0 \end{cases} \tag{16}$$

$$\begin{bmatrix} \lambda_3 (I \otimes Q_3) - (I \otimes W_2) & -(I \otimes W_2) h_3 \\ -h_3^\dagger (I \otimes W_2) & \gamma_3 \sigma_3^2 - \lambda_3 \theta_3 - h_3^\dagger (I \otimes Q_3) h_3 \end{bmatrix} \succeq 0 \tag{17}$$

where $\lambda_3 >= 0$ is a slack variable, $h_3 = vec(H_3)$.

The problem is recast as

$$\max_{\{W_1,W_2,\gamma_1,\gamma_2,\phi_1,\phi_2,\lambda_3\}} \quad \log_2(1+\gamma_1^1) + \log_2(1+\gamma_2^2)$$

$$-\log_2(1+\gamma_3^{1,2}) \tag{18a}$$

$$\text{s.t.} \quad Tr(DW_1) = \gamma, \tag{18b}$$

$$Tr(AW_2) \geq \gamma_1\sigma_1^2, \tag{18c}$$

$$Tr(BW_2) \geq \gamma_2\sigma_2^2, \tag{18d}$$

$$Tr(W_1) \leq \frac{1}{\tilde{a}}, Tr(W_2) \leq \frac{1}{\tilde{a}}, \tag{18e}$$

$$Tr(AW_2) \geqslant Tr(BW_2), \tag{18f}$$

$$Tr(BW_2) \geqslant Tr(CW_2), \tag{18g}$$

$$(14b)-(14e),(17). \tag{18h}$$

where γ_0 denotes the minimum SNR requirement of the weak user U_2; $W_1 = F_1^H F_1$, $W_2 = F_2^H F_2$, $A = H_1^H H_1$, $B = H_2^H H_2$, $C = H_3^H H_3$, $D = H_r^H H_r$. Because the objective and constraints are non-convex, problem (18) can not be directly solved by convex method. As far as we know, (18) has no global optimal solution, so we will use an effective method to attain the local optimal solution.

By introducing slack variables $\{a, b, c\}$ such that $\gamma_1^1 \geq a$, $\gamma_2^2 \geq b$, $\gamma_3^{1,2} \leq c$, (18) is equivalently recast as

$$\max_{\{W_1,W_2,\gamma_1,\gamma_2,\phi_1,\phi_2,a,b,c\}} \quad \log_2(1+a) + \log_2(1+b)$$

$$-\log_2(1+c) \tag{19a}$$

$$\text{s.t.} \quad \gamma_2^2 \geq \gamma_0, \tag{19b}$$

$$\gamma_1^1 \geq a, \tag{19c}$$

$$\gamma_2^2 \geq b, \tag{19d}$$

$$\gamma_3^{1,2} \leq c, \tag{19e}$$

$$(18b)-(18h). \tag{19f}$$

Based on (12), (19e) is expressed as

$$\frac{1}{\gamma_3} + \frac{1}{\gamma} + \frac{1}{\gamma_3\gamma} \geq \frac{P_s}{cK} \tag{20a}$$

$$\tau_3 + \tau_{min} + t_3 \geq \frac{P_s}{cK} \tag{20b}$$

$$\tau_3\gamma_3 \leq 1 \tag{20c}$$

$$\tau_{min}\gamma \leq 1 \tag{20d}$$

$$t_3\gamma_3\gamma \leq 1 \tag{20e}$$

Equation (20e) is expressed as

$$\begin{bmatrix} \mu_3 & t_3 \\ \gamma & \mu_3 \end{bmatrix} \succeq 0 \tag{21a}$$

$$\begin{bmatrix} \mu_{3,3} & 1 \\ 1 & \mu_{3,3} \end{bmatrix} \succeq 0 \tag{21b}$$

$$\mu_3 \mu_{3,3} \leq 1 \tag{21c}$$

Similarly, (19b), (19c), (19d) are expressed as

$$\frac{1}{\gamma_1} + \frac{1}{\gamma} + \frac{1}{\gamma_1 \gamma} \leq \frac{\phi_1 P_s}{aK} \tag{22a}$$

$$\tau_1 + \tau_{max} + t_1 \leq \frac{\phi_1 P_s}{aK} \tag{22b}$$

$$\frac{1}{\gamma_2} + \frac{1}{\gamma} + \frac{1}{\gamma_2 \gamma} \leq \frac{1}{b - \frac{\phi_1}{\phi_2}} \frac{\phi_2 P_s}{K} \tag{22c}$$

$$\tau_2 + \tau_{max} + t_2 \leq \frac{1}{b - \frac{\phi_1}{\phi_2}} \frac{\phi_2 P_s}{K} \tag{22d}$$

$$\frac{1}{\gamma_2} + \frac{1}{\gamma} + \frac{1}{\gamma_2 \gamma} \leq (\frac{1}{\gamma_0} - \frac{\phi_1}{\phi_2}) \frac{\phi_2 P_s}{K} \tag{22e}$$

$$\tau_2 + \tau_{max} + t_2 \leq (\frac{1}{\gamma_0} - \frac{\phi_1}{\phi_2}) \frac{\phi_2 P_s}{K} \tag{22f}$$

$$\tau_i \gamma_i \geq 1 \tag{22g}$$

$$\tau_{max} \gamma \geq 1 \tag{22h}$$

$$t_i \gamma_i \gamma \geq 1 \tag{22i}$$

Hyperbolic constraints (22g) and (22h) can be converted into convex forms

$$\tau_i + \gamma_i \geq \|[\sqrt{2}, \tau_i, \gamma_i]^T\| \tag{23a}$$

$$\tau_{max} + \gamma \geq \|[\sqrt{2}, \tau_{max}, \gamma]^T\| \tag{23b}$$

Equation (22i) can be expressed as

$$\begin{bmatrix} t_i & \mu_i \\ \mu_i & \gamma \end{bmatrix} \succeq 0 \tag{24a}$$

$$\begin{bmatrix} 1 & \mu_{i,i} \\ \mu_{i,i} & \gamma_i \end{bmatrix} \succeq 0 \tag{24b}$$

$$\mu_i + \mu_{i,i} \geq \|[\sqrt{2}, \mu_i, \mu_{i,i}]^T\| \tag{24c}$$

where $i \in \{1, 2\}$.

Because of the non-convexity of these functions: $-\log_2(1 + c)$, (20c), (20d) and (21c), we can employ the first-order Taylor method to recast them, which are expressed as $f(c, \bar{c})$, $f(\tau_3, \bar{\tau}_3)$, $f(\tau_{min}, \bar{\tau}_{min})$, $f(\mu_{3,3}, \bar{\mu}_{3,3})$, respectively.

Then problem (18) can be rewritten as

$$\max_{\{W_1, W_2, \gamma_1, \gamma_2, \phi_1, \phi_2, a, b, c, \mu, \tau, t\}} \log_2(1+a) + \log_2(1+b)$$

$$- f(c, \bar{c}) \tag{25a}$$

$$\text{s.t.} \quad (22b), (22d), (22f), (23), \tag{25b}$$

$$(24), (20b), \tag{25c}$$

$$(21a), (21b), \tag{25d}$$

$$(18b) - (18h) \tag{25e}$$

$$f(\tau_3, \bar{\tau}_3), f(\tau_{min}, \bar{\tau}_{min}), f(\mu_{3,3}, \bar{\mu}_{3,3}). \tag{25f}$$

To solve problem (25), we employ AO method which alternatively optimizes ϕ_1 and $(W_1, W_2, a, b, c, \mu, \tau, t)$. Given ϕ_1, problem (25) is convex with respect to $(W_1, W_2, a, b, c, \mu, \tau, t)$, which can be solved by interior point method [6].

With given $(W_1, W_2, a, b, c, \mu, \tau, t)$, we can obtain ϕ_1 as follow. Based on $\gamma_2^2 \geq \gamma_0$, the objective function is an increasing function with respect to ϕ_1, then the closed-form solution of ϕ_1 can be achieved as

$$\phi_1 = \frac{1}{1+M} - M \frac{1 + Tr(W_1 D) + Tr(W_2 A)}{0.5 P_s (1 + M) Tr(W_2 B) + Tr(W_1 D)}. \tag{26}$$

The proposed AO method is shown in Algorithm 1 as below.

Algorithm 1 The Proposed AO Method to Solve Problem (21)

1: **Initialization:** $n = 0$ and $\phi_1^{(0)}$ and an accuracy parameter ϵ;
2: **Repeat:**
   Solve problem (21) to update $(W_1, W_2, a, b, c, \mu, \tau, t)^{(n)}$, with $\phi_1^{(n)}$ fixed;
   Use (26) to update $\phi_1^{(n+1)}$ with $(W_1, W_2, a, b, c, \mu, \tau, t)^{(n)}$;
   Update $n = n + 1$;
3: **Until:** $|R^{(n)} - R^{(n-1)}| \leq \epsilon$, where $R^{(n)}$ is the objective value in the nth iteration.

By solving problem (25), we get the optimal solution $(W_1, W_2, a, b, c, \mu, \tau, t)^{(o)}$. If the rank of $\widehat{W}^o$ equals to one, the optimal F^o can be easily found. Otherwise, we can generate a suboptimal solution through the Gaussian random method in [9].

4   Simulation Results

Here, the simulation results are provided to validate the proposed robust precoding schemes with NOMA OSTBC are more effective than other cases.

In Fig. 2, we present the average achievable secrecy sum rates achieved by the proposed scheme, equal power allocation scheme and the conventional orthogonal multiple access scheme both with OSTBC (denoted as "OSTBC+NOMA",

"OSTBC+EP" "OSTBC+OMA", respectively), versus the transit power P_s with setting $M_s = M_r = M_1 = M_2 = M_3 = 2$, $\tilde{a} = 1$ and the minimum SNR requirement of the weak user $\gamma_0 = 0.10$ dB.

It can be observed that our proposed "NOMA+OSTBC" scheme achieves better performance than the other two schemes for different P_s. This is because the superposition of signals to different users and SIC cause the bandwidth is explored more efficiently.

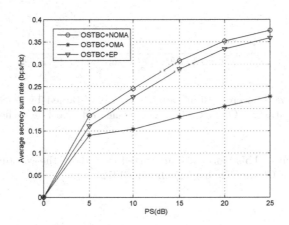

Fig. 2. The achievable secrecy sum rate versus P_s, obtained by "OSTBC+NOMA" scheme, the conventional "OSTBC+OMA" and "OSTBC+EP" scheme.

In Fig. 3, we show the effect of the power allocation factors of U_1, i.e.,ϕ_1, where no eavesdropper situation (denoted as "No Eve") is also compared. From Fig. 3, we find that "NOMA+OSTBC" scheme outperforms the "OSTBC+OMA" scheme with different ϕ_1, for NOMA can achieve higher spectral efficiency and better user fairness than conventional OMA. Besides, larger ϕ_1 leads to higher secrecy sum rate. As ϕ_1 continues to increase, the performance gab between the "OSTBC+NOMA" and "OSTBC+OMA" becomes larger.

5 Conclusion

In this paper, the precoders and power allocation design for MIMO NOMA robust system with OSTBC is studied, where source and relay have imperfect CSI of the eavesdropper. To tackle the non-convexity of the optimization problem, we employ the CCCP-based algorithm and AO method to optimize the precoders and power allocation. The simulation results demonstrate the superiority of our proposed scheme.

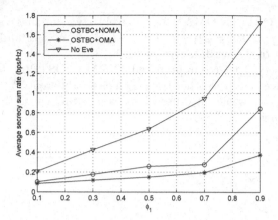

Fig. 3. The achievable secrecy sum rate versus ϕ_1, obtained by "OSTBC+NOMA" scheme, the conventional "OSTBC+OMA" scheme and "No Eve" scheme.

Acknowledgments. This work was supported in part by the National Natural Science Foundation of China (No. 61672549, No. 61472458).

References

1. Ding, Z., Adachi, F., Poor, H.V.: The application of MIMO to nonorthogonal multiple access. IEEE Trans. Wirel. Commun. **15**(1), 537–552 (2016)
2. Kader, M.F., Shin, S.Y.: Cooperative relaying using space-time block coded non-orthogonal multiple access. IEEE Vehicular Technology Society, vol. 66, pp. 5894–5903 (2017)
3. Kader, M.F., Shin, S.Y.: Cooperative spectrum sharing with space time block coding and non-orthogonal multiple access. In: ICUFN 2016, pp. 490–494 (2016)
4. Tian, M. et al.: Secrecy sum rate optimization for downlink MIMO non-orthogonal multiple access systems. IEEE Signal Process. Lett. **24**(8), 1113–1117 (2017)
5. Hjorungnes, A., Gesbert, D.: Precoding of orthogonal-space time block codes in arbitrarily correlated MIMO channels: iterative and closed-form solutions. IEEE Trans. Wirel. Commun. **6**(3), 1072–1082 (2007)
6. Boyd, S., Vandenberghe, L.: Convex Optimization. Cambridge University Press, Cambridge (2004)
7. Horst, R., Thoai, N.V.: DC programming: overview. J. Optim. Theory Appl. **103**(1), 1–43 (1999)
8. Charnes, A., Cooper, W.W.: Programming with linear fractional functionals. Nav. Res. Logist. Quart. **9**(3/4), 181–186 (1962)
9. Karipidis, E., Sidiropoulos, N.D., Luo, Z.-Q.: Quality of service and max-min fair transmit beamforming to multiple cochannel multicastgroups. IEEE Trans. Signal Process. **56**(3), 1268–1279 (2008)

Network-coding-based Cooperative V2V Communication in Vehicular Cloud Networks

Rui Chen[1], Weijun Xing[1], Chao Wang[1,2(✉)], Ping Wang[1], Fuqiang Liu[1,3], and Yusheng Ji[4]

[1] College of Electronics and Information Engineering, Tongji University, Shanghai, China
{214chenrui,1210487,chaowang,liufuqiang}@tongji.edu.cn
[2] Department of Computer Science, University of Exeter, Exeter, UK
[3] College of Design and Innovation, Tongji University, Shanghai, China
[4] National Institute of Informatics, Graduate University for Advanced Studies, Hayama, Japan

Abstract. We investigate the potential of applying cooperative relaying and network coding techniques to support vehicle-to-vehicle (V2V) communication in vehicular cloud networks (VCN). A reuse-mode MIMO content distribution system with multiple sources, multiple relays, and multiple destinations under Nakagami-m fading is considered. We apply a class of finite field network codes in the relays to achieve high spatial diversity in an efficient manner and derive the system communication error probability that the destinations fail to recover the desired source messages. The results show that our method can improve the performance over conventional data transmission solutions.

Keywords: Vehicular cloud networks · Network coding · Cooperative relaying

1 Introduction

In the past decades, the ever-increasing numbers of vehicles have caused serious problems, such as congestion, accidents and pollution, all over the world. There is a strong desire for an efficient, safe, and clean road traffic system. The concept of intelligent transportation system (ITS) enabled by information and communication technology (ICT) has been accepted as the most promising solution [1]. As a key element in ITS, "smart vehicles" equipped with advanced on-board sensing and computation devices are designed to help reducing accidents caused by human errors, through the advance driving assistance system (ADAS) or even self-driving functions. Google's self-driving car and Baidu's Apollo platform are examples of autonomous driving vehicle and developing platform prototypes.

© ICST Institute for Computer Sciences, Social Informatics and Telecommunications Engineering 2019
Published by Springer Nature Switzerland AG 2019. All Rights Reserved
X. Liu et al. (Eds.): ChinaCom 2018, LNICST 262, pp. 315–324, 2019.
https://doi.org/10.1007/978-3-030-06161-6_31

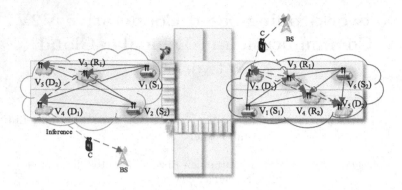

Fig. 1. System model.

One issue with most today's smart vehicle prototypes is that they normally work individually. Even though they have advanced sensors, their sensing range and reliability are relatively limited in complex driving conditions. Allowing vehicles to interact through wireless communications is seen as an effective solution. Since the application scenario is very different from traditional mobile Internet/wireless LAN, a number of wireless technologies dedicated for establishing the Internet of vehicles or V2X (vehicle to everything) communications have been developed in recent years, including e.g., the IEEE 802.11p Dedicated Short Range Communication (DSRC) technology and some LTE-based solutions [2]. More importantly, the current 5G research campaign also considers vehicular communication environment as a key application scenario [3].

With the support of 5G and other advanced V2X technologies, the data exchanged among vehicles and road-side units would not be limited to conventional small-size heart-beat or event-driven messages. Sharing content-rich sensing data will be feasible, which triggers a promising concept *Vehicular Cloud Network* (VCN) [4,5]. VCN stems from the idea of cloud computing and is envisioned as a framework where multiple vehicles share their sensing, communication, computation, and storage resources to realize functions that each individual vehicle cannot. For example, sharing sensing data allows each vehicle to access the others' sensing resources to extend its own ability of perceiving the surrounding environment. Sharing communication resources (e.g., bandwidth and power) potentially enhances the data exchange capability within the network.

Two example application scenarios of VCN are shown in Fig. 1. The left hand side (LHS) illustrates a VCN with five vehicles conducting cooperative objective detection in complicated driving conditions such as road intersections [6]. V_1 and V_2 have advantageous locations to obtain accurate sensing results regarding their surrounding environment, e.g., the existence, position and movement of obstacles near the intersection corner. Sharing their sensing data with those who relatively far away from the traffic scene (i.e., V_4 and V_5) enables greatly improved sensing range. The right hand side (RHS) of Fig. 1 shows a VCN formed by six autonomous vehicles operating in a cooperative driving scenario

[7]. Now V_1 and V_6 act as sensing information sources and desire to distribute their data to V_2 and V_5. These illustrate the concept of sensing resource sharing.

However, due to fast changing network topology and complicated signal propagation environment, the quality of message sharing is often hard to guarantee. In this case, cooperative relaying can serve to significantly improve performance [8]. For example, in the LHS scenario of Fig. 1, V_3 shares its own communication resources with others and serves as a relay to guarantee the sensing data to be delivered from V_1 and V_2 to V_4 and V_5 with low probability of failure. In the RHS scenario, the message sharing can be helped by V_3 and V_4 acting as relays.

The conventional relay forwarding operation is to directly repeat received signals. In multi-source scenarios, such a repetition-coding based relaying strategy does not efficiently use the available channel resources. In this paper, we study applying network coding techniques to handle this issue. Specifically, we consider a reuse-mode multi-source multi-relay multi-destination cooperative MIMO V2V communication scenario in VCN. A class of maximum distance separable finite field network codes (MDS-FFNC) [9–11] is applied in the relays to attain high spatial diversity in an efficient fashion. We derive the system communication error probability, in a Nakagami-m fading environment. It is shown that our method can improve performance over conventional direct transmission and repetition-coding based relaying schemes.

2   System Model

We consider a VCN scenario with a total of U member vehicles $V_1, \cdots, V_U$, among which M vehicles (denoted as sources $S_1, \cdots, S_M$) have attained some sensing data from their own sensors, represented respectively by independent messages $I_{S_1}, \cdots, I_{S_M}$, and intend to share them (i.e., sensing resource sharing) with N other members (denoted as destinations $D_1, \cdots, D_N$). To ensure the quality of the desired message sharing, K vehicles (denoted as relays $R_1, \cdots, R_K$) can serve as cooperative relays (i.e., communication resource sharing). Practical applications abstracted by such a multi-source multi-relay multi-destination content distribution network are illustrated in Fig. 1.

The set of the M source messages to be shared within the VCN is denoted as $\mathcal{I} = \{I_{S_1}, \cdots, I_{S_M}\}$. We consider the case that every destination D_i is interested in attaining the whole source message set $\mathcal{I}$ in order to reinforce its own sensing capability. Therefore, the communication in the considered VCN is successful only if all $D_1, \cdots, D_N$ correctly recover $\mathcal{I}$. Otherwise, if any of them is not capable of completely recovering $\mathcal{I}$, a system communication error is reported.

In addition, in order to efficiently utilize available channel resources, we consider that the V2V transmissions in our VCN can be potentially operated in a reuse mode. In other words, all the aforementioned message delivery process can coexist with the uplink transmission of a certain cellular user, denoted as C. A proper cellular user scheduling strategy is assumed such as that C is chosen to be close to its serving base station. The V2V transmission power is limited, since message sharing in the considered VCN occurs locally in a limited area, to avoid introducing intolerable interference to the base station. But the reception of each member in the VCN may experience interference from C.

Every node a in the considered system can potentially have $A_a \geq 1$ antennas. All message transmissions are conducted in a narrow-band Nakagami-m slow fading environment. The channel fading matrix between transmitting node a and receiving node b is denoted by an $A_b \times A_a$ matrix $\mathbf{H}_{b,a}$. The element on the jth ($j \in \{1, \cdots, A_b\}$) row and ith column ($i \in \{1, \cdots, A_a\}$) of $\mathbf{H}_{b,a}$, denoted by h_{b^j,a^i}, represents the fading coefficient between the ith antenna of a and jth antenna of b. All the coefficients in $\mathbf{H}_{b,a}$ are modeled by independently and identically distributed (i.i.d.) random variables with absolute values following a Nakagami distribution with parameters $m_{b,a}$ (Nakagami-m fading parameter) and $\hat{\Omega}_{b,a}$ (expected channel power gain) i.e., $|h_{b^j,a^i}| \sim \text{Nakagami}(m_{b,a}, \hat{\Omega}_{b,a})$. This means that the channel power gain $|h_{b^j,a^i}|^2$ follows a Gamma distribution with shape parameter $\alpha_{b,a} = m_{b,a}$ and rate parameter $\beta_{b,a} = \frac{m_{b,a}}{\hat{\Omega}_{b,a}}$ [12].

The message exchange in the considered VCN is conducted in a slotted fashion. Each message is encoded using a capacity-achieving Gaussian random codeword, so that focus of performance analysis can be mainly put on the impact of channel variation. Every codeword is transmitted using a unit-bandwidth time-division-multiple-access (TDMA) time slot. The channel fading coefficients remain unchanged during the whole period of transmission and are known only at the associated receivers. Since transmitter-side channel knowledge is not available, dynamic power control and rate adaptation are not performed. Therefore, all the messages $I_{S_1}, \cdots, I_{S_M}$ have the same data rate R bits per codeword. A transmitter a evenly spreads it power P_a across its A_a antennas and applies orthogonal space-time block coding (OSTBC) to broadcast its message.

In the next section, we will elaborate the message transmission process.

3   Message Transmission Process

The complete transmission of the message set $\mathcal{I}$ from the M sources to the N destinations, through the help of K relays, is carried out using $M + K$ individual time slots, each of which is allocated to a transmitting vehicle. Instead of demanding the relays to simply repeat the source messages, we consider applying MDS-FFNC [9–11] at the relays. In particular, each relay R_k ($k \in \{1, \cdots, K\}$) is assigned with a set of MDS-FFNC coding coefficients $\omega_k^{[1]}, \cdots, \omega_k^{[M]}$. All the coding coefficients within the network are properly constructed in advance to guarantee the transfer matrix $\mathbf{G} = [\mathbf{I}_M \ \mathbf{W}]^T$ to be non-singular, where $\mathbf{I}_M$ is an $M \times M$ identity matrix and $\mathbf{W}$ is the $M \times K$ coefficient matrix whose kth column elements are $\omega_k^{[1]}, \cdots, \omega_k^{[M]}$. Therefore, any M elements of the output of the encoding process $\mathbf{G} \cdot [I_{S_1} \ \cdots \ I_{S_M}]^T$, where summations are taken in a certain finite field, are sufficient to recover $I_{S_1}, \cdots, I_{S_M}$. Using these coefficients, if relay R_k correctly attains the complete message set $\mathcal{I}$, it can re-encode the source messages to a new message using $I_{R_k} = \sum_{i=1}^M \omega_k^{[i]} I_{S_i}$. Knowing any M messages among $I_{S_1}, \cdots, I_{S_M}, I_{R_1}, \cdots, I_{R_K}$ is sufficient to recover $\mathcal{I}$.

The detailed message transmission process can be described as follows. The first M time slots are allocated to S_1, $\cdots$, S_M (i.e., during the slot i ($i \in \{1, \cdots, M\}$), the source S_i broadcasts its message I_{S_i} to all relays and destinations). After all sources complete transmission, the next K time slots are reserved for the relays R_1, $\cdots$, R_K respectively. All the relays also listen to each other in order to recover the source messages. During time slot $M + k$, if R_k obtains all the messages in $\mathcal{I}$ from its received signals in the previous $M + k - 1$ time slots (i.e., it can correctly recover at least M messages among $I_{S_1}, \cdots, I_{S_M}, I_{R_1}, \cdots, I_{R_{k-1}}$), it re-encodes the source messages into I_{R_k} and forwards it to other relays and the destinations. Otherwise, R_k keeps silent.

Clearly, during each of the $M + K$ time slots, if a transmitting vehicle a ($a \in \{S_1, \cdots, S_M, R_1, \cdots, R_K\}$) is actually activated as being scheduled, the received signal at any receiving vehicle b ($b \in \{R_1, \cdots, R_K, D_1, \cdots, D_N\}$) can be expressed as the following general form:

$$y_{b,a} = \sqrt{\frac{P_a}{A_a}} H_{b,a} x_a + \sqrt{\frac{P_C}{A_C}} H_{b,C} x_C + n_b, \tag{1}$$

where $y_{b,a}$ is the A_b-dimensional received signal, x_a is the transmitted signal from a, x_C denotes the co-channel interference from cellular user C, and n_b represents additive white Gaussian noise with total power N_0.

Since each vehicle uses OSTBC to transmit its message in the VCN, the received signal-to-interference-plus-noise-ratio (SINR) can be expressed as [13]

$$\gamma_{b,a} = \frac{\|H_{b,a}\|_F^2 \cdot \rho_a}{\|H_{b,C}\|_F^2 \cdot \rho_C + 1}, \tag{2}$$

where $\| \cdot \|_F$ represents the Frobenius norm, and $\rho_a = \frac{P_a}{A_a N_0}$ and $\rho_C = \frac{P_C}{A_C N_0}$ represent transmitter-side signal-to-noise ratio (SNR).

As we mentioned earlier, The probability of occurring system communication error event can be expressed as

$$P_{\text{err}} = P_r\{D_i \text{ cannot recover } \mathcal{I}, \exists i \in \{1, \cdots, N\}\}. \tag{3}$$

We will provide the method to derive P_{err} in the following section.

4   System Error Performance Analysis

Due to the facts that the probability of activation of each relay depends on the other relays that are scheduled before it, and that the decoding events at different destinations are related (in fact, they are conditionally independent given the decoding set of the relays), the derivation of P_{err} is actually quite involved. In this paper, we decouple the calculation of P_{err} into three elementary probability expressions that would be easier to tackle.

The first is the link decoding probability $\mathcal{P}_{b,a}^L$ that, during each time slot, a receiving vehicle b correctly decodes the transmit signal of the activated vehicle

a. We consider Gaussian random coding so that this probability can be found by the probability that the SINR expressed by (2) is larger than the message data rate R, i.e., $\mathcal{P}_{b,a}^L = Pr\{\log_2(1 + \gamma_{b,a}) > R\}$.

The second elementary probability is the probability that a particular set of activated relays occurs. Let us use $\Delta = [\delta_1, \cdots, \delta_K]$ to denote the activation behaviors of the K relays after the scheduled $M + K$ time slots, where $\delta_k = 1$ denotes R_k is able to recover $\mathcal{I}$ from its received signals and $\delta_k = 0$ denotes the opposite situation. The second elementary probability is denoted by $\mathcal{P}_\Delta$.

The third, $\mathcal{P}_{D_n|\Delta}$ ($n \in \{1, \cdots, N\}$), is the conditional probability that the destination D_n can fully recover all the source messages given a particular relay activation behavior Δ. Since the destinations' decoding capabilities are conditionally independent, the system's error probability P_{err} can be calculated by

$$P_{err} = 1 - \sum_{\Delta} \left(\mathcal{P}_\Delta \prod_{n=1}^{N} \mathcal{P}_{D_n|\Delta} \right), \tag{4}$$

where the summation is taken for all $\delta_k \in \{0,1\}, k \in \{1, \cdots, K\}$, and both $\mathcal{P}_\Delta$ and $\mathcal{P}_{D_n|\Delta}$ are functions of $\mathcal{P}_{b,a}^L$.

(1) *Link decoding probability* $\mathcal{P}_{b,a}^L$:

The link decoding probability $\mathcal{P}_{b,a}^L$ can also be written as $Pr\{\gamma_{b,a} > 2^R - 1\}$. To simplify presentation, we define equivalent received SNR $\hat{\gamma}_{b,a} = \|\mathbf{H}_{b,a}\|_F^2 \cdot \rho_a$, $\hat{\gamma}_{b,C} = \|\mathbf{H}_{b,C}\|_F^2 \cdot \rho_C$, and $\gamma_{th} = 2^R - 1$. Now $\mathcal{P}_{b,a}^L = Pr\left\{ \frac{\hat{\gamma}_{b,a}}{\hat{\gamma}_{b,C}+1} > \gamma_{th} \right\}$.

It is easy to see that $\|\mathbf{H}_{b,a}\|_F^2$ (resp. $\|\mathbf{H}_{b,C}\|_F^2$) is the sum of $A_b \times A_a$ (resp. $A_b \times A_C$) i.i.d. Gamma random variables with shape parameter $m_{b,a}$ and rate parameter $\frac{m_{b,a}}{\hat{\Omega}_{b,a}}$ (resp. $m_{b,C}$ and $\frac{m_{b,C}}{\hat{\Omega}_{b,C}}$) [13]. The sum of n i.i.d. Gamma random variables with shape parameter α and rate parameter β is Gamma distributed with parameters $n\alpha$ and β. Thus the pdf of $\hat{\gamma}_{b,a}$ is

$$f_{\hat{\gamma}_{b,a}}(x) = \left(\frac{m_{b,a}}{\rho_a \hat{\Omega}_{b,a}} \right)^{m_{b,a} A_b A_a} \cdot \frac{x^{m_{b,a} A_b A_a - 1}}{\Gamma(m_{b,a} A_b A_a)} \cdot \exp \left(-\frac{m_{b,a} x}{\rho_a \hat{\Omega}_{b,a}} \right). \tag{5}$$

The pdf of $\hat{\gamma}_{b,C}$ can be expressed similarly, by replacing $m_{b,a}$, A_a, $\hat{\Omega}_{b,a}$, ρ_a with $m_{b,C}$, A_C, $\hat{\Omega}_{b,C}$, ρ_C respectively. As a result, we can derive $\mathcal{P}_{b,a}^L$ as

$$\mathcal{P}_{b,a}^L = \left\{ \frac{\hat{\gamma}_{b,a}}{\hat{\gamma}_{b,C}+1} \geq \gamma_{th} \right\} = \int_0^\infty Pr\{\hat{\gamma}_{b,a} \geq \gamma_{th}(x+1)\} f_{\hat{\gamma}_{b,C}}(x) dx. \tag{6}$$

Denote $\Omega_{b,a} = \rho_a \hat{\Omega}_{b,a}$ and $\Omega_{b,C} = \rho_C \hat{\Omega}_{b,C}$. Using the CDF of Gamma distribution, we have

$$Pr\{\hat{\gamma}_{b,a} \geq \gamma_{th}(x+1)\} = \exp \left(-\frac{m_{b,a} \gamma_{th}(x+1)}{\Omega_{b,a}} \right) \sum_{k=0}^{m_{b,a} A_b A_a - 1} \frac{1}{k!} \left(\frac{m_{b,a} \gamma_{th}}{\Omega_{b,a}} (x+1) \right)^k \tag{7}$$

Now we can substitute (5) and (7) into (6) and take integration

$$\mathcal{P}_{b,a}^L = \exp\left(-\frac{m_{b,a}\gamma_{th}}{\Omega_{b,a}}\right)\left(\frac{m_{b,C}}{\Omega_{b,C}}\right)^{A_b A_C m_{b,C}} \frac{1}{\Gamma(A_b A_C m_{b,C})} \sum_{k=0}^{A_b A_a m_{b,a}-1} \sum_{n=0}^{k} \binom{k}{n} \frac{1}{k!}$$

$$\cdot \left(\frac{m_{b,a}\gamma_{th}}{\Omega_{b,a}}\right)^k \cdot \int_0^\infty \exp\left(-\left(\frac{m_{b,a}\gamma_{th}}{\Omega_{b,a}} + \frac{m_{b,C}}{\Omega_{b,C}}\right)x\right) x^{n+A_b A_C m_{b,C}-1} dx$$

$$= \exp\left(-\frac{m_{b,a}\gamma_{th}}{\Omega_{b,a}}\right)\left(\frac{m_{b,C}}{\Omega_{b,C}}\right)^{A_b A_C m_{b,C}} \frac{1}{\Gamma(A_b A_C m_{b,C})} \cdot \sum_{k=0}^{m_{b,a}A_b A_a-1} \sum_{n=0}^{k} \binom{k}{n} \frac{1}{k!}$$

$$\cdot \left(\frac{m_{b,a}\gamma_{th}}{\Omega_{b,a}}\right)^k \left(\frac{m_{b,a}\gamma_{th}}{\Omega_{b,a}} + \frac{m_{b,C}}{\Omega_{b,C}}\right)^{-(n+A_b A_C m_{b,C})} \cdot (n + A_b A_C m_{b,C} - 1)! \quad (8)$$

(2) *Relay activation probability* $\mathcal{P}_\Delta$:
$\mathcal{P}_\Delta = P_r\{\delta_1, \cdots, \delta_K\}$ denotes the probability that a particular relay activation sequence occurs. Whether a relay can be activated is related to the activities of those relays scheduled prior to it. Hence we can decompose $\mathcal{P}_\Delta$ using the chain rule $\mathcal{P}_\Delta = P_r\{\delta_1\}P_r\{\delta_2|\delta_1\}P_r\{\delta_3|\delta_1,\delta_2\}\cdots P_r\{\delta_K|\delta_1,\cdots,\delta_{K-1}\}$. In what follows, we will focus on two expressions $P_r\{\delta_1\}$ and $P_r\{\delta_k|\delta_1,\cdots,\delta_{k-1}\}$.

First, $P_r\{\delta_1 = 1\}$ denotes the probability that R_1 can fully recover $\mathcal{I}$. This event happens only if R_1 successfully decodes the signals transmitted from all the sources. Using the link decoding probability $\mathcal{P}_{b,a}^L$ (8), we can directly have

$$P_r\{\delta_1 = 1\} = \prod_{s=1}^M \mathcal{P}_{R_1,S_s}^L \quad \text{and} \quad P_r\{\delta_1 = 0\} = 1 - \prod_{s=1}^M \mathcal{P}_{R_1,S_s}^L. \quad (9)$$

To derive $P_r\{\delta_k|\delta_1,\cdots,\delta_{k-1}\}$, we define a decoding relay set $\mathcal{R}_{k-1}$ after the $(M+k-1)$th time slot. This set contains the relays corresponding to $\delta_i = 1$ among $R_1, \cdots, R_{k-1}$, and therefore has size $\sum_{i=1}^{k-1}\delta_i$. Thus $P_r\{\delta_k = 1|\delta_1,\cdots,\delta_{k-1}\}$ is the probability that R_k can correctly decode at least M out of the $M + \sum_{i=1}^{k-1}\delta_i$ signals it received from the M sources and the $\sum_{i=1}^{k-1}\delta_i$ relays in $\mathcal{R}_{k-1}$. We further define an $(M + \sum_{i=1}^{k-1}\delta_i) \times 1$ indicator vector $\kappa^{[k-1]} = [\kappa_1^{[k-1]}, \cdots, \kappa_{M+\sum_{i=1}^{k-1}\delta_i}^{[k-1]}]$, where each element $\kappa_j^{[k-1]} \in \{0,1\}$ denotes whether R_k can decode the jth node in $\mathcal{S} \cup \mathcal{R}_{k-1}$, and $\mathcal{S}$ is the set of all sources. Now

$$P_r\{\delta_k = 1|\delta_1,\cdots,\delta_{k-1}\} = \sum_{\sum_j \kappa_j^{[k-1]} \geq M} \left(\prod_{a_i \in \mathcal{S}\cup\mathcal{R}_{k-1}} \left(\mathcal{P}_{R_k,a_i}^L\right)^{\kappa_i^{[k-1]}} \left(1 - \mathcal{P}_{R_k,a_i}^L\right)^{1-\kappa_i^{[k-1]}}\right)$$

$$(10)$$

Again, $P_r\{\delta_k = 0|\delta_1,\cdots,\delta_{k-1}\} = 1 - P_r\{\delta_k = 1|\delta_1,\cdots,\delta_{k-1}\}$. Now, for each $\Delta = [\delta_1,\cdots,\delta_K]$, we substitute (9) and (10) into the chain rule to attain $\mathcal{P}_\Delta$.

(3) *Destination conditional decoding probability* $\mathcal{P}_{\mathrm{D}_n|\Delta}$:

We follow the logic presented in the above subsection to calculate $\mathcal{P}_{\mathrm{D}_n|\Delta}$. In particular, define $\mathcal{R}_K$ to be the decoding relay set after all the relays finish forwarding source messages. This set contains the relays corresponding to $\delta_i = 1$ among $R_1, \cdots, R_K$, and has size $\sum_{i=1}^{K} \delta_i$. Thus $\mathcal{P}_{\mathrm{D}_n|\Delta}$ is the probability that D_n can correctly decode at least M out of the $M + \sum_{i=1}^{K} \delta_i$ signals it received from the M sources and all activated relays. Again, we define an $(M + \sum_{i=1}^{K} \delta_i) \times 1$ indicator vector $\pi^{[n]} = [\pi_1^{[n]}, \cdots, \pi_{M+\sum_{i=1}^{K} \delta_i}^{[n]}]$, where each element $\pi_i^{[n]} \in \{0,1\}$ denotes whether D_n can decode the ith node in $\mathcal{S} \cup \mathcal{R}_K$. We have

$$\mathcal{P}_{\mathrm{D}_n|\Delta} = \sum_{\sum_j \pi_j^{[n]} \geq M} \left(\prod_{a_i \in \mathcal{S} \cup \mathcal{R}_K} \left(\mathcal{P}_{\mathrm{D}_n, a_i}^L\right)^{\pi_i^{[n]}} \left(1 - \mathcal{P}_{\mathrm{D}_n, a_i}^L\right)^{1 - \pi_i^{[n]}} \right). \tag{11}$$

Substituting (10) and (11) into (4) leads to the system error probability.

Remark: The above derivation is complicated because the large-scale path loss, transmit power and antenna numbers can be different for different V2V links, which makes the decoding probability of each link unique. If the considered VCN system has the same fading and path loss factors for all V2V links (i.e., $m_{b,a} = m$ and $\Omega_{b,a} = \Omega$), and that between C and vehicles ($m_{b,C} = m_C$ and $\Omega_{b,C} = \Omega_C$), and has the same transmit power and antennas at each vehicle (i.e., $P_a = P_V$ and $A_a = A$), we have the same link decoding probability $\mathcal{P}_{b,a}^L$ for all V2V pairs, which can be denoted by $\mathcal{P}^L$. We have simplified expressions

$$Pr\{\delta_k = 1|\delta_1, \cdots \delta_{k-1}\} = \sum_{j=M}^{M+\sum_{i=1}^{k-1} \delta_i} \binom{M + \sum_{i=1}^{k-1} \delta_i}{j} \left(\mathcal{P}^L\right)^j \left(1 - \mathcal{P}^L\right)^{M+\sum_{i=1}^{k-1} \delta_i - j}$$

$$\mathcal{P}_{\mathrm{D}_n|\Delta} = \sum_{j=M}^{M+\sum_{i=1}^{K} \delta_i} \binom{M + \sum_{i=1}^{K} \delta_i}{j} \left(\mathcal{P}^L\right)^j \left(1 - \mathcal{P}^L\right)^{M+\sum_{i=1}^{K} \delta_i - j}.$$

5   Numerical Results

In what follows, we use numerical results to demonstrate the advantage of the proposed scheme. We consider an example network with $M = 2$ sources, $K = 2$ relays and $N = 2$ or $N = 10$ destinations. Each node has 2 antennas. We demand all the vehicles to transmit with the same power P_a. The relationship between P_a and P_C is set to be $\frac{P_a}{P_C} = \tau P_a^{\theta} = P_a^{0.8}$, so that we can scale them together by changing P_a. The system bandwidth is set as $W = 100\,\mathrm{kHz}$, and the noise density is $-174\,\mathrm{dBm/Hz}$. The Nakagami fading parameters are set to $m_{b,a} = 2$, $m_{b,C} = 4$. $\Omega_{b,a}$ and $\Omega_{b,C}$ take into account the path loss as $\Omega_{b,a} = \frac{10^{0.1 \cdot (10\log_{10}(1000 P_a) - PL(d_{b,a}))}}{1000 N_0 A_a}$, $\Omega_{b,C} = \frac{10^{0.1 \cdot (10\log_{10}(1000 P_C) - PL(d_{b,C}))}}{1000 N_0 A_C}$, $PL(d_{b,a}) = 103.4 + 24.2\log_{10} d_{b,a}$ and $PL(d_{b,C}) = 127 + 30\log_{10} d_{b,C}$, where $d_{b,a}$ denotes the distances between a V2V pair a and b, $d_{b,C}$ denotes that between C

and b, and the path loss setting follows [14] for V2V and C2V communication environments. All distances $d_{b,a}$ $(d_{b,C})$ are generated from Gaussian distribution with mean 50 m (200 m) and standard deviation 2 m (5 m).

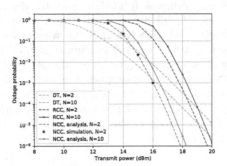

Fig. 2. System error performance comparison

In addition to our scheme (termed NCC, networking coding cooperation), we also consider the direct transmission (DT) and the repetition-coding based cooperation (RCC) scheme. For fair comparison, we define the average transmission rate of each scheme to be $\bar{R} = 0.1$ bps/Hz. Figure 2 shows that the analytical error performance of our NCC scheme is in line with results attained through simulation. This demonstrates the accuracy of our derivation. The DT scheme ($R = \bar{R}MW$ bits) has lower diversity gain due to the lack of the assistance of relays (i.e., communication resource sharing in VCN) to combat fading and interference. Higher power has to be consumed to guarantee sufficiently good performance. The RCC scheme attains the same diversity as NCC. But since it has to use a high data rate ($R = \bar{R}M(K+1)W$ bits) to compensate inefficient channel usage, which causes difficulties in the decoding process at the relays and destinations, especially when the received SNR is relative low. Hence the advantages of cooperative transmission emerge only when the transmit power of each node is sufficiently large. Finally, our NCC ($R = \bar{R}(M+K)W$ bits) scheme adopts the efficient network coding concept and properly chooses the coding structure suitable for multi-source multi-relay networks. Although it uses more time slots than the DT scheme to conduct transmissions, the increase in channel consumption is limited. This fact balances the decoding errors due to fading and high data rate. The situation will become severe if more information sources and destinations are involved in the VCN, because it is easier to make errors. But high diversity still remains. These clearly show the benefits of our scheme.

6   Conclusion

We have studied applying cooperative relaying and network coding techniques to support V2V communication in VCN. We have considered a reuse-mode MIMO multi-sources, multi-relay, and multi-destination network under Nakagami-m fading. A class of MDS-FFNC is applied and the closed-form system error probability has been derived. Our results have demonstrated the potential of combining relaying and network coding techniques in future VCN.

Acknowledgement. This work was funded in part by the National Natural Science Foundation of China (61771343 and 61331009), the EU PF7 QUICK project (PIRESES-GA- 2013-612652) and JSPS KAKENHI Project Number JP16H02817. This is also a part of a project that has received funding from the European Unions Horizon 2020 research and innovation programme under the Marie Sklodowska-Curie grant agreement No 752797. R. Chen's work is supported in part by the international internship program of the National Institute of Informatics. It reflects only the authors view and the Research Executive Agency and the European Commission are not responsible for any use that may be made of the information it contains. C. Wang is the correspondence author.

References

1. Papadimitratos, P., Fortelle, A., Evenssen, K., Brignolo, R., Cosenza, S.: Vehicular communication systems: Enabling technologies, applications, and future outlook on intelligent transportation. IEEE Commun. Mag. **47**(11), 84–95 (2009)
2. Ahmed, E., Gharavi, H.: Cooperative vehicular networking: a survey. IEEE Trans. Intell. Transp. Syst. **19**(3), 996–1014 (2018)
3. Andrews, J.G., et al.: What will 5G be? IEEE J. Sel. Areas Commun. **32**(6), 1065–1082 (2014)
4. Lee, E., Lee, E.K., Gerla, M., Oh, S.Y.: Vehicular cloud networking: architecture and design principles. IEEE Commun. Mag. **52**(2), 148–155 (2014)
5. Whaiduzzaman, M., Sookhak, M., Gani, A., Buyya, R.: A survey on vehicular cloud computing. J. Netw. Comput. Appl. **40**, 325–344 (2014)
6. Kim, S.W., et al.: Multivehicle cooperative driving using cooperative perception: Design and experimental validation. IEEE Trans. Intell. Transp. Syst. **16**(2), 663–680 (2015)
7. Dolk, V., et al.: Cooperative automated driving for various traffic scenarios: experimental validation in the GCDC 2016. IEEE Trans. Intell. Transp. Syst. **19**(4), 1308–1321 (2018)
8. Li, M., Yang, Z., Lou, W.: CodeOn: Cooperative popular content distribution for vehicular networks using symbol level network coding. IEEE J. Sel. Areas Commun. **29**(1), 223–235 (2011)
9. Xiao, M., Kliewer, J., Skoglund, M.: Design of network codes for multiple-user multiple-relay wireless networks. IEEE Trans. Commun. **60**(12), 3755–3766 (2012)
10. Wang, C., Xiao, M., Skoglund, M.: Diversity-multiplexing tradeoff analysis of coded multi-user relay networks. IEEE Trans. Commun. **59**(7), 1995–2005 (2011)
11. Xing, W., Liu, F., Wang, C., Xiao, M., Wang, P.: Multi-source network-coded D2D cooperative content distribution systems. J. Commun. Netw. **20**(1), 69–84 (2018)
12. Soleimani-Nasab, E., Matthaiou, M., Ardebilipour, M.: Multi-relay MIMO systems with OSTBC over Nakagami-m fading channels. IEEE Trans. Veh. Technol. **62**(8), 3721–3736 (2013)
13. Chalise, B.K., Czylwik, A.: Exact outage probability analysis for a multiuser MIMO wireless communication system with space-time block coding. IEEE Trans. Veh. Technol. **57**(3), 1502–1512 (2008)
14. 3GPP TR 36.814: Further advancements for E-UTRA physical layer aspects (2009)

Cluster-Based Dynamic FBSs On/Off Scheme in Heterogeneous Cellular Networks

Xiaoge Huang[✉], She Tang, Dongyu Zhang, and Qianbin Chen

School of Communication and Information Engineering, Chongqing University
of Posts and Telecommunications, Chongqing 400065, China
{Huangxg,Chenqb}@cqupt.edu.cn
{tangshedhl,zhangdongyu}@outlook.com

Abstract. Recent years, with the explosive growth of mobile data traffic, cellular communication system is faced with enormous challenges. The ultra-dense deployment of small cells will increase the network capacity while increasing the energy consumption. In this paper, we study a cluster-based dynamic FBSs on/off scheme in heterogeneous cellular networks, where the overall objective is to maximize the network energy efficiency by optimizing jointly the cell association, the base station on/off strategies and the cluster division, taking into account the load balancing and the QoS requirement of heterogenous cellular networks. The optimization problem is divided into three processes: the base station and the user equipment (UE) association scheme, the femtocell base station (FBS) clustering, and the FBS on/off scheme according to the current traffic load. A cluster-based dynamic FBSs on/off scheme is proposed to improve EE in HCNs while ensuring the load balancing, the probability of outage, and the communication requirement of UEs in the core area. Simulation result shows that the proposed algorithm could achieve significant improvement of the network energy efficiency in all aspects than comparison algorithms in literature.

Keywords: Heterogeneous cellular networks · Energy efficiency
Femtocell base station · Cluster

1 Introduction

Recently, in order to deal with the explosive increment of demand in high speed mobile communication traffic, the network operator urge to seek for various means to increase network capacity. In the traditional cellular network,

This work is supported by the National Natural Science Foundation of China (NSFC) (61401053), and Innovation Project of the Common Key Technology of Chongqing Science and Technology Industry (Grant no. cstc2015zdcyztzx40008).

© ICST Institute for Computer Sciences, Social Informatics and Telecommunications Engineering 2019
Published by Springer Nature Switzerland AG 2019. All Rights Reserved
X. Liu et al. (Eds.): ChinaCom 2018, LNICST 262, pp. 325–335, 2019.
https://doi.org/10.1007/978-3-030-06161-6_32

macrocell base stations (MBSs) are used to provide communication service in a wide area. However, the user equipment (UE) located in the marginal of the area will get a low signal-to-interference-plus noise ratio (SINR) and suffer from low communication quality. Accordingly, Small cell is suggested as a kind of local BSs with lower transmission power, which include the microcell base station, the picocell base station (PBS) and the femtocell base station (FBS), while deploying in abundance under the coverage of the MBS to provide higher SINR for cell-edge UEs and further increase capacity of the network. Ultra-dense heterogeneous cellular networks play a critical role in 5G communication system. However, the large amount of small cells could lead to increasing of the extra electric energy consumption. In addition, technologies to enhance network energy efficiency (EE) have become a critical design due to increasing energy price and growing attention toward environmental factors.

There are many challenges in deploying small cell base stations (SBSs) under the coverage of MBS, and various resource allocation schemes have been designed for HCNs [1–4]. Normally, frequency reuse technologies among small cells and macro cells could improve the spectrum efficiency and the capacity of HCNs. In [1], the author proposed a joint subcarrier assignment and power allocation scheme to optimize spectrum efficiency in the downlink transmission of HCNs. Due to the large number of deployed SBSs, a serious co-layer interference will occur in the network. In [2], the author introduced a greedy algorithm to allocate resource blocks under the QoS constraint to reduce interference among FBSs. Especially, it is a feasible method to eliminate interference by clustering. In [3], the author designed clusters through graph-theoretic approaches to eliminate interference in HCNs. In [4], the author applied a modify cluster algorithm and Stackelberg game for resource allocation to improve network throughput.

Moreover, in order to improve EE in the HCN, various schemes have been proposed in [5–7]. In [5], the author designed a dynamic gNB (the name of the BS in 5G) on/off strategy to improve EE, while considering the quality of service (QoS) constraint of UEs and the load balancing among gNBs. The author attempted to explore the relationship of energy efficiency, transmission power and the number of SBSs in [6]. In [7], water-filling method was used to structure the optimal solution of the power allocation problem to further improve EE.

In this paper, we aim to improve EE by clustering with dynamic on/off strategies of FBSs according to the current traffic load, while achieving the optimal tradeoff between the UE SINR threshold and the outage probability in HCNs. The main contributions are summarized as follows:

Firstly, in order to optimize network EE while taking into account the UE SINR requirement, the network outage probability and the load balancing, we propose an optimal UE-FBS association load balancing (UFALB) algorithm.

Secondly, in order to eliminate intra-cluster interference, FBSs are divided into several clusters based on EE and geographic location of FBSs.

Finally, in order to further improve EE, cluster-based FBSs on/off (CBFOO) algorithm is proposed. Notice that, to ensure the communication requirement of UEs in the core transmission area, the FBS can not be turned off if there are UEs in the core transmission area.

The rest of the paper is organized as follows. The system model is presented in Sect. 2. Section 3 provides the cluster formulation and the UE-FBS association strategy. In Sect. 4, the CBFOO algorithm is proposed to improve EE. Simulation results are discussed in Sect. 5. Section 6 draws the conclusion.

2   System Model

2.1   Network Scenario

In this paper, we consider a two-ties HCN which includes the MBS and the FBS for each tier, respectively, as shown in Fig. 1. FBSs are randomly located in the coverage of the MBS. The transmission power of the MBS and the FBS are denoted as P_m and P_f, respectively. The UE will connect with the BS according to UFALB algorithm to obtain the maximum EE. In our scenario, FBSs will be divided into different clusters based on the reactive distance to other FBSs and the different contribution to EE. Accordingly, every cluster will select the cluster head FBS (H-FBS) and member FBSs (M-FBS). The H-FBS could collect all UEs information in the cluster and forward the signaling message to turn on/off M-FBSs to improve NEE. Let $\mathbf{M}$ denote the set of the MBS, $\mathbf{N}$ represent the set of the FBS, $\mathbf{K}$ indicate the set of UEs and $\mathbf{C}$ denote the set of clusters.

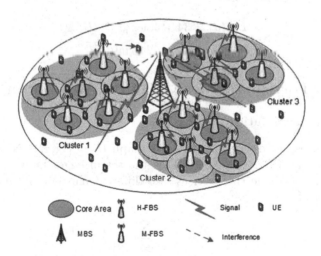

Fig. 1. Network scenario

2.2   Power Consumption and Path Loss Model

A typical FBS hardware model consists of the microprocessor module, the power amplifier (PA) module, the radio frequency (RF) module and the field-programmable gate array (FPGA) module [8]. The power consumption of these modules are represented by P_{mic}, P_{pa}, P_{rf}, and P_{fpga} respectively. Hence, the total power consumption of a FBS is expressed as:

$$P_{total} = P_{mic} + P_{pa} + P_{rf} + P_{fpga} \tag{1}$$

We assume a FBS has two operation modes: "ON" mode and "OFF" mode.

- "ON" mode: the FBS is in full operation, the power consumption is P_{total}.
- "OFF" mode: the FBS is turned off, but it still consumes a little power to maintain the wake-up function.

In this paper, transmission channels are assumed to be time-invariant with slow fading. Thus, the received SINR γ_k^n of UE k from FBS n is calculated by:

$$\gamma_k^n = \frac{P_n L(d_{k,n}) \lambda_n}{\sum\limits_{n' \in I} P_{n'} L(d_{k,n'}) \lambda_{n'} + N_0 B} \tag{2}$$

where I is the set of BSs (e.g. includes the FBS and the MBS) which could generate interference to UE k. P_n is the transmission power of BS n ($n \in I$). $d_{k,n}$ is the distance between UE k ($k \in K$) and BS n and B is the bandwidth of the sub-channel. $\lambda_n \in \{0,1\}$ indicates the operation mode of FBS n, $\lambda_n = 1$ indicates FBS n is in full operation mode, otherwise, $\lambda_n = 0$ indicates FBS n is turned off. $L(\cdot)$ is the path loss function. The path loss model of the MBS and the FBS are expressed as follow:

$$L_M(d) = 34 + 40\log_{10}(d)dB$$
$$L_F(d) = 37 + 30\log_{10}(d)dB \tag{3}$$

3   Cluster Formulation and UE-FBS Association Strategy

3.1   Cluster Formulation

The HCN is considered as an undirected graph $G = \{N, E\}$. N is the set of vertices (regard as FBSs in HCNs). E is the set of edges between two FBSs. Our purpose is to divide the vertices into different clusters and maximize the following object function [9]:

$$\sum_{(i,j) \in E_1(N)} \beta w_{i,j}^+ + \sum_{(i,j) \in E_2(N)} (1 - \beta) w_{i,j}^- \tag{4}$$

where $E_1(N)$ represents the set of edges whose vertices are in the same cluster and $E_2(N)$ represents the set of edges whose vertices are in the different cluster. $w_{i,j}^+ = |EE_i - EE_j|$ is the similarity degree between FBS i and FBS j, the greater the EE difference between FBSs, the higher probability they will be divided into the same cluster. $w_{i,j}^- = D_{i,j}$ is the difference degree between FBS i and FBS j, which indicates that the smaller the distance between FBSs, the higher probability they be divided into the same cluster. $\beta(0 < \beta < 1)$ represents the weight of w^+. According to the above description, the original problem (5) can be expressed as following:

$$\max_{X} \sum_{i,j\in\mathbf{N}} \beta\,|EE_i - EE_j|\,x_{i,j} + (1-\beta)D_{i,j}(1 - x_{i,j})$$

$$\begin{aligned}
\text{s.t.}\quad &\text{C1}: \quad x_{i,j} = x_{j,i}, \forall i,j \in \mathbf{N}\\
&\text{C2}: \quad x_{i,i} = 1, \forall i \in \mathbf{N}\\
&\text{C3}: \quad x_{i,j} + x_{j,l} + x_{l,i} \le 1, \forall i,j,l \in \mathbf{N}\\
&\text{C4}: \quad \sum_{j\in\mathbf{N}} x_{i,j} \le M, \forall i \in \mathbf{N}\\
&\text{C5}: \quad X = (x_{i,j}) \in \{0,1\}, \forall i,j \in \mathbf{N}
\end{aligned} \tag{5}$$

C1 denotes that FBS i and j are in the same cluster. C2 represents that a FBS only belongs to one cluster. C3 means if FBS i and j are in the same cluster, meanwhile, FBS j and l are in the same cluster, thus FBS i, j, l are divided into the same cluster. C4 is cluster size constraint. C5 indicates whether FBS i and j are in the same cluster.

There are several approaches to solve the optimization problem (5), such as the traversal search, the Branch and Bound (BnB) scheme, the semidefinite programming (SDP) and so on. However, for ultra-dense networks with a large number of FBSs, the computation complex of the traversal search is too high. Thus, in order to reduce the computational complexity, the SDP-based correlation clustering algorithm [10] is used to obtain the solution:

$$\max_{X} \sum_{i,j\in\mathbf{N}} \beta\,|EE_i - EE_j|\,x_{i,j} + (1-\beta)D_{i,j}(1 - x_{i,j})$$

$$\begin{aligned}
\text{s.t.}\quad &\text{C1}: \quad x_{i,i} = 1, \forall i \in \mathbf{N}\\
&\text{C2}: \quad x_{i,j} + x_{j,l} + x_{l,i} \le 1, \forall i,j,l \in \mathbf{N}\\
&\text{C3}: \quad \sum_{j\in\mathbf{N}} x_{i,j} \le M, \forall i \in \mathbf{N}\\
&\text{C4}: \quad x_{i,j} \ge 0, \forall i,j \in \mathbf{N}\\
&\text{C5}: \quad X = (x_{i,j}) \succeq 0, \forall i,j \in \mathbf{N}
\end{aligned} \tag{6}$$

The constraint C4 indicates that $x_{i,j}$ is slack. In constraint C5, X is a positive semidefinite matrix which is symmetry and non-negative, which can be written as $X = B^T B$, where $B = \{\mathbf{b}_1, \mathbf{b}_2, \cdots, \mathbf{b}_N\}$. Notice that $x_{i,j} = \mathbf{b}_i^T \mathbf{b}_j = \mathbf{b}_i \cdot \mathbf{b}_j$.

Based on the slack solution X of (6), we can obtain a integer solution X_{int} by iterations. Firstly, we obtain L random hyperplane with independent random vectors $\mathbf{r}_y = \{r_{y1}, r_{y2}, \cdots, r_{yN}\}$, $1 \le y \le L$, $2^L \ge N$ as their normals. We can divide vertices into the following sets:

$$\begin{aligned}
\mathbf{C}_1 &= \{i \in \mathbf{N} : \mathbf{r}_1 \cdot \mathbf{b}_i \ge 0, \cdots, \mathbf{r}_L \cdot \mathbf{b}_i \ge 0\}\\
\mathbf{C}_2 &= \{i \in \mathbf{N} : \mathbf{r}_1 \cdot \mathbf{b}_i \ge 0, \cdots, \mathbf{r}_L \cdot \mathbf{b}_i < 0\}\\
&\cdots\\
\mathbf{C}_{2^L} &= \{i \in \mathbf{N} : \mathbf{r}_1 \cdot \mathbf{b}_i < 0, \cdots, \mathbf{r}_L \cdot \mathbf{b}_i < 0\}
\end{aligned} \tag{7}$$

Accordingly, we can obtain 2^L clusters, such as $\mathbf{C} = \{\mathbf{C}_1, \mathbf{C}_2, \cdots, \mathbf{C}_{2^L}\}$. Thus, we could obtain the optimal solution ZZ_R with X_{int} in (6). Note that $ZZ_R \ge$

$\alpha ZZ_{opt}(\alpha < 1)$, ZZ_{opt} is the optimal solution in (5) and α is related to L. The SDP-based correlation clustering algorithm is summarized as follows.

3.2   UE-FBS Association Strategy

We aim to maximize network energy efficiency of HCNs, while ensuring the outage probability, the load balancing and the UE target-SINR. Therefore, the UE-FBS association load balancing (UFALB) algorithm [9] is introduced to optimize the cell association between FBSs and UEs. The utility function of UE k is composed of access factor θ_k^n and the UE EE_k^n, which is given by:

$$\omega_k^n = \theta_k^n \cdot EE_k^n \tag{8}$$

Algorithm 1 SDP-Based Correlation Clustering Algorithm

1: Solve X in (6) by CVX
2: Calculate $B = \{\mathbf{b}_1, \mathbf{b}_2, \cdots, \mathbf{b}_V\}$ by $X = B^T B$
3: **for** $t = 1 : t_{\max}$ **do**
4:     Generate the independent random vectors $\mathbf{r}_i = \{r_{i1}, r_{i2}, \cdots, r_{iV}\}$, $1 \leq i \leq L$
5:     Calculate $\mathbf{C} = \{\mathbf{C}_1, \mathbf{C}_2, \cdots, \mathbf{C}_{2L}\}$ according to (7)
6:     Map $\mathbf{C}$ into the solution X_{int}
7:     Calculate ZZ_R in (6) by X_{int}
8: **end for**
9: Find the largest $ZZ_R(ZZ_R \geq \alpha ZZ_{opt})$ and its corresponding cluster $\mathbf{C}$
10: **return** $\mathbf{C}$

---

where θ_k^n is the access factor, which indicates the probability of UE k successfully access to FBS n, which can be formulated as:

$$\theta_k^n = \begin{cases} \frac{L_n^{max} - L_n}{L_n^{max}} & \text{if} \quad L_n < L_n^0 \\ \frac{L_n^{max} - L_n}{L_n^{max}} \cdot \frac{L_n^0}{L_n} & \text{if} \quad L_n \geq L_n^0 \end{cases} \tag{9}$$

where L_n^{max} is the maximum acceptable traffic load of FBS n in a subframe. Aa subframe is divided into two slots: $L_n^0 = L_n^{max}/2$ by Rounding Robin Scheduling. L_n is the current traffic load of FBS n. The expression of EE_k^n is given as follows:

$$EE_k^n = \frac{R_k^n}{P_n} = \frac{B \cdot \log_2(1 + \gamma_k^n)}{P_n} \tag{10}$$

4   Cluster-Based FBSs On/Off Algorithm

Based on the previous analysis, we could obtain the optimal association scheme for UEs and FBSs. To further improve EE, a cluster-based FBSs on/off (CBFOO) algorithm is proposed in this section, which jointly consider the load balancing, the intra-cluster interference and the transmission performance of UEs in the core area. The optimization problem with the load-balancing and the target-SINR threshold constraints can be formulated as:

$$\max_{\lambda, S} \quad EE = \frac{\sum_{k \in \mathbf{K}} \sum_{n \in \{\mathbf{M}, \mathbf{C}\}} B \log 2(1 + S_k^n \gamma_k^n)}{\sum_{n \in \{\mathbf{M}, \mathbf{C}\}} P(\lambda_n)}$$

s.t. $C1 : S_k^n \in (0, 1), \forall k \in \mathbf{K}, n \in \{\mathbf{M}, \mathbf{C}\}$

$C2 : \sum_{n \in \{\mathbf{M}, \mathbf{C}\}} S_k^n \leq 1, \forall k \in \mathbf{K}$

$C3 : \sum_{k \in \mathbf{K}} S_k^n \leq L_n^{max}, \forall n \in \{\mathbf{M}, \mathbf{C}\}$

$C4 : P_{out} \leq \rho$

$C5 : SINR_k^n \geq SINR_k^{th}, \forall k \in \mathbf{K}, n \in \{\mathbf{M}, \mathbf{C}\}$

$$C6 : \gamma_k^n = \frac{P_n L(d_{k,n}) \lambda_n}{\sum_{n' \in \mathbf{I} \setminus \mathbf{C}_z} P_{n'} L(d_{k,n'}) \lambda_{n'} + N_0 B} \tag{11}$$

where C1 is the connection indicator of UE k and FBS n or not. C2 indicates that a UE can only associate with one FBS at a time. C3 ensures the load balancing of FBSs. C4 shows the total outage probability of HCNs should not exceed the predetermined threshold. C5 is the target-SINR constraint of UE k. C6 indicates that there is no intra-cell interference, where $\mathbf{C}_z$ is the cluster that UE k belongs to. Therefore, the CBFOO algorithm is described in Algorithm 2.

Algorithm 2 Cluster-Based FBSs on/off (CBFOO) Algorithm

1: Initialization: $t = 0$, $\lambda = 1$, $\mathbf{C} = \{\mathbf{C}_1, \mathbf{C}_2, \cdots, \mathbf{C}_Z\}$, $\Phi = \{EE^{\mathbf{C}_1}, EE^{\mathbf{C}_2} \cdots, EE^{\mathbf{C}_Z}\}$
2: Associate UEs and BSs by the UFALB algorithm
3: Generate clusters by algorithm 1 and calculate $EE(t)$ by (11)
4: **for** $z = 1 : Z$ **do**
5:     Calculate energy efficiency of cluster $\mathbf{C}_z$, $EE^{\mathbf{C}_z}$
6: **end for**
7: **repeat**
8:     Find the cluster $\mathbf{C}_z$ with the minimum EE, $\mathbf{C}_z = arg \min \Phi$
9:     **repeat**
10:        Find FBS n of cluster $\mathbf{C}_z$ with the minimum EE, $n = arg \min EE^{\mathbf{C}_z}$
11:        **if** no UEs locate in the core area of FBS n **then**
12:            $\lambda_n = 0$ and calculate $EE(t+1)$ by (11)
13:            **if** $EE(t+1) \leq EE(t)$ or $P_{out} > \rho$ **then**
14:                $\lambda_n = 1$
15:                Break
16:            **end if**
17:            $EE(t) = EE(t+1)$
18:            $t = t + 1$
19:        **end if**
20:        Delete FBS n from $\mathbf{C}_z$
21:    **until** $\mathbf{C}_z$ is empty
22:    Delete $EE^{\mathbf{C}_z}$ from Φ
23:    Delete $\mathbf{C}_z$ from $\mathbf{C}$
24: **until** $\mathbf{C}$ is empty
25: **return** $\lambda, EE(t)$

---

5   Simulation Results

In this section, we evaluate the performance of the proposed CBFOO algorithm. In the scenario, FBSs and UEs are randomly distributed in the coverage of the MBS. In the simulation, a comparison between the proposed algorithm, the MAX-SINR algorithm, the UE-FBS association load balancing (UFALB) algorithm and the random cluster-based (RCB) algorithm is adopted to evaluate the performance of the proposed algorithm in various aspects. The simulation parameters are shown in Table 1.

Table 1. Simulation parameters

Parameter	Value
Macro base station radius	200 m
MBS transmission power	46 dBm
FBS transmission power	20 dBm
Number of FBSs	25
Maximum acceptable load of a MBS	120
Maximum acceptable load of a FBS	8
Outage probability threshold ρ	0.01
Noise power N_0	-174 dBm/Hz
Maximum size of a cluster	5

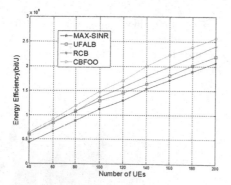

Fig. 2. EE versus number of UEs for different algorithms

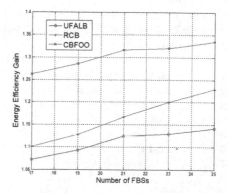

Fig. 3. EE gain versus the number of FBSs for different algorithms

Figure 2 shows that EE of the four algorithms gradually increases with the number of UEs increasing. In the MAX-SINR algorithm, UEs will select FBSs

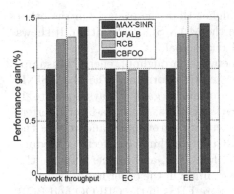

Fig. 4. Performance gain for different algorithms

Fig. 5. Impact of β on the cluster size/cluster number

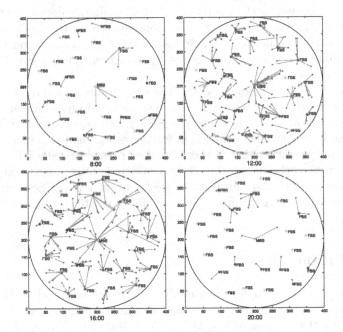

Fig. 6. UE-FBS association and FBSs on/off strategy at time 8:00 a.m, 12:00 a.m, 16:00 p.m, and 20:00 p.m

with the largest received SINR. The UFALB algorithm can choose a suitable association manner between UEs and FBSs, while taking into account the load balancing and the network outage probability. Based on the UFALB algorithm, the RCB algorithm randomly divide FBSs into different clusters. Additionally, the proposed CBFOO algorithm will turn off parts of FBSs based on the current traffic load to further improve EE.

In Fig. 3, compared with the MAX-SINR algorithm, the EE gain of three algorithms significantly raises with the number of FBSs increasing. Moreover, the EE gain of the CBFOO algorithm is higher than the others. On the other hand, the UFALB algorithm is higher than the MAX-SINR algorithm. It shows that the CBFOO algorithm can achieve ultimate goal.

Figure 4 illustrates the performance of network throughput, energy consumption (EC), and energy efficiency (EE) for different algorithms. The MAX-SINR algorithm is the basic algorithm whose performance gains always equal to 1. Since the others algorithms can turn off parts of FBSs, the intra-layer interference is smaller, which could increase the network throughput and reduce the total EC. Furthermore, the CBFOO and RCB algorithm can effectively reduce the intra-cluster interference, increase SINR for UE k, and further improve the throughput compared with the UFALB algorithm. On the other hand, the strict constraints restrain HCNs from turning off more FBSs in the CBFOO and RCB algorithms, thus the network energy consumption in these two algorithms is larger than that in the UFALB algorithm.

In Fig. 5, β represents the weight of w^+, i.e., similarity degree. From (5), we can see that when β is small, there are more clusters in the HCN in order to maximize the object function. As β increasing, the cluster size and the cluster number gradually reach a plateau, because the maximum cluster size M restrains more FBSs join a cluster.

Figure 6 shows the dynamic BS-UE association during a day at time 8 : 00 a.m, 12 : 00 a.m, 16 : 00 p.m, and 20 : 00 p.m. respectively. The star represents the MBS, the yellow points indicate the FBS which is turned off, whereas the blue points indicate the FBS which is in full operation mode, and the red points indicate the UE. Mention that if the FBS does connect to any UEs, it will be turned off. One the other hand, the UE is considered outage user when it doesn't connect with a BS. Through this picture, we can observe that FBSs can achieve a load balancing with a large number of UEs while guaranteeing outage probability in HCNs.

6   Conclusion

In this paper, we studied a cluster-based dynamic FBSs on/off scheme to improve EE in HCNs. The proposed cluster-based dynamic FBSs on/off scheme could jointly consider the load balancing, Qos requirement of HCNs, and EE improvement. Specially, the SDP-based Correlation Clustering algorithm with low computational complexity was introduced to obtain good correlation clustering, and the CBFOO algorithms is used to improve EE observably. Our work had significant contributions for future EE research.

References

1. Tang, J., So, D.K.C., Alsusa, E., Hamdi, K.A., Shojaeifard, A., Wong, K.K.: Energy-efficient heterogeneous cellular networks with spectrum underlay and overlay access. IEEE Trans. Veh. Technol. **67**(3), 2439–2453 (2018)

2. Aghababaiyan, K., Maham, B.: QoS-aware downlink radio resource management in OFDMA-based small cells networks. IET Commun. **12**(4), 441–448 3 6 (2018,)
3. Li, Y., Niu, C., Ye, F.: Graph-based femtocell enhanced universal resource allocation strategy for LTE-A HetNets. In: 2017 Progress in Electromagnetics Research Symposium - Fall (PIERS - FALL), pp. 3073–3078. Singapore (2017)
4. Liu, Y., Wang, Y., Zhang, Y., Sun, R., Jiang, L.: Game-theoretic hierarchical resource allocation in ultra-dense networks. In: 2016 IEEE 27th Annual International Symposium on Personal, Indoor, and Mobile Radio Communications (PIMRC), pp. 1–6. Valencia (2016)
5. Huang, X., Tang, S., Zheng, Q., Zhang, D., Chen, Q.: Dynamic femtocell gNB on/off strategies and seamless dual connectivity in 5G heterogeneous cellular networks. IEEE Access **6**, 21359–21368 (2018)
6. Liu, J., Sun, S.: Energy efficiency analysis of cache-enabled cooperative dense small cell networks. IET Commun. **11**(4), 477–482 3 9 (2017)
7. He, P., Zhang, S., Zhao, L., Shen, X.: Multi-channel power allocation for maximizing energy efficiency in wireless networks. IEEE Trans. Veh. Technol. **99**, 1–14 (2018)
8. Claussen, H., Ashraf, I., Ho, L.T.W.: Dynamic idle mode procedures for femtocells. Bell Labs Tech. J. **15**(2), 95–116 (2010)
9. Vazirani, V.V.: Approximation Algorithms. Springer, Berlin (2001)
10. Abdelnasser, A., Hossain, E., Kim, D.I.: Clustering and resource allocation for dense femtocells in a two-tier cellular OFDMA network. IEEE Trans. Wirel. Commun. **13**(3), 1628–1641 (2014)

Application Identification for Virtual Reality Video with Feature Analysis and Machine Learning Technique

Xiaoyu Liu[(✉)], Xinyu Chen, Yumei Wang, and Yu Liu

School of Information and Communication Engineering, Beijing University of Posts and Telecommunications, Beijing 100876, China
{liuxiaoy, chenxinyu, ymwang, liuy}@bupt.edu.cn

Abstract. Immersive media services such as Virtual Reality (VR) video have attracted more and more attention in recent years. They are applications that typically require large bandwidth, low latency, and low packet loss ratio. With limited network resources in wireless network, video application identification is crucial for optimized network resource allocation, Quality of Service (QoS) assurance, and security management. In this paper, we propose a set of statistical features that can be used to distinguish VR video from ordinary video. Six supervised machine learning (ML) algorithms are explored to verify the identification performance for VR video application using these features. Experimental results indicate that the proposed features combined with C4.5 Decision Tree algorithm can achieve an accuracy of 98.6% for VR video application identification. In addition, considering the requirement of real-time traffic identification, we further make two improvements to the statistical features and training set. One is the feature selection algorithm to improve the computational performance, and the other is the study of the overall accuracy in respect to training set size to obtain the minimum training set size.

Keywords: Application identification · Statistical feature · Machine learning VR video application

1 Introduction

Nowadays, online video has become one of the most popular network services, and video traffic is increasing on a large scale. For ordinary video, delay or stalling will reduce the Quality of Experience (QoE) of users. Delay or stalling can even cause users' physiological discomfort for Virtual Reality (VR) video. Therefore, it is necessary to establish an effective identification system for VR video application to manage network resources. To the best of our knowledge, there are few studies related to VR video traffic identification. Therefore, we survey several popular methods of network traffic identification and analyze their identification performance for VR video.

Typically, there are four different kinds of methods for network traffic identification, i.e., port-based, host-behavior-based, payload-based, and machine learning (ML) -based. The *port-based* method checks the port number of each packet and compares it with the Internet Assigned Numbers Authority (IANA) list [1]. The IANA

© ICST Institute for Computer Sciences, Social Informatics and Telecommunications Engineering 2019
Published by Springer Nature Switzerland AG 2019. All Rights Reserved
X. Liu et al. (Eds.): ChinaCom 2018, LNICST 262, pp. 336–346, 2019.
https://doi.org/10.1007/978-3-030-06161-6_33

list characterizes the one-to-one relationship between the port number and application. The *host-behavior-based* approach analyzes the host-behavior pattern of the transport layer and then associates the host-behavior with one or more application types [2]. Kim et al. [3] proved that the method based on port and host-behavior were not suitable for video identification.

Initially, we try to identify VR video with the *payload-based* method. Payload-based method checks if the payload of the packet contains a pre-registered special application sequence which is associated with one or more application types. We attempt to search a special application sequence that can distinguish VR video from ordinary video from the following three aspects, i.e., the specific host domain name included in the request packet, the specific video extension name, and the specific content type in the reply packet. However, there are no new discoveries. Therefore, we determine to distinguish VR video from ordinary video with a *ML-based* method. ML can classify each traffic flow by using its statistical features. We use analysis and traffic capture method to obtain statistical features, e.g., average packet size, throughput, packet arrival interval, which can be used to distinguish VR video from ordinary video. Experiments show that using these features, VR video can be well identified. To improve identification speed, we make two improvements to the statistical features and training set, i.e., reducing feature numbers and minimizing training set size.

The remainder of this paper is structured as follows. Section 2 presents background. Section 3 proposes VR video application identification system. And our experimental results are presented in Sect. 4. Finally, Sect. 5 concludes the paper.

2 Background

2.1 Related Work

As far as we know, currently there are few papers related to VR video traffic identification. Most of the work is the categorization of various types of network traffic. For example, karagiannis et al. [2] classified traffic into Web, News, Streaming, Gaming, etc. Moreover, authors in [4, 5] divided the flow data into video and non-video. They compared the performance obtained by Random Forest and AdaBoost, respectively. The results showed that, ignoring the classification speed of the model, the two algorithms could achieve similar classification accuracy (about 93%). Random Forest could guarantee a smaller model while ensuring classification accuracy, leading to faster classification speed. This is also what we obtain during our experiments.

Moore et al. [6] proposed the definition and the calculation of 249 flow features. Later researchers who use statistical features to classify traffic flow will generally adopt a subset of these flow features. We also use 37 of them in this study. Authors in [7] did not divide applications into categories like Streaming, Email, etc. Nevertheless, they considered the importance of application classification for network security and trend analysis, and divided applications into popular end-user applications such as Facebook, Skype, etc. The idea of categorizing end-user applications is applied in this paper.

The problem of ML algorithm is large training time which makes it ineffective of real time traffic classification. Solution of this problem is to reduce the number of features that represent the application type. The work of paper [8] shows that the feature selection algorithm can reduce the training time of the Bayes Net algorithm, making the Bayes Net classifier more suitable for real-time and online IP traffic classification. Williams [9] certified that feature selection could improve computational performance without sacrificing classification accuracy. In our study, we classify popular end-user applications such as iQiyi, iQiyiVR, Youku, YoukuVR and non-video.

2.2   Brief Introduction of ML

In this paper, we evaluate video application identification performance with six commonly used ML algorithms. Next, we briefly introduce the basic concepts of these six algorithms.

(1) **Naïve Bayes** classifiers are a family of simple "probabilistic classifiers" based on Bayes theorem. Naïve Bayes assumes strong independence between features [8]. The probability that an instance x belongs to a class c can be expressed as:

$$P(C = c|X = x) = \frac{P(C = c) \prod_i P(X_i = x_i | C = c)}{P(X = x)} \tag{1}$$

Where X is a vector of instances where each instance is described by features $\{X_1, X_2, \cdots X_k\}$, and C is the class of an instance [9].
We evaluate Naïve Bayes with discretization (NBD) which converts successive features into discrete features in this paper.

(2) **Bayesian Network** is a directed acyclic graph model [10]. The nodes of the model represent features or classes, and the links between nodes represent their probabilistic relationship.

(3) **K-Nearest Neighbors (KNN)** calculates the Euclidean distance from each test instance to the k nearest neighbors [11]. The k nearest neighbors vote to determine the class of test instance.

(4) **AdaBoost** is a meta-learning algorithm, which is built from a linear combination of simple classifiers. AdaBoost uses several classification models to decide the class label of an instance [12].

(5) **C4.5 Decision Tree** is a tree structure (a binary tree or a non-binary tree). Each non-leaf node indicates a test on features. Each branch indicates the output of the feature in a range of values, and each leaf node stores a category. In order to determine the class of a test instance, C4.5 Decision Tree starts testing the feature attributes corresponding to the test instance from the root node. Then this algorithm selects the output branch according to the value of the feature attribute. C4.5 Decision Tree repeats this process until it reaches the leaf node [13]. The category stored in the leaf node is the decision result.

(6) **Random Forest** is a classifier that contains multiple classification trees. All trees in the forest have the same distribution. The output category is determined by the mode of the individual tree's output [14].

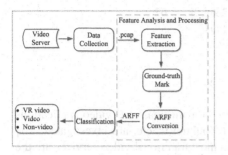

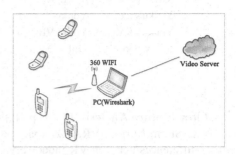

Fig. 1. The outline of identification system

Fig. 2. The data collection system for dataset gathering

3   VR Video Application Identification System

We present the outline of our identification system in Fig. 1, which consists of three parts: (1) Data collection, (2) Feature analysis and processing module for flow feature extraction, ground-truth mark, and ARFF conversion, (3) Classification module. The definition of a flow based on 5-tuple (source IP address, destination IP address, source port, destination port, and protocol) is adopted in this paper.

3.1   Data Collection

Our dataset is gathered via Wireshark [15]. Traffic is collected with five categories, e.g., iQiyiVideo, iQiyiVRVideo, YoukuVideo, YoukuVRVideo and non-video. Among these categories, a total of seven applications are applied. More details are given in Table 1.

As most VR video applications, such as 3D broadcast, iQiyiVR, and YoukuVR, are running on mobile devices, we collect mobile traffic with smartphones and iPad via WIFI access. At the same time, the computer runs Wireshark to collect traffic, which is stored in.pcap format for subsequent processing. The architecture of the data collection system is depicted in Fig. 2.

3.2   Feature Analysis and Processing

Appropriate flow feature acquisition is the premise of using ML algorithms to classify network traffic. In this section, we firstly analyze the different statistical features between VR video and ordinary video in detail. Then we introduce the further processing of these features.

Table 1. Categories, applications and the number of instances in our dataset

Category	Application	Number	Percentage of total (%)
iQiyiVideo	iQiyi	6993	36.74
iQiyiVRVideo	iQiyiVR	3800	19.96
YoukuVideo	Youku	2669	14.02
YoukuVRVideo	YoukuVR	1494	7.85
Non-video	Zhihu, Mail, Taobao	4078	21.43

Flow Feature Analysis. In this part, the current mainstream VR video applications such as Storm Mirrors VR, 3D broadcast, and Youku VR, etc. are investigated. The main differences between VR video service and most traditional video services are the process of increasing multi-camera video splicing and 360° video projection before video coding. Nevertheless, almost all current VR videos are still encoded in H264, which is the same as ordinary video. Therefore, the main coding related parameters for VR video are still resolution, bit rate, frame rate, etc.

However, due to the characteristics of VR video, higher requirements are placed on these video parameters. As seen from Fig. 3, the Field of View (FOV) in VR video is only part of the entire video. In order to achieve the appropriate resolution for the FOV, the entire VR video requires very high resolution. Take a 4 K (3840 × 1920) VR video as an example. Assuming that the HMD's angle of view is 90° in both directions of the horizontal α and the vertical β, the video resolution in the FOV is only 960 × 480, which is far away from the near-future 4 K video requirement. In order to improve users' experience, VR video requires even higher resolution.

High-resolution video requires a higher bit rate. In order to save the packet packaging cost, each packet size of VR video will be larger than that of ordinary video. Therefore, the average packet throughput and average byte throughput of VR video will be larger than that of ordinary video. We exploit traffic capture and analysis to obtain the differences of flow features between VR video and ordinary video.

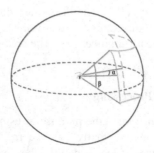

Fig. 3. VR video spherical projection and the FOV

We capture traffic and save them in.pcap format files, namely youkuvr.pcap and youku.pcap, for VR video and ordinary video, respectively. Each.pcap file has about 180,000 packets, which is equal to 124 MB. However, it takes 325 s to capture VR video and 770 s to capture ordinary video. Thus, VR video is two times the average packet throughput and byte throughput of the ordinary video. On the other hand, we analyze the difference of packet arrival interval between ordinary video and VR video. Packet arrival interval of them are given in Fig. 4a, b, respectively. There is a big difference between them. In terms of the packet arrival interval of ordinary video, the overall trend is relatively flat, and there are some protrusions in short time (about 30 s). However, for VR video, there are some large protrusions in a relatively long period of time (about 40 s). The maximum number of packets arriving per second is also dif-

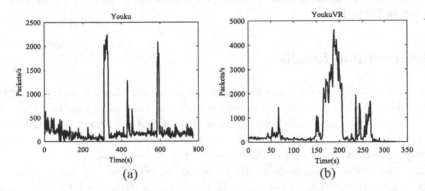

Fig. 4. **a.** Packet arrival interval of ordinary video **b.** Packet arrival interval of VR video

Table 2. Statistical features in this paper

	Before feature selection (37)	After feature selection (22)
Features	# Protocol, source and destination ports	# Protocol, source and destination ports
	# The number of packets/bytes	# The number of packets/bytes
	# The number of packets without Layer 4 payload	# The number of packets without Layer 4 payload
	# Start time, end time, duration	# Start time, end time, duration
	# Average packet throughput, average byte throughput	# Average packet throughput, average byte throughput
	# Max/min/average/standard deviation of packet sizes and inter-arrival times	# Max/min/average/standard deviation of packet sizes and inter-arrival times
	# Number of TCP packets with FIN, SYN, RSTS, PUSH, ACK, URG, CWE, ECE flags set (all zero for UDP packets)	# Number of TCP packets with FIN, SYN, RSTS(all zero for UDP packets).
	# The size of the first ten packets.	

ferent, which are 2200 and 4500, respectively. We also find that the packet sizes of them are different. In summary, the average packet/byte throughput, packet arrival interval and packet size will be representative features. Initially, we select 37 unidirectional flow features from those in [3] according to the findings made in above data analysis. The 37 features are shown in Table 2 (the column on the left).

Ground-truth Mark. In our study, each traffic from the video/VR video application is labeled as video or VR video. So, the classifiers will model some non-video features as video features, such as the features of ads and others. Yet the goal in this study is to ensure the overall experience of people using video applications.

ARFF Format Conversion. We make use of the PostgreSQL database to store data so that we can conveniently convert.pcap format into ARFF format. In the ARFF file, each flow is considered as an instance. The number of instances for each category is shown in Table 1.

4   Experimental Results

Through extensive experiments, we try to observe: (a) the identification performance for VR video application using the proposed statistical features; (b) the best algorithm for VR video application identification, considering both accuracy and build time; (c) the effect of feature selection on algorithm performance; (d) the change of overall accuracy in respect to the size of the training set.

The classification module is mainly composed of six most often-used supervised ML algorithms from WEKA [16]: Naïve Bayes, Bayesian Network, KNN (k is chosen as 1 in our experimental setup), AdaBoost (J48 is the Base classifier), J48 (C4.5 Decision Tree in WEKA), and Random Forest. The ML algorithms applied in this study are all implemented using the WEKA tool and only a few parameters are adjusted. Our test option is set to 10-fold cross validation, by which we gain the best overall accuracy during the entire experiments. We test 5-fold, 10-fold, 15-fold and 20-fold in our experiments.

4.1   Performance Metrics

To evaluate the performance of the six ML algorithms using the proposed statistical features, we use three metrics: overall accuracy (Acc), F-measure (F1) and build time.

First, we introduce the definition of True Positive (TP), False Positive (FP), True Negative (TN), and False Negative (FN). TP means that the forecast is positive and actually positive. FP means that the forecast is positive but actually negative. TN means that the forecast is negative and actually negative. FN means that the forecast is negative but actually positive.

$$Acc = \frac{TP + TN}{TP + TN + FP + FN} \tag{2}$$

$$F1 = \frac{2 * Precision * Recall}{Pr\,ecision + Re\,call} \tag{3}$$

where Precision is $\frac{TP}{TP+FP}$, and Recall is $\frac{TP}{TP+FN}$.

Acc is applied to measure the accuracy of an algorithm on the whole dataset. F1 is to evaluate the identification performance for each category. Build time is the time taken to create an identification model given a training set.

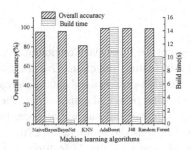

Fig. 5. Overall accuracy and the build time using 37 features

4.2 Classification Results Using 37 Features

The results of our dataset are given in Fig. 5. The Acc and the build time of each algorithm are shown in Table 3. The evaluation criteria for each strategy is F1, given in Sect. 4.1.

The results show an overall accuracy of 81% to 98.7%. In addition to KNN algorithm, the accuracy of other algorithms is above 95%. This phenomenon indicates that we can effectively distinguish VR video from ordinary video using the proposed statistical features. At the same time, it indicates that the proposed statistical features have good performance on various ML algorithms, and they are universal. These statistical features can also be used to distinguish video from non-video.

As shown in Fig. 5, AdaBoost algorithm gives the best accuracy and the longest build time. In addition, the accuracy of J48 algorithm is similar to that of AdaBoost algorithm, but J48 algorithm builds model in a shorter period of time, only one second. This is explained by the fact that J48 algorithm has less training demands, and that it has lower complexity than AdaBoost algorithm.

In general, the proposed features combined with J48 algorithm can achieve good performance for VR video application identification. They can achieve an overall accuracy of 98.6%, and the build time is about one second.

4.3 Further Discussions

Considering the requirement of real-time network traffic identification, we make two improvements to the experiments, i.e., study feature selection and the change of the overall accuracy in respect to training set size.

Table 3. The accuracy (%) and the build time(s) of each category

	NaiveBayes	BayesNet	KNN	AdaBoost	J48	Random Forest
iQiyi	95.7	96.0	87.3	98.9	98.8	98.6
iQiyiVR	95.9	96.4	81.1	98.9	99.0	99.1
Youku	93.4	93.7	75.6	97.9	97.7	97.4
YoukuVR	93.2	94.2	63.0	98.1	98.2	97.7
Non-video	95.7	96.2	79.5	98.9	98.9	98.9
Acc	95.3	95.7	81.0	98.7	98.6	98.5
Time	1.01	0.6	0.01	14.53	1	10.1

Feature Selection. We adopt the Principal Components Analysis (PCA), one of the most well-known algorithms in feature selection. This algorithm looks for a series of projection directions. After the high-dimensional data are projected in these directions, the variance is maximized. The first principal component is the largest variance, and the second principal component is the second largest variance. PCA algorithm selects the first 22 features of the 37 features, as shown in Table 2 (the column on the right). The overall accuracy and the build time with the 22 features and all features are given in Fig. 6a, b, respectively. After feature selection, the overall accuracy is hardly changed

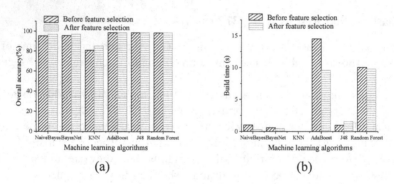

(a)   (b)

Fig. 6. **a**. Overall accuracy of six algorithms using all features and selected features **b**. The build time of six algorithms using all features and selected features

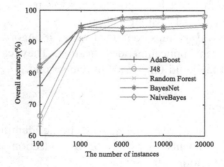

Fig. 7. The change of the overall accuracy in respect to training set size

and even some algorithms have slight accuracy improvement. In addition, the build time is greatly shortened so that it is qualitatively consistent with the result in [9]. Therefore, the 22 features in Table 2 (the column on the right) are more suitable for VR video identification.

The Change of the Overall Accuracy in Respect to Training Set Size. We consider 19,034 network flows in our study, which are too large in real-time traffic identification. Therefore, we study the change of the overall accuracy in respect to training set size. The details are given in Fig. 7. Here we ignore KNN algorithm. We can find that when the number of flows exceeds 1000, AdaBoost can always provide the best performance, followed by J48 (and it is quite fast to train) and Random Forest. Due to the scarcity of training data in real-time network traffic identification, it is very exciting that 6000 network flows can provide good identification results.

5 Conclusions

In this study, we proposed 22 statistical features that can well represent VR video application. Classification strategies such as iQiyiVideo, iQiyiVRVideo, Youku-Video, YoukuVRVideo and non-video, etc. were adopted, and we evaluated and obtained the C4.5 Decision Tree algorithm which performed the best in terms of overall accuracy and the build time. These 22 main statistical features combined with C4.5 Decision Tree algorithm could achieve an accuracy of 98.6% for VR video application identification while maintaining high computational performance. In addition, this paper proved that as long as the training set exceeds 6,000 flows, high accuracy could be achieved, which makes it possible to identify real-time video application. Our work can effectively distinguish VR video from ordinary video, which provides a good foundation for other works such as resource scheduling.

Acknowledgements. This work has been sponsored by Huawei Research Fund (grant No. YBN2016110032) and National Science Foundation of China (No. 61201149). The authors would also like to thank the reviewers for their constructive comments.

References

1. IANA, http://www.iana.org/assignments/port-numbers. Accessed 06 June 2018
2. Karagiannis, T., Papagiannaki, K., Faloutsos, M.: BLINC: multilevel traffic classification in the dark. In: ACM SIGCOMM Computer Communication Review, Pennsylvania, pp. 229–240 (2005)
3. Kim, H., Claffy, K.C., Fomenkov, M., Barman, D., Faloutsos, M., Lee, K.: Internet traffic classification demystified: myths, caveats, and the best practices. In: Proceedings of the 2008 ACM CoNEXT Conference, Spain, p. 11 (2008)
4. Andersson, R.: Classification of video traffic: an evaluation of video traffic classification using random forests and gradient boosted trees. Digitala Vetenskapliga Arkivet. **83** (2017)
5. Västlund, F.: Video flow classification: a runtime performance study. Digitala Vetenskapliga Arkivet. **68** (2017)
6. Moore, A., Zuev, D.,Crogan, M.: Discriminators for use in flow-based classification. (2013)

7. Yamansavascilar, B., Guvensan, M. A., Yavuz, A. G., Karsligil, M. E.: Application identification via network traffic classification. In: International Conference on ICNC, Santa Clara, pp. 843–848(2017)

8. Aggarwal, R., Singh, N.: A new hybrid approach for network traffic classification using SVM and Naïve Bayes algorithm. Int. J. Comput. Sci. Mobile Comput. **6**, 168–174 (2017)

9. Williams, N., Zander, S., Armitage, G.: A preliminary performance comparison of five machine learning algorithms for practical IP traffic flow classification. ACM SIGCOMM Comput. Commun. Rev. **36**(5), 5–16 (2006)

10. Williams, N., Zander, S.: Evaluating machine learning algorithms for automated network application identification. (2006)

11. Chen, Z., Chen, R., Zhang, Y., Zhang, J., Xu, J.: A Statistical-Feature ML Approach to IP Traffic Classification Based on CUDA. In: IEEE Trustcom/BigDataSE/ISPA, Tianjin, pp. 2235–2239 (2017)

12. Datta, J., Kataria, N., Hubballi, N.: Network traffic classification in encrypted environment: a case study of google hangout. In: Twenty First National Conference on Communications (NCC), pp. 1–6 Mumbai, India (2015)

13. Munther, A., Alalousi, A., Nizam, S., Othman, R. R., Anbar, M.: Network traffic classification-A comparative study of two common decision tree methods: C4.5 and Random forest. In: International Conference on Electronic Design, pp. 210–214 Penang (2014)

14. Wang, C., Xu, T., Qin, X.: Network traffic classification with improved random forest. In: International Conference on Computational Intelligence and Security, pp. 78–81 Shenzhen (2015)

15. Wireshark, https://www.wireshark.org/. Accessed 22 May 2018

16. WEKA: Data Mining Software in Java. https://www.cs.waikato.ac.nz/ml/weka. Accessed 26 May 2018

28-GHz RoF Link Employing Optical Remote Heterodyne Techniques with Kramers–Kronig Receiver

Yuancheng Cai, Xiang Gao, Yun Ling, Bo Xu[(✉)], and Kun Qiu

School of Information and Communication Engineering, Key Laboratory of
Optical Fiber Sensing and Communications (Education Ministry of China),
University of Electronic Science and Technology of China, Chengdu 611731,
Sichuan, China
xubo@uestc.edu.cn

Abstract. We propose and demonstrate a 28-GHz optical remote heterodyne
RoF link using KK receiver for the first time. An optical SSB modulated signal
is obtained utilizing an IQ modulator and two free-running lasers. Due to its
minimum phase property, KK algorithm can be adopted to reconstruct the
complex 16-QAM signal from the received intensity signal. This scheme is
effective in eliminating the SSBI penalty introduced by square-law detection.
Through the use of the KK receiver, the power penalty caused by the 80 km
SSMF transmission is found to be less than 1 dB with digital CDC post pro
cessing. The KK-based receiver can also provide about 2 dB advantage over the
traditional receiver at the 7% HD-FEC threshold in the case of 28 GBaud rate
transmission over 80 km fiber. Furthermore, as the baud rate increases, the
benefit of KK receiving scheme is more obvious and superior than that of the
traditional receiving scheme.

Keywords: Millimeter-wave · Radio-over-fiber (RoF) · Optical heterodyning
Fiber optics communication

1 Introduction

Radio over fiber (RoF) technique can offer a strong and cost-effective solution on
enhancing the system capacity and mobility of wireless links. It can make the network
structure more flexible by using fibers to connect the central office (CO) with numerous
simplified base stations (BSs). Among different schemes, optical remote heterodyne
technique provides a promising and low-cost solution for RoF transmission [1–3]. In
optical remote heterodyne, the modulated signal light is coupled with an optical local
oscillator (LO) at the centralized CO (not at the BSs) and each BS is only responsible
for O/E, E/O, filtering and amplification of the signal.

At the receiving side of an optical remote heterodyne direct-detection (DD) RoF
system, a conventional receiver generally requires a microwave LO and an electric
mixer for down conversion. In addition, this solution cannot eliminate the DD-induced
signal-to-signal interference (SSBI) without extra signal processing techniques [4, 5].
In contrast, the recently proposed approach named Kramers–Kronig (KK) algorithm

© ICST Institute for Computer Sciences, Social Informatics and Telecommunications Engineering 2019
Published by Springer Nature Switzerland AG 2019. All Rights Reserved
X. Liu et al. (Eds.): ChinaCom 2018, LNICST 262, pp. 347–352, 2019.
https://doi.org/10.1007/978-3-030-06161-6_34

can fully reconstruct the complex field signal from the detected photocurrent amplitude waveform with a low digital signal processing (DSP) complexity, assuming that the signal is a single-sideband and minimum phase signal [6–9]. As a result, the microwave LO and electric mixer can be avoided at the receiver. Moreover, the KK-based approach is able to alleviate the SSBI very well [10].

In this paper, we report and demonstrate a 28-GHz RoF link employing optical remote heterodyne techniques with Kramers–Kronig receiver. It is worth highlighting that the 28-GHz is considered in this paper instead of the widely investigated 60-GHz because higher distances can be achieved even in nonline-of-sight (NLOS) transmission [2] for 28-GHz systems plus better choices on available devices for future 5G operating frequency. Optical remote heterodyne techniques and single-ended photodiode (PD) with KK reception scheme for 16-quadrature amplitude modulation (QAM) signal reconstruction are combined to achieve the low-cost and efficient scheme for 28-GHz RoF link. Furthermore, to the best of our knowledge, this is for the first time to adopt KK reception scheme in ROF systems. The paper is organized as follows. Section 2 explains the fundamental principles of the proposed scheme and the simulation results are presented and discussed in Sect. 3. Section 4 gives the conclusion.

2 Scheme of Proposed Remote Heterodyne RoF Link with Kramers–Kronig Receiver

The schematic of the proposed remote heterodyne RoF link with KK receiver is shown in Fig. 1. At the CO transmitter, a suppressed carrier double sideband (SC-DSB) signal is generated by modulating 16-QAM baseband data onto a continuous wave (CW) laser utilizing an IQ modulator. Afterwards, the SC-DSB signal light is coupled with a carrier from the LO, then an optical single sideband signal (OSSB) is obtained in this case. By controlling the carrier signal power ratio (CSPR) and adjusting the signal bandwidth (proportional to baud rate) and RF frequency (namely the frequency difference between the CW laser and the LO), we can make the SSB signal satisfy the minimum phase signal condition, which is an indispensable condition for adopting the KK algorithm.

At the BS, the photocurrent signal (see inset (iii) in Fig. 1) which is obtained using a single-ended PD for heterodyne detection, is wireless transmitted via a pair of antennas after amplification. At the receiver, the received signal is converted into digital signal by an analog-to-digital converter (ADC), and is fed to the DSP for offline processing finally.

Suppose that the complex signal $s(t)$ is a conventional bandwidth-limited 16-QAM small signal with a bandwidth of B, then the SSB signal can be described as

$$y(t) = A + s(t) \exp(j2\pi f_{RF} t) \tag{1}$$

where A is constant and represents the amplitude of the carrier, $f_{RF} = f_C - f_{LO}$ is the radio frequency. It can be shown that $y(t)$ is a minimum phase signal when $f_{RF} \geq B$ and $|A|$ is large enough compared to $|s(t)|$ [7].

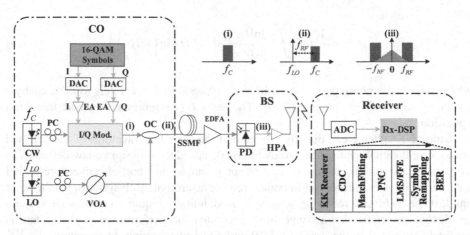

Fig. 1. Schematic of system setup. Insets: (i) modulated SC-DSB signal. (ii) OSSB minimum phase signal generated by coupling the SC-DSB signal with a carrier from the LO. (iii) photocurrent signal with DC and SSBI after square-law detection. PC: polarization controller. EA: electric amplifier. OC: optical coupler. EDFA: Erbium-doped optical fiber amplifier. HPA: high-power amplifier.

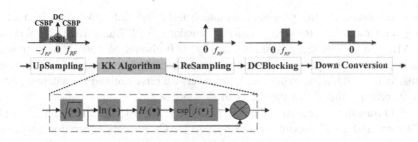

Fig. 2. DSP flow of KK reception scheme

After transmission and square-law detection, the photocurrent signal can be written as

$$I(t) = |A|^2 + 2\Re e[s(t) \exp(j2\pi f_{RF} t)] + |s(t)|^2 \tag{2}$$

where $\Re e[x]$ stands for the real part of x. In the Eq. (2), the first and second terms are the direct current (DC) and the desired carrier-signal beating products (CSBP), the third term is the SSBI.

The process of DSP of KK receiver is shown in Fig. 2. KK algorithm reconstructs the complex field signal from its detected amplitude

$$E_s(t) = \left\{ \sqrt{I(t)} \exp[j\varphi(t)] - A \right\} \exp(-j2\pi f_{RF} t) \tag{3}$$

$$\varphi(t) = \frac{1}{2}p.v. \int\limits_{-\infty}^{+\infty} \frac{\ln[I(t')]}{\pi(t-t')}dt' = H\left\{\ln\left[\sqrt{I(t)}\right]\right\} \tag{4}$$

where $\varphi(t)$ is the phase part of the minimum phase signal, $p.v.$ refers to the Cauchy's principal value of the integral [7, 9]. The term $H\{\bullet\}$ represents Hilbert transform operation.

Since KK scheme can directly reconstruct the complex waveform of the detected photocurrent signal, and hence avoids the SSBI introduced by square-law detection as mentioned above. Therefore, we can adopt a simple and cost-effective pure-digital receiver solution at the receiver in optical remote heterodyne RoF systems. On the other hand, if the traditional receiving scheme is used, it firstly requires a microwave LO and an electric mixer for down conversion. Secondly, the performance of the system is seriously degraded in the case of CSBP and SSBI overlapping. In summary, the KK reception scheme has a significant advantage from either the overall cost or system performance when compared to the conventional reception solution.

3   System Setup and Results

We have demonstrated the proposed remote heterodyne RoF link with KK receiver utilizing co-simulation through industry standard VPI-Transmission Maker and MATLAB. The system setup is shown in Fig. 1. It should be noted that in practical applications, the signal obtained from PD should be wireless transmitted via a pair of antennas. In our simulation experiment, however, we have omitted the antenna part for simplicity because this is not the focus of this article.

The CO transmitter consists of a CW laser at 1552.524 nm, a free-running LO at 1552.749 nm and an IQ modulator fed by two DACs for data generation and modulation. Both lasers have a linewidth of 1 MHz and 11 dBm output optical power, and their wavelength difference corresponds to 28-GHz in frequency, as shown in Fig. 1 inset (ii). The modulation format is 16-QAM at 4 samples-per-symbol. Root-raised cosine (RRC) filters with a roll-off factor of 0.1 are used for Nyquist pulse-shaping. The IQ modulator is biased at its transmission null point and generates an optical SC-DSB signal. Afterwards, we obtain the OSSB by coupling with the carrier generated by the LO. In order to prove the effectiveness of the KK receiver, we consider an 80 km link over standard single mode fiber (SSMF). The attenuation coefficient of the SSMF is set at 0.2 dB/km. Other parameters of SSMF include dispersion coefficient D of 16 ps/nm/km, and nonlinear index of $2.6 \times 10^{-20} \mathrm{m^2/W}$.

At the BS, we use a single-ended PD which has a responsivity of 0.84 A/W and dark current of 0.43 nA to conduct heterodyne beating detection. A variable optical attenuator (VOA) is placed before the PD in order to adjust the received optical power (ROP). Then the obtained photocurrent is fed to the DSP for offline processing after ADC at the receiver. Firstly, a KK-based receiver reconstructs the complex field signal from the photocurrent amplitude directly, then chromatic dispersion compensation (CDC) is carried out by digital post-processing. It is necessary and indispensable for laser phase noise compensation (PNC) since both the CW laser and LO have a line-width of 1 MHz. After matched filtering, channel equalization based on least-mean-square (LMS) with feed forward equalizer (FFE) are applied. Finally, the bit error rate (BER) is calculated following symbol decision and remapping.

In order to ensure good system performance, the CSPR is set to about 9 dBm. Figure 3a shows the BER curves of the 28-GHz 16-QAM SSB signal with 28 GBaud symbol rate for two cases in RoF optical remote heterodyne system. One is optical back-to-back (B2B) transmission and the other is 80 km SSMF transmission. It can be seen that the hardware-decision pre-forward error correction (HD-FEC) BER lower than 3.8×10^{-3} can be achieved at a ROP of −13.3 dBm for B2B case and −12.5 dBm for 80 km transmission case. Obviously, the power penalty caused by the 80 km SSMF is less than 1 dB.

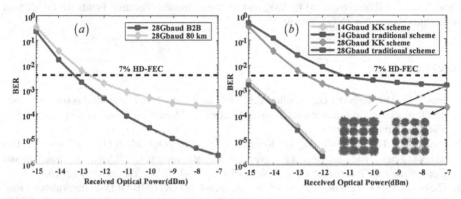

Fig. 3. The BER performance of 28-GHz 16-QAM RoF system. (a) 28 GBaud for B2B and 80 km SSMF. (b) 80 km SSMF transmission in contrast of KK receiver and traditional receiver for 14 GBaud and 28 GBaud 16-QAM signal.

Figure 3b compares the BER performance between the KK reception scheme and the conventional reception scheme at different baud rates after 80 km SSMF transmission. In the case of 14 GBaud rate, the BER results of two receiving schemes are almost the same, because CSBP and SSBI have no overlap in this case. On the other hand, the KK scheme has about 2 dB advantage over the traditional scheme at 28 GBaud rate case, which exists a partial overlap between CSBP and SSBI. The insets in Fig. 3b are shown the constellation diagrams of two reception schemes at a ROP of —7 dBm, respectively. Qualitative comparison between these two figures shows that KK receiving solution brings significant improvement in the signal quality. Moreover, as the signal's baud rate increases, there exists more serious signal degradation caused by SSBI and the KK receiving scheme shows higher performance gain than that of the traditional receiving scheme.

4   Conclusions

In this paper, we propose and demonstrate a 28-GHz RoF link employing optical remote heterodyne techniques with KK Receiver. The simulation results reveal that the power penalty caused by the 80 km SSMF can be reduced to less than 1 dB with digital CDC post processing. Moreover, when compared with traditional receiver, the receiver sensitivity has improved about 2 dB at the 7% HD-FEC threshold for 28 GBaud 16-QAM signal transmission over 80 km SSMF by using KK receiver. In summary, the KK reception scheme has a significant advantage from either the overall cost or system performance compared to the conventional reception solution. It is thus a promising candidate for the future 5G millimeter wave radio access network applications.

Acknowledgements. This work is supported by the National Natural Science Foundation of China (No. 61420106011 and 61471088) and the Fundamental Research Funds for the Central Universities (Grant No.ZYGX2016J014).

References

1. Insua, I.G., Plettemeier, D., Schäffer, C.G.: Simple remote heterodyne RoF system for Gbps wireless access. In: International Topical Meeting on Microwave Photonics, pp. 1–4. IEEE, Valencia, Spain (2009)
2. Latunde, A.T., Milosavljevic, M., Kourtessis, P., et al.: OQAM-OFDM RoF with IM-DD remote heterodyne 28 GHz upconversion for 5G millimeter RANs. In: International Conference on Transparent Optical Networks, pp. 1–4. IEEE, Trento, Italy (2016)
3. Omomukuyo, O., Thakur, M.P., Mitchell, J.E.: Simple 60-GHz MB-OFDM Ultrawideband RoF System Based on Remote Heterodyning. IEEE Photonics Technol. Lett. **25**(3), 268–271 (2013)
4. Lin, C.T., Chiang, S.C., Li, C.H., et al.: V-band gapless OFDM RoF system with power detector down-conversion and novel Volterra nonlinear filtering. Opt. Lett. **42**(2), 207–210 (2017)
5. Li, Z., Erkilinc, M.S., Galdino, L., et al.: Comparison of digital signal-signal beat interference compensation techniques in direct-detection subcarrier modulation systems. Opt. Express **24**(25), 29176–29189 (2017)
6. Chen, X., Antonelli, C., Chandrasekha, S., et al.: 218-Gb/s single-wavelength, single-polarization, single-photodiode transmission over 125-km of standard singlemode fiber using Kramers-Kronig detection. In: Optical Fiber Communications Conference and Exhibition, pp. 1–3. Optical Society of America, Los Angeles, USA (2017)
7. Mecozzi, A., Antonelli, C., Shtaif, M.: Kramers-Kronig Coherent Receiver. Opt. **3**(11), 1220 (2016)
8. Shu, L., Li, J., Wan, Z., et al.: Single-Lane 112-Gbit/s SSB-PAM4 Transmission With Dual-Drive MZM and Kramers-Kronig Detection Over 80-km SSMF. IEEE Photonics J. **9**(6), 1–9 (2017)
9. Fan, S.J., Zhuge, Q.B., et al.: Twin-SSB direct detection transmission over 80 km SSMF using Kramers-Kronig Receiver. In: European Conference on Optical Communication, pp. 1–3. IEEE, Gothenburg, Sweden (2017)
10. Li, Z., Erkilinc, M.S., Shi, K., et al.: SSBI Mitigation and the Kramers-Kronig Scheme in Single-Sideband Direct-Detection Transmission With Receiver-Based Electronic Dispersion Compensation. J. Lightwave Technol. **35**(10), 1887–1893 (2017)

Fairness-Based Distributed Resource Allocation in Cognitive Small Cell Networks

Xiaoge Huang$^{(\boxtimes)}$, Dongyu Zhang, She Tang, and Qianbin Chen

School of Communication and Information Engineering, Chongqing
University of Posts and Telecommunications, Chongqing, China
{Huangxg, Chenqb}@cqupt.edu.cn, {zhangdongyu, tangshedhl}@outlook.com

Abstract. In this paper, we aim to maximize the total throughput of
the cognitive small cell networks by jointly considering interference management, fairness-based resource allocation, average outage probability
and channel reuse radius. In order to make the optimization problem
tractable, we decompose the original problem into three sub-problems.
Firstly, we derive the average outage probability function of the system
with respect to the channel reuse radius. With a given outage probability
threshold, the associated range of the channel reuse radius is obtained. In
addition, a fairness-based distributed resource allocation (FDRA) algorithm is proposed to guarantee the fairness among cognitive small cell
base stations (CSBSs). Finally, based on the channel reuse range we
could find the maximum throughput of the small cell network tire. Simulation results demonstrate that the proposed FDRA algorithm could
achieve a considerable performance improvement relative to the schemes
in literature, while providing a better fairness among CSBSs.

Keywords: Cognitive small cell · Resource allocation · Channel reuse
radius · Fairness

1 Introduction

In the future, 5G networks are moving in the direction of diversification, broad-
bandization, integration and intelligence, which lead to a huge demand in the
traffic load and spectrum resources. HetNets, consisting of macro cells and small
cells, provide a cost-effective and flexible solution to satisfy the ever increasing
demand for network capacity. However, the coexistence between small cells is
very challenging due to their lack of coordination and random locations.

To provide effective resource allocation, several mechanisms have already
been proposed. In [1,2], the authors used game-based methods to assignment

This work is supported by the National Natural Science Foundation of China (NSFC)
(61401053), and Innovation Project of the Common Key Technology of Chongqing
Science and Technology Industry (Grant no. cstc2015zdcyztzx40008).

X. Liu et al. (Eds.): ChinaCom 2018, LNICST 262, pp. 353–362, 2019.
https://doi.org/10.1007/978-3-030-06161-6_35

channels effectively. The authors used cluster to address the resource allocation problem for ultra-dense networks and proposed a K-means clustering algorithm to divide the small cell base stations into different cluster in [3]. Based on this algorithm, the clusters could adjusted dynamic to adapt to the changing network topology.

In the aspect of interference management, in paper [4–8], the authors used the advantages of cognitive radio to solve the interference problem in small cell networks. The authors in [9] proposed a centralized user-centric merge-and-split (MAS) rule based coalition formation game, which could utilize users' information to estimate inter-user interference and mitigate interference accurately and effectively.

However, resource allocation for two-tire HetNets jointly considering interference management, fairness resource allocation, average outage probability and spectrum reuse radius has not been investigated in previous works. In this paper, firstly, we model the two-tire HetNet by stochastic geometry which considers the differences between small cells and macro cells. Secondly, the average outage probability of the two-tire HetNet is discussed. Finally, a fairness-based distributed resource allocation (FDRA) algorithm is proposed to guarantee the fairness between small cells and maximize the total data rate under the certain constraints.

The rest of this paper is organized as follows. We present the system model in Sect. 2. The optimization problem is formulated in Sect. 3. The detail procedure of the FDRA algorithm is described in Sect. 4. The simulation results are discussed in Sect. 5. Finally, Sect. 6 draws the conclusion.

2   System Model

2.1   Network Topology

We consider a two-tire HetNet consisting of macro base stations (MBSs) and cognitive small cell base stations (CSBSs). The network topology is modeled by stochastic geometry. Each CSUE will associate with either a MBS or a CSBS in the two-tire HetNet. In this paper, we assume that CSUEs associate with the network entity which provide the highest receive signal strength.

2.2   Channel Model

Consider a downlink OFDMA transmission system, which consists of M MBSs, F CSBSs and K CSUEs. The total system bandwidth is B_w which is divided into N channels. We denote the set of channels as $\mathcal{N} = \{n_1, n_2, n_3, \cdots, n_N\}$. We assume that the MBS, the CSBS and the CSUE in the system are independent and each follows an independent homogeneous Poisson Point Process (PPP) distribution with intensity $\mathcal{B}$, $\mathcal{A}$ and $\mathcal{U}$, respectively. The MBSs and CSBSs use the same channel and each CSBS will sense the channel periodically and opportunistically access the free channel. Let $g_{i,k,n}$ denote the channel gain of the kth CSUE in CSBS $i \in \psi_a$ on channel $n \in \mathcal{N}$. Hence, the received signal to interference plus noise ratio (SINR) of CSUE k in CSBS i on channel n is given as

$$SINR_{i,k,n} = \frac{p_{i,k,n} g_{i,k,n}}{I + N_0} \tag{1}$$

where $p_{i,k,n}$ denotes the transmission power of CSBS i to CSUE k on channel n; $I = \sum_{j=1, j \neq i}^{\psi_a} p_{j,k,n} g_{j,k,n}$ is the received interference from other CSBSs; N_0 denotes the noise power. Hence, the throughput of CSUE k in CSBS i on channel n can be denoted as

$$R_{i,k,n} = \frac{B_w}{N} \log_2 \left(1 + SINR_{i,k,n}\right). \tag{2}$$

3   Problem Formulation

3.1   Tire Connection Probability

The probability of a user in the two-tire HetNet connecting to the small cell network tire is given by [10]

$$\xi_a = 1 - \int_0^\infty \frac{\mathcal{B}}{(1+h)^2 \left(\left(\frac{P_a}{P_b} h\right)^{\frac{2}{\eta}} \mathcal{A} + \mathcal{B} \right)} dh \tag{3}$$

where P_a denotes the transmission power of CSBSs and P_b denotes the transmission power of MBSs; η is the path loss factor. The probability of a user in the two-tire HetNet connecting to the macro cell network tire is given by

$$\xi_b = 1 - \xi_a. \tag{4}$$

Hence, based on (3) and (4), we can obtain the PPP intensity of the set of users connecting to CSBSs is $\mathcal{U}_a = \mathcal{U}\xi_a$ and the PPP intensity of the set of users associating with MBSs is $\mathcal{U}_b = \mathcal{U}\xi_b$.

3.2   Average Outage Probability of the Macro User Equipment

A macro user equipment (MUE) can successfully decode a signal if and only if the SINR of the signal higher than a threshold β. The average outage probability of the MUE can be expressed as

$$O_b = P\left\{\text{SINR} < \beta\right\} = 1 - \int_0^\infty 2\pi \mathcal{B} r e^{-\mathcal{B}\pi r^2} e^{-\pi \mathcal{B}_{ac} r^2 \sqrt{\beta} \arctan \sqrt{\beta}} dr \tag{5}$$

where $\mathcal{B}_{ac} = \left(1 - \left(\frac{\mathcal{B}c}{\mathcal{B}c + \mathcal{U}_b}\right)^c\right)$, $\mathcal{B}$ denotes the PPP intensity of MBS transmitting on channel n_1; $c = 3.575$ is a constant for Voronoi tessellation.

3.3　Average Outage Probability of the CSUE

The average outage probability of a CSUE since the received SINR less than the threshold β can be expressed as

$$
\begin{aligned}
O_a^{SINR}(N_i) = 1 - \int_0^\infty & 2\pi \mathcal{A}r \exp\left\{-\mathcal{A}\pi r^2\right. \\
& - \mathcal{B}_{in}(N_i)\,\pi r^2 \sqrt{\frac{\beta}{p}}\,\arctan\left(\frac{r^2\sqrt{\beta/p}}{(r_{sb}^2 - r)^2}\right) \\
& - \mathcal{A}_{n_{N-(N_i-1)}}\,\pi r^2 \sqrt{\beta}\,\arctan\left(\frac{r^2\sqrt{\beta}}{(r_{sa} - r)^2}\right) \\
& \left. - \frac{\beta r^n(\sigma^2)}{P_a}\right\}\,dr
\end{aligned}
\tag{6}
$$

where r_{sa} and r_{sb} are the channel reuse radius; N_i represents the number of available channels to CSBS i within the range r_{sb}, that is, the number of channels not used by the MBSs; $\mathcal{B}_{in}(N_i)$ is the PPP intensity of MBSs which are using the channel $n_{N-(N_i-1)}$; $\mathcal{A}_{n_{N-(N_i-1)}}$ is the intensity of CSBSs which are using the channel $n_{N-(N_i-1)}$; $p = P_a/P_b$; σ is the noise power. The average outage probability of a CSUE since the channel is unavailable can be expressed as

$$
O_a^{access}(N_i) = 1 - P_{ac}(N_i)
\tag{7}
$$

where $P_{ac}(N_i)$ is the probability that CSBS i successfully access the free channels. Hence, the average outage probability of the CSUE is given by

$$
O_a = \sum_{n=0}^{N} P\{N_i = n\}\left[(1 - O_a^{access}(N_i))O_a^{SINR}(N_i) + O_a^{access}(N_i)\right]
\tag{8}
$$

According to (5) and (8), the average outage probability of the system is given by

$$
O_t = \sum_{n=0}^{N} P\{N_i = n\}\left[\xi_a O_a(N_i) + \xi_b O_b(N_i)\right]
\tag{9}
$$

3.4　Optimization Problem

Our aim is to maximize the total throughput of the small cell networks tire while ensuring the transmission performance of the CSUE and the average outage probability of the system. Hence, the optimization problem can be expressed as follows

$$\max_{\tau_{i,k,n},p_{i,k,n}} \sum_{i=1}^{F}\sum_{k=1}^{K_i}\sum_{n=1}^{N_i} \tau_{i,k,n}R_{i,k,n}$$

s.t.   C1: $p_{i,k,n} \geq 0 \quad \forall i,k,n$

C2: $\sum_{k=1}^{K_i}\sum_{n=1}^{N_i} p_{i,k,n} \leq p_{\max} \quad \forall i$

C3: $\sum_{n=1}^{N_i} \tau_{i,k,n}R_{i,k,n} \geq R_{i,k}^0 \quad \forall i,k$

C4: $\tau_{i,k,n} \in \{0,1\} \quad \forall i,k,n$

C5: $\sum_{k=1}^{K_i} \tau_{i,k,n} \leq 1 \quad \forall i,n$

C6: $O_t \leq \varepsilon$ \hfill (10)

where $\tau_{i,k,n} = 1$ or 0 indicates whether channel n is allocated to CSUE k in CSBS i or not; $p_{\max}$ is the maximum transmission power of a CSBS; $R_{i,k}^0$ is the minimum throughput requirement of CSUE k in CSBS i.

In the above optimization problem, C1 and C2 are transmission power constraints which represent the transmission power of a CSBS should less than the maximum transmission power $p_{\max}$. C3 is the minimum throughput requirement for CSUEs. C4 and C5 are the channel allocation constraints, representing that channel n can not be allocated to two different CSUEs in the same small cell simultaneously. C6 is the average outage probability constraint of the system. Via C6, the average outage probability of the system will be limited below the threshold ε.

4   Distributed Resource Allocation in HetNets

4.1   Channel Reuse Radius

A CSBS opportunistically uses a available channel within its channel reuse radius (CRR). Due to the different transmission power among CSBSs and MBSs, as well as the unified spectrum reuse threshold, each CSBS has two different CCR, namely, the small CRR and the macro CRR. The spectrum reuse threshold defines the channel reuse radius of CSBSs. That is, a CSBS can only use the channels which are not used by the MBSs in the macro CRR and the channels which are not used by other CSBSs in the small CRR.

The small CRR and the macro CRR can be expressed as

$$r_{sa} = (P_a/\gamma)^{\frac{1}{\eta}} \hfill (11)$$

$$r_{sb} = (P_b/\gamma)^{\frac{1}{\eta}} \hfill (12)$$

where r_{sa} denotes the small CRR and the r_{sb} denotes the macro CRR; γ denotes the spectrum reuse threshold.

4.2   Channel Quality Estimation Table and Status Information Table

For CSBS i, the total interference on channel n is the sum interference from other cognitive small cells and the noise power, which can be represented as

$$TI_n^i = N_0 + \sum_{j=1, j\neq i}^{T} I_n^{j,i} \tag{13}$$

where N_0 is the noise power; T is the number of CSBSs which are using channel n; $I_n^{j,i}$ is the interference to the users of CSBS i from CSBS j on channel n, which can be expressed as

$$I_n^{j,i} = p_n^j \bar{h}_n^{j,i} \tag{14}$$

where p_n^j is the transmission power of CSBS j on channel n; $\bar{h}_n^{j,i}$ is the average channel gain between CSBS j and users of CSBS i, which can be obtained from the arithmetic mean of channel gains between CSBS j and all users of CSBS i. In order to evaluate each RB's channel condition, we use the reciprocal of TI_n^i as the channel quality of CSBS i for channel n, which is denoted as

$$Q_n^i = \frac{1}{TI_n^i}. \tag{15}$$

Let N_i be the number of unoccupied channel out of $\mathcal{N}$ within the r_{sb}, and we can obtain the probability mass function of N_i according to the network model by using the stochastic geometry theory. Furthermore, we randomly generate the number of unoccupied channels of CSBS i within the r_{sb} based on this function. Then, we calculate the channel quality of each free channel of CSBS i, and sort them in descending order. Table 1 is an example of the Channel Quality Estimation Table (CQET).

In the proposed resource allocation scheme, each CSBS also need to create a state information table (SIT), as shown in Table 2. PL_A^i and PL_B^i are two selection priorities of the CSBS, which is used to guarantee the fairness among CSBSs.

Table 1. CQET

SN	Channel ID	Channel quality
1	N	Q_1^i
2	$N-1$	Q_2^i
3	$N-2$	Q_3^i
...	...	...
K_i	$N-K_i+1$	$Q_{K_i}^i$

Table 2. SIT

Parameters	Description
N_D^i	No. of total required channels
N_{Init}^i	No. of initial service channels
N_N^i	No. of still needed channels
N_A^i	No. of available channels
N_S^i	No. of current serving channels
K_i	No. of current serving users
PL_A^i	N_D^i/N_S^i
PL_B^i	N_D^i/K_i

Algorithm 1 Fairness-based Distributed Resource Allocation Algorithm

---

1: Let $\mathcal{G} = \{1, 2, \ldots, G\}$ denotes the set of CSBSs involved in the algorithm.
2: **for** all CSBSs within $\mathcal{G}$ **do**
3:    **if** CSBS i $N_S^i = 0, N_A^i = 1, K_i \neq 0$ **then**
4:       Select the first rank channel in the CQET of CSBS i
5:       Update CQET and SIT for the CSBS in $\mathcal{G}$
6:    **else**
7:       Select the CSBS $i = argmax[PL_A^i]$
8:       **if** There are multiple CSBSs with the same PL_A^i **then**
9:          Select the CSBS $i = argmax[PL_B^i]$
10:         **if** There are multiple CSBSs with the same PL_B^i **then**
11:            Select the CSBS $i = argmax[Q_1^i]$
12:            Select the first rank channel in the CQET of CSBS i
13:            Update CQET and SIT for the CSBS in $\mathcal{G}$
14:         **else**
15:            Select the first rank channel in the CQET of CSBS i
16:            Update CQET and SIT for the CSBS in $\mathcal{G}$
17:         **end if**
18:       **else**
19:          Select the first rank channel in the CQET of CSBS i
20:          Update CQET and SIT for the CSBS in $\mathcal{G}$
21:       **end if**
22:    **end if**
23: **end for**
24: **output:** Channel allocation results $\left\{\tau_{i,k,n}^*\right\}$.

In addition, each CSBS needs to establish its own neighboring CSBS list which can be obtained according to the reference signal received power (RSRP) from other CSBSs.

4.3 Fairness-Based Distributed Resource Allocation Algorithm

The proposed fairness-based distributed resource allocation (FDRA) algorithm is performed in distribute manner and each CSBS can only select one channel at each round. Once a CSBS requests to re-initialize, the resource allocation process will be triggered. All the CSBSs within the coverage of the triggered CSBS will participate in the resource allocation. Each CSBS creates its own CQET and SIT, and forward them to its neighboring CSBSs. In the resource allocation process, the CSBS with $N_S^i = 0, N_A^i = 1, K_i \neq 0$ will be selected firstly. Secondly, the CSBS according to the PL_A and PL_B will be selected, which could guarantee the fairness among CSBSs. Until the requirements of all the CSBSs in the group have been satisfied or there are no free channels, the FDRA algorithm will stop. We outline the main procedure of the FDRA algorithm in Algorithm 1. Furthermore, the CQET and SIT update procedure is shown in Algorithm 2.

Algorithm 2 Update Procedure of CQET and SIT

1: **if** Channel n is selected by CSBS i **then**
2:    Delete channel n from the CQET of CSBS i
3:    Update the SIT of CSBS i: $N_N^i = N_N^i - 1, N_A^i = N_A^i - 1, N_S^i = N_S^i + 1, PL_A^i = N_D^i/N_S^i$
4:    **for** CSBS $j(j \in \mathcal{G}, j \neq i)$, j is the neighboring CSBS of CSBS i **do**
5:       **if** Channel n is one of the available channel of CSBS j **then**
6:          Delete channel n from CQET of CSBS j
7:          Update the SIT of CSBS j: $N_A^j = N_A^j - 1$
8:       **end if**
9:    **end for**
10: **end if**

---

5   Simulation Results

We consider MBSs, CSBSs and CSUEs are randomly distributed in a range of $500\,\mathrm{m} \times 500\,\mathrm{m}$. $P_a = 20\,\mathrm{dBm}$, $P_b = 46\,\mathrm{dBm}$, $B_w = 40\,\mathrm{kHz}$, $\eta = 4$, $\mathcal{B} = 2\,\mathrm{MBS/km^2}$ and $\beta = 2$. In the simulation, we compare two existing algorithms with the proposed FDRA algorithm, named RRA algorithm and CSRA algorithm, respectively. In the RRA algorithm, each CSBS randomly takes the required amount of channels from its CQET without considering neighbor CSBSs. The CSRA algorithm only allocates channels based on their channel state, in which a channel is allocated to the CSBS with the highest channel quality among the neighboring CSBSs.

Figure 1 shows the effect of the CSBS intensity and the spectrum reuse threshold on the average outage probability of the system when $N = 25, \mathcal{U} = 3(\mathcal{A}+\mathcal{B})$. From the figure, we can see that the greater the density of the CSBS, the greater the optimal spectrum reuse threshold. Moreover, at the begin, the average outage probability is relatively large. That is because the lower rate of channel reuse results in a higher outage probability. With the increase of γ, the channel reuse rate increases and the outage probability decreases, but at the same time the interference increases, so, the outage caused by the strong interference dominates the average outage probability at the end.

Figure 2 shows the fairness of the proposed FDRA algorithm and the CSRA algorithm. The horizontal axis is the CSBSs ID in the resource allocation group. The black part of the histogram represents the percentage of the number of initial service channels to the number of total required channels before resource allocation. The gray part represents the percentage of the number of channels allocated by the algorithm to the number of total required channels and the white part represents the percentage of the number of still needed channels to the number of total required channels after resource allocation. It can be seen that, the proposed FDRA algorithm can guarantee the fairness among CSBSs than the CSRA algorithm.

Figure 3 shows the effect of the number of channels on the satisfaction degree variance among CSBSs. The satisfaction degree of a CSBS is defined by the ratio of the number of current serving channels to the number of total needed channels. From the figure, we can see that the satisfaction degree variance decrease with the number of channels increases. This is because the more channels can be allocated to each CSBS. Specially, the RRA algorithm is based on requirement, the variance is always zero when the number of channels reaches a certain value. In addition, Fig. 3 illustrates that the proposed FDRA algorithm performs better in term of the satisfaction degree than the CSRA algorithm.

Figure 4 shows the effect of the channel reuse radius on the total throughput of the system for different CSBS density. Form the figure, we can see higher CSBS density results in larger the network throughput. At the beginning, the channel reuse radius is vary small, lead to a higher channel reuse ratio and larger interference. The gain of the channel reuse to the total throughput is less than the effect of interference, and the total throughput decreases with the decrease of channel reuse radius at the beginning. With the increase of channel reuse radius, the interference becomes smaller, meanwhile, the spectrum efficiency becomes lower. Consequently, the total throughput decreases with the channel reuse radius after a certain channel reuse radius value.

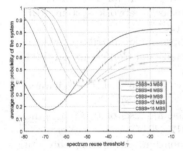

Fig. 1. The average outage probability versus γ

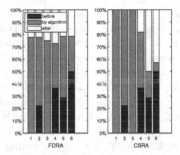

Fig. 2. Resource allocation among CSBSs for different algorithms

6   Conclusion

In this paper, we studied the resource allocation problem in cognitive small cell networks. The original optimization problem is intractable, which could be decomposed into three sub-problems. The proposed fairness-based distributed resource allocation algorithm not only achieves a better fairness among cognitive small cell base stations but also guarantees the throughput of the system. By introducing the channel reuse radius parameter, we obtain the maximum throughput of the cognitive small cell network under the average outage probability constraint. Numerical results demonstrated the effectiveness of the proposed algorithm.

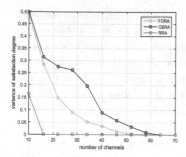

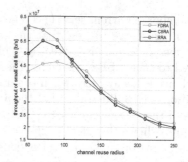

Fig. 3. Satisfaction degree variance versus numbers of channels for different algorithms

Fig. 4. Network throughput versus the channel reuse radius for different CSBS density

References

1. Yu, J., Han, S., Li, X.: A Robust Game-Based Algorithm for Downlink Joint Resource allocation in hierarchical OFDMA Femtocell network system. IEEE Trans. Syst. Man Cybern. Syst. **2168–2216**, 1–11 (2018)
2. Bui, K.N., Jung, J.J.: Cooperative game theoretic approach for distributed resource allocation in heterogeneous network. In: 2017 International Conference on Intelligent Environments (IE), Seoul, pp. 168–171 (2017)
3. Li, W., Zhang, J.: Cluster-based resource allocation scheme with QoS guarantee in ultra-dense networks. IET Commun. **12**(7), 861–867 (2018)
4. Yan, Z., Zhou, W., Chen, S., Liu, H.: Modeling and analysis of two-tier hetnets with cognitive small cells. IEEE Access **5**, 2904–2912 (2017)
5. Zhang, H., Nie, Y., Cheng, J., Leung, V.C.M., Nallanathan, A.: Sensing time optimization and power control for energy efficient cognitive small cell with imperfect hybrid spectrum sensing. IEEE Trans. Wirel. Commun. **16**(2), 730–743 (2017)
6. Zhao, M., Guo, C., Feng, C., Chen, S.: Consistent-estimated eigenvalues based cooperative spectrum sensing for dense cognitive small cell network. In: 2017 IEEE International Conference on Communications Workshops (ICC Workshops), Paris, pp. 510–515 (2017)
7. Kuang, Q., Utschick, W.: Energy management in heterogeneous networks with cell activation, user association, and interference coordination. IEEE Trans. Wirel. Commun. **15**(6), 3868–3879 (2016)
8. Tung, L., Wang, L., Chen, K.: An interference-aware small cell on/off mechanism in hyper dense small cell networks. In: 2017 International Conference on Computing, Networking and Communications (ICNC), Santa Clara, CA, pp. 767–771 (2017)
9. Cao, J., Peng, T., Qi, Z., Duan, R., Yuan, Y., Wang, W.: interference management in ultra-dense networks: a user-centric coalition formation game approach. IEEE Trans. Veh. Technol. **67**(6), 5188–5202 (2018)
10. ElSawy, H., Hossain, E.: Two-tier HetNets with cognitive femtocells: downlink performance modeling and analysis in a multichannel environment. IEEE Trans. Mob. Comput. **13**(3), 649–663 (2014)

A Distributed Self-healing Mechanism Based on Cognitive Radio and AP Cooperation in UDN

Zhongming Gao$^{(\boxtimes)}$, Xi Li, Hong Ji, and Heli Zhang

Key Laboratory of Universal Wireless Communications, Ministry of Education,
Beijing University of Posts and Telecommunications, Beijing,
People's Republic of China
{gaozhongming,lixi,jihong,zhangheli}@bupt.edu.cn

Abstract. Self-healing is considered as an indispensable function to achieve intelligent network management in future wireless communication systems. However, in ultra-dense networks (UDNs), it's a great challenge to realize efficient self-healing due to the massive and diverse network nodes, as well as complex transmission environment. The failed network access point (AP) may result in sudden traffic outage and severe user service degrading. In this paper, we propose an effective self-healing mechanism for UDNs with complete procedure of intelligent failure detection, diagnosis and recovery. Cognitive technology has been introduced to realize the effective detection of the AP working status. Then the processed information are analyzed based on multi-armed bandit model for possible AP failure judgement. After it is confirmed that an AP is failed, the impacted users, which are served originally by the failed AP, would be accessed to the proper neighbor APs. Furthermore, the corresponding resource allocation based on Non-Orthogonal Multiple Access (NOMA) is proposed. Simulation results show that the proposed mechanism could detect the AP failure effectively and realize quick self-healing for the network.

Keywords: Ultra-dense network · Self-healing · Failure detection · Resource allocation

1 Introduction

Recently, with the rapid development of wireless communications, ultra-dense networks (UDNs) have been considered as an inspiring approach to meet the huge traffic requirements, nearly 1 ms latency and massive devices access in typical scenarios [1]. With densely deployed network access points (APs), traditional manual or semi-automatic network management methods are inefficient

© ICST Institute for Computer Sciences, Social Informatics and Telecommunications Engineering 2019
Published by Springer Nature Switzerland AG 2019. All Rights Reserved
X. Liu et al. (Eds.): ChinaCom 2018, LNICST 262, pp. 363–373, 2019.
https://doi.org/10.1007/978-3-030-06161-6_36

and costly [2]. Therefore, self-organizing networking (SON) is introduced into UDN to realize the intelligent network parameters self-optimization, possible failure self-healing, and entities self-deployment [3,4]. Relative fields have attracted huge research interests and still need further investigation.

As one of the key technologies of SON, self-healing could detect the AP failure in time, and then efficiently provide service for the impacted users (served originally by the failed AP) automatically, thereby, prevent the network performance degradation. Generally, it contains three parts: failure detection, diagnosis, and recovery [4]. Currently, there are some existing works in relative fields. In [5], a self-healing framework based on cognitive learning is proposed for failure detection and compensation. In [6], the authors propose a self-healing algorithm with the water ripple algorithm and variable transmission power to provide seamless and reliable service despite AP failures.

However, self-healing in UDN is quite different from the traditional conditions and still an open problem. On one hand, the performance of self-healing usually has sensitive time constraint for failure detection. Existing centralized failure detection algorithms are too complicated to be implemented for UDN with a lot of APs. On the other hand, due to the short distance between neighbor APs, the simple increasing transmission power to cover the impacted users may cause severe interference in UDN. Therefore, the traditional failure recovery resource allocation algorithms are not suitable to be adopted in UDN directly.

Recently, there are some self-healing mechanism in small cell networks. In [7], a optimization algorithm considering APs selection and resource allocation is proposed to guarantee the reliable and seamless service for the impacted users. Reference [8] explores a hidden Markov model to automatically capture current states of the BSs and probabilistically estimate a cell outage. In [9], the authors propose a cell outage detection architecture based on the handover statistics in a two-tier heterogeneous network. Reference [10] presents a novel cell outage management framework for heterogeneous networks with split control and data planes. However, these detection methods require the network nodes to send report data to APs frequently, and there are huge costs. Moreover, the possible serious interference among APs are not considered.

Therefore, the self-healing in UDN is an important problem and needs further discussion. In this paper, we propose a self-healing mechanism consisting of intelligent failure detection, diagnosis and recovery. APs are divided into clusters and a leading AP (L-AP) is responsible for detecting their working information by cognitive technology in each cluster. Then these information are analyzed based on multi-armed Bandit Model to judge possible AP failure. Once an AP is confirmed to be failed, the impacted users would be connected to the optimal APs in the cluster. In addition, the corresponding resource allocation based on AP selection and Non-Orthogonal multiple Access (NOMA) is developed to reduce interference.

The remainder of our work is organized as follows. Section 2 gives the system model. In Sect. 3, we introduce the proposed self-healing mechanism. Simulation results and discussions are given in Sect. 4. Finally, we conclude this paper and present some future work in Sect. 5.

2   System Model

2.1   Topology Scene

We consider a typical scenario in UDN, where a series of APs are deployed by network operator with fixed positions. Considering the large number of APs, centralized algorithms require intolerable computation for failure detection, which couldn't meet the sensitive time constraint. Thus, we propose a distributed self-healing mechanism and it's suitable for the clusters that take spatial location as main similarity feature.

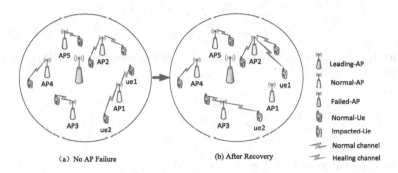

Fig. 1. System model of self-healing mechanism in UDN

In each cluster, the total system bandwidth is equally divided by APs, each AP occupies a sub-band, and the AP would serve different users based on NOMA. Each cluster has a L-AP to monitor the work of other APs by cognitive radio technology (CR). The L-AP has a list that records the users connection tables of all APs, and it would be updated synchronously when users access or disconnect from an AP, therefore, it works well when users move. Then it could schedule the APs and users in the cluster. In addition, we assume that the wireless channel would not vary during the transmission of a packet and the perfect channel quality information (CQI) is available by APs. The detailed system model is shown in Fig. 1.

In a cluster, $\mathbb{M}$ is the set of total APs, $\mathbb{M}_N$ and $\mathbb{M}_F$ are the sets of normal and failed APs respectively. Therefore, $\mathbb{M} \triangleq \mathbb{M}_N \cup \mathbb{M}_F$. Similarly, $\mathbb{U}$ is the set of total users, while $\mathbb{U}_N$ and $\mathbb{U}_F$ are the sets of normal users and impacted users respectively; here $\mathbb{U} \triangleq \mathbb{U}_N \cup \mathbb{U}_F$. $\mathbb{N}_i$ represents the users set of i-th AP. B_i represents the bandwidth allocated to the i-th AP. p_{ij} is the power allocated to the j-th user in the i-th AP. h_{ij} denotes the j-th user's channel gain in the i-th AP. n_0 is the power spectral density of Additive White Gaussian Noise (AWGN), $p_{i,max}$ denotes the maximum transmission power of the i-th AP, $p_{i,min}$ denotes the minimum power of each sub-channel. $r_{j,min}$ indicates the rate requirement of the j-th user, and C_{max} is the constraint of maximum connections of APs.

2.2   Downlink NOMA Channel Model

In this NOMA systems, each AP's spectrum band is orthogonal and no longer divided, then multiple users are served by power-domain NOMA. In addition, successive interference cancelation (SIC) is adopted to decode the different users' information in the same sub-channel. There are N_i users ($N_i \leq C_{max}$) in the AP ($\forall i \in M$). For each user in the same AP, user k is able to correctly decode then remove the interfering signals of other users $g \in N_i \backslash \{k\}$ with $h_{ig} < h_{ik}$ and treat the interfering signals of other users $g \in N_i \backslash \{k\}$ with $h_{ig} > h_{ik}$ as interference. Therefore, the interference after SIC for user k could be expressed as [11]:

$$\sum_{\substack{g \in N_i \backslash \{k\} \\ h_{ig} > h_{ik}}} p_{ig}|h_{ik}|^2, \quad \forall k \in U \tag{1}$$

Then the signal to interference noise ratio (SINR) of user k in the AP is described as following

$$SINR_k = \frac{p_{ik}|h_{ik}|^2}{\sum\limits_{\substack{g \in N_i \backslash \{k\} \\ h_{ig} > h_{ik}}} p_{ig}|h_{ik}|^2 + n_0 B_i}, \quad \forall k \in U \tag{2}$$

Thus, the achievable rate of user k in i-th AP is

$$R_k = B_i \log_2(1 + SINR_k) \tag{3}$$

3   Algorithm Design

Our proposed self-healing mechanism includes the complete procedure of failure detection, diagnosis and recovery. Firstly, a distributed failure detection algorithm is developed based on CR to reduce the detection time. Then the intelligent judgement of whether an AP is normal would be decided by our proposed algorithm based on multi-armed bandit model. Finally, we optimize the system energy consumption and consider possible inter-interference during recovery and resource allocation.

3.1   Failure Detection Based on CR

In the aspect of failure detection, the traditional methods require AP to report their work status periodically. Due to large number of APs in UDN, this might lead to "signal storm", resulting in degradation of network performance. Therefore, we set up a L-AP in each cluster, and adopt CR to sense other APs' working status in the cluster, as shown in Fig. 1(a). Every L-AP have a detection cycle T, which is determined by the number of APs (i.e. $\|M\|$) and its computation ability. Meanwhile, L-AP builds a vector $V = \{v_1, v_2 \ldots v_m\}$, where $v_i \in \{0, 1\}, \forall i \in M$ represents whether i-th AP's spectrum is occupied. An AP is considered as working when $v_i = 1$, otherwise it's idle or failed. If AP's spectrum isn't occupied, we will use the multi-armed Bandit Model to judge the AP's status according to collected information, then return the judgment result. If the result is idle, no further action needs to be taken. Otherwise, the L-AP needs to send a inquiry signaling to this AP. If the reply comes back in time, this AP is judged as normal, otherwise it is considered as failed.

3.2 Diagnostic Model Based on Multi-armed Bandit

In order to analyze the actual status of the APs, we introduced multi-armed bandit model to set up the proposed diagnose model. The multi-armed Bandit Model comes from a realistic problem of maximizing revenue [12]. There are K rocker arms in a gambling machine, μ_1, μ_2...μ_K represent the return function of each arm. The gambler presses one of the rocker arms after puts a coin, then this rocker gives the corresponding revenue, and the gambler's goal is to get the best benefit.

In this paper, we set the number of arms as 2, representing the idle or failed statuses respectively. Considering that the probability of an AP failure is usually low, we use the Upper Confidence Bound Algorithm (UCB) to achieve this model. This algorithm not only focuses on the reward of each arm, but also considers the number that each arm is chosen. Then we need to record the information of each arm, with the following format:

$$S = (m, c, v) \tag{4}$$

Here, m is the identification of each arm, c is the number that each arm is selected and v is the average reward value of this arm.

The understanding degree of an arm is denoted as:

$$bonus = \frac{\sqrt{2 \times \ln(tc)}}{S_i.c} \tag{5}$$

tc is the sum that all arms are selected, and $S_i.c$ is the number that i-th arm is selected. $bonus$ is a indicator showing the degree that an arm is understood. If we have little knowledge about the arm, its v has a low confidence at this time, and we need to choose the arm to get more information. So its $bonus$ is big, then we are more probably to select this arm.

In the algorithm, if there is an arm with its c as 0, that is, the arm has never been selected, then choose it first (ie, each arm will be selected once at the beginning). If all arms have been selected, this algorithm will calculate the sum of $bonus$ and v for each arm, and selects the arm with maximum sum. Then, the reward of the selected arm is updated. For example, i-th arm is selected, and its v is updated as follows:

$$S_i.v = \frac{S_i.v \times S_i.c + res}{S_i.c + 1} \tag{6}$$

res is the reward after choosing the i-th arm.

The details are shown in Algorithm 1, and it will be triggered when an AP's spectrum isn't occupied.

3.3 Recovery Model Based on Optimizing Energy Consumption

When an AP is detected to be failed, the L-AP acquires the failed AP's user connection table, and manages these users to access nearby normal APs. The L-AP informs other normal APs about this failure, and every AP sends a beacon frame to detect the impacted users near it. Then the L-AP asks each AP to report the CQI between itself and these impacted users, and it verifies whether all the users are found. If not, it notifies the nearby APs to increase the detection range until all impacted users are found. After that, every user is accessed to the AP that its CQI is optimal under the scheduling of L-AP, and if this AP has reached the maximum connections, impacted users are scheduled to access the suboptimal AP, and so on. As shown in Fig. 1(b), when AP1 failed, the impacted users UE1 and UE2 are connected to AP2 and AP3 respectively.

Algorithm 1 Multi-armed Bandit Model Based on UCB Algorithm

1: Initialization:
2: (a) Create a record structure for each AP, containing the following data:
   - S defined in Eq. (4): the two S are marked with 0 and 1 respectively, their v and c are set to 0.
   - tc is set to 0.
   - Set the reward function for each state.
   (b) $Id = 0$, $maxB = 0$
3: **for** $i = 1, 2$ **do**
4:     **if** $S_i.c == 0$ **then**
5:         Set $Id = i$ then jump to line 11
6:     **else**
7:         Calculate this state's *bonus* according to Eq. (5)
8:         **if** $S_i.v + bonus \geq maxB$ **then**
9:             $maxB = S_i.v + bonus$; $Id = i$
10: Update $S_{Id}.v$ based on this state's return function
11: $S_{Id}.c = S_{Id}.c + 1$
12: $tc = tc + 1$
13: **if** $S_{Id}.m == 0$ **then**
14:     L-AP send inquiry signaling to this AP
15:     **if** Receive response from this AP **then**
16:         The judgement result is incorrect, and punish this state (i.e. Cut down its v).
17:     **else**
18:         This AP has failed, and call Algorithm 2.

When all impacted users are connected to the assigned AP, we begin to allocate power. Firstly, the AP sorts user connection table in the descending order of CQI, then calculates the power that should be assigned for each channel to meet the user's rate requirements. For example, the demanding power of user k in the i-th AP is denoted as:

$$p_{ik} = \frac{(2^{\frac{B_i}{r_{k,min}}} - 1)(\sum_{g\in\mathbb{N}_i\setminus\{k\};h_{ig}>h_{ik}} p_{ig}|h_{ik}|^2 + n_0 B_i)}{|h_{ik}|^2} \tag{7}$$

We need to guarantee that the allocated sum power is less than the maximum power of this AP. If this AP couldn't meet the power requirements of some users, other AP would be assigned for these users.

Hence the system energy consumption is formulated as

$$\min_{p_{ij}} \sum_{i\in\mathbb{M}_N} \sum_{j\in N_i} p_{ij} + (M - M_F) \cdot P_C$$

$$\text{s.t.} \quad C1 : \sum_{j\in\mathbb{N}_i} p_{ij} \leq p_{i,max} \quad \forall i \in \mathbb{M}_N, j \in \mathbb{N}_i \tag{8}$$

$$C2 : p_{ij} \geq p_{i,min} \quad \forall i \in \mathbb{M}_N, j \in \mathbb{N}_i$$

$$C3 : 0 \leq j \leq C_{max} \quad \forall i \in \mathbb{M}_N, j \in \mathbb{N}_i$$

Here, P_C is the circuit power consumption at each AP.

The detailed steps are shown in Algorithm 2.

Algorithm 2 Self-healing Power Allocation Algorithm

1: Initialization:
2: Create a record structure for each AP, containing the following data:
3:   • A *list* records the users' CQI connecting to this AP. $list = [h_{i1}, h_{i2}...h_{i,C_{max}}]$
      • A *Plist* records the power allocated to these users, and the initial values are set to 0. $Plist = \{p_{i1}, p_{i2}...p_{i,C_{max}}\}$
      • Set $P_{i,max}$ for i-th AP.
4: L-AP asks normal APs to detect $\mathbb{U}_F$
5: **while** $\mathbb{U}_N \cup \mathbb{U}_F < \mathbb{U}$ **do**
6:     All APs increase the detection range
7: L-AP acquires the CQI between APs and impacted users, then determine the optimal AP (i.e. i-th AP) for each user by comparing CQI
8: **if** length($AP_i.list$)$< C_{max}$ **then**
9:     L-AP schedules this user to access the AP
10: **else**
11:     L-AP selects a sub-optimal AP for this user, then jump to line 8
12: *list* is sorted in the descending order of channel quality
13: Calculate the subchannel power p_{ik} assigned for each user according to Eqs. (1), (7)
14: **if** $p_{i,max} - \sum Plist \leq p_{ik}$ **then**
15:     The user will access a new AP that can provide service

---

In this algorithm, the self-healing mechanism is achieved without affecting the operation of other normal APs in the cluster as far as possible, and it is compatible with manual maintenance. If the AP fails, then its spectrum is allocated, other normal APs need to divide the spectrum again. And the failed AP will not have available spectrum when it is repaired. Therefore, the corresponding spectrum of failed AP is no longer allocated.

4   Simulation Results and Discussions

In this section, simulation results are presented to illustrate the performance of the proposed mechanism. We consider a circular area with a radius of 26 m in the square. The L-AP is deployed in the center of this area, and there are 4 ~ 10 APs as well as 16 users. The wireless channel is modeled as rayleigh fading channel including pathloss, where the channel coefficient is $h_{ij}^2 = h_{0j}^2 L_{ij}^{-\kappa}$, in which L_{ij} is the distance between AP i and user j. h_{0j} is the complex Gaussian channel coefficient [7].

Figure 2 illustrates the average time to find AP failure as AP number changes in a cluster. In order to ensure that the result is more accurate, we take the average of multiple outcomes, and find that detection time for each AP failure is small. As the number of APs increases in the cluster, the detection time becomes longer, this is because the L-AP needs to detect more APs each cycle and performs more operations (Table 1).

Figures 3, 4, 5 and 6 are from the same simulation. Figure 3 shows the APs' throughput changes in this simulation. When an AP failed, it can't continue to provide services and its throughput (i.e. AP3) drop to 0. After a period of time, L-AP finds this failure, and schedules neighboring APs service for these impacted users. In order to save the transmission power, these impacted users are accessed to the AP with the best channel gain. In this simulation, AP3 serves two users originally, and the users are connected to AP1 and AP4 respectively after AP3 fails, therefore the throughput of AP1 and AP4 increases. And the added throughput of AP1 and AP4 is exactly equal to the original throughput of AP3.

Table 1. The simulation parameters

Simulation parameters	Value
Carrier center frequency	2.5 GHz
The system bandwidth	$W = 30$ MHz
AP radius	10 m
Path loss exponent κ	4
Power spectral density of noise	-100 dBm/Hz
The maximum transmission power of AP	5 W
The detection threshold at SIC receiver	10 dBm
Users' minimum rate	5 ~ 15 Mbps
Monitoring time interval of L-AP (Iteration time interval)	10 ms

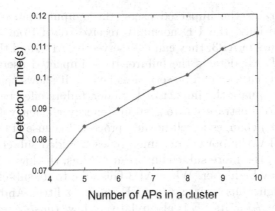

Fig. 2. Detection time with different number of APs

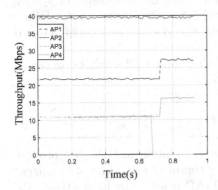

Fig. 3. APs' throughput change in self-healing process

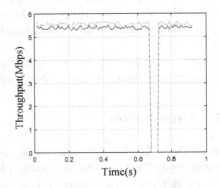

Fig. 4. UEs' throughput change in self-healing process

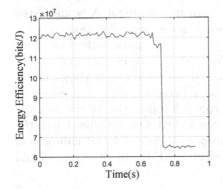

Fig. 5. System energy efficiency change

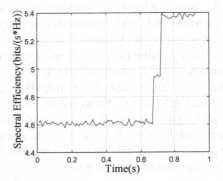

Fig. 6. System spectral efficiency change

Figure 4 illustrates the impacted users' throughput change in the process of self-healing. Initially, the UE normally receives data from AP3. When an abrupt failure occurs in AP3, they can't be served by AP3, and their throughput drop to 0. After L-AP detects the failure, these impacted users access nearby normal APs, and each user can be served as before. It shows that our proposed mechanism can eliminate the impact of a sudden failure effectively.

Figures 5 and 6 illustrate entire system's energy efficiency (EE) and spectrum efficiency (SE) changes in self-healing process. Because of the sub-channels allocated by NOMA, the power consumed by each added sub-channel increases significantly. Thus, the more sub-channels an AP has, the lower its EE is generally, but its SE may be higher. When AP3 fails, due to its sub-channels less, its EE is relatively high, thus entire system EE reduces a little. And its throughput is lower than other APs, its SE is also relatively low, thus system SE increases after AP3 fails. When these impacted users are connected to nearby APs, they require greater transmission power to achieve the original rate, thus the system EE further decreases. However, the system SE increases because an AP failure results in a reduction in system bandwidth yet the system throughput remains unchanged.

We simulate the channels' changes in reality, and CQI can't be obtained immediately, thus simulation results fluctuate.

5   Conclusion

In order to facilitate management and reduce calculation in UDN, we propose a distributed self-healing mechanism including complete procedure of intelligent failure detection, diagnosis and recovery. A L-AP is set to monitor other APs' status in each cluster, and their working information are collected by cognitive technology. When some spectrum is perceived to be unoccupied, the processed information of corresponding APs are analyzed based on multi-armed Bandit Model for possible failure AP judgement. If an AP is confirmed to be failed, the impacted users would be accessed to the optimal neighbor APs under the scheduling of L-AP. Furthermore, the corresponding resource allocation based on AP selection and NOMA is proposed, which considers inter-interference. Simulation results prove this mechanism can find AP failure quickly and realize effective self-healing for the network. However, there are some shortages in the mechanism. On the one hand, the current failure detection algorithm considers less about the AP status parameters, so that couldn't reflect the AP status accurately, thus more AP information should be considered. On the other hand, the proposed self-healing mechanism doesn't consider the case of L-AP failure, and this will reduce the detection sensitivity, therefore L-AP failure detection should be considered in the future.

Acknowledgements. This paper is sponsored by National Natural Science Foundation of China (Grant 61771070 and 61671088).

References

1. Rakshit, S.M., Banerjee, S., Hempel, M., Sharif, H.: Towards an integrated approach for distributed 5G cell association in UDN under interference and mobility. In: 2018 International Conference on Computing, Networking and Communications (ICNC), March 2018, pp. 810–814 (2018)
2. Jiang, W., Strufe, M., Schotten, H.D.: Intelligent network management for 5G systems: the SELFNET approach. In: 2017 European Conference on Networks and Communications (EuCNC), June 2017, pp. 1–5 (2017)
3. Nagarajan, D.R., Thiagarajah, S.P., Alias, M.Y.: Robust son system with enhanced handover performance system. In: 2017 IEEE 13th Malaysia International Conference on Communications (MICC), Nov 2017, pp. 276–281 (2017)
4. Moysen, J., Giupponi, L.: A reinforcement learning based solution for self-healing in LTE networks. In: IEEE 80th Vehicular Technology Conference (VTC2014-Fall), Sept 2014, pp. 1–6 (2014)
5. Chernogorov, F., Repo, I., Räisänen, V., Nihtilä, T., Kurjenniemi, J.: Cognitive self-healing system for future mobile networks. In: 2015 International Wireless Communications and Mobile Computing Conference (IWCMC), Aug 2015, pp. 628–633 (2015)
6. Lin, F. Y.-S., Tsai, M., Wen, Y., Hsiao, C.: Adaptive power ranges and associations for self-healing in multiple types of Wi-Fi networks. In: 2017 13th International Wireless Communications and Mobile Computing Conference (IWCMC), June 2017, pp. 1084–1089 (2017)
7. Liu, Y., Li, X., Ji, H., Wang, K., Zhang, H.: Joint APS selection and resource allocation for self-healing in ultra dense network, July 2016, pp. 1–5 (2016)
8. Alias, M., Saxena, N., Roy, A.: Efficient cell outage detection in 5G HetNets using hidden Markov model. IEEE Commun. Lett. **20**(3), 562–565 (2016)
9. Zhang, T., Feng, L., Yu, P., Guo, S., Li, W., Qiu, X.: A handover statistics based approach for cell outage detection in self-organized heterogeneous networks. In: 2017 IFIP/IEEE Symposium on Integrated Network and Service Management (IM), May 2017, pp. 628–631 (2017)
10. Onireti, O., et al.: A cell outage management framework for dense heterogeneous networks. IEEE Trans. Veh. Technol. **65**(4), 2097–2113 (2016)
11. Lei, L., Yuan, D., Ho, C.K., Sun, S.: Joint optimization of power and channel allocation with non-orthogonal multiple access for 5G cellular systems. In: 2015 IEEE Global Communications Conference (GLOBECOM), Dec 2015, pp. 1–6 (2015)
12. Zhou, Z.: Machine Learning. Tsinghua University Press (2016)

A High-Speed Large-Capacity Packet Buffer Scheme for High-Bandwidth Switches and Routers

Ling Zheng[1], Zhiliang Qiu[1], Weitao Pan[1(✉)], and Ya Gao[2]

[1] State Key Laboratory of Integrated Services Networks, Xidian University,
Xi'an, China
wtpan@mail.xidian.edu.cn
[2] School of Internet of Things Technology, Wuxi Institute of Technology, Wuxi,
China

Abstract. Today's switches and routers require high-speed and large-capacity packet buffers to guarantee a line rate up to 100 Gbps as well as more fine-grained quality of service. For this, this paper proposes an efficient parallel hybrid SRAM/DRAM architecture for high-bandwidth switches and routers. Tail SRAM and head SRAM are used for guaranteeing the middle DRAMs are accessed in a larger granularity to improve the bandwidth utilization. Then, a simple yet efficient memory management algorithm is designed. The memory space is dynamically allocated when a flow arrives, and a hard timeout is assigned for each queue. Hence, the SRAM space is utilized more efficiently. A queueing system is used to model the proposed method, and theoretical analysis is performed to optimize the timeout value. Simulation shows that the proposed architecture can reduce packet loss rate significantly compared with previous solutions with the same SRAM capacity.

Keywords: Switching system · Packet buffer · SRAM · DRAM
Queueing system

1 Introduction

The rapid development of the Internet as well as appearance of the 40/100GE standard have placed higher demands on network switches and routers. With the emergence of new generation data centers and various killer applications such as cloud service, high-definition video streaming and social media, it requires the switching devices to develop towards larger capacity, higher bandwidth and more sophisticated Quality of Service (QoS) guarantees to meet a variety of demands of new applications.

Most current packets switches deploy Store-and-Forward architecture. The incoming packets are first stored in the packet buffers and then forwarded to the destination ports. The packet buffers are critical for network congestion control. The access time and capacity of the switch memory have significant influences on the performance of a switch [1, 2]. On one hand, the ever-increasing bandwidth requires the memory to operate extremely fast to keep up with the line rate of the input port. On the other hand, the novel network techniques, such as software-defined networking

© ICST Institute for Computer Sciences, Social Informatics and Telecommunications Engineering 2019
Published by Springer Nature Switzerland AG 2019. All Rights Reserved
X. Liu et al. (Eds.): ChinaCom 2018, LNICST 262, pp. 374–383, 2019.
https://doi.org/10.1007/978-3-030-06161-6_37

[3] and OpenFlow [4], need fine-grained flow classifications for sophisticated QoS. As a result, the switch memory often maintains several thousands of queues. Therefore, a large amount of memory space is commonly required. Current memory technologies, such as Static Random Access Memory (SRAM) or Dynamic Random Access Memory (DRAM), cannot simultaneously satisfy both the speed and capacity requirements.

To simultaneously meet these requirements, hybrid SRAM/DRAM (HSD) architecture has been proposed [5]. Fast but small-size SRAM meets the random access time and large but slow DRAM can satisfy the capacity requirement. The main idea of HSD is to use two SRAM caches. The tail SRAM is used to store packets at the tail of each FIFO queue, while the head SRAM holds packets at the head. The bulk DRAM maintains the large middle part of the queues. Both SRAM and DRAM maintain Q separate flow queues, and each queue store the input packet from a certain flow. The incoming packets are first stored in the tail SRAM. A memory management algorithm (MMA) is responsible for the transmission between Tail/Head SRAM and DRAM. The basic HSD architecture is shown in Fig. 1.

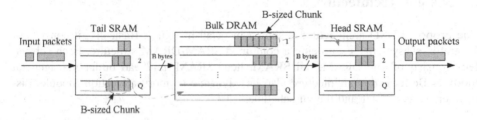

Fig. 1. Basic hybrid SRAM/DRAM architecture

The key function of MMA is to aggregate the input packets into B-sized chunks, and transfer a chunk from tail SRAM to DRAM each time. Similarly, the MMA always read a chunk of B bytes from DRAM and then send it to the head SRAM. When a SRAM queue accumulates B bytes of data, it is transferred to DRAM through a single write. The chunk size B is usually a constant and generally can be determined by $B = 2RA$, where R is the input line rate and A is the random access time of DRAM. Reference [6] proved that the required tail and head SRAM size is $Q(B-1)$ bytes. However, this buffer size is too large for current SRAM chip, due to the fact that today's data center switch may maintain from hundreds to tens of thousands of flow queues [7] and the input line rate is upgrading towards 100/400G. Besides, today's DRAM device cannot provide such a bit width of B bytes for a single data transfer.

To reduce the SRAM capacity requirement and the chunk size, researchers introduced interleaved DRAM [8, 9] and parallel DRAMs architectures [10, 12]. However, the interleaved DRAMs need additional effort for DRAM bank management and the modification of DRAM controller. This task may be too complex for real time processing. In addition, in the parallel DRAM architecture, the bandwidth utilization of each DRAM is not optimized because it often use single FIFO queue in the tail SRAM [11]. To reduce the chunk size to 64 bytes, we need $k = 20$ parallel DRAMs for $R = 100$ Gbps line rate processing [13], which cannot be implemented now and in near future.

In this paper, we introduce a high-speed large-capacity packet buffer scheme for high-bandwidth switches and routers. Parallel DRAMs are used to build the bulk memory, and SRAMs are used as caches for tail and head packets. Per-flow data aggregation is employed to achieve sequential access to DRAMs, so that the effective bandwidth can be improved and in-sequence packet service is guaranteed. Besides, we design a simple yet efficient MMA. The main idea is that the memory space of tail SRAM is only allocated for the most recent active flows, increasing the memory utilization. Theoretical analysis is performed and the optimal timeout value can be obtained. Numerical results show that the proposed architecture can reduce the overall packet loss rate at lower hardware costs compared with existing methods.

The rest of the paper is organized as follows. In Sect. 2, we introduce the proposed HSD architecture. In Sect. 3, theoretical analysis is performed, then the closed-form formulas is derived for analysis and optimization. Performance simulation are presented in Sect. 4. Finally, the conclusions are given in Sect. 5.

2 System Architecture

The proposed HSD architecture is shown in detail in Fig. 2. Basically, it consists of k parallel subsystems, each being a combined SRAM/DRAM structure. Each subsystem consists of one DRAM, tail SRAM, head SRAM, tail/head buffer management module. Besides, this architecture also has dynamic memory allocation module, dispatcher, re-assembler and output scheduler.

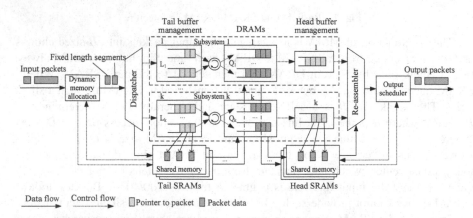

Fig. 2. Proposed HSD system architecture

Dynamic memory allocation: When a packet comes into the switch, the dynamic memory allocation module request free memory space form the tail SRAM, then pointers to the free memory are returned if the request is successful. The memory space in tail SRAM and head SRAM is managed by fixed length buffer, for purpose of fully utilize the limited SRAM size. So the incoming variable length packet are segmented into fixed length segments, typically 64 bytes.

Dispatcher and tail buffer management: The dispatcher module distributes the fixed length segments to their destination queue according to the flow classification results. The tail buffer management module maintains k queue groups, one group for one DRAM. The i-th queue group maintains L_i logic queues, $i = 1, \cdots, k$. The total number of logic queues is $L = \sum_{i=1}^{k} L_i$. These logic queues are actually linked lists of pointers. The packet data is physically stored in the SRAM. The tail SRAM is used for the aggregation of packet data *per-flow*, and the transmission to DRAM in granularity of chunks.

Parallel DRAMs: There are k parallel DRAMs, and each DRAM provides $1/k$ of the required bandwidth. The i-th DRAM contains Q_i physical queue for *per-flow* buffering, where $i = 1, \cdots, k$. Let Q denote to the number of flows the switch can classify, then $Q = \sum_{i=1}^{k} Q_i$. This Q queues should be uniformly distributed in this k DRAMs. Due to the fact that the DRAMs have large enough memory space, the Q physical queues is pre-allocated for each flow. The input packets are transmitted to DRAM in granularity of B-sized chunks to improve the effective bandwidth.

Head buffer management and output scheduler: The output scheduler polls each flow queue according to some schedule algorithm, and fetch the packet data from it to the head SRAM. The read granularity of the DRAM is B-sized chunks too. Similar to the tail buffer management, the head buffer management also uses linked list to manage the logic queue, and the packet is physically stored in the head SRAM. The difference is that the head buffer management only manage k logic queue, one queue per DRAM, because there is no need for per-flow data aggregation in the output side. The outgoing fixed length segments are re-assembled into variable length packet before output.

Memory Management Algorithm: The MMA works as follows. When a flow f arrives, the MMA requests a chunk of B bytes from tail SRAM for data buffering of this flow. If there is no enough memory space, this flow is blocked and dropped. If successfully allocated, a pointer to the allocated memory space is returned and the subsequent packets of this flow are stored in the memory as a queue for data aggregation. Meanwhile, a countdown timer with a default timeout T is assigned to the queue. The timer starts count down once it is enabled. When it reaches 0, a timeout signal is generated. The subsequent packets of flow f are stored in the allocated queue. When data in the queue reaches to B bytes, the aggregated data is transferred into DRAM with sequential address access. If a timeout signal of a queue is received, the data in this queue is transferred to DRAM, even if the queue length is smaller than the chunk size B. Then the memory space of the queue is freed and the timer is disabled. The MMA is simple enough to be implemented in high-speed switches and routers, because it only needs several timers to identity the timeout.

3 Performance Analysis

In the memory management algorithm, the timeout value T should be set a proper value. If T is too long, more memory space is cost for stale flows, and the blocking probability of newly arrive flow will increase. If T is too short, we cannot aggregate enough data for sequential DRAM access, leading to low DRAM transfer speed. In this section, we perform theoretical analysis and wish to obtain the optimal value of T.

Since the k subsystems operate completely parallel and the input traffic is uniformly distributed in the k subsystem, the k subsystem can be regarded as equally. We only analyze a behavior of a certain subsystem. The system analysis model is shown in Fig. 3.

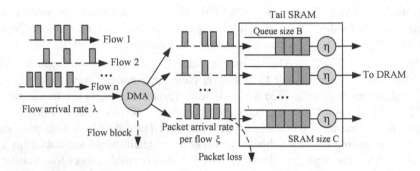

Fig. 3. System analysis model

The traffic applied to the subsystem is composed of different network flows, and the stochastic process of each flow is independent and identically distributed. The behavior of tail SRAM can be regarded as a queueing system where the arriving flows are customers and the memory space allocated for each flow is service station [14]. When the memory space is dynamically allocated for an arriving flow, a customer gets service in the queueing system. When the memory space is freed, this customer leaves the system. The arriving process of flows can be modeled as Poisson process with parameter λ, and the flow durations are exponentially distributed with parameter μ [15]. Let random variable W denote the flow size (number of packets) of each flow, and R denote the total packet arrival rate.

Within each flow, the packet arrival also can be modeled as Poisson process with parameter ξ, and the service time of each packet is exponentially distributed with parameter $1/\eta$. Denote variable X to express the flow duration. Let T denote the timeout value, and the total service time of a flow is denoted as Y. When a new flow come but there is no enough free memory space to be allocated, this flow is *blocked* and dropped. The flow blocking probability is denoted as P_C and C is the total size of the tail SRAM (the number of queues the SRAM can accommodate). The overall packet loss rate is $P_{overall}$.

First, at the view of flow arrival, we analyze the flow blocking probability. If a flow's duration is less than T, it will get a service time of T. If the duration is larger than T, a new buffer will be allocated for this flow and the flow will get another service time of T. The flow duration is exponentially distributed, so that the cumulative distribution function (CDF) is:

$$F(x) = 1 - e^{-\mu x} \tag{1}$$

And the expectation of X is $E[X] = 1/\mu$. Then we have the probability distribution of Y as follows

$$P(Y = iT) = P\{(i-1)T < X \leq iT\} = e^{-(i-1)\mu T} - e^{-i\mu T}, \quad i = 1, 2, \cdots \tag{2}$$

And the expectation of Y is

$$E[Y] = \sum_{i=1}^{\infty} iT \times P\{Y = iT\} = T(1 + e^{-\mu T} + e^{-2\mu T} + \cdots) \tag{3}$$

$$= T \times \lim_{n \to \infty} \frac{1 - e^{-n\mu T}}{1 - e^{-\mu T}} = \frac{1}{1 - e^{-\mu T}}$$

The above queueing system is an M/G/C/C system, where the flow arrival rate is λ and the service rate is $1/E[Y]$. So that the blocking probability can be obtained by the Erlang-B formula

$$P_C = \frac{\rho^C}{C!} \left[\sum_{i=0}^{C} \frac{\rho^i}{i!} \right]^{-1} \tag{4}$$

where $\rho = \lambda E[Y]$.

Next, at the view of packet arrival per flow, we study the packet loss rate within a flow queue in tail SRAM. Once a flow arrives and memory space is allocated for it, the flow queue can be modeled as an M/M/1/B queue, where B is the allocated memory space for each queue.

According to the characteristic of compound Poisson process, we have the following relation

$$R = \lambda E[W] \tag{5}$$

The average packet arrive rate per flow can be obtained by the flow size divided by the flow duration, which is

$$\xi = \frac{E[W]}{E[X]} = \frac{\mu R}{\lambda} \tag{6}$$

And the service rate of this M/M/1/B queue depends on the transmission speed of DRAM and average number of queues. Let random variable N denote the number of queues in the system, then the expectation of N can be obtained as follows

$$E[N] = \sum_{n=0}^{C} n p_n = \sum_{n=0}^{C} \left(n \times \frac{\rho^n}{n!} \left[\sum_{i=0}^{C} \frac{\rho^i}{i!} \right]^{-1} \right)$$

$$= \frac{\rho \left(\sum_{n=0}^{C-1} \frac{\rho^n}{n!} \right)}{\sum_{i=0}^{C} \frac{\rho^i}{i!}} = \rho (1 - P_C) \tag{7}$$

It can be considered that the total bandwidth of DRAM is equally shared by these N queues. As mentioned above, the more data aggregated in the SRAM queue, the higher speed of DRAM. In this study, we assume that the speed of DRAM has a linear relationship with the aggregated queue length. According to Little's theorem, the average queue length equals to packet arrival rate multiplied by the timeout value, so we have the service rate of each M/M/1/B queue by the following equation

$$\eta = \begin{cases} \frac{S_{\min} + \alpha \xi T}{E[N]}, & 0 < T \le \frac{S_{\max} - S_{\min}}{\alpha \xi} \\ \frac{S_{\max}}{E[N]}, & T > \frac{S_{\max} - S_{\min}}{\alpha \xi} \end{cases} \tag{8}$$

where $S_{\max}$ and $S_{\min}$ is the maximum and minimum speed of DRAM, and α is a constant. When $T > (S_{\max} - S_{\min})/(\alpha \xi)$, there is no improvement in DRAM speed since the DRAM can operate at its peak bandwidth, so we only consider the situation where $0 < T \le (S_{\max} - S_{\min})/(\alpha \xi)$.

Therefore, in the M/M/1/B queue, the packet loss rate can be obtained by

$$P_B = \frac{(\xi/\eta)^B (1 - \xi/\eta)}{1 - (\xi/\eta)^{B+1}} \tag{9}$$

Together with (4), (5), (6), (7), (8) and (9), we have the overall packet loss rate as follows

$$P_{overall} = P_C + (1 - P_C) P_B$$

$$= \frac{\rho^C}{C!} \left[\sum_{i=0}^{C} \frac{\rho^i}{i!} \right]^{-1} + \left(1 - \frac{\rho^C}{C!} \left[\sum_{i=0}^{C} \frac{\rho^i}{i!} \right]^{-1} \right) \frac{(\xi/\eta)^B (1 - \xi/\eta)}{1 - (\xi/\eta)^{B+1}} \tag{10}$$

The optimal value of T can be obtained by solving the optimization problem

$$T* = \arg \min_{0 < T < (S_{\max} - S_{\min})/(\alpha \xi)} P_{overall} \tag{11}$$

4 Simulation

In this section, we evaluate the proposed architecture through software simulation. The following parameters are assumed throughout the simulations. We assume that the input packets have a fixed length and time is split into discrete time slots. The SRAM size is 1600 packets, and the chunk size allocated for each arriving flow is 16 packets. The total packet arrive rate R is fixed to 100, i.e. 100 packets arrive in the system each time slot on average. Each simulation is performed using 10^5 time slots.

In the first simulation, we consider two traffic scenario: "light flow arrival with large flow size" and "heavy flow arrival with small flow size". We use triple $(R, \lambda, 1/\mu)$ to describe traffic condition, then the first traffic scenario is express by $(100, 8, 5)$, while the second one is $(100, 18, 5)$. The results are shown in Table 1 and the following results can be found. First, in the light flow arrival traffic pattern, a relatively long timeout ($T = 6$) should be set to perform better data aggregation, so that the DRAM's access speed can be improved. Second, the heavy flow arrival traffic will cause a higher packet loss rate. This is because heavy flow arrive rate will cause higher blocking probability, and data aggregation is more difficult. But we can choose a relatively small timeout ($T = 2$) to let the SRAM to accommodate more flows. In addition, we use Distributed Packet Buffers (DPB) [11] and Semi-Parallel Hybrid SRAM/DRAM (SPHSD) [13] as the benchmarking methods. It is found that overall packet loss rates of DPB and SPHSD are independent of T. Under the same traffic scenario, if a proper T is set, the proposed method has a better performance in packet loss rate compared with DPB and SPHSD.

Table 1. Results of simulation-1, comparison of different traffic scenario.

Traffic Scenario		Overall packet loss rate		
		Proposed	DPB [11]	SPHSD [13]
(100, 8, 5)	T = 1	0.11058	0.111438	0.361268
light flow arrival with large flow size	T = 2	0.0310028	0.109163	0.363204
	T = 4	0.0304534	0.107947	0.358306
	T = 6	0.0297913	0.108581	0.361994
	T = 8	0.0398018	0.111425	0.360386
	T = 10	0.0919443	0.109131	0.358593
(100, 18, 5)	T = 1	0.377449	0.32771	0.358723
heavy flow arrival with small flow size	T = 2	0.263501	0.327208	0.358788
	T = 4	0.30571	0.325895	0.357888
	T = 6	0.412023	0.328218	0.362129
	T = 8	0.496174	0.328878	0.359489
	T = 10	0.559625	0.326562	0.362606

In the second simulation, different HSD methods is compared under different flow arrive rate. The results as shown in Fig. 4. For all given λ, the proposed method achieves better performance in packet loss rate compared with DPB and SPHSD. When

$\lambda = 16$, the packet loss rate performance is improved by about 50% compared with DPB and about 60% compared with SPHSD.

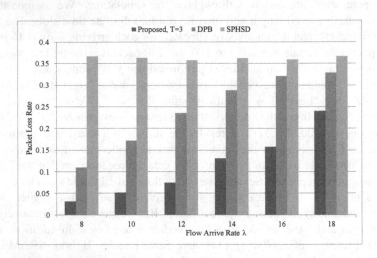

Fig. 4. Result of simulation-2, comparison of different HSD methods.

5   Conclusion

Today's most switches and routers employ Store-and-Forward architecture, and packets need to be stored in memory before sent to the destination ports. To simultaneously meet the high-speed and large-capacity requirements for packet buffers, this paper proposes an efficient parallel hybrid SRAM/DRAM architecture for high-bandwidth switches and routers. Tail SRAMs and head SRAMs are used for guaranteeing the middle DRAMs are accessed in a larger granularity. To improve the SRAM utilization, a simple yet efficient memory management algorithm is designed. When a flow arrives, memory space is dynamically allocated for it, and a hard timeout is assigned for the allocated queue. After the specific period, the memory space is freed, in order to accommodate the most recent active flows. A queueing system is used to model the proposed method, and theoretical analysis is performed to optimize the timeout value. Simulation results indicate that with a proper timeout value, packet loss rate can be significantly reduced compared with existing methods with the same SRAM capacity.

In the future work, we will develop a prototype of the proposed architecture and MMA based on Field Programmable Gate Array (FPGA) platform. Then further evaluate its performance using real traffic loads, making the proposal more practical.

Acknowledgements. This work was supported in part by the project of Science and Technology on Information Transmission and Dissemination in Communication Networks Laboratory (KX152600010/ITD-U15001), the National Natural Science Foundation of China (61502204, 61306047), the Fundamental Research Funds for the Central Universities (JB140112), and the Qing Lan Project of Jiangsu.

References

1. Wang, F., Hamdi, M.: Memory subsystems in high-end routers. IEEE Micro **29**(3), 52–63 (2009)
2. Ganjali, Y., McKeown, N.: Update on buffer sizing in internet routers. ACM SIGCOMM Comput. Commun. Rev. **36**(5), 67–70 (2006)
3. Kreutz, D., Ramos, F.M.V., Verissimo, P.E., et al.: Software-Defined networking: a comprehensive survey. Proc. IEEE **103**(1), 14–76 (2014)
4. Kao, S.C., Lee, D.Y., Chen, T.S., Wu, A.Y.: Dynamically updatable ternary segmented aging bloom filter for OpenFlow-compliant low-power packet processing, IEEE/ACM Trans. Netw. **26**(2), 1004–1017 (2018)
5. Iyer, S., Kompella, R., Mckeown, N.: Analysis of a memory architecture for fast packet buffers. In: Proceedings of IEEE International Conference on High Performance Switching and Routing (HPSR), pp. 368–373. Dallas, TX, USA (2001)
6. Iyer, S., Kompella, R., McKeown, N.: Designing packet buffers for router linecards. IEEE/ACM Trans. Netw. **16**(3), 705–717 (2008)
7. Juniper E Series Router (2011). http://juniper.net/products/eseries/
8. Garcia, J., March, M., Cerda, L., Corbal, J., Valero, M.: A DRAM/SRAM memory scheme for fast packet buffers. IEEE Trans. Comput. **55**(5), 588–602 (2006)
9. Wang, F., Hamdi, M.: Scalable router memory architecture based on interleaved DRAM: analysis and numerical studies. In: Proceedings of IEEE International Conference on Communications (ICC), pp. 6380–6385. Glasgow, UK (2007)
10. Lin, D., Hamdi, M., Muppala, J.: Designing packet buffers using random round robin. In: Proceedings of IEEE Global Telecommunications Conference (GLOBECOM), pp. 1–5. Miami, FL, USA (2010)
11. Lin, D., Hamdi, M., Muppala, J.: Distributed packet buffers for high-bandwidth switches and routers. IEEE Trans. Parallel Distrib. Syst. **23**(7), 1178–1192 (2012)
12. Wang, F., Hamdi, M., Muppala, J.: Using parallel DRAM to scale router buffers. IEEE Trans. Parallel Distrib. Syst. **20**(5), 710–724 (2009)
13. Mutter, A.: A novel hybrid memory architecture with parallel DRAM for fast packet buffers. In: Proceedings of IEEE International Conference on High Performance Switching and Routing (HPSR), pp. 44–51. Richardson, TX, USA (2010)
14. Zhang, L., Lin, R., Xu, S., Wang, S.: AHTM: achieving efficient flow table utilization in software defined networks. In: Proceedings of IEEE Global Telecommunications Conference (GLOBECOM), pp. 1897–1902. Austin, TX, USA (2014)
15. Rai, I.A., Urvoy-Keller, G., Biersack, E.W.: Analysis of LAS scheduling for job size distributions with high variance. In: Proceedings of ACM International Conference on Measurement and Modeling of Computer Systems (SIGMETRICS), pp. 2018–228. San Diego, CA, USA (2003)

Non-stationary Characteristics for Indoor Massive MIMO Channels

Qi Wang[✉] [ID], Jiadong Du, and Yuanyuan Cui

China Academy of Information and Communications Technology, Beijing 100191,
China
qiwang.609@gmail.com

Abstract. Massive Multiple Input Multiple Output (MIMO) has been
widely considered as one of the most promising technologies for the fifth-
generation (5G) wireless communication. In massive MIMO system, the
research on channel characteristics is important. In this paper, the characteristics for massive MIMO channels at both 2 GHz and 6 GHz are
investigated. Based on the real-world measurements, the channel parameters in the delay and frequency domains are extracted to show the
non-stationary phenomenon over the large-scale antenna array. Furthermore, the characteristics of the angular parameters extracted by space-
alternating generalized expectation-maximization (SAGE) algorithm are
investigated and the fluctuations are modeled. The results for different
frequencies are useful for deep understanding of massive MIMO channels
in the future.

Keywords: Massive MIMO · Channel characteristics · Angular
parameter

1 Introduction

Nowadays, the fifth-generation (5G) wireless communication is a very hot topic
in both academic and industry fields [1]. The key capabilities of 5G systems
will dramatically outperform the previous generation systems. In general, 5G is
expected to have more than 5 times improvement on spectrum efficiency and
more than 100 times improvement on energy and cost efficiency [2]. To fulfill the
requirements of 5G, massive Multiple Input Multiple Output (MIMO) technology has been introduced and regarded as one of the most important technologies
in 5G system.

This work was supported by the Science and Technology Foundation of the State Grid
Corporation of China: Adaptability research and application between LTE wireless
private network and electric power service.

© ICST Institute for Computer Sciences, Social Informatics and Telecommunications Engineering 2019
Published by Springer Nature Switzerland AG 2019. All Rights Reserved
X. Liu et al. (Eds.): ChinaCom 2018, LNICST 262, pp. 384–393, 2019.
https://doi.org/10.1007/978-3-030-06161-6_38

Some key issues for massive MIMO have already been figured out in recent years. As the foundation of wireless communication, the deep understanding of propagation channel is of importance. However, the researches on channel characteristics and modeling approaches are still open issues for massive MIMO. In a massive MIMO system, a large number of antennas will be equipped at the base stations [3]. Therefore, the massive MIMO channel characteristics should be obviously different from the traditional MIMO channels. Channel measurements for massive MIMO have been performed in various of scenarios, such as outdoor campus [4,5], indoor scenario [6], and indoor venue [7]. Different antenna array shapes have been adopted for massive MIMO measurements, which include 128-element linear and cylindrical array [8,9], 112-element scalable virtual antenna array [10], and 64-element switch array with different shapes [6], respectively. Based on the measurement results, channel characteristics such as channel gain, delay spread, Ricean K-factor, and angular power spectrum are analyzed, and the non-stationary phenomenon over the large-scale antenna array has been discovered [8]. However, the differences between channel characteristics at different frequencies in massive MIMO channels are not considered in the existing measurements. Moreover, the channel behaviors in some typical 5G scenario are not fully addressed.

Hence, channel measurements at both 2 GHz and 6 GHz in an indoor hall scenario have been briefly introduced in this paper. The non-stationary phenomenon over the large-scale antenna array has been presented by the obtained statistical parameters in the delay and frequency domains, and these behaviors at different frequencies are further explained. Finally, the characteristics of the channel angular parameters are investigated, and the non-stationarity of these parameters is modeled.

The rest of this paper is organized as follows. The massive MIMO channel characteristics at different central frequencies are presented in Sect. 2. Section 3 describes the characteristics and modeling of angular parameters. Finally, Sect. 4 concludes the paper.

2   Massive MIMO Channel Characteristics

2.1   Measurement Description

The measurement environment was selected in an indoor hall scenario, which was expected to have ultra-high connections during the events. The transmitter (Tx) antenna was a 64-element virtual linear antenna array generated by using a high-accuracy automatic turntable, thus the Tx array could be regarded as a large-scale antenna array. At the receiver (Rx) side, a 4-element linear antenna array was constituted by moving one single antenna manually. The central frequency of the measurements was selected as 2 GHz or 6 GHz with a bandwidth of 200 MHz, so as to compare the channel characteristics at different propagation frequencies. Note that both 2 GHz and 6 GHz are considered as the alternative frequencies for the massive MIMO system in 5G. During the measurements, both the line-of-sight (LOS) and the non-LOS (NLOS) cases were considered, and the NLOS case was realized by placing the Rx antenna array below the seats. Moreover, a multi-carrier signal in the frequency domain was utilized to measure the channels by the self-designed channel sounder system. Thus, the wideband channel transfer function (CTF) between the m-th Tx antenna and the n-th Rx antenna can be obtained from the recorded IQ data, which can be expressed as $H(m, n, \tau)$, and τ is the multipath delay. Finally, all the details of the measurement scenario and system can be found in [11].

2.2 Power Delay Profile

The channel impulse response (CIR) $h(m, n, \tau)$ can be obtained from the inverse Fourier transform of the CTF $H(m, n, \tau)$. The power delay profile (PDP) and the corresponding average PDP (APDP) along the Tx antenna array APDP (m, τ) can then be calculated according to the method in [11]. It is noted that the effective multipath components (MPCs) are extracted by using the power threshold selected as the noise floor of measurement system plus 6 dB [12].

The root mean square (RMS) delay spreads (DSs) can therefore be computed based on APDPs

$$\sigma_\tau = \sqrt{\frac{\int_0^\infty \tau^2 \text{APDP}(m, \tau) d\tau}{\int_0^\infty \text{APDP}(m, \tau) d\tau} - \overline{\tau}^2}, \tag{1}$$

where $\overline{\tau}$ is the mean delay, and can be expressed as

$$\overline{\tau} = \frac{\int_0^\infty \tau \text{APDP}(m, \tau) d\tau}{\int_0^\infty \text{APDP}(m, \tau) d\tau}. \tag{2}$$

The RMSDSs for different conditions are illustrated in Fig. 1. It is noticed that the RMSDSs in the LOS condition are roughly smaller than the values in the NLOS condition across the large-scale array at both 2 GHz and 6 GHz. However, the RMSDSs in the NLOS case are just slightly larger than the values in the LOS case at 2 GHz, which can be explained by a combination of smaller free space path loss and penetration loss at the lower frequency, i.e., the NLOS condition in such an indoor hall scenario is more like an obstructed LOS condition at 2 GHz. One can observe that the received power difference between LOS and NLOS cases is larger in the 6 GHz case than in the 2 GHz case. Diffraction over the seats will also be somewhat stronger at the lower frequency. Worth noting is the unexpected and fairly large variation of the LOS component in the 2 GHz LOS case. At present we do not have a good explanation for this phenomenon. One hypothesis is a stronger two-ray condition from the floor reflection at 2 GHz, yielding lower received power around the center of the array. Finally, the statistics of RMSDS are summarized in Table 1.

2.3 Frequency Correlation Function

The frequency correlation function (FCF) of the m-th Tx position, which can be obtained from the Fourier transform of the APDP of the m-th Tx position, is defined as [13]

$$R_H(m, \Delta f) = \int\limits_{-\infty}^{+\infty} \mathrm{APDP}(m, \tau) \cdot e^{-j2\pi\Delta f\tau} d\tau, \tag{3}$$

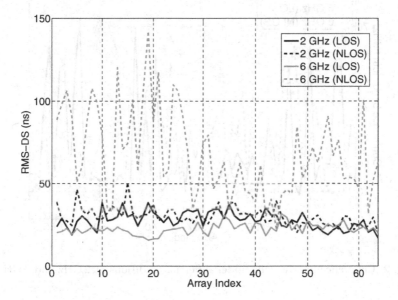

Fig. 1. RMSDSs under different conditions at 2 GHz and 6 GHz.

Table 1. Statistics of RMSDSs in nanoseconds under different conditions at 2 GHz and 6 GHz

Condition	Minimum value	Maximum value	Mean value	Standard deviation (Std)
2 GHz (LOS)	16.99	38.42	27.71	5.17
2 GHz (NLOS)	20.05	50.51	30.15	5.59
6 GHz (LOS)	15.64	31.56	22.48	3.32
6 GHz (NLOS)	22.82	142.2	68.13	26.31

where the Δf is the frequency difference. The measured FCFs at 2 GHz and 6 GHz across the Tx array are therefore obtained. The coherence bandwidth B_{coh} can then be defined as a bandwidth at which the correlation drops to a value of

$1/e$. Here also we might draw the same conclusion that non-stationarity exists for both 2 GHz and 6 GHz with different extents. Figure 2 gives the coherence bandwidth across the Tx array. As expected, the NLOS values of coherence bandwidth are generally smaller than those for the LOS case at 6 GHz, but there is no clear dependency at 2 GHz. Also, it can be observed that the values vary significantly over the large-scale antenna array at both center frequencies. Finally, the statistics of coherence bandwidths are summarized in Table 2.

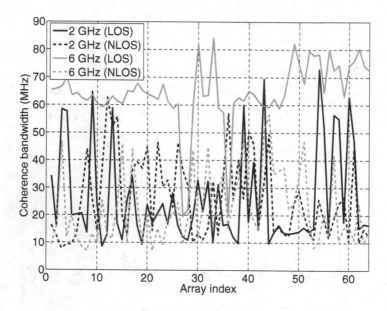

Fig. 2. Coherence bandwidths under different conditions at 2 GHz and 6 GHz.

Table 2. Statistics of coherence bandwidths in MHz under different conditions at 2 GHz and 6 GHz

Condition	Minimum value	Maximum value	Mean value	Standard deviation (Std)
2 GHz (LOS)	8.58	72.90	25.96	17.80
2 GHz (NLOS)	7.80	62.77	26.27	14.51
6 GHz (LOS)	19.49	84.21	64.42	12.00
6 GHz (NLOS)	7.02	56.53	22.64	13.16

3 Non-stationary Characteristics of Angular Parameters

In this section, the non-stationarity on the angular characteristics over the large-scale antenna array at different frequencies is further investigated, and the fluctuations of different statistical parameters in the angular domain are also modeled. Note that the characteristics in the delay domain can be found in [14].

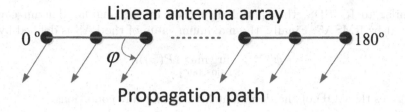

Fig. 3. Definition of the propagation path at the Tx side.

To obtain the angular parameters of MPCs, the space-alternating general-ized expectation-maximization (SAGE) algorithm is adopted. The number of MPCs in SAGE is chosen as 80, and the delay resolution is 5 ns, which is the same with the measurement system. As the linear antenna array at the TX side is used in our measurements, the azimuth angle of departure (AOD) over the large-scale antenna array can be estimated in SAGE, which ranges from 0° to 180° as defined in Fig. 3. Therefore, the MPCs with the same propagation delay can be further distinguished as different MPCs based on the estimated AODs, and the CIR of each MPC can be expressed as $h(n, m, \tau, \varphi)$, compared to the CIR of MPC $h(n, m, \tau)$ without the angular information. The φ is the azimuth AOD estimated for each MPC in the SAGE algorithm. The power azimuth spec-trum (PAS) of each sub-channel between the m-th Tx antenna and the n-th Rx antenna can therefore be computed as

$$P(\varphi) = \sum_{l=1}^{L} \delta(\varphi - \varphi_l)|h(n, m, \tau, \varphi)|^2, \tag{4}$$

where L is the number of MPCs, which equals to 80. By using the $P(\varphi)$, the statistical parameters in the angular domain can be obtained.

The root mean square azimuth spread (RMSAS) at each Tx position is defined by

$$\sigma_\varphi = \sqrt{\frac{\sum\limits_{\varphi=0°}^{180°} |\varphi - \mu_\varphi|^2 P(\varphi)}{\sum\limits_{\varphi=0°}^{180°} P(\varphi)}}, \tag{5}$$

where μ_φ is the mean angular value of the PAS at each TX position, and can be calculated as

$$\mu_\varphi = \frac{\sum\limits_{\varphi=0°}^{180°} \varphi P(\varphi)}{\sum\limits_{\varphi=0°}^{180°} P(\varphi)}. \tag{6}$$

Similar to RMSDS, the spatial dispersion in the angular domain can be described by RMSAS. Finally, the maximum value of the PAS is defined by

$$\varphi^{\max} = \underset{\varphi \in [0°,180°]}{\arg\max} \, (P(\varphi)), \tag{7}$$

which gives the AOD of the dominant MPC under any conditions.

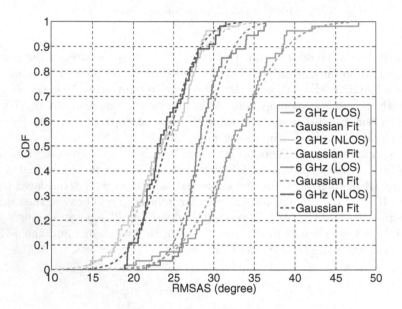

Fig. 4. CDFs of the RMSAS under different conditions at 2 GHz and 6 GHz.

The fluctuations of RMSASs across the large-scale antenna array at different frequencies are aimed to be modeled, so as to characterize the differences of non-stationarity between different frequencies. However, no clear trends can be discovered for the angular parameters over the large-scale antenna array. Thus, it can be concluded that the statistical angular parameters change irregularly for different antenna positions, and the cumulative distribution functions (CDFs) of the RMSASs under different conditions at both 2 GHz and 6 GHz are plotted in Fig. 4. It can be observed that the range of RMSASs at 2 GHz in the LOS condition is larger than the ranges in other three cases, which agrees with the result in Sect. 2.3 where this case was found to have the largest degree of non-stationarity. We can also see that RMSASs in the LOS conditions are larger than the values in the NLOS conditions at both 2 GHz and 6 GHz. This can be explained that the NLOS condition in our measurement is more like the obstructed LOS condition due to the limited measurement spaces, and most of the MPCs with large AODs are blocked by the seats. According to the results in [11], the dominant MPCs in the NLOS conditions are the ground reflection path and the diffracted path over the seats, and the AODs of these paths regarding to the AOD of LOS path are very small. With larger path loss and penetration loss, this phenomenon is more distinct at 6 GHz.

As also shown in Fig. 4, the CDFs of RMSASs under different conditions at both 2 GHz and 6 GHz can all be well modeled by the Gaussian distributions. By adopting the Kolmogorov–Smirnov (KS) testing, the goodness of all these fittings have been tested. The measured values and Gaussian fit values of RMSASs are further summarized in Table 3. From the results of modeled standard deviation, one can also draw a conclusion that the fluctuations of RMSASs are positively correlated to the non-stationarity characterized by collinearity [11] in different conditions.

Table 3. Statistics of RMSASs under different conditions at 2 GHz and 6 GHz

Condition	Measured values		Gaussian fit values	
	Median	90th percentile	Mean	Std
2 GHz (LOS)	32.26°	38.58°	31.89°	9.09°
2 GHz (NLOS)	23.52°	28.45°	23.41°	7.94°
6 GHz (LOS)	27.89°	33.97°	28.19°	5.34°
6 GHz (NLOS)	22.99°	29.48°	23.80°	6.21°

As the strongest path in each condition, the maximum value $\varphi^{\max}$ of the PAS over the large-scale antenna array is investigated. Obviously, the maximum value $\varphi^{\max}$ in the LOS condition can be expressed as

$$\varphi^{\max} = \varphi_{LOS}, \tag{8}$$

and φ_{LOS} is the AOD for the LOS path, which can be simply calculated by the geometry. However, $\varphi^{\max}$ in the NLOS condition is more difficult to determine. Figure 5 gives the maximum value $\varphi^{\max}$ of the PAS over the large-scale antenna array in the NLOS conditions. As discussed above, a potentially strong diffracted path (and possibly a floor reflected path) still exists in some sub-channels, and these paths have similar AOD to that of the LOS path in this position. Thus, based on the results in Fig. 5, the maximum value $\varphi^{\max}$ in the NLOS condition is modeled as

$$\varphi^{\max} = \varphi_{LOS} + \sigma_\varphi, \tag{9}$$

where σ_φ is the standard deviation of the $\varphi^{\max}$ at each center frequency. The values of σ_φ are calculated and listed in Table 4.

Table 4. Standard deviation of the value σ_φ in the NLOS condition

	2 GHz	6 GHz
σ_φ	5.84°	4.01°

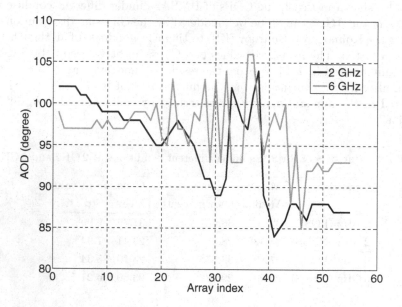

Fig. 5. Maximum value of the PAS over the large-scale antenna array in the NLOS condition.

4   Conclusion

The massive MIMO channel characteristics at both 2 GHz and 6 GHz have been presented in this paper. The non-stationary phenomenon across the large-scale antenna array were investigated by the obtained statistical parameters in the delay and frequency domains. It was found that the RMSDSs and the coherence bandwidths over the large-scale antenna array were clearly different in different conditions, and the statistical values were also provided in this paper. Furthermore, the RMSAS and the maximum value of the PAS over the large-scale antenna array were investigated, and the median values of RMSAS ranged from 22.99° to 32.26° under different cases. Finally, the non-stationarity models of these angular parameters were further established. These results should be useful for deep understanding of massive MIMO channels in the future.

References

1. Gohil, A., Modi, H., Patel, S.: 5G technology of mobile communication: a survey. In: 2013 International Conference on Intelligent Systems and Signal Processing (ISSP), pp. 288–292. IEEE, Gujarat, India (2013)

2. IMT-2020 Promotion Group: 5G vision and requirements. http://www.imt-2020. org.cn/en. Accessed May 2014
3. Jungnickel, V., et al.: The role of small cells, coordinated multipoint, and massive MIMO in 5G. IEEE Commun. Mag. **52**(5), 44–51 (2014)
4. Payami, S., Tufvesson, F.: Channel measurements and analysis for very large array systems at 2.6 GHz. In: 2012 6th European Conference on Antennas and Propagation (EUCAP), pp. 433–437. IEEE, Prague, Czech Republic (2012)
5. Payami, S., Tufvesson, F.: Delay spread properties in a measured massive MIMO system at 2.6 GHz. In: 2013 IEEE 24th Annual International Symposium on Personal, Indoor, and Mobile Radio Communications (PIMRC), pp. 53–57. IEEE, London, UK (2013)
6. Gauger, M., Hoydis, J., Hoek, C., Schlesinger, H., Pascht, A., Brink, S.: Channel measurements with different antenna array geometries for massive MIMO systems. In: 10th International ITG Conference on Systems, Communications and Coding (SCC), pp. 1–6. IEEE, Hamburg, Germany (2015)
7. Martinez, A., Carvalho, E., Nielsen, J.: Towards very large aperture massive MIMO: a measurement based study. In: 2014 IEEE Globecom Workshops (GC Wkshps), pp. 281–286. IEEE, Austin, USA (2014)
8. Gao, X., Edfors, O., Rusek, F., Tufvesson, F.: Massive MIMO performance evaluation based on measured propagation data. IEEE Trans. Wirel. Commun. **14**(7), 3899–3911 (2015)
9. Gao, X., Edfors, O., Tufvesson, F., Larsson, E.: Massive MIMO in real propagation environments: do all antennas contribute equally? IEEE Trans. Commun. **63**(11), 3917–3928 (2015)
10. Hoydis, J., Hoek, C., Wild, T., Brink, S.: Channel measurements for large antenna arrays. In: 2012 International Symposium on Wireless Communication Systems (ISWCS), pp. 811–815. IEEE, Paris, France (2012)
11. Wang, Q., et al.: Spatial variation analysis for measured indoor massive MIMO channels. IEEE Access **5**, 20828–20840 (2017)
12. Molisch, A., Steinbauer, M.: Condensed parameters for characterizing wideband mobile radio channels. Int. J. Wirel. Inf. Netw. **6**(3), 133–154 (1999)
13. Molisch, A.: Wireless Communications, 2nd edn. Wiley, Hoboken (2010)
14. Li, J., et al.: Measurement-based characterizations of indoor massive MIMO channels at 2 GHz, 4 GHz, and 6 GHz frequency bands. In: 2016 IEEE 83rd Vehicular Technology Conference (VTC Spring), pp. 1–5. IEEE, Nanjing, China (2016)

DSP Implementation and Optimization of Pseudo Analog Video Transmission Algorithm

Chengcheng Wang, Pengfei Xia[✉], Haoqi Ren, Jun Wu,
and Zhifeng Zhang

College of Electronics and Information Engineering, Tongji University,
Shanghai, China
1631731@tongji.edu.cn, pengfei.xia@gmail.com

Abstract. With the development of wireless video technology and embedded technology, a dedicated digital signal processor (DSP) can achieve the video transmission stably and flexibly. Some existing wireless video transmission algorithms do not perform well in response to complex channel environments. A pseudo-analog video algorithm that can be run in a dedicated instruction set was proposed. At the transmitter, the image data which are removed spatially redundant are divided into L-shaped blocks for power allocation, and the digital signal are sent to CRC and Turbo coding. Finally, the modulated digital signal and the pseudo-analog data after power allocation are sent to framing. The receiver includes channel estimation and de-framing, recovers digital signal and pseudo-analog signal through error detection and decoding. We have optimized the algorithm at the assembly level, so that the entire system is more flexible. The entire transfer system will run on the FPGA and hardware DSP boards for debugging.

Keywords: DSP · DMA · Dedicated instruction set · Power allocation
Turbo encoding · Framing

1 Introduction

With the rapid development of wireless communication technologies, various smart devices have rapidly become popular. Emerging application platforms such as drones and smart wearable devices have emerged, which have increased the demand for various applications of wireless video transmission. The current video transmission scheme can not match the video quality with the channel quality [1], thus it is not optimal from the perspective of network information theory [2]. Someone designed a set of pseudo-analog video transmission system, which is based on SoftCast [3, 4]. SoftCast transmits video in a way of linear transformation, so that guarantees the linear relationship between video signal and image pixels. This relationship will not be changed by the noise or interference [2]. SoftCast is a video multicast method based on real-number transmission, which greatly reduces the redundancy of video frames. Pseudo-analog system has a low delay and eliminates the cliff effect in a certain extent [5], and provides different video quality for different users in different channel environments.

© ICST Institute for Computer Sciences, Social Informatics and Telecommunications Engineering 2019
Published by Springer Nature Switzerland AG 2019. All Rights Reserved
X. Liu et al. (Eds.): ChinaCom 2018, LNICST 262, pp. 394–404, 2019.
https://doi.org/10.1007/978-3-030-06161-6_39

1.1   Structure of Transmitter and a Receiver

The entire implementation of Softcast includes a transmitter and a receiver. As shown in Fig. 1, the transmitter's data are sent to the encoder through the HDMI interface. Encoder executes discrete cosine transform (DCT) on the data firstly. The DCT transformation can remove the intra correlations of pixel values [6]. Then, the power allocation module calculates the power allocation factor. Power allocation also sends the average power of each transform domain as a digital signal to encoding and modulation. In the digital signal part, CRC and Turbo coding are involved to improve the system's error correction capability. In framing module, in order to combat frequency-selective fading in digital part code blocks, the modulated digital signal and the pseudo-analog signal after power allocation will be interleaved. The synchronization data and pilot data are inserted into the OFDM symbols in the radio frame.

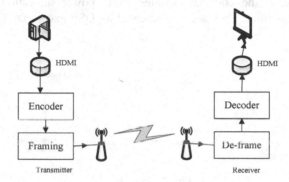

Fig. 1. The pseudo analog video transmission system

1.2   Advantages of DSP Implementation

If each module is implemented in hardware, the entire system would have a much lower flexibility. Although the DSP algorithm is more complex and less efficient, it is more flexible to implement the scheme by DSP programming. An assembly-based power allocation, CRC and Turbo algorithm are designed. Due to the uniqueness of the frame format, a framing algorithm which is multiple look-up tables with linear time complexity is designed. DSP will process the channel estimation section, the de-framing section, and the linear least-squares estimation (LLSE) in the receiver. Digital signals and pseudo-analog signals can be separated by de-framing algorithms. The digital signal can be demodulated based on the result of the channel estimation.

1.3   Dedicated DSP Features

This pseudo-analog video algorithm was ported to a dedicated digital signal processor. A digital signal processor (DSP) is a specialized microprocessor, which is Harvard structure [7]. A specific instruction set is designed for this DSP processor. To improve its performance, single instruction multiple data (SIMD) [8] and very long instruction word (VLIW) [9, 10] are widely used in DSP design. DSP has huge capacity in data processing, as well as powerful memory access(DMA) channels. By configuring the corresponding source address register, target address register, and transmission counter to access data. CPU core and the DMA may operate in parallel. Therefore, the efficiency of the entire computing system is greatly improved.

The rest of this paper is organized as follows: Sect. 2, introduce the structure of the pseudo-analog transmission system. Section 3, discuss the DSP implementation of pseudo-analog algorithm. Section 4, analyze the DSP performance. Section 5, draw the conclusion.

2   Architecture Design

As shown in Fig. 2, the encoder includes DCT, power allocation, CRC, Turbo encoding and modulation. They are implemented by DSP except DCT.

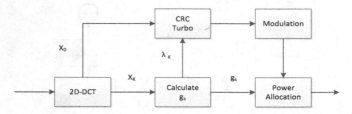

Fig. 2. Encoder

2.1   Power Allocation Algorithm

Firstly, the image data will be converted by 2D-DCT. The DCT transform can eliminate the spatial redundancy of the image data, and can transform the image from the spatial domain to the frequency domain without damaging the image information. The DCT result is a number of matrices (32*32). According to the characteristics of the DCT operation, the matrix stores the DC component and the low frequency coefficient which record most of the information of the image in the top left corner. So the first data stored in each row of each matrix is sent to the Turbo encoding. Turbo encoding has strong anti-interference and anti-fading capabilities. CRC is added before Turbo encoding to ensure error checking at the receiver.

As shown in Fig. 3, during the process of power allocation, the data matrix is divided into L-shaped blocks. There are 15 L-shaped blocks of each matrix. Each L-shaped block corresponding to the corresponding average power λ_k (k = 1, 2... 15), λ_k is the weighted sum of the squares of the corresponding block data. 30 matrices are divided into groups and obtain the average power $\bar{\lambda}_k$. This can ensure that the average power obtained for different coefficients has universality and stability.

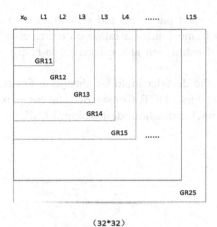

Fig. 3. L-shaped block

The formula for λ_k is as follows:

$$\lambda_k = \frac{\sum_{x \in L_k} x^2}{N} \tag{1}$$

L_k is the data set of the L-shaped blocks, N is the number of the data set L_k. After calculating the corresponding average power for each matrix, the average power $\bar{\lambda}_i$ of the corresponding blocks of the 30 matrices should be calculated, According to the average power $\bar{\lambda}_i$, the power allocation factor g_k can be calculated by the following formula:

$$g_k = \bar{\lambda}_k^{-1/4} \sqrt{\frac{P}{\sum_{k=1}^{15} \sqrt{\bar{\lambda}_k}}} \tag{2}$$

P is the total power [3]. Then we can get the pseudo-analog data Y_K to be transmitted, $Y_K = X_k * g_k$.

2.2 Turbo Encoding and Framing

As already mentioned above, the first data of each matrix is sent to the CRC and Turbo encoding as a digital signal, actually average power is added to it too. Then, the coded digital part code block is subjected to 16QAM constellation point modulation according to the normalized energy. The modulated digital data and the pseudo-analog data are framing together. This algorithm proposes a set of schemes that can be compatible with the traditional physical layer. The real-number signal generated by the encoder on the transmitter does not perform traditional error correction, but directly maps the data to OFDM symbols. A radio frame contains 32 OFDM symbols with 2048 subcarriers per OFDM symbol, each subcarrier can carry 4 bytes of data.

Some OFDM symbols contain pilot data and synchronization data. We use synchronization data to perform frame synchronization on the received frame data. The pilot data are used for channel estimation at the receiver to help restore the digital signal data.

As shown in Fig. 4, the decoder includes channel estimation, de-framing, LLSE, demodulation, Turbo decoding, IDCT. Considering the complexity of related modules, DSP only processes channel estimation, de-framing, LLSE.

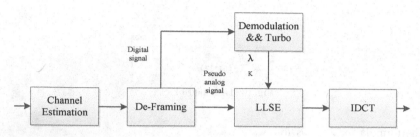

Fig. 4. Decoder

2.3   Channel Estimation

The receiver sends the received frame data to the channel estimation module, and the channel estimation is divided into two parts: one is the least square (LS) channel estimation which estimates the pilot positions only. The advantage of the LS is that the calculation is small and the structure is simple. However, when the channel noise is large, its accuracy will be reduced. Another is minimum mean square error (MMSE) which estimates the position of other subcarriers outside the pilot.

2.4   LLSE Algorithm

The digital signal is used for soft demodulation and Turbo decoding, both are achieved by hardware. Linear least squares estimation (LLSE) is executed on the pseudo-analog data. The average power is got based on the result of the decoding of the digital part. According to the power allocation algorithm, the data received can be described like this: $Y = g_k * X + sigma$. The LLSE calculation formula as follows:

$$X_{llse} = \frac{\bar{\lambda}_k * g_k}{\bar{\lambda}_k * g_k * \bar{\lambda}_k + sigma^2} * Y \tag{3}$$

Y is the data received by the receiver; g_k is the corresponding the power allocation factor; $\bar{\lambda}_k$ is the average power; sigma is the Gaussian white noise signal matrix.

3 DSP Implementation and Optimization

3.1 Swift DSP

A digital signal processor named Swift [11] is designed for the wireless communications. It has a high-performance 32-bit fixed-point arithmetic unit, a high-performance 160-bit SIMD unit, and supports four 40-bit, eight 20-bit, or 16 10-bit vector operations. The instruction set includes 11 control instructions, 19 data transfer instructions, 54 load/store instructions, 49 scalar operation instructions, and 60 vector operation instructions. At the same time, there is a powerful DMA module to transfer data from external storage to RAM.

3.2 Power Allocation and LLSE Implementation

Firstly, the DMA transfers the result of the DCT operation to the RAM and hands it to the DSP for power allocation. The problem that power allocation ported on the DSP platform needs to solve is its unique L-shaped data block. Since the result of the DCT output is a 32*32 matrix data block stored in a row, the DSP needs to read the data which are not sequential storage when calculating the average power. We have designed an algorithm that uses scalar instructions to configure the scalar load's addressing address in advance. This algorithm needs to configure the memory address for each instruction, resulting in a huge amount of code, and a lot of instruction cycles. The number of cycles is roughly 4 times that of the algorithm implemented by the later optimized algorithm. So an algorithm that can use the vector instruction sequential addressing dislocation addition to obtain the power sum is proposed. As shown in Fig. 3, general registers are allocated for each L data block to hold the current power sum. The vector instructions are used to read data by row. The vector instructions need to be completed in multiple cycles. The data processing of the upper and lower rows are made into the instruction pipeline and are relatively parallel.

LLSE is similar to the power allocation. The power allocation factor g_k should be calculated firstly. Because of the result of decoder, the average power λ_i can be obtained directly. According to the formula (2), then the power allocation factor g_k can be calculated easily.

3.3 Turbo Encoding and Channel Estimation

Turbo encoding in high-level language can be achieved through the finite state machine. But it is very complicated to carry on the conditional judgment in DSP. So as shown in Table 1, a Turbo encoding algorithm that sets general register as flag register to realize convolution is designed. The running time of the algorithm has been tested far less than the running time of hardware acceleration.

The LS channel estimation is mainly the multiplication of the second-order complex matrix. Each second-order complex matrix was respectively sent to the specified memory area through DSP. As shown in Fig. 5, After the hardware module calculation is completed, the result is stored into the sequential address space. The basic operation of MMSE channel estimation is similar to LS.

Table 1. Turbo

Algorithm Turbo Convolutional coding   32-bit register

```
1:Load 32-bit data to GR1
2:State{GR2,GR3,GR4};
3:For i in range(31)
4:   GR5 = GR1 && 0x01 , then  GR1 >>1
5:   GR6 = GR5^GR2, GR7 = GR5^GR3
6:   GR6 = GR6^GR3, GR7 = GR7^GR4
7:   Store GR6 as the result of  the convolution
8:   GR4 = GR3, GR3 = GR2, GR2 = GR7
9:end For
```

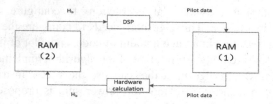

Fig. 5. Channel estimation

3.4 Framing and De-Framing

A single physical frame contains 32 OFDM symbols. One OFDM symbol contains 2048 subcarriers, 1296 effective subcarriers. Every 4 OFDM symbols have 2 OFDM symbols storing pilots. The second and third OFDM symbols store synchronization data. Because the location of the pilot data and synchronization data are fixed, these data are stored into RAM before the project is run. Framing and De-framing need to store or read valid data into the specified RAM space. Firstly, all symbols are divided into two kinds of symbols according to the synchronization pilot data OFDM and normal OFDM symbols. The former is considered to be a scalar symbol and the latter is a vector symbol. The difference between the two symbols is that the framing target address of the scalar symbol is not in order. After data are stored into the target address, scalar symbol needs look up the table to calculate the next target address. The target address of the vector symbol is in order. After the target data stored, the target address can add a fixed number to get the next storage target address.

Since the framing and de-framing algorithms are similar, the framing algorithm will be introduced. In the beginning, a large number of judgment statements are added to distinguish different symbols and memory addresses. The result of this is that the execution cycle is too long. Therefore, an algorithm using multiple lookup tables to obtain the target address of the framing is designed. The precondition of the algorithm is that the source data address is in order. After each fetch, the address can be obtained by adding a fixed number. As shown in Fig. 6, the target address was divided into 32 memory regions corresponding to 32 OFDM symbols. The rules of address changes of these 32 memory regions are set into 32 tables that are stored sequentially. The flag in Fig. 6A is set to let the program distinguish between two symbols. The framing algorithm execution parameters for different symbols are different.

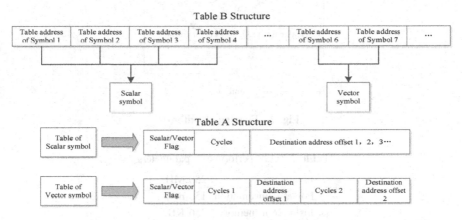

Fig. 6. Table structure

The Table A can configure the framing algorithm for the number of cycles and the target address offset. As shown in Fig. 6B, Table B is the address index of Table A. As flow chart is shown in Fig. 7, obtain the address of the current framing symbol parameter table through Table B. After obtaining the address of Table A, reading the contents of Table A to configure the framing parameters. Transfer data and complete the data loading of the symbol. After 32 cycles, the framing of one radio frame is completed.

4   Performance Analysis

The implementation of the pseudo-analog video transmission algorithm is based on the DSP and the corresponding hardware accelerators. There is a hardware system that contains a DSP. The DSP performance parameters are shown in Table 2. The real video data are used to test the entire system. The resolution of the video is 960*640, Y : U : V is 4:2:0. The video data are sent to the FPGA(Xilinx'-s Virtex − 7V x 475T) board through the HDMI interface. The DSP is connected to the FPGA board through the FMC interface. Furthermore, there is a hardware monitoring module which is used for calculating the operating cycles. Besides this, there is a fully hardware designed Softcast system based on the same FPGA for comparison.

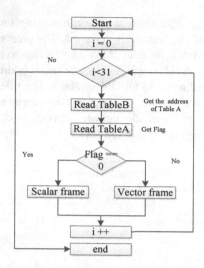

Fig. 7. Framing algorithm

Table 2. DSP performance parameters

Frequency	250 MHz
Power Consumption	150 mW
Instruction memory	256 KB
Data Memory	1 MB

As shown in Table 3, the number of receiver LLSE cycles is less than the number of transmitter PA cycles, because the average power at the receiver can be obtained according to the digital data. Framing and de-framing take more cycles because DSP involves a lot of data transferring. In fact, DMA is not completely parallel to DSP, DMA also consumes cycles.

Table 3. The average number of cycles for each module

Transmitter		Receiver	
Modules	Cycles	Modules	Cycles
PA	1,584,000	LLSE	1,068,000
Turbo	716,000	Channel estimation	721,000
Framing	2,542,000	De-framing	2,244,000
Modulation	520,000	–	–
DMA	64,1000	DMA	64,2000
Total	6,003,000	Total	4,675,000

According to the above table data, when the DSP clock frequency is 200 MHz, the video can reach 30FPS, while the fully hardware designed system performs 30 FPS at 150 MHz. However, the DSP based design has more flexibility, that the algorithm of encoder and decoder can be easily modified as applications demand, and the frame structure can be easily changed as well. It is maintainable and can be optimized. The DSP can also run at higher frequencies with better silicon process, to further improves the processing video resolution.

5 Conclusion

This paper focuses on the design and the DSP implementation of pseudo-analog video transmission algorithm. The video transmission efficiency can be improved through optimizing the assembly algorithm later. Increasing the frequency of the DSP can run the higher resolution video. In the future research, we will continue to improve the system efficiency and optimize DSP instruction set, making the system more stable and efficient.

Acknowledgment. This work was supported in part by the National Natural Science Foundation of China (Nos.61631017 and 61502341); National Science and Technology Support Plan (Grant no. 2012BAH15F03).

References

1. Fan, X., Wu, F., Zhao, D.: D-cast: Dsc based soft mobile video broadcast. In: ACM International Conference on Mobile and Ubiquitous Multimedia (2011)
2. Ding, Z., Wu, J., Yu, W., Han, Y.Q., Chen, X.H.: Pseudo analog video transmission based on LTE physical layer. In: IEEE/CIC International Conference on Communications in China (ICCC) (2016)
3. Jakubczak, S., Katabi, D.: Softcast:one-size-fits-all wireless video. In: ACM SIGCOMM 2010 Conference on Applications, Technologies, Architectures, and Protocols for Computer Communications (2010)
4. Zheng, S., Antonini, M., Cagnazzo, M., Guerrieri, L.: Softcast with per-carrier powerconstrained channels. In: IEEE International Conference on Image Processing (2016)
5. Jiang, J., Xia, P.F., Wu, J., Chen, S., Zhang, B.Y.: Pseudo-analog wireless stereo video transmission in hardware acceleration. In: 2017 9th International Conference on Wireless Communications and Signal Processing (2017)
6. Borkar, S., Chien, A.: The future of microprocessors: communications of the ACM, **54**(67–77) (2011)
7. Zhao, C.X., Wu, J., Chen, X.: Design and implementation of a memory architecture in DSP for wireless communication. In: 2015 10th International Conference on Communications and Networking in China (2015)
8. Derby, J.H., Moreno, J.: A high-performance embedded DSP core with novel SIMD features. In: Acoustics, Speech, and Signal Processing, pp. 301–304 (2003)

9.
  Anderson, T., Bui, D., et al.: A 1.5 GHz VLIW DSP CPU with integrated floating point and fixed point instructions in 40 nm CMOS. In: 2011 20th IEEE Symposium on Computer Arithmetic (ARITH), pp. 82–86 (2011)
10. Fridman, J., Greenfield, Z.: The TigerSHARC DSP Architecture. IEEE Micro 20, 66–76 (2000)
11. Ren, H.Q., Zhang, Z.F., Wu, J.: SWIFT: A computationally-intensive DSP architecture for communication applications. Mob. Netw. Appl. 21(6), 974–982 (2016)

Spectrum Modulation of Smart-Surfaces for Ultra High Frequency Radars

Kai Liu, Yang Wang[✉], Qilong Song, and Xi Liao

School of Communication and Information Engineering,
Chongqing University of Posts and Telecommunications, Chongqing, China
wangyang@cqupt.edu.cn

Abstract. Smart surfaces are reconfigurable meta-materials whose electromagnetic characteristics can be altered for applications such as remote identification, stealth, etc. This paper introduces the spectrum modulation of smart-surfaces for long range radars. Applying controlling signals onto the tunable lumped elements loaded on smart-surfaces, modulations can be achieved on reflecting signals illuminating on smart surfaces. Changing the spectral characteristics of the modulated signals, radar receivers can only detect the limited information of the target. This paper introduces the operation mechanism of smart surfaces and analyzes two specific modulating signals, the square wave signal and the pseudo-random Gaussian white noise signal. The spectrum of reflecting signals will change accordingly, making it difficult for the radar receiver to detect. Simulations and results show that the proposed method can change the reflecting echo of the radar and reduce the probability of the target being detected.

Keywords: Radar Cross-Section (RCS) · Spectrum modulation · Metamaterial · Remote identifications

1 Introduction

The rapid development of radar technology poses a great need in remote identifications of various civil and military objectives. The main objectives of the remote identifications are to reduce technique of the radar cross section (RCS) and to eliminate the possibility of detection, identification and tracking. The radar cross section is related to the shape, size, structure and material of the target, as well as the frequency, polarization and angle of incidence of the incident electromagnetic wave [1]. The traditional method of reducing RCS mainly includes two categories. One is to design a special shape to make the target-reflected radar wave deviate from the radar emission direction [2], but this method will affect the objects aerodynamic performance; another is Radar Absorbing Materials (RAM)

© ICST Institute for Computer Sciences, Social Informatics and Telecommunications Engineering 2019
Published by Springer Nature Switzerland AG 2019. All Rights Reserved
X. Liu et al. (Eds.): ChinaCom 2018, LNICST 262, pp. 405–413, 2019.
https://doi.org/10.1007/978-3-030-06161-6_40

[3,4], which uses coating or structural materials to absorb incident electromagnetic waves. Consequently, the electric field strength of the target scattering can be decreased and the RCS will be reduced. Obviously, this method has good performance in absorbing the incident wave. Unfortunately, it is sensitive to the frequency of incident waves and cannot cope with the complicated electromagnetic environment.

Ultra-High Frequency (UHF, 300 MHz−3000 MHz) radar is mainly used to detect the speed and angle of long-distance moving targets due to its relatively long wavelength. Radars in this frequency band have flexible working modes, large bandwidth (>20%), and are likely to adopt multi-frequency or Linear Frequency Modulation (LFM) waveforms, which poses a considerable challenge for designing RCS reduction materials. In recent years, artificial materials have been used to manufacture ultra-thin and wide-band radar absorbers [5,6]. Electromagnetic metamaterials are a new type of artificially composite electromagnetic materials which can controlled dielectric constants and magnetic permeability [7]. The frequency response of metamaterials is affected by cell size, arrangement, dielectric substrate, incident angle and other factors [8]. A variety of radar absorber structures based on Metamaterials have been proposed in the literatures, such as single-layer resistive reflection absorbers [9], multilayer metal metamaterial absorbers [10], non-magnetic broadband radar absorbers [11]. These structures have their own advantages and disadvantages, each with different absorbing properties and are suitable for a specific operating frequency and working bandwidth. The electromagnetic characteristics will be fixed once the radar absorbing surface based on metamaterials is designed and manufactured. However, smart-surfaces [12] can dynamically adjust their electromagnetic characteristics, and the surface does not only absorb the wave but also changes the electromagnetic characteristics of the electromagnetic wave irradiated on the surface, such as the amplitude, the phase, the direction of the reflected wave. It will affect the judgment of the radar receiver and reduce the possibility that the target is detected, therefore the function of RCS reduction can be realized.

In our previous contribution [13], we introduce active meta-surfaces to mitigate the windmill clutters. This paper studies the operation of smart surfaces for UHF radars. The active-surfaces consist of reflecting elements loaded with voltage-controlled elements. Controlling the voltage applied on the elements, the impedance of surfaces could be ideally switched between zero and extreme high impedance. If selecting appropriate modulating signals, we can design the signals arrived at the radar receivers. The content focus on the spectrum of the reflection signal modulated by the square wave signal and the pseudo-random Gaussian white noise signal. Actively modulating to the incident signal, the spectral characteristics of the reflected signal are changed. Not only the power of the reflection signal within the radar operation frequency band will be reduced, but also the power outside the frequency will be altered. This are possible to provide a new dimension in radar coating technology such as smart friend and foe identification, full-band RCS reduction, remote sensing, etc.

The paper is organized as follows. Section 2 introduces the principle of spectrum modulation based on the smart surfaces, including numerical model and transmission line analysis of Smart Meta-surfaces. Section 3 analyzes the performance of reflecting signals using periodic square wave and Gaussian white noise. Finally, Sect. 4 is the conclusion.

2 Spectrum Modulation of Smart Surfaces

2.1 Numerical Model of Smart Surfaces

Generally, spectrum modulation of smart surfaces is the way that modulating the incident radar signal to make the reflecting signal is different from the original incident signal in spectral characteristics, consequently, reducing the power of radar reception and processing. The modulated signal tends to change in characteristics such as spectral bandwidth, amplitude, and frequency.

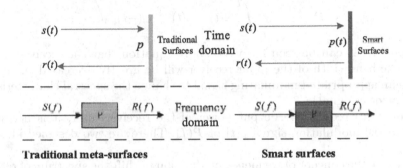

Fig. 1. Principle of traditional meta-surfaces and smart surfaces

Shown in Fig. 1, conventional meta-surfaces reduce the power of reflection signals using a radio wave absorbing structure, which approximately equals to multiply a static coefficient before the reflecting signals. The proposed smart surfaces can change its reflection coefficient dynamically that spectrum modulation can be applied on the reflecting signals. When the incident wave signal $s(t)$ irradiates onto the smart surface, the incident signal will interact with the smart surface and be modulated by the modulating signal $p(t)$ to form a reflecting signal with expected spectral characteristics. The relationship between the three kinds of signals at the time domain can be expressed as:

$$r(t) = s(t) \times p(t) \tag{1}$$

According to the Fourier transform, their relationship at the frequency domain can be depicted as:

$$R(f) = S(f) \otimes P(f) \tag{2}$$

where $s(t)$, $p(t)$, and $r(t)$ represent the frequency response of the incident signal, the modulating signal and the reflecting signal, respectively. $\otimes$ indicates a convolution operation.

For radar receivers, it is assumed that the widely used matched filter is adopted which can be expressed as $w(t) = s^*(-t)$. Then, the incident signal can reach the radar receiver without loss in an ideal situation, therefore $r(t) = s(t)$, and the output signal power of the radar receiver is expressed as:

$$P = \frac{1}{T}|\sigma_0| |\int_0^T [s(t) \otimes s^*(-t)]^2 dt| \tag{3}$$

where σ_0 denotes a time-invariant coefficient which represents the combined effect of the gains obtained in various stages of signal transmission, propagation, reflection and reception. After passing through the modulation board, the output signal of radar receiver is first multiplied by the modulating signal. Then the Eq. (3) becomes:

$$P = \frac{1}{T}|\sigma_0| |\int_0^T [s(t) \times p(t) \otimes s^*(-t)]^2 dt| \tag{4}$$

Therefore, it can be seen from the above equation that the received power within the bandwidth of the radar receiver will be greatly decreased as long as there is an appropriated modulating signal to change the spectral characteristics of the incident signal.

In order to reduce the output power of the receiver filter, it is necessary to design the modulating signal $p(t)$ or $P(f)$. There are two designed ideas as follows:

(1) Using the concept of frequency modulation to shift the spectrum of reflecting waves to other frequencies.

From the traditional analog modulation technique, it is known that the modulating signal will have a shifting effect on the spectrum when one signal is used to modulate another signal. Based on the similar principle, we can design an appropriated $p(t)$ so that the incident signal $s(t)$ can be shifted in the spectrum.

Assume that p(t) is a single frequency signal:

$$p(t) = exp(2\pi f_1 t) \tag{5}$$

Its spectrum $P(f)$ is equivalent to an impulse response at the corresponding frequency f_1. Then, after the incident signal is modulated, the echo signal spectrum is shifting f_1 on the basis of the incident signal frequency. The frequency of the reflecting signal can be shifted outside the bandwidth of the radar receiver when f_1 is designed reasonable so that the output power of the receiver is minimized.

(2) Using the idea of a filter to design a set of modulating signals, making the power spectrum of the reflected wave approximates Gaussian white noise.

The design of the modulating signal to convert $R(f)$ into a white noise-like spectrum and make R(f) does not contain the original incident signal, as shown in the following equation:

$$R(f) = \xi_1 + j\xi_2 \tag{6}$$

$$p(t) = \frac{r(t)}{s(t)} = \frac{if ft[R(f)]}{s(t)} \tag{7}$$

where ξ_1 and ξ_2 are both mutually independent, zero-mean normal distribution variables.

Thus, the information of $s(t)$ has disappeared in $r(t)$, therefore the radar receiver cannot recover the information of the incident signal and consequently reduce the probability of the target being found.

2.2 Transmission Line Analysis of Smart Meta-Surfaces

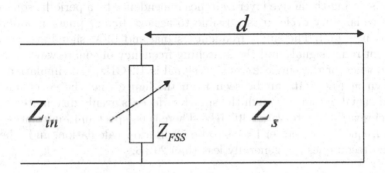

Fig. 2. Transmission line model of smart meta-surfaces

Similar to the conventional modulated surface, the proposed smart meta-surface consists of periodic structures. However, active components are introduced in modulated signals to manage the impedance of the array. The arrays consist one layer of active wideband frequency selective surfaces (FSS) and a perfect electric conductor (PEC) ground plane. Shown in Fig. 2, assume the active FSS and the ground plane are distanced by d, the transmission line model of the active layer can be described using the time varying impendence of FSS $Z_{FSS}(t)$, the medium impendence Z_s, and the propagation constant β, across whose input terminals is placed a time-varying admittance $Z_{in}(t)$, where $Z_{in}(t)$ defined as Z_{in}

$$Z_{in}(t) = \frac{jZ_{FSS}(t)Z_s tan(\beta d)}{Z_{FSS}(t) + jZ_s tan(\beta d)} \tag{8}$$

The results in reflection coefficient, which is related to $Z_{in}(t)$ by

$$\rho(t) = \frac{Z_{in} - Z}{Z_{in}(t) + Z} \tag{9}$$

where $Z = Z_0 \cos \theta$ or $Z = Z_0/ \cos \theta$ for parallel or perpendicular polarization, respectively, and Z_0 represents the impedance of free space. Since $Z_{FSS}(t)$ is arbitrary and dependent on the status of active controlling components, then $\rho(t)$ is generally complex number whose absolute value is from 0 to 1. Assume one radar signal $s(t)$ arrive at the smart-surfaces, the reflected radar signal can be described by

$$r(t) = s(t) \times \rho(t) \tag{10}$$

If $\rho(t)$ can be controlled by the periodic square wave or designed in the form of Gaussian white noise, the reflecting signal will difficult to be detected by the radar receiver.

3   Simulation and Result Analysis

3.1   Binary Phase Modulation

First considering the active layer switches impendence by a periodic square wave with an equal duty cycle, it is possible to achieve binary phase modulation of the incident signal. The single frequency signal and LFM signal are considered as incident radar signals and the switching frequency of square wave is 20 MHz. The frequency of the single frequency signal is 1.2 GHz, the simulation results are shown in Fig. 3. It can be seen from the figure that the spectrum of the reflected signal passing through the smart surfaces is evenly distributed to other frequencies with an interval of 40 MHz. There is no spectrum information at the original frequency, its power loss is to be 52 dB after calculation, and other RCS reduction techniques are generally less than 20 dB.

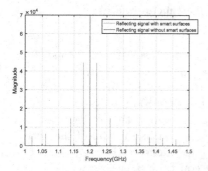
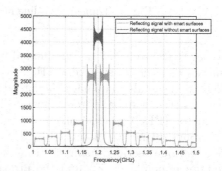

Fig. 3. Single frequency signal   **Fig. 4.**   LFM   signal(switching frequency is 20 MHz)

Figure 4 shows the spectral variation of the reflected signal when the incident signal is the LFM signal. The center frequency is 1.2 GHz and the bandwidth is 30 MHz of the incident signal, the switching frequency of the square wave is still 20 MHz. It can be seen from the simulation results that reflected signal produces a lot of the same reflected signal spectrum distributed near the original signal, and there is still have spectrum of the reflected signal within the bandwidth of the original signal, the power loss is only 5 dB. In this case, false echo information appears around the original spectrum, which also affects the judgment of the radar receiver.

When the switching frequency is increased to 30 MHz, and the simulation result is shown in Fig. 5. It is obvious that the reflecting signal with smart surfaces not only has the decreased magnitude but also the spectrum transferred to other frequencies, and the spectrum has no component at frequency of incident signal. In this case, the power loss of the reflected signal is increased to 24 dB. Consequently, proposed method can get better performance when the switching frequency is greater than the bandwidth of LFM signal owing to the reflecting signal can be moved outside the bandwidth of the radar receiver.

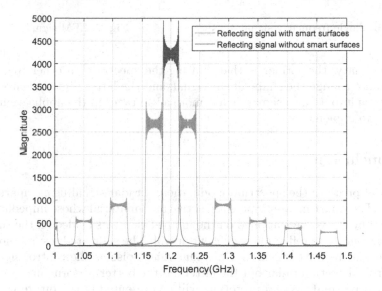

Fig. 5. LFM signal(switching frequency is 30 MHz)

3.2   Pseudo-Random Gaussian Modulation

If $\rho(t)$ can be designed in the form of Gaussian white noise, another way to modulate the incident signal can be obtained. The simulation results are as follows. Figure 6 shows the case that the incident signal is single frequency. After passing through the smart-surfaces, the reflected signal becomes the noise signal and the signal energy is evenly distributed within the frequency band. When such the form of reflected signal arrives at the radar receiver, it will be ignored as the noise signal. And the probability of the target is detected is greatly reduced. Similarly, when the LFM signal as the incident signal irradiates on the smart surfaces, the reflecting signal also becomes a form of a noise signal. The simulation result is shown in Fig. 7. The radar receiver does not detect echo information about the target.

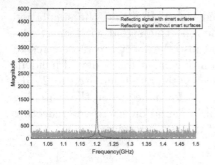

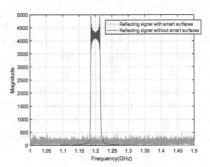

Fig. 6. Single frequency signal   **Fig. 7.** LFM signal

In summary, the former is that shifting the spectrum to other frequencies but does not change the shape of the signal; and the latter directly converts the radar signal into a noise signal, it has more advantages in the application but it is harder to implement.

4   Conclusion

This paper presents the spectrum modulation of radar signal using smart meta-surfaces. The smart meta-surfaces are active metamaterial whose impedance can controlled by the voltage applied on the lumped elements loaded on the surfaces, so we can control its reflection coefficient at will. The simulated results show that using the square wave signal or Gaussian white noise as the control signal can escape from detection radar but the latter has the better performance. The more comprehensive analysis and prototype will be presented in the future work.

References

1. Mahafza, B.R.: Radar Systems Analysis and Design Using MATLAB[M], 3rd edn. CRC Press, Inc. (2013)
2. Knott, E.F., Tuley, M.T., Shaeffer, J.F.: Radar Cross Section[M], 2nd edn. Norwood, Artech House (1993)
3. Fante, R.L., Mccormack, M.T.: Reflection properties of the Salisbury screen[J]. Antennas Propag. IEEE Trans. **36**(10), 1443–1454 (1988)
4. Tennant, A., Chambers, B.: A single-layer tunable microwave absorber using an active FSS[J]. Microw. Wirel. Compon. Lett. IEEE **14**(1), 46–47 (2004)
5. Paquay, M., Iriarte, J.C., Ederra, I., et al.: Thin AMC structure for radar cross-section reduction[J]. IEEE Trans. Antennas Propag. **55**(12), 3630–3638 (2007)

6. Pang, Y., Cheng, H., Zhou, Y., et al.: Ultrathin and broadband high impedance surface absorbers based on metamaterial substrates[J]. Opt. Express **20**(11), 12515–12520 (2012). https://doi.org/10.1364/OE.20.012515
7. Calaoz, C., Itoh, T.: Electromagnetic Metamaterials: Transmission Line Theory and Microwave Applications[M]. Wiley (2006)
8. Bozzi, M., Perrengrini, L., Weinzierl, J., et al.: Design, fabrication and measurement of frequency-selective surfaces[J]. Opt. Eng. **39**(8), 2263–2269 (2000)
9. Li, W., Chen, M., Zeng, Z., et al.: Broadband composite radar absorbing structures with resistive frequency selective surface: optimal design, manufacturing and characterization[J]
10. Xu, H., Bie, S., Xu, Y., et al.: Broad bandwidth of thin composite radar absorbing structures embedded with frequency selective surfaces[J]. Compos. Part A: Appl. Sci. Manuf. **80**, 111–117 (2016)
11. Wang, C., Chen, M., Lei, H., et al.: Radar stealth and mechanical properties of a broadband radar absorbing structure[J]. Compos. Part B: Eng. **123**, 19–27 (2017)
12. Martinez, I., Panaretos, A., Werner, D.: Reconfigurable ultrathin beam redirecting metasurfaces for RCS reduction[J]. IEEE Antennas Wirel. Propag. Lett. **99**, 1 (2017)
13. Wang, Y., Yun, M., Lin, F., et al.: Windmill clutter mitigation using active metasurfaces[C]. In: IEEE International Symposium on Antennas and Propagation and USNC/URSI National Radio Science Meeting, pp. 1495–1496. IEEE (2017)

Two Stage Detection for Uplink Massive MIMO MU-SCMA Systems

Cuitao Zhu[⊠], Ning Wei, Zhongjie Li, and Hanxin Wang

Hubei Key Laboratory of Intelligent Wireless Communication, College of
Electronics and Information Engineering, South-Central University for
Nationalities, Wuhan, China
Cuitaozhu@mail.scuec.edu.cn

Abstract. In this paper, we propose a two stage multiuser detection scheme: a linear pre-filtering and iteration removal based message passing algorithm (RM-MPA). As the first stage of the proposed detection, a linear pre-filtering based on Richardson method is proposed to avoid the complicated matrix inversion in an iterative way. Meanwhile, we also present a sub-optimum relaxation parameter to Richardson for lower-complexity. Then the RM-MPA is used for multiuser decoding, which compared the decoding advantages of users and sorted users according to decoding advantages. After the each iteration, the users with higher decoding advantages directly are decoded and removed. The removed users do not participate in the subsequent iterations, therefore, the complexity of subsequent iterations decrease gradually. Simulation results show that the proposed two stages multiuser detection can significantly reduce the computational complexity with better symbol error rate performance.

Keywords: Massive MIMO · Non-orthogonal multiple access
Message passing algorithm

1 Introduction

Future fifth generation (5G) wireless networks are expected to support massive number of connected devices, non-orthogonal multiple access (NOMA) has attracted much attention as an enabling technology [1]. SCMA, as one of the competitive NOMA schemes, has been proposed to address the above requirement. To further improve the spectral efficiency, SCMA can be combined with Massive MIMO technology [2]. However, low complexity and efficient multiuser (MU) detection is one of the vital issues for combining SCMA with Massive MIMO in 5G, which need to be further addressed.

Due to the sparse structure of SCMA spreading signature, several iterative multiuser detection schemes for uplink SCMA systems [3, 4], which are based on message passing algorithm (MPA), were proposed to efficiently approximate to the maximum a

This work was partially supported by the National Natural Science Foundation of China under grant No. 61671483 and the Natural Science Foundation of Hubei province under grant No. 2016CFA089.

X. Liu et al. (Eds.): ChinaCom 2018, LNICST 262, pp. 414–424, 2019.
https://doi.org/10.1007/978-3-030-06161-6_41

posteriori (MAP). Message passing schedule strategy plays an important role in MPA-based detection schemes, which influences the convergence rate. References [5–7] proposed various strategies to further reduce the complexity of the MPA detector. Reference [8] proposed a resource-selection based MPA detector, in which resources with well-conditioned channels are selected to perform the jointly Gaussian algorithm in order to achieve satisfactory performance with low complexity. However, the complexity of the improved methods based on MPA is still high for SCMA with massive MIMO.

To address this issue, we propose an iteration removal based message passing algorithm with linear pre-filtering, which included the sequentially operating two stages. The first is a linear pre-filtering based on Richardson method to avoid the complicated matrix inversion in each iteration. We also propose a suboptimum relaxation parameter to Richardson for lower-complexity. After linear pre-filtering, RM-MPA was used for multiuser decoding based on the filtered signals. We propose a simple and novel method to compare the decoding advantages of users and sort the users according to decoding advantages. Then, n users with higher decoding advantages were directly decoded and removed after each iteration. Moreover, the removed users did not participate in the subsequent iteration, so the complexity of subsequent iterations decreased gradually.

2 System Model

For an uplink multiuser massive MIMO-NOMA system, in which a single base station with a uniform linear array of N_r antennas serves J users with single-antenna, J users share K physical resource elements, the j-th user transmits binary bits are mapped into a K dimensional complex codeword x_j, $X_j = \left[x_{1,j}, x_{2,j}, \ldots, x_{K,j}\right]^T$, for an SCMA encoder, it selected from the corresponding SCMA codebook χ_j. We assume that the power of a codeword is normalized to be 1 for all users, i.e., $\|x_j\|^2 = 1$. The system model is depicted in Fig. 1.

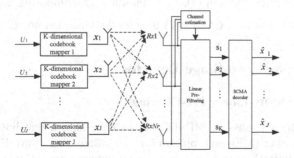

Fig. 1. System model

Since the SCMA codewords have sparse nature, the structure of SCMA encoder can be interpreted by an indicator matrix $F = [f_1, f_2, \ldots, f_J]$, in which $f_j = [f_{1,j}, f_{2,j}, \ldots, f_{K,j}]^T$ is the encoder indicator vector of user j, if $f_{k,j} = 1$, indicates k - th resource occupied by user j. Let d_f denotes the number of nonzero entries in each row of matrix F, i.e. the number of users sharing same resource, d_j denotes the number of nonzero entries in each column of matrix F, which corresponds to the number of resources allocated to the j-th user. In addition, let $\varsigma_j = \{k | f_{k,j} = 1\}$ be the set of resource, which indicates the resources allocated to the j - th user, and $\zeta_k = \{j | f_{k,j} = 1\}$ be the set of users, which indicates the users sharing the k-th resource, respectively. The K-dimensional codewords are transmitted to the receiver through K resources. The received signal at the k - th resource $y_k \in C^{N_r \times 1}$ can be expressed as follows:

$$y_k = H_k x_{k,[d_f]} + n_k, \quad k = 1, 2, \ldots, K \tag{1}$$

Where,

$$y_k = \left[y_k^1, y_k^2, \ldots, y_k^{N_r} \right]^T \tag{2}$$

$$H_k = \begin{bmatrix} h_{k,j_1}^1 & h_{k,j_2}^1 & \cdots & h_{k,j_{df}}^1 \\ h_{k,j_1}^2 & h_{k,j_2}^2 & \cdots & h_{k,j_{df}}^2 \\ & & \cdots & \\ h_{k,j_1}^{N_r} & h_{k,j_2}^{N_r} & \cdots & h_{k,j_{df}}^{N_r} \end{bmatrix} \tag{3}$$

$$x_{k,[df]} = \left[x_{k,j_1}, x_{k,j_2}, \ldots, x_{k,j_{df}} \right]^T \tag{4}$$

$$n_k = \left[n_k^1, n_k^2, \ldots, n_k^{N_r} \right]^T \tag{5}$$

In (2), the element y_k^p denotes the received signal by the p - th antenna from the k - th resource. In (3), the element $h_{k,j}^p$ is the channel gain between user j and the p - th antenna at the k - th resource, In (4), $x_{k,[d_f]}$ indicates superimposing users codeword at the k - th resource, and $n_k \sim CN(0, \sigma^2 I)$ denotes the Gaussian noise.

3   The Proposed Two Stage Detector

3.1   Low-Complexity Linear Pre-Filtering

Theoretical results for massive MIMO systems have shown that linear pre-filtering, such as matched filter (MF), and minimum mean square error (MMSE) filter, are able to achieve near-optimal performance in the large-antenna limit. However, these linear pre-filters involve troublesome matrix inversion. For this reason, we propose an approach based on Richardson method for a linear pre-filtering, and present a simple approach to determine the suboptimum relaxation parameter.

Let W_k be a linear filtering weighting matrix of size $N_r \times d_f, 1 \leq k \leq K$, where the entries depend on the perfectly known channel gains at the receiver. The weighting matrix of MMSE defined as follow:

$$W_k = \left(H_k^H H_k + \frac{d_f}{SNR} I_{d_f} \right)^{-1} H_k^H \tag{6}$$

In (6), I_{d_f} is an identity matrix of size $d_f \times d_f$. MMSE based linear filtering matrix was applied to (1), filtered signal model can be derived as

$$s_k = \left(H_k^H H_k + \frac{d_f}{SNR} I_{d_f} \right)^{-1} H_k^H y_k = P^{-1} \hat{y}_k \tag{7}$$

where, S_k is the filtered signal from k - th resource, $P = \left(H_k^H H_k + \frac{d_f}{SNR} I_{d_f} \right)$ is Hermitian positive definite matrix. Owing to the direct computation of the inverse P^{-1} requires high complexity, so we propose the Richardson method [9] to efficiently solve (7) to reduce the complexity. The method using the Richardson method to solve (7) is given by

$$s_k^{(i+1)} = s_k^{(i)} + \gamma \left(\hat{y}_k - P s_k^{(i)} \right) \tag{8}$$

Where, $s_k^{(i)}$ denotes the solutions at the i - th iteration, γ is the relaxation parameter, which plays an important role in the convergence and convergence rate.

3.2   Selection of the Suboptimal Relaxation Parameter

In this section, we will deduce the suboptimum relaxation parameter based on approximation error. By formula (7) and (8), we can have the following equation:

$$\begin{aligned} s_k^{(i+1)} - s_k &= s_k^{(i)} + \gamma \left(\hat{y}_k - P s_k^{(i)} \right) - s_k \\ &= (I_k - \gamma P) \left(s_k^{(i)} - s_k \right) \\ &= \cdots \\ &= B^{i+1} \left(s_k^{(0)} - s_k \right) \end{aligned} \tag{9}$$

Where, $B = I_K - \gamma P$ is the iteration matrix of Richardson algorithm, from (9) we can derive the approximation error induced by the Richardson algorithm, which can be evaluated as the following:

$$\left\| s_k^{(i+1)} - s_k \right\|_2 = \left\| B^{i+1} \right\|_F \left\| s_k^{(0)} - s_k \right\|_2 \tag{10}$$

From (10), the approximation error induced by Richardson algorithm is mainly affected by iteration matrix B and initial solution $S_k^{(0)}$. The relaxation parameter γ only affected by matrix B, if $\left\|B^{i+1}\right\|_F$ was minimized, the value of γ is optimum. However, the direct computation of $\left\|B^{i+1}\right\|_F$ is complicated, since the iteration matrix B is a random matrix and it is hard to obtain the joint distribution of all the elements. Fortunately, it has been proved in [9] that when i goes infinity, we have the following equation

$$\lim_{i\to\infty}\left\|B^{i+1}\right\|_F^{1/(i+1)} = \rho(B) \tag{11}$$

Where, $\rho(B)$ is the spectral radius of iteration matrix B. From (11), we can regard $\rho(B)$ as the asymptotic convergence rate of Richardson algorithm, and a smaller $\rho(B)$ will lead to a faster convergence rate. Let $\lambda_{\max}$ and $\lambda_{\min}$ denote the largest and the smallest eigenvalues of matrix P. According to the definition $B = I_K - \gamma P$, the spectral radius $\rho(B)$ will depend on γ as

$$\rho(B) = \max(|1 - \gamma\lambda_{\max}|,\ |1 - \gamma\lambda_{\min}|) \tag{12}$$

From (12), when $|1 - \gamma\lambda_{\min}| = |1 - \gamma\lambda_{\max}|$, the smallest spectral radius can be achieved. Since γ should also satisfy $0 < \gamma < 2/\lambda_{\max}$, to guarantee the convergence of Richardson algorithm. So, we can conclude that when $\gamma\lambda_{\max} - 1 = 1 - \gamma\lambda_{\min}$, the smallest spectral radius $\rho(B)$ can be achieved and the fastest convergence rate, which means the optimal relaxation parameter γ_{opt} is

$$\gamma_{opt} = \frac{2}{\lambda_{\max} + \lambda_{\min}} \tag{13}$$

From (13), we can see that to obtain the optimal relaxation parameter γ_{opt}, we need to know a priori $\lambda_{\max}$ and $\lambda_{\min}$, which in practice is difficult. Therefore, directly using (13) to determine the optimum relaxation parameter γ_{opt} is very sophisticated in practical massive MIMO systems. To address this issue, we propose a suboptimal relaxation parameter γ'_{opt} with a negligible performance loss.

In massive MIMO systems, the elements of channel matrix $H_k \in C^{N_r \times d_f}$ are independent and identically distributed complex Gaussian random variables, the sample covariance matrix Φ is [10]

$$\Phi = \frac{1}{d_f}\sum_{v=1}^{d_f} h_v h_v^H = \frac{1}{d_f} H_k H_k^H \tag{14}$$

Where, h_v denotes the v column vector of matrix H_k, when $\frac{N_r}{d_f} \to \beta \in (0,\ \infty)$, according to the law of Bai-Yin, the largest and the smallest eigenvalues of matrix Φ meet as

$$\lim_{N_r \to \infty} \lambda'_{\max}(\Phi) = \sigma^2 \left(1 + \sqrt{\beta}\right)^2 \tag{15}$$

$$\lim_{N_r \to \infty} \lambda'_{\min}(\Phi) = \sigma^2 \left(1 - \sqrt{\beta}\right)^2 \tag{16}$$

Where, σ_z^2 is the noise power. In massive MIMO systems, the number of antennas N_r is very large, the smallest and the largest eigenvalues of matrix P will converge to the deterministic values as

$$\lambda_{\max} \approx d_f \left(1 + \sqrt{\beta}\right)^2 + \frac{d_f}{SNR} \tag{17}$$

$$\lambda_{\min} \approx d_f \left(1 - \sqrt{\beta}\right)^2 + \frac{d_f}{SNR} \tag{18}$$

By replacing (17) and (18) in (13), the suboptimum relaxation parameter γ'_{opt} is given by

$$\gamma'_{opt} = \frac{2}{d_f \left[(1 + \sqrt{\beta})^2 + (1 - \sqrt{\beta})^2\right] + 2\frac{d_f}{SNR}} \tag{19}$$

From (19), we can observe that γ'_{opt} depends on the number of receive antennas N_r and d_f, which are deterministic and known after the massive MIMO configuration has been fixed. Thus, we do not need to re-compute γ'_{opt} as H varies. Furthermore, γ'_{opt} does not need to know a priori $\lambda_{\max}$ and $\lambda_{\min}$ in comparison to γ_{opt} given by (13).

4   Low Complexity SCMA Decoder Based on Iteration Removal

4.1   Decoding Advantage Analysis

The procedure of RM-MPA can be explained by the factor graph, which is a bipartite graph including resource nodes r_k and user nodes u_j. In general, MPA consists of the exchange of messages between the nodes of a factor graph. There are two approaches for scheduling messages in MPA, i.e., parallel schedule strategy and serial schedule strategy. In this paper, we tackle the problem of how to schedule message for the serial schedule strategy that results in effectively reduced complexity and the best convergence rate.

Let $I_{r_k \to u_j}(x_j)$ and $I_{u_j \to r_k}(x_j)$ be the message propagated along branch from r_k to u_j, and from u_j to r_k, respectively. In the each iteration, messages are first sent from user nodes to resource nodes. Each resource nodes then computes extrinsic messages and sends back to the user nodes based on the previously received information. These user nodes-to-resource nodes messages will then be used to calculate the new resource nodes-to-user nodes messages in the next iteration. Thus, the two messages generating functions are defined as follows:

$$
I_{r_k \to u_j}^t (x_j) = \sum_{\sim x_j} \left\{ \frac{1}{\sqrt{2\pi}\sigma} \exp\left(-\frac{1}{2\sigma^2} \left\| y_k - \sum_{m \in \zeta_k} h_{k,m} x_{k,m} \right\|^2 \right) \right.
$$

$$
\left. \times \prod_{\substack{l \in \zeta_k/j \\ l < j}} I_{u_l \to r_k}^t (x_l) \prod_{\substack{l \in \zeta_k/j \\ l > j}} I_{u_l \to r_k}^{t-1} (x_l) \right\}
\tag{20}
$$

$$
I_{u_j \to r_k}^t (x_j) = normalize \left(\prod_{m \in \varsigma_j/k} I_{r_m \to u_j}^t (x_j) \right)
\tag{21}
$$

Where, $l < j$ and $l > j$ represent the user l before and after the user j, respectively. The $\zeta_{k/j}$ denotes to remove the user j from the set ζ_k, t is the iteration index, at the maximum iterations $t_{\max}$, the decoding soft output can be expressed as

$$
Q(x_j) = \prod_{k \in \varsigma_j} I_{r_k \to u_j}^{t_{\max}} (x_j)
\tag{22}
$$

As mentioned above, the computational complexities of the original MPA based on serial schedule strategy, relate to $t_{\max} \cdot K M^{d_f}$. Furthermore, for original MPA, we can see that codewords of all users are decoded together until the maximum itcrations $t_{\max}$ is reached. Thus, the schedule strategy for message update has a disadvantage that all users participate in the process of the each iteration, and the parameter d_f is same in the each iteration, this leads to high computational complexity in the each iteration. To tackle this issue, we present a low complexity MPA detector based on iteration removed.

Owing to the proper property of message passing serial schedule strategy, RM-MPA based on serial schedule strategy can decode and remove the users with higher decoding advantages after the first iteration. Furthermore, when the resources are handled later, the users have the higher reliability of the message carried by them. Thus, later users occupy resources, which have the higher reliability of the decoding soft output in the each iteration. So, there is an inherent decoding advantage between users in RM-MPA based multiuser detection schemes. Let $e_{k,j}^t(x_j)$ denotes the number of external message used by the user j, which occupied the k-th resource update message at the t iterations. That can be defined as follow

$$
e_{k,j}^t(x_j) = \left\{ n = \sum f_{l,j} \mid f_{k,j} = 1 \,\& f_{l,j} = 1 \,, \, j' \neq j, l < k \right\}
\tag{23}
$$

From (23), the user j utilizes the total number of external messages produced in this iteration at the t iterations.

$$e_j^t(x_j) = \sum_{k \in \varsigma_j} e_{k,j}^t(x_j) \tag{24}$$

From (24), the bigger $e_j^t(x_j)$, the more number of external messages utilized by user j in this iteration. Then, user j has higher decoding reliability of after this iteration, and has higher decoding advantage to other users. According to the sequence of resources being processed in the RM-MPA, we will define an advantage level for each resource. The higher level, the more decoding advantage, decoding reliability is greater. For instance, resource k, the corresponding advantage level is l_k, then, the decoding advantage level of user j defined as

$$a_j = \sum_{k \in \varsigma_j} f_{k,j} \cdot l_k \tag{25}$$

According to (25), users sorted by decoding advantage in descending order, then the order of decoding reliability of users can be obtained after the cach iteration, and decode and remove the decoding reliable users, then do next iteration.

4.2   Iteration Removed Strategy

(1)   Iteration Removed Strategies Based on Decoding Advantages of Users

As mentioned above, greater decoding advantage of users, the higher decoding reliability in RM-MPA-based detection schemes. Therefore, users can be sorted according to decoding advantage from high to low. After each iteration, the n users with higher decoding advantages were directly decoded and removed, and these users no longer to participate in the next iteration. Thus, after the each iteration the number of users is cut down n, $1 \leq n \leq K/N$. After t_{max} iterations, no matter how much remaining user, all decoded together. Thus, it can be seen that more users removed, the lower complexity of decoding algorithm after the each iteration.

(2)   Iteration Removed Strategies Based on Orthogonal User Grouped

In the system with SCMA, if each user occupies a data stream layer, then there are maximum users C_K^N, meanwhile, there are maximum C_K^N indicator vectors f_j in the indicator matrix F. If indicator vectors are grouped according to mutually orthogonal, all indicator vectors can be divided into $C_K^N/(K/N) = C_{K-1}^{N-1}$ groups. In this paper, we assume that K can be divisible by N. There are K/N indicator vectors in each group, and the indicator vector f_j of users is mutually orthogonal in a group. If there are same number of users in each group, and users occupy all resources, this grouping is called an orthogonal complete user group, otherwise, called an orthogonal non-complete.

On former orthogonal users grouped, iteration decoding and removal based on Orthogonal users grouped describe as follow: optionally selecting a group of orthogonal user after the each iteration, the users in a group are directly decoded and removed, and these users no longer to participate in the next iteration. To an orthogonal complete user group, selected randomly an orthogonal users group, then the users are decoded and removed.

5 Numerical Results

This section presents the performances of the proposed linear pre-filtering and multiuser decoding schemes based on decoding advantages of users (RM-MPA-URS) and Orthogonal user grouped (RM-MPA-OUGR) for uplink massive MIMO MU-SCMA systems.

Figure 2 shows the Symbol Error Rate (SER) performance of the MMSE linear pre-filtering based on Richardson algorithm and Neumann series, and exact matrix inversion method over Rayleigh fading channels. The SCMA system is with 16-QAM, $\beta = 8$, $N_r = 64$, $d_f = 8$.

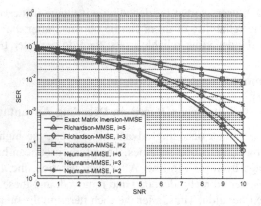

Fig. 2. SER performance comparison

We can see from the Fig. 2 that the SER performance of the MMSE linear pre-filtering based on Richardson algorithm outperforms Neumann series with 5 iterations, 2 iterations and 3 iterations, and the more advantage with increase of SNR. On the other hand, the SER performance of the MMSE linear pre-filtering based on Richardson algorithm and exact matrix inversion method have become much closer with the increase of iterations, however, the computational cost of the exact matrix inversion method is higher than that in Richardson algorithm.

Figure 3 shows the SER performance of the two stages detections: MMSE based on exact matrix inversion method with PMPA, and MMSE based on Richardson with PMPA, and MMSE based on Richardson with RM-MPA. The SCMA system with 16-QAM, $N_r = 64$, $J = 48$, over Rayleigh fading channels.

We can see from the Fig. 3 that the SER performance of the exact matrix inversion-MMSE-PMPA is the best with 10 iterations. However, the exact matrix inversion-

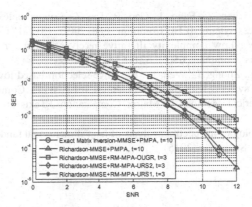

Fig. 3. SER performance comparison of two stages detections

MMSE-PMPA requires higher computational costs. The SER performance of Richardson-MMSE-PMPA is lower than that obtained by the exact matrix inversion-MMSE-PMPA. Meanwhile, for the SER performance comparison of Richardson-MMSE-RM-MPA with two removed strategies with 10 iterations, Fig. 3 shows the SER performance of the Richardson-MMSE-RM-MPA-URS is the best. However, Richardson-MMSE-RM-MPA-OUGR requires the lowest computational cost.

6 Conclusions

In this paper, we proposed a novel Richardson-MMSE-RM-MPA detector for uplink SCMA systems with massive MIMO. In order to detect the transmitted signals of users with lower computational cost, the proposed detection sequentially employs a linear pre-filtering and iteration removal based on Message Passing Algorithm. Simulation results show that the Richardson-MMSE-RM-MPA-OUGR detector can observably reduce the computational complexity with the performance degraded unnoticeably.

References

1. Dai, L., Wang, B., Yuan, Y., Han, S., Chih-Lin, I., Wang, Z.: Non-orthogonal multiple access for 5G: solutions, challenges, opportunities, and future research trends. IEEE Commun. Mag. **53**(9), 74–81 (2015)
2. Liu, T.T., Li, X.M., Qiu, L.: Capacity for downlink massive MIMO MU-SCMA system. In: *Proceedings of the* IEEE WCSP, pp. 1–5. Nanjing, China (2015)
3. Du, Y., Dong, B.H., Chen, Z., Fang, J., Wang, X.J.: A fast convergence multiuser detection scheme for uplink SCMA systems. IEEE Wirel. Commun. Lett. **5**(4), 388–391 (2016)
4. Du, Y., et al.: Low-complexity detector in sparse code multiple access systems. IEEE Commun. Lett. **20**(9), 1812–1815 (2016)
5. Du, Y., Dong, B., Chen, Z., Fang, J., Yang, L.: Shuffled multiuser detection schemes for uplink sparse code multiple access systems. IEEE Commun. Lett. **20**(6), 1231–1234 (2016)

6. Du, Y., Dong, B., Chen, Z., Fang, J., Gao, P., Liu, Z.: Low-complexity detector in sparse code multiple access sytems. IEEE Commun. Lett. **20**(9), 1812–1815 (2016)
7. Wei, F., Chen, W.: Low complexity ilterative receiver design for sparse code multiple access. IEEE Trans. Commun. **65**(2), 621–634 (2016)
8. Tian, L., Zhao, M., Zhong, J., Wen, L.: Resource-selection based low complexity detector for uplink SCMA systems with multiple antennas. IEEE Wirel. Commun. Lett. (2017)
9. Bjorck, A.: Numerical Methods in Matrix Computations. Texts in Applied Mathematics (2015)
10. Bai Z.D.: Methodologies in spectral analysis of large dimensional random matrices, a review [m]// Advances In Statistics. (2015)

Robust Spectrum Sensing for Cognitive Radio with Impulsive Noise

Liping Luo[✉]

College of Information Science and Engineering,
Guangxi University for Nationalities, Nanning, China
lping.luo@gmail.com

Abstract. Spectrum sensing plays an important role in cognitive radio. In this paper, a robust spectrum sensing method via empirical characteristic function based on goodness-of-fit testing is proposed, named as ECF detector. The test statistic is derived from the empirical characteristic function of the observed samples, thus the secondary users do not require any prior knowledge of the primary signal and the noise distribution. Extensive simulations are performed and compared with the existing spectrum sensing methods, such as energy detector, eigenvalue-based detector, AD detector and KS detector. The results show that, the proposed ECF detector can offer superior detection performance under both the Gaussian noise and the impulsive noise environments.

Keywords: Cognitive radio · Empirical characteristic function Goodness-of-fit testing · Impulsive noise · Spectrum sensing

1 Introduction

Cognitive radio is a spectrum shared technology to alleviate the spectrum shortage problem and to improve the spectrum utilization. In cognitive radio networks, the secondary users are able to access the licensed spectrum without causing interference to the primary users. Spectrum sensing plays an important role to detect the presence of the primary user.

Based on the local observations, a variety of spectrum sensing methods have been proposed in [1–6]. The energy detector [1,2] is one of the most commonly employed spectrum sensing schemes, since it does not require any prior information of the primary signal. The problem of energy detector is that it requires the

This research was supported by the Natural Science Foundation of China under Grant No.61762011, Guangxi Natural Science Foundation under Grant No.2016GXNSFAA380091, Guangxi One Thousand Young and Middle-Aged College and University Backbone Teachers Cultivation Program.

X. Liu et al. (Eds.): ChinaCom 2018, LNICST 262, pp. 425–437, 2019.
https://doi.org/10.1007/978-3-030-06161-6_42

knowledge of the noise variance which is estimated by some estimation procedure. The energy detector is fairly sensitive to the estimated error, named noise uncertainty. To circumvent this difficulty, assuming no prior knowledge of the primary signal and noise variance at the secondary users, some eigenvalue-based spectrum sensing methods based on the generalized likelihood ratio test (GLRT) paradigm [3–6] have been proposed which utilize the eigenvalues of the sample covariance matrix of the received signal vector. However, the aforementioned spectrum sensing methods are developed under the Gaussian noise assumption. Their performance degrades substantially in the presence of non-Gaussian noise.

Although it is common to justify the Gaussian assumption on noise with the central-limit theorem, it also frequently deals with noise environments where the non-Gaussian (impulsive or heavy-tailed) nature of noise prevails in the system. For instance, car ignition noise, moving vehicles, electromagnetic interference, man-made noise, and arc generating circuit components are impulsive noise sources, which are encountered in metropolitan areas [7]. In indoor wireless communication, devices with electromechanical switches such as electrical motors in elevators refrigerators units and printers are also considered as impulsive noise. Furthermore, microwave ovens, cash register receipt printers, gas-powered engines produce impulsive noise on frequency bands which coincide with the operating frequencies of current cellular and wireless local area networks [8,9]. Under such impulsive noise circumstances, the spectrum sensing algorithms developed under Gaussian noise may be highly susceptible to a severe degradation of the performance.

To cope with the impulsive noise, using the goodness of fit testing, robust spectrum sensing methods have been proposed in [10–14]. They consider the spectrum sensing as a nonparametric hypothesis testing problem. When there is no primary signal, the local observations are a sequence of samples drawn independently from the noise distribution. To detect the presence of the primary user, it is equivalently to test whether the observations are drawn from the noise distribution. Depending on how to measure the distance between the sample distribution and noise distribution, Anderson-Darling(AD) detector [10,11] and Kolmogorov-Smirnov(KS) detector [12–14] are developed for spectrum sensing. Although they can work under both the Gaussian and the impulsive noise environments, the performance of the AD detector degrades significantly when uses the empirical cumulative distribution function (CDF) to instead of the real CDF. In addition, the performance of the KS detector depends on the number of noise-only samples and observations.

The motivation of this work is to provide a robust spectrum sensing method for cognitive radio under both Gaussian noise and impulsive noise environments. The secondary users do not require any prior knowledge of the primary signal and the noise distribution. A common model that is symmetric α-stable ($S\alpha S$) distribution is used for the impulsive noise with Gaussian noise as special case. In this paper, another goodness-of-fit testing method based on empirical characteristic function (c.f.) is applied, then an ECF detector is proposed, which is available under both Gaussian noise and impulsive noise environments. Moreover, an ECF-based moment estimator is employed to estimate the noise parameters. Thus, the ECF detector does not require the prior information of primary signal and noise parameters. The performance of the method is evaluated through Monte Carlo simulations. It is shown that the proposed spectrum sensing method outperforms the exist detectors.

The remainder of this paper is organized as follows. In Sect. 2, the spectrum sensing problem and the $S\alpha S$ distribution for impulsive noise model are introduced. In Sect. 3, an ECF detector is proposed for spectrum sensing, and a moment estimator based on ECF is developed to estimate the impulsive noise parameters. In Sect. 4, simulation results are illustrated to compare the proposed ECF detector with some existing spectrum sensing methods. Finally, the paper is concluded in Sect. 5.

2 System Model and Preliminary Knowledge

2.1 Spectrum Sensing Problem

In cognitive radio, the secondary users require to detect whether the primary user exists or not based on the local observed samples. The spectrum sensing problem can be formulated as the following binary hypothesis test:

$$
\begin{aligned}
H_0 : \quad & y(n) = \nu(n) \\
H_1 : \quad & y(n) = hs(n) + \nu(n), \qquad n = 0, 1, ..., N-1
\end{aligned}
\tag{1}
$$

where $y(n)$ is the observed samples at the secondary user. N is the number of the observations. $s(n)$ denotes the primary signal, $\nu(n)$ is a class of impulsive noise including Gaussian noise as a special case. Without loss of generality, the signal and the noise are assumed to be complex-valued. h denotes the channel coefficient between the primary user and the secondary user, which is assumed to be constant during the sensing interval.

2.2 $S\alpha S$ Distribution

For the impulsive noise, the $S\alpha S$ distribution, which is a generalization of Cauchy, Lévy and Gaussian distribution, has been proved to be the most accurate model [15]. A real-valued $S\alpha S$ random variable with zero mean, denoted by $S_\alpha(\gamma, 0)$, has a characteristic function given by [16]:

$$
\phi_{\alpha,\gamma}(\omega) = e^{-\gamma|\omega|^\alpha},
\tag{2}
$$

where α is the characteristic exponent, and γ is a quantity analogous to variance called dispersion. The characteristic exponent α in (2) controls the heaviness of the pdf tails ($0 < \alpha \leq 2$), a small positive value of α indicates severe impulsiveness, while a value of α close to 2 indicates a more Gaussian type of behavior. Although the characteristic function of $S\alpha S$ has a simple form, there are only two distributions-Gaussian ($\alpha = 2$) and Cauchy ($\alpha = 1$)-for which the probability density function (pdf) can be expressed in terms of elementary functions. For all other $\alpha's$, the pdf does not have a closed form.

For complex-valued $S\alpha S$ random variables, the original definition can be found in [16,17]. For simplicity, the equivalent and explicitly expression is given by

$$\nu = \nu_R + j\nu_I \tag{3}$$

where ν_R and ν_I are independent and identically distributed (i.i.d.) random variables with $S_\alpha(\frac{\gamma}{2}, 0)$ distribution. Then, ν follows complex $S\alpha S$ distribution denoted by $\nu \sim \mathcal{CS}_\alpha(\gamma, 0)$. The noise parameters $\boldsymbol{\theta} = (\alpha, \gamma)$ are not known prior for the secondary users, thus are required to estimate for spectrum sensing.

2.3   Goodness-of-Fit Testing

From a mathematical statistics point of view, the classical detection algorithms such as energy detector, matched filter detector and cyclostationarity feature detector fall into the category of parametric hypothesis testing. If the assumption about the parameters related to the known patterns is invalid or not accurate, their performance will deteriorate. Thus, to improve the detection performance, goodness-of-fit testing, a nonparametric hypothesis testing method, is employed for spectrum sensing [10–14].

EDF-Based Goodness-of-Fit Testing Empirical distribution function (EDF) test is a widely used goodness-of-fit testing in statistics. EDF test measures the distance between two distributions $F_Y(y)$ and $F_0(y)$, which are CDF of the observations and the noise respectively. Some EDF-based goodness-of-fit tests have been proposed in the literature of mathematical statistics, including the AD test and KS test.

The AD test is a generalization of the Cramer-von Mises test and defined by

$$D_Y^{AD} = N \int_{-\infty}^{+\infty} (F_Y(y) - F_0(y))^2 \Phi(F_0(y)) dF_0(y) \tag{4}$$

where $\Phi(F_0(y))$ is a nonnegative weight function given by $\Phi(F_0(y)) = (F_0(y)(1 - F_0(y)))^{-1}$. In [10,11], an AD detector is proposed based on the AD test, the test statistic is

$$A_c^2 = -\frac{\sum\limits_{n=1}^{N} (2n-1)(\ln Z_n + \ln(1 - Z_{N+1-n}))}{N} - N \tag{5}$$

where $Z_n = F_0(y_n)$, y_n is the observed sample at the secondary user. From (5), it is seen that the secondary user requires the closed form of noise CDF $F_0(y)$ for AD detector. However, for impulsive noise, the $S\alpha S$ distribution does not have a closed form of the CDF except for Gaussian and Cauchy distribution. To make the AD detector be available, it has to use the empirical CDF instead of the real CDF, i.e., $Z_i = \hat{F}_0(y_i)$.

The KS test first forms the empirical CDF from $(z_1, z_2, ..., z_N)$ and the noise-only samples $(\nu_1, \nu_2, ..., \nu_{N_0})$ as follows,

$$\hat{F}_1(z) \triangleq \frac{1}{N} \sum_{n=1}^{N} \mathbb{I}(z_n \leq z)$$

$$\hat{F}_0(\nu) \triangleq \frac{1}{N_0} \sum_{n=1}^{N_0} \mathbb{I}(\nu_n \leq \nu) \tag{6}$$

where z_n is the function of the observed samples y_n.

The KS test statistics is the largest absolute difference between the two CDFs given by

$$D_Y^{KS} = \max |\hat{F}_1(z_n) - \hat{F}_0(z_n)| \tag{7}$$

In [12], two types of KS detector are proposed for spectrum sensing. One is the KS-mag detector, in which z_n is the magnitude of the observations, i.e., $z_n = |y_n|$. The other is KS-qua detector, in which z_n is formed by the real part and the imaginary part of y_n, i.e., $z_i = \Re[y_n], z_{N+n} = \Im[y_n]$.

ECF-Based Goodness-of-Fit Testing Similar to EDF tests, empirical c.f. (ECF) tests measure the distance between the empirical c.f. of the observations and the noise c.f.. The advantages of ECF-based goodness-of-fit testing includes the mathematical tractability of the $S\alpha S$ distribution and favorable properties such as strong consistency and asymptotic normality [18]. Thus, an ECF detector will be proposed according to the ECF-based goodness-of-fit testing in this paper.

3   Proposed ECF Detector for Spectrum Sensing

3.1   ECF Detector

According to the ECF-based goodness-of-fit testing, the spectrum sensing problem in (1) can be reformulated as:

$$\begin{aligned} H_0 : & \quad \phi_y(\omega; \boldsymbol{\theta}) = \phi_\nu(\omega; \boldsymbol{\theta}) \\ H_1 : & \quad \phi_y(\omega; \boldsymbol{\theta}) \neq \phi_\nu(\omega; \boldsymbol{\theta}) \end{aligned} \tag{8}$$

where $\phi_y(\omega; \boldsymbol{\theta})$ and $\phi_\nu(\omega; \boldsymbol{\theta})$ represent the characteristic functions of the observations and the noise respectively. For a complex-valued y and $\omega = \omega_R + j\omega_I$, the characteristic function of the observations is defined by [19]

$$\phi_y(\omega; \boldsymbol{\theta}) = \mathbb{E}\{e^{j\Re[\bar{\omega}y]}\} = \mathbb{E}\{e^{j(\omega_R \Re[y] + \omega_I \Im[y])}\} \triangleq C(\omega; \boldsymbol{\theta}) + jS(\omega; \boldsymbol{\theta}) \tag{9}$$

where $\bar{\omega}$ is the conjugate of ω. Similarly, the characteristic function of the noise is

$$\phi_\nu(\omega; \boldsymbol{\theta}) = \mathbb{E}\{e^{j(\omega_R \nu_R + \omega_I \nu_I)}\} \triangleq C_\nu(\omega; \boldsymbol{\theta}) + jS_\nu(\omega; \boldsymbol{\theta}) \tag{10}$$

where $\Re(\cdot)$ and $\Im(\cdot)$ represent the real and imaginary part of y. $C(\omega; \boldsymbol{\theta})$ and $S(\omega; \boldsymbol{\theta})$ denote the real and imaginary part of the c.f.

For N i.i.d. observations $y_1, ..., y_N$, the empirical c.f. is

$$\hat{\phi}_y(\omega) = \frac{1}{N} \sum_{n=1}^{N} e^{j\Re[\bar{\omega}y_n]} = \frac{1}{N} \sum_{n=1}^{N} e^{j(\omega_R \Re[y_n] + \omega_I \Im[y_n])} \triangleq C_N(\omega) + jS_N(\omega) \quad (11)$$

where

$$C_N(\omega) = \Re[\hat{\phi}_y(\omega)] = \frac{1}{N} \sum_{n=1}^{N} \cos(\omega_R \Re[y_n] + \omega_I \Im[y_n]),$$

$$S_N(\omega) = \Re[\hat{\phi}_y(\omega)] = \frac{1}{N} \sum_{n=1}^{N} \sin(\omega_R \Re[y_n] + \omega_I \Im[y_n]).$$

Since $\hat{\phi}_y(\omega)$ is the consistent estimate of $\phi_y(\omega; \boldsymbol{\theta})$, it holds that $\mathbb{E}[C_N(\omega)] = C(\omega; \boldsymbol{\theta}), \mathbb{E}[S_N(\omega)] = S(\omega; \boldsymbol{\theta})$.

For m points $\bar{\omega} = [\omega_1, ..., \omega_m]$, according to (10) and (11), we define

$$\boldsymbol{\xi}_0(\boldsymbol{\theta})^T = [C_\nu(\omega_1; \boldsymbol{\theta}), ..., C_\nu(\omega_m; \boldsymbol{\theta}), S_\nu(\omega_1; \boldsymbol{\theta}), ..., S_\nu(\omega_m; \boldsymbol{\theta})]$$
$$\boldsymbol{\xi}_N^T = [C_N(\omega_1), ..., C_N(\omega_m), S_N(\omega_1), ..., S_N(\omega_m)] \quad (12)$$

Then

$$\boldsymbol{\xi}_N - \boldsymbol{\xi}_0(\boldsymbol{\theta}) = \begin{bmatrix} C_N(\omega_1) - C_\nu(\omega_1; \boldsymbol{\theta}) \\ \vdots \\ C_N(\omega_m) - C_\nu(\omega_m; \boldsymbol{\theta}) \\ S_N(\omega_1) - S_\nu(\omega_1; \boldsymbol{\theta}) \\ \vdots \\ S_N(\omega_m) - S_\nu(\omega_m; \boldsymbol{\theta}) \end{bmatrix} \quad (13)$$

Let $\boldsymbol{\Omega}(\bar{\omega})$ be the covariance matrix of $\sqrt{2N}(\boldsymbol{\xi}_N - \boldsymbol{\xi}_0(\boldsymbol{\theta}))$, it is derived that $\boldsymbol{\Omega}(\bar{\omega})$ contains the following elements,

$$\Omega_{jk}(\omega, \boldsymbol{\theta}) = \begin{cases} C(\omega_j + \omega_k; \boldsymbol{\theta}) + C(\omega_j - \omega_k; \boldsymbol{\theta}) - 2C(\omega_j; \boldsymbol{\theta})C(\omega_k; \boldsymbol{\theta}) & (1 \leq j, k \leq m) \\ C(\omega_j - \omega_k; \boldsymbol{\theta}) - C(\omega_j + \omega_k; \boldsymbol{\theta}) - 2S(\omega_j; \boldsymbol{\theta})S(\omega_k; \boldsymbol{\theta}) & (m+1 \leq j, k \leq 2m) \\ S(\omega_j + \omega_k; \boldsymbol{\theta}) - S(\omega_j - \omega_k; \boldsymbol{\theta}) - 2C(\omega_j; \boldsymbol{\theta})S(\omega_k; \boldsymbol{\theta}) & (1 \leq j \leq m, m+1 \leq k \leq 2m) \end{cases}$$
$$(14)$$

where $\omega_j = \omega_{j-m}$ for $m+1 \leq j \leq 2m$. Since $C_N(\omega)$ and $S_N(\omega)$ are consistent estimate of $C(\omega; \boldsymbol{\theta})$ and $S(\omega; \boldsymbol{\theta})$, $\Omega_{jk}(\omega, \boldsymbol{\theta})$ can be replaced by $\hat{\Omega}_{jk}(\omega)$ which is defined in terms of $C_N(\omega)$ and $S_N(\omega)$, that is

$$\hat{\Omega}_{jk}(\omega) = \begin{cases} C_N(\omega_j + \omega_k) + C_N(\omega_j - \omega_k) - 2C_N(\omega_j)C_N(\omega_k) & (1 \leq j, k \leq m) \\ C_N(\omega_j - \omega_k) - C_N(\omega_j + \omega_k) - 2S_N(\omega_j)S_N(\omega_k) & (m+1 \leq j, k \leq 2m) \\ S_N(\omega_j + w_k) - S_N(\omega_j - w_k) - 2C_N(\omega_j)S_N(\omega_k) & (1 \leq j \leq m, m+1 \leq k \leq 2m) \end{cases}$$
$$(15)$$

According to [20], an ECF detector is proposed, the test statistic is given by the following quadratic form:

$$T_N = 2N(\boldsymbol{\xi}_N - \boldsymbol{\xi}_0(\boldsymbol{\theta}))^T \boldsymbol{\Omega}^{-1}(\bar{\omega})(\boldsymbol{\xi}_N - \boldsymbol{\xi}_0(\boldsymbol{\theta})) \overset{H_1}{\underset{H_0}{\gtrless}} \tau \qquad (16)$$

where τ is the threshold selected according to the given false alarm probability η.

$$\Pr[T_N > \tau | H_0] = \eta \qquad (17)$$

However, it is observed that $\boldsymbol{\xi}_0(\boldsymbol{\theta})$ is dependent on the unknown noise parameters. Thus, it is required to develop an estimation procedure to obtain the information of the noise parameters before spectrum sensing.

3.2 Noise Parameters Estimation Based on the Empirical c.f.

To estimate the noise parameters, it requires a sequence of noise-only samples. Essentially, this is the same requirement as the energy detector and the KS detector. For the energy detector, it needs to estimate the nose variance. Since the pdf of $S\alpha S$ is not expressible in closed form, the conventional methods such as the maximum likelihood estimation (MLE) cannot be applied. Based on the empirical c.f., some methods was proposed in mathematical literatures [21–23], of which Press's method, named as moment estimator, can offer an explicit estimator while only need minimal computation. In this paper, the Press's method is extended to the complex $S\alpha S$ random variables.

Assume that N_0 independent noise-only samples $\{\nu_i\}_{i=1}^{N_0}$, the empirical c.f. is given by

$$\hat{\phi}_\nu(\omega) = \frac{1}{N_0} \sum_{i=1}^{N_0} e^{j\Re[\bar{\omega}\nu_i]} = \frac{1}{N} \sum_{i=1}^{N_0} e^{j(\omega_R \Re[\nu_i] + \omega_I \Im[\nu_i])} \qquad (18)$$

Note that for any ω, $|\hat{\phi}_\nu(\omega)|$ is bounded above by unity. Hence, all moments of $|\hat{\phi}_\nu(\omega)|$ are finite. Moreover, for any fixed ω, $\hat{\phi}_\nu(\omega)$ is the sample average of i.i.d. random variables. Thus, by the law of large numbers, $\hat{\phi}_\nu(\omega)$ is a consistent estimator of $\phi_\nu(\omega, \boldsymbol{\theta})$. Based on the empirical c.f., consistent estimator can be developed to estimate the noise parameters $\boldsymbol{\theta}$.

For all α, γ, $\log|\phi_\nu(\omega, \boldsymbol{\theta})| = -\gamma|\omega|^\alpha$. Choose two different nonzero values ω_a, ω_b,

$$-\gamma|\omega_a|^\alpha = \log|\phi_\nu(\omega_a, \boldsymbol{\theta})|$$
$$-\gamma|\omega_b|^\alpha = \log|\phi_\nu(\omega_b, \boldsymbol{\theta})| \qquad (19)$$

Since $\hat{\phi}_\nu(\omega)$ is consistent estimate of $\phi_\nu(\omega; \boldsymbol{\theta})$, it can use $\hat{\phi}_\nu(\omega_a), \hat{\phi}_\nu(\omega_b)$ to replace $\phi_\nu(\omega_a, \boldsymbol{\theta})$ and $\phi_\nu(\omega_b, \boldsymbol{\theta})$ respectively. Solving these two equations simultaneously for α and γ, it gives

$$\hat{\alpha} = \frac{\log \left| \frac{\log |\hat{\phi}_\nu(\omega_a)|}{\log |\hat{\phi}_\nu(\omega_b)|} \right|}{\log \left| \frac{|\omega_a|}{|\omega_b|} \right|} \tag{20}$$

$$\hat{\gamma} = e^{\log(-\log|\hat{\phi}_\nu(\omega_a)|) - \hat{\alpha}\log|\omega_a|} \quad \text{or} \quad \hat{\gamma} = e^{\log(-\log|\hat{\phi}_\nu(\omega_b)|) - \hat{\alpha}\log|\omega_b|} \tag{21}$$

In order to improve the accuracy of the estimation, one can choose multiple couples of (ω_a, ω_b), by averaging, a more accurate estimation value $\hat{\boldsymbol{\theta}} = (\hat{\alpha}, \hat{\gamma})$ can be obtained. Then, $\boldsymbol{\xi}_0(\boldsymbol{\theta})$ in (16) can be calculated by $\boldsymbol{\xi}_0(\hat{\boldsymbol{\theta}})$.

Therefore, the ECF detector involves the following two phases.

Estimation phase: the secondary user employs N_0 independent noise-only samples $(\nu_1, ..., \nu_{N_0})$ to estimate the noise parameters (α, γ) using (20) and (21).

Spectrum sensing: the secondary user collects N observed samples $(y_1, ..., y_N)$, computes $\boldsymbol{\xi}_N - \boldsymbol{\xi}_0(\hat{\boldsymbol{\theta}})$ in (13) and $\boldsymbol{\Omega}^{-1}(\bar{\omega})$ in (15), then forms the corresponding test statistics according to (16). The threshold τ is determined by (17). If $T_N > \tau$, it declares the primary users' presence; otherwise no primary user is present.

4 Simulation Results and Discussion

In this section, the performance of the proposed ECF detector is presented and compared with the energy detector, eigenvalue-based detector, AD detector and KS detector under both the Gaussian noise and the impulsive noise environments.

Simulation parameters setup: In the sequel, the parameters of the complex impulsive $\alpha-$stable noise are set to $\gamma = 1, \alpha = 1.5$, the complex Gaussian noise is zero mean unit variance, which is equivalent to $\gamma = \frac{\sqrt{2}}{2}, \alpha = 2$. The noise parameters are not known prior and required to estimate. The desired false alarm probability is fixed to $\eta = 0.05$. The primary users employ 16-QAM modulated signal. For the impulsive noise, the average SNR is defined as the ratio of the transmit power of the signal to the dispersion of the impulsive noise, i.e. $SNR = \frac{P_s}{\gamma}$ [24]. The independent flat Rayleigh fading channels are simulated between the transmitter-receiver pairs.

The choice of $\bar{\omega}$: In [23, 25, 26], it is shown that the estimate accuracy and detection performance are dependent on the choices of ω. In [26], the authors have demonstrated that the optimal choice of ω is $(0.8, 0.9, 0.85, 0.95)$ by simulations. Thus, we also choose the complex value of $\bar{\omega} = (0.8 + j0.8, 0.9 + j0.9, 0.85 + j0.85, 0.95 + j0.95)$.

Performance analysis of the ECF detector under impulsive noise: In Figs. 1 and 2, the detection performance of the proposed ECF detector are demonstrated under nonfading and Rayleigh fading scenarios. The numbers of the noise-only samples for estimation and the observations for detection are $N_0 = 500, N = 500$. The average SNR is -9dB. Figure 1 shows the ROC curves (P_d versus P_f). It is seen from the simulation results, the sensing performance over Rayleigh fading channel is worse than that over nonfading environment as expected.

The detection performance of the ECF detector versus exponential parameter α are shown in Fig. 2. It is observed that the detection probability becomes larger as α increasing. Since α characterizes the impulsiveness, α close to 2 indicates a more Gaussian of behavior. This implies that the ECF detector can achieve better detection performance under Gaussian noise, while worse performance for severe impulsive noise.

Performance comparison with other methods under impulsive noise: In Fig. 3, the detection performance of the proposed ECF detector is compared with the energy detector, eigenvalue-based detector and AD detector under both the Gaussian noise and the impulsive $S\alpha S$ noise environments. In order to make a fair com-

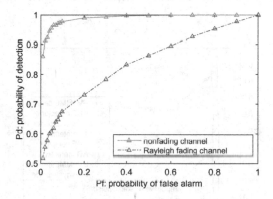

Fig. 1. ROC curves of the proposed ECF detector over nonfading and Rayleigh fading channel, with $N_0 = 500$ noise-only samples and $N = 500$ observations, the average SNR is $= -9dB$, impulsive noise parameters $\gamma = 1, \alpha = 1.5$

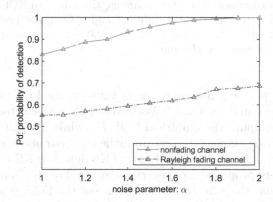

Fig. 2. Detection probability of the ECF detector versus noise parameter α over nonfading and Rayleigh fading channel, with $\gamma = 1, N_0 = 500$ noise-only samples and $N = 500$ observations, $P_f = 0.05$, SNR$= -9dB$, 16-QAM modulated signal.

parison, the ED is performed based on the estimated noise parameters. For the eigenvalue-based detector, the noise-only samples are also employed for detection. The observations $y_1, ..., y_N$ are divided into L groups with M samples. Assume that $N_0/M = \delta$ is an integer, all of the noise-only samples and the observations can form a $(L + \delta) \times M$-Dimension signal matrix. Then making eigen-decomposition on the sample covariance matrix and computing the ratio of the maximum eigenvalue to the minimum eigenvalue, the test statistic of the detector, denoted by EV-MME detector, is obtained. As shown, although the EV-MME detector outperforms the ECF detector under the Gaussian noise, while the ED and the EV-MME detector exhibit a severe degradation of performance, even become too weak to detect the primary signal when the noise is impulsive. Moreover, for the AD detector, it is also inferior to the ECF detector in performance under both Gaussian noise and impulsive noise. Thus, the ECF detector is more robust than the above methods under the impulsive environment.

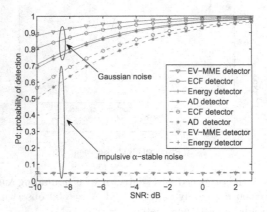

Fig. 3. Detection performance comparison among the proposed ECF detector, energy detector, EV-MME detector and AD detector under Gaussian noise and impulsive noise, with parameters $\gamma = 1, \alpha = 1.5$, with $N_0 = 500$ noise-only samples and $N = 500$ observations, 16-QAM modulated signal

The detection performance comparison between the ECF detector and the KS detector is shown in Figs. 4 and 5. For two sample KS detector, it needs noise-only samples to compute the empirical CDF $\hat{F}_0$, while for ECF detector, these noise-only samples are employed to estimate the noise parameters. Since 16-QAM signal is complex-valued, in [12], two kinds of KS detector: KS-mag detector and KS-qua detector are proposed for spectrum sensing. For the Gaussian noise, it is seen from Fig. 4 that the detection performance of the ECF detector is between the KS-mag detector and the KS-qua detector when the observed samples is $N = 500$. As the increasing of the samples, the detection performance is improved. When $N = 1500$, the ECF detector outperforms the two KS detectors, which implies that the detection performance of the ECF detector is improved more quickly than the KS detectors as the increasing of the observed samples. For the impulsive noise, as shown in Fig. 5, the detection probability of the ECF detector is also higher than those of the KS detectors when $N = 1500$. Therefore, the ECF detector is better than the KS detector with large number of samples. Moreover, the threshold of the ECF detector can be easily calculated from P_f, thus more extensive Monte Carlo simulations are avoided.

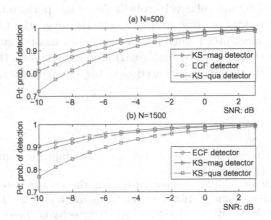

Fig. 4. Detection performance comparison between the proposed ECF detector and KS detector with $N_0 = 500$ independent noise-only samples under Gaussian noise environment. Observation samples: (a)$N = 500$, (b)$N = 1500$

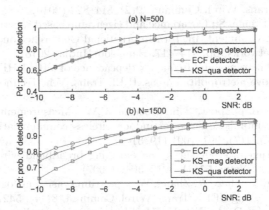

Fig. 5. Detection performance comparison between the proposed ECF detector and KS detector under impulsive noise environment, with $N_0 = 500$ independent noise-only samples, (a)$N = 500$, (b)$N = 1500$ observation samples, impulsive noise parameters are $\gamma = 1, \alpha = 1.5$.

5   Conclusion

In this paper, a robust ECF detector is proposed in the presence of impulsive noise. Extensive simulations are performed and compared with other methods. Among the comparisons between the ECF detector and other detectors, it is shown that the eigenvalue-based detector and the energy detector which are proposed under Gaussian noise cannot be available under the impulsive noise environment. Using goodness-of-fit testing, the AD detector, KS detector and ECF detector can provide relatively robust detection performance under both the Gaussian and impulsive noise environments. However, the ECF detector has strong advantages including higher performance and mathematical tractability of the impulsive noise modeled by $S\alpha S$ distribution. Therefore, the ECF detector are more powerful than the EDF based detector involved the AD and the KS detector.

References

1. Urkowitz, H.: Energy detection of unknown deterministic signals. Proc. IEEE **55**(4), 523–531 (1967)
2. Digham, F., Alouini, M.-S., Simon, M.K.: On the energy detectionof unknown signals over fading channels. IEEE Trans. Commun. **55**(1), 21–24 (2007)
3. Zeng, Y., Liang, Y.-C.: Eigenvalue-based spectrum sensing algorithms for cognitive radio. IEEE Trans. Commun. **57**(6), 1784–1793 (2009)
4. Wang, Pu, Fang, Jun, Li, Hongbin: Multiantenna-assisted spectrum sensing for cognitive radio. IEEE Trans. Veh. Technol. **59**(4), 1791–1800 (2010)
5. Abbas, T., Masoumeh, N.-K.: Multiple antenna spectrum sensing in cognitive radios. IEEE Trans. Wirel. Commun. **9**(2), 814–823 (2010)
6. Althaf, C.I.M., Prema, S.: Covariance and eigenvalue based spectrum sensing using USRP in real environment. In: 10th International Conference on Communications Systems and Networks, pp. 414–417. Bangalore, India (2018)
7. Skomal, E.N.: The range and frequency dependence of VHF-UHF man-made radio noise in and above metropolitan areas. IEEE Trans. Veh. Technol. **19**(2), 213–221 (1970)
8. Blackard, K.L., Rappaport, T.S., Bostian, C.W.: Measurements and models of radio frequency impulsive noise for indoor wireless communication. IEEE J. Sel. Areas Commun. **11**(7), 991–1001 (1993)
9. Kuran, M.S., Tugcu, T.: A survey on emerging broadband wireless access technologies. Comput. Netw. **51**(1), 3013–3046 (2007)
10. Wang, H., Yang, E., Zhao, Z., Zhang, W.: Spectrum sensing in cognitive radio using goodness of fit testing. IEEE Trans. Wirel. Commun. **8**(11), 5427–5430 (2009)
11. Sheers, B., Teguig, D., Le Nir, V.: Modified Anderson-Darling detector for spectrum sensing. Electron. Lett. **15**(25), 2156–2158 (2015)
12. Zhang, G., Wang, X., Liang, Y.-C., Liu, J.: Fast and robust spectrum sensing via Kolmogorov-Smirnov test. IEEE Trans. Commun. **58**(12), 3410–3416 (2010)
13. Arshad, K., Moessner, K.: Robust spectrum sensing based on statistical tests. IET Commun. **7**(9), 808–817 (2013)
14. Lekomtcev, D., Marsalek, R.: Spectrum sensing under transmitter front-end constraints. In: 23rd International Conference on Systems, Signals and Image Processing, pp. 1–4. Bratislava, Slovakia (2016)

15. Shao, M., Nikias, C.: Signal processing with fractional lower order moments: stable processes and their applications. Proc. IEEE **81**(7), 986–1010 (1993)
16. Nikias, C., Shao, M.: Sibgnal Processing with Alpha-stable Distributions and Applications. Wiley, New York (1995)
17. Rajan, A., Tepedelenlioglu, C.: Diversity combining over Rayleigh fading channels with symmetric alpha-stable noise. IEEE Trans. Wirel. Commun. **9**(9), 2968–2976 (2010)
18. Brcich, R.F., Iskander, D.R., Zoubir, A.M.: The stability test for symmetric alpha-stable distributions. IEEE Trans. Signal Process. **53**(3), 977–986 (2005)
19. Andersen, H.H, Hoejbjerre, M., Soerensen, D., et al.: Linear and Graphical Models: For the Multivariate Complex Normal Distribution. Springer, New York (1995)
20. Fan, Y.: Goodness-of-fit tests for a multivariate distribution by the empirical characteristic function. J. Multivar. Anal. **62**, 36–63 (1997)
21. Press, S.J.: Estimation in univariate and multivariate stable distributions. J. Am. Stat. Assoc. **67**(340), 842–846 (1972)
22. Koutrouvelis, I.A.: Regression-type estimation of the parameters of stable laws. J. Am. Stat. Assoc. **75**(372), 918–928 (1980)
23. Feuerverger, A., Mcdunnough, P.: On the efficiency of empirical characteristic function procedures. J. R. Stat. Soc. **43**(1), 20–27 (1981)
24. Tsihrintzis, G., Nikias, C.: Performance of optimum and suboptimum receivers in the presence of impulsive noise modeled as an alpha-stable process. IEEE Trans. Commun. **43**(4), 904–914 (1995)
25. Koutrouvelis, I.A.: A goodness-of-fit test of simple hypotheses based on the empirical characteristic function. Biometrika **67**(1), 238–240 (1980)
26. Ilow, J., Hatzinakos, D.: Applications of the empirical characteristic function to estimation and detection problems. Elsevier Signal Process. **65**(2), 199–219 (1998)

Resource Allocation for Mobile Data Offloading Through Third-Party Cognitive Small Cells

Qun Li, Zheng Yin, and Ding Xu[✉]

Nanjing University of Posts and Telecommunications, Nanjing, China
xuding@ieee.org

Abstract. Mobile data offloading is considered as an effective way to solve the network overloading issue. In this paper, we study the mobile data offloading problem through a third-party cognitive small cell providing data offloading service to a macrocell. Particularly, four scenarios, namely, successive interference cancellation (SIC) available at neither the macrocell base station (MBS) nor the small cell BS (SBS), SIC available at both the MBS and the SBS, SIC available at only the MBS, and SIC available at only the SBS are considered. For all the four scenarios, iterative optimization based data offloading schemes are proposed. We show that the proposed data offloading schemes outperform the corresponding schemes without data offloading. We also show that equipping SIC at the SBS is more beneficial compared to equipping SIC at the MBS.

Keywords: Cognitive radio · Mobile data offloading
Successive interference cancellation

1 Introduction

Mobile phones and wireless mobile communications are developing very rapidly in recent years. The unprecedented increase in mobile data traffic has created many challenges for cellular networks, such as the network overloading issue. Mobile users in overloaded cellular networks will undergo degraded mobile services, such as high call blocking probability and low data rate. In this respect, mobile data offloading is an effective method to solve the network overloading issue by offloading part of the data traffic load off the main cellular networks [5].

This work was supported by the National Science and Technology Major Project of China (Grant No. 2017ZX03001008), the Postdoctoral Research Plan of Jiangsu Province (Grant No. 1701167B), the Postdoctoral Science Foundation of China (Grant No. 2017M621795), the NUPTSF (Grant Nos. NY218007 and NY218026), and the NSF of Jiangsu Province (Grant No. BK20160900).

X. Liu et al. (Eds.): ChinaCom 2018, LNICST 262, pp. 438–447, 2019.
https://doi.org/10.1007/978-3-030-06161-6_43

So far, small cells and WiFi are the preferred candidates for data offloading and have attracted a lot of attention [1]. For data offloading through small cells, the work in [7] proposed a two-level offloading scheme that takes the network load and interference conditions into account in small cell networks, the work in [8] proposed a learning mechanism based fair auction scheme for data offloading in small cell networks, and the work in [9] proposed an optimal energy efficient offloading scheme based on the auction theory. For data offloading through WiFI, the work in [3] proposed a network-assisted user-centric WiFi offloading scheme in a heterogeneous network, the work in [6] analyzed the efficiency of the opportunistic and the delayed WiFi offloading schemes, and the work in [4] jointly considered the problem of base station (BS) switching, resource allocation and data offloading.

Meanwhile, the concept of cognitive radio (CR) has been proposed to address the conflict between spectrum scarcity and low spectrum utilization [2]. The CR allows the secondary users with no licensed spectrum band to access the spectrum band licensed to the primary users under the condition that the QoS of the primary users is guaranteed. By adopting the concept of CR, we propose to use third-party cognitive small cells with no licensed spectrum band to offload data traffic from macrocells with licensed spectrum band and at the same time gain transmission opportunities for cognitive small cells. The advantages of data offloading through third-party cognitive small cells are three fold: (1) There is no extra cost for building small cell infrastructures to support data offloading as these small cells are third-party; (2) The throughput of the macrocells can be improved as long as the QoS of data offloading is guaranteed by the cognitive small cells; (3) The third-party cognitive small cells can use the remaining resources from the macrocells to let their own users transmit information.

Therefore, this paper considers the data offloading scenario where a third-party cognitive small cell provides data offloading service to a macrocell. The transmission time is assumed to consist of two slots, where the first time slot is for macrocell user transmission and the second time slot is for small cell user transmission. The resource allocation problem for data offloading through such a third-party cognitive small cell is investigated. Particularly, we consider four scenarios, namely, successive interference cancellation (SIC) available at neither the macrocell BS (MBS) nor the small cell BS (SBS), SIC available at both the MBS and the SBS, SIC available at only the MBS, and SIC available at only the SBS. For all the four scenarios, we propose iterative optimization based data offloading schemes to maximize the sum rate of the small cell UEs (SUEs) subject to the required minimum sum rate of the macrocell UEs (MUEs). Simulation results are given to verify our proposed data offloading schemes.

2 System Model

This paper considers an uplink macrocell network with M MUEs served by a MBS, which is licensed with a narrow spectrum band for data communication. We assume that there is an uplink small cell network with K SUEs served by a SBS, which is in the coverage area of the MBS and is not licensed with any spectrum band for data communication. Since some MUEs may be near the SBS and far from the MBS, it is better to direct these MUEs to be served by the SBS. To reward the small cell network for data offloading, it can use the spectrum licensed to the macrocell network for its own purpose under the condition that the performance of the macrocell network is guaranteed.

We assume that all the channels are block-fading, i.e., the channel power gains are constant in each transmission block and change independently. The channel power gains from the MUE m to the MBS, from the SUE k to the SBS, and from the MUE m to the SBS are denoted by h_m^p, h_k^s and h_m^{ps}, respectively. The transmission time for each transmission block is denoted by T. We assume that the total transmission time is divided into two slots. The first time slot is for MUE data communication with time τ_1, while the second time slot is for SUE data communication with time τ_2. Thus, we have $\tau_1 + \tau_2 \leq T$. Let p_m^p and p_k^s denote the transmit powers of the MUE m and the SUE k, respectively. The transmit powers of the MUEs and the SUEs are restricted as $p_m^p \leq P_{max}^p$ and $p_k^s \leq P_{max}^s$, respectively for $m = 1, \ldots, M$, $k = 1, \ldots, K$. Let $\alpha_m \in \{0, 1\}$ and $\beta_m \in \{0, 1\}$ denote whether the MUE m is connected to the MBS and the SBS, respectively. Specifically, $\alpha_m = 1$ denotes that the MUE m is connected to the MBS and vice versa, while $\beta_m = 1$ denotes that the MUE m is connected to the SBS and vice versa. Since each MUE is assumed to be able to connect either the MBS or the SBS, we have $\alpha_m + \beta_m \leq 1$, for $m = 1, \ldots, M$.

In this paper, we assume that the performance of the macrocell network is guaranteed by satisfying the required minimum sum rate of the MUEs, given by $R_p(\tau_1, \boldsymbol{\alpha}, \boldsymbol{p}^p) + R_{ps}(\tau_1, \boldsymbol{\beta}, \boldsymbol{p}^p) \geq R_{min}$, where $R_p(\tau_1, \boldsymbol{\alpha}, \boldsymbol{p}^p)$ is the sum rate of the MUEs connected to the MBS, $R_{ps}(\tau_1, \boldsymbol{\beta}, \boldsymbol{p}^p)$ is the sum rate of the MUEs offloaded to the SBS, $\boldsymbol{\alpha} = [\alpha_1, \ldots, \alpha_M]^T$, $\boldsymbol{\beta} = [\beta_1, \ldots, \beta_M]^T$, and $\boldsymbol{p}^p = [p_1^p, \ldots, p_M^p]^T$. Our aim is to maximize the sum rate at the SBS denoted by $R_s(\tau_2, \boldsymbol{p}^s)$, where $\boldsymbol{p}^s = [p_1^s, \ldots, p_K^s]^T$. The exact expression of the sum rate depends on whether SIC decoder is available at the MBS or the SBS. If SIC decoder is not available at the MBS, the sum rate of the MUEs connected to the MBS can be written as

$$R_p^{NSIC}(\tau_1, \boldsymbol{\alpha}, \boldsymbol{p}^p) = \frac{\tau_1}{T} \sum_{m=1}^{M} \ln\left(1 + \frac{\alpha_m p_m^p h_m^p}{\sigma^2 + \sum_{m'=1, m' \neq m}^{M} \alpha_{m'} p_{m'}^p h_{m'}^p}\right), \quad (1)$$

where σ^2 is the background noise power. If SIC decoder is not available at the SBS, the sum rate of the MUEs connected to the SBS and the sum rate of the SUEs can be written as

$$R_{ps}^{NSIC}(\tau_1, \boldsymbol{\beta}, \boldsymbol{p}^p) = \frac{\tau_1}{T} \sum_{m=1}^{M} \ln\left(1 + \frac{\beta_m p_m^p h_m^{ps}}{\sigma^2 + \sum_{m'=1, m' \neq m}^{M} \beta_{m'} p_{m'}^p h_{m'}^{ps}}\right), \quad (2)$$

and

$$R_s^{NSIC}(\tau_2, \boldsymbol{p}^s) = \frac{\tau_2}{T} \sum_{k=1}^{K} \ln\left(1 + \frac{p_k^s h_k^s}{\sigma^2 + \sum_{k'=1, k' \neq k}^{K} p_{k'}^s h_{k'}^s}\right), \quad (3)$$

respectively. If SIC decoder is available at the MBS and the SBS, the sum rate of the MUEs connected to the MBS, the sum rate of the MUEs connected to the SBS and the sum rate of the SUEs can be written respectively as

$$R_p^{SIC}(\tau_1, \boldsymbol{\alpha}, \boldsymbol{p}^p) = \frac{\tau_1}{T} \ln\left(1 + \frac{\sum_{m=1}^{M} \alpha_m p_m^p h_m^p}{\sigma^2}\right), \tag{4}$$

$$R_{ps}^{SIC}(\tau_1, \boldsymbol{\beta}, \boldsymbol{p}^p) = \frac{\tau_1}{T} \ln\left(1 + \frac{\sum_{m=1}^{M} \beta_m p_m^p h_m^{ps}}{\sigma^2}\right), \tag{5}$$

$$R_s^{SIC}(\tau_2, \boldsymbol{p}^s) = \frac{\tau_2}{T} \ln\left(1 + \frac{\sum_{k=1}^{K} p_k^s h_k^s}{\sigma^2}\right). \tag{6}$$

Depending on whether SIC decoder is available at the MBS or the SBS, we can study the resource allocation for data offloading in four cases.

3   Resource Allocation Schemes

3.1   Without SIC Decoders at the MBS and the SBS

In this subsection, we investigate the case when SIC decoders are not available at the MBS and the SBS. The optimization problem is formulated as

$$\max_{\tau_1, \tau_2, \boldsymbol{\alpha}, \boldsymbol{\beta}, \boldsymbol{p}^p, \boldsymbol{p}^s} \quad R_s^{NSIC}(\tau_2, \boldsymbol{p}^s) \tag{7}$$

$$\text{s.t. } \tau_1 + \tau_2 \leq T, \tau_1 \geq 0, \tau_2 \geq 0, \tag{8}$$

$$\alpha_m \in \{0, 1\}, \beta_m \in \{0, 1\}, m = 1, \ldots, M, \tag{9}$$

$$\alpha_m + \beta_m \leq 1, m = 1, \ldots, M, \tag{10}$$

$$0 \leq p_m^p \leq P_{max}^p, m = 1, \ldots, M, \tag{11}$$

$$0 \leq p_k^s \leq P_{max}^s, k = 1, \ldots, K, \tag{12}$$

$$R_p^{NSIC}(\tau_1, \boldsymbol{\alpha}, \boldsymbol{p}^p) + R_{ps}^{NSIC}(\tau_1, \boldsymbol{\beta}, \boldsymbol{p}^p) \geq R_{min}. \tag{13}$$

The problem in (7) is highly nonlinear and nonconvex. We solve the problem in (7) by optimizing τ_1, τ_2 with given $\boldsymbol{\alpha}, \boldsymbol{\beta}, \boldsymbol{p}^p, \boldsymbol{p}^s$, optimizing $\boldsymbol{\alpha}, \boldsymbol{\beta}$ with given $\tau_1, \tau_2, \boldsymbol{p}^p, \boldsymbol{p}^s$, and optimizing $\boldsymbol{p}^p, \boldsymbol{p}^s$ with given $\tau_1, \tau_2, \boldsymbol{\alpha}, \boldsymbol{\beta}$.

With given $\boldsymbol{\alpha}, \boldsymbol{\beta}, \boldsymbol{p}^p, \boldsymbol{p}^s$, we optimize τ_1, τ_2 by maximizing $R_s^{NSIC}(\tau_2, \boldsymbol{p}^s)$ subject to the constraints (8) and (13). It is easy to observe that a larger value of τ_2 can lead to a larger objective function value. Thus, according to the constraints (8) and (13), the optimal τ_2 is given by $\tau_2^* = T - \tau_1^*$, where the value of τ_1^* is achieved when the constraint (13) is satisfied at equality and is given by

$$\tau_1^* = R_{min} T \left(\sum_{m=1}^{M} \ln\left(1 + \frac{\alpha_m p_m^p h_m^p}{\sigma^2 + \sum_{m'=1, m' \neq m}^{M} \alpha_{m'} p_{m'}^p h_{m'}^p}\right) \right.$$
$$\left. + \ln\left(1 + \frac{\beta_m p_m^p h_m^{ps}}{\sigma^2 + \sum_{m'=1, m' \neq m}^{M} \beta_{m'} p_{m'}^p h_{m'}^{ps}}\right) \right)^{-1}. \tag{14}$$

With given $\tau_1, \tau_2, \boldsymbol{p}^p, \boldsymbol{p}^s$, we optimize $\boldsymbol{\alpha}, \boldsymbol{\beta}$ by maximizing $R_s^{NSIC}(\tau_2, \boldsymbol{p}^s)$ subject to the constraints (9), (10) and (13). It is observed that $R_s^{NSIC}(\tau_2, \boldsymbol{p}^s)$ does not depend on $\boldsymbol{\alpha}, \boldsymbol{\beta}$. Considering the fact that a smaller τ_1 leads to a higher $R_s^{NSIC}(\tau_2, \boldsymbol{p}^s)$, we optimize $\boldsymbol{\alpha}, \boldsymbol{\beta}$ by maximizing $R_p^{NSIC}(\tau_1, \boldsymbol{\alpha}, \boldsymbol{p}^p) + R_{ps}^{NSIC}(\tau_1, \boldsymbol{\beta}, \boldsymbol{p}^p)$ subject to the constraints (9) and (10). If the obtained $R_p^{NSIC}(\tau_1, \boldsymbol{\alpha}, \boldsymbol{p}^p) + R_{ps}^{NSIC}(\tau_1, \boldsymbol{\beta}, \boldsymbol{p}^p)$ is smaller than R_{min}, then the original problem is infeasible. Since such problem is hard to be solved, we propose a heuristic scheme to solve it. The values of $\boldsymbol{\alpha}, \boldsymbol{\beta}$ are initialized as $\alpha_m = 0, \beta_m = 0$ for all $m = 1, \ldots, M$. Then, we sequentially set $\alpha_m = 1$ or $\beta_m = 1$ by selecting the one that provides higher value of $R_s^{NSIC}(\tau_2, \boldsymbol{p}^s)$.

With given $\tau_1, \tau_2, \boldsymbol{\alpha}, \boldsymbol{\beta}$, we optimizing $\boldsymbol{p}^p, \boldsymbol{p}^s$ by maximizing $R_s^{NSIC}(\tau_2, \boldsymbol{p}^s)$ subject to the constraints (11), (12) and (13). Such problem can be solved by solving the following two subproblems as given by

$$\max_{\boldsymbol{p}^s} R_s^{NSIC}(\tau_2, \boldsymbol{p}^s) \tag{15}$$

s.t. constraint (12),

and

$$\max_{\boldsymbol{p}^p} R_p^{NSIC}(\tau_1, \boldsymbol{\alpha}, \boldsymbol{p}^p) + R_{ps}^{NSIC}(\tau_1, \boldsymbol{\beta}, \boldsymbol{p}^p) \tag{16}$$

s.t. constraint (11).

If the obtained $R_p^{NSIC}(\tau_1, \boldsymbol{\alpha}, \boldsymbol{p}^p) + R_{ps}^{NSIC}(\tau_1, \boldsymbol{\beta}, \boldsymbol{p}^p)$ from solving the problem in (16) is smaller than R_{min}, then the original problem is infeasible. We solve the problems in (15) and (16) by iteratively optimizing one variable with other variables being fixed. First, we solve the problem in (15). With given $p_1^s, \ldots, p_{k-1}^s, p_{k+1}^s, \ldots, p_K^s$, the variable p_k^s is optimized by solving the problem

$$\max_{0 \le p_k^s \le P_{max}^s} f_k(p_k^s), \tag{17}$$

where

$$f_k(p_k^s) = \ln\left(1 + \frac{p_k^s h_k^s}{\sigma^2 + \sum_{l=1, l \ne k}^K p_l^s h_l^s}\right)$$
$$+ \sum_{k'=1, k' \ne k}^K \ln\left(1 + \frac{p_{k'}^s h_{k'}^s}{\sigma^2 + \sum_{l=1, l \ne k'}^K p_l^s h_l^s}\right). \tag{18}$$

The first derivative of $f_k(p_k^s)$ can be obtained as

$$\frac{df_k(p_k^s)}{dp_k^s} = \frac{h_k^s}{\sigma^2 + \sum_{l=1}^K p_l^s h_l^s}\left(1 - \sum_{k'=1, k' \ne k}^K \frac{p_{k'}^s h_{k'}^s}{\sigma^2 + \sum_{l=1, l \ne k'}^K p_l^s h_l^s}\right). \tag{19}$$

It is seen that the first part of the above expression is positive and the second part is a strictly increasing function of p_k^s. Thus, the solution to $\frac{df_k(p_k^s)}{dp_k^s} = 0$ is unique and is denoted by $p_k^s = x_k$. If $x_k \leq 0$, then $f_k(p_k^s)$ is a monotonically increasing function of p_k^s and the solution to the problem in (17) is thus $p_k^s = P_{max}^s$. If $x_k \geq P_{max}^s$, then $f_k(p_k^s)$ is a monotonically decreasing function of p_k^s and the solution to the problem in (17) is thus $p_k^s = 0$. If $0 < x_k < P_{max}^s$, then $f_k(p_k^s)$ first decreases as p_k^s increases and turns to increase when p_k^s is beyond x_k, and the solution to the problem in (17) is thus $p_k^s = 0$ if $f_k(0) > f_k(P_{max}^s)$ and is $p_k^s = P_{max}^s$ otherwise. The problem in (16) can be solved similarly as the problem in (15) and we omit here for brevity.

3.2 With SIC Decoders at the MBS and the SBS

In this subsection, we investigate the case when SIC decoders are available at both the MBS and the SBS. The optimization problem is formulated as

$$\max_{\tau_1, \tau_2, \alpha, \beta, p^p, p^s} R_s^{SIC}(\tau_2, p^s) \tag{20}$$

$$\text{s.t. } R_p^{SIC}(\tau_1, \alpha, p^p) + R_{ps}^{SIC}(\tau_1, \beta, p^p) \geq R_{min}. \tag{21}$$

$$\text{and constraints } (8)-(12)$$

Similar to the problem in (7), we solve the problem in (20) by optimizing τ_1, τ_2 with given α, β, p^p, p^s, optimizing α, β with given τ_1, τ_2, p^p, p^s, and optimizing p^p, p^s with given $\tau_1, \tau_2, \alpha, \beta$.

With given α, β, p^p, p^s, the variables τ_1, τ_2 are optimized by maximizing $R_s^{SIC}(\tau_2, p^s)$ subject to the constraints (8) and (21). The optimal τ_2 is easily obtained as $\tau_2^* = T - \tau_1^*$, where the value of τ_1^* can be obtained as

$$\tau_1^* = \frac{R_{min}T}{\ln\left(1 + \frac{\sum_{m=1}^M \alpha_m p_m^p h_m^p}{\sigma^2}\right) + \ln\left(1 + \frac{\sum_{m=1}^M \beta_m p_m^p h_m^{ps}}{\sigma^2}\right)}. \tag{22}$$

With given τ_1, τ_2, p^p, p^s, the variables α, β are optimized by maximizing $R_s^{SIC}(\tau_2, p^s)$ subject to the constraints (9), (10) and (21). We solve the problem by maximizing $R_p^{SIC}(\tau_1, \alpha, p^p) + R_{ps}^{SIC}(\tau_1, \beta, p^p)$ subject to the constraints (9) and (10). If the obtained $R_p^{SIC}(\tau_1, \alpha, p^p) + R_{ps}^{SIC}(\tau_1, \beta, p^p)$ is smaller than R_{min}, then the problem is infeasible. Similar to Sect. 3.1, a heuristic scheme can be proposed to optimize α, β and we omit it here for brevity.

With given $\tau_1, \tau_2, \alpha, \beta$, the variables p^p, p^s are optimized by maximizing $R_s^{SIC}(\tau_2, p^s)$ subject to the constraints (11), (12) and (21). It is observed that the problem can be solved by solving the following two subproblems as given by

$$\max_{p^s} R_s^{SIC}(\tau_2, p^s) \tag{23}$$

$$\text{s.t. constraint (12),}$$

and

$$\max_{p^p} R_p^{SIC}(\tau_1, \alpha, p^p) + R_{ps}^{SIC}(\tau_1, \beta, p^p) \tag{24}$$

$$\text{s.t. constraint (11).}$$

It is noted that the original problem is infeasible if the obtained $R_p^{SIC}(\tau_1, \boldsymbol{\alpha}, \boldsymbol{p}^p) + R_{ps}^{SIC}(\tau_1, \boldsymbol{\beta}, \boldsymbol{p}^p)$ from solving the problem in (24) is smaller than R_{min}. It can be verified that the objective functions in (23) and (24) are increasing functions of $\boldsymbol{p}^s$ and $\boldsymbol{p}^p$, respectively. Thus, the optimal solutions to the problems in (23) and (24) are $p_k^s = P_{max}^s$ and $p_m^p = P_{max}^p$, for $m = 1, \ldots, M$, $k = 1, \ldots, K$.

3.3 With SIC Decoder at the MBS

In this subsection, we investigate the case when SIC decoder is available only at the MBS. The optimization problem is formulated as

$$\max_{\tau_1, \tau_2, \boldsymbol{\alpha}, \boldsymbol{\beta}, \boldsymbol{p}^p, \boldsymbol{p}^s} R_s^{NSIC}(\tau_2, \boldsymbol{p}^s) \tag{25}$$

$$\text{s.t. } R_p^{SIC}(\tau_1, \boldsymbol{\alpha}, \boldsymbol{p}^p) + R_{ps}^{NSIC}(\tau_1, \boldsymbol{\beta}, \boldsymbol{p}^p) \geq R_{min}. \tag{26}$$

$$\text{and constraints } (8)-(12)$$

Similar to the problem in (7), we solve the problem in (25) by optimizing τ_1, τ_2 with given $\boldsymbol{\alpha}, \boldsymbol{\beta}, \boldsymbol{p}^p, \boldsymbol{p}^s$, optimizing $\boldsymbol{\alpha}, \boldsymbol{\beta}$ with given $\tau_1, \tau_2, \boldsymbol{p}^p, \boldsymbol{p}^s$, and optimizing $\boldsymbol{p}^p, \boldsymbol{p}^s$ with given $\tau_1, \tau_2, \boldsymbol{\alpha}, \boldsymbol{\beta}$.

With given $\boldsymbol{\alpha}, \boldsymbol{\beta}, \boldsymbol{p}^p, \boldsymbol{p}^s$, the optimal τ_2 can be obtained as $\tau_2^* = T - \tau_1^*$, where the value of τ_1^* is obtained by

$$\tau_1^* = \frac{R_{min}T}{\ln\left(1 + \frac{\sum_{m=1}^M \alpha_m p_m^p h_m^p}{\sigma^2}\right) + \sum_{m=1}^M \ln\left(1 + \frac{\beta_m p_m^p h_m^{ps}}{\sigma^2 + \sum_{m'=1, m' \neq m}^M \beta_{m'} p_{m'}^p h_{m'}^{ps}}\right)}. \tag{27}$$

With given $\tau_1, \tau_2, \boldsymbol{p}^p, \boldsymbol{p}^s$, the variables $\boldsymbol{\alpha}, \boldsymbol{\beta}$ are optimized by maximizing $R_s^{NSIC}(\tau_2, \boldsymbol{p}^s)$ subject to the constraints (9), (10) and (26). We solve the problem by maximizing $R_p^{SIC}(\tau_1, \boldsymbol{\alpha}, \boldsymbol{p}^p) + R_{ps}^{NSIC}(\tau_1, \boldsymbol{\beta}, \boldsymbol{p}^p)$ subject to the constraints (9) and (10), and if the obtained $R_p^{SIC}(\tau_1, \boldsymbol{\alpha}, \boldsymbol{p}^p) + R_{ps}^{NSIC}(\tau_1, \boldsymbol{\beta}, \boldsymbol{p}^p)$ is smaller than R_{min}, then the problem is infeasible. Similar to Sect. 3.1, a heuristic scheme can be proposed to solve the problem and we omit it here for brevity.

With given $\tau_1, \tau_2, \boldsymbol{\alpha}, \boldsymbol{\beta}$, the variables $\boldsymbol{p}^p, \boldsymbol{p}^s$ are optimized by maximizing $R_s^{NSIC}(\tau_2, \boldsymbol{p}^s)$ subject to the constraints (11), (12) and (26). The problem can be solved by solving the following two subproblems as given by

$$\max_{\boldsymbol{p}^s} R_s^{NSIC}(\tau_2, \boldsymbol{p}^s) \tag{28}$$

$$\text{s.t. constraint } (12),$$

and

$$\max_{\boldsymbol{p}^p} R_p^{SIC}(\tau_1, \boldsymbol{\alpha}, \boldsymbol{p}^p) + R_{ps}^{NSIC}(\tau_1, \boldsymbol{\beta}, \boldsymbol{p}^p) \tag{29}$$

$$\text{s.t. constraint } (11).$$

It is noted that the original problem is infeasible if the obtained $R_p^{SIC}(\tau_1, \boldsymbol{\alpha}, \boldsymbol{p}^p) + R_{ps}^{NSIC}(\tau_1, \boldsymbol{\beta}, \boldsymbol{p}^p)$ from solving the problem in (29) is smaller than R_{min}. Since

$R_p^{SIC}(\tau_1, \boldsymbol{\alpha}, \boldsymbol{p}^p)$ is an increasing functions of $\boldsymbol{p}^p$. Thus, the optimal p_m^p is given as $p_m^p = P_{max}^p$ for $m \in \{m | \alpha_m = 1, m = 1, \ldots, M\}$. For the optimal $\{p_k^s\}$ of the problem in (28) and the optimal $\{p_m^p, m \in \{m | \beta_m = 1, m = 1, \ldots, M\}\}$ of the problem in (29), we iteratively optimize one variable with other variables being fixed similar to that in Sect. 3.1, which we omit here for brevity.

3.4   With SIC Decoder at the SBS

In this subsection, we investigate the case when SIC decoder is available only at the SBS. The optimization problem is formulated as

$$\max_{\tau_1, \tau_2, \boldsymbol{\alpha}, \boldsymbol{\beta}, \boldsymbol{p}^p, \boldsymbol{p}^s} R_s^{SIC}(\tau_2, \boldsymbol{p}^s) \tag{30}$$

$$\text{s.t.} \; R_p^{NSIC}(\tau_1, \boldsymbol{\alpha}, \boldsymbol{p}^p) + R_{ps}^{SIC}(\tau_1, \boldsymbol{\beta}, \boldsymbol{p}^p) \geq R_{min}. \tag{31}$$

$$\text{and constraints } (8)-(12)$$

Similar to the problem in (7), we solve the problem in (30) by optimizing τ_1, τ_2 with given $\boldsymbol{\alpha}, \boldsymbol{\beta}, \boldsymbol{p}^p, \boldsymbol{p}^s$, optimizing $\boldsymbol{\alpha}, \boldsymbol{\beta}$ with given $\tau_1, \tau_2, \boldsymbol{p}^p, \boldsymbol{p}^s$, and optimizing $\boldsymbol{p}^p, \boldsymbol{p}^s$ with given $\tau_1, \tau_2, \boldsymbol{\alpha}, \boldsymbol{\beta}$.

With given $\boldsymbol{\alpha}, \boldsymbol{\beta}, \boldsymbol{p}^p, \boldsymbol{p}^s$, the optimal τ_2 can be obtained as $\tau_2^* = T - \tau_1^*$, where the value of τ_1^* is obtained by

$$\tau_1^* = \frac{R_{min}T}{\sum_{m=1}^M \ln\left(1 + \frac{\alpha_m p_m^p h_m^p}{\sigma^2 + \sum_{m'=1, m' \neq m}^M \alpha_{m'} p_{m'}^p h_{m'}^p}\right) + \ln\left(1 + \frac{\sum_{m=1}^M \beta_m p_m^p h_m^{ps}}{\sigma^2}\right)}. \tag{32}$$

With given $\tau_1, \tau_2, \boldsymbol{p}^p, \boldsymbol{p}^s$, the variables $\boldsymbol{\alpha}, \boldsymbol{\beta}$ are optimized by maximizing $R_s^{SIC}(\tau_2, \boldsymbol{p}^s)$ subject to the constraints (9), (10) and (31). The problem is solved by maximizing $R_p^{NSIC}(\tau_1, \boldsymbol{\alpha}, \boldsymbol{p}^p) + R_{ps}^{SIC}(\tau_1, \boldsymbol{\beta}, \boldsymbol{p}^p)$ subject to the constraints (9) and (10), and if the obtained $R_p^{NSIC}(\tau_1, \boldsymbol{\alpha}, \boldsymbol{p}^p) + R_{ps}^{SIC}(\tau_1, \boldsymbol{\beta}, \boldsymbol{p}^p)$ is smaller than R_{min}, then the problem is infeasible. Similar to Sect. 3.1, a heuristic scheme can be proposed to solve the problem and we omit it here for brevity.

With given $\tau_1, \tau_2, \boldsymbol{\alpha}, \boldsymbol{\beta}$, the variables $\boldsymbol{p}^p, \boldsymbol{p}^s$ are optimized by maximizing $R_s^{SIC}(\tau_2, \boldsymbol{p}^s)$ subject to the constraints (11), (12) and (31). The problem can be solved by solving the following two subproblems as given by

$$\max_{\boldsymbol{p}^s} R_s^{SIC}(\tau_2, \boldsymbol{p}^s) \tag{33}$$

$$\text{s.t. constraint } (12),$$

and

$$\max_{\boldsymbol{p}^p} R_p^{NSIC}(\tau_1, \boldsymbol{\alpha}, \boldsymbol{p}^p) + R_{ps}^{SIC}(\tau_1, \boldsymbol{\beta}, \boldsymbol{p}^p) \tag{34}$$

$$\text{s.t. constraint } (11).$$

Noted that the original problem is infeasible if the obtained $R_p^{NSIC}(\tau_1, \boldsymbol{\alpha}, \boldsymbol{p}^p) + R_{ps}^{SIC}(\tau_1, \boldsymbol{\beta}, \boldsymbol{p}^p)$ from solving the problem in (34) is smaller than R_{min}. Since

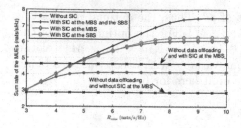 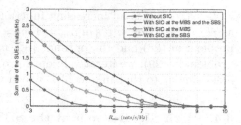

Fig. 1. Sum rate of the MUEs against R_{min}.

Fig. 2. Sum rate of the SUEs against R_{min}.

$R_s^{SIC}(\tau_2, \boldsymbol{p}^s)$ and $R_{ps}^{SIC}(\tau_1, \boldsymbol{\beta}, \boldsymbol{p}^p)$ are increasing functions of $\boldsymbol{p}^s$ and $\boldsymbol{p}^p$, respectively. Thus, the optimal p_k^s is given as $p_k^s = p_{max}^s$ and the optimal p_m^p is given as $p_m^p = P_{max}^p$ for $m \in \{m | \beta_m = 1, m = 1, \ldots, M\}$. For the optimal $\{p_m^p, m \in \{m | \alpha_m = 1, m = 1, \ldots, M\}\}$ of the problem in (34), we iteratively optimize one variable with other variables being fixed similar to that in Sect. 3.1, which we omit here for brevity.

4   Simulation Results

In this section, we verify the performance of the proposed data offloading schemes. The channels involved are assumed to follow Rayleigh fading with unit mean. In the following results, we set $\sigma^2 = 1$, $T = 1$, $M = 10$, $K = 10$, $P_{max}^p = 10\,\mathrm{W}$ and $P_{max}^s = 10\,\mathrm{W}$.

In Fig. 1, we illustrate the sum rate of the MUEs against the required minimum sum rate of the MUEs R_{min} for different data offloading schemes with or without SIC decoders at the MBS and/or the SBS. For the purpose of comparison, the results obtained from the schemes without data offloading are also given. It is observed that the proposed data offloading schemes can achieve the sum rate of the MUEs equal to R_{min} when R_{min} is not too high. When R_{min} is high, the achieved sum rates of the MUEs by the proposed schemes gradually saturate as R_{min} increases further. This is because that when R_{min} is high, the required minimum sum rate of the MUEs R_{min} will not be supported even if all the available time T is allocated for the MUE data communication with data offloading to the SBS. It is also observed that by choosing a proper large value of R_{min}, the proposed data offloading schemes can achieve much higher sum rate of the MUEs compared to the schemes without data offloading. In addition, it is observed that the sum rate of the MUEs achieved by the data offloading scheme with SIC at the SBS is higher than that achieved by the data offloading scheme with SIC at the MBS when R_{min} is high. This indicates that equipping a SIC decoder at the SBS is more beneficial for the MUEs compared to equipping a SIC decoder at the MBS. It is also observed that the sum rates of the MUEs achieved by the data offloading schemes with SIC at the MBS (or SBS) are always higher than that achieved by the data offloading schemes without SIC at the MBS (or SBS).

In Fig. 2, we illustrate the sum rate of the SUEs against the required minimum sum rate of the MUEs R_{min} for different data offloading schemes with or without SIC decoders at the MBS and/or the SBS. It is observed that the data offloading scheme without SIC achieves the lowest sum rate of the SUEs among the four data offloading schemes, while the data offloading scheme with SIC at the MBS and the SBS achieves the highest sum rate of the SUEs among the four data offloading schemes. It is also observed that the data offloading scheme with SIC at the SBS achieves higher sum rate of the SBS than that achieved by the data offloading scheme with SIC at the MBS. This indicates that equipping a SIC decoder at the SBS is more beneficial for the SUEs compared to equipping a SIC decoder at the MBS.

5   Conclusions

This paper investigates the mobile data offloading problem through a third-party cognitive small cell for a macrocell. By considering whether SIC is available at the MBS and/or the SBS, four scenarios are considered. For each scenario, iterative optimization based data offloading scheme is proposed. It is shown that the proposed data offloading schemes outperform the corresponding schemes without data offloading. It is also shown that equipping SIC at the SBS is more beneficial compared to equipping SIC at the MBS.

References

1. Aijaz, A., Aghvami, H., Amani, M.: A survey on mobile data offloading: technical and business perspectives. IEEE Wirel. Commun. **20**(2), 104–112 (2013)
2. Haykin, S.: Cognitive radio: brain-empowered wireless communications. IEEE J. Sel. Areas Commun. **23**(2), 201–220 (2005)
3. Jung, B.H., Song, N.O., Sung, D.K.: A network-assisted user-centric wifi-offloading model for maximizing per-user throughput in a heterogeneous network. IEEE Trans. Veh. Technol. **63**(4), 1940–1945 (2014)
4. Luong, P., Nguyen, T.M., Le, L.B., Dao, N.D., Hossain, E.: Energy-efficient wifi offloading and network management in heterogeneous wireless networks. IEEE Access **4**, 10210–10227 (2016)
5. Rebecchi, F., de Amorim, M.D., Conan, V., Passarella, A., Bruno, R., Conti, M.: Data offloading techniques in cellular networks: a survey. IEEE Commun. Surv. Tutor. **17**(2), 580–603 (2015)
6. Suh, D., Ko, H., Pack, S.: Efficiency analysis of wifi offloading techniques. IEEE Trans. Veh. Technol. **65**(5), 3813–3817 (2016)
7. Sun, Y., Xu, X., Zhang, R., Gao, R.: Offloading based load balancing for the small cell heterogeneous network. In: Proceedings of the WPMC, pp. 288–293 (2014)
8. Trakas, P., Adelantado, F., Verikoukis, C.: A novel learning mechanism for traffic offloading with small cell as a service. In: Proceedings of the IEEE ICC, pp. 6893–6898 (2015)
9. Xu, X., Zhang, H., Dai, X., Tao, X.: Optimal energy efficient offloading in small cell HetNet with auction. In: Proceedings of the CHINACOM, pp. 335–340 (2014)

Performance Analysis of Non-coherent Massive SIMO Systems with Antenna Correlation

Weiyang Xu$^{(\boxtimes)}$ (iD), Huiqiang Xie, and Shengbo Xu

Chongqing University, No. 174 Shazhengjie, Shapingba, Chongqing 400044, China
{weiyangxu,huiqiangxie,shengboxu}@cqu.edu.cn

Abstract. Recently, energy detection (ED) has been investigated in massive single-input multiple-output (SIMO) systems, where transmit symbols can be decoded by averaging the received power across all receive antennas. In this paper, we concentrate on the performance of non-coherent massive SIMO in the presence of antenna correlation. Specifically, closed-form expressions of symbol error rate (SER) and achievable rate are derived. Furthermore, asymptotic behaviors of SER and achievable rate in regimes of a large number of receive antennas, high antenna correlation and large signal-to-noise ratio (SNR) are investigated. Interestingly, the results show that antenna correlation poses a great impact to SER, but has little effect on the achievable rate. Numerical results are presented to verify our analytical results.

Keywords: Energy detection · Performance analysis
Spatially correlated channel
Massive single-input multiple-output (SIMO)

1 Introduction

Massive multiple-input multiple-out (MIMO) systems, which deploy a large number of antennas at base station (BS) to serve a relatively small number of users, has become a promising technology due to its increased degrees of freedom [1,2]. Besides, massive MIMO is energy efficient since the transmit power scales down with the number of antennas at BSs. However, non-orthogonal pilots among adjacent cells would deteriorate the system performance as channel estimates obtained in a given cell will be corrupted by pilots transmitted by users in the other cells.

Non-coherent communications systems based on energy detection (ED), which require no knowledge of instantaneous channel state information (CSI) at either the transmitter or receiver, have attracted a great attention [3,4]. In spite of a sub-optimal performance, non-coherent receivers enjoy the benefits

© ICST Institute for Computer Sciences, Social Informatics and Telecommunications Engineering 2019
Published by Springer Nature Switzerland AG 2019. All Rights Reserved
X. Liu et al. (Eds.): ChinaCom 2018, LNICST 262, pp. 448–457, 2019.
https://doi.org/10.1007/978-3-030-06161-6_44

of low complexity, low power consumption and simple structures compared to coherent communications systems [5]. Specifically, for an ED-based non-coherent massive single-input multiple-output (SIMO) system, the average symbol-error-rate (SER) is derived with channel statistics, based on which a minimum distance constellation is presented in [6]. An asymptotically optimal constellation is proposed with varying levels of uncertainty in channel statistics [7]. Also, it is proved that non-coherent massive SIMO system satisfies the same scaling law as its coherent counterpart [8]. More importantly, given that the number of receive antennas is asymptotically infinite, the ED-based non-coherent massive SIMO system can provide the same error performance as that of the coherent system.

In real applications, deploying a large number of antennas leads to inadequate antenna separation. Thus, a new challenge emerges as the correlation between antennas could adversely affects the communications systems performance and capacity. The impact of antenna correlation on conventional MIMO has been investigated thoroughly. In [9] and [10], the effects of spatial correlation and mutual antenna coupling are studied when an increasing number of antennas is fitted in a fixed physical space. Furthermore, it is shown that energy efficiency does not increase unboundedly in massive MIMO system when antennas are to be accommodated within a fixed physical space [11]. The analysis of antenna correlation is not restricted to the popular separable correlation model, but rather it embraces a more general representation [12] and closed-form expressions for the capacity of correlative channel based on the eigenvalues of input covariance and channel matrix are proposed in [13].

The aforementioned studies validate that antenna correlation has an adverse impact in coherent MIMO systems. However, for non-coherent massive SIMO systems, whether the antenna correlation influences the capacity or error performance is still not clear. Inspired by this, this paper presents a thorough performance analysis of non-coherent massive SIMO systems with ED-based receivers. In this work, we derive analytical expressions of ergodic rate and the SER for non-coherent massive SIMO systems with receive antenna correlation. For the SER, when antenna correlation is large enough, increasing the number of antennas cannot further reduce the error probability. Conversely, antenna correlation has little impact on the achievable rate.

2   System Model

We consider a massive SIMO configuration with one transmit antenna and a large number of receive antennas. The flat-fading channels of different transmit-receive pairs are assumed to be mutually independent. The received signal vector is represented by

$$\mathbf{y} = \mathbf{h}x + \mathbf{n} \tag{1}$$

where $\mathbf{y} \in \mathbb{C}^{M \times 1}$ is the received signal at the multi-antenna receiver, $\mathbf{n} \in \mathbb{C}^{M \times 1}$ indicates a complex Gaussian noise vector with elements $n_i \sim \mathcal{CN}(0, \sigma_n^2)$, $\mathbf{h} \in \mathbb{C}^{M \times 1}$ refers to the channel realization with $h_i \sim \mathcal{CN}(0, \sigma_h^2)$, x denotes the transmit symbol drawn from a certain non-negative constellation $\mathcal{P} = \{\sqrt{p_1}, \sqrt{p_2}, \ldots, \sqrt{p_K}\}$, K indicates the constellation size and M the number of receive antennas. The channel statistics is supposed to be known to the receiver instead of the instantaneous CSI.

In the case of one transmit antenna, the spatially correlated channel can be characterized by the well-known Kronecker model [14]

$$\mathbf{h} = \mathbf{\Phi}_r^{1/2}\mathbf{g} \tag{2}$$

where $\mathbf{g} \in \mathbb{C}^{M \times 1}$ is an uncorrelated complex channel vector whose entries are independent identically distributed (i.i.d.) with $g_i \sim \mathcal{CN}(0,1)$. $\mathbf{\Phi}_r^{1/2}$ indicates the deterministic receive correlation matrix, which depends on the angle spread, antenna beamwidth and antenna spacing.

For the structure of $\mathbf{\Phi}_r$, the exponential correlation model are often utilized to quantify the level of spatial correlation [14]. Specifically, according to the exponential model, the receive correlation matrix can be constructed utilizing a single coefficient $\rho \in \mathbb{C}$, namely

$$\Phi_{ij} = \begin{cases} \rho^{|j-i|}, & i \leqslant j \\ \left(\rho^{|j-i|}\right)^*, & j < i \end{cases} \tag{3}$$

where $|\cdot|$ denotes the absolute value operation and Φ_{ij} the $(i,j)^{th}$ entry of $\mathbf{\Phi}_r$, $\rho = ae^{j\theta}$ is the correlation coefficient with $0 \leq a < 1$. Note that the eigenvalues of $\mathbf{\Phi}_r$ only depend on a, while θ decides the eigenvectors of $\mathbf{\Phi}_r$. Because only the eigenvalues of $\mathbf{\Phi}_r$ will be used in the following analysis, we assume $\rho = a$ throughout this paper. Also, $\mathbf{\Phi}_r$ is supposed to be known as a prior, since it is supposed to be less frequently varying than the channel matrix.

3   ED-Based Receiver Using a Finite Number of Antennas

Based on the ED principle, after the received signal having been filtered, squared and integrated, the decision metric for symbol decoding can be written as

$$z = \frac{\|\mathbf{y}\|_2^2}{M}. \tag{4}$$

We assume that the knowledge of channel and noise statistics is available at the receiver, this is achieved by sending a sequence of training symbols before data transmission [9]. First, the decision metric with a finite number of antennas can be expanded as

$$z = \frac{1}{M}[\mathbf{h}^H\mathbf{\Phi}_r\mathbf{h}]x^2 + \frac{1}{M}\mathbf{n}^H\mathbf{n} + \frac{2}{M}\Re\left(\mathbf{n}^H\mathbf{\Phi}_r^{1/2}\mathbf{h}\right)x. \tag{5}$$

From [15], the first component of (5) can be expressed as

$$\frac{1}{M}[\mathbf{h}^H\mathbf{\Phi}_r\mathbf{h}]x^2 = \frac{1}{M}[\mathbf{h}^H\mathbf{U}\mathbf{\Lambda}\mathbf{U}^H\mathbf{h}]x^2$$

$$= \frac{1}{M}[\mathbf{v}^H\mathbf{\Lambda}\mathbf{v}]x^2 \tag{6}$$

where the eigendecomposition is employed to translate $\boldsymbol{\Phi}_r$ into $\boldsymbol{\Phi}_r = \mathbf{U}\boldsymbol{\Lambda}\mathbf{U}^H$, $\boldsymbol{\Lambda}$ is a eigenvalue diagonal matrix and $\mathbf{U}$ is a unitary matrix consisting of corresponding eigenvectors. $\mathbf{v} = \mathbf{U}^H\mathbf{h}$ follows the identical distribution with $\mathbf{U}^H$ and the entries of $\mathbf{v}$ are mutually independent [15].

The third component in (5) can be expanded in the same way. Therefore, the decision metric is transformed into

$$
\begin{aligned}
z &= \frac{1}{M}[\mathbf{v}^H\boldsymbol{\Lambda}\mathbf{v}]x^2 + \frac{1}{M}\mathbf{n}^H\mathbf{n} + \frac{2}{M}\Re\left(\mathbf{q}^H\boldsymbol{\Lambda}^{1/2}\mathbf{v}\right)x \\
&= \frac{x^2}{M}\sum_{i=1}^{M}\lambda_i|v_i|^2 + \frac{1}{M}\sum_{i=1}^{M}|n_i|^2 + \frac{2x}{M}\sum_{i=1}^{M}\lambda_i^{\frac{1}{2}}\Re(q_iv_i)
\end{aligned}
\tag{7}
$$

where $\mathbf{q} = \mathbf{U}^H\mathbf{n}$, $2|v_i|^2$ and $\frac{2}{\sigma_n^2}|n_i|^2$ are chi-square variables with 2 degrees of freedom, q_iv_i is a product of two Gaussian variables. Although $\frac{2}{M}\Re\left(\mathbf{q}^H\boldsymbol{\Lambda}^{1/2}\mathbf{v}\right)$, $\frac{1}{M}\mathbf{n}^H\mathbf{n}$ and $\frac{1}{M}[\mathbf{v}^H\boldsymbol{\Lambda}\mathbf{v}]$ are not mutually independent, the asymptotic independence can be validated among them [16]. Thus, it is assumed the elements in (7) are mutually independent in the following analysis.

Lemma 1. *If the number of antennas M grows large, the following approximations are attainable thanks to Lyapunov Central Limit Theorem (CLT).*

$$
\sum_{i=1}^{M}\lambda_i|v_i|^2 \sim \mathcal{N}\left(\sum_{i=1}^{M}\lambda_i, \sum_{i=1}^{M}\lambda_i^2\right),
$$

$$
\sum_{i=1}^{M}|n_i|^2 \sim \mathcal{N}\left(M\sigma_n^2, M\sigma_n^4\right), \quad \sum_{i=1}^{M}\lambda_i^{\frac{1}{2}}\Re(q_iv_i) \sim \mathcal{N}\left(0, \frac{\sigma_n^2}{2}\sum_{i=1}^{M}\lambda_i\right)
\tag{8}
$$

where

$$
\sum_{i=1}^{M}\lambda_i = M, \quad \sum_{i=1}^{M}\lambda_i^2 = M + 2\sum_{i=1}^{M-1}(M-i)\rho^{2i} = M + f(\rho)
\tag{9}
$$

with

$$
f(\rho) = 2\frac{\rho^{2M+2} + M(\rho^2 - \rho^4) - \rho^2}{(1-\rho^2)^2}, \quad 0 \le \rho < 1.
\tag{10}
$$

Since $0 \le \rho < 1$, the above equation is further simplified as

$$
f(\rho) = \frac{2M(\rho^2 - \rho^4) - 2\rho^2}{(1-\rho^2)^2}, \quad 0 \le \rho < 1.
\tag{11}
$$

where $0 \le f(\rho) < M^2 - M$, $\sum_{i=1}^{M}\lambda_i$ is equal to the trace of $\boldsymbol{\Phi}_r$ and $\sum_{i=1}^{M}\lambda_i^2$ is the trace of $\boldsymbol{\Phi}_r^2$. Obviously, $f(\rho)$ is an increasing function of ρ.

Applying Lemma 1 and with some straightforward mathematical manipulations, it is shown that the decision metric z follows a real Gaussian distribution, namely $z \sim \mathcal{N}(\mu_z, \sigma_z^2)$. The corresponding mean and variance are given below

$$
\mu_z = x^2 + \sigma_n^2 \to \mu(p_k) = p_k + \sigma_n^2
$$

$$
\sigma_z^2 = \frac{1}{M}\left(x^2 + \sigma_n^2\right)^2 + \frac{f(\rho)}{M^2} \to \sigma^2(p_k) = \frac{1}{M}\left(p_k + \sigma_n^2\right)^2 + \frac{f(\rho)}{M^2}
\tag{12}
$$

where $\mu(p_k)$ and $\sigma^2(p_k)$ are mean and variance of z when $\sqrt{p_k}$ is the transmit symbol.

4   Performance Analysis

In this section, a closed-form expression of the SER is presented. The asymptotic behaviors of infinite number of antennas and high SNRs are taken into consideration. Afterwards, a closed-form expression of the achievable rate is given.

4.1   SER Analysis

Given multiple decoding regions $\{d_k\}_{k=1}^{K-1}$, z can be decoded by

$$\widehat{x} = \sqrt{p_k} : d_{k-1} \leqslant z < d_k. \tag{13}$$

Proposition 1. *With a finite number of receive antennas, the SER of the ED-based massive SIMO system with antenna correlation is given by*

$$P_e = 1 - \frac{1}{K} \sum_{k=1}^{K} P(p_k)$$

$$= 1 - \frac{1}{2K} \sum_{k=1}^{K} \left(\text{erf} \left(\frac{\Delta_{k,L}}{\sqrt{2}\sigma(p_k)} \right) + \text{erf} \left(\frac{\Delta_{k,R}}{\sqrt{2}\sigma(p_k)} \right) \right) \tag{14}$$

where $\Delta_{k,L} = \mu(p_k) - d_{k-1}$ and $\Delta_{k,R} = d_k - \mu(p_k)$.

Proof. Since z is a Gaussian variable that has been proved, the correct probability of each p_k can be obtained as follows

$$P(p_k) = \Pr(d_{k-1} \leqslant z < d_k)$$

$$= \frac{1}{2} \left(\text{erf} \left(\frac{\mu(p_k) - d_{k-1}}{\sqrt{2}\sigma(p_k)} \right) + \text{erf} \left(\frac{d_k - \mu(p_k)}{\sqrt{2}\sigma(p_k)} \right) \right). \tag{15}$$

The error probability is $P_e(p_k) = 1 - P(p_k)$, thus the average error probability, P_e, is equal to $\frac{1}{K} \sum_{k=1}^{K} P_e(p_k)$, Proposition 1 is proved.

It is worth noting that the expression in (14) is a generalized result suitable for a variety of non-negative constellations. Given variance $\sigma(p_k)$ and decoding regions, one can obtain the error probability. Moreover, the result in (14) reveals how antenna correlation affects the error performance. When M, SNR and constellation size are fixed, $\sigma^2(p_k)$ grows with a larger ρ. Since SER is an increasing function of $\sigma^2(p_k)$, the error probability will increase if channels of different transmit-receive pairs are more correlated. In the limit of $\rho \to 1$ and SNR $\to \infty$, the following results is obtained

$$\lim_{\rho \to 1} \frac{\Delta_{k,R}^2}{\sigma^2(p_k)} = \frac{(d_k - \mu(p_k))^2}{2}. \tag{16}$$

$$\lim_{\sigma_n^2 \to 0} \sigma^2(p_k) = \frac{1}{M} p_k^2 + \frac{f(\rho)}{M^2}. \tag{17}$$

It is readily observed from (16) that no matter how large the number of receive antennas is, it will not be helpful to reduce SER. On the other hand, as long as ρ is not that large, increasing M can reduce error rate. From (17), it can be found that $\sigma^2(p_k)$ will not converge to zero even if SNR $\to \infty$, which means that an error floor appears in high SNR regions.

4.2 Achievable Rate Analysis

Proposition 2. *The SNR of received signal at BS can be represented as*

$$\gamma \sim \frac{X_1}{X_2} \tag{18}$$

where X_1 and X_2 are independent real Gaussian random variables, namely

$$X_1 \sim \mathcal{N}\left(\mu_{X_1}, \sigma_{X_1}^2\right), X_2 \sim \mathcal{N}\left(\mu_{X_2}, \sigma_{X_2}^2\right) \tag{19}$$

where

$$\mu_{X_1} = p_k^2 \left(1 + \frac{M + f(\rho)}{M^2}\right), \mu_{X_2} = \sigma_n^4 \left(1 + \frac{1}{M}\right) + \frac{2p_k \sigma_n^2}{M},$$

$$\sigma_{X_1}^2 = p_k^4 \left(\frac{2 + M + f(\rho)}{M^2} + \frac{4Mf(\rho) + 2f^2(\rho)}{M^4}\right), \sigma_{X_2}^2 = 2\sigma_n^8 \left(\frac{1 + 2M}{M^2}\right) + \frac{8p_k^2 \sigma_n^4}{M^2}.$$

Proof. From (7), the SNR of received signal at BS is defined as

$$\gamma = \frac{\left(\frac{1}{M} \sum_{i=1}^{M} \lambda_i |v_i|^2\right)^2 x^4}{\left(\frac{1}{M} \sum_{i=1}^{M} |n_i|^2\right)^2 + \left(\frac{2}{M} \sum_{i=1}^{M} \lambda_i^{\frac{1}{2}} \Re(q_i v_i)\right)^2 x^2}. \tag{20}$$

At first, $\frac{1}{M} \sum_{i=1}^{M} \lambda_i |v_i|^2$ follows a non-zero mean Gaussian distribution when M is large according to Lemma 1. Therefore, the numerator of (20) is a non-central Chi-square random variable. When the variance of a non-central Chi-square distribution is small enough, it can also be approximated as a Gaussian distribution [9]. In the same way, the denominator of (20) is able to be considered as a Gaussian variable too.

From Proposition 2, the achievable rate with respect to the kth constellation point is able to be computed by averaging over X_1 and X_2

$$R_k = \mathbb{E}_{X_1, X_2} \left\{ \log_2 \left(1 + \frac{X_1}{X_2}\right) \right\}. \tag{21}$$

Proposition 3. *In the presence of antenna correlation, the achievable rate when* $\sqrt{p_k}$ *is transmitted is given by*

$$
\begin{aligned}
R_k = {} & \frac{\log_2 e}{\sqrt{\pi}} \sum_{i=0}^{n} W_i \ln(1 + v_i) K(v_i) \\
& - \frac{\log_2 e}{m\sqrt{\pi}} \sum_{i=1}^{m-1} \ln\left(1 + \frac{i}{m}\right) K\left(\frac{i}{m}\right) - \frac{1}{2m\sqrt{\pi}} K(1) \\
& + \frac{\log_2 e}{\sqrt{\pi}} \sum_{i=0}^{n} \frac{A_i}{2\sqrt{s_i}} \ln\left(\frac{\mu_z \mu_{X_2} + \sqrt{2}\mu_{X_2}\sigma_z\sqrt{s_i}}{\mu_z \mu_{X_2} + \sqrt{2}\mu_z\sigma_{X_2}\sqrt{s_i}}\right) \\
& + \frac{\log_2 \mu_z}{2} \mathrm{erfc}\left(-\frac{\mu_z}{\sqrt{2}\sigma_z}\right) - \frac{\log_2 \mu_{X_2}}{2} \mathrm{erfc}\left(-\frac{\mu_{X_2}}{\sqrt{2}\sigma_{X_2}}\right)
\end{aligned}
\tag{22}
$$

where

$$
\mu_z = \mu_{X_1} + \mu_{X_2}, \quad \sigma_z^2 = \sigma_{X_1}^2 + \sigma_{X_2}^2 \tag{23}
$$

$$
K(x) = \frac{\mu_z}{\sqrt{2}\sigma_Z} e^{-\left(\frac{\mu_z}{\sqrt{2}\sigma_z}x\right)^2} - \frac{\mu_{X_2}}{\sqrt{2}\sigma_{X_2}} e^{-\left(\frac{\mu_{X_2}}{\sqrt{2}\sigma_{X_2}}x\right)^2} \tag{24}
$$

and the value of W_i *and* v_i *are derived from Gauss–Legendre quadrature formula, the value of* A_i *and* s_i *is derived from Gauss–Laguerre quadrature formula.* $1/m$ *is the step in compound trapezoid formula.*

Proof. The proof is omitted because of the length constraint.

The average achievable rate can be simply calculated by $R = \frac{1}{K} \sum_{k=1}^{K} R_k$.

5   Numberical Results

Monte Carlo simulations are performed to illustrate the effect of antenna correlation and verify our analysis. We assume that the non-negative PAM is employed and the channel is Rayleigh fading with a correlation matrix $\mathbf{\Phi}_r$.

Figure 1 shows SER versus SNR for different numbers of receive antennas, where $K = 4$ and $\rho = 0.5$. As expected, when M increases, the SER decreases as a consequence. However, there exists a distinct discrepancy between simulation and analytical results. This is attributed to the CLT approximation, where the tail of Gaussian distribution shows a slight difference with actual distribution. Although it is small in absolute value, the logarithmic representation in Fig. 1 will amplify this difference. However, the tendency of simulation and analytical curves is quite similar. Beside, P_e will converge to a non-zero error floor with SNR growing. Generally, there are two approaches to reduce the error floor, one is to employ more the number of antennas, and the other is constellation optimization.

The impact of antenna correlation on the error rate can be further verified in Fig. 2, where $K = 4$ and SNR = 6 dB. This figure clearly demonstrates the

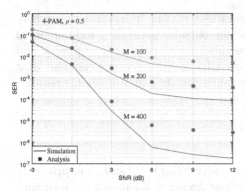

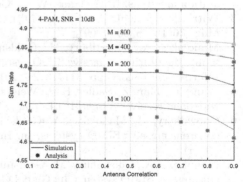

Fig. 1. SER versus SNR for different numbers of receive antennas, where $K = 4$ and $\rho = 0.5$

Fig. 2. SER versus antenna correlation with various number of antennas, where $K = 4$ and SNR $= 6$ dB.

adverse effect of antenna correlation on error performance. Meanwhile, the performance gain provided by massive antenna array would be counteracted by spatially correlated channels.

Figure 3 plots the relationship between the achievable rate and SNR in the presence of antenna correlation, where $K = 4$ and $\rho = 0.5$. The numerical results are obtained by performing simulation using (20), while analytical results are computed with (22). Unlike the situation of SER comparison, the numerical and analytical results of achievable rate fit each other very well. Furthermore, since there is no interference in the considered system model, the achievable rate increases unboundedly with growing SNRs.

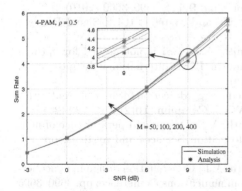

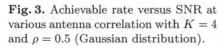

Fig. 3. Achievable rate versus SNR at various antenna correlation with $K = 4$ and $\rho = 0.5$ (Gaussian distribution).

Fig. 4. Achievable rate versus antenna correlation at various number of antennas with $K = 4$ and SNR $= 10$ dB (Gaussian distribution).

Figure 4 shows how the achievable rate varies with antenna correlation, where $K = 4$ and SNR $= 10$ dB. The remarkable gap between analytical and numerical results at $M = 100$ arises because the number of antennas is insufficient and resulting Gaussian approximation by using CLT is not accurate enough. Most importantly, for a large range of antenna correlation, the sum rate almost remains unchanged, especially when $M > 200$.

6 Conclusion

Non-coherent receivers are attractive in massive SIMO systems, due primarily to their low complexity and cost. This paper presents a through performance analysis of non-coherent massive SIMO systems over spatially correlated channels.

We have derived the approximated analytical closed-form expression of the average SER based on CLT. Simultaneously, the achievable rate is given according to Gaussian distribution approximation. Both analytical and numerical results indicate an error floor of P_e will appear at high SNRs, which can be reduced by constellation optimization or increasing the number of receive antennas. Interestingly, the simulation results report that the antenna correlation has far less impact on the achievable rate than the error probability.

References

1. Lu, L., Li, G.Y., Swindlehurst, A.L., Ashikhmin, A., Zhang, R.: An overview of massive MIMO: benefits and challenges. IEEE J. Sel. Top. Signal Process. **8**(5), 742–758 (2014)
2. Marzetta, T.L.: Noncooperative cellular wireless with unlimited numbers of base station antennas. IEEE Trans. Wirel. Commun. **9**(11), 3590–3600 (2010)
3. Witrisal, K., et al.: Noncoherent ultra-wideband systems. IEEE Signal Process. Mag. **26**(4), 48–66 (2009)
4. Manolakos, A., Chowdhury, M., Goldsmith, A.J.: CSI is not needed for optimal scaling in multiuser massive SIMO systems. In: IEEE International Symposium on Information Theory, pp. 3117–3121 (2014)
5. Wang, F., Tian, Z., Sadler, B.M.: Weighted energy detection for noncoherent ultra-wideband receiver design. IEEE Trans. Wirel. Commun. **10**(2), 710–720 (2011)
6. Chowdhury, M., Manolakos, A., Goldsmith, A.J.: Design and performance of non-coherent massive SIMO systems. In: Information Sciences and Systems, pp. 1–6 (2014)
7. Manolakos, A., Chowdhury, M., Goldsmith, A.J.: Constellation design in noncoherent massive SIMO systems. In: Global Communications Conference, pp. 3690–3695 (2015)
8. Manolakos, A., Chowdhury, M., Goldsmith, A.: Energy-based modulation for noncoherent massive SIMO systems. IEEE Trans. Wirel. Commun. **15**(11), 7831–7846 (2015)
9. Jing, L., Carvalho, E.D., Popovski, P., Martnez, A,O.: Design and performance analysis of noncoherent detection systems with massive receiver arrays. IEEE Trans. Signal Process. **64**(19), 5000–5010 (2016)

10. Biswas, S., Masouros, C., Ratnarajah, T.: Performance analysis of large multiuser MIMO systems with space-constrained 2-d antenna arrays. IEEE Trans. Wirel. Commun. **15**(5), 3492–3505 (2016)
11. Masouros, C., Sellathurai, M., Ratnarajah, T.: Large-scale MIMO transmitters in fixed physical spaces: the effect of transmit correlation and mutual coupling. IEEE Trans. Commun. **61**(7), 2794–2804 (2013)
12. Castro-Ors, I.D., et al.: Impact of antenna correlation on the capacity of multi-antenna channels. IEEE Trans. Inf. Theory **51**(7), 2491–2509 (2005)
13. Alfano, G., Tulino, A.M., Lozano, A., Verdu, S.: Capacity of MIMO channels with one-sided correlation. In: IEEE Eighth International Symposium on Spread Spectrum Techniques and Applications, pp. 515–519 (2004)
14. Chatzinotas, S., Imran, M.A., Hoshyar, R.: On the multicell processing capacity of the cellular MIMO uplink channel in correlated Rayleigh fading environment. IEEE Trans. Wirel. Commun. **8**(7), 3704–3715 (2009)
15. Park, D., Park, S.: Performance analysis of multiuser diversity under transmit antenna correlation. IEEE Trans. Commun. **56**(4), 666–674 (2008)
16. Hudson, W.N., Tucker, H.G.: Asymptotic independence in the multivariate central limit theorem. Ann. Probab. **7**(4), 662–671 (1979)

Coalition Formation Game Based Energy Efficiency Oriented Cooperative Caching Scheme in UUDN

Yu Li[✉], Heli Zhang, Hong Ji, and Xi Li

Key Laboratory of Universal Wireless Communications, Ministry of Education,
Beijing University of Posts and Telecommunications, Beijing, China
{liyu1,zhangheli,jihong,lixi}@bupt.edu.cn

Abstract. It is generally considered that Ultra-Dense Network (UDN) is a promising solution for 5G and the network is going to turn into user centric. Caching popular contents at the edge of network is an efficient way to reduce the energy consumption and data traffic of backhaul link. But most of current researches on caching in UDN fail to take into account of user centric and energy efficiency performance during caching files delivery process. In this paper, we consider an User-centric Ultra-Dense Network (UUDN) with cache-enabled Small Base Stations (SBSs) and investigate the energy efficiency of cooperative caching in UUDN. In order to achieve energy efficiency during delivery, we design a novel SBS grouping rule and a cooperative caching scheme based fragmentation with the consideration of user mobility. We formulate an energy optimization problem on caching and introduce coalition formation game to simplify and solve our optimization objective. Then we analyze the impacts of system parameters on the overall performance and compare our scheme to some other schemes. Numerical results demonstrate our scheme is energy efficient and outperforms the others.

Keywords: Energy efficiency · Cooperative cache · UUDN · Content fragmentation · Coalition game

1 Introduction

Cache popular contents at the wireless edge of network is an extensively accepted technology in 5G [6]. Cache can offload backhaul burden and improve energy efficient of the network by reducing duplicate downloads effectively [5]. UUDN aims at making every user feel like a network is always following it via very high data rate and intelligent service [3]. Thus it is imperative to deploy cache in UUDN to provide high-quality intelligent service.

© ICST Institute for Computer Sciences, Social Informatics and Telecommunications Engineering 2019
Published by Springer Nature Switzerland AG 2019. All Rights Reserved
X. Liu et al. (Eds.): ChinaCom 2018, LNICST 262, pp. 458–468, 2019.
https://doi.org/10.1007/978-3-030-06161-6_45

Recently, more and more researches focus on caching at SBSs, e.g. femto-cell or picocell. Reference [6] demonstrates that energy efficient can benefit from caching and investigates the key parameters and locations that influence energy efficiency of caching significantly. However, considering the characters of SBSs, there are some obstacles on deploying cache in UDN [13]. First, the storage capacity of a single SBS is too limited to cache enough popular contents especially the large volume of multimedia contents. And the number of users under each SBS is too small to reflect the content aggregation effect [12]. Jointing SBSs to cache cooperatively is an effective solution to these problems and can improve network performance in multiple aspects in comparison with the non-cooperative one. In [7], the potential of energy efficiency in cache-enabled cooperative dense small cell networks is explored based on affinity propagation-based clustering. Reference [4] designs a combined cooperative caching and transmission policy in cluster-centric small cell network. Reference [2] investigate the problem of caching placement on SBS leveraging user mobility, aiming to maximize the cache hit ratio. In [3], UUDN has been defined and the authors analyze challenges and requirements of UUDN. Reference [8] investigate the problem of dynamic access point grouping in UUDN. However, most of the researches on caching fail to consider the characteristics of UUDN or ignore the delivery energy efficiency of caching files in UUDN.

In this paper, we investigate the issue of cooperative caching in UUDN in order to optimize the delivery energy efficient of caching files. We consider SBSs can form an SBS group (SBSG) to serve an User Equipment (UE) cooperatively in our UUDN. SBSG cache files according to the UE's preference. We divide the files into several fragments to cache in different SBSs. The reason of fragmentation is the burden of SBSs can be decentralized and multiple SBSs share a transmission task is more stable. In addition, we formulate an energy optimization problem on the basis of our caching scheme. Then we introduce coalition formation game to settle the optimization objective. Coalition games prove to be a very powerful tool for designing fair, practical and efficient cooperation strategies in communication networks [11]. And in this work, coalition formation game simplify and solve our problem effectively. Numerical results demonstrate that our scheme and proposed algorithm can improve the delivery energy efficiency to a great degree. The major contributions of this paper are summarized as follows:

- We study energy efficiency of cooperative caching and UUDN architecture. Then we formulate an cache-enabled UUDN model basing the previous study.
- We propose a novel SBS grouping rule and a fragmentation caching scheme with the goal of improving energy efficiency. We formulate an optimization objective on reducing energy consumption in order to improve energy efficiency.
- We introduce coalition formation game and evolve it to adapt to and solve our objective problem. Numerical results demonstrate that our work is of vital benefit and outperforms the others.

The remainder of this paper is organized as follows. Section 2 gives the system model and problem formulation. In Sect. 3, coalition formation game is introduced and evolve to solve our problem. Numerical results are given in Sect. 4. Finally, we conclude this paper in Sect. 5.

2   System Model and Problem Formulation

In this section, we introduce the system model and formulate the optimization problem.

2.1   System Model

(a) Network model: As illustrated in Fig. 1, we consider an user-centric ultra-dense wireless small cell network consisting of cache-enabled SBSs. When an UE joins the network, the SBSs around the UE would form an SBSG to serve it cooperatively. In the SBSG, SBS can communicate with the UE directly or through multi-hops. For notation ease, we denote $\mathbb{M} = \{1, 2, \cdots, M\}$ as the SBSG set and m as index for the m-th SBS in a SBSG where $m \in \mathbb{M}$. In each SBSG, an enhanced SBS (ESBS) is selected dynamically to control and manage the others.

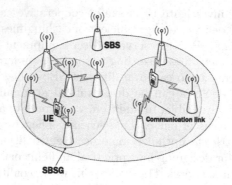

Fig. 1. Network structure

(b) Request model: We consider that all the requested files are ranked into a requesting library $\mathbb{F} = \{f_1, f_2, \ldots, f_{N_L}\}$ with all the files having same size q, where f_i is the i-th most popular file of the library and N_L is the total number of files in the library. We assume that the request probability distribution $\mathbb{P} = \{p_1, p_2, \ldots, p_{N_L}\}$ follows Zipf's distribution [1], thus the request probability of f_i is calculated as:

$$p_i = \frac{i^{-\gamma}}{\sum_{j=1}^{N_L} j^{-\gamma}} \tag{1}$$

where γ is the skewness parameter of which typical value is between 0.5 and 1.0 reflecting different levels of skewness of the distribution.

(c) Caching model: In our UUDN, each SBS is equipped with a cache which capacity is C. Files are cached according to the order of requesting library and we regard an SBSG as a cache entirety. For simplification, we assume an SBS only belong to one single SBSG. Thus the first N_C files in the requesting library are cached in the SBSG. Each caching file is divided into same-sized fragments to cache in different SBSs respectively. Denote $\mathbb{L} = \{l_1, l_2, \ldots, l_{N_C}\}$ as the fragmentation matrix in which l_i represents how many fragments f_i is divided into. We denote $f_i^{(j)}$ as the j-th fragment of file f_i, where $j \in \mathbb{Z}^+$ is between 1 and l_i. And $\mu_m^{i,j}$ is the cache coefficient denoting whether the m-th SBS in the SBSG caches $f_i^{(j)}$. ESBS has the information of $\mathbb{L}$ and each $\mu_m^{i,j}$.

(d) Energy model: For wireless links, we adopt Rayleigh fading model. Hence the achievable data transmission rate among SBSs is formulated as:

$$r_{m_1,m_2} = Wlog_2(1 + \frac{pd_{m_1,m_2}^{-\alpha}}{\beta pI_{m_2} + \sigma^2}) \tag{2}$$

where $m_1, m_2 \in \mathbb{M}$ and $m_1 \neq m_2$, W and $d_{m_1m_2}$ is the transmission bandwidth and the distance between m_1 and m_2 respectively, α is the path-loss exponent of small-scale Rayleigh fading channel, I_{m_2} is the power of ICI (inter-cell interference) at the SBS m_2 normalized by the transmit power p, σ^2 is the variance of the white Gaussian noise. Besides, $\beta \in [0, 1]$ reflects the percentage of how much of ICI can be eliminated by interference management techniques [6], i.e. $\beta = 0$ reflects the optimistic condition in which all ICIs are assumed to be removed and $\beta = 1$ represents the pessimistic case.

Similarly, the achievable throughput from SBS m to UE can be calculated by:

$$r_m = Wlog_2(1 + \frac{pd_m^{-\alpha}}{\beta pI + \sigma^2}) \tag{3}$$

where $m \in \mathbb{M}$ and d_m is the distance between the SBS m and the UE, I is the power of ICI at the UE normalized by the transmit power p.

We denote w_{m_1,m_2} as the energy consumption coefficient between SBS m_1 and m_2. They can be calculated by:

$$w_{m_1,m_2} = \frac{p}{r_{m_1,m_2}} \tag{4}$$

Similarly, for the SBS communicating with the UE directly, the energy consumption coefficient is:

$$w_m = \frac{p}{r_m} \tag{5}$$

For the SBS communicating with the UE through multi-hops, the transmission path and energy consumption coefficient with the UE can be calculated by Dijkstra algorithm and denoted as w_m as well. We set w_0 as a threshold that all the SBS with the w_m less than w_0 can join the SBSG of the UE.

2.2   Problem Formulation

We formulate the main optimization objective investigated throughout the paper in this section. In our research, we aim at maximize the average energy efficiency during the content delivery process. As q is constant, our optimization problem is to minimize the average content transmission energy consumption as below:

$$\min_{l_i, \mu_m^{i,j}} \sum_{i=1}^{N_C} \frac{p_i}{\sum_{i=1}^{N_C} p_i} \left(\sum_{j=1}^{l_i} \sum_{m=1}^{M} \mu_m^{i,j} w_m \frac{q}{l_i} \right)$$

$$s.t. \quad C1 : 1 \leq l_i \leq M$$

$$C2 : N_C \leq \frac{MC}{q}$$

$$C3 : \mu_m^{i,j} = \{0,1\} \tag{6}$$

$$C4 : \sum_{m=1}^{M} \mu_m^{i,j} = 1$$

$$C5 : \sum_{j=1}^{l_i} \sum_{m=1}^{M} \mu_m^{i,j} = l_i$$

$f_1{}^1$	$f_1{}^2$	$f_1{}^3$	$f_1{}^4$
$f_2{}^1$	$f_2{}^2$	$f_2{}^3$	$f_2{}^4$
$f_3{}^1$	$f_3{}^2$	$f_3{}^3$	$f_3{}^4$
SBS1	SBS2	SBS3	SBS4

Coalition 1	Coalition 2	Coalition 3
SBS1,SBS5	SBS3	SBS2,SBS4,SBS6
File1~File8	File9~File12	File13~File24

(a) The coalition consists of SBS1, SBS2, SBS3 and SBS4 and file 1,2,3 need to be cached in the coalition.

(b) Coalition 1 with two SBSs caches the first 8 files, Coalition 2 with one SBS caches 9th-12th files and Coalition 3 with three SBSs caches 13th-24th files.

Fig. 2. Allocation rule

Here $C1$ is the fragmentation constraint, $C2$ is the caching capacity constraint, $C4$ means each fragment is only cached at a single SBS and $C5$ means the SBSG cache the whole f_i.

3   Proposed Solution

From the above description, we can know that several SBSs cache different parts of a file cooperatively in an SBSG. Thus we use coalition formation game to divide the SBSG into several small coalitions. SBSs in the same coalition cache different fragments of same files cooperatively. Coalition formation game usually aim to seek cooperative group where network structure and cost for cooperation play a major role. The following of this section will introduce the algorithm based coalition formation game in detail.

In order to solve the objective properly, we make some adaptive settings on the optimal objective. As previously mentioned, several SBSs composing a coalition cache different part of same files. In the coalition, each SBS is numbered according to their w_* in the order of small to large. The number of fragments the caching files divided into is the same with the size of their caching coalition and the sequence numbers of the SBSs and fragments are corresponding. As is described in Fig. 2(a), the blocks in four colors represent the caching space of the four SBSs and they are signed by the fragments cached inside them. In addition, the files are cached in the order of coalition. As illustrated in Fig. 2(b), Coalition 1 caches File 1- File 8, Coalition 2 caches File 9- File 12 and so forth. The ranking rule of coalitions will be described in the following part.

We denote $\mathbb{S} = \{S_1, S_2, \cdots, S_J\}$ as the coalition set, in which S_j is the j-th coalition in the SBSG, $\bigcup_{j=1}^{j=J} S_j = \mathbb{M}$ and $S_i \bigcap S_j = 0$ for any $i \neq j$. A coalition set is usually regarded as a partition. We define the utility function to evaluate the energy efficiency capacity of coalition $\mathbb{S}$ as following:

$$v(\mathbb{S}) = \sum_j P_j \overline{w_j} \tag{7}$$

where P_j and w_j are the total file request probability and the average of energy consumption coefficient in the coalition S_j respectively. They can be calculated by:

$$P_j = \sum_{\sum_{i=1}^{j-1} \alpha_i + 1}^{\sum_{i=1}^{j} \alpha_i} P_i \tag{8}$$

$$\overline{w_j} = \frac{\sum_k w_k}{\alpha_j} \tag{9}$$

in which SBS k belong to the coalition S_j. α_j represents the total cache capacity of S_j i.e. how many files can be cached in S_j in all calculated by:

$$\alpha_j = \lfloor \frac{|S_j|C}{q} \rfloor \tag{10}$$

where $|S_j|$ is the number of SBSs in coalition S_j.

Algorithm 1 Coalition formation game based algorithm

1: Initialization:
   Set each SBS form a coalition itself in the SBSG and this partition is the initial $\mathbb{S}$.
2: Calculate the initial $\mathbb{S}$ according to (8).
3: **repeat**
4:     Arbitrary merging and splitting in $\mathbb{S}$ to form a temporary $\mathbb{S}'$.
5:     Calculate the utility value $v(\mathbb{S}')$ according to (8).
6:     **if** $v(\mathbb{S}') < v(\mathbb{S})$ **then**
7:         $\mathbb{S} = \mathbb{S}'$
8:     **end if**
9: **until** $\mathbb{S}$ converge to a $\mathbb{D}_{hp}$-stable partition

---

In our coalition formation game based algorithm, there are several key operations described as follows:

1. Ranking: Calculate all the $\overline{w_j}$s in $\mathbb{S}$ and rerank the coalitions in the order of small to large.
2. File allocation: Assign files to each coalition. The coalitions with smaller numbers cache the top files in the requesting library, as described in Fig. 2b. Thus the files with high requesting probability can be transmitted in a low energy consumption.
3. Merging: Merge any set of coalitions $\{S_1, \cdots, S_k\}$ to form a new coalition S_j thereby composing a new partition $\mathbb{S}'$.
4. Splitting: Split any coalition S_j into several coalitions $\{S_1, \cdots, S_k\}$ to compose a new partition $\mathbb{S}'$ where $\bigcup_{i=1}^{i=k} S_i = S_j$.

The ultimate propose of our algorithm is to converge to a final stable partition by any arbitrary sequence of merging and splitting. According to [10], every partition resulting from our proposed merging and splitting way is $\mathbb{D}_{hp}$-stable. In a $\mathbb{D}_{hp}$-stable partition $\mathbb{S}$, no players in $\mathbb{S}$ are interested in leaving their coalitions through merging and splitting to form other partitions. The specific process of our proposed algorithm is summarized in Algorithm 1.

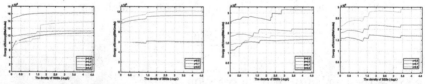

(a) UE1: Energy ef-(b) UE1: Energy ef-(c) UE2: Energy ef-(d) UE2: Energy efficiency versus SBS ficiency versus SBS ficiency versus SBS ficiency versus SBS
density with different density with different γ density with different density with different γ
β β

Fig. 3. Impact of SBS density λ

4   Numerical Results

In this section, simulation results of the proposed coalition formation game based algorithm are presented to discuss the energy efficiency performance and the impacts of different parameters in our UUDN. In this work, we consider the SBSs in the network are uniform distribution and the density of SBSs is λ. But we must emphasize that our caching scheme and proposed algorithm can be further generalized to any network topology. The simulation parameters are described as follows. The transmission power of SBS is 0.1 W. The background noise is -95 dBm. The path-loss exponent α of small-scale Rayleigh fading channel is 3. The transmission bandwidth of SBSs is 200 kHz. There are 1000 files in the requesting library and the size q of each file is 50 Mb. The impacts of SBS density λ, threshold of SBSG w_0, SBS's storage capacity C, interference elimination coefficient β and skewness parameter of Zipf's distribution γ are investigated in the next part of this section.

(a) *Impact of SBS density*: In Fig. 3, we first investigate the impact of SBS density λ with $C = 500$ Mb and $w_0 = 10^{-5}$ Joule/bit. After extensive simulation, we find that the energy efficiency is related to the UEs' location with different SBS density. So we pick two represented UEs to investigate the impact of SBS density. Figure 3(a) and (b) is about UE1 and Fig. 3(c) and (b) is about UE2. In order to decentralize the observation point, we set the x axis as $-log^\lambda$ and λ's unit is m^{-2}. All of the four figures has similar tendency that is rising a little. As $-log^\lambda$ is a monotone decreasing function, the λ is smaller, the $-log^\lambda$ is larger. Thus the curves on the rise mean that the energy efficiency is descending lightly with the density increasing. And when the distribution of SBSs is relatively dense, the energy efficiency is fluctuant a little with λ change. The fluctuation is result from the member of SBSG change frequently when the λ is larger. In addition, the interference elimination coefficient β and skewness parameter γ have obvious influence on the energy efficiency. Lower β or higher γ can generate higher energy efficiency. And β can influence the overall trend to varying degrees while γ can not. This is because β changes the $SINR$ which influences the SBSG while γ change the requesting probability distribution which influence the caching files assignment.

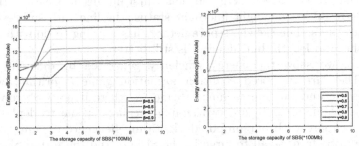

(a) Energy efficiency versus storage capacity with different β

(b) Energy efficiency versus storage capacity with different γ

Fig. 4. Impact of SBS storage capacity C

(b) *Impact of storage capacity*: In Fig. 4, we plot the energy efficiency with respect to the SBS storage capacity C with $\lambda = 0.04\,\mathrm{m}^{-2}$. β and γ are serving as the other parameters in (a) and (b) respectively as well. In Fig. 4(a), all of the four tracks are increasing rapidly first and then turn the rapidly increasing into gently increasing. Smaller β has bigger slope and more increase. However, when the β is large, there is a slow rising before the rapidly increasing. In Fig. 4(b), the tracks of $\gamma = 0.5$, $\gamma = 0.6$, $\gamma = 0.8$ and $\gamma = 0.9$ have similar tendency that is the energy efficiency rises gently with the storage capacity getting larger. However, there is a sharp rise on the track of $\gamma = 0.7$ reaping from the low energy efficiency group to the high energy efficiency group. Thus $\gamma = 0.7$ can be seen as a watershed.

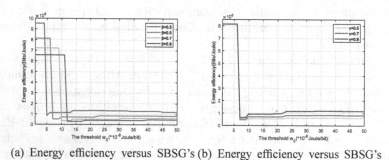

(a) Energy efficiency versus SBSG's threshold with different β (b) Energy efficiency versus SBSG's threshold with different γ

Fig. 5. Impact of SBSG's threshold w_0

(c) *Impact of SBSG's threshold*: Fig. 5 depicts the energy efficiency with different SBSG's threshold w_0 under different β and γ respectively in (a) and (b). Their overall trends are similar that are the energy efficiency keeps a stable status first, then declines dramatically and enters a slight fluctuant status. All the decline slopes are the same. And the sharp decline result from SBSs with higher w_* join in the SBSG then the average of energy costs coefficient is improved. The interference elimination coefficient β can affect the first status's value and length. Smaller β has higher energy efficiency first but declines earlier. However, when they are all in the slight fluctuant status, the line with smallest β still exceeds the others. The skewness parameter γ has nothing to do with the first status's value and length, but higher γ is still better.

In Fig. 6, we show the energy efficiency generated by different caching schemes. They are proposed coalition formation game algorithm, sequence caching scheme and random caching scheme. As there is no work solving our proposed problem, we only use two simple caching scheme to deploy in our UUDN to compare with our proposed algorithm. Sequence caching scheme is sorting the SBSs from small to large in the SBSG. And the sorted SBSs cache files after the previous one filled up with files. Random caching scheme is the files in the requesting library are cached in random SBS in order. We notice

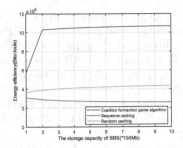

Fig. 6. Energy efficient performance comparison

that the energy efficiency of our proposed algorithm is higher than the other two schemes obviously. So our fragmentation scheme and coalition formation game are benefit.

5   Conclusion

In this paper, we concentrate on the issue of energy efficient cooperative caching in UUDN. We define a novel but simple SBS grouping rule basing the energy consumption coefficient to form SBSG. Considering the situation that multiple SBSs serve one UE cooperatively in UUDN, we divide each file into several fragments to cache in different SBSs. For solving the optimizing problem effectively, coalition formation game is introduced and promoted. Simulation results demonstrate that our proposed scheme and algorithm is energy efficient and outperforms the others. For future work, we could investigate the tradeoff between service quality and energy costs or deployment costs. Or we could turn our focus into investigating the dynamic contents by online caching.

Acknowledgement. This paper is sponsored by the National Science and Technology Major Project of China (Grant No.2017ZX03001014).

References

1. Breslau, L., Cao, P., Fan, L., Phillips, G., Shenker, S.: Web caching and Zipf-like distributions: evidence and implications. In: INFOCOM '99, Eighteenth Annual Joint Conference of the IEEE Computer and Communications Societies. Proceedings, vol. 1, pp. 126–134. IEEE (1999). https://doi.org/10.1109/INFCOM.1999. 749260
2. Chen, M., Hao, Y., Hu, L., Huang, K., Lau, V.K.N.: Green and mobility-aware caching in 5G networks. IEEE Trans. Wirel. Commun. **16**(12), 8347–8361 (2017). https://doi.org/10.1109/TWC.2017.2760830
3. Chen, S., Qin, F., Hu, B., Li, X., Chen, Z.: User-centric ultra-dense networks for 5G: challenges, methodologies, and directions. IEEE Wirel. Commun. **23**(2), 78–85 (2016). https://doi.org/10.1109/MWC.2016.7462488

4. Chen, Z., Lee, J., Quek, T.Q.S., Kountouris, M.: Cooperative caching and transmission design in cluster-centric small cell networks. IEEE Trans. Wirel. Commun. **16**(5), 3401–3415 (2017). https://doi.org/10.1109/TWC.2017.2682240

5. Kamel, M., Hamouda, W., Youssef, A.: Ultra-dense networks: a survey. IEEE Commun. Surv. Tutor. **18**(4), 2522–2545 (Fourthquarter 2016). https://doi.org/10.1109/COMST.2016.2571730

6. Liu, D., Yang, C.: Energy efficiency of downlink networks with caching at base stations. IEEE J. Sel. Areas Commun. **34**(4), 907–922 (2016). https://doi.org/10.1109/JSAC.2016.2549398

7. Liu, J., Sun, S.: Energy efficiency analysis of cache-enabled cooperative dense small cell networks. IET Commun. **11**(4), 477–482 (2017). https://doi.org/10.1049/iet-com.2016.0680

8. Liu, Y., Li, X., Yu, F.R., Ji, H., Zhang, H., Leung, V.C.M.: Grouping and cooperating among access points in user-centric ultra-dense networks with non-orthogonal multiple access. IEEE J. Sel. Areas Commun. **35**(10), 2295–2311 (2017). https://doi.org/10.1109/JSAC.2017.2724680

9. Ozfatura, E., Gndz, D.: Mobility and popularity-aware coded small-cell caching. IEEE Commun. Lett. **22**(2), 288–291 (2018). https://doi.org/10.1109/LCOMM.2017.2774799

10. Saad, W., Han, Z., Debbah, M., Hjorungnes, A.: A distributed merge and split algorithm for fair cooperation in wireless networks. In: ICC Workshops - 2008 IEEE International Conference on Communications Workshops, pp. 311–315 (May 2008). https://doi.org/10.1109/ICCW.2008.65

11. Saad, W., Han, Z., Debbah, M., Hjorungnes, A., Basar, T.: Coalitional game theory for communication networks. IEEE Signal Process. Mag. **26**(5), 77–97 (2009). https://doi.org/10.1109/MSP.2009.000000

12. Wang, X., Chen, M., Taleb, T., Ksentini, A., Leung, V.C.M.: Cache in the air: exploiting content caching and delivery techniques for 5g systems. IEEE Commun. Mag. **52**(2), 131–139 (2014). https://doi.org/10.1109/MCOM.2014.6736753

13. Zhou, Y., Zhao, Z., Li, R., Zhang, H., Louet, Y.: Cooperation-based probabilistic caching strategy in clustered cellular networks. IEEE Commun. Lett. **21**(9), 2029–2032 (2017). https://doi.org/10.1109/LCOMM.2017.2717398

A Joint Frequency Offset Estimation Method Based on CP and CRS

Xiaoling Hu$^{(\boxtimes)}$, Zhizhong Zhang, and Yajing Zhang

School of Communication and Information Engineering, Chongqing University of Posts and Telecommunications, Chongqing, China
`hu-xiaoling@foxmail.com`

Abstract. In order to solve the problem that Fraction Frequency Offset (FFO) estimation algorithm has the problem of low estimation precision, small range and high occupancy rate of spectrum resource, this paper proposed a FFO estimation method based on the combination of Cyclic Prefix (CP) and Cell-specific reference signals (CRS). First, judging the range of the true frequency offset value according to the results of the frequency offset estimation algorithm based on CP. Then the possible true frequency offset value obtained by adding value calculated by frequency offset estimation algorithm based on CRS and the possible rotation value of 2000 nHz. Finally, comparison the results of the frequency offset estimation algorithm based on CP and the possible true frequency offset, the minimum deviation is its true. The accuracy is the same as that frequency offset estimation algorithm that based on CRS. The range is the same as frequency offset estimation algorithm based on CP, which is [−7500 Hz, 7500 Hz]. The principle of the algorithm is simple and does not occupy additional bandwidth resources.

Keywords: CP · CRS · Combination · FFO

1 Introduction

As a key technology in wireless communication, OFDM has the advantages of high spectrum resource utilization and strong anti-fading ability, but its disadvantage is strict requirements on the orthogonality among sub- carriers. Carrier frequency offset can destroy the orthogonality between subcarriers, leading to Inter-Channel Interference (ICI), which severely degrades receiver performance. Therefore, the frequency offset must be estimated and compensated at the receive side.

This work is supported by the National Science and Technology Major Project of the Ministry of Science and Technology of China (2015ZX03001013), the Generic Technology Innovation Project of Key Industries in Chongqing (cstc2017zdcy-zdzx0030), the Innovation Team of University in Chongqing (KJTD201312).

X. Liu et al. (Eds.): ChinaCom 2018, LNICST 262, pp. 469–479, 2019.
https://doi.org/10.1007/978-3-030-06161-6_46

Frequency offset estimation methods include Integer Frequency Offset (IFO) methods and Fractional Frequency Offset (FFO) methods. IFO causes a cyclic shift in the received data, making the demodulated data completely wrong; the FFO destroys the orthogonality among the subcarriers and seriously affects the performance of the receiver. The research on integer frequency multiplication bias estimation algorithm is already mature [1–3], but the fractional frequency multiplication frequency estimation algorithm still has the problems of low estimation accuracy, small estimation range and high utilization of spectrum resources [4]. Especially In high-speed scenarios, it is difficult to achieve a balance in the frequency offset estimation range, the frequency offset estimation accuracy, and the spectrum resource utilization. Fractional frequency offset estimation methods mainly include frequency offset estimation based on Cyclic Prefix (CP) [5,6], frequency offset estimation based on Reference Signal (RS) [7], and training based on Sequence frequency offset estimation algorithm [8,9], or combination of the above algorithms for frequency offset estimation [10,11].

In this paper, Combination the advantages of estimation range based on CP is large, and the estimation accuracy based on cell-specific reference signals (CRS) is high, and the two algorithms are simple and do not occupy additional frequency offset resources. A fractional frequency offset estimation method based on the combination of CP and CRS is proposed, and the frequency offset estimation range is extended from $[-1000\,\text{Hz}, 1000\,\text{Hz}]$ to $[-7500\,\text{Hz}, 7500\,\text{Hz}]$; the frequency offset estimation accuracy and CRS-based frequency offset estimation are extended. The accuracy of the algorithm is the same; it does not occupy additional frequency band resources; it has little impact on noise and multipath interference. It has very practical in communication engineering.

2   Traditional Frequency Offset Estimation Method

2.1   Frequency Offset Estimation Algorithm Based on CP

The cyclic prefix structure shown in Fig. 1, which can be used to eliminate Inter Symbol Interference (ISI), and it can be used to estimate frequency offset. The CP-based frequency offset estimation algorithm uses a strong correlation between the cyclic prefix part of the OFDM symbol and the end of the signal to calculate the frequency offset. In the CP-based frequency offset estimation algorithm, the performance of the bias estimation algorithm based on one CP is poor [5]. Jointly multiple CPs of OFDM symbols to perform frequency offset estimation can obtain significant performance improvements. Therefore, in order to improve the accuracy of CP-based frequency offset estimation algorithm, this paper combines the information of multiple OFDM symbols to estimate the frequency offset [6]. The CP-based frequency offset estimation algorithm described in paper [6]. Its frequency deviation is:

$$\hat{\varepsilon}_{CP} = -\frac{1}{2\pi} \arg \left\{ \sum_{m=1}^{N_R} \sum_{l=1}^{N_f} \sum_{n=n_1}^{n_2} y_{l,n}^{(m)} y_{l,n+N}^{(m)*} \right\} \tag{1}$$

where $y_{l,n}^{(m)}$ is the received signal includes the CP in time domain. $m \in [1, N_R]$ is the received antenna index. N represents the number of FFT points. N_R is calculating the number of relevant receiving antennas. l is used to calculate the position of the related OFDM symbol in the subframe. $l \in [1, N_f]$ is the OFDM symbol index, and N_f represents the number of OFDM symbols in a subframe used for calculating the correlation; $n \in [n_1, n_2]$ denotes the time index within one OFDM symbol, n_1 denotes the starting position for calculating the relevant cyclic prefix, and n_2 denotes the ending position for calculating the relevant cyclic prefix.

In order to obtain the performance of the CP-based frequency offset estimation algorithm, the CP-based frequency offset estimation algorithm is simulated in the AWGN channel, multipath fading channel, and Rice channel (for simulation parameter settings, see the fourth Section).

Figure 1 shows the performance simulation of the frequency offset estimation algorithm based on CP. In the simulation diagram, the MSE of the frequency offset estimation algorithm is less than 10^{-3}. When the SNR is less than 10 dB, the performance of the frequency offset estimation algorithm in the multipath fading channel and Rice channel is slightly better than that in the AWGN channel. In multipath fading channels and Rician channels, the performance of the frequency offset estimation algorithm is less affected by SNR when the SNR more than 10 dB.

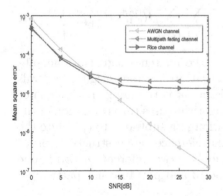

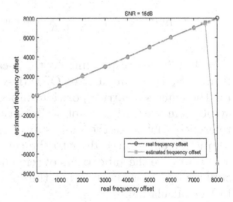

Fig. 1. Accuracy of frequency offset estimation algorithm based on CP in different channels

Fig. 2. Relationship between actual frequency offset and estimated frequency offset in AWGN channel

Figure 2 shows the relationship between the actual frequency offset and the estimated frequency offset under the $SNR = 15$ dB and AWGN channel. From the simulation results, when the value is less than 7500 Hz, the CP-based frequency offset estimation algorithm can obtain better estimation performance, but when the value is greater than 7500 Hz, the frequency offset estimation panning performance drops sharply. Therefore, the estimation range of this frequency offset estimation algorithm is $[-7500\,\text{Hz}, 7500\,\text{Hz}]$.

2.2 Frequency Offset Estimation Algorithm Based on CRS

In paper [7], the frequency offset estimation algorithm based on CRS uses the correlation of the reference symbols of the same subcarrier position of two OFDM symbols in one subframe to calculate the frequency offset.

First, cross-correlate the transmitted CRS reference signal with the CRS reference signal of the receiving end to obtain a channel value for removing noise and multipath interference. Assume that the frequency domain transmit signal is $X_{l,k}^{(m,p)}$ and the frequency domain receive signal is $Y_{l,k}^{(m,p)}$. p is the transmit antenna port index value. then:

$$H_{l,k}^{(m,p)} = Y_{l,k}^{(m,p)} X_{l,k}^{(m,p)*} \tag{2}$$

Then, to do a cross-correlation operation to calculate the frequency offset with two channel values of reference signal of the OFDM symbol in the same sub-carrier position.

$$\hat{\varepsilon}_{CRS} = -\frac{1}{2\pi} \frac{N}{N_s(N + N_{CP})} * \arg\left\{ \sum_{m=1}^{N_R} \sum_{l'=1,2} \sum_{k=1}^{\kappa} H_{l',k}^{(m,p)} H_{l'+1,k}^{(m,p)*} \right\} \tag{3}$$

$l' = 1, 2, 3, 4$ corresponding to the location of the reference signal in one subframe. N_s is the number of OFDM symbols that are reference signal intervals for the same subcarrier position in one subframe. κ is the number of reference symbols in one OFDM symbol. Because the algorithm uses the correlation of the reference symbols of the same subcarrier position in a subframe to calculate the frequency offset, according to the cell-specific reference signal structure, port3 and port4 are the subcarriers of reference symbols in two different OFDM in one subframe. The location is not the same, so port3 and port4 do not participate in the calculation.

Figure 3 shows the accuracy simulation of the frequency offset estimation algorithm based on CRS in different channels. In the simulation results, the frequency offset estimation algorithm based on CRS can obtain good performance in AWGN channel, multipath fading channel and Rice channel, the maximum MSE is less than 10-4. Among them, the performance is best under the AWGN channel. In Rician channel and multi-path fading channel, the signal-to-noise ratio has little effect on the performance of this frequency offset estimation algorithm.

Figure 4 shows the relationship between the actual frequency offset and the estimated frequency offset under a fixed SNR = 15 dB, AWGN channel condition. From the simulation results, we can see that when the value is less than 1000 Hz, the frequency offset estimation algorithm based on CRS can obtain good estimation performance. When the value is greater than 1000 Hz, the accuracy of the estimation method sharply decreases.

2.3 Traditional Joint Frequency Offset Estimation Method

The existing joint frequency offset estimation algorithm is mainly to increase the frequency offset estimation range by increasing symbol estimation. For example, in paper [10,11], the symbol (positive and negative) of frequency offset is judged by using the polarity parameters λ in CP.

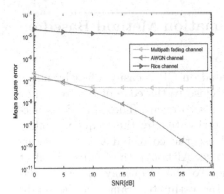 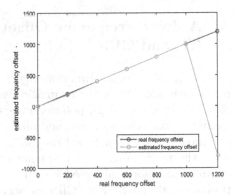

Fig. 3. Accuracy of frequency offset estimation algorithm based on CRS

Fig. 4. The relationship between the actual frequency offset and the estimated frequency offset in the AWGN channel

$$\lambda = sign(\hat{\varepsilon}_{CP}) = \begin{cases} 1 & 0 < \hat{\varepsilon}_{CP} < 7500 \\ 0 & \hat{\varepsilon}_{CP} = 0 \\ -1 & -7500 < \hat{\varepsilon}_{CP} < 0 \end{cases} \tag{4}$$

Sign is used to determine the polarity of $\hat{\varepsilon}_{CP}$.

$$\varepsilon_{CRS} = \begin{cases} \hat{\varepsilon}_{CRS} + 2000 & \lambda > 0 \& \hat{\varepsilon}_{CRS} < 0 \\ \hat{\varepsilon}_{CRS} & others \\ \hat{\varepsilon}_{CRS} - 2000 & \lambda < 0 \& \hat{\varepsilon}_{CRS} > 0 \end{cases} \tag{5}$$

The frequency offset estimation method extends the frequency offset estimation range of $\hat{\varepsilon}_{CRS}$ to $[-2000\,\text{Hz}, 2000\,\text{Hz}]$ in theoretically.

However, this method of calculating frequency deviation has the following problems:

(1) The frequency offset symbol judgment is crucial to the final frequency offset estimation value. If the symbol is judged incorrectly, the frequency offset estimation result will differ by 2π. The frequency offset estimation algorithm based on CP is easily affected by noise, fading, etc. And when the frequency offset value is close to the edge of the frequency offset estimation range, the sign of the frequency offset value is easily reversed, resulting in incorrect polarity parameter λ determination.

(2) The above frequency offset estimation algorithm only doubles the frequency offset estimation range based on the original frequency offset estimation range.

3   A Joint Frequency Offset Estimation Method Based on CP and CRS

The frequency offset estimation algorithm based on CP is easily affect by noise, multipath, etc. When the estimation value is close to the edge of the frequency offset estimation range, polarity inversion tends to occur. Therefore, it is necessary to correct the frequency offset value estimated by the frequency offset estimation algorithm based on the CP, and use the corrected value to determine the range of the true frequency offset value. Although the frequency offset estimation algorithm based on CP is inferior to the frequency offset estimation algorithm based on CRS in accuracy, the estimated range of the modified frequency offset estimation algorithm based on CP is still accurate. When the true frequency offset value is greater than 1000 Hz or less than -1000 Hz, the deviation between the value estimated by the frequency offset estimation algorithm based on the CRS and the true frequency offset value is 2000 nHz ($n = \ldots - 3, -2, -1, 1, 2, 3 \ldots$), so the modified CP-based frequency offset is needed. The estimation value estimated by the estimation algorithm corrects the value calculated by the frequency offset estimation algorithm based on CRS. The overall algorithm flow is as follows (Fig. 5):

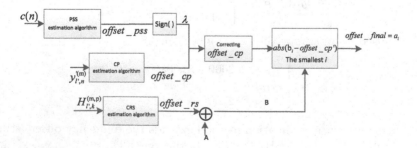

Fig. 5. Algorithm flow

3.1 Polarity Correction of Frequency Offset Estimation Algorithm Based on CP

The polarity correction method of the frequency offset estimation algorithm based on CP is: using the mirror symmetry property of PSS to give the polarity of the frequency offset value, and then using the CP-based frequency offset estimation algorithm to calculate the absolute value of the frequency offset value [6]. Finally, the absolute value of the value calculated by the frequency offset estimation algorithm based on the CP is polarity-maintained with λ.

$$\lambda = sign(\hat{\varepsilon}_{PSS}) = \begin{cases} 1 & 0 < \hat{\varepsilon}_{PSS} < 7500 \\ 0 & \hat{\varepsilon}_{PSS} = 0 \\ -1 & -7500 < \hat{\varepsilon}_{PSS} < 0 \end{cases} \tag{6}$$

$$\hat{\varepsilon}_{cp} = \lambda |\hat{\varepsilon}_{cp}| \tag{7}$$

3.2 Frequency Offset Correction of Frequency Offset Estimation Algorithm Based on CRS

The frequency offset estimation algorithm based on CP and the frequency offset estimation algorithm based on CRS are all based on the sequence phase difference introduced by the frequency offset, and the frequency offset value is obtained by tangent cutting the phase difference. However, the scope of this method is only that $[-\pi, \pi]$, it constrains the frequency offset estimation range, resulting in the frequency offset estimation range of the frequency offset estimation algorithm based on CRS being $[-1000\,\text{Hz}, 1000\,\text{Hz}]$, and phase angle rotation will occur if this frequency offset estimation range is exceeded. For every rotation 2π of the phase angle, the value is differs 2000 Hz. the rotation of the phase angle 2π, the corresponding deviation value of the frequency deviation is 2000 Hz; the rotation of the phase angle 4π, corresponds to a deviation of the frequency deviation of 4000 Hz; In this way, if you know how many weeks the phase angle rotates, you can expand the frequency offset estimation range.

First, calculate frequency offset value of based on CRS. Compare the estimation value and the real value, the rotation of phase angle $2n\pi$ ($n = \ldots -3, -2, -1, 1, 2, 3\ldots$), the corresponding value is different by 2000 nHz. Using the frequency offset estimate $\hat{\varepsilon}_{CRS}$ plus its possible rotation frequency offset gives its possible true frequency offset value ε_{CRS} ($\varepsilon_{CRS} = \hat{\varepsilon}_{CRS} + 2000n$). Finally, compare the CP-based frequency offset estimation value $\hat{\varepsilon}_{CP}$ and the CRS value ε_{CRS}, and the smallest difference is the true frequency offset value. In the frequency offset estimation method, because the estimated range of the CP-based frequency offset estimation algorithm is $[-7500\,\text{Hz}, 7500\,\text{Hz}]$, if the frequency offset value exceeds 7500 Hz, the CP-based frequency offset estimation algorithm will not be able to compare the true frequency offset value. The range is effectively judged so that the frequency offset estimation range based on CRS is limited to $[-7500\,\text{Hz}, 7500\,\text{Hz}]$. The specific steps of correction based on the CRS frequency offset estimation algorithm are as follows.

(1) First set up an A sequence, which is the frequency offset it may rotate:

$$A = [a_1, a_2, a_3, a_4, a_5, a_6, a_7, a_8, a_9]$$
$$= [-8000, -6000, -4000, -2000, 0, 2000, 4000, 6000, 8000] \qquad (8)$$

(2) $\hat{\varepsilon}_{CRS}$ add each value in the A sequence separately to get its possible true frequency offset value.

$$B = [a_1, a_2, a_3, a_4, a_5, a_6, a_7, a_8, a_9]$$
$$= [b_1, b_2, b_3, b_4, b_5, b_6, b_7, b_8, b_9] \qquad (9)$$

(3) Find the smallest $abs(b_i - \hat{\varepsilon}_{CP})$ of i, then the real frequency offset value: offset_final $= b_i$.

4   Analysis of Simulation Results

In order to verify the performance of the joint frequency offset estimation algorithm, this method, the frequency offset estimation algorithm based on CP, and the frequency offset estimation algorithm based on CRS are simulated under the AWGN channel, multipath fading channel and Rice channel, and the simulation parameter settings are as follows (Table 1).

Table 1. Simulation parameters setting

Parameter name	Value
System channel bandwidth	20 MHz
Subcarrier interval	15 kHz
FFT points	2048
CP length	144
The number of antenna ports	1 × 4
Subcarrier number	1200
Channel type	AWGN
Doppler shift	AWGN channel
modulation mode	64QAM
Number of users	1
transmission mode	Open loop spatial multiplexing

Figure 6 shows the relationship between the actual frequency offset and the estimated frequency offset under a fixed $SNR = 30$ dB condition. As can be seen from the figure, the estimation effect is no error.

Figure 7 simulates the frequency offset estimation range of the three algorithms under $SNR = 15$ dB. We can see from the figure that the deviation estimation method rapidly increases when the estimation value is greater than 7500 Hz. Therefore, the estimation range of this frequency offset estimation algorithm is [−7500 Hz, 7500 Hz], which is the same as the estimation range of the frequency offset estimation algorithm based on CP. Therefore, this algorithm extends the frequency offset estimation range based on the frequency offset estimation algorithm based on CRS.

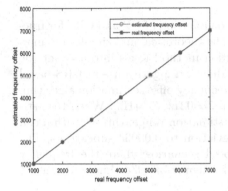

Fig. 6. The relationship between the actual and the estimated frequency offset

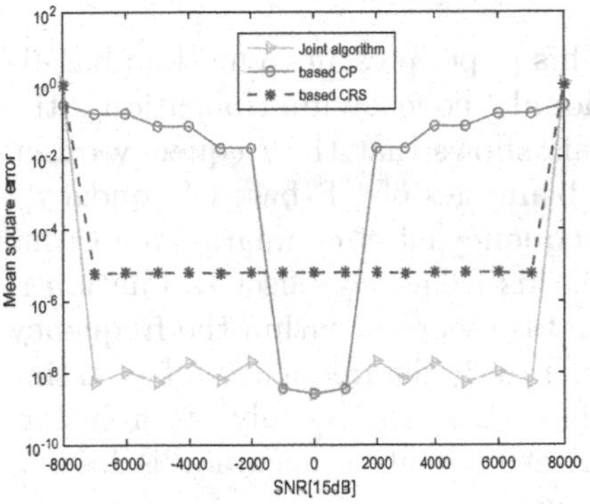

Fig. 7. The range of frequency offset estimation

Figure 8 shows the relationship between system SNR and root-mean-squared error for a fixed frequency offset of 500 Hz, AWGN channel, multi-path fading channel, and Rice channel. It can be seen from the figure that the frequency offset estimation accuracy curve of the present algorithm coincides with the frequency offset estimation accuracy curve of the frequency offset estimation algorithm

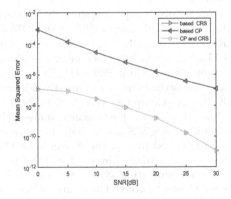

Fig. 8. The accuracy of frequency offset estimation

based on CRS, and the root mean square error thereof under the condition of low signal to noise ratio or high signal to noise ratio. Both are much smaller than the root mean square error of the CP-based frequency offset estimation algorithm. Therefore, the accuracy of the frequency offset estimation method is as high as that of the CRS-based frequency offset estimation algorithm.

5   Conclusion

This paper presents a method based on the combination of CP and CRS for fractional frequency multiplication estimation. The theoretical and simulation analysis shows that the frequency offset estimation method takes into account the advantages of CP-based frequency offset estimation algorithm and CRS-based frequency offset estimation algorithm. The frequency offset estimation algorithm of this frequency offset estimation range is $[-7500\,\text{Hz}, 7500\,\text{Hz}]$. When the estimation value is within the frequency offset estimation range, this algorithm can well limit the frequency offset estimation deviation to 0.00034 subcarriers. The algorithm is especially suitable for high-speed scenarios where the frequency offset estimation accuracy is high and the frequency offset estimation range is large.

References

1. Qiu, W., Schmidl, T.M.: Circuits and methods for frequency offset estimation in FSK communications
2. Huang, Z.: Frequency offset estimation method based on internal penalty function and gradient iteration. In: Modern Navigation (2017)
3. Bhowmick, S., Vasudevan, K.: An improved differential correlation based integer frequency offset estimation method for OFDM signals. In: IEEE, Uttar Pradesh, pp. 334–339. IEEE (2017)
4. Minn, H., Fu, X., Bhargava, V.K.: Optimal periodic training signal for frequency offset estimation in frequency -selective fading channels. IEEE Trans. Commun. **54**(6), 1081–1096 (2006)
5. Moose, P.H.: A technique for orthogonal frequency division multiplexing frequency offset correction. IEEE Trans. Commun. **42**(10), 2908–2914 (1994)
6. Wang, Q., Mehlführer, C., Rupp, M.: Carrier frequency synchronization in the downlink of 3GPP LTE. In: IEEE, International Symposium on Personal Indoor and Mobile Radio Communications, pp. 939–944. IEEE (2010)
7. Moose, P.H.: Technique for orthogonal frequency division multiplexing frequency offset correction. IEEE Trans. Commun. **42**(10), 2908–2914 (1994)
8. Song, J.H., Ra, S.J., Jung, J.Y., et al.: Performance analysis and residual frequency offset estimation method in DOCSIS 3.1 System. In: Symposium of the Korean Institute of Communications and Information Sciences (2017)
9. Wang, Y., Li, L., Li, R., et al.: An improvement method of frequency offset estimation performance in communication system carrier synchronization. In: International Conference on Cyber-Enabled Distributed Computing and Knowledge Discovery, pp. 454–458. IEEE (2017)

10. Tang, Y., Fu, Y., Zhang, H., et al.: Device and method for frequency offset estimation (2017)
11. Chong, D.H., Kim, H.I., Yeo, S.B., et al.: Method and apparatus for performing cell search and frequency offset estimation (2018)

Mobility-Aware Caching Specific to Video Services in Hyper-Dense Heterogeneous Networks

Zhenya Liu[✉], Xi Li, Hong Ji, and Heli Zhang

Key Laboratory of Universal Wireless Communications, Ministry of Education,
Beijing University of Posts and Telecommunications,
Beijing, People's Republic of China
{liuzhenya,lixi,jihong,zhangheli}@bupt.edu.cn

Abstract. Caching at the network edge has emerged as a promising technique to cope with the dramatic increase of mobile data traffic. It is noted that different types of video applications on mobile devices have different requirements for cached contents, thus corresponding caching policies should be developed accordingly. In hyper-dense heterogeneous networks, due to the user mobility and limited connection duration, the user often could not download the complete cached contents from an associated SBS before it moves away, which makes the design of caching strategy more challenging. In this paper, we propose two different caching strategies to adapt to multimedia applications of different video contents. For ordinary network video files, coded caching is used to increase the efficiency of content access. The caching problem is formulated as an optimization problem to minimize the average transmission cost of cached contents. We first present an optimal caching strategy based on the critical value of validity period of user requests. Then, for the validity period greater than its critical value, an iterative optimization on the basis of the above optimal solution is performed. For typical streaming video, uncoded video fragments is considered to be stored in the caches to meet the needs of online viewing. The principle of the proposed caching scheme is to cache data chunks in advance according to the sequences of SBSs passed by the user based on the mobility prediction results. Simulation results indicate that the proposed mobility-based caching performs better than the existing popularity-based caching scheme.

Keywords: Mobility-aware caching · Video services · Transmission cost · Heterogeneous networks (HetNets)

1 Introduction

With the popularization of smart terminals and the diversification of multimedia applications, mobile data traffic has exhibited unprecedented growth worldwide. This traffic growth in mobile networks will result in higher transmission

X. Liu et al. (Eds.): ChinaCom 2018, LNICST 262, pp. 480–490, 2019.
https://doi.org/10.1007/978-3-030-06161-6_47

delay and energy consumption. To cope with the stern challenge, researchers have proposed to deploy SBSs together with macro-cell base stations (MBSs) in existing networks to boost the network capacity. However, the overloaded and costly backhaul links connecting the SBSs with the core network becomes the bottleneck in improving network performance.

Caching popular contents at the SBSs equipped with storage facility has been proposed to relieve congestion in the backhaul links [1]. By storing the frequently requested contents in SBSs cache in advance, this technique not only avoids redundant file retrieval over the backhaul links, but reduces the user experienced delay. Motivated by this, content caching have been widely studied in small cell networks. However, many works do not consider user mobility [2, 3]. In realistic environments, the users are not stationary and their association with the SBSs may change during data transfer, which makes the design of caching strategy more intractable.

In hyper-dense HetNets, the association between the user and the SBSs will change more frequently. Reference [4] has shown that analyzing and exploiting the mobility patterns of users, the mobility-aware caching can markedly improve the efficiency of caching. In [5], the authors formulate the content caching in small-cell networks as an optimization problem, with the goal of maximize the caching utility. In [6], the authors first model the user sojourn time as a random variable that obeys the exponential distribution, and then propose a file alloca-tion strategy based on coded caching. In [7], the authors research the impact of user mobility on content caching aimed at minimizing the load of the macrocells. In [8], the authors develop a novel algorithm for content placement at the cache based on estimated popularity.

Although these works have taken user mobility into account, most of them assume that all data of requested file stored in the cache of connected SBS can be downloaded, once the user established a connection with it. In practice, the user may only get parts of the cached contents from the associated SBS during each connection due to the limit of connection duration. Moreover, in recent years, various multimedia applications are emerging endlessly to meet the different user needs. We notice that different types of video applications on user terminals have different requirements for cached contents. Therefore, corresponding caching policies should also be developed.

In this paper, we propose two different caching strategies specific to different video services. For ordinary network video files, coded caching is introduced to increase the efficiency of content access. The caching problem is formulated as an optimization problem to minimize the average transmission cost of cached contents. We first present an optimal caching strategy at the critical value of validity period of user requests. Then, for the validity period greater than its critical value, an iterative optimization based on the aforementioned optimal result is implemented to maximize the average amount of coded data delivered by local caches. For typical streaming video, storing uncoded video segments at SBSs is considered to meet the needs of online viewing. The idea of the proposed caching scheme is to cache data chunks in advance according to the sequences of SBSs traversed by the user. Simulation results show that our proposed mobility-based caching performs better than maximum popularity caching.

2   System Model

2.1   Network Model

Consider a heterogeneous network for video delivery like the one depicted in Fig. 1, which consists of one MBS and a set $\mathcal{N}$ of N SBSs deployed in a macro-cell. Each SBS_n, $n \in \mathcal{N}$ is equipped with a cache of storage size C_n (bytes). The coverage areas of the SBSs may overlap each other, and a user may be concurrently covered by multiple SBSs. Since the SBS coverage area is relatively small, the user may repeatedly move in and out of the small cells and thus connect to different SBSs at different times.

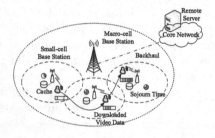

Fig. 1. Graphical illustration of heterogeneous network and user trajectory.

User preference to content files can be learned by analyzing previous statistics of user requests, and it is assumed to be known and fixed within a time period. We consider that each user independently requests a item from the content library $\mathcal{F}$ consisting of F video files, and file popularity follows Zipf distribution [9].

2.2   Video Service Model

Given the different requirements of multimedia applications, the cache space of each SBS is divided into two areas to store different contents, thus providing mobile users with different video services. For ordinary network video files, storing the encoded data of video files is considered to increase the efficiency of content access. By appropriate coding, the requested file can be successfully recovered when the total amount of downloaded coded data in any order is at least the size of the original file [10]. For typical streaming video, in-order packet delivery should be guaranteed. Since the storage buffers at the SBSs specific to such applications is limited, only some video fragments are expected to be placed in the cache to improve cache hit probability. We consider splitting each item into some data chunks with the same size, each of which is identified by a sequence number. The user collects data chunks sequentially from the encountered SBSs to meet the needs of viewing streaming video online.

2.3   Mobility Model

The user mobility is modeled from two dimensions of time and space. We first divide time into identical time intervals, and each time interval corresponds to the shortest duration that the SBS is accessed. Then, several important locations that are frequently visited are identified in the macro-cell, e.g., crowded cross-road, shopping center, stadium, etc. These important locations can be extracted by using clustering algorithms from previous trajectories of the user, and each location may be covered by multiple SBSs. Besides, there is also a non-important location covered by all remaining SBSs.

In the coded caching, we consider delayed offloading scheme [11]. To meet QoS requirements, we associate each user request with a period of validity. That is, each request must be completely served within T time intervals by the encountered SBSs once it is initiated. We refer to the sequence of visited locations during the validity period as movement pattern of the user, i.e. $r_w = \{v_1, v_2, \cdots, v_T\}$, where v_i represents the location visited at i-th interval. We denote the set of all possible movement patterns with $\mathcal{W}$. The probability that a user takes r_w, $w \in \mathcal{W}$ can be derived as follows:

$$q_w = p(v_1) \prod_{i=1}^{T-1} p(v_{i+1}|v_i) \tag{1}$$

where $p(v_1)$ denotes the probability that the user appears in location v_1 when initiating a video request, and $p(v_{i+1}|v_i)$ denotes the transition probability between v_i and v_{i+1}. These probabilities can be estimated by leveraging previous time statistics.

In the chunk-based uncoded caching, our work focuses on utilizing either the specified or the predicted mobility information to facilitate effective content placement at SBSs. We refer to the sequence of SBSs accessed within the coverage of the identified locations as movement trajectory of the user. We assume that the prediction of user trajectory is performed by the mobility prediction entity deployed at the MBS. In the following, two different kinds of caching strategies are presented.

3   Proposed Coded Caching Strategy

3.1   Problem Formulation

We use $\mathcal{X} = \{x_{n,i}|n \in \mathcal{N}, i \in \mathcal{F}\}$ to denote the caching strategy, where $x_{n,i}$ indicates the amount of coded data of video file f_i stored in the cache of SBS_n. Considering that there are differences between SBSs in the deployed bandwidth and the average workload, B_n is used to represent the average amount of data that SBS_n can deliver to a user within a time interval. In a movement pattern r_w, a user may be connected to the same SBS multiple times. Let $\mathcal{N}_w$ represent the subset of SBSs that are encountered in r_w, and $a_{w,n}$ is introduced to denote the number of time intervals that $\text{SBS}_n \in \mathcal{N}_w$ is accessed in r_w. During the j-th connection with SBS_n, the non-redundant amount of coded data of file f_i which can be downloaded within this interval is given by

$$d_{n,i}^{(j)} = \max\{\min\{x_{n,i} - (j-1)B_n, B_n\}, 0\} \tag{2}$$

For a user requesting file f_i and taking movement pattern r_w, the total amount of coded data downloaded from the local encountered SBSs can be expressed as follows:

$$u_{i,w} = \min\{\sum_{n\in\mathcal{N}_w}\sum_{j=1}^{a_{w,n}} d_{n,i}^{(j)}, s_i\} = \min\{\sum_{n\in\mathcal{N}} \min\{x_{n,i}, B_n a_{w,n}\}, s_i\} \tag{3}$$

where s_i is the size of file f_i. If $u_{i,w}$ is less than s_i, the remaining video segments need to be downloaded from the remote server. Then, the amount of coded data downloaded from the core network over backhaul link for file f_i is equal to $s_i - u_{i,w}$.

Obviously, serving a user request from the SBSs cache and the remote server will incur different levels of transmission costs to the operator, such as the energy consumed at the SBSs or the traffic generated in the backhaul network. Suppose that the cost of transmitting the unit of coded data volume from the cache of the encountered SBSs is ω_0, and the cost from the core network over backhaul link is ω_1 ($\omega_1 > \omega_0$). Then, in the case that the user takes the movement pattern r_w, the download cost of requested video file f_i is given by $(u_{i,w}\omega_0 + (s_i - u_{i,w})\omega_1)$.

Our goal is to find the optimal caching strategy to minimize the average transmission cost of the requested items. The optimization problem can be formulated as follows:

$$\min_{\mathcal{X}} \Omega(\mathcal{X}) = \sum_{w\in\mathcal{W}} q_w \sum_{i\in\mathcal{F}} p_i \cdot (s_i\omega_1 - u_{i,w}(\omega_1 - \omega_0))$$
$$= \omega_1 \sum_{i\in\mathcal{F}} p_i s_i - (\omega_1 - \omega_0) \sum_{w\in\mathcal{W}}\sum_{i\in\mathcal{F}} q_w p_i u_{i,w} \tag{4}$$

$$s.t. \quad x_{n,i} \in [0, s_i], \forall n \in \mathcal{N}, \forall i \in \mathcal{F}; \quad \sum_{i\in\mathcal{F}} x_{n,i} \leq C_n', \forall n \in \mathcal{N}. \tag{5}$$

where C_n' denotes the capacity used to store MDS-encoded data of ordinary video files in the cache of SBS_n, satisfying $C_n' < C_n$. Let $\Phi(\mathcal{X})$ denote the average amount of coded data delivered by the encountered SBSs, which is derived as follows:

$$\Phi(\mathcal{X}) = \sum_{w\in\mathcal{W}} q_w \sum_{i\in\mathcal{F}} p_i u_{i,w} \tag{6}$$

Since $\omega_1 > \omega_0$, minimizing the transmission cost of file download is equivalent to maximizing $\Phi(\mathcal{X})$.

3.2   Distributed Approximate Solution

Before solving the above optimization problem, we first discuss the period of validity T of user requests. The critical value T_c of validity period is defined as

s_{min}/B_{max}. In the case of $T < T_c$, it means that no matter which movement pattern is taken, and no matter which video file is requested, successful recovery of requested videos cannot occur. That is, there are hardly user requests that can be completely served by the local SBSs, which is clearly not what the operator would like to see. Thus, in this paper, we consider that the validity period of user requests satisfies $T \geq T_c$.

Next, we first present the optimal caching strategy when $T = T_c$, and then the iterative optimization on the basis of this result is implemented to maximize the average amount of coded data downloaded from local SBSs, thereby minimizing the transmission cost. In the case of $T = T_c$, Eq. (3) can be simplified as follows:

$$u_{i,w} = \sum_{n \in \mathcal{N}} \min\{x_{n,i}, B_n a_{w,n}\} \tag{7}$$

The Eq. (6) can be rewritten as follows:

$$\Phi(\mathcal{X}) = \sum_{n \in \mathcal{N}} \sum_{w \in \mathcal{W}} q_w \sum_{i \in \mathcal{F}} p_i \cdot \min\{x_{n,i}, B_n a_{w,n}\} \tag{8}$$

Algorithm 1 Mobility-based Optimal Caching Algorithm $(T = T_c)$

1: **Input:** B_n, C_n', $\lambda_{n,i}^k$, s_i. **Output:** The optimal solution $\mathcal{X}_n^*$.
2: $val_{n,i}^k \leftarrow \lambda_{n,i}^k$, $wgt_{n,i}^k \leftarrow s_i$, $i \in \mathcal{F}$, $k \in \{1, \cdots, T\}$; $x_{n,i} \leftarrow 0$, $i \in \mathcal{F}$;
3: $D \leftarrow \{1, \cdots, F\} \times \{1, \cdots, T\}$;
4: **for** $i = 1, 2, \cdots, F \cdot T$ **do**
5:     **while** $C_n > 0$ **do**
6:         $(i^*, k^*) = \arg\max\limits_{(i,k) \in D} \dfrac{val_{n,i}^k}{wgt_{n,i}^k}$; $x_{n,i^*} \leftarrow x_{n,i^*} + \min\{B_n, C_n'\}$;
7:         $D \leftarrow D \setminus (i^*, k^*)$; $C_n' \leftarrow C_n' - B_n$;
8:     **end while**
9: **end for**

From the structure of $\Phi(\mathcal{X})$ and the aforementioned constraints (5), we can observe that the caching strategy at one SBS does not affect the other SBSs. Therefore, we can decompose this problem into N independent sub-problems and solve them in a distributed way. For SBS$_n$, the sub-problem $\mathcal{P}_n$ can be expressed as follows:

$$\max_{\mathcal{X}_n} \sum_{w \in \mathcal{W}} q_w \sum_{i \in \mathcal{F}} p_i \cdot \min\{x_{n,i}, B_n a_{w,n}\}$$

$$= \sum_{i \in \mathcal{F}} \sum_{k=1}^{T} \sum_{w \in \mathcal{W}_{n,k+}} p_i q_w \cdot d_{n,i}^{(k)} = \sum_{i \in \mathcal{F}} \sum_{k=1}^{T} \lambda_{n,i}^k \cdot d_{n,i}^{(k)} \tag{9}$$

$$s.t. \ x_{n,i} \in [0, s_i], \ \forall i \in \mathcal{F}; \ \sum_{i \in \mathcal{F}} x_{n,i} \leq C_n'. \tag{10}$$

where $\mathcal{W}_{n,k}$ and $\mathcal{W}_{n,k+}$ denotes the subset of movement patterns where the number of time intervals that SBS_n is accessed is equal to k and not less than k, respectively. Thereinto, $\lambda_{n,i}^k = \sum_{w \in \mathcal{W}_{n,k+}} p_i q_w$. It can be observed that the objective function in $\mathcal{P}_n$ is a superposition of F monotonically increasing piecewise linear functions.

With respect to the optimization variables $x_{n,i}(i \in \mathcal{F})$, the objective function and the constraints of this problem are linear, and thus it can be solved by using linear optimization techniques. Specifically, we classify problem $\mathcal{P}_n$ as a class of knapsack problems. The optimal solution of this knapsack type problem can be obtained by the following scheme, that is, iteratively placing the item with the highest ratio of value to weight into the knapsack until there is no space left. The specific procedure is summarized in Algorithm 1.

When $T > T_c$, the amount of coded data $u_{i,w}$ downloaded from the SBSs cache can't be reduced to the form of Eq. (7), so we cannot obtain the optimal caching strategy. Thus, an iterative optimization on the basis of the optimal caching strategy $\mathcal{X}_n^*$ is implemented to minimize the transmission cost of cached contents. The criterion of cache optimization is mainly based on the popularity of files. That is, if there are several video files that are stored with the same amount of coded data, then we can consider decreasing the amount of coded data corresponding to the less popular files, while increasing the amount of coded data corresponding to the more popular files. Let $V_{in}(i)$ and $V_{de}(i)$ represent the changes in the average transmission cost $\Omega(\mathcal{X})$ when the amount of cached data of the file f_i is increased or decreased by B_n. When $|V_{in}(i^+)| > V_{de}(i^-)$, the cache optimization can be performed. The specific procedure is described in Algorithm 2.

Algorithm 2 Mobility-based Approximate Caching Algorithm $(T > T_c)$

1: **Input:** B_n, $\mathcal{X}_n^*$. **Output:** The approximate solution $\mathcal{X}_n$.
2: **while** *true* **do**
3:     Initialize $\mathcal{F}_{in} \leftarrow \emptyset$, $\mathcal{F}_{de} \leftarrow \emptyset$, $V_{in}(i) \leftarrow 0$, $V_{de}(i) \leftarrow 0$, $i \in \mathcal{F}$;
4:     $x_0 = \max\{x_{n,i}, i \in \mathcal{F}\}$;
5:     **while** $x_0 \geq 0$ **do**
6:         $i^- = \max\{i|x_{n,i} \geq x_0, i \in \mathcal{F}\}$;
7:         **if** $x_{n,i^-} > 0$ **then**
8:             $\mathcal{F}_{de} \leftarrow \mathcal{F}_{de} \cup \{i^-\}$; calculate the variation in average cost $V_{de}(i^-)$;
9:         **end if**
10:        $i^+ = \min\{i|x_{n,i} \geq x_0, i \in \mathcal{F}\}$; $\mathcal{F}_{in} \leftarrow \mathcal{F}_{in} \cup \{i^+\}$;
11:        calculate the variation in average cost $V_{in}(i^+)$; $x_0 \leftarrow x_0 - B_n$;
12:     **end while**
13:     **if** $|\min\{V_{in}(i^+)\}| > \min\{V_{de}(i^-)\}$ **then**
14:         $i_m^+ = \arg\min_{i^+ \in \mathcal{F}_{in}} V_{in}(i^+)$; $x_{n,i_m^+} \leftarrow \min\{x_{n,i_m^+} + B_n, s_{i_m^+}\}$;
15:         $i_m^- = \arg\min_{i^- \in \mathcal{F}_{de}} V_{de}(i^-)$; $x_{n,i_m^-} \leftarrow \max\{x_{n,i_m^-} - B_n, 0\}$;
16:     **else**
17:         break;
18:     **end if**
19: **end while**

---

4   Chunk-Based Uncoded Caching Strategy

Once a user initiates a video request, the associated SBS immediately informs the MPE to forecast possible movement trajectories within the coverage of the current location. Based on the predicted trajectories and corresponding sojourn time (intervals), the MBS determines the set of SBSs where the cached contents should be updated and the video fragments that they respectively need to store. Upon the user leaves current location and enters the coverage of adjacent location, the above steps are implemented again until the requested file is completely downloaded. The principle of the proposed caching strategy is to store in advance the data chunks in the likely sequences of SBSs that are accessed by the user at a higher probability, and these chunks are most probably downloaded at each small-cell traversed.

The detailed caching scheme is presented as follows. Firstly, the first data chunks of each video file are placed in the SBSs cache at the beginning. The size of cached video fragments of file f_i in the cache of SBS_n is given by $\min\{p_i C_n^{''}, s_i\}$, where $C_n^{''}$ denotes the capacity used to store uncoded video fragments in the SBS_n cache, satisfying $C_n^{''} < C_n$. Secondly, when a video file is requested by the user, the MBS sorts the possible movement trajectories predicted by MPE in descending order of occurrence probability, and then selects the most probable trajectories so that the sum of their probabilities is not less than the threshold τ. Finally, according to the predicted SBSs sequences and the sojourn time in the corresponding small-cell, the data chunks of requested file that should be stored in the cache of each SBS are determined. At the time of performing caching at each SBS, the previous chunk and the latter chunk (if exist) of the decision result should also be placed in the cache, so that the data can also be downloaded from the local cache rather than remote server in the case that the movement speed of the user changes slightly.

Considering that the predicted results cannot be completely accurate no matter what kind of mobility prediction model is adopted, so the several trajectories is chosen instead of the most likely one. Because SBSs have limited storage space, a cache replacement strategy is necessary to determine which data chunks should be removed, in the case that the cache is full and new chunks should be cached. Obviously, the video fragments that have already been delivered are the ones to be evicted.

5   Simulation Results

In this section, we evaluate the performance of the proposed caching strategies. The performance criterion we use is the average transmission cost of the requested files denoted with Ω. In the case of $T \geq T_c$, we compare the performance of the approximate caching strategy in which iterative optimization is implemented with the maximum popularity caching strategy and the optimal caching strategy ($T = T_c$).

In our simulation, we consider that there are $F = 100$ files with a size of 60 MB in the video library, and their popularity follows a Zipf distribution with exponent $\gamma = 1.0$. There are $N = 7$ SBSs with the same cache capacity $C_n' = 20\%$ of the video library size used to store coded data in a macro cell to achieve seamless coverage, and we consider that these SBSs respectively represent different locations. During the movement, the user moves to the adjacent locations with equal transition probabilities. The validity period of user requests and the amount of data delivered by SBS_n are set to $T = 5$ time intervals and $B_n = 20$ MB, respectively. In addition, we assume that the costs of serving a file request from SBSs cache and remote server are 0 and 100.

Impact of Cache Capacity. We first investigate the impact of cache capacity on the average transmission cost incurred by the presented algorithms, as depicted in Fig. 2. In this experiment, the cache capacity of SBSs span a wide range, from 10% to 50% of the entire video files library size. As expected, increasing the cache capacity reduces the average transmission cost. Furthermore, we can observe that the approximate caching algorithm always outperforms the optimal caching algorithm, and they perform significantly better than the maximum popularity caching. This can be explained from the fact that maximum popularity caching policy takes caching decisions considering only user demand, ignoring the movement patterns of users.

Impact of Zipf Exponent. Figure 3 examines the relation between the average transmission cost and the exponent of Zipf distribution γ. We notice that the approximate caching algorithm that performs iterative optimization consistently outperforms the other algorithms, and the gap between their performances diminishes with increasing the parameter of Zipf. As we know, the exponent γ characterizes the correlation level of user requests. The larger the value of it, the more the user preferences are concentrated on a few most popular video files. That is, increasing the value of γ accordingly enhances the probability of requesting files in the SBSs cache, thereby reducing the average transmission cost of the requested contents.

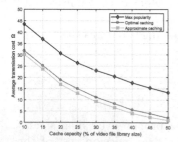

Fig. 2. Average transmission cost Ω versus cache capacity C.

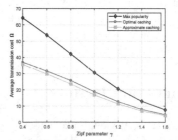

Fig. 3. Average transmission cost Ω versus Zipf exponent γ.

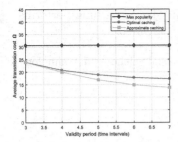

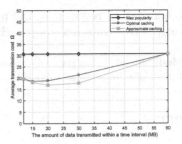

Fig. 4. Average transmission cost Ω versus validity period T.

Fig. 5. Average transmission cost Ω versus amount of data delivered within a time interval B.

Impact of Validity Period. We explore how the period of validity of user requests impacts the results in Fig. 4. In this experiment, the validity period T varies from 3 time intervals (corresponding to the critical value T_c) to 7 time intervals. It can be observed that the average transmission cost gradually decreases for the proposed caching algorithms, since the user has more opportunities to contact with different SBSs. The amount of coded data downloaded from the remote server over backhaul link is reduced. It is worth mentioning that the performance of the maximum popularity caching remains unchanged, because the most popular files have been completely replicated in the cache. The change in T has no impact on the amount of data delivered by SBSs.

Impact of Transmission Rate. In Fig. 5, we analyze the impact of the amount of data delivered by SBSs within a time interval on the performance of caching algorithms. Specifically, the transmission rate B varies in $\{12, 15, 20, 30, 60\}$ MB per interval. As the transmission rate increases, it is expected that the SBSs can transmit more data during connection with the user, thus reducing the transmission cost. However, with the data rate continuous increasing, the average cost no longer decreases, but instead starts to increase. In the case of higher transmission rate, we need to store more coded data for each of the most popular files that should be cached. Due to the limited cache capacity, this may cut down the number of requests directly served by the caches.

6   Conclusion

In this paper, due to the different requirements of multimedia applications on user terminals for cached contents, we present two different kinds of caching policies. For ordinary network video files, the caching problem is formulated as an optimization problem to minimize the average transmission cost of cached contents. For typical streaming video, we consider storing uncoded video fragments in the caches. The proposed caching scheme is to cache data chunks in advance at SBSs passed by the user based on the mobility prediction results. The results of simulation reveal that the proposed mobility-based caching performs significantly better than max-popularity caching.

Acknowledgement. This paper is sponsored by the National Science and Technology Major Project of China (Grant No.2017ZX03001014).

References

1. Golrezaei, N., Shanmugam, K., Dimakis, A.G., Molisch, A.F., Caire, G.: Femto-Caching: wireless video content delivery through distributed caching helpers. In: 2012 Proceedings IEEE INFOCOM, March 2012, pp. 1107–1115 (2012)
2. Poularakis, K., Iosifidis, G., Tassiulas, L.: Approximation algorithms for mobile data caching in small cell networks. IEEE Trans. Commun. **62**(10), 3665–3677 (2014)
3. Chae, S.H., Choi, W.: Caching placement in stochastic wireless caching helper networks: channel selection diversity via caching. IEEE Trans. Wirel. Commun. **15**(10), 6626–6637 (2016)
4. Wang, R., Peng, X., Zhang, J., Letaief, K.B.: Mobility-aware caching for content-centric wireless networks: modeling and methodology. IEEE Commun. Mag. **54**(8), 77–83 (2016)
5. Guan, Y., Xiao, Y., Feng, H., Shen, C.C., Cimini, L.J.: MobiCacher: mobility-aware content caching in small-cell networks. In: 2014 IEEE Global Communications Conference, Dec 2014, pp. 4537–4542 (2014)
6. Liu, T., Zhou, S., Tsinghua, Z.N.: Mobility-aware coded-caching scheme for small cell network. In: 2017 IEEE International Conference on Communications, May 2017, pp. 1–6 (2017)
7. Poularakis, K., Tassiulas, L.: Code, cache and deliver on the move: a novel caching paradigm in hyper-dense small-cell networks. IEEE Trans. Mob. Comput. **16**(3), 675–687 (2017)
8. Mishra, S.K., Pandey, P., Arya, P., Jain, A.: Efficient proactive caching in storage constrained 5G small cells. In: 2018 10th International Conference on Communication Systems Networks (COMSNETS), Jan 2018, pp. 291–296 (2018)
9. Cha, M., Kwak, H., Rodriguez, P., Ahn, Y.-Y., Moon, S.: I tube, you tube, everybody tubes: analyzing the world's largest user generated content video system. In: Proceedings of the 7th ACM SIGCOMM Conference on Internet Measurement, 2007, pp. 1–14 (2007)
10. Leong, D., Dimakis, A.G., Ho, T.: Distributed storage allocations. IEEE Trans. Inf. Theory **58**(7), 4733–4752 (2012)
11. Lee, K., Lee, J., Yi, Y., Rhee, I., Chong, S.: Mobile data offloading: how much can wifi deliver? IEEE/ACM Trans. Netw. **21**(2), 536–550 (2013)

Long-Reach PON Based on SSB Modulated Frequency-Shifted QAM and Low-Cost Direct-Detection Receiver with Kramers–Kronig Scheme

Xiang Gao, Bo Xu[✉], Yuancheng Cai, Mingyue Zhu, Jing Zhang,
and Kun Qiu

School of Information and Communication Engineering, Key Laboratory of
Optical Fiber Sensing and Communications (Education Ministry of China),
University of Electronic Science and Technology of China, Chengdu 611731,
Sichuan, China
xubo@uestc.edu.cn

Abstract. As PON systems move towards terabit/s aggregated data rates with longer transmission distance, optical coherent receivers become preferred due to their high tolerance to power fading from fiber transmission. To solve the high complexity and high cost problems of optical coherent receivers, a scheme for complex QAM signal transmission with simple direct detection is recommended in this paper. The scheme based on optical SSB modulation with frequency-shifted QAM signals and low-cost single-ended PD provides an efficient low-cost solution for long reach coherent PON. Due to its minimum phase property of the optical SSB modulated signal, Kramers-Kronig scheme can be used to reconstruct the complex QAM signal from the received intensity signal. The efficiency of the proposed scheme is validated by both numerical simulations and experiments for both QPSK and 16-QAM modulated signals. By using standard commercially available components, the experiments demonstrated that the combination of SSB modulation of frequency-shifted QAM signal and its single-ended PD receiver with KK scheme can support SSMF transmission over 75 km for both QPSK and 16-QAM signals with receiver optical power penalty less than 1.5 dB.

Keywords: Fiber optics communication · Modulation ·
Optical communications

1 Introduction

Recently, the live-streaming of high definition video multimedia and cloud computing using remote storage of data are evolving steadily. As a result, the demand for high volumes of data keeps increasing the transmission speed requirements in both short- and long-reach optical access network [1]. As the expanding coverage of modern metropolitan areas leads to longer transmission distance, the Long-reach passive optical network (LR-PON) is proposed. LR-PON is a promising solution of optical access

© ICST Institute for Computer Sciences, Social Informatics and Telecommunications
Engineering 2019 Published by Springer Nature Switzerland AG 2019. All Rights Reserved
X. Liu et al. (Eds.): ChinaCom 2018, LNICST 262, pp. 491–500, 2019.
https://doi.org/10.1007/978-3-030-06161-6_48

networks and is able to combine the capacity of metro and access networks by extending the coverage span of the transmission [2]. Direct detection (DD) has been the main technology for LR-PON because of its low cost and simple structure. However, as PON moves towards terabit/s aggregated data rates, several shortcomings have come out including power fading due to chromatic dispersion (CD) at high symbol rates and long transmission distance plus limited modulation formats.

To solve the problems of PON transmission with DD, coherent PON has been proposed which achieve information modulation on both amplitude and phase [3–5]. A much higher tolerance on power fading from fiber dispersion is also possible with coherent PON and digital optical coherent receiver with electrical dispersion compensation. Traditional optical coherent receivers require one local oscillator (LO) laser, one 90°optical hybrid, and two balanced detectors with four photodiodes (PDs) in total. Even for some simplified coherent receivers where the LO laser is replaced by optical carrier sent from the source [6], coherent receivers are still high cost due to their high implementation complexity for PON applications [7].

Single-side band (SSB) signal with pulse amplitude modulation (PAM) is another known way to overcome power fading problem from DD and it can also double the optical spectrum efficiency [8–10]. However, as the fiber CD accumulates for LR-PON, the influence of signal-to-signal beating interference (SSBI) from square-law detection of PD cannot be neglected. Different digital signal processor (DSP) techniques for SSBI mitigation have been proposed in [11, 12]. Kramers-Kronig (KK) scheme with DD is able to alleviate the SSBI very well by fully reconstructing the optical field signal from the detected photocurrent signal [13–15].

Unlike SSB-PAM which is limited to amplitude modulation, quadrature amplitude modulation (QAM) can provide much more choices on modulation schemes. In this paper, we recommend a scheme for complex 16-array QAM signal transmission with simple DD for LR-PON. Optical SSB modulation of frequency-shifted QAM signal and single-ended PD with KK scheme for QAM signal reconstruction are combined to achieve the low-cost and efficient scheme for coherent LR-PON. The paper is organized as follows. Section 2 explains the fundamental principles of the proposed schemes. Numerical simulations are then used to validate the proposed scheme in Sect. 3 followed by experimental results in Sect. 4. Section 5 concludes the paper.

2   Principles of the Scheme

2.1   Principle of SSB Modulation of the Frequency-Shifted QAM Signal

Suppose that the complex signal $m(t)$ is a conventional bandwidth-limited QAM signal, its spectrum is shown Fig. 1(a) where B is the signal's bandwidth. To modulate the signal onto optical carrier, an IQ modulator based on two intensity modulating Mach-Zehnder modulators (MZMs) is commonly used. A relative carrier phase difference exists between the upper and lower arm of the IQ modulator, and the real part and imaginary part of the signal $m(t)$ are used to drive each one of the two MZMs.

Different from conventional IQ modulation for QAM signal, we propose in this paper to use a single dual-drive MZM (DDMZM) to modulate the frequency-shifted

QAM signal as a SSB signal. To do this, we first achieve frequency shifting of the QAM signal as shown in Eq. (1) where B is the signal's bandwidth. The spectrum of the frequency-shifted QAM signal $s(t)$ is shown in Fig. 1(b). If the signal $s(t)$ is written in the form of its real part and imaginary part, it can be shown that $s_i(t)$ is actually the Hilbert transform of $s_r(t)$, which is the same as required for SSB modulation [13].

$$s(t) = m(t) \exp(j2\pi Bt)$$
$$= s_r(t) + js_i(t) \tag{1}$$

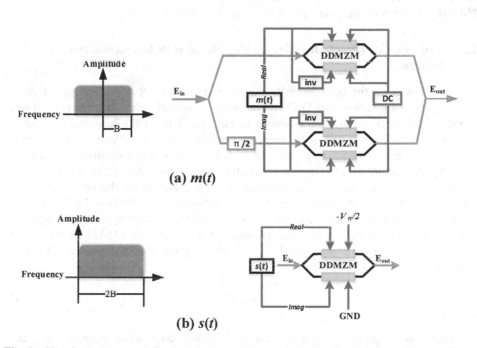

(a) $m(t)$

(b) $s(t)$

Fig. 1. Signal spectrum and its modulation scheme comparison for the conventional QAM signal (a) and the frequency-shifted QAM signal (b).

Figure 1(b) also shows how to achieve the SSB modulation of the frequency-shifted QAM signal by using a DDMZM. The DDMZM consists of two parallel phase modulators with separate radio frequency (RF) input ports and direct current (DC) bias ports. The DDMZM output optical field can be written as [10]

$$E_{out}(t) = E_{in}\left[\exp(j\pi\frac{V_{r1}(t) + V_{b1}}{V_\pi}) + \exp(j\pi\frac{V_{r2}(t) + V_{b2}}{V_\pi})\right] \tag{2}$$

Where E_{in} stands for the input optical field, $V_{r1}(t)$ and $V_{r2}(t)$ are the radio frequency signals applied to each one of the two branches, and V_{b1} and V_{b2} are the DC voltage bias which will induce a constant phase shift between the two branches. V_π is the half-wave voltage parameter of the MZM. To obtain an optical SSB signal, the real part and imaginary part of the signal $s(t)$ are used as the two radio frequency signals, that is

$V_{r1}(t) = s_r(t)$, $V_{r2}(t) = s_i(t)$. Moreover, it is important to set the two DC bias voltages as $V_{b1} = -V_\pi/2$, $V_{b2} = 0$. Under the assumption that the $s_r(t)$ and $s_i(t)$ are small signals, the output of DDMZM can be found as

$$E_{out}(t) = E_{in}\left[1 - j + \frac{\pi}{V_\pi}(s_r(t) + js_i(t))\right] \qquad (3)$$

In the above equation, $E_{in}(1-j)$ is the optical carrier signal. Since $s_i(t)$ is actually the Hilbert transform of $s_r(t)$, an optical SSB signal is obtained.

2.2   Principles of Direct Detection of SSB Signal with Kramers-Kronig Scheme

Figure 2 compares the conventional optical coherent receiver and direct detection of SSB signal. For conventional QAM modulation, an optical coherent receiver with two balanced detectors is required as shown in Fig. 2(a). One of the balanced detector is used to recover the real part of the received complex QAM signal and the other one is used to recover its imaginary part. Since each balanced detector consists of two PDs, four PDs are required for the optical coherent receiver totally. Moreover, a LO working at the same wavelength as the transmitter must be available for the optical coherent receiver which might be expensive for optical network unit (ONU) of PON systems.

Instead using an optical coherent receiver to recover the complex QAM signal, a simple single-ended PD for the SSB modulated frequency-shifted QAM signal is used in the proposed system as shown in Fig. 2(b). The output of the single-ended PD can be written as

$$I(t) = \eta|E_{out}(t)|^2 + n(t) \qquad (4)$$

Where $I(t)$ is the photocurrent produced by the PD, η denotes the responsivity of the PD, and $n(t)$ is the noise of the PD. It is clear from the equation that the phase information of the transmitted signal will be lost due to the square-law detection nature of a PD. To recover the complex signal of $s(t)$ from the received real signal of $I(t)$, KK scheme is used in the following. Compared with the optical coherent receiver with two balanced detector, the proposed single-ended PD receiver is much cheaper as no local laser diode is required at the receiver side.

The key of KK scheme relies on whether the received signal is of minimum phase [13]. For SSB modulated optical signal with an optical carrier of sufficient power, the signal can be written as

$$b(t) = A + m(t)\exp(j2\pi Bt) \qquad (5)$$

Where A is constant. It can be shown that $b(t)$ is a minimum phase signal when $|A|$ is large enough. As described in [13], under this condition, the KK scheme can be used to reconstruct the signal $E_m(t)$ from the intensity signal of the minimum phase signal as follow

$$\phi(t) = \frac{1}{2\pi} p.v. \int_{-\infty}^{+\infty} dt' \frac{\log[I(t')]}{t - t'} \tag{6}$$

$$E_m(t) = \left\{ \sqrt{I(t)} \exp[j\phi(t)] - E_0 \right\} \exp(-j2\pi Bt) \tag{7}$$

Where $\phi(t)$ is the phase part of the minimum phase signal, $I(t)$ is the photocurrent produced by the PD which is proportional to the field intensity as in Eq. (4), $p.v.$ denotes the Cauchy principal value of the integral [16]. $E_m(t)$ is the reconstructed complex field signal transmitted through fiber.

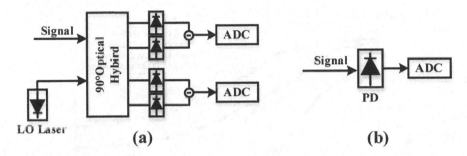

Fig. 2. Comparison of conventional optical coherent receiver (a) and direct detection of SSB signal with single-ended PD (b).

After using the KK scheme to reconstruct the field signal, we can obtain the QAM signal by doing an inverse frequency shifting of Eq. (1). Then, the QAM signal can be further processed with conventional digital signal processing functions as used in digital optical coherent receiver.

3 Simulation Results

In order to validate the proposed scheme, numerical simulations based on VPI Transmission Maker and MATLAB are implemented. The schematic of the simulated system is a simplified version of the experiment setup shown in Fig. 4 to be included in the next section. In the simulations, standard single-mode fiber (SSMF) with a length of 75 km is assumed as for LR-PON. The attenuation coefficient of the SSMF is set at 0.2 dB/km. Other parameters of SSMF include dispersion coefficient D of 16 ps/nm/km, and nonlinear index of 2.6×10^{-20} m^2/W. At the transmitter side, a pseudo random bit sequence (PRBS) with 2^{17} bits is used for QAM modulation. Root raised cosine (RRC) filter with a roll-off factor of 0.1 is applied for pulse shaping. After the QAM signal is shifted in frequency, the real part and the imaginary part of the frequency shifted signal are fed into the two arms of DDMZM. The launch optical power is fixed at 0 dBm. A tunable optical attenuator before the PD is used for bit error rate test under different receiver optical power. The received signal is firstly proceed by the KK scheme to reconstruct the optical field. After the inverse frequency shifting for getting the base band signal, phase correcting and least mean squares (LMS) based feed forward equalizer (FFE) are applied before symbol decision making. Finally, the number of bit error is counted for BER computation.

In the simulations, 25 GBaud is assumed for transmission with both quadrature phase shift keying (QPSK) and 16-QAM. The net bit rate for the QPSK and 16-QAM signals are thus 50 Gbps and 100 Gbps respectively. As the KK scheme can help to recover the complex signal $s(t)$ from a single-ended PD, digital dispersion compensation can thus be applied instead of using dispersion compensation fiber (DCF). The linewidth of laser is set to 1 MHz.

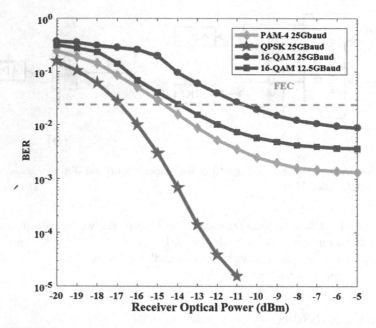

Fig. 3. BER performance comparison among different modulation schemes as the ROP varies.

Figure 3 shows the simulation results with the KK-based receiver for both QPSK (4-QAM) and 16-QAM. SSB modulated PAM-4 modulation scheme is also included in the simulation comparison. As expected, QPSK has the best performance among these modulations because there are only two levels on the real part and imaginary part of QPSK modulation. From the simulation results, it can be seen that the pre-forward error correction (FEC) BER lower than 2.4×10^{-2} can be achieved at a receiver optical power of -17 dBm for QPSK modulation.

For SSB-modulated PAM-4 signal, about 3 dB receiver optical power (ROP) penalty is observed as compared with the QPSK modulation. This ROP penalty is used to obtain a doubled spectrum efficiency as SSB-modulated PAM-4 can achieve the same transmission bit rate with only half of the signal bandwidth as compared with the QPSK signal.

Among the different modulation schemes, the SSB modulated frequency-shifted 16-QAM signal has the highest transmission rate, but about 6 dB ROP penalty is found as compared with the QPSK modulation at the same baud rate. However, if we reduce the baud rate of the 16-QAM signal to 12.5 GBaud, then a similar performance is found between the lower-rate 16-QAM signal and the SSB-PAM-4 signal at the same bit rate.

4   Experimental Setup and Results

The experimental setup for the optical SSB modulation of the frequency-shifted QAM signal system is shown in Fig. 4. A PRBS with 2^{18} bits is used for QPSK and 16-QAM modulation and up-sampled to 4 samples-per-symbol at the transmitter side. An RRC filter with a roll-off factor of 0.1 is applied to generate the QAM modulated signal sequence. Then the signal is frequency shifted using Eq. (1). Arbitrary wavelength generator (AWG) of AWG700002A is used to generate the real part and the imaginary part of the signal, $s_r(t)$ and $s_i(t)$, at the highest sampling rate of 25GS/s for AWG700002A. Thus, the bit rates are 12.5 Gbps and 25 Gbps for QPSK and 16-QAM modulated signals respectively. A DDMZM biased at its quadrature point is used to generate the optical SSB signal. As the carrier to signal power ratio (CSPR) is an important parameter to optimize the system's performance, a pair of attenuators are used on the AWG's output signal to optimize the CSPR. The optical source used in the experiment has an output optical power about 12 dBm at 1550 nm wavelength. The optical signal power to be launched into the transmission fiber is fixed at 0 dBm and

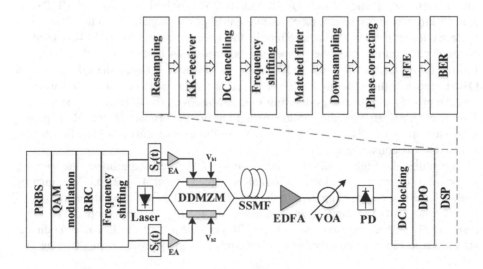

Fig. 4. Schematic of experimental setup. DPO: digital-processing oscilloscope.

75 km SSMF with an attention coefficient of 0.2 dB/km is used for transmission. At the receiver side, an Erbium-doped optical fiber amplifier (EDFA) and a variable optical attenuator (VOA) are used to compensate the transmission loss plus varying the ROP during the experiments.

To check whether the KK scheme can successfully reconstruct the complex QAM signal from single-ended PD with optical SSB modulation of frequency-shifted QAM signal, Fig. 5(a) shows the constellation diagram of the 16-QAM signal after the KK scheme. With a relatively high ROP, a clear 16-point constellation diagram is observed from Fig. 5 which proves the validation of the KK scheme with the proposed system. After phase compensation, the conventional rectangular 16-QAM constellation diagram is obtained as shown in Fig. 5(b).

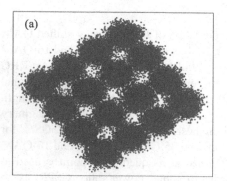 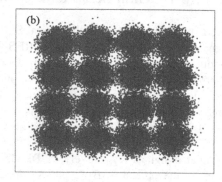

Fig. 5. The constellation diagram of the reconstructed 16-QAM signal after the KK scheme before phase compensation (a) and after phase compensation (b).

Figure 6(a) gives the experimental results on the BER performance of the SSB modulation of frequency-shifted QPSK and 16-QAM for both back-to-back (B2B) case and 75 km fiber transmission case. As the optical power launched into the fiber is low, there exists little fiber nonlinear effect in the system. Only a linear FFE is used for channel equalization like fiber dispersion compensation before the BER calculation. From Fig. 6(a), about 5–6 dB ROP penalty is observed between the QPSK and 16-QAM signal in the experiments when we consider the BER of 1.0×10^{-2} which is consistent with simulation results from Fig. 3. Moreover, the ROP penalty is less than 1 dB for QPSK signals after 75 km fiber transmission. However, this ROP penalty increases up to 1.5 dB for 16-QAM signals in the experiments if a pre-FEC BER of 2.4×10^{-2} is considered.

To utilize the highest sampling rate of our AWG in the experiment, the samples-per-symbol of its output electrical signal is reduced from 4 to 2.76. Consequently, higher baud rate can be supported in the experiment and the bit rates of the transmitted QPSK and 16-QAM signals are increased up to 18.125 Gbps and 36.25 Gbps, respectively. Figure 6(b) compares the BER performance of the low-rate modulated signals as in Fig. 6(a) and the high-rate modulated signals from down sampling after

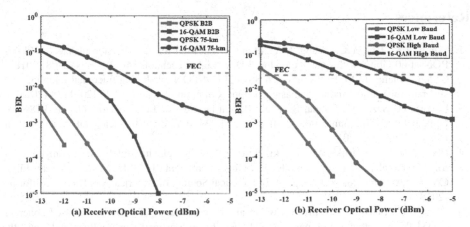

Fig. 6. BER performance comparisons from experimental results on the SSB modulation of frequency-shifted QPSK and 16-QAM between B2B and fiber transmission cases (a) and between high-rate and low-rate cases (b).

75 km fiber transmission in the experiments. At the pre-FEC BER threshold, both the QPSK and 16-QAM signals shows an ROP penalty about 1.5 dB.

5   Conclusions

Optical SSB modulation with frequency-shifted QAM signals with low-cost single-ended PD is recommend in this paper as an efficient low-cost solution for long reach coherent PON system. After simple frequency shifting on the QAM signal at the transmitter side, a single DDMZM can be used to generate the SSB modulated signal. At the receiver side, a single-ended PD without local laser diode can replace the expensive two balanced detectors from conventional optical coherent receiver. Since the optical SSB modulated signal is of minimum phase, KK scheme is available to reconstruct the complex QAM signal from the received intensity signal. The efficiency of the proposed scheme is validated by both numerical simulations and experiments for both QPSK and 16-QAM modulated signals for LR-PON communications. By using standard commercially available components, the experiments demonstrated that the combination of SSB modulation of frequency-shifted QAM signal and its single-ended PD receiver with KK scheme can support SSMF transmission over 75 km for both QPSK and 16-QAM signals with ROP penalty less than 1.5 dB.

Acknowledgements. This work is supported by the National Natural Science Foundation of China (#61471088).

References

1. Weng, Z.K., Chi, Y.C., Wang, H.Y., et al.: 75-km long reach dispersion managed OFDM-PON at 60 Gbit/s with quasi-color-free LD. J. Light. Technol. **36**(12), 2394–2408 (2018)
2. Guo, C., Liang, J., Li, R.: Long-reach SSB-OFDM-PON employing fractional sampling and super-nyquist image induced aliasing. J. Opt. Commun. Netw. **7**(12), 1120–1125 (2015)
3. Lavery, D., Sezer, E., Polina, B., et al.: Recent progress and outlook for coherent PON. In: Optical Fiber Communication Conference, pp. M3B.1. Optical Society of America, San Diego, California (2018)
4. Matsumoto, R., Keisuke, M., Suzuki, N.: Fast, Low-complexity widely-linear compensation for IQ imbalance in burst-mode 100-Gb/s/λ coherent TDM-PON. In: Optical Fiber Communication Conference, pp. M3B.2. Optical Society of America, San Diego, California (2018)
5. Suzuki, N., Satoshi, Y., Hiroshi, M., et al.: Demonstration of 100-Gb/s/λ-based coherent WDM-PON system using new AGC EDFA based upstream preamplifier and optically superimposed AMCC function. J. Light. Technol. **35**(8), 1415–1421 (2017)
6. Kim, D., Kim, B.G., Kim, H.: 60-km transmission of 28-Gb/s QPSK upstream signal in RSOA-based WDM PON using SBS suppression technique. In: Optical Fiber Communication Conference, pp. W4G.4. Optical Society of America, San Diego, California (2018)
7. Zhang, M., Xu, B., Guo, Q., et al.: Study and comparison of low cost intensity modulation direct detection radio over fiber systems. In: Asia Communications and Photonics Conference 2016, pp. AF2A.86. Optical Society of America, Wuhan (2016)
8. Lin, B.J., Li, J.H., Yang, H., et al.: Comparison of DSB and SSB transmission for OFDM-PON. J. Opt. Commun. Netw. **4**(11), B94–B100 (2012)
9. Zhang, X.L., Zhang, C.F., Chen, C., et al.: Non-optical carrier SSB-OFDM PONs with the improved receiver sensitivity and potential transmission nonlinearity tolerance. IEEE Photonics J. **9**(1), 1–10 (2017)
10. Zhu, M.Y., Zhang, J., Yi, X.W., et al.: Hilbert superposition and modified signal-to-signal beating interference cancellation for single side-band optical NPAM-4 direct-detection system. Opt. Express **25**(11), 12622–12631 (2017)
11. Li, Z., Erkilinc, M.S., Galdino, L., et al.: Comparison of digital signal-signal beat interference compensation techniques in direct-detection subcarrier modulation systems. Opt. Express **24**(25), 29176–29189 (2017)
12. Zhu, M.Y., Zhang, J., Yi, X.W., et al.: Optical single side-band Nyquist PAM-4 transmission using DDMZM modulation and direct detection. Opt. Express **26**(6), 6629–6638 (2018)
13. Mecozzi, A., Antonelli, C., Shtaif, M.: Kramers-Kronig coherent receiver. Optica **3**(11), 1220 (2016)
14. Li, Z., Erkilinc, M.S., Shi, K., et al.: SSBI mitigation and the Kramers-Kronig scheme in single-sideband direct-detection transmission with receiver-based electronic dispersion compensation. J. Light. Technol. **35**(10), 1887–1893 (2017)
15. Antonelli, C., Mecozzi, A., Shtaif, M., et al.: Polarization multiplexing with the Kramers-Kronig receiver. J. Light. Technol. **35**(24), 5418–5424 (2017)
16. Mecozzi, A.: Retrieving the full optical response from amplitude data by Hilbert transform. Opt. Commun. **282**(20), 4183–4187 (2009)

Two-Layer FoV Prediction Model
for Viewport Dependent Streaming
of 360-Degree Videos

Yunqiao Li, Yiling Xu$^{(\boxtimes)}$, Shaowei Xie, Liangji Ma, and Jun Sun

Shanghai Jiao Tong University, Shanghai, China
{liyunqiao,yl.xu,junsun}@sjtu.edu.cn, {xswjsnj,maliangji2012}@163.com
http://cmic.sjtu.edu.cn/CN/Default.aspx

Abstract. As the representative and most widely used content form of Virtual Reality (VR) application, omnidirectional videos provide immersive experience for users with 360-degree scenes rendered. Since only part of the omnidirectional video can be viewed at a time due to human's eye characteristics, field of view (FoV) based transmission has been proposed by ensuring high quality in the FoV while reducing the quality out of that to lower the amount of transmission data. In this case, transient content quality reduction will occur when the user's FoV changes, which can be improved by predicting the FoV beforehand. In this paper, we propose a two-layer model for FoV prediction. The first layer detects the heat maps of content in offline process, while the second layer predicts the FoV of a specific user online during his/her viewing period. We utilize a LSTM model to calculate the viewing probability of each region given the results from the first layer, the user's previous orientations and the navigation speed. In addition, we set up a correction model to check and correct the unreasonable results. The performance evaluation shows that our model obtains higher accuracy and less undulation compared with widely used approaches.

Keywords: Omnidirectional video · Field of view prediction
FoV-based transmission

1 Introduction

VR immerses the user into the virtual world by realizing free interaction, which reshapes the consumption experience for users. Intensive attention is being attracted from the industry and the academia [1]. Omnidirectional video is one of the most typical application format in VR. Obviously, to get the same visual perception quality as traditional videos, the whole omnidirectional videos covering 360-degree scenes will contain much more amount of data. However, due to the restriction of human visual system (HVS), users can only see the content

© ICST Institute for Computer Sciences, Social Informatics and Telecommunications Engineering 2019
Published by Springer Nature Switzerland AG 2019. All Rights Reserved
X. Liu et al. (Eds.): ChinaCom 2018, LNICST 262, pp. 501–509, 2019.
https://doi.org/10.1007/978-3-030-06161-6_49

within their FoV at a time, and the size of FoV equipped by the head mounted display (HMD) of HTC Vive system is about 110-degree in horizontal range.

To save the bandwidth while not sacrificing the quality of experience (QoE), FoV-based transmission [9] of omnidirectional videos is widely recognized as a effective scheme. The principle of this scheme is to transmit the content that covers user's FoV with high quality, while the other regions with lower quality to avoid the blank screen [3]. Studies about FoV adaptation model finds that the refinement duration affects the QoE of the user significantly [15]. After the user switching the viewport by turning head, the quality within the new FoV will decrease before the high quality content is received covering the new FoV. While, this can be improved by FoV prefetching which has to be motivated by accurate FoV prediction beforehand.

Recently, some researchers have investigated the very short-term FoV prediction. Feng Qian proposed to utilize weighted linear regression algorithm in the prediction based on previous viewing orientation information [2,12], and this approach has been widely adopted in transmission improvement thanks to its simplicity. Meanwhile, as a state-of-art technology, Convolutional Neural Network (CNN) is also applied to improve the prediction [16]. Furthermore, content features are being considered in prediction [5,13], while still limited to short-term.

The navigation trajectory of a user keeps close relevance to the personalized FoV prediction. We introduce the orientation and navigation speed which reflects the user's viewing status in our prediction model. In addition, content attractiveness detection helps to build reliable prediction as users are often stimulated to switch the viewport for certain objects. We propose a two-layer model to work on the FoV prediction. The first layer is designed as an offline process to detect the heat value of each region in the omnidirectional videos. Motion features of the content is paid more emphasis, which proves to better comply with this problem. In the second layer, FoV of the user is predicted in real time. The previous orientation and navigation speed as part of the inputs illustrates the short-term viewing trend and status. What's more, we input the results from the first layer related to short-term and long-term step into the second layer to provide distinguish guidance to the prediction. In this layer, a LSTM module is trained to learn the intricate connections in this problem, and a correction module is designed to check and correct the unreasonable results at last.

This paper is organized as follows. In Sect. 2, we describe the overall framework and concrete details of each step in our model. Experiment setup and results are presented in Sect. 3. Finally, in Sect. 4 we provide conclusions about our research and look into the future work related.

2   Prediction Model

To predict the FoV of a specific user, we investigate the viewing behavior of users. According to the research of Sitzmann, the viewing status can be divided into focusing and browsing status [13]. Under the focusing status, users are attracted by the specific content in the video. They switch the viewport to follow the targeted content or pay attention to the extended part of it. In this status, the navigation speed appears to be lower than the average, and the heat region of the content close to current FoV has a strong correlation with the viewport switching. When the user is under browsing status, their movement shows great irregularity, for they do not pay real attention to the content in omnidirectional video, and the head rotation speed appears to be higher than the average. In this status, the prediction should mainly focus on heat region of the video, which may get the users attention later and change one's viewing state into gazing status, with wider range than the previous circumstances. Therefore, we propose a two-layer FoV prediction model as shown in Fig. 1 considering above key factors to adapt to different viewing status.

In order to save the computing resources and to predict faster during the viewing process, the detection of the heat region of the video is completed offline before viewed, which refers to the first layer in our model. By detecting the saliency feature of each static frame and the moving object between frames, two main kinds of feature maps are obtained. As people tend to focus on and follow the moving objects when immerses in an omnidirectional video, we place extra emphasis on the motion features when combine two features into heat maps.

Besides the heat value already detected, user's orientation and navigation speed on previous frames are tracked and feeded back into the data-driven LSTM architecture during the user' consumption. Furthermore, to make up the oversight of the FoV characteristics, a correction module is added as the final part in second layer.

We adopt the tile based transmission scheme in this paper [8], where a full omnidirectional video is partitioned into many coding independent rectangular tiles, so that each tile can be transmitted independently [6]. Our prediction model is established on this transmission scheme as well.

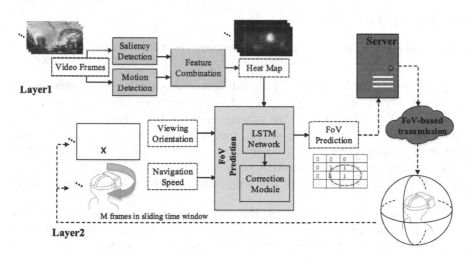

Fig. 1. Overview of the proposed two-layer FoV prediction model. The example of a tile-based streaming scheme is shown.

2.1   Layer1- User Attention Prediction

This layer is an offline process to detect the attractive regions in the content with higher probability to be viewed. The heat value $HT_n, n \in N$ of N tiles are obtained in this layer.

Saliency Detection. Image saliency map is obtained by detecting the objects that show distinct differences in features of colors, textures, etc. from the surroundings. The Fused Saliency Map (FSM) [4] is used for detection for it adapt current detection models to omnidirectional image characteristics.

Motion Detection. People tend to be attracted by moving objects such as moving animals or athletes in a sports game. This can be detected by analyzing the features between consecutive frames. We utilize the Lucas-Kanade optical flow approach to detect pixel-level motion features [14].

Content Feature Combination. We convolute the motion maps with a 2D Gaussian filter so that the motion features can present overall impact on the attractive region detection results. Pixel-wise heat maps are calculated as:

$$HP_k = \begin{cases} MP_k, & MP_k \geq SP_k \\ \frac{MP_k + SP_k}{2}, & MP_k < SP_k \end{cases} \tag{1}$$

HP_k is the heat value of kth pixel, SP_k and MP_k are saliency value and filtered motion value. Then tile-wise heat value HT is obtained by calculating and normalizing the summation of the pixel-wise heat values HP within each tile.

2.2   Layer2- FoV Prediction

The second layer is an online process to predict the tiles to be viewed next.

LSTM Prediction Module. This module is to predict the probability of each tile to be viewed by a specific user, which is based on the recurrent neural network (RNN) architecture. This scheme presents high efficiency in temporal sequential problems as realizing weight sharing in time domain. We choose Long short-term memory(LSTM) [7,10] model replacing the ordinary node in network by the memory cell to avoid gradient vanishing and explosion by adding a forget gate towards the cell to rectify the long-term and short-term memory of the node.

As the impact of inputs towards the prediction gradually changes (the reliability of the head orientation in the sliding window gradually decreases), the output of the prediction should also reflect this trend. When omnidirectional videos are partitioned into N tiles, our model provides short-term prediction results $Pred_{short} = \{p\}_1^N$ and additional long-term prediction results $Pred_{long} = \{p\}_1^N$ to adapt to the variation of user's FoV in the prediction window.

The input to our LSTM model consists of the tile-wise heat maps of next $2 * M$ frames including $\{H\}_{t+1}^{t+M}$ related to short-term content and $\{H\}_{t+M+1}^{t+2*M}$ related to long-term content, along with user's orientation and navigation speed of previous M frames $\{O\}_{t-M}^{t}$, $\{S\}_{t-M}^{t}$. The head orientation information $O(x, y, z)$ is expressed in the form of the position on the x, y, z coordinates; and the navigation speed on sphere surface $S(yaw, pitch, raw)$ is calculated as the orientation change compared with the previous frame in rotation coordinates $yaw, pitch, raw$.

We adopt two layer LSTM model with 256 neurons each layer, which presents better performance compared with the other parameters we have tried. The prediction results are obtained at the last time step, while inner results obtained in each time step is retained in the form of state information and becomes the input into the same neuron at the next time step to establish the association in the time domain as shown in Fig. 2.

When training the LSTM module, learning rate is set to be 0.01, and a dropout layers with 50% drop rate is added to prevent the overfitting. We adopt the Adam Optimizer to minimize the cross-entropy loss. The sliding window is set to be 1 second, as well as the size of short-term and long-term prediction window. A down-sampling process is performed on frame rate of the input by 2 times to shorten the input length which accelerates the prediction as the content and viewing features of adjacent frames has minor changes. We divide the dataset into 80% and 20% for training and validation respectively. Different videos are trained and validated separately, so that a unique prediction network is trained for each video.

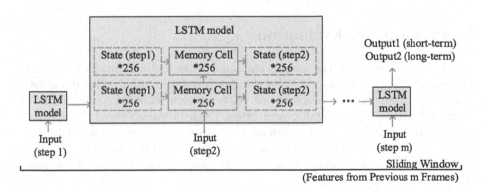

Fig. 2. LSTM prediction module in our approach.

Correction Module. By setting an appropriate threshold, tiles are decided whether to be covered by FoV. However, the separately judged tile-wise prediction results cannot always compose a reasonable FoV region. We propose a correction module with two steps to make sure the predicted FoV as an integrated region with reasonable size.

The first step is to make up the omitting error that a few tiles are misjudged to be negative, while their surroundings are all judged to be covered by FoV. This violate the integrity of FoV. When there is a wide range of misjudged tiles, we believe that this situation does not belong to omitting error, but the drawback of the prediction model which has been avoided in our model. We scan the prediction results for each row of all tiles. When the predicted FoV is partitioned to be at two side of the projected plane, we correct the negative predicted tiles among the range of the frame edge and the farmost positive predicted tiles away from the frame edge. Otherwise we find the edge tiles that predicted to be viewed and make sure each misjudged tile among them is reset as the positive tiles.

The second step is to determine the minimum range that the positive judged tiles should cover. Equirectangular projection is widely used to project the omni-directional video from spherical 3-Dimention (3D) to 2D plane, sampling uniformly on latitude. However, this approach causes the distortion of the content near two poles. Equally partitioned tiles on projected 2D frames correspond to spherical tiles with larger size near the equator and smaller size when approaching poles. Therefore, during the playback of an omnidirectional video that has been rendered as a spherical three-dimension video, the range of tiles falling into the user's FoV is not always the same.So that the FoV shows a changing shape and size on the projected 2D plane video. We locate the center of the predicted FoV by averaging the position of all tiles. The shape and range of the FoV on the 2D projected video related to the above calculated center can be obtained according to the projection relationship. Then we complement the negative judged tiles around the center to fill in the range of FoV.

3   Performance Evaluation

The LSTM network of the prediction model has to be trained by the omni-directional video dataset [11]. The dataset has ten omnidirectional videos all lasting one minutes. All videos are of 4K resolution with 30fps. The videos are projected with equirectangular projection. Each video is viewed by 30 viewers wearing HMD and the orientation of each frame has been collected by Open Track and presented in both Cartesian coordinates and Euler angles. The omni-directional video is partitioned into 10?20 tiles, which are labeled whether is viewed by the user.

We evaluate the performance of our two-layer model of FoV prediction given the pre-detected heat map and continuously collected user's head movement information related to the previous frames in sliding window.

We compare the performance of our model with the typical and widely used prediction approach of weighted linear regression [12].

Table 1 provides the specific measurement of accuracy and F-score of the prediction on each video. Accuracy presents the primary performance of the prediction model by calculating the ratio of tiles correctly classified to all tiles. F-score is the weighted mean of the precision and recall (precision is the ratio of

Table 1. Performance of the model on each video.

Videos	Training set		Validation set	
	Accuracy	F-score	Accuracy	F-score
Mega coaster	89.53%	0.7312	88.90%	0.7011
Roller coaster	90.08%	0.7270	88.28%	0.6693
Driving with the 360	79.08%	0.3719	77.34%	0.2821
Shark shipwreck	83.46%	0.5097	83.26%	0.5259
Perils panel	90.16%	0.7408	89.12%	0.7246
Kangaroo Island	84.53%	0.5735	82.62%	0.5308
SFR spore	92.37%	0.8014	89.29%	0.7282
Hog rider	80.29%	0.3544	77.09%	0.2742
Pac-Man	88.74%	0.7151	87.45%	0.6826
Chariot race	88.26%	0.6818	87.79%	0.6040
Average(ours)	86.65%	0.6207	85.11%	0.5722
Average(w-reg)	73.00%	0.5352	72.54 %	0.5155

tiles classified to be positive correctly to all positive classified tiles, and recall is the ratio of tiles classified to be positive correctly to all viewed tiles).

The accuracy presents the primary performance of the prediction model. Our model outperforms over the weighted linear regression approach with a considerably increase of accuracy. Our prediction approach obtains balance between precision and recall over most video sequences. The performance improvement caused by our model is probably because (a) the newly introduced navigation speed leads the model to be more sensible to different viewing status, (b) correction module in the second layer improves the prediction to be more reasonable, and (c) our model predict both short-term FoV and long-term FoV to match the user's viewport switching in longer time (d) we synthetically utilize the content features and viewing movement information with our two-layer model.

Moreover, we evaluate the fluctuation of the prediction approach on each video sequence. As shown in the Fig. 3, our model obtains great stability compared with the weighted linear regression approach. Maintaining the prediction at a higher accuracy all the time during the user's viewing period still proves to be a great challenge. In the process of viewing most video sequences, the prediction accuracy of our model is stable near the average value in spite of the changeable viewing status of users. However, when our prediction model achieves higher accuracy, the stability prediction reduces. In contrast, the weighted linear regression approach not only obtains low average accuracy, but also shows strong instability. In other words, our approach capture the regularity of the FoV switching better, and is able to be adaptive to various users and videos.

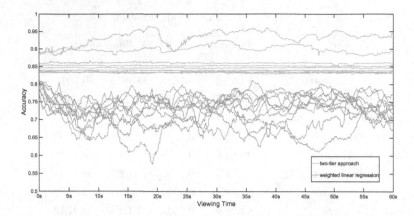

Fig. 3. The prediction performance comparison of our approach against weighted linear regression ap-proach on ten videos during the whole consumption period.

4   Conclusion

In this paper, we propose a two-layer model to predict user's FoV on omnidirectional video during one's viewing. The user's navigation speed is considered for the first time with user's viewing orientation and content features. LSTM model is trained and used to predict the probability of each region to be viewed, which is further optimized by a correction module. The experimental results show that our prediction approach is superior to other conventional approaches, obtaining great accuracy and stability improvement under different viewing status. In the future, we will utilize our prediction model in our realistic multi-network system to further evaluate the performance in real application.

Acknowledgments. This paper is supported in part by National Natural Science Foundation of China (61650101), Scientific Research Plan of the Science and Technology Commission of Shanghai Municipality (16511104203), in part by the 111 Program (B07022).

References

1. Virtual and augmented reality: Understanding the race for the next computing platform. Technical report. The Goldman Sachs Group Inc. (2016)
2. Bao, Y., Wu, H., Zhang, T., Ramli, A.A., Liu, X.: Shooting a moving target: motion-prediction-based transmission for 360-degree videos. In: 2016 IEEE International Conference on Big Data (Big Data), pp. 1161–1170. IEEE (2016)
3. Corbillon, X., Simon, G., Devlic, A., Chakareski, J.: Viewport-adaptive navigable 360-degree video delivery. In: 2017 IEEE International Conference on Communications (ICC), pp. 1–7. IEEE (2017)

4. De Abreu, A., Ozcinar, C., Smolic, A.: Look around you: Saliency maps for omni-directional images in vr applications. In: 2017 Ninth International Conference on Quality of Multimedia Experience (QoMEX), pp. 1–6. IEEE (2017)
5. Fan, C.L., Lee, J., Lo, W.C., Huang, C.Y., Chen, K.T., Hsu, C.H.: Fixation prediction for 360 video streaming in head-mounted virtual reality. In: Proceedings of the 27th Workshop on Network and Operating Systems Support for Digital Audio and Video, pp. 67–72. ACM (2017)
6. Graf, M., Timmerer, C., Mueller, C.: Towards bandwidth efficient adaptive streaming of omnidirectional video over http: Design, implementation, and evaluation. In: Proceedings of the 8th ACM on Multimedia Systems Conference, pp. 261–271. ACM (2017)
7. Hochreiter, Sepp, Schmidhuber, Jürgen: Long short-term memory. Neural Comput. 9(8), 1735–1780 (1997)
8. Mohammad H., Swaminathan, V.: Adaptive 360 vr video streaming: Divide and conquer. In: 2016 IEEE International Symposium on Multimedia (ISM), pp. 107–110. IEEE (2016)
9. Hu, Y., Xie, S., Xu, Y., Sun, J.: Dynamic VR live streaming over MMT. In: 2017 IEEE International Symposium on Broadband Multimedia Systems and Broadcasting (BMSB), pp. 1–4. IEEE (2017)
10. Lipton, Z.C., Berkowitz, J., Elkan, C.: A critical review of recurrent neural networks for sequence learning (2015). arXiv:1506.00019
11. Lo, W.C., Fan, C.L., Lee, J., Huang, C.Y., Chen, K.T., Hsu, C.H.: 360 video viewing dataset in head-mounted virtual reality. In: Proceedings of the 8th ACM on Multimedia Systems Conference, pp. 211–216 (2017)
12. Qian, F., Ji, L., Han, B., Gopalakrishnan, V.: Optimizing 360 video delivery over cellular networks. In: Proceedings of the 5th Workshop on All Things Cellular: Operations, Applications and Challenges, pp. 1–6 (2016)
13. Sitzmann, Vincent, Serrano, Ana, Pavel, Amy, Agrawala, Maneesh, Gutierrez, Diego, Masia, Belen, Wetzstein, Gordon: Saliency in vr: How do people explore virtual environments? IEEE Trans. Vis. Comput. Graph. 24(4), 1633–1642 (2018)
14. Wu, Z., Su, L., Huang, Q., Wu, B., Li, J., Li, G.: Video saliency prediction with optimized optical flow and gravity center bias. In: 2016 IEEE International Conference on Multimedia and Expo (ICME), pp. 1–6. IEEE (2016)
15. Xie, S., Xu, Y., Qian, Q., Shen, Q. Ma, Z., Zhang, W.: Modeling the perceptual impact of viewport adaptation for immersive video. In: 2018 IEEE International Symposium on Circuits and Systems (ISCAS), pp. 1–5. IEEE (2018)
16. Xu, M., Song, Y., Wang, J., Qiao, M., Huo, L., Wang, Z.: Modeling Attention in Panoramic Video: A Deep Reinforcement Learning Approach (2017)

Energy Efficient Caching and Sharing Policy in Multihop Device-to-Device Networks

Yuling Zuo[✉], Heli Zhang, Hong Ji, and Xi Li

Key Laboratory of Universal Wireless Communications, Ministry of Education,
Beijing University of Posts and Telecommunications, Beijing,
People's Republic of China
{zuoyuling,zhangheli,jihong,lixi}@bupt.edu.cn

Abstract. Caching content at the user device and sharing files via multihop Device-to-Device link can offload the traffic from the Base Station, which is inevitable to consume the user's energy. But most works usually assume that the battery capacity is implicitly infinite and rarely consider the impact of the user's remaining battery energy on the file transmission. In fact, the user device has limited battery capacity and the transmission may be not completed due to the insufficient battery energy. So it is important to utilize the limited battery energy to ensure more successful transmission and traffic offloading. In this paper, we firstly optimize the caching policy and obtain the minimum energy cost of cache-enabled multihop D2D communications. For this purpose, we classify users into different clusters and use a weighted undirected graph to represent the topological relationship of users in one cluster. Then, we propose a novel algorithm to find the optimal path to transmit files via multihop D2D link. Finally, we obtain the minimum energy cost and optimal caching policy. Simulation results show that the proposed caching policy performs better than other general caching strategies in terms of energy conservation.

Keywords: Caching policy · Energy cost · Multihop D2D
Undirected graph

1 Introduction

With the development and popularization of mobile devices, such as smartphones and tablets, the demand for mobile data is increasing rapidly [7]. At the same time, research shows that the majority of users' requests for content are often asynchronous and duplicated [2], which increases the load on base stations heavily. Therefore, caching content at the edge of the network and sharing content via D2D link become a trend of content distribution, which not only can offload traffic from the base station, but also can provide lower latency and higher quality user experience [1].

© ICST Institute for Computer Sciences, Social Informatics and Telecommunications Engineering 2019
Published by Springer Nature Switzerland AG 2019. All Rights Reserved
X. Liu et al. (Eds.): ChinaCom 2018, LNICST 262, pp. 510–520, 2019.
https://doi.org/10.1007/978-3-030-06161-6_50

Recent researches about D2D cache network are almost focus on the performance of caching policy. In [9], the probabilistic caching policy was proved to outperform the universal caching strategy due to its promising performance and feasibility in practical use. However, the authors of [8] showed that the optimal caching scheme for D2D content distribution was similar to the demand distribution, which can be modeled as a Zipf distribution. In fact, sharing content has to consume transmitter's energy, nevertheless, the user device has limited battery capacity and may be not willing to consume its power to transmit files to other users. The authors of [12] proposed an energy-aware incentive mechanism to motivate mobile users to participate in D2D content sharing. In addition, taking the user's battery capacity into account contributes to a reduction in the possibility of offloading traffic. In [4], the author investigated the relationship between offloading gain and energy cost of each helper user and introduced a user centric protocol to control the energy cost for a helper user to transmit the file. In [10], a new definition, energy-consumption-ratio was proposed, which was used to measure the energy efficiency that caching scheme can achieve. In these previous research [4,8,10,12], users obtained the required file from other users only via one-hop transmission, not consider multihop transmission. Therefore, in [6], the author extended the one-hop D2D network and considered multihop transmission, and in [5] a sequence of adjacent nodes were modeled as a linear topology to investigate the performances of both local and distributed cooperative caching policies. In practice, however, the users are randomly distributed and not always be represented by linear topology.

In the previous researches for caching in D2D network [8 10], for simplicity, the user device is assumed to have enough battery energy to transmit files. However, in fact, the user device has the limited battery energy. There are few papers consider the impact of the user's remaining battery energy on content delivery via multihop D2D link. Once the transmission fails, the energy consumed is wasted. In this paper, we take the user's battery capacity into account to make sure successful transmission. In addition, in order to better reflect the users location in practice, we consider a weighted undirected graph to represent the topology among users in a cluster. We aim to optimize caching policy to minimize the energy consumption by helper users via multihop D2D link. To acquire the minimum energy cost, we first derive the expression of successful transmission probability, defined as the probability that user's battery power and transmission rate are no less than the predicted threshold. Then, we proposed a novel algorithm to calculate the optimal transmission path between any two users via multihop D2D link and obtain the optimal caching scheme. Finally, The proposed caching policy is proved to outperform other general caching strategies in terms of energy conservation by simulation results.

The remainder of the paper is organized as follows. The Sect. 2 elaborates the system model and problem formulation. In Sect. 3, we introduce the energy efficient caching and sharing policy. Simulation results and discussions are given in Sect. 4. Finally, we conclude this paper in Sect. 5.

2   System Model and Problem Formulation

We consider a single cell cellular network with numberable user devices, which are modeled as homogeneous Poisson point process (PPP) with density λ_u denoted as Φ_u. According to their physical location, users are clustered with M users in one cluster. The topological relationship among users in each cluster is represented by a random geometric graph $G = (V, E)$, where V is the users set and E represents the relationship among users. When the distance between two users is no more than the D2D communication distance R_c, there is one edge between this two users, facilitating a one-hop D2D link. Each user is equipped with a cache memory. For the sake of simplicity, assume that each user can only cache one file. Users can download the file to their devices during the off-peak time, such as midnight. When a user requests the file, users cached the file within communication distance R_c denoted as D2D transmitter can transmit the file to the D2D receiver via one-hop D2D link, otherwise, via multi-hop D2D link. When a user is not connected to any user in the cluster, it can request the file from the BS. Assume that Each cache-enabled user is equipped with one transmit antenna with the same transmission power and each requester has one receive antenna. The BS is aware of the files cached in user devices and D2D communication.

2.1   Content Popularity and Caching Placement

We consider a finite file library $F = [f_1, f_2, ...f_N]$ in one cluster, where N is the library size and f_k is sorted in descending order of the files' popularity. Each content has the same size of S bits, assuming that the large files can be divided into several the same size files. Then, the probability that the kth most popular file is requested follows the Zipf distribution:

$$p_k = \frac{1/k^\beta}{\sum_{t=1}^{N} 1/t^\beta} \tag{1}$$

where β is the Zipf parameter reflecting how skewed the popularity distribution is. Large β means that a few files are responsible for the majority of requests [3].

We consider the probability caching strategy which each user randomly caches file f_k according to probability c_k, where c_k is the probability that user caches the file f_k. Then, the users caching the file f_k follow a PPP Φ_c with density $\lambda_c = \lambda_u c_k$. Since each user only caches one file, so c_k must meet the following conditions:

$$0 \le c_k \le 1, k = 1, 2, ...N \tag{2}$$

$$\sum_{k=1}^{N} c_k = 1 \tag{3}$$

2.2 Channel Model and Energy Analysis

In this paper, we consider that D2D users work in Underlay Mode and reuse uplink resource of cellular network, because battery supplies energy for transmitter, which is of great significance for improving energy efficiency. Assume that each cellular user can only be reused by one D2D user. In this way, there is no interference between D2D users, as shown in Fig. 1.

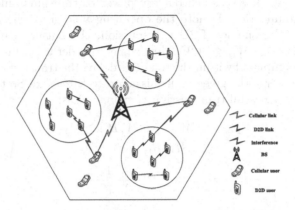

Fig. 1. System model of cellular and D2D network

The general channel model considers large-scale fading channels and small-scale fading channels. Large-scale fading channels are generally modeled by the standard path loss propagation model $r^{-\alpha}$, where r is the communication distance and α is the path loss exponent. Small-scale fading channels are usually modeled by Rayleigh fading channels whose channel gain h follows the exponential distribution with unit mean. In this paper, we consider D2D users reuse the cellular users which frequency is orthogonal and a cellular user is assumed to only be reused by one D2D user. Thus, when user i transmits a file to user j through one-hop D2D, the signal to interference plus noise ratio (SINR) is

$$SINR_{ij} = \frac{P_d h r_{ij}^{-\alpha}}{P_u h_u r_u^{-\alpha} + \sigma^2} \qquad (4)$$

where P_d and P_u is respectively the transmit power at each D2D user and each cellular user. h and h_u is respectively the Rayleigh fading channel gain of D2D user and cellular user. r_{ij} is the distance between user i and user j. σ^2 is the variance of Gaussian white noise.

Therefore,the transmission rate between user i and user j is $R_{ij} = B\log_2(1 + SINR_{ij})$. The energy consumed by D2D transmitter which transmits file f_k to D2D receiver through one-hop D2D link directly is:

$$W_{ij} = Pt = \frac{P_d S}{R_{ij}} = \frac{P_d S}{B\log_2(1 + SINR_{ij})} \qquad (5)$$

Here, we don't consider the circuit power.

2.3   Problem Formulation

We assume that each device have the same battery capacity of C, and only a fraction η of what can be used to transmit a file requested by other user devices. We make this assumption according to the following reasons: on one hand, when a user device transmits the file, it may do other communication businesses, which also consume energy; on the other hand, we must make sure that the user device has so much battery power to transmit a file on the basis of normal communication. Denote the operating voltage of the user device as V. To complete the one-hop D2D transmission, the battery capacity of D2D transmitter must meet $W_{ij} \leq \eta CV$. Meanwhile, in order to ensure that the D2D transmission is completed within the tolerable delay, the transmit rate must meet $R_{ij} \geq R_0$, where R_0 is the given transmission rate threshold. So the probability of one-hop D2D successful transmission is expressed as:

$$
\begin{aligned}
p_{ts} &= P(W_{ij} \leq \eta CV, R_{ij} \geq R_0) \\
&= P(SINR_{ij} \geq \tau) \\
&= e^{\frac{-\tau\sigma^2 r_{ij}^\alpha}{P_d}} \times \frac{1}{bP_u r_u^{-\alpha} + 1}
\end{aligned}
\tag{6}
$$

where $\tau = e^{\max(\frac{P_d S \ln 2}{\eta CVB}, \frac{R_0 \ln 2}{B})} - 1$ and $b = \frac{\tau r_{ij}^\alpha}{P_d}$. According to the expression in (6), when $\eta \leq \frac{P_d S}{R_0 CV}$, p_{ts} is related to the remaining battery ratio η. That is, when the remaining power is too low, the user device's battery capacity limits p_{ts}. while $\eta \geq \frac{P_d S}{R_0 CV}$, that is, user has enough power, p_{ts} is limited by the data rate threshold R_0.

According to the above analysis, the total energy cost that the request user j get the desired file f_k from the cache user i in cluster (one-hop D2D or multihop D2D) is:

$$
E = \sum_{i=1}^{M} \sum_{j=1}^{M} P_{ts} E_{ij}
\tag{7}
$$

where E_{ij} denotes the total energy consumed by user i to transmit file f_k to user j via multihop D2D link, and it is the sum of W_{ij} in (5). P_{ts} is the probability that user i transmit file f_k to user j successfully and is product of several p_{ts}.

Due to the user's limited battery capacity, we must minimize the energy cost for transmitting file via multihop D2D link in (7). At the same time, reducing the energy cost for the network should not compromise the hit probability denoted as the requested file is cached by at least one user in the cluster. Therefore, we optimize the caching policy to maximize the hit probability firstly. Then, we minimize the energy cost for a given optimal caching policy. Because the users caching the file f_k follow a PPP Φ_c with density $\lambda_c = \lambda_u c_k$, So the probability that the requested file f_k is cached in the cluster with radius R_{clu} is:

$$
P_{ck} = 1 - e^{-\lambda_c \pi R_{clu}^2}
\tag{8}
$$

then, the hit probability can be expressed as:

$$P_{hit} = \sum_{k=1}^{N} P_k P_{ck} \qquad (9)$$

According to the above analysis, we firstly maximize the hit probability for the optimal caching policy. Then, based on the acquired caching policy, we can minimize the total energy cost E for transmitting the file via multihop D2D link in (7) by solve the following problem:

$$\min E = \sum_{i=1}^{M} \sum_{j=1}^{M} P_{ts} E_{ij} \qquad (10)$$

3 The Optimization Algorithm for Minimum Energy Cost

To solve the optimization problem (10), we must evaluate the E_{ij} and P_{ts} in expression (7) at first. Since E_{ij} denotes the energy consumed by user i to transmit file f_k to user j via multihop D2D link, it is of great significance to find the optimal path between user i and user j in graph $G = (V, E)$. Once we obtain the optimal path, the E_{ij} and P_{ts} can be calculated. To find the optimal path for transmitting files, we propose a novel algorithm to evaluate the minimum E_{ij} based on the Dijkstra algorithm in undirected graph. The algorithm includes the following two steps:

(1) Initialization:
  (a) Determine the number of users M in one cluster, define the mark-matrix $final_{M \times M}$ to judge $E(i, j)$ need be updated or not.
  (b) Calculate the distance d_{ij} between user i and user i, initialize the energy cost matrix $E_{M \times M}$ and $P_{ts M \times M}$. When $d_{ij} < R_c$, we consider that user i and user j can transmit files via one-hop D2D link directly and the default $E(i, j)$ is minimum. We don't consider the situation that the energy cost of multihop D2D link is less than that of one-hop D2D link. So we can calculate the $E(i, j)$ and $P_{ts}(i, j)$ according to (5) and (6) directly, also set $final(i, j) = 1$. when $d_{ij} > R_c$, we suppose the user i and user j are unreachable and set $E(i, j) = infinite$, $P_{ts}(i, j) = 0$ and $final(i, j) = 0$.
(2) Find the shortest path:
  (a) when $d_{ij} > R_c$, for user i, we find the user mid who consume the least energy among the users connected to user i. Similarly, for user mid, we find the the user who consume the least energy among the users connected to user mid, and so forth until to user j.
  (b) If there is no path between user i and user j, we set $E(i, j) = infinite$, $P_{ts}(i, j) = 0$ and $final(i, j) = 1$. For each user, we perform the above process to find the shortest path to other users. If a user is not connected to any other users, it can get the required files from the Base Station. The detailed process is shown in Algorithm 1.

4   Simulation Results and Discussions

In this section, some simulation results are presented to illustrate the performance of the proposed algorithm and caching scheme. We consider a single cellular cell with only one base station and radius of 500 m. The location of users follows the PPP with $\lambda_u = 0.01$. There are 15 users randomly distributed in one cluster, that is square region with side length of 100 m, as shown in Fig. 3. The other simulation parameter setting are showed in Table 1. The variance of

---

Algorithm 1. Shortest Path for Minimum Energy Based on Dijkstra Algorithm

---

1: Initialization:
 (a)The number of users M in one cluster and mark-matrix $final_{M \times M}$ to judge $E(i,j)$ be updated or not.
 (b)Calculate d_{ij}, if $i = j$, then $d_{ij} = 0$.
2: **for** $i = 1, \cdots, M$ **do**
3:   **for** $j = 1, \cdots, M$ **do**
4:    **if** $i = j$ **then**
5:     $E(i,j) = 0$, $P_{ts}(i,j) = 1$ and $final(i,j) = 1$
6:    **else**
7:     **if** $d_{ij} \leq R_c$ **then**
8:      Calculate $E(i,j)$ according to (5) and $P_{ts}(i,j)$ according to (6), set $final(i,j) = 1$
9:     **else**
10:      $E(i,j) = \inf$ and $P_{ts}(i,j) = 0$
11:     **end if**
12:    **end if**
13:   **end for**
14: **end for**
15: **for** $k = 1, \cdots, M$ **do**
16:   **for** $i = 1, \cdots, M$ **do**
17:    **for** $j = 1, \cdots, M$ **do**
18:     Find the minimum value min of the ith row of matrix E and record the $P_{ts}(k,j)$ and j
19:     $min = E(k,j)$
20:     $pmin = P_{ts}(k,j)$
21:     $minnum = j$
22:    **end for**
23:    $final(k, minnum) = 1$
24:    **for** $w = 1, \cdots, M$ **do**
25:     **if** $final(k, w) = 0$ && $(min + E(minnum, w)) < E(k, w)$ **then**
26:      Update $E(k, w) = min + E(minnum, w)$
27:      Update $P_{ts}(k, w) = pmin * P_{ts}(minnum, w)$
28:     **end if**
29:    **end for**
30:   **end for**
31: **end for**
32: Output the matrix E and matrix P_{ts}

Gaussian white noise is adopted from [11]. We compare the performance of the following caching policy with the proposed caching policy in this simulation:

- Uniform caching policy: Each user caches any file of the file library according to uniform probability
- Popularity based caching policy: Each user caches the file of the file library according to content popularity.

Table 1. The simulation parameters

Simulation parameters	Value
The size of the file library in one cluster N	500
The size of each file S	50 MBytes
The exponent of file popularity distribution β	0.7
The communication distance of D2D link R_c	30 m
The transmit power of D2D user P_d and P_u	200 mw
The variance of Gaussian white noise σ^2	-174 dBm
The path loss exponent α	4
The bandwidth of D2D B	20 MHz
The battery capacity C	3000 mAh
The operating voltage of the user device V	4 V

In Fig. 2, we show the probability of transmission successful for different transmission rate threshold R_0 and remaining battery power ratio η. In Fig. 2(a), it can be observed that at the beginning, as the remaining power increases, the probability of successful transmission gradually increases. When the remaining

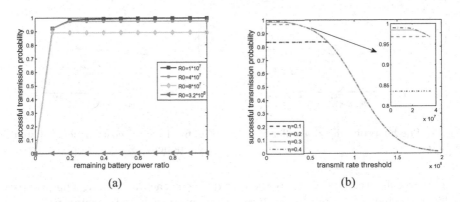

Fig. 2. Impact of remaining battery power ratio and transmission rate threshold on successful transmission probability

power is greater than a certain value, the probability of successful transmission almost keeps constant. At the same time, there are different constant values for different R_0. This is because when the user's battery power is poor, the probability of successful transmission is mainly influenced by the the user's battery power. whereas the battery power is so much to transmit a file completely, the transmission rate threshold is limiting factor. Also, in Fig. 2(b), when the transmission rate threshold is small, it is easy to transmit a file successfully. So the probability of successful transmission is limited by the battery power. However, when the transmission rate threshold is too large, it is difficult to transmit completely. These simulation results are consistent with the analysis of (6) in the above.

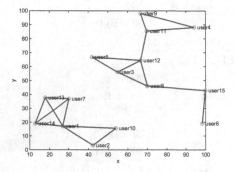

Fig. 3. The topology of users in one cluster

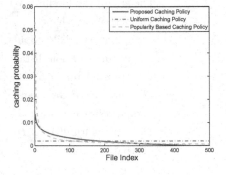

Fig. 4. Comparison of different cache policies

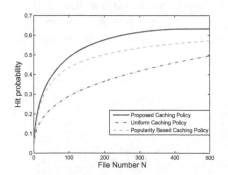

Fig. 5. The hit probability of different cache policies

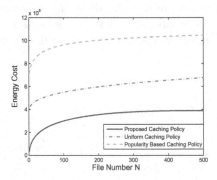

Fig. 6. Energy cost under different cache policies

Figure 4 illustrates the comparison of different caching policy. The proposed caching scheme is a compromise between popularity based caching policy and uniform caching policy, and when the file's popularity is too low, the caching probability of file is almost zero. In Fig. 5, it is showed that the proposed caching policy has higher hit probability than other two caching scheme. Figure 6 reflects the energy cost under different caching policy and it is obvious that the proposed caching policy can reduce energy cost, compared with uniform caching policy and popularity based caching policy.

5   Conclusion

In this paper, we consider the impact of users' battery energy on the content transmission in the caching network with D2D communication. We focus on obtaining minimize the energy cost of delivering a file via multihop D2D link. In addition, reducing the energy cost should not compromise the caching hit probability. To solve this problem, we classify users into different clusters and model users' topology relation as an undirected weighted graph in one cluster. Then, we determine the caching policy for making sure the hit probability. Based on the caching policy, to obtain the minimum energy cost, a novel algorithm is proposed to find the optimal path between any two users to transmit files. Moreover, we also take the battery capacity into account to make sure successful transmission. Simulation results show that the proposed caching policy outperforms other general caching strategies for less energy cost.

Acknowledgements. This paper is sponsored by National Natural Science Foundation of China (Grant 61671088 and 61771070).

References

1. Asadi, A., Wang, Q., Mancuso, V.: A survey on device-to-device communication in cellular networks. J. Guilin Univ. Electron. Technol. **16**(4), 1801–1819 (2014)
2. Blaszczyszyn, B., Giovanidis, A.: Optimal geographic caching in cellular networks. In: IEEE International Conference on Communications, pp. 3358–3363 (2015)
3. Breslau, L., Cao, P., Fan, L., Phillips, G., Shenker, S.: Web caching and zipf-like distributions: evidence and implications. In: INFOCOM 1999. Eighteenth Annual Joint Conference of the IEEE Computer and Communications Societies. Proceedings, vol. 1, pp. 126–134. IEEE (1999). https://doi.org/10.1109/INFCOM.1999.749260
4. Chen, B., Yang, C., Molisch, A.F.: Cache-enabled device-to-device communications: offloading gain and energy cost. IEEE Trans. Wirel. Commun. **PP**(99), 1–1 (2017)
5. Iqbal, J., Giaccone, P., Rossi, C.: Local cooperative caching policies in multi-hop d2d networks. In: IEEE International Conference on Wireless and Mobile Computing, NETWORKING and Communications, pp. 245–250 (2014)
6. Jeon, S.W., Hong, S.N., Ji, M., Caire, G.: Caching in wireless multihop device-to-device networks. In: IEEE International Conference on Communications, pp. 6732–6737 (2015)
7. Long, Y., Wu, D., Cai, Y., Qu, J.: Joint cache policy and transmit power for cache-enabled d2d networks. IET Commun. **11**(16), 2498–2506 (2017). https://doi.org/10.1049/iet-com.2017.0025

8. Malak, D., Al-Shalash, M.: Optimal caching for device-to-device content distribution in 5g networks. In: GLOBECOM Workshops, pp. 863–868 (2015)
9. Tarnoi, S., Suksomboon, K., Kumwilaisak, W., Ji, Y.: Performance of probabilistic caching and cache replacement policies for content-centric networks, pp. 99–106 (2014)
10. Yang, C., Chen, Z., Yao, Y., Xia, B., Liu, H.: Energy efficiency in wireless cooperative caching networks. In: IEEE International Conference on Communications, pp. 4975–4980 (2014)
11. Zhao, Y., Li, Y., Zhang, H., Ge, N., Lu, J.: Fundamental tradeoffs on energy-aware d2d communication underlaying cellular networks: a dynamic graph approach. IEEE J. Sel. Areas Commun. **34**(4), 864–882 (2016)
12. Zhu, H., Cao, Y., Liu, B., Jiang, T.: Energy-aware incentive mechanism for content sharing through device-to-device communications. In: Global Communications Conference (2017)

A Task Scheduling Algorithm Based on Q-Learning for WSNs

Benhong Zhang[1], Wensheng Wu[2], Xiang Bi[1（✉）], and Yiming Wang[1]

[1] School of Computer Science and Information Engineering,
Hefei University of Technology, Hefei 230009, Anhui, China
bixiang@hfut.edu.cn

[2] Intelligent Manufacturing Institute, Hefei University of Technology,
Hefei 230009, Anhui, China

Abstract. In industrial Wireless Sensor Networks (WSNs), the transmission of packets usually have strict deadline limitation and the problem of task scheduling has always been an important issue. The problem of task scheduling in WSNs has been proved to be an NP-hard problem, which is usually scheduled using a heuristic algorithm. In this paper, we propose a task scheduling algorithm based on Q-Learning for WSNs called Q-Learning Scheduling on Time Division Multiple Access (QS-TDMA). The algorithm considers the packet priority in combination with the total number of hops and the initial deadline. Moreover, according to the change of the transmission state of packets, QS-TDMA designs the packet transmission constraint and considers the real time change of packets in WSNs to improve the performance of the scheduling algorithm. Simulation results demonstrate that QS-TDMA is an approximate optimal task scheduling algorithm and can improve the reliability and real-time performance of WSNs.

Keywords: Wireless sensor networks · Q-Learning · Task scheduling

1 Introduction

In recent years, the flexibility and cost efficiency of wireless networks have become the main motivations for adopting wireless communications in industrial environments. Various network specifications such as WIA-PA [16], WirelessHART and ISA 100.11a [13] have been used to meet strict industrial requirements like real-time and reliability.

One of the major approaches to improve network performance is to use TDMA-based scheduling algorithm. The classic scheduling algorithms of WSNs based on TDMA are mainly the Earliest Deadline First (EDF) [15] and the

This paper is supported by the National Key Research Development Program of China (2016YFC0801800), Anhui Postdoctoral Research Funding Project (2017B144) and the National Natural Science Foundation of China (61501161).

X. Liu et al. (Eds.): ChinaCom 2018, LNICST 262, pp. 521–530, 2019.
https://doi.org/10.1007/978-3-030-06161-6_51

improvement of EDF algorithm [9]. But the problem of packet transmission scheduling in WSNs has been proved to be an NP-hard problem [4,10]. It is difficult to achieve an optimal or approximate optimal scheduling scheme by using the traditional methods.

In this context, many researchers have turned their attention to the field of machine learning [1]. Many novel methods based on machine learning have been applied to many aspects of scheduling tasks. The role-free clustering with Q-Learning for Wireless Sensor Networks (CLIQUE) is introduced in [6]. CLIQUE allows each node to investigate its capabilities as a cluster head node by combining the Q-Learning algorithm with some dynamic network parameters. A task scheduling algorithm for wireless sensor networks based on Q-Learning and sharing value function(QS) to solve the problem of frequent exchange of cooperative information in WSNs is introduced in [14]. QS can ensure that the nodes complete the application functional requirements while performing good cooperative learning. Considering that the real-time scheduling problem in multi-hop wireless network, a markov decision process of the packet transmission is proposed in [8]. All the above papers proved superiority of solving problems related to WSNs based on machine learning. However, few of them research on scheduling algorithms that improve the real-time and reliability of the network.

In this paper, We propose QS-TDMA algorithm for real-time scheduling of WSNs with strict deadline limitation in the form of flow. The algorithm uses Q-Learning [5] to achieve an approximate optimal scheduling scheme and improves the real-time and reliability of WSNs under the packet transmission constraint.

2   System Model

2.1   Data Flow Model

The transmission process of the generated packet sent from the source node to the destination node defined as a data flow. We consider that a set of M flows $F = \{f_1, f_2, \cdots, f_M\}$ are arranged in a single frequency band, The flow f_i is defined as follows:

$$f_i = (T_i, D_i, \phi_i, H_i) \tag{1}$$

Where $1 \leq i \leq M$, T_i represents the packet generation period of flow f_i; positive discrete variable D_i represents the deadline for each packet generated from the source, and $D_i \subset \{h_i^*, h_i^* + 1, \cdots, D_i^*\}$, h_i^* denotes the total number of hops required for the flow f_i to be transmitted from the source node to the destination node, and the value of D_i^* is large enough; ϕ_i represents the route of the flow f_i; H_i represents the length of f_i.

The packet transmission state on the flow f_n is defined as (t_n, h_n), t_n denotes the remaining deadline of packet transmission to the destination node, and h_n denotes the remaining hops of packet transmission from the current node to the destination node. If $t_n > h_n$, it means that the packet can be transmitted to the destination theoretically. If f_n is assigned to a time slot, the current transmission state of f_n is changed to $(t_n - 1, h_n - 1)$, and the transmission state on the other flows changes to $(t_n - 1, h_n)$. If $t_n = 0$, a new packet will be generated and entered into the flow to wait for transmission scheduling, and the packet state will be changed to the initial state (D_n, h_n).

2.2 Q-Learning Model

Reinforcement learning is a branch of machine learning. By identifying optimal strategies, each state is mapped to actions that the system should take in these states in order to maximize the numerical target reward over time [2]. Figure 1 illustrates the reinforcement learning model for task scheduling. It performs actions in WSNs and uses the reward feedback of a specific environment as a new learning process for the next experience.

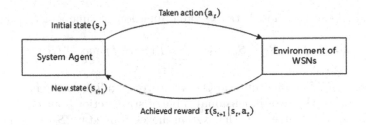

Fig. 1. Reinforcement learning model

As a popular method of reinforcement learning, Q-Learning can learn the usefulness of each task in the system and the benefit value of task execution in a certain environment over time to achieve the best adaptation to the current environment. Before the packet transmission, the system will give an immediate reward for the current action and make an evaluation. Before the end of the current transmission, the cumulative reward in the finite-state space is calculated by the value function and the Q-value evaluation is given. In the process of system learning, considering the flow transmission and constraints in the actual application scenario, we make the following assumptions:

- Only one flow per slot performs packet transmission;
- The packet generated on the flow has a strict deadline, assuming that the packet generation period is equal to the deadline;
- The probability of successful transmission from node i to node j through wireless communication will be affected by many physical factors such as transmission power, coding method, and modulation scheme. In this paper, we only consider the effect of the scheduling of node time slots, the probability that the node successfully transmitted to the next node $p = 1$.

Q-Learning algorithm can be seen as a random expression of the value iterative algorithm. The value iteration can be expressed by the action value function [12], and $V^\pi(S)$ represents that under the strategy π, the system performs the action f with the probability $P(S'|S, f)$ from the state S to the next state S'. The action value function of the state S is defined as follows:

$$V^\pi(S) = \max_{f\in F}[R(S'|S, f) + \gamma \sum_{S'\in S} P(S'|S, f)V^\pi(S')] \tag{2}$$

Where $P(S'|S, f)$ represents transition probability of the system when the agent select the flow f to perform from state S to state S'; $R(S'|S, f)$ represents the average reward for state transitions and γ is the discount factor, $\gamma \in (0, 1)$. The optimal strategy is to obtain the execution action that maximizes the value function. The optimal strategy $\pi^*(S)$ in the state S defined as follows:

$$\pi^*(S) = \arg V^{*(\pi)}(S) = \arg \max_{f\in F}[R(S'|S, f) + \gamma \sum_{S'\in S} P(S'|S, f)V^\pi(S')] \tag{3}$$

In our Q-Learning model, the Q-value function of time slot t is defined as follows:

$$Q_t(S_t, f) = R(S'|S, f) + \gamma \sum_{S'\in S} P(S'|S, f) \max_{f\in F} Q_t(S_t, f) \tag{4}$$

Where $Q_t(S_t, f)$ represents the Q-value corresponding to the flow f selected by the state S in the two-dimensional table of state actions; $\max Q_t(S_t, f)$ represents that at the t time slot, the system moves from state S to the next state S' of all flows that may perform packet transmission tasks and selects one of the actions that maximizes its Q-value. The Q-value update of the system is defined as follows:

$$Q_t(S_t, f) \leftarrow (1 - \alpha)Q_t(S_t, f) + \alpha[R(S'|S, f) + \gamma \sum_{S'\in S} P(S'|S, f) \max_{f\in F} Q_t(S'|S, f)] \tag{5}$$

Where $\alpha \in (0, 1)$ represents the learning rate factor. The larger α is, the more the system learning process relies on reward function and value function, the smaller α is, the more the system relies on accumulated learning experience and the slower the learning rate is.

3   QS-TDMA Scheduling Algorithm

3.1   System Space

In order to implement the task scheduling problem for M flows, the system needs to select a flow to perform the task in a hyper-period scheduling table. Hyper-period is usually defined in the industrial environment as the least common multiple of the packet generation periods of the field devices [15]. In this paper,we define the hyper-period as the least common multiple of the time slots to the deadlines of all flows. Therefore, the action space of the system is to determine which flow is assigned at each time slot, that is, the action space is $A = \{f_1, f_2, \cdots, f_M\}$. Each time slot of the hyper-period is mapped to the state space of the system, the state space is $S = \{1, 2, \cdots, T\}$.

3.2   Reward Function

The reward function reflects the value of rewards and punishments for the execution of the task, including two ways proposed in [11,14], respectively. Literature [11] adopts the mutative reward mechanism to achieve the applicability prediction to control the task executions during the learning process, while Literature [14] defines different fixed reward value for each task according to the priority of task execution in the application. In this paper, we consider the immediate reward of real-time flows allocated to the time slots, represented by r, and the influence of other flows not assigned to the time slots, represented by R_L. The combination of two factors reward function is defined as follows:

$$R(S, f) = r + R_L \tag{6}$$

The immediate reward r is composed of the total number of hops of the flow and the initial deadline. The smaller the total number of hops, the longer the initial deadline, and the smaller the value, the lower the priority of the current flow. The immediate reward r defined as follows:

$$r = k_1 \frac{h}{t} + k_2 \frac{1}{t - h + 1} \tag{7}$$

Where $\frac{h}{t}$ reflects the urgency of the packet; $t-h$ reflects the effect of the actual remaining time, which is not reflected in $\frac{h}{t}$; $t \geq h$, k_1, k_2 satisfy $0 < k_1, k_2 < 1$ and $k_1 + k_2 = 1$.

R_L denotes the feedback of the action, which reflects the negative reward. When the system in the state S_i and select f^i to perform in slot i, assume that there are L_{i0} flows in all flows satisfies $t_i - h_i = -1$, L_{i1} flows satisfies $t_i - h_i = 0$, and L_{i2} flows satisfies $t_i - h_i = 1$ before entering the next state S_{i+1}. Then we can define R_L as follows:

$$R_L = -(\rho_1 L_{i0} + \rho_2 L_{i1} + \rho_2 L_{i2}) \tag{8}$$

Where ρ_1, ρ_2, ρ_3 is the relevant discount parameters, $0 < \rho_1, \rho_2, \rho_3 < 1, \rho_1 > \rho_2 > \rho_3$, and $\rho_1 + \rho_2 + \rho_3 = 1$. Combining Eqs. (6)–(8) to get final reward function as follows:

$$R(s, f) = k_1 \frac{h}{t} + k_2 \frac{1}{t - h + 1} - (\rho_1 L_{i0} + \rho_2 L_{i1} + \rho_2 L_{i2}) \tag{9}$$

Local separation and combination of reward parameters and reward factors allow the reward function to be adjusted for external weights. And hence the behavior of the overall system is up to the initial states and the reward feedback of all flows.

3.3   Exploration-Exploitation Policy

In the trial and error process of the system, the relationship between exploration and exploitation needs to be balanced. The general $\varepsilon - greedy$ strategy is prone to converge rapidly. Developing in a situation where exploration is not adequate can result in a short learning process and serious learning biases. In this paper, we introduce the Metropolis Criterion (MC) in the Simulated Annealing [3,7] method into the flow selection in our exploration and exploitation, which can better solve the problem of excessive convergence and the balance between exploration and exploitation. The exploration probability ε_t is defined as follows:

$$\begin{cases} \varepsilon_p = \exp[-|(Q(S, a_r) - Q(S, a_o))|/KT_k] \\ \varepsilon_t = \max\{\varepsilon_{min}, \varepsilon_p\} \end{cases} \tag{10}$$

Where $Q(S, a_r)$ represents the Q-value that randomly selects an action in the state S; $Q(S, a_o)$ represents the Q-value that selects an optimal action in the state S; T_k is a fixed value, K is a coefficient, and $K = \lambda^e$, decline factor $\lambda \in (0, 1)$, e is the number of learning, as e increases, the value of ε_p will become smaller and smaller, and the entire exploration process will become stable; ε_t is the exploration probability based on MC, which is maximum value of ε_p and ε_{min}; ε_{min} is the minimum exploration probability given, which is the lower bound of exploration.

3.4   Q-Value Function Update

The definition of Q-value function update can be known from formula (5). In the formula, $P(S'|S, f)$ denotes the probability that the system selects flow f from the state S to the next state S'. $P(S'|S, f)$ is usually unknown, but the Q-Learning algorithm is obtained by replacing $R(S,f) + \sum_{S' \in S} P(S'|S,f) \max_{f \in F} Q_t(S'|S,f)$ by its simplest unbiased estimator built from the current transition $R_{t+1} + \max_{f \in F} Q_t(S'|S, f)$, In this way, the final Q-Learning algorithm Q-value function update formula is obtained:

$$Q_t(S_t, f) \leftarrow (1 - \alpha)Q_t(S_t, f) + \alpha[R(S'|S, f) + \gamma \max_{f \in F} Q_t(S'|S, f)] \tag{11}$$

3.5   Algorithm Description

The process description of our Q-Learning task scheduling algorithm can be obtained from the above two parts, the system model and the QS-TDMA scheduling algorithm. But in industrial environment, we have to consider the packet transmission scheduling process of WSNs is subject to a strict deadline limitation. For packets whose remaining time is less than the remaining transmission hops, it is impossible to be sent to the destination node theoretically and the packet will be lost. If we continue to allocate time slots for these flows, it is undoubtedly a waste of resources, and it will lengthen the entire process of learning and exploring. Therefore, we add selection constraints in the process of trial and error, and we do not allocate time slots for the flows that lose the meaning of theoretical transmission. By this way, the efficiency of the exploration-exploitation policy and the accuracy of system convergence can be improved.

In the process of learning, the system always has MDP feature in the face of external environment, and gradually rewards the flow with a small number of packet losses to perform packet transmission task, and finally obtains an approximate optimal scheduling algorithm. Specific algorithm description is shown in Table 1.

Table 1. QS-TDMA algorithm description

Algorithm 1 QS-TDMA algorithm description

1.	Initialize $Qmatrix$, T, α etc;
2.	Episode start;
	Initialize state $S_i = 1$, $i = 1$;
3.	If the number of episode reaches the requirement, repeat 7, otherwise select the task f^t to be performed according to the exploration-exploitation strategy in time slot i;
	3.1: Produce a random number $a(a \in (0,1))$;
	3.2: Select the optimal task f_p;
	3.3: Produce a random task f_r, calculate ε_t;
	3.4: If $a < \varepsilon_t$, $f^t = f_r$, otherwise $f^t = f_p$;
	3.5: If the state of the packet f^t is $t_{f^t} < h_{f^t}$, repeat 3.1;
4.	Execute task f^t, obtain reward R_t;
5.	According to (9) and (12), update the Q-value;
6.	If $S_t < T$, the system goes to the next state S_{t+1}, $i = i + 1$, repeat 3, otherwise repeat 2;
7.	End.

4   Simulation Analysis

In this paper, we simulate the performance of the algorithm and give the experimental results. The basic parameters are set as follows: the learning rate and discount factor $\alpha = \gamma = 0.9$; the weight of the two influencing factors of the immediate reward function $k_1 = k_2 = 0.5$; the influence parameters of flows that are not assigned to time slots $\rho_1 = 0.5, \rho_2 = 0.4, \rho_3 = 0.1$; The constant term in the exploration-exploitation policy $T_k = 1000$, $\lambda = 0.9$ and $\varepsilon_{min} = 0.01$.

We use the number of lost packets in a hyper-period as the criterion to measure the performance of the scheduling algorithms. We assume that the 'Optimal' algorithm (OP) in literature [8] is achievable and we compare QS-TDMA with other three strategies, respectively, as OP, EADF and RB. Analyze the network performance of QS-TDMA and other three algorithms in the number of different flows and different deadlines.

In Table 2, we consider two flows like [8], the total number of hops for the two flows with $H_1 = H_2 = 2$, the deadlines with $D_1 = 2, D_2 = 6$ and the hyper-period with $T = 1000$. The hyper-period takes a long enough value to offset the effect of randomness. In the following experiment, the hyper-period take the same value. Although the result of the QS-TDMA is good enough, but the result of learning has minimal fluctuations. The simulation data here take the average of 10 results, later comparisons will also be compared in the same way. As shown in the Table 2, the results of the QS-TDMA are close to the OP and the performance is much better than EADF and the RB. The scheduling of the RB is the worst in the feasible algorithms.

Table 2. The number of lost packets in the scheduling algorithm

Algorithm	QS-TDMA	OP	EADF	RB
Number of lost packets	176	166	251	500
Number of successful packets	490	500	415	166
Loss rate (%)	26.4	24.9	37.7	75.1

In Fig. 2, we consider the change in the number of lost packets for QS-TDMA, EADF, RB and OP as the flows increase. The number of selected flows increase from 2 to 6 in sequence. We consider three sets of symmetric flows. The deadline of the flow is randomly generated, and the flow deadline is one to three times the total hop count. The total number of hops for the flow is 2, 2, 3, 3, 5, 5, and the corresponding deadline is 2, 6, 6, 9, 10, and 15. As we can see in the figure, with the number of flows increase, the average number of lost packets for the four methods is increasing. Considering that only one flow can be sent for each time slot in a single frequency band, this result is in line with the actual situation. At the same time, it can be found that the OP is the optimal scheduling with the least theoretical packets loss. The QS-TDMA is close to OP, EADF is the second, and RB loses most packets. Moreover, with the increase in the number of flows, the result of RB is worse.

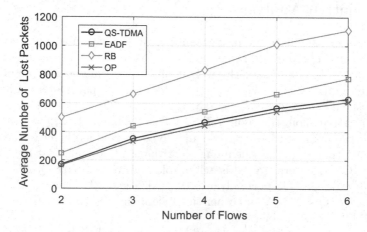

Fig. 2. The total number of lost packets with the increase of flows

In Fig. 3, we consider three flows with a total number of hops of $H_1 = 1$, $H_2 = H_3 = 3$, and the deadline for the three flows is the same. We analyze the number of lost packets for the four algorithms when the deadline increases from 3 to 8. It can be seen from Fig. 3 that with the increase of the deadline, the OP has always been the optimal result. When the deadline increases from 3 to 6, EADF and RB have the same number of lost packets, and performance of the EADF is worse than that of the RB when the deadline reaches 7. This is because the EADF is scheduled based on the principle of minimum average deadline first. The performance of EADF decreases with the same deadline for each flow. The overall performance of QS-TDMA is stable with the increase of deadline, and the performance is better than EADF and RB.

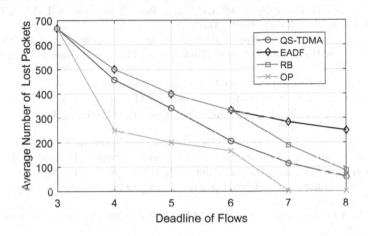

Fig. 3. The total number of lost packets with the increase of deadlines

5   Conclusion

In this paper, we proposed a TDMA-based task scheduling algorithm for wireless sensor networks. The algorithm combined the two factors of priority and real-time change of the packets. Under the given packet transmission constraints, an approximate optimal scheduling scheme is achieved by rewarding feedback and value iteration of system scheduling. The next step of the paper, we will consider how to improve the real-time and reliability of WSNs in the case of allowing concurrent data.

References

1. Abu Alsheikh, M., Lin, S., Niyato, D., Tan, H.P.: Machine learning in wireless sensor networks: algorithms, strategies, and applications. Commun. Surv. Tutor. IEEE **16**(4), 1996–2018 (2015)

2. Arnold, B.: Reinforcement learning: an introduction (adaptive computation and machine learning). IEEE Trans. Neural Netw. **9**(5), 1054 (1998)
3. Chen, S.L., Wu, H.Z., Xiao, L., Zhu, Y.Q.: Metropolis policy-based multi-step Q learning algorithm and performance simulation. J. Syst. Simul. **19**(6), 1284–1287 (2007)
4. Choi, H., Wang, J., Hughes, E.A.: Scheduling on sensor hybrid network. In: International Conference on Computer Communications and Networks, 2005. ICCCN 2005. Proceedings, pp. 503–508 (2013)
5. Watkins, C.J.C.H.: Q-learning. Mach. Learn. **8**, 279–292 (1992)
6. Forster, A., Murphy, A.L.: Clique: role-free clustering with Q-learning for wireless sensor networks. In: IEEE International Conference on Distributed Computing Systems, pp. 441–449 (2009)
7. Guo, M., Liu, Y., Malec, J.: A new Q-learning algorithm based on the metropolis criterion. IEEE Trans. Syst. Man Cybern. Part B Cybern. A Publ. IEEE Syst. Man Cybern. Soc. **34**(5), 2140 (2004)
8. Kashef, M., Moayeri, N.: Real-time scheduling for wireless networks with random deadlines. In: IEEE International Workshop on Factory Communication Systems, pp. 1–9 (2017)
9. Li, Q., Ba, W.: Two improved EDF dynamic scheduling algorithms in soft real-time systems. Chin. J. Comput. **34**(5), 943–950 (2011)
10. Saifullah, A., Xu, Y., Lu, C., Chen, Y.: Real-time scheduling for WirelessHART networks, pp. 150–159 (2010)
11. Shah, K., Kumar, M.: Distributed independent reinforcement learning (DIRL) approach to resource management in wireless sensor networks. In: IEEE International Conference on Mobile Adhoc and Sensor Systems, pp. 1–9 (2008)
12. Sigaud, O., Buffet, O.: Markov Decision Processes in Artificial Intelligence. ISTE, New York (2010)
13. Stig, P.: A comparison of WirelessHART and ISA100.11a for wireless instrumentation (2014)
14. Wei, Z., Zhang, Y., Xu, X., Shi, L., Feng, L.: A task scheduling algorithm based on Q-learning and shared value function for WSNs. Comput. Netw. **126**, 141–149 (2017)
15. Wu, C., Sha, M., Gunatilaka, D., Saifullah, A.: Analysis of EDF scheduling for wireless sensor-actuator networks. In: Quality of Service, pp. 31–40 (2014)
16. Yu, P.: Analysis on features of industrial wireless standard WIA-PA and the application in prospect. Process Autom. Instrum. **31**(1), 1–4 (2010)

Performance Analysis of Task Offloading in Double-Edge Satellite-Terrestrial Networks

Peng Wang[✉], Xing Zhang, Jiaxin Zhang, and Zhi Wang

Wireless Signal Processing and Networks Lab (WSPN),
Key Laboratory of Universal Wireless Communication,
Beijing University of Posts and Telecommunications (BUPT), Beijing, China
PWang@bupt.edu.cn

Abstract. With the rapid development of wireless networks, the growing number of mobile applications results in massive computation task to be processed. Multi-access edge computing (MEC) can efficiently minimize computational latency, reduce response time, and improve quality of service (QoS) by offloading tasks in the access network. Although lots of MEC task offloading schemes have been proposed in terrestrial networks, the integrated satellite-terrestrial communication, as an emerging trend for the next generation communication, has not taken MEC offloading into consideration. In this paper, we proposed a cooperative offloading scheme in a double-edge satellite-terrestrial (DESTN) network. Performance of offloading efficiency and energy consumption are derived analytically. Simulations show that the proposed offloading scheme in the double-edge satellite-terrestrial outperforms the traditional terrestrial-only offloading scheme by approximately 18.7%. Our research provides an insight for following studies in task offloading of double-edge satellite-terrestrial networks.

Keywords: Satellite-terrestrial network · Offloading scheme · Edge computing

1 Introduction

Recent years have witnessed the explosive growth of communication traffic, and a new communication network architecture is required to provide continuous geographic coverage for services and break the limitation of shortage in local communication and computation resources, saving up energy that services cost, connecting the global especially the remote area and meanwhile, reducing the average service time cycle [1]. To meet the demand resulted from this dramatic change, more effective resource cooperation deployments and distributed task offloading schemes should be proposed [2].

© ICST Institute for Computer Sciences, Social Informatics and Telecommunications Engineering 2019
Published by Springer Nature Switzerland AG 2019. All Rights Reserved
X. Liu et al. (Eds.): ChinaCom 2018, LNICST 262, pp. 531–540, 2019.
https://doi.org/10.1007/978-3-030-06161-6_52

Satellite communication systems, with the merits of wide coverage and flexible multiple-link capability, break the limitations of terrestrial network and deliver resilient and high-speed connectivity across the globe [3]. Scholars have already envisaged the idea of constructing an integrated satellite-terrestrial communication system to make full use of different resources [4–6]. Many new innovations have already been deployed in the integrated satellite-terrestrial communication, providing higher spectral efficiency of the overall system. Software Defined Networking (SDN) and Network Function Virtualization (NFV) technologies is also used for the realization of End-to-End Traffic Engineering in a combined satellite-terrestrial network used for mobile backhauling [7] and for supporting flexible and customized resource scheduling [8]. Recently there are some explorations on satellite task processing. Challenges and opportunities of onboard computers for small satellites are presented in [9] and the writer focuses upon ideas of hybrid computing and reconfigurable computing. While the emerging technologies mentioned above start to play vital roles in the new 5G system architecture, multi-access edge computing, named by the European Telecommunications Standards Institute (ETSI) [10], has emerged to provides cloud computing capabilities and IT service environments at the edge of the access network. New applications, e.g., interactive gaming, virtual reality, natural language processing and the Internet of Vehicles (IoV), put forward more requirements on the manageability and scalability of the system. Specifically, mobile edge networks provide computing and caching capabilities at the edge of cellular networks [11].

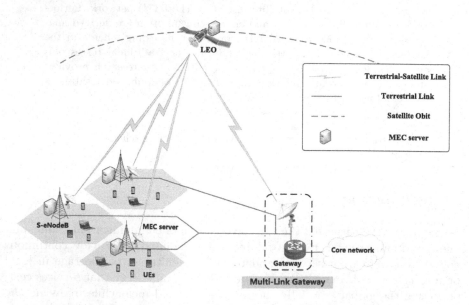

Fig. 1. Double-edge satellite-terrestrial network architecture

Taking advantages of computing ability, multi-access edge computing is a deployment to sink communication resources in the proximity of mobile user equipment (UE), at the same time, allowing users offloading computation tasks to MEC servers and thus enhancing the services experience, especially decreasing the service respond time. Currently there are lots of research works on offloading computation tasks to MEC in communication system. In [12], MEC network is designed to offload the traffic and tackle the backhaul congestion. In [13], each user equipment (UE) offloads its computation tasks to the MEC or cloud server for better performance. In [14], the communication, computing and caching resources are optimized together. Considering small cell network architecture, authors in [15] proposed an offloading scheme by solving energy optimization problem.Although many advancing technologies are deployed in integrated satellite-terrestrial network, few researchers take MEC computing and task offloading into consideration. As MEC server provides extra computation resources, in this paper, we analyze the task offloading efficiency of double-edge satellite-terrestrial networks and evaluate the energy consumption of our proposed scheme in the novel network architecture.

The rest of this paper is organized as follows. The system model of double-edge satellite-terrestrial network is described in Sect. 2. In Sect. 3, we proposed a task offloading scheme according to the new network architecture. Offloading efficiency and energy consumption are derived in Sect. 4. Numerical results are obtained and analyzed in Sect. 5. Finally, we conclude the paper in Sect. 6.

2 System Model

In this section, the double-edge satellite-terrestrial network architecture and the system model are introduced to describe the probability of the number of UEs that a randomly chosen satellite eNodeB (S-eNodeB) has.

2.1 Double-Edge Satellite-Terrestrial Architecture

The proposed double-edge satellite-terrestrial architecture is depicted in Fig. 1. Satellites, equipped with MEC server, form a space network edge of caching and computation resources. Covering the terrestrial base stations from a high altitude, they leverage distributed spot beams offering a continuous service to the specified area. On the terrestrial network edge, each S-eNodeB has a bidirectional link to communicate with satellites, connecting the space and terrestrial parts as an integrated one. Therefore, the proposed architecture has the double-edge network structure to meet computing and caching requirements of various services. Multi-link gateway is utilized in order to connect satellite and terrestrial backhaul for acquiring resources from evolved packet core (EPC) network.

Employing multiple beams over different geographical regions, satellites make use of space division multiplexing effectively for continuous handover. When UEs offload tasks to S-eNodeB, MEC servers distributed in double-edge satellite-terrestrial network will provide cooperative computation capability to reduce service latency.

2.2   System Model

In this section, we consider a double-edge satellite-terrestrial network wherein M S-eNodeBs are located in the coverage of a spotbeam of a satellite, each S-eNodeB has K UEs, which are denoted by $\mathcal{U} = \{1, \ldots, K\}$, we apply CDMA access scheme so that UEs share the same spectrum. Assuming that UE i has a task $Task_i = \{s_i, r_i, t_i^d\}$, $i \in \mathcal{U}$, as notations listed in Table 1. Typically, we assume that in the coverage of one satellite, both S-eNodeB and UE are distributed as classical homogeneous Spatial Poisson Point Processes (SPPP) distribution F. So the probability that a random chosen S-eNodeB have UEs access at the same time is [16]

$$P[N = n] = \frac{3.5^{3.5} \Gamma(n + 3.5)(\lambda_u/\lambda_b)^n}{\Gamma(3.5)n!(\lambda_u/\lambda_b + 3.5)^{n+3.5}}, \tag{1}$$

where $\Gamma(x) = \int_0^\infty \exp(-t)t^{x-1}dt$ is the gamma function, λ_b and λ_u refer to the density of S-eNodeB and UE, respectively.

Table 1. Notation summary.

Notation	Description
$\mathcal{U}$	UE set
s_i	Data size of UE i's computation task (bits)
r_i	Computation resources needed for UE i's task (CPU cycles)
t_i^d	Tolerable delay upper bound of the task (s)
λ_b	Density of S-eNodeB $(1/\text{m}^2)$
λ_u	Density of UE $(1/\text{km}^2)$

Tasks from UEs should be processed by the cooperation of satellite and terrestrial MEC servers, initially it should be transmitted to terrestrial MEC server located in S-eNodeB, the data rate of UE i received by the S-eNodeB is derived as

$$R_i^t = W_t \log_2 \left(1 + \frac{p_{i,m} g_{i,m}}{\sigma^2 + \sum\limits_{l=1}^{M} \sum\limits_{j=1, j \neq i}^{K} p_{j,l} g_{j,m}} \right), \tag{2}$$

where W_t is the wireless channel bandwidth, $p_{i,m}$ refers to transmission power of UE i to the S-eNodeB and $g_{i,m}$ the channel gain, σ^2 is the Gaussian noise.

Similarly, the data rate of terrestrial-satellite uplink m is

$$R_m^{su} = W_s \log_2 \left(1 + \frac{p_{m,s} g_{m,s}}{\sigma^2 + \sum\limits_{l=1, l \neq m}^{M} p_{m,s} g_{m,s}} \right), \tag{3}$$

where W_s is the channel bandwidth between satellite and terrestrial S-eNodeB, $p_{m,s}$ refers to transmission power of S-eNodeB m to the satellite and $g_{m,s}$ the channel gain.

3   Offloading Scheme

The offloading procedure can be divided into two phases: task transmission and task computation. In the double-edge satellite-terrestrial architecture, we propose a cooperative offloading scheme for more effective and reasonable task computation. As Fig. 2. shows, UEs transmit all tasks to the terrestrial MEC server in S-eNodeB, S-eNodeB calculate the priority index of the tasks and decide the task size to be processed in distributed MEC servers, tasks computation operate parallelly at terra and satellite. On the satellite, tasks queue and wait for computing, after finishing the task computation, the results will be converged back to S-eNodeB and are delivered to UEs. Taking up less computation resources, the tasks which have smaller size and are more sensitive to time delay should be served in the first place, at the same time, satellite with computation ability help compute part of tasks. When S-eNodeB receives the tasks from UEs, it sorts the tasks according to priority index β_i, which is denoted as

$$\beta_i = t_i^d / (r_i \cdot s_i) \tag{4}$$

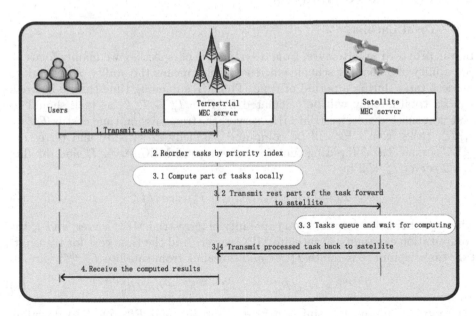

Fig. 2. Offloading scheme of double-edge satellite-terrestrial networks

In the proposed offloading scheme, each task should be divided into two parts according to parameter α. Here we define

$$\alpha = \frac{\beta_i - \beta_{min}}{\beta_{max} - \beta_{min}} \tag{5}$$

Then α percentage of the task should be transmitted and computed at satellite MEC server, remaining the rest of the task to be processed locally. When UE i's task priority index is the minimum β_{min}, then $\alpha = 0$, that means the task should be processed in terrestrial MEC server directly. For UE i's task priority index is the β_{max}, that means the task can tolerate the largest time delay, have the biggest file size and need the most computation resources, we can derive that $\alpha = 1$, then the whole task should be transmitted to MEC server on satellite.

The total time cost for task of UE i can be denoted by

$$T_i = T_i^P + T_i^{TR} + T_q, \tag{6}$$

where T_i^P refers to time cost of UE's task for being completely computed T_i^{TR} denotes transmission time which need to finish the task, and T_q is the time delay when task waits in the queue, if $T_i > t_i^d$, then UE's task will be abandoned.

Besides, we define task need to be computed in $T_i^c = r_i/f_0$, where f_0 is the computation capability of MEC server.

4   Performance Analysis

4.1   Total Latency

In our proposed architecture, from a systematic perspective, we mainly focus on the ability of offloading scheme efficiency, which means the ability of network to process tasks during a period of time. Thus UE's queuing time can be ignored, UE i's total latency will be calculated by $T_i = T_i^P + T_i^{TR}$, as task should be first transmitted to terrestrial MEC server, the transmission time cost is $T_i^{TR} = s_i/R_i^t$, tasks from UEs will be computed parallelly at satellite and terrestrial MEC servers, the computing time cost in terrestrial MEC server T_t^c and satellite MEC server T_s^c will be

$$T_t^c = (1-\alpha)r_i/f_t \qquad T_s^c = \alpha r_i/f_s, \tag{7}$$

where f_t stands for computation capability of terrestrial MEC server, and f_s the computation capability of satellite MEC server. And the time cost for transmitting tasks uplink to satellite $T_s^{tr\_up}$ and downlink from satellite $T_s^{tr\_down}$ are

$$T_s^{tr\_up} = \alpha s_i/R_m^{su} \qquad T_s^{tr\_down} = \alpha s_i/R_m^{sd}, \tag{8}$$

where R_m^{su} is uplink transmission rate of satellite and R_m^{sd} refers to downlink transmission rate. Finally, the time cost for task is calculated as

$$T_i^P = max\left\{ T_t^c, T_s^{tr\_up} + T_s^{tr\_down} + T_s^c \right\} \tag{9}$$

In the proposed scheme, tasks from UEs are divided into two parts, one is for terrestrial MEC servers, the other is for satellite MEC server. UE's whole task time will be

$$T_{total}^c = max\left\{ \sum_{i=1}^K \frac{(1-\alpha_i)r_i}{E\{f_t\}} , \sum_{i=1}^K \left(\frac{\alpha_i r_i}{E\{f_s\}} + \frac{\alpha_i s_i}{R_m^{su}} + \frac{\alpha_i s_i}{R_m^{sd}} \right) \right\}, \tag{10}$$

the first part is the time cost of MEC server on terra and the second part refers to time cost on satellite, therefore, the total time for the task is

$$T_{total} = E\{ \sum_{i=1}^K T_i^{TR} \} + T_{total}^c = P[N=K] \sum_{i=1}^K T_i^{TR} + \sum_{i=1}^K T_i^P$$

$$= \frac{3.5^{3.5}\Gamma(K+3.5)(\lambda_u/\lambda_b)^K}{\Gamma(3.5)K!(\lambda_u/\lambda_b + 3.5)^{K+3.5}} \sum_{i=1}^K T_i^{TR} + \sum_{i=1}^K T_i^P \tag{11}$$

To evaluate the performance of offloading scheme, offloading efficiency is defined as

$$\eta = r_{total}/T_{total} = \sum_{i=1}^K r_i/T_{total} \tag{12}$$

4.2   Energy Consumption

The total energy consumption is comprised of energy cost for transmitting tasks and the energy for task computation in MEC servers, for UE i, energy can be derived as:

$$e_i = p_{i,m}T_i^{TR} + p_{m,s}T_s^{tr} + p_c r_i, \tag{13}$$

where p_c is the power of MEC server when computing the tasks. The total energy consumption in the system is the sum of each UE's energy consumption e_i

$$E = \sum_{i=1}^K e_i \tag{14}$$

5   Numerical Results

In this section, simulation results are proposed in comparison with conventional terrestrial architecture which takes no satellite MEC server into consideration. The simulation is based on Matlab R2016b. In Table 2, we list important parameters. For simplicity, we consider 1 satellite MEC server, 1 terrestrial MEC server. λ_b and λ_u are $1/km^2$ and $2/km^2$, respectively. The total wireless bandwidth is 5 MHz, and the transmission power is ranging from 50 mW to 100 mW randomly. CPU Cycles required by UE's task is randomly distributed between 0.1 G and 2 G, we assume that terrestrial and satellite MEC computing capability is $f_t, f_s \in (1, 10)$, respectively. For LEO satellite uplink, Effective isotropic radiated power (EIRP) and bandwidth is 54.4 dBW and 3 MHz while the downlink

Table 2. Simulations parameter values [15].

Parameter	Min	Default	Max
UEs access to S-eNodeB	5	–	35
Task size	1 Mb	–	30 Mb
Background noise	–	100 dBm	–
Pathloss factor	–	4	–
Energy cost of MEC server	–	4 J/Hz	–

40.2 dB and 5 MHz, taking Eutelsat KA-SAT for reference, the downlink transmission rate is 20 Mbits/s, the wavelength of the carrier is 137.3 mm and the satellite operating altitude is 800 km, here we ignore the rain attenuation.

Figure 3 illustrates the offloading performance between the offloading scheme which simply leverage terrestrial MEC server and cooperative scheme we proposed in the double-edge satellite terrestrial architecture. The UE's task sizes are randomly distributed between 1 to 5 Mbits, 5 to 15 Mbits and 15 to 35 Mbits. The results show that the relationship between the number of UE and the offloading performance, from the results, we can make a conclusion that the offloading scheme performance which we proposed can improve approximately 18.7% than terrestrial MEC server only offloading scheme. It also illustrates that when number of UEs is small, transmission rate is high, so tasks with bigger data size have a better offloading performance. As the number of UEs and the amount of task size become larger, transmission bit rate on backhauls become lower thus transmission time would be the mainly part of the whole-time cost, although it can affect offloading efficiency, under this circumstance, the proposed scheme still gets a better performance.

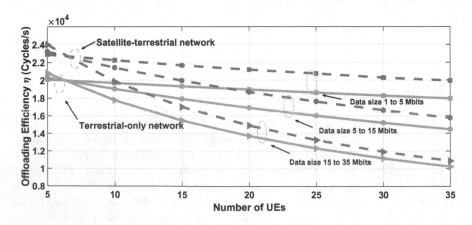

Fig. 3. Comparison of offloading efficiency η between terrestrial-only offloading scheme and double-edge satellite-terrestrial offloading scheme with different numbers of UE

In Fig. 4., we also evaluate the energy consumption in our proposed scheme versus number of UE. When the size of the task becomes larger, the energy consumption also increases, the reason of this phenomenon main lies in the transmission cost and computation cost.

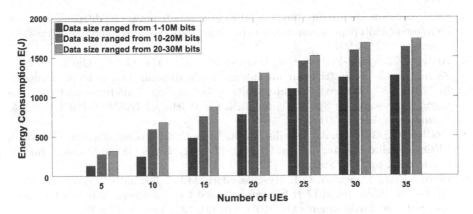

Fig. 4. Energy consumption versus the size of the offloading data in double-edge satellite-terrestrial networks

In summary, compared with the traditional terrestrial-only offloading scheme, the offloading scheme we proposed in the double-edge satellite and terrestrial architecture could improve the average offloading efficiency by nearly 18.7%, as the number of UE and the tasks' size increase, the energy consumption of the proposed offload-ing scheme will also become larger.

6   Conclusion

In this paper, under the background of a new architecture of DESTN, by using terrestrial and satellite MEC server, we analyze two key performance indicators of offloading scheme considering satellite and terrestrial resource cooperation. The simulation results show that the proposed scheme in satellite-terrestrial network can improve the average offloading efficiency by nearly 18.7% than terrestrial-only scheme and the energy consumption of the proposed offloading scheme will also become larger. In future, more complicated cooperative task offloading schemes between satellite constellations and terrestrial networks will be proposed.

Acknowledgement. This work is supported by the Beijing Municipal Science and Technology Commission Research under Project Z171100005217001 and by "the Fundamental Research Funds for the Central Universities" 500418765.

References

1. 5GPPP Homepage. https://5g-ppp.eu/. Accessed 15 May 2018
2. E-inclusion: satellites are the answer. https://www.ses.com/news/whitepapers/ses-delivers-innovative-connectivity-improve-and-save-lives. Accessed 15 May 2018
3. SES's mission to produce positive change through networks. https://www.ses.com/news/whitepapers/sess-mission-produce-positive-change-through-networks. Accessed 15 May 2018
4. Artiga, X., Nunez-Martinez, J., Perez-Neira, A., Vela, G.J.L., Garcia, J.M.F., Ziaragkas, G.: Terrestrial-satellite integration in dynamic 5G backhaul networks. In: 2016 8th Advanced Satellite Multimedia Systems Conference and the 14th Signal Processing for Space Communications Workshop (ASMS/SPSC), pp. 1–6. Palma de Mallorca (2016)
5. Kodheli, O., Guidotti, A., Vanelli-Coralli, A.: Integration of satellites in 5G through LEO constellations. In: GLOBECOM 2017 - 2017 IEEE Global Communications Conference, pp. 1–6. Singapore (2017)
6. Birrane, E.J., Copeland, D.J., Ryschkewitsch, M.G.: The path to space-terrestrial inter-networking. In: 2017 IEEE International Conference on Wireless for Space and Extreme Environments (WiSEE), pp. 134–139. Montreal (2017)
7. Ferrus, R., Sallent, O., Ahmed, T., Fedrizzi, R.: Towards SDN/NFV-enabled satellite ground segment systems: end-to-end traffic engineering use case. In: 2017 IEEE International Conference on Communications Workshops (ICC Workshops), pp. 888–893. Paris (2017)
8. Zhang, J., Zhang, X., Imran, M.A., Evans, B., Zhang, Y., Wang, W.: Energy efficient hybrid satellite terrestrial 5G networks with software defined features. J. Commun. Netw. **19**(2), 147–161 (2017)
9. George, A.D., Wilson, C.M.: Onboard processing with hybrid and reconfigurable computing on small satellites. Proc. IEEE **106**(3), 458–470 (2018)
10. ETSI.: Mobile-edge computing. In: Introductory Technical White Paper, pp. 1–36 (2014)
11. Wang, S., Zhang, X., Zhang, Y., Wang, L., Yang, J., Wang, W.: A survey on mobile edge networks: convergence of computing, caching and communications. IEEE Access **5**, 6757–6779 (2017)
12. Zhang, J., Zhang, X., Wang, W.: Cache-enabled software defined heterogeneous networks for green and flexible 5G networks. IEEE Access **2**(99), 1–1 (2016)
13. Chen, X., Jiao, L., Li, W., Fu, X.: Efficient multi-user computation offloading for mobile-edge cloud computing. IEEE/ACM Trans. Netw. **24**(5), 2795–2808 (2016)
14. Zhou, Y., Yu, F.R., Chen, J., Kuo, Y.: Resource allocation for information-centric virtualized heterogeneous networks with in network caching and mobile edge computing. IEEE Trans. Veh. Technol. **PP**(99), 1–1 (2017)
15. Yang, L., Zhang, H., Li, M., Guo, J., Ji, H.: Mobile edge computing empowered energy efficient task offloading in 5G. IEEE Trans. Veh. Technol. **67**(7), 6398–6409 (2018)
16. Singh, S., Andrews, J.G.: Joint resource partitioning and offloading in heterogeneous cellular networks. IEEE Trans. Wirel. Commun. **13**(2), 888–901 (2014)

Performance Analysis of Relay-Aided D2D Communications with Traffic Model

Jun Huang$^{(\boxtimes)}$, Yong Liao, and Yide Zhou

Chongqing University of Posts and Telecommunications, Chongqing 40065, China
xiaoniuadmin@gmail.com

Abstract. In this paper, we consider a communication scenario where relay users assist nearby a pair of D2D users underlaying cellular network. In our communication scenario, we analyze not only fading channel model but also different traffic models. In order to jointly consider the impact of interference level and network traffic condition, the packet loss probability (PLP) of D2D link is carefully orchestrated from two perspectives, i.e., link outage probability and packet delivery failure probability. The closed-form expressions of them are respectively obtained based on a Rician–Rayleigh fading model and different traffic models, and then the performance of our relay-aided D2D communication scenario is evaluated by the PLP of D2D link. Finally, the PLP of D2D link with three representative traffic models including Pareto, FBM, and Poisson traffic models are compared, respectively. We believe that the proposed analytical approach can provide a useful insight into the application of traffic model in relay-aided D2D communications.

Keywords: Relay-aided D2D communication · Traffic model
Packet loss probability

1 Introduction

Due to the demand for Internet access is increasing dramatically with the increment of mobile users, device-to-device (D2D) communication is proposed to address this issue. The close-range D2D communication underlaying cellular network has been considered as an effective way to improve transmission rate, reduce transmission latency and power consumption, and enhance spectrum efficiency. However, the direct D2D communication can only work with a very limited distance, but not apply to all communication scenarios. Therefore, a relay-aided D2D communication is considered to be indispensable in expanding D2D coverage [1].

The relay-aided D2D communication utilizes relay users to forward data packets. On the one hand, it can expand the signal communication range, which makes D2D communication adaptable to complex and diversified environments. On the other hand, it can also shorten the distance of per-hop D2D link, which

© ICST Institute for Computer Sciences, Social Informatics and Telecommunications Engineering 2019
Published by Springer Nature Switzerland AG 2019. All Rights Reserved
X. Liu et al. (Eds.): ChinaCom 2018, LNICST 262, pp. 541–550, 2019.
https://doi.org/10.1007/978-3-030-06161-6_53

eventually reduces the transmit power and energy consumption. Many efforts have been made to analyze the performance of relay-aided D2D communication. Wei et al. investigated a multi-hop D2D communication scenario where relay nodes assist to exchange information with PNC, and then analyzed the average energy efficiency and spectral efficiency under Rayleigh fading channel [2]. Hasan et al. proposed a robust distributed solution for resource allocation with a view to maximizing network rate when the interference from other relay nodes and the link gains are uncertain [3]. The authors of [4] proposed a game-theoretic model for the compensation power acquisition of D2D link transmitters underlaying cellular system. However, when the process of loss packet occurs at the per-hop D2D link, the receiver can not be able to correctly receive the desired data packets. Therefore, the packet loss probability (PLP) of D2D link is a key performance metric for relay-aided D2D communications [5].

In the most of related works, the Rayleigh fading model which ignores line-of-sight (LoS) signal components is adopted. But in fact, a close-range communication often leads to the existence of dominant LoS signal components in the received desired signals [6,7]. When multiple D2D users reuse a cellular uplink channel, they can be subject to interfering signals from cellular users. The larger interference can result in the higher outage probability of D2D link, which has an impact on the successful reception of the desired signals. In addition, the relay users in multi-hop D2D communication often consider limited queue capacity and service capability [8,9]. Therefore, how to select fading channel model and traffic model for multi-hop D2D communication scenario is of great significance. Our contribution is summarized in the following aspects. First, a general multi-hop D2D communication underlaying cellular network is introduced, followed by the formulation of corresponding fading channel model and traffic model. Second, in order to jointly consider the impact of interference level and network traffic condition, the PLP of D2D link is carefully orchestrated from two perspectives, i.e., the outage probability and packet delivery failure probability. Meanwhile, the closed-form expressions of them are respectively obtained based on fading channel model and traffic model.

The rest of this paper is organized as follows. In Sect. 2, a general relay-aided D2D communication underlaying cellular network is introduced, followed by the formulation of corresponding channel model and traffic model. In Sect. 3, the packet loss probability of D2D link is analyzed. In next section, simulation results verifying the link outage probability with fading channel model and the PLP of D2D link under different traffic models are provided, respectively. Finally, we conclude the paper in Sect. 5.

2   System Model

2.1   Network Model

In this paper, we consider a general relay-aided D2D communication underlaying cellular network, as shown in Fig. 1. There are some randomly distributed cellular user equipment (CUE) and a pair of D2D user equipment (DUE_1 and DUE_2) in this cell. This pair of D2D UE has the desire to establish D2D links to exchange some multimedia content like pictures, live video, or interactive games. It is assumed that the distance between DUE_1 and DUE_2 is too long to directly communicate with each other using the traditional one-hop D2D link. Therefore, the relay user equipment (RUE) is required to establish multi-hop D2D links. In this paper, we consider that the number of RUE is denoted by M and all of them are seen essentially as D2D UE. The scheduling and resource allocation for all D2D UE $(DUE_1, DUE_2$ and $RUE)$ can be done by the eNodeB. In addition, it is considered that all D2D UE reuse the same cellular uplink channel that is assigned to the l-th cellular UE (CUE_l). Meanwhile, the eNodeB has the control over transmission power of the CUE_l and all D2D UE to reduce interference level. Figure 1 shows that there are $M + 2$ user equipment including one cellular user equipment and $M + 1$ D2D user equipment. For simplicity, we name DUE_1 as node 0, RUE_k as node k $(k \in \{1, 2, \cdots, M\})$, DUE_2 as node $M + 1$, and CUE_l as node $M + 2$. The notation (i, j) is used to represent the transmission link from node i to node j.

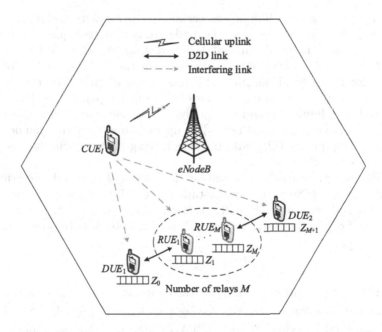

Fig. 1. A general relay-aided D2D communication underlaying cellular network

2.2   Channel Model

The path loss channel model and additive white Gaussian noise are considered in our channel model. The large-scale fading is determined by the Euclidean distance $d_{i,j}$ between node i and node j and the path loss exponent α which is relevant to the communication environment. Then, the small-scale fading of the link (i,j) is captured by a Rayleigh or Rician random variable $f_{i,j}$. Therefore, the instantaneous SINR at node j is given by

$$\gamma_j = \frac{P_{T_i} \cdot d_{i,j}^{-a} \cdot F_{i,j}}{P_{T_l} \cdot d_{l,j}^{-a} \cdot F_{l,j} + \sigma^2} = \frac{F_{i,j}}{b \cdot c \cdot F_{l,j} + b \cdot \sigma^2}, \tag{1}$$

where P_{T_i} is the transmit power of node i, $d_{i,j}^{-a}$ is the path loss of the link (i,j) and $F_{i,j} = |f_{i,j}|^2$ is the channel gain of the link (i,j). Similarly, $F_{l,j}$ is the interfering power used by node l. The symbols $d_{l,j}^{-a}$ and $F_{l,j} = |f_{l,j}|^2$ are the path loss and channel gain of the link (l,j), respectively. It is assumed that all nodes suffer the same additive white Gaussian power σ^2. For simplicity, we define $b = \frac{1}{P_{T_i} \cdot d_{i,j}^{-a}}$ and $c = P_{T_l} \cdot d_{l,j}^{-a}$.

2.3   Traffic Model

As illustrated in Fig. 1, each node has certain queue capacity and service capacity. In this paper, we assume that the node j has a limited queue capacity Z_j (packets) to store dynamically arrival data packets, and each packet has an equal length of b (bits). Moreover, the transmission in the time is slot-by-slot based and each slot has a fixed duration ΔT (ms). In each time slot, the spectrum resource can be allocated to one or more D2D links, depending on the resource sharing and scheduling strategies. Finally, during each time slot, the average service rate at node j, μ_j (packets/ms), is upper bounded by the channel capacity of the corresponding D2D link. In the following part, we will discuss several typical network traffic models.

The Pareto traffic model is a self-similar model that mainly describes the traffic inter-arrival time with a heavy-tailed probability density function (pdf). The inter-arrival time of such traffic is independent and identical distributed. Then, the pdf of the inter-arrival time X is the Pareto distribution as follows:

$$f(x) = \frac{S\beta^S}{x^{S+1}}, x \geq \beta, \tag{2}$$

where S is the shape parameter and β is the minimum value of inter-arrival time. The mean and variance are respectively $\frac{S\beta}{S-1}$ and $\frac{S\beta^2}{(S-1)^2(S-2)}$, $1 < S < 2$.

The fractional Brownian motion (FBM) traffic model is an important self-similar model that mainly describes the cumulative arrival amount of traffic and models the variation of connectionless traffic with a self-similar Gaussian process. A standard FBM random process $Y(t)$ with Hurst parameter $H \in [0.5, 1)$ is an essential Gaussian process with the zero mean and the variance of $|t|^{2H}$. Therefore, during each time slot, the cumulative arrival amount at the node j, $A_j(\Delta T)$, that satisfies the self-similar FBM input traffic model is expressed as [10]

$$A_j(\Delta T) = \lambda_j \Delta T + \sqrt{\eta \cdot \lambda_j} Y(\Delta T), \tag{3}$$

where η is a variance coefficient, and λ_j is the average packet arrival rate at node j.

The Poisson traffic model, unlike the above presentation, is a very attractive memoryless model (future behavior has no link to past behavior), which means that it is easy to analyze but cannot effectively reflect the burstiness nature of the traffic. During each time slot, the cumulative arrival amount $A_j(\Delta T)$ is a Poisson process with a parameter $\lambda_j \Delta T$, which is given by

$$\Pr\{A_j(\Delta T) = n\} = e^{-\lambda_j \Delta T} \frac{(\lambda_j \Delta T)^n}{n!}, \qquad (4)$$

where the expectation of $A_j(\Delta T)$ is equal to $\lambda_j \Delta T$, means $E(A_j(\Delta T)) = \lambda_j \Delta T$. Meanwhile, the packet inter-arrival time has an exponential distribution with mean $1/\lambda_j$.

3   The Packet Loss Probability of D2D Link

In Fig. 1, when the node 0, node $M + 1$ and node k forming D2D links reuse the same cellular uplink channel, these nodes will be subject to interference from node $M + 2$. However, the larger interference can result in the higher outage probability of D2D links, which has an impact on the successful reception of the desired signals. In addition, network traffic is an important factor in D2D links. It is found that each node has own limitation on queue capacity and service capability. When network traffic exceeds a certain level, the queue capacity and service capability cannot be sufficient, which eventually causes network congestion and compromise the packet delivery process in D2D links. Therefore, in order to jointly consider the impact of interference level and network traffic condition, the packet loss probability of D2D link proposed in [5] is adopted in this paper as follows

$$\Pr_{i,j}^l = \hat{\Pr}_{i,j} \cdot \Pr_{i,j}^o = \left(1 - \bar{\Pr}_{i,j}\right) \cdot \Pr_{i,j}^o, \qquad (5)$$

where $\Pr_{i,j}^l$ is the packet loss probability of link (i, j). $\hat{\Pr}_{i,j}$ is the packet delivery probability defined to reflect traffic conditions over the link (i, j). $\bar{\Pr}_{i,j}$ is the packet delivery failure probability. $\Pr_{i,j}^o$ is the outage probability of link (i, j).

3.1   The Link Outage Probability

The outage probability of transmission link needs to be analyzed based on its fading channel model. In previous research, most of the papers consider the desired signals and interfering signals of the receiving terminal from the perspective of Rayleigh fading. However, in the close-range D2D communication, the desired signals at the receiving terminal also contain the dominant LoS signal components in addition to the scattered signal components. Therefore, in this paper, the Rician fading is considered to model the channel gain of the per-hop D2D link, while for the interfering links with long distance, the Rayleigh fading is adopted. Such simulation is called a Rician–Rayleigh fading channel model.

In the Rician–Rayleigh fading channel model, the link outage probability is defined as the probability that the instantaneous SINR at node j expressed by Eq. (1) is less than the SINR threshold γ_0, that is

$$
\mathrm{Pr}_{i,j}^o = \frac{(K+1)\gamma_0 \left(1 - \frac{\sigma^2}{c \cdot \omega + \sigma^2}\right)}{\gamma_a + (K+1)\gamma_0 \left(1 - \frac{\sigma^2}{c \cdot \omega + \sigma^2}\right)} e^{-\frac{K \cdot \gamma_a}{\gamma_a + (K+1)\gamma_0 \left(1 - \frac{\sigma^2}{c \cdot \omega + \sigma^2}\right)} + \frac{\sigma^2}{c \cdot \omega}}, \tag{6}
$$

where $\gamma_a = \frac{\Omega}{b \cdot c \cdot \omega + b \cdot \sigma^2}$ is the average SINR at node j, K is the Rican factor and Ω is the expected receiving power. Note that the Eq. (6) can also express the link outage probability under the Rayleigh–Rayleigh fading channel model when $K = 0$. When the mean interfering power $c \cdot \omega$ is much larger than white Gaussian power σ^2, the Eq. (6) can be shown that

$$
\mathrm{Pr}_{i,j}^o \approx \frac{(K+1)\gamma_0}{\gamma_a + (K+1)\gamma_0} e^{-\frac{K \cdot \gamma_a}{\gamma_a + (K+1)\gamma_0}}. \tag{7}
$$

It is found that this closed-form approximation is quite similar to the result of Yao and Sheikh [[11], Eq. (7)].

3.2 The Packet Delivery Failure Probability

The packet delivery failure probability is a network layer parameter that indicates the network congestion caused by the link traffic. When the queue length of relay nodes reaches queue capacity, the subsequent arriving data packets are dropped, which can lead to packet delivery failure. Let $Q_j(t)$ be the queue length (packets) of node j at the beginning of t-th time slot, which is also classified as either a continuous-time or a discrete-time queue length. In this paper, it is only considered to be a continuous-time queue length because equivalent results can also be obtained for the discrete-time case. During the t-th time slot, the queue length of $Q_j(t+1)$, the cumulative arrival amount of data packets placed in the queue capacity, expressed by

$$
Q_j(t+1) = \min\left\{Z_j, \max\left\{0, Q_j(t) - \mu_j \Delta T + A_j(\Delta T)\right\}\right\}. \tag{8}
$$

It is assumed that each node needs to send control signaling for requesting data forwarding to the base station before each time slot. Then, the arriving data packets during each time slot can only be transmitted at the next time slot. Finally, the base station achieves the sequence of the data forwarding. This assumption is based on the following aspects. First, the existing literature has little research on the sequence of multi-hop D2D data forwarding, which leads to having no uniform standard; Second, it is convenient that the base station can control the sequence of data forwarding, which can effectively reduce the complexity. According to the above assumption, it can be seen that the data packets placed in queue capacity at the beginning of $(t + 1)$-th time slot have been transmitted, means $Q_j(t) = 0$.

In this paper, when the queue length of the node j, $Q_j(t+1)$, reaches its queue capacity Z_j, the packet delivery failure probability over the link (i, j) is defined as follows

$$\bar{\text{Pr}}_{i,j} = \text{Pr}\{Q_j(t+1) = Z_j\} = \text{Pr}\{A_j(\Delta T) \geq Z_j + \mu_j \Delta T\}. \tag{9}$$

In the following part, the closed-form expressions $\bar{\text{Pr}}_{i,j}$ based on different traffic models are obtained, respectively.

The FBM traffic model can model the variation of connectionless traffic with a self-similar Gaussian process. According to substitute Eq. (3) to (9), the packet failure delivery probability over the link (i, j) is presented as

$$\bar{\text{Pr}}_{i,j} = 1 - \Phi \left(\frac{(\mu_j - \lambda_j)\Delta T + Z_j}{\sqrt{\eta \cdot \lambda_j \Delta T^H}} \right), \tag{10}$$

where $\Phi(\cdot)$ is the cumulative distribution function of the standard Gaussian distribution. Similarly, the packet failure delivery probability based on the Poisson traffic model, substituting Eq. (4) to (9), can be calculated by

$$\bar{\text{Pr}}_{i,j} = 1 - e^{-\lambda_j \Delta T} \sum_{n=0}^{Z_j + \mu_j \Delta T} \frac{(\lambda_j \Delta T)^n}{n!}. \tag{11}$$

On the other hand, due to the Pareto traffic model mainly analyzes the packet inter-arrival time, we consider a Pareto/M/1 queue model. The packet failure delivery probability is given by [12]

$$\bar{\text{Pr}}_{i,j} = \left[1 - \frac{S(S-1)}{\rho} M^{\frac{S}{2}-1} e^{M/2} \left(\sqrt{M} W_{-\frac{S+1}{2}, -\frac{S}{2}}(M) - W_{-\frac{S}{2}, -\frac{1-S}{2}}(M) \right) \right] \sigma^{Z_j} \tag{12}$$

where $\rho = \frac{\lambda_j}{\mu_j}$ is the service utilization, $W_{\eta,\xi}(\phi)$ is Whittakers function, M is equal to $\frac{(S-1)(1-\sigma)}{\rho}$ and $\sigma = \alpha M^{\frac{\alpha-1}{2}} e^{M/2} W_{-\frac{(\alpha+1)}{2}, -\frac{\alpha}{2}}(M)$ is a geometric parameter.

4 Numerical Result

In this section, we evaluate the performance of our relay-aided D2D communication scenario using the packet loss probability (PLP) of D2D link. Table 1 summarizes the list of main simulation parameters and their default values.

As shown in Fig. 2, we compare the link outage probability under different fading models when the range of SINR is appropriately chosen. For simplicity, we randomly select two SINR thresholds γ_0, -5 dB and 0 dB, respectively. According to numerically calculate Eq. (7), it is found that the link outage probability decreases with the increasing SINR under a fixed γ_0. In addition, the $\gamma_0 = 0$

Table 1. Parameters settings

Parameter description	Value
Reuse uplink bandwidth B	2 MHz
Path loss exponent α	4
Rician factor K	7 dB
The time slot duration ΔT	1 ms
The coefficient of variance η	1
Hurst parameter H	0.7
Pareto shape parameter S	1.5

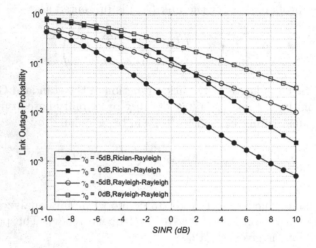

Fig. 2. The link outage probability with different SINR thresholds and fading channel models

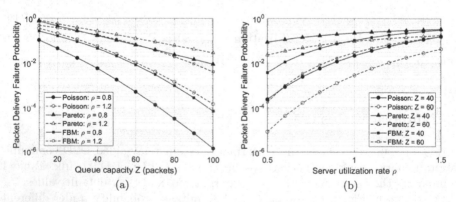

Fig. 3. The packet delivery failure probability with different traffic models

leads to much larger link outage probability than the $\gamma_0 = -5$. Meanwhile, the Rician–Rayleigh fading model which takes the LoS signal components into account leads to a lower link outage probability than the Rayleigh–Rayleigh fading model. In order to ensure a small link outage probability, we choose the value of γ_0 to be -5 dB rather than 0 dB in the subsequent analysis.

In order to characterize traffic conditions, three representative traffic models, i.e., Pareto, FBM, and Poisson traffic models, are adopted to represent $\bar{Pr}_{i,j}$ in the Eqs. (10), (11) and (12). Figure 3(a) and (b) illustrate the packet delivery failure probability with different parameter configurations. It is seen that the smaller queue capacity is, the more likely the packet delivery fails. Moreover, the $\rho > 1$ is more likely to cause packet delivery failure than the $\rho < 1$. Finally, it is found that the packet delivery failure probabilities of the Pareto and FBM traffic models which consider the burstiness and self-similarity of network traffic are higher than that of the Poisson traffic model.

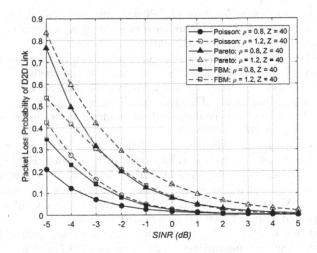

Fig. 4. The packet loss probability of D2D link with different traffic models

Figure 4 presents the PLP of D2D link expressed by Eq. (5) with the above traffic models. It is concluded that the PLP of Poisson traffic model is smaller than that of the other traffic models. When the node 0 and node $M + 1$ send a small number of data packets to each other at the communication scenario of Fig. 1, the D2D links usually have less traffic burstiness. The PLP of Poisson traffic model is more suitable than other traffic models.

5   Conclusion

In this paper, we consider a general relay-aided D2D communication underlaying cellular network. In our communication scenario, we consider not only fading channel model but also different traffic models. It is verified that the Rician–Rayleigh fading model which takes the LoS signal components into account leads to a lower link outage probability than Rayleigh-Rayleigh fading model. Then, the packet delivery failure probabilities of the Pareto and FBM traffic models which consider the burstiness and self-similarity of network traffic are higher than that of the Poisson traffic model. Finally, it is concluded that the PLP of Poisson traffic model is smaller than that of the other traffic models. In short, the insight is expected to shed light on the application of traffic model in relay-aided D2D communications.

References

1. Liu, J., Kato, N., Ma, J., Kadowaki, N.: Device-to-device communication in LTE-advanced networks: a survey. IEEE Commun. Surv. Tutor. **17**(4), 1923–1940 (Fourthquarter 2015)
2. Wei, L., Hu, R.Q., Qian, Y., Wu, G.: Energy efficiency and spectrum efficiency of multihop device-to-device communications underlaying cellular networks. IEEE Trans. Veh. Technol. **65**(1), 367–380 (2016)
3. Hasan, M., Hossain, E., Kim, D.I.: Resource allocation under channel uncertainties for relay-aided device-to-device communication underlaying LTE-A cellular networks. IEEE Trans. Wirel. Commun. **13**(4), 2322–2338 (2014)
4. Huang, J., Huang, S., Xing, C.C., Qian, Y.: Game-theoretic power control mechanisms for device-to-device communications underlaying cellular system. IEEE Trans. Veh. Technol. **67**, 1–1 (2018)
5. Huang, J., Gharavi, H.: Performance analysis of relay-based two-way D2D communications with network coding. IEEE Trans. Veh. Technol. **PP**(99), 1–1 (2018)
6. Penda, D.D., Risuleo, R.S., Valenzuela, P.E., Johansson, M.: Optimal power control for D2D communications under Rician fading: a risk theoretical approach. In: GLOBECOM 2017 - 2017 IEEE Global Communications Conference, pp. 1–6 (2017). https://doi.org/10.1109/GLOCOM.2017.8254838
7. Lin, M., Ouyang, J., Zhu, W.P.: Joint beamforming and power control for device-to-device communications underlaying cellular networks. IEEE J. Sel. Areas Commun. **34**(1), 138–150 (2016)
8. Kim, J., Kim, S., Bang, J., Hong, D.: Adaptive mode selection in D2D communications considering the bursty traffic model. IEEE Commun. Lett. **20**(4), 712–715 (2016)
9. Huang, S., Liang, B., Li, J.: Distributed interference and delay aware design for D2D communication in large wireless networks with adaptive interference estimation. IEEE Trans. Wirel. Commun. **16**(6), 3924–3939 (2017)
10. Norros, I.: On the use of fractional Brownian motion in the theory of connectionless networks. IEEE J. Sel. Areas Commun. **13**(6), 953–962 (1995)
11. Yao, Y.D., Sheikh, A.U.H.: Outage probability analysis for microcell mobile radio systems with cochannel interferers in Rician/Rayleigh fading environment. Electron. Lett. **26**(13), 864–866 (1990)
12. Rodrfguez-Dagnino, R.M.: Some remarks regarding asymptotic packet loss in the Pareto/M/1/K queueing system. IEEE Commun. Lett. **9**(10), 927–929 (2005)

Phase Noise Estimation and Compensation Algorithms for 5G Systems

Shuangshuang Gu[✉], Hang Long, and Qian Li

Wireless Signal Processing and Network Lab, Key Laboratory of Universal Wireless Communication, Ministry of Education, Beijing University of Posts and Telecommunications, Beijing, China
gss@bupt.edu.cn

Abstract. In the 5G system, orthogonal frequency division multiplexing (OFDM) waveform survived for its superior performance. As is well known, OFDM systems are sensitive to the phase noise introduced by local oscillators and it may get worse for likely higher carrier frequency in 5G systems. There are two aspects of the impact of phase noise, namely the common phase error (CPE) and the inter-carrier-interference (ICI). In this paper, first, we propose a more accurate way to estimate CPE. Then, we focus on ICI cancellation. To simplify the ICI model, we only consider the interference from adjacent sub-carriers. Based on the simplified model, we propose two schemes to estimate ICI. The performance of phase noise compensation algorithm we proposed is presented. Simulation results show that the algorithm we proposed can significantly reduce the impact of phase noise and improve the throughput of 5G systems.

Keywords: Phase noise · CPE estimation · ICI cancellation

1 Introduction

As a new generation of mobile communication technology, 5G has great innovations on data transmission rate, system bandwidth, carrier frequency, etc. Different from long term evolution (LTE), 5G new radio (NR) may support higher carrier frequency, such as 28 GHz and 40 GHz. Generally, the higher the carrier frequency, the stricter the requirements on the radio frequency (RF) circuits, and the more serious the phase noise. By default, for per decade increase in carrier frequency, power spectral density (PSD) increases by 20 dBc/Hz [1]. 3GPP working group clearly stipulates that when the carrier frequency is greater than 6 GHz, 5G systems must consider the impact of phase noise, and hence introduces a new phase tracking reference signal (PT-RS) for phase noise compensation [2].

Phase noise has always been a problem in high-frequency communication systems and there have been many studies. Reference [3] estimates phase noise and

Supported by China Unicom Network Technology Research Institute and project 61302088 which was supported by National Natural Science Foundation of China.

X. Liu et al. (Eds.): ChinaCom 2018, LNICST 262, pp. 551–561, 2019.
https://doi.org/10.1007/978-3-030-06161-6_54

channel simultaneously. Although the channel and phase noise models have been simplified, the algorithm still requires a large number of pilots, and the iterative process is complex. Reference [4] selects orthogonal discrete cosine transform as the basis function to model the phase noise, and obtain the data and phase noise parameters iteratively according to the maximum likelihood criterion. There is a certain model error and multiple iterations are needed to obtain better results. Reference [5] uses a filter with a certain number of coefficients for extraction of phase noise and use scattered pilots to estimate coefficients. This method approximates the phase noise on condition that phase noise is very small, when the phase noise is large, this approximate method is not applicable. Reference [6] estimates the fast fourier transform (FFT) components of the current phase noise realization and suppress the inter-carrier-interference (ICI) by performing a deconvolution in the frequency domain. This method requires the correlation between ICI are known. Reference [7] proposes two phase noise cancellation schemes. The first one is based on linear interpolation of the common phase error (CPE) values over adjacent orthogonal frequency division multiplexing (OFDM) symbols. The second one improves the ICI estimation based on the methods presented in [8] by improving the accuracy of phase noise estimation at symbol boundaries using proper interpolation. Based on complex models and iterative operations, existing algorithms can eliminate the effects of phase noise including CPE and most of the ICI successfully. In practical systems, ICI introduced by phase noise is not very serious and does not need to be completely eliminated. Therefore, effective ICI cancellation schemes with lower complexity is needed.

In this paper, we propose an advanced CPE estimation method which takes advantage of a better merger scheme. Next, based on a simplified model which only consider adjacent sub-carrier interference, we propose an ICI cancellation algorithm. In this algorithm, we propose two solutions for ICI estimation and they are independent estimation and joint estimation respectively. Simulation results show that eliminating only the interference from adjacent subcarriers can greatly improve the performance of the system, especially when the phase noise is large. Simultaneously, the algorithm proposed in this paper is based on the pilot structure of the 5G standard and is easy to be extended and implemented in 5G systems.

This paper is organized as follows. Section 2 introduces the system model. Section 3 introduces a CPE estimation method and proposes an enhanced one. Section 4 explores the ICI cancellation algorithm with two solutions. Finally, simulation results are presented for comparison.

2   System Model

In OFDM systems, if only consider the phase noise at the receiving end, the baseband time domain receiving signal can be given as

$$y[n] = (x[n] \otimes h[n])e^{j\phi[n]} + w[n], \tag{1}$$

where $x[n]$ denotes the time domain transmitting sequence, $h[n]$ indicates the time domain channel, $e^{j\phi[n]}$ refers to the phase noise, $w[n]$ is the additive noise and $\otimes$ represents circular convolution. Denoting the FFT of transmitted signal, channel, phase noise and additive noise on subcarrier k ($0 \leq k \leq N-1$) as X_k, H_k, Φ_k and W_k, the received signal in the frequency domain can be given as

$$Y_k = X_k H_k \underbrace{\Phi_0}_{CPE} + \underbrace{\sum_{i=0, i \neq k}^{N-1} X_i H_i \Phi_{(k-i)_N}}_{ICI} + W_k, \tag{2}$$

where

$$\Phi_k = \frac{1}{N} \sum_{n=0}^{N-1} e^{j\phi[n]} e^{-j\frac{2\pi}{N}nk}, \tag{3}$$

N denotes the number of system FFT points and $(\cdot)_N$ represents modulo N. As shown in (2), the phase noise has two main impacts: 1. CPE, which causes common phase rotation in constellations of received symbols. Because the magnitude of CPE is close to 1, we define it as $e^{j\theta}$; and 2. ICI, which breaks the orthogonality of OFDM waveform.

3   CPE Estimation

To estimate CPE, 5G systems introduce a new reference signal PT-RS [2]. An example of reference signals' structure in a resource block (RB) is illustrated in Fig. 1.

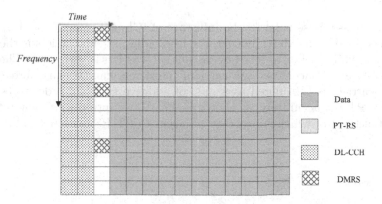

Fig. 1. Example of structure of reference signals.

In Fig. 1, dedicated demodulation reference signals (DM-RS) used for channel estimation are placed at the third OFDM symbol. Meanwhile, there are one PT-RS in each OFDM of one RB. Next, we will study CPE estimation scheme based on the structure of reference signals illustrated in Fig. 1.

According to (2) and assuming the channel is invariant during a slot, the received signals for sub-carrier k in the third and l-th symbol ($l > 3$) can be obtained as

$$Y_{k,3} = X_{k,3} H_{k,3} e^{j\theta_3} + W_{k,3} \tag{4}$$

$$Y_{k,l} = X_{k,l}H_{k,3}e^{j\theta_l} + W_{k,l}, \tag{5}$$

where ICI is considered as a part of additive noise. Denoting the total number of PT-RS in frequency domain M, the difference of CPE in symbol 3 and l ($l > 3$) can be obtained as [9]

$$\theta_\tau = \frac{1}{M}\sum_{k=1}^{M} \text{angle}\left\{Y_{k,3}^* Y_{k,l}\right\}, \tag{6}$$

where Y^* represents the conjugation of Y. Through rotating phase angle θ_τ based on $H_{k,3}e^{j\theta_3}$ which obtained by channel estimation, we can get the channel affected by phase noise of all data symbols. In case the channel is additive white gaussian noise (AWGN), we ought to calculate CPE of every symbol alone.

In (6), each PT-RS in same symbols estimates one CPE difference and θ_τ is the average value of all. This combing method do not consider the amplitude of vector $Y_{k,3}^* Y_{k,l}$. In fact, the larger the amplitude of the vector, the smaller the impact of additive noise relatively and as a result, the CPE difference is more accuracy. Thus, we give a better scheme to obtain θ_τ

$$\theta_\tau = \text{angle}\left\{\sum_{k=1}^{M} Y_{k,3}^* Y_{k,l}\right\}. \tag{7}$$

4   ICI Cancellation

When the phase noise is large, the influence of ICI cannot be ignored. Like ICI introduced by doppler shift, interference from adjacent sub-carriers is the maximum. Meanwhile, ICI introduced by distant sub-carriers is very small. Therefore, we just estimate adjacent two sub-carriers' interference and the interference from rest sub-carriers is considered as a part of additive noise. In order to estimate ICI using PT-RS accurately, as shown in Fig. 2, we introduce a new reference signal structure, in which subcarriers adjacent to PT-RS are left blank.

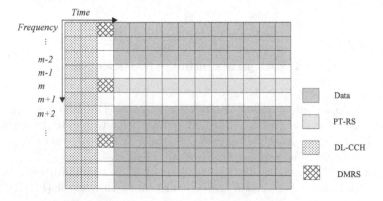

Fig. 2. Structure of reference signal for ICI reduction.

As shown in Fig. 2, we assume the PT-RS is placed at subcarrier m and the index of vacant subcarriers is $m-1$ and $m+1$ consequently. For the subcarrier $m-1$, since it has no data of its own, the signals it receives are composed of interference from subcarriers $m-2$, m and additive noise. So, according to (2), the received signals in vacant subcarriers in one OFDM symbol can be given as

$$Y_{m-1} = X_{m-2}H_{m-2}\Phi_1 + X_m H_m \Phi_{N-1} + W_{m-1}, \tag{8}$$

$$Y_{m+1} = X_m H_m \Phi_1 + X_{m+2}H_{m+2}\Phi_{N-1} + W_{m+1}. \tag{9}$$

When the channel is AWGN, Eqs. (8) and (9) becomes

$$Y_{m-1} = X_{m-2}\Phi_1 + X_m \Phi_{N-1}+W_{m-1}, \tag{10}$$

$$Y_{m+1} = X_m \Phi_1 + X_{m+2}\Phi_{N-1} + W_{m+1}. \tag{11}$$

To solve Φ_1 and Φ_{N-1}, in addition to the known PT-RS X_m, data X_{m-2} and X_{m+2} are also needed. Thus, in this paper, the ICI estimation and cancellation process are performed after the soft demodulation of receiver end. When an OFDM symbol does not configure PT-RS, both CPE and ICI of this symbol can be obtained by interpolation.

When the channel is not AWGN, through channel estimation and CPE compensation, we can obtain $H_{m-2}\Phi_0$, $H_m\Phi_0$ and $H_{m+2}\Phi_0$. Equations (8) and (9) now can be rewritten as

$$Y_{m-1} = X_{m-2}H_{m-2}\Phi_0\Phi_1/\Phi_0 + X_m H_m \Phi_0\Phi_{N-1}/\Phi_0 + W_{m-1}, \tag{12}$$

$$Y_{m+1} = X_m H_m \Phi_0\Phi_1/\Phi_0 + X_{m+2}H_{m+2}\Phi_0\Phi_{N-1}/\Phi_0 + W_{m+1}. \tag{13}$$

Thus, when considering real channel, just estimate Φ_1/Φ_0, Φ_{N-1}/Φ_0 instead of Φ_1, Φ_{N-1} and the remaining steps remain unchanged.

Of course, the structure of reference signals is not only one form of Fig. 2. Through double the number of PT-RS or just keep it like Fig. 1, ICI can also be calculated without leaving blank. The method of leaving blank increases the accuracy of ICI estimation at the expense of system throughput and others are the opposite. In this paper, we focus on how to estimate ICI more accurately, and two solutions under AWGN channel are introduced next in detail.

4.1 Calculate Φ_1 and Φ_{N-1} Independently

When there is only one subcarrier to place PT-RS in the frequency domain, namely $M = 1$, we define the received signals in (10) and (11) as $\mathbf{Y}$ and it can be obtained by

$$\begin{bmatrix} Y_{m-1} \\ Y_{m+1} \end{bmatrix} = \mathbf{A} \begin{bmatrix} \Phi_1 \\ \Phi_{N-1} \end{bmatrix} + \begin{bmatrix} W_{m-1} \\ W_{m+1} \end{bmatrix}, \tag{14}$$

where

$$\mathbf{A} = \begin{bmatrix} X_{m-2} & X_m \\ X_m & X_{m+2} \end{bmatrix}. \tag{15}$$

Here, we define the average power of W is σ^2 and assume Φ_1 and Φ_{N-1} are independent and their average power both are P. A standard minimum mean square error (MMSE) estimation can be utilized

$$\begin{bmatrix} \Phi_1 \\ \Phi_{N-1} \end{bmatrix} = \mathbf{V}_{\text{MMSE}} \mathbf{Y}, \tag{16}$$

where

$$\mathbf{V}_{\text{MMSE}} = \left(\mathbf{A}^H \mathbf{A} + \frac{\sigma^2}{P} \mathbf{I} \right)^{-1} \mathbf{A}^H. \tag{17}$$

Since the average power of PT-RS and data is both 1, the inverse of $\mathbf{A}^H \mathbf{A}$ may not exist and this method effectively avoid this problem.

When $M > 1$, according to (16), we can obtain M pair of Φ_1 and Φ_{N-1}. Because averaging complex number directly makes no sense, it is not suggested to calculate ICI by each PT-RS separately and then average them as the final result. In this paper, we propose to merge items obtained by all PT-RS in the same OFDM symbol firstly and get one final result. Namely, the dimension of $\mathbf{A}$ and $\mathbf{Y}$ is extended to be $2M \times 2$ and $2M \times 1$. Since all PT-RS are merged, the value of $\frac{\sigma^2}{P}\mathbf{I}$ is very small relative to $\mathbf{A}^H \mathbf{A}$ and the probability of $\mathbf{A}^H \mathbf{A}$ irreversible is almost zero. In practical situations, Φ_1 and Φ_{N-1} are necessarily independent of each other. Therefore, we suggest use zero forcing (ZF) scheme in place of MMSE scheme in this condition, namely $\mathbf{V}_{\text{ZF}} = \left(\mathbf{A}^H \mathbf{A} \right)^{-1} \mathbf{A}^H$. The subsequent simulations of this paper also use ZF scheme.

In this section, we propose a method to calculate ICI independently and discuss the situations under different PT-RS numbers. This method is simple and unaffected by the size of the phase noise.

4.2   Calculate Φ_1 and Φ_{N-1} Jointly

The actual phase noise consists of CPE and the variable part relative to it. Hence, the phase noise in an OFDM symbol can be represented as

$$\phi[n] = \text{angle}\,(\Phi_0) + \Delta\phi\,[n], \tag{18}$$

and further we can get

$$e^{j\phi[n]} = e^{j(\text{angle}(\Phi_0) + \Delta\phi[n])} \approx \Phi_0 \left(1 + j\Delta\phi\,[n] \right). \tag{19}$$

This approximation method is more accurate than that in [5], which consider $e^{i\phi[n]} \approx 1 + j\Delta\phi\,[n]$, especially when the phase noise is large. According to the definition of Φ, we can get

$$\begin{aligned}
\Phi_1 &= \frac{1}{N} \sum_{n=0}^{N-1} \Phi_0 \left(1 + j\Delta\phi[n] \right) e^{-j\frac{2\pi}{N}n} \\
&= \frac{1}{N} \sum_{n=0}^{N-1} \Phi_0 e^{-j\frac{2\pi}{N}n} + j\frac{1}{N}\Phi_0 \sum_{n=0}^{N-1} \Delta\phi[n] e^{-j\frac{2\pi}{N}n} \\
&= j\frac{1}{N}\Phi_0 \sum_{n=0}^{N-1} \Delta\phi[n] e^{-j\frac{2\pi}{N}n},
\end{aligned} \tag{20}$$

$$\Phi_{N-1} = \frac{1}{N} \sum_{n=0}^{N-1} \Phi_0 \left(1+j\Delta\phi[n]\right) e^{j\frac{2\pi}{N}n}$$

$$= \frac{1}{N} \sum_{n=0}^{N-1} \Phi_0 e^{j\frac{2\pi}{N}n} + j\frac{1}{N}\Phi_0 \sum_{n=0}^{N-1} \Delta\phi[n] e^{j\frac{2\pi}{N}n}$$

$$= j\frac{1}{N}\Phi_0 \sum_{n=0}^{N-1} \Delta\phi[n] e^{j\frac{2\pi}{N}n}. \tag{21}$$

By comparing (20) and (21), the relationship between Φ_1 and Φ_{N-1} can be obtained as

$$\Phi_{N-1} = -\Phi_0^2 \cdot conj(\Phi_1). \tag{22}$$

Using the relationship between Φ_1 and Φ_{N-1}, Eqs. (10) and (11) can solve a result separately. In detail, when there is only one subcarrier to place PT-RS in the frequency domain, we define $\Phi_0 = c + dj$, $\Phi_1 = a + bj$, $\Phi_{N-1} = -\Phi_0^2(a - bj)$, $X_k = x_k^r + x_k^i j$, $Y_k = y_k^r + y_k^i j$ and $W_k = w_k^r + w_k^i j$. By solving (10), we get $\mathbf{Y}'$

$$\begin{bmatrix} y_{m-1}^r \\ y_{m-1}^i \\ y_{m+1}^r \\ y_{m+1}^i \end{bmatrix} = \mathbf{A}' \begin{bmatrix} a \\ b \end{bmatrix} + \begin{bmatrix} w_{m-1}^r \\ w_{m-1}^i \\ w_{m+1}^r \\ w_{m+1}^i \end{bmatrix}, \tag{23}$$

where

$$\mathbf{A}' = \begin{bmatrix} x_{m-2}^r - (c^2 - d^2) x_m^r + 2cdx_m^i & -x_{m-2}^i - (c^2 - d^2) x_m^i - 2cdx_m^r \\ x_{m-2}^i - (c^2 - d^2) x_m^i - 2cdx_m^r & x_{m-2}^r + (c^2 - d^2) x_m^r - 2cdx_m^i \\ x_m^r - (c^2 - d^2) x_{m+2}^r + 2cdx_{m+2}^i & -x_m^i - (c^2 - d^2) x_{m+2}^i - 2cdx_{m+2}^r \\ x_m^i - (c^2 - d^2) x_{m+2}^i - 2cdx_{m+2}^r & x_m^r + (c^2 - d^2) x_{m+2}^r - 2cdx_{m+2}^i \end{bmatrix} \tag{24}$$

Same as Sect. 4.1, we can solve a and b by the MMSE criterion. When $M > 1$, also same as Sect. 4.1, merge all PT-RS firstly and the dimension of $\mathbf{A}'$ and $\mathbf{Y}'$ is extended to be $4M \times 2$ and $4M \times 1$ respectively.

In this section, by seeking the relationship between Φ_1 and Φ_{N-1}, we estimate ICI jointly. Compared with estimating ICI independently, the items can be used to merge is double. However, the relationship between Φ_1 and Φ_{N-1} relied on by the joint estimation is obtained by approximation. There is a certain model error when estimating ICI jointly, especially when the phase noise is large, and the independent estimation method has no such problem.

5   Simulations and Performance Analysis

In this section, we make simulations for comparison and analysis. Main assumptions are shown in Table 1.

Table 1. Simulation assumption.

Parameters	Assumptions
Carrier frequency	30 GHz
Subcarrier spacing	120 kHz
Allocated bandwidth	100e6 Hz
UE speed	0 km/h
Coding scheme	Turbo

5.1   CPE Estimation

In this part, we compare the CPE estimation scheme in [9] and proposed in this paper with scheduled bandwidth 32 RBs under CDL-B channel. The PSD of phase noise is -70, -70, -140 and -140 dBc/Hz in frequency 0, 10e3, 1e6 and 9e9 Hz corresponding. Specific definition of phase noise PSD can refer to [10]. The modulation method is 256QAM and the code rate is 0.75. With the PT-RS FD of every 4th RB and TD of every 1 symbol, the simulation results are given as follow.

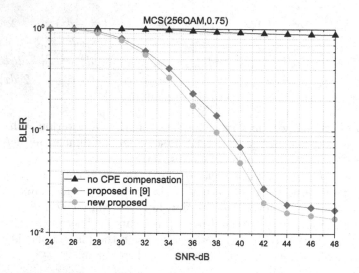

Fig. 3. Simulation results of CPE estimation.

In Fig. 3, no CPE compensation represents the system does not configure PT-PS. As can be seen from Fig. 3, the method we proposed can decrease the block error ratio (BLER) about 1 dB compared with the method in [9]. High-order modulation is very sensitive to phase rotation, so no CPE compensation leads to BLER remain high. The impact of ICI on high-order modulation systems is relatively large, and therefore, in Fig. 3, when the SNR is high, the BLER without ICI cancellation tends to be flat.

5.2 ICI Cancellation

In order to highlight the effect of ICI cancellation, we increase the ICI of the system by changing the PSD of phase noise. In detail, the frequencies are adjusted to 0, 10e4, 10e6, 9e9 Hz and 0, 50e3, 1e6, 9e9 Hz in AWGN and CDL-B channel respectively, and other configurations remain unchanged. When the channel is CDL-B, in order to exclude the effect of channel estimation error on the result of ICI cancellation, we perform simulations under both ideal channel estimation and exponential power delay profile channel estimation. Simulation results under AWGN and CDL-B channel are given as follows.

As can be seen from the Figs. 4 and 5, the two ICI estimation solutions proposed in this paper both are effective, and the second solution is about 2 dB better than the first in AWGN channel. The results indicate that, compared with estimating ICI independently, though estimating ICI jointly has model error, double the number of items using to merge make the ICI estimation more accurate and ultimately lead to a better result.

In Figs. 4 and 5(b), when the SNR is high, due to the impact of the remaining ICI, BLER tends to be flat. When the modulation mode changes from 256QAM to 64QAM, as shown in Fig. 5(a), BLER can be reduced to 0 and smaller performance gains from ICI elimination. Therefore, it can be concluded that when the modulation order is high, only eliminating adjacent subcarriers' interference can significantly improve the system performance, but the remaining ICI still impair performance. When the modulation order is low, the impact of ICI is relatively small and the effects of distant ICI can be ignored. According to several experiments, increasing the number of iterations does not make the results better, so only one iteration is sufficient.

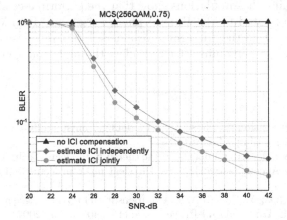

Fig. 4. Simulation of ICI cancellation under AWGN with code rate 0.75.

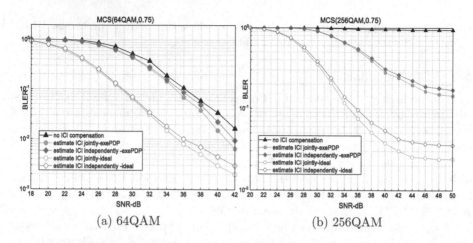

(a) 64QAM   (b) 256QAM

Fig. 5. Simulations of ICI cancellation under CDL-B channel with code rate 0.75.

6   Conclusion

In this paper, based on the method proposed in [9], we proposed a new method to estimate CPE. Simulation Results indicate that when there are multiple subcarriers in the frequency domain placing PT-RS, considering the influence of vector's amplitude makes CPE estimation more accurate.

Based on the CPE estimation, we propose an ICI cancellation algorithm with low complexity. In this algorithm, we use a simplified ICI model which only consider the interference from adjacent subcarriers and propose two solutions to estimation ICI. Simulations show that make use of the relationship between ICI is better than estimate them independently and hence, it can be concluded that the number of the items used to merge has a great influence on the ICI estimation. Meanwhile, the simulations show that the performance of the algorithm we proposed are related to the PSD of phase noise and MCS. When the MCS is low, only eliminating the adjacent subcarriers' interference is sufficient. When the MCS is high, the remaining ICI will affect the performance of the system.

References

1. R1-163984, Discussion on phase noise modeling. Samsung, 3GPP TSG RAN WG1 Meeting #85, Nanjing, China (2016)
2. 3GPP TS 38.214, 3rd Generation Partnership Project, Technical Specification Group Radio Access Network, NR, Physical layer procedures for data, Release 15 (2018)
3. Zou, Q., Tarighat, A., Sayed, A.H.: Compensation of phase noise in OFDM wireless systems. IEEE Trans. Signal Process. **55**(11), 5407–5424 (2007)
4. Bhatti, J., Noels, N., Moeneclaey, M.: Phase noise estimation and compensation for OFDM systems: a DCT-based approach. In: 2010 IEEE Conferences on Spread Spectrum Techniques and Applications, Taichung, Taiwan, pp. 93–97 (2010)

5. Gholami, M.R., Nader-Esfahani, S., Eftekhar, A.A.: A new method of phase noise compensation in OFDM. In: IEEE Conferences on Communications, Anchorage, AK, USA, vol. 5, pp. 3443–3446 (2003)
6. Petrovic, D., Rave, W., Fettweis, G.: Effects of phase noise on OFDM systems with and without PLL: characterization and compensation. IEEE Trans. Commun. **55**(8), 1607–1616 (2007)
7. Syrjala, V., Valkama, M., Tchamov, N.N., Rinne, J.: Phase noise modelling and mitigation techniques in OFDM communications systems. In: IEEE Conferences on Wireless Telecommunications Symposium, Prague, Czech Republic, pp. 1–7 (2009)
8. Bittner, S., Zimmermann, E., Fettweis, G.: Exploiting phase noise properties in the design of MIMO-OFDM receivers. In: IEEE Conference on Wireless Communications and Networking, Las Vegas, NV, USA, pp. 940–945 (2008)
9. R1-1611981, On phase tracking for NR, 3GPP TSG-RAN WG1 #87, Intel Corporation, Reno, USA (2016)
10. R1-162885, On the phase noise model for 5G New Radio evaluation, Nokia, Alcatel-Lucent Shanghai Bell, 3GPP TSG-RAN WG1 Meeting #84bis, Busan, Korea (2016)

A Machine Learning Based Temporary Base Station (BS) Placement Scheme in Booming Customers Circumstance

Qinglong Dai[1(✉)], Li Zhu[2,3], Peng Wang[1], Guodong Li[1], and Jianjun Chen[1]

[1] China Academy of Electronics and Information Technology, Beijing 100041, China
qldaisd@126.com
[2] China Transport Telecommunications and Information Center, Beijing 100011, China
jolie.zhl@hotmail.com
[3] People's Public Security University of China, Beijing 100038, China

Abstract. Explosive increase of terminal users and the amount of data traffic give a great challenge for Internet service providers (ISPs). At the same time, this big data also brings an opportunity for ISPs. How to solve network planning problem in emergency or clogging situation, based on big data? In this paper, we try to realize effective and flexible temporary base station (BS) placement through machine learning in a booming customers situation, with ISPs' massive data. A machine learning based temporary BS placement scheme is presented. A K-means based model training algorithm is put forward, as a vital part of machine learning based temporary BS placement scheme. K-means algorithm is selected as a representative example of machine learning algorithm. The performances of BS position with random starting point, BS position iteration, average path length with different parameters, are conducted to prove the availability of our work.

Keywords: Network planning · Machine learning · Big data
K-means algorithm

1 Introduction

Nowadays, with the explosive increase of terminal users and the amount of data traffic, Internet service providers (ISPs) seems not equal to this trend. In China, until Dec 2017, the number of Internet users via cellphone is 753 million, accounting for 97.5% in Internet users. In 2017, the amount of data consumed by Chinese Internet users is 21.21 billion GB [2]. As for 2017, according to ITU-T's statistics, the number of cellphone in the world is 7.74 billion [1]. There will be a promis-

Supported by key special project of National Key Research and Development Program (Grant No. 2017YFC0803900)

X. Liu et al. (Eds.): ChinaCom 2018, LNICST 262, pp. 562–572, 2019.
https://doi.org/10.1007/978-3-030-06161-6_55

ing future for mobile internet usage, as global mobile data traffic is projected to increase nearly sevenfold between 2016 and 2021. In 2016, global mobile data traffic amounted to 84 EB [3]. With the development of Internet of things, industry 4.0 and Internet of vehicles, a mass of sensors are applied in varied scenarios [7,16]. The communications between these sensors and backup data centers may take up a large amount of bandwidth. These make ISPs' situation worse.

Everything has two sides. Increasing terminals and data traffic also bring a huge opportunity for ISPs, because of the emergence of big data. Big data encompasses unstructured, semi-structured and structured data and represents the information assets characterized by so-called 4 V, i.e., high volume, high variety, high velocity and low veracity [10]. Based on the numerous terminals and varied services, ISPs could acquire generous customer and service data. From these considerable data, some hidden knowledge or regular patterns about users, like shopping habit, preference, trip information and etc., can be obtained via a certain method. The common method that is used to extract knowledge or pattern from big data is machine learning.

Machine learning is widely known as a technique of artificial intelligence [4]. Machine learning is a field of computer science that gives computer systems the ability to learn (i.e. progressively improve performance on a specific task) with data, without being explicitly programmed [18]. With the increase of data scale, the complexity of machine learning algorithm becomes the important factor that should be considered. The application of machine learning based on big data would be more effective, compared with that based on traditional data. The specific algorithms or methods in machine learning usually come from many diverse existing fields, including statistics, probability theory, neural network and etc. Machine learning can be applied in many areas such as speech recognition, image processing and fraud detection [5].

Network planning is always a significant direction in communication area. Simply, network planning determines where to place the base stations (BSs) and how to connect them. Through BSs, users' terminals including cellphone, laptop, tablet PC and so on, can access to Internet service. BSs are normally hexagonal honeycomb distributed. However, in some special situations with booming customers in a brief time, such as concert or football match, extra surge terminal access and service data exceed the capability of ordinary deployed BSs. Temporary BSs are necessary. In traditional network planning, temporary BSs placement is on the basis of experience and estimation. With the help of machine learning, temporary BSs placement would be more effective and flexible.

The contributions of this paper are listed as follows:

1. A machine learning based temporary BS placement scheme is presented, in which ISPs' operation data about users is exploited as dataset. Proposed temporary BS placement scheme scheme is classified into two phases: training phase and verification phase. Each component in this scheme is also explained in detail.
2. A K-means based model training algorithm is put forward, as a vital part of machine learning based temporary BS placement scheme. K-means algorithm

is selected as a representative example of machine learning algorithm, because of its low complexity.

3. Numerical simulation is conducted. Typical performances including BS position with random starting point, BS position iteration, average path length with different parameters, are demonstrated, to prove the availability of proposed scheme.

The rest of this paper is organized as follows. Section 2 reviews related work on machine learning. Section 3 proposes a machine learning based temporary BS placement scheme. A K-means based model training algorithm is put forward in Sect. 4. Section 5 represents the numerical simulation. Finally, Sect. 6 concludes the paper.

2   Related Work

Around machine learning, a great number of works have been done.

Gregory Piatetsky-Shapiro, the co-founder of knowledge discovery and data mining conferences, deemed that data science or data mining, big data and machine learning were related. The relationship of them is shown in Fig.1. Any one of data science, machine learning and clustering algorithm had overlap regions with others. And clustering algorithm could be seen as a part of machine learning [19].

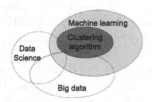

Fig. 1. The relationship of data science, big data, machine learning and clustering algorithm

Mohammadi et al. focused on the challenge of the big data generated by smart cities from a machine learning perspective. In their framework, machine learning was used to meet the cognitive services of smart city, based on the big data that generated by smart city sensors [15].

Most machine learning algorithms are categorized into supervised, unsupervised and reinforcement learning. Supervised learning deals with a labeled dataset, to build the relationship of input, output and other parameters. Unsupervised learning is provided with an un-labeled dataset, to classify data into different groups or clusters. In reinforcement learning, the agent learns by interacting with environment [5,9].

To detect mobile botnets and minimize the threat, a machine learning algorithm, exactly classification algorithm based approach was presented to identify anomalous behaviors in statistical features extracted from system calls by da Costa et al. [6]. The core part of their approach was a classifier. This classifier was responsible to classify normal or mobile botnet activity.

Li et al. extensively applied supervised machine learning in spam emails classification and evaluated different classifiers in three environments with 1000 users. Some supervised machine learning classifiers such as decision tree and support vector machine (SVM) were acceptable in real emails classification [11].

Liu et al. adopted deep learning to investigate the latent relationship between flow information and link usage, and then classified the used and unused links. Through this, the scale of network optimization problems was reduced [13]. Similarly, Rottondi et al. also introduced machine learning into the quality of transmission, as a novel way to achieve pre-deployment estimation [17].

As for video service accounting for large wireless traffic, Lin et al. applied supervised machine learning and SVM to forecast video starvation events, like the number of users existing in cell. They demonstrated the correlation of video starvation and recorded users' features for streaming with diverse characteristics [12].

Dargie et al. came up with a feasible way to locate wireless sensor network node in a 3D environment, using supervised neural networks with four input measurements, i.e., signal strength indicator, time of arrival, time difference of arrival and angle [8].

3   A Machine Learning Based BS Placement Scheme

In order to place temporary BSs more effectively, avoiding the uncertainty involved by experience and estimation, clustering algorithm is selected as a feasible method. Clustering algorithm is one of the most common machine learning applications. In clustering algorithm, the typical algorithm is K-means algorithm.

The K-means algorithm is used to recognize data with no labels into different classes. In here, no labels means that there is no output vector or non-judgmental property, like *yes* or *no*, *good* or *bad*. Each class is called as a cluster. The data in the same cluster have high similarity, meanwhile the data in different clusters have low similarity. Due to K-means algorithm's linear complexity, simple implementation and the data with no labels that collected by ISPs, it is selected by us.

We give up a machine learning based temporary BS placement scheme, as shown in Fig. 2. There are two phases in this scheme: training phase and verification phase. In training phase, based on a portion of the data from ISPs, all the terminals are divided into several different clusters. This division process is called model. To verify the rationality of this model, the other portion of the data from ISPs, is used to input the model. If the model output of verification data accords with already obtained result, the model is rational. Otherwise, the model is unqualified. And a feedback new model should be generated based on new data.

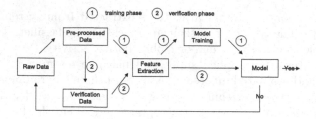

Fig. 2. A machine learning based temporary BS placement scheme

The components in our proposed scheme are interpreted as follows:

1. Raw data. The original data from ISPs' database. Raw data is generated by users requesting ISPs service. Under the large volume and chaotic data, the habit and preference of users can be gained. Because of the diversity of services and the existence of over the top (OTT) services from different companies, there are a lot of redundant, repeated, default and blank fields in raw data. Therefore, pre-process for raw data is necessary.
2. Pre-processed data. Through pre-process, raw data becomes pre-processed data. The redundant field is simplified. The repeated field is deleted. The default field is filled up with default value. The blank field is replenished by random value. Pre-processed data is split into two parts. One part of pre-processed data is used for model training, the others is used for the rationality verification of trained model.
3. Feature extraction. In machine learning, the scale of data is too large to be processed and it is suspected to be redundant (e.g., the same record in different sheets), then the data can be transformed into a reduced set of feature or a featured vector. Feature extraction facilitates the subsequent model training.
4. Model training. By import extracted data record successively, a general pattern is optimized in a iterated way, by the manner of confirming different parameters' coefficients.
5. Model. The model is the final result of model training, based on the aforementioned part of pre-processed data that is used in training phase. Note that the model is relaying on model training input data. In other words, the model is limited. As it were, all models are wrong, but some are useful.
6. Verification data. Verification data is used to verify the rationality of trained model. If the number of inapplicable data for trained model is inferior to a pre-set threshold value, the trained model is the final model; Otherwise, the trained model is unqualified. A brand new raw data for model retraining is necessary.

Note that the data pre-process is strongly related to machine learning purpose. A sample sheet in pre-processed data is shown in Fig. 3. For different machine learning purposes, some fields in sheet can be deleted. For instance, for a machine learning predicting users' connection duration, the information of terminal position is not needed.

Algorithm 1 K-means based Model Training Algorithm

Input:

    Dataset of previous terminal communications, $\mathbf{A}$, in which the jth entry is denoted as a_j;

    K initial temporary BS location: (lon_1, lat_1), $(lon_2, lat_2), \cdots, (lon_i, lat_i), \cdots, (lon_K, lat_K)$;

    Maximum distance of connection establishment, D;

Output:

1: **for** each entry in dataset $\mathbf{A}$, i.e., $Terminal_j$ **do**

2:    **for** each initial BS location, i.e. BS_i **do**

3:       Compute the distance between $Terminal_j$ and BS_i, i.e., D_{ij};

4:       **if** $D_{ij} \leq D$ **then**

5:          Replace (lon_i, lat_i) with $((lon_i + lon_j)/2, (lat_i + lat_j)/2)$;

6:       **else**

7:          Continue;

8:       **end if**

9:    **end for**

10: **end for**

11: **return** K final BS location, i.e., (lon_1, lat_1), $(lon_2, lat_2), \cdots, (lon_i, lat_i), \cdots, (lon_K, lat_K)$;

No.	Terminal ID	BS ID	Terminal longitude	Terminal latitude	BS longitude	BS latitude	Connection Duration
1	Tid_003	BSid_7	E116°20'13.53"	N39°54'39.35"	E116°20'13.57"	N39°54'39.33"	337s
2	Tid_013	BSid_7	E116°20'13.48"	N39°54'39.30"	E116°20'13.57"	N39°54'39.33"	0.6s
3	Tid_011	BSid_4	E116°20'47.57"	N39°54'36.86"	E116°20'47.51"	N39°54'36.91"	23s
...	...	...	...	...	...	...	...

Fig. 3. A sample sheet in pre-processed data

4   K-means based Model Training Algorithm

In this section we select a typical clustering algorithm, i.e., K-means algorithm, to accomplish temporary BS placement, i.e., the model training in our proposed scheme. For a booming customers circumstance, a targeted model training algorithm is raised in Algorithm 1.

    Model training algorithm is started, based on the dataset of previous terminal communications, initial temporary BSs locations and pre-set maximum distance of connection establishment. Each entry in the dataset is applied to training model. K initial temporary BSs locations are seen as the initial centroids for different clusters. Compute the distance between each terminal and each BS. Compare each time computation result with pre-set maximum distance of connection establishment. If computation result is smaller, replace corresponding BS location with the middle point location of this BS location and terminal location. Otherwise, continue the computation of next pair of BS and terminal. Finally, the ultimate BSs locations are training result.

In our proposed scheme, the location is recoded by geographic coordinate system. Therefore, for distance comparison conveniently, geographic coordinate has to be transformed into distance, according to Eq. 1. In Eq. 1, R is the radius of the earth, lat is short for latitude, lon is short for longitude, sin is sine function, cos is cosine function and $arcsin$ is arc-sin function.

$$D_{ij} = 2R * \arcsin\left(\sqrt{sin^2\left(\frac{lat_i - lat_j}{2}\right) + cos\left(lat_i\right) * cos\left(lat_j\right) * sin^2\left(\frac{lon_i - lon_j}{2}\right)}\right)$$

$$(1)$$

5   Numerical Results

The performances of proposed BS placement scheme are analyzed via simulation using $MATLAB$. The BSs placement in [14] is selected as a baseline method. With varied parameters, the performances of proposed BS placement scheme are depicted in detail.

In our simulation, unless explicitly stated, otherwise, the simulation parameters shown in Table 1 are used. The network area of terminals and BSs is a $100\,\text{km} \times 100\,\text{km}$ area. The terminal number in this simulation is 120. And these terminals are randomly distributed in network area. These are all demonstrated in Fig. 4. Based on these conditions and parameters, a series of evaluations are performed in different scenarios.

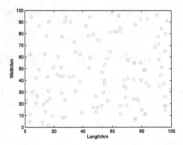

Fig. 4. User distribution in $100\,\text{km} \times 100\,\text{km}$ area

Figure 5 is the temporary BS placement with different random starting points, namely, (100, 0), (100, 100), (0, 100) and (0, 0). The red rectangle in each sub-fig is the final BS placement. The maximum distance of connection establishment is 20 km. It is obviously that the temporary BS placement has a strongly correlation with starting point. Normally, from different starting points, different temporary BS placements are obtained. However, every result of different starting point is the best BS placement for the terminals within its coverage

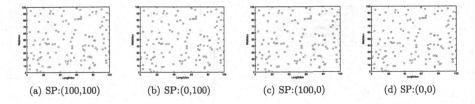

(a) SP:(100,100)   (b) SP:(0,100)   (c) SP:(100,0)   (d) SP:(0,0)

Fig. 5. Base position with random starting point (SP)

with distance D. Therefore, to cover all the terminals, when D is fixed, increased starting points is a feasible solution.

In the process of BS placement, the BS position is varied after each iteration. The BS position iteration with starting point (0, 100) is shown in Fig. 6. Each rectangle in Fig. 6 is the result of a position re-calculation. Through N position changes, BS position moves from (0, 100) to final position. According to Algorithm 1, once there is one or more terminal emerging in BS's coverage, position re-calculation happens, until all the terminals in network area are considered. These lead to the BS movement.

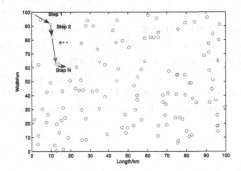

Fig. 6. BS position iteration

Average path length is negatively related to communication distance, D, as shown in Fig. 7. The number of added temporary BS is 1. With the increase of communication distance, average path length decreases, no matter proposed algorithm or baseline method. Greater communication distance means that more terminals can connected to a BS, becoming a cluster. Similarly, different clusters are linked together, like the way of terminals connecting to BS. The greater communication distance, the less cluster, and then the smaller average path length. Besides, no matter which one the starting point is, the average path lengths of baseline method with different starting points are almost the same. They are greater than the average path length of proposed algorithm all the time.

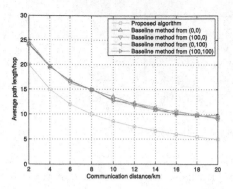

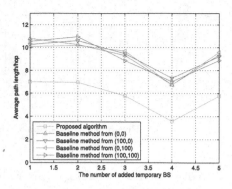

Fig. 7. World Map

Fig. 8. APL vs. communication distance

In Fig. 8, the relationship of average path length and the number of added temporary BS is demonstrated. The distance of connection establishment is set as 10 km, using the average distance. Average path length of proposed algorithm is always smaller than those of baseline methods with different starting points. When the number of added temporary BS grows, average path length firstly decreases to a minimum value, and then increases. The connections among different clusters are considered in average path length computation. More added temporary BSs means more clusters. More clusters make an extra burden for average path length, if added temporary BSs number beyond a specific value. In this specific simulation for Fig. 8, the appropriate value of added temporary BS is 4.

6   Conclusion

With explosive increase of terminal users and the amount of data traffic, in order to place temporary BS effectively and flexibly in a booming customers situation. Based on ISPs collected big data, a machine learning based temporary BS placement scheme was presented by us. A K-means based model training algorithm was also put forward. K-means algorithm was selected as a representative example of machine learning algorithm. Through simulation, propose scheme was superior to compared work. In our future work, service-oriented machine learning application and the complexity analysis of different machine learning algorithms will be key emphases.

References

1. Key ict indicators for developed and developing countries and the world (totals and penetration rates). Technical report, ITU-T, Geneva, Switzerland (2017)
2. The 41st china statistical report on internet development. Technical report, China Internet Network Information Center, Beijing, China (2018)
3. Mobile internet usage worldwide - statistics and facts. Technical report, Statista, Hamburg, Germany (2018)
4. Alpaydin, E.: Introduction to Machine Learning, 2nd edn. The MIT Press, USA (2009)
5. Alsheikh, M.A., Lin, S., Niyato, D., Tan, H.P.: Machine learning in wireless sensor networks: algorithms, strategies, and applications. IEEE Commun. Surv. Tutor. 16(4), 1996–2018 (2014). https://doi.org/10.1109/OMST.2014.2320099. Fourthquarter
6. da Costa, V.G.T., Barbon, S., Miani, R.S., Rodrigues, J.J.P.C., Zarpelao, B.B.: Detecting mobile botnets through machine learning and system calls analysis. In: 2017 IEEE International Conference on Communications (ICC). pp. 1–6 (2017). https://doi.org/10.1109/ICC.2017.7997390
7. Curry, E., Hasan, S., Kouroupetroglou, C., Fabritius, W., ul Hassan, U., Derguech, W.: Internet of things enhanced user experience for smart water and energy management. IEEE Internet Comput. 22(1), 18–28 (2018). https://doi.org/10.1109/MIC.2018.011581514
8. Dargie, W., Poellabauer, C.: Localization. Wiley, New York (2010)
9. Klaine, P.V., Imran, M.A., Onireti, O., Souza, R.D.: A survey of machine learning techniques applied to self-organizing cellular networks. IEEE Commun. Surv. Tutor. 19(4), 2392–2431 (2017). https://doi.org/10.1109/COMST.2017.2727878. Fourthquarter
10. LHeureux, A., Grolinger, K., Elyamany, H.F., Capretz, M.A.M.: Challenges and approaches. IEEE Access 5, 7776–7797 (2017). https://doi.org/10.1109/ACCESS.2017.2696365
11. Li, W., Meng, W.: An empirical study on email classification using supervised machine learning in real environments. In: 2015 IEEE International Conference on Communications (ICC), pp. 7438–7443 (2015). https://doi.org/10.1109/ICC.2015.7249515
12. Lin, Y.T., Oliveira, E.M.R., Jemaa, S.B., Elayoubi, S.E.: Machine learning for predicting qoe of video streaming in mobile networks. In: 2017 IEEE International Conference on Communications (ICC), pp. 1–6 (2017). https://doi.org/10.1109/CC.2017.7996604
13. Liu, L., Cheng, Y., Cai, L., Zhou, S., Niu, Z.: Deep learning based optimization in wireless network. In: 2017 IEEE International Conference on Communications (ICC), pp. 1–6 (2017). https://doi.org/10.1109/ICC.2017.7996587
14. Liu, Y., Zhou, C., Cheng, Y.: Integrated bs/onu placement in hybrid epon-wimax access networks. In: GLOBECOM 2009–2009 IEEE Global Telecommunications Conference, pp. 1–6 (2009). https://doi.org/10.1109/GLOCOM.2009.5425770
15. Mohammadi, M., Al-Fuqaha, A.: Enabling cognitive smart cities using big data and machine learning: Approaches and challenges. IEEE Commun. Mag. 56(2), 94–101 (2018). https://doi.org/10.1109/MCOM.2018.1700298
16. Nguyen, T.T.T., Armitage, G.: A survey of techniques for internet traffic classification using machine learning. IEEE Commun. Surv. Tutor. 10(4), 56–76 (2008). https://doi.org/10.1109/SURV.2008.080406. Fourth

17. Rottondi, C., Barletta, L., Giusti, A., Tornatore, M.: Machine-learning method for quality of transmission prediction of unestablished lightpaths. IEEE/OSA J. Opt. Commun. Netw. **10**(2), A286–A297 (2018). https://doi.org/10.1364/JOCN. 10.00A286
18. Samuel, A.L.: Some studies in machine learning using the game of checkers. Ibm J. Res. Dev. **3**(3), 210–229 (1959)
19. Taylor, D.: Battle of the data science venn diagrams (2016)

An Improved Preamble Detection Method for LTE-A PRACH Based on Doppler Frequency Offset Correction

Yajing Zhang[1(✉)], Zhizhong Zhang[2], and Xiaoling Hu[1,2]

[1] Chongqing University of Posts and Telecommunications, Chongqing, China
zhangyajing2969@163.com
[2] Test Engineering Research Center of Communication Networks, Chongqing, China

Abstract. In the random access process of the Long Term Evolution Advanced (LTE-A) system, the Doppler shift influences the detection of the Physical Random Access Channel (PRACH) signal, resulting in the appearance of the pseudo correlation peaks at the receiving end. In the 3GPP protocol, the frequency offset in the mid-speed and low-speed modes is not processed, and the frequency offset processing algorithm in the high-speed mode only applies to the case where the Doppler frequency offset does not exceed the unit sub-carrier. For solve the problem, a three-step improvement method is proposed. The first step is to perform the maximum likelihood (ML) offset estimation to do the frequency offset correction; the second step is to perform the sliding average filter processing to eliminate the influence of multipath; the third step is to use multiple sliding window peak detection algorithm. Compared with the traditional algorithm, the performance of the proposed method is better. And the false alarm performance under the AWGN channel is at least 3.8 dB better, and the false alarm performance under ETU channel is at least 1 dB better.

Keywords: Preamble detection · Frequency offset correction · Sliding average filter processing · Multiple sliding window

1 Introduction

The performance evaluation of random access in the Long Term Evolution Advanced (LTE-A) system has become an important research topic in recent years, because the quality of random access will play an important role in the future 5G [1]. The most prominent feature of 5G technology is unmatched speed, which requires shorter access delay and higher random access success rate. The

This work is supported by the National Science and Technology Major Project of the Ministry of Science and Technology of China (2015ZX03001013), the Generic Technology Innovation Project of Key Industries in Chongqing (cstc2017zdcy-zdzx0030), the Innovation Team of University in Chongqing (KJTD201312).

X. Liu et al. (Eds.): ChinaCom 2018, LNICST 262, pp. 573–582, 2019.
https://doi.org/10.1007/978-3-030-06161-6_56

first and most important step in the uplink random access process of the LTE-A system is the successful transmission and correct resolution of the Physical Random Access Channel (PRACH) preamble signal [2]. At the receiving end, the preamble ID and the timing advance (TA) [3] can be correctly parsed by the preamble detection.

The sub-carrier spacing of the random access channel of the LTE-A system is very narrow, It is more susceptible to the influence of frequency offset. In the 3GPP protocol, the frequency offset in the mid-speed and low-speed modes is not processed, and the frequency offset processing algorithm in the high-speed mode only applies to the case where the Doppler frequency offset does not exceed the unit sub-carrier. Therefore, to ensure correct analysis of the random access detection signal, Doppler shift estimation and compensation need to be performed on the uplink.

With regard to the PRACH signal detection technology, experts and scholars at home and abroad have done a lot of research. For example, the preamble detection algorithm for large transmission delay proposed in [4], based on the strong correlation of sequences as an enhanced version of the algorithm to overcome the propagation delay; The ZC sequence grouping and in-group peak sliding detection algorithm proposed in [5], which has efficient hardware implementation by Discrete Fourier Transform (DFT) and large-point Fast Fourier Transform (FFT) algorithm. This type of algorithms do not analyze the increasingly serious effects of frequency offset. Other existing documents do not discuss the effect of preamble detection after frequency offset correction. For example, [6] only studied the frequency offset estimation method of PRACH signal.

Compared with the traditional frequency correlation detection method, this paper mainly completes the detection of the preamble ID signal of random access through Doppler frequency offset correction, sliding average filter processing, multiple sliding window peak detection. And the performance of the proposed method is better.

The remainder of this article is organized as follows. The PRACH signal detection process and effect of Doppler frequency offset on random access signal detection are detailed in Sect. 2. The improved detection method is detailed in Sect. 3. Performance evaluation results that demonstrate the efficiency of the traditional and proposed algorithms are presented in Sect. 4. Conclusions are drawn in Sect. 5.

2   Signal Detection and Effect of Frequency Offset

2.1   PRACH Signal Detection Process

The random access preamble sequence is generated by the cyclic shift of the Zadoff–Chu (ZC) sequence [2]. Since the ZC sequence just satisfies the good correlation characteristics required for the random access preamble sequence, the definition of ZC sequence is

$$x_u(n) = e^{-j\frac{\pi u n(n+1)}{N_{zc}}}, 0 \le n \le N_{zc} - 1 \tag{1}$$

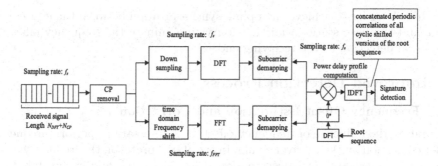

Fig. 1. Random access to the receiving process.

Where N_{zc} is the length of the ZC sequence, u is physical root sequence.

The random access preambles with zero correlation zones based on the uth root ZC sequence with length of $N_{cs} - 1$ are defined by cyclic shift according to

$$x_{u,v}(n) = x_u((n + C_v) \bmod N_{zc}) \tag{2}$$

Where C_v is the cyclic shift value and N_{cs} is the length of cyclic shift.

The random access process eNodeB receiving end [7] is shown in Fig. 1. The preamble signal is synchronized with the local ZC root sequence after CP synchronization, (Cyclic Prefix) CP removal, down sampling, DFT, and subcarrier selection to obtain correlation values in the frequency domain correlation. And then Inverse Discrete Fourier Transform (IDFT) and modular-squares are performed on correlation values to obtain time-domain correlation energy power delay profile (PDP) sequences. The correlation peak value of PDP sequence is searched by the detection threshold, to judge whether there is random access. If there is random access, preamble ID and timing advance (TA) are calculated.

2.2 Effect of Doppler Frequency Offset

When the user equipment (UE) moves fast, a large Doppler shift occurs, affecting the zero auto-correlation of the preamble sequence. Assuming Δf is the Doppler shift, and the define of preamble sequence with containing frequency offset is

$$\begin{aligned} x_u(n, \Delta f) &= e^{-j\frac{\pi u n(n+1)}{N_{zc}}} \cdot e^{j2\pi n \frac{\Delta f}{f_s}} \\ &= x_u(n - d_u) \cdot e^{j2\pi n \frac{\Delta f \cdot T_{seq} - k}{f_s}} \cdot e^{j\phi} \end{aligned} \tag{3}$$

Where T_{seq} is the last time of preamble sequence, f_s is the sampling frequency, $e^{j\phi}$ is a phase rotation constant with modulus 1 and n independent. According to formula 3, preamble sequence has changed with containing Doppler shift. Assuming $\Delta f = k f_{RA} = k \frac{1}{T_{seq}}$, the value of displacement is d_u, d_u satisfies $(d_u \cdot u) \bmod N_{zc} = 1$, the original main peak completely disappears and the position transfers to $C_v \pm (k + 1)d_u$.

Thus, In order to achieve the uplink synchronization between the user equipment and the base station, it is necessary to eliminate the frequency offset on the signal detection at the receiving end.

3   Improved Detection Process

3.1   Frequency Offset Estimation and Correction

To analyze the influence of frequency offset, it is necessary to perform frequency offset correction on the receiver signal. In the 3GPP protocol, the frequency offset in the mid-speed and low-speed modes is not processed, and cyclic shift limit set algorithm is used by default in high-speed mode. But frequency offsets in mid-speed and low-speed modes sometimes lead to severe false alarm rates. The cyclic shift limit set algorithm in high-speed mode is computationally complex [8], takes a long time, requires cooperation from the upper layer, and only applies to the case where the Doppler frequency offset does not exceed the unit subcarrier. Therefore, without considering the UE moving speed, it is better to use the frequency offset estimation method to correct the frequency offset.

The CP in the preamble signal is copied from the last part of the useful signal, and the length of CP is known. According this redundant information, the maximum likelihood (ML) frequency offset estimation method is used to estimate the time and frequency offsets. The signal received from the PRACH channel is modeled from both time and frequency domain, to analyze the time and frequency shift of the signal at the receiving end.

First, the received signal is modeled. The received signal contains the time offset and frequency offset. The time shift is represented by the channel impulse response $h(n - d)$. The frequency shift is expressed by multiplying a rotation factor. Thus, the definite of the received signal is

$$r(n) = x(n - d) \cdot e^{j \cdot \frac{2\pi n \Delta f}{N_{zc}}} + w(n) \tag{4}$$

Assuming L is the length of CP, T_{GP} is ignored, so the length of the received signal is $N_{zc} + L$. Assuming that the time offset and frequency offset are determined, then the probability density function of the $2N_{zc} + L$ sampling point for the log likelihood function is

$$\wedge (d, \Delta f) = \log f(r(n)|d, \Delta f) \tag{5}$$

Through collection selection and related calculations, all constants and factors that have no effect on the results of the maximum likelihood function are eliminated as useless factors. So the Eq. 5 is simplified to

$$\wedge (d, \Delta f) = \gamma(d) \cos(2\pi \Delta f + \angle \gamma(d)) - \beta \phi(d) \tag{6}$$

where

$$\gamma(d) = \sum_{n-d}^{n-d+L-1} r(n)r^*(n + N)$$

$$\phi(d) = \frac{1}{2} \sum_{n-d}^{n-d+L-1} (|r(n)|^2 + |r(n + N)|^2) \tag{7}$$

$$\beta = \frac{SNR}{SNR+1}$$

Equations 6 and 7 indicate that the unknown parameters are d and Δf. The likelihood function gets the maximum score in two steps. First, let d be a constant and find the maximum value of the likelihood function corresponding to Δf. When the case is $\cos(2\pi\Delta f + \angle\gamma(d)) = 1$, we obtain the maximum value.

$$2\pi\Delta f + \angle\gamma(d) = 2k\pi$$
$$\Delta\hat{f}_{ML}(d) = -\frac{1}{2\pi}\angle\gamma(d) + k \tag{8}$$

When the frequency offset does not exceed the unit sub-carrier, then $k = 0$, and when the frequency offset exceeds the unit sub-carrier, then k value needs to be calculated by trial. Second, the likelihood function is $\wedge(d, \Delta\hat{f}_{ML}) = |\gamma(d)| - |\beta\phi(d)|$. When the likelihood function gets maximum value, d is estimated. Finally, the d and Δf are

$$\hat{d}_{ML} = \arg\max_{d}\{|\angle\gamma(d)| - |\beta\phi(d)|\}$$
$$\Delta\hat{f}_{ML}(d) = -\frac{1}{2\pi}\angle\gamma(d) + k \tag{9}$$

Doppler frequency offset is corrected based on frequency offset value by multiplying frequency deviation factor $\exp(\frac{j2\pi n\Delta\hat{f}_{ML}}{N_{zc}})$. TA is calculated by time deviation factor, but base station signal processing complexity is increased and will not be discussed here.

3.2   Sliding Average Filter Processing

The sliding average filter is a Finite Impulse Response (FIR) filter with a coefficient of 1. The peak value in the window is processed by the sliding average filter to eliminate the effect of multipath, enhance the intensity of the main peak, and reduce the false alarm probability.

The frequency offset-corrected received signal is multiplied by the local ZC root sequence in the frequency domain, and the result of the time domain correlation is obtained by the IFFT. Then the time domain results of the multiple antennas at the receiving end are summed to obtain the energy-combined PDP spectrum sequence $p(k)$.

According to the cyclic shift value and the number of multipaths in different channel models, the value of the MA_value is set, $MA\_value = MA\_order(k)$, where $k = zeroCorrelationZoneConfig + 1$, values of $zeroCorrelationZoneConfig$ and MA_order are listed in Table 1. The value of moving average group delay is calculated by formula $\lfloor MA\_value/2 \rfloor$, and the N_{shift_num} values at the end of $p(k)$ are advanced, the end of $p(k)$ is filled with 0 and the number of 0 is the value of moving average group delay. When the format of preamble is 0 3, the value of N_{shift_num} is 3; when the format of preamble is 4, the value of N_{shift_num} is 0.

FIR filtering is performed according to the value of MA_value, and then the data of the group delay of $p(k)$ is removed. a new energy-combined PDP spectrum sequence $p'(k)$ is obtained, since time domain peak values are corrected by value of cal_value, the value of cal_value is listed in Table 1.

Table 1. Different format parameters configuration.

Format 0–3 parameters configuration			
zeroCorrelationZoneConfig	N_{cs}	MA_order	cal_value
0	0	7	3
1	13	7	3
2	15	7	3
3	18	7	3
4	22	7	3
...	...	...	...
Format 4 parameters configuration			
zeroCorrelationZoneConfig	N_{cs}	MA_order	cal_value
0	2	1	1
1	4	1	1
2	6	3	1
3	8	5	1
4	10	5	1
5	12	5	2

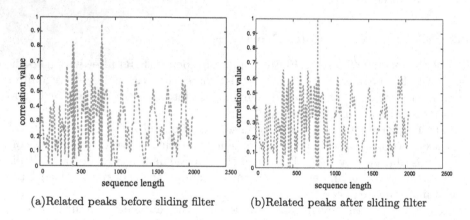

(a)Related peaks before sliding filter   (b)Related peaks after sliding filter

Fig. 2. Related peaks change with sliding average filtering.

When the relevant signal output performance deteriorates, the peak energy is not obvious due to noise, and it is not easy to do peak detection. Figure 2 manifest that it is improved by sliding average filter processing to eliminate the effect of multipath, enhance the main peak intensity, and increase the probability of successful detection.

3.3   Multiple Sliding Window Peak Detection

On the basis of time-domain peak correction and main peak enhancement, multiple sliding window peak detection is performed, which can further reduce the false alarm rate and increase the success rate of random access.

The noise power is calculated according to the new energy-combined PDP spectrum sequence $p'(k)$, and then according to the noise power and protocol requirements that the false alarm rate should be less than 0.1%, the absolute threshold $Thre\_A$ and the detection threshold $Thre\_B$ are calculated.

The detection window is divided. According to the new energy-combined PDP spectrum sequence $p'(k)$, the detection window is divided by formula 10, and finally divided into 64 detection windows.

$$
\begin{cases}
main\_win = (2048 * mDetect)/839 \\
mDetect = 839 - C_v \\
win\_length = 2048 * N_{cs}/839
\end{cases}
\tag{10}
$$

Where $main\_win$ is the position of main window, win_length is the length of detection window. Taking format 0 as an example, the preamble sequence length N_{zc} is 839, and the cyclic shift value N_{cs} is 13, so a root sequence can generate $\lfloor 839/13 \rfloor = 64$ preamble sequences, the win_length is 31.

Searching for the maximum value of the main detection window through three kinds of rectangular windows with a multiple relationship, the peak value and the position are detected. Using the first sliding rectangular window of length N_{cs} to slide within each main detection window (N_{cs} is the minimum non-zero value of the limit set, 13 at low speed, and 15 at high speed), calculate the total energy of the data within the window, search for the peak $MaxValue\_win1$ of the window and the corresponding starting position; using the second sliding rectangular window of length $\lfloor \frac{N_{cs}}{2} \rfloor$ to slide within the first sliding rectangular window where the peak is detected, search for the peak $MaxValue\_win2$ of the window and the corresponding starting position; using the third sliding rectangular window of length $\lfloor \frac{N_{cs}}{4} \rfloor$ to slide within the second sliding rectangular window where the peak is detected, search for the peak $MaxValue\_win3$ of the window and the corresponding starting position.

Finally, preamble ID and timing advance are calculated according to the value $MaxValue\_win3$ and the corresponding starting position.

4   Simulation Results

The performance of the preamble detection in LTE-A system is measured by false alarm probability and successful detection probability [9]. According to the protocol requirements of [10], the false alarm probability shall be less than or equal to 0.1%. The probability of successful detection shall be equal to or exceed 99%. According to the noise power and false alarm probability protocol requirements, the random access preamble detection thresholds are obtained by a large number of simulations. The PRACH parameters used in simulation are listed in Table 2.

Figure 3 shows that when the frequency offset is 600 HZ, the proposed method performance is better than traditional method whether it is AWGN channel or ETU channel. Due to the influence of frequency offset, the false alarm rate rises and the detection performance decrease. In this case, the detection threshold needs to be raised. But the proposed method does not need to increase the detection threshold. Under the condition of the false alarm probability is equal to 0.1%, the SNR of the proposed method is about 20 dB when the channel condition is AWGN and the SNR of the proposed method is about 17.8 dB when the channel condition is ETU.

Table 2. Simulation parameters.

System bandwidth (MHz)	20
PRACH bandwidth (MHz)	1.08
Sample frequency (MHz)	30.72
FFT point	2048
N_{zc}	839
N_{cs_index}	4
Preamble format	0
Subcarrier spacing	1.25
Channel models	AWGN/ETU

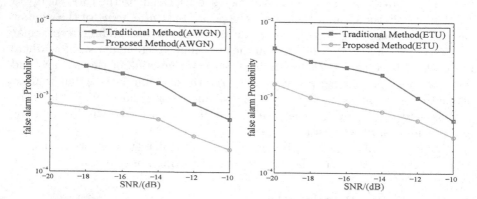

Fig. 3. Performance of different methods for 600 Hz frequency offset.

Figure 4 shows that the false alarm probability gradually decreases as the SNR increases. With the increase of SNR in AWGN channel, the reduction of false alarm rate of traditional detection algorithm becomes more and more slowly. However, with the increase of SNR, the reduction of false alarm rate of proposed detection algorithm becomes more increase. The false alarm probability performance is 3.8 dB better than the protocol requirement in AWGN channel and the

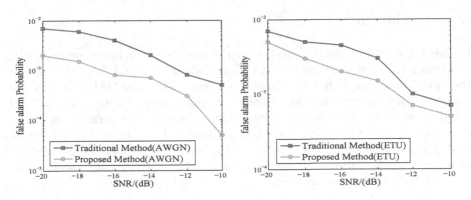

Fig. 4. Performance of different methods for 900 Hz frequency offset.

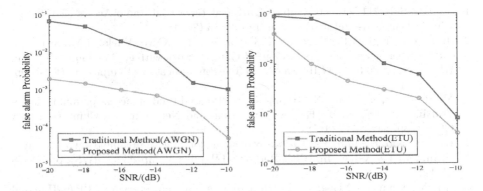

Fig. 5. Performance of different methods for 1500 Hz frequency offset.

false alarm probability performance is 1 dB better than the protocol requirement in ETU channel.

Figure 5 shows the false alarm probability is very large when the frequency offset exceeds 1250 Hz unit sub-carrier. Under the condition of the false alarm probability is equal to 0.1%, the SNR of the proposed method is about 14.7 dB and the SNR of the traditional method is about 10 dB when the channel condition is AWGN. The false alarm probability performance is 1.2 dB better than the traditional method in ETU channel.

In a word, the proposed method has different degrees of performance improvement for different frequency offset. We can draw a conclusion that the false alarm performance under the AWGN channel is at least 3.8 dB better, and the false alarm performance under ETU channel is at least 1 dB better.

5   Conclusion

In this paper, we formulated and studied a mixed-combinatorial programming problem for EE optimization in a downlink multiuser OFDM DASs. We divided the optimization problem into two sub-problems, which are DAU selection, and subcarrier assignment and power allocation optimization. Firstly, an energy efficient resource allocation scheme is presented. Then, we transform the optimization problem into a subtractive form, and solved by Lagrangian dual decomposition. From simulation results, we can see the better performance of the proposed algorithm. In the future, the inter-cell interference and cell-edge users' performance will be taken into consideration to how to design EE optimization scheme.

References

1. 3GPP TS 36.211 V12.3.0 Evolved Universal Terrestrial Radio Access (E-UTRA), Physical channels and modulation, pp. 46–55 (2014)
2. de Figueiredo, F.A.P., Mathilde, F.S., Cardoso, F.A.C.M., Vilela, R.M., Miranda, J.P.: Efficient frequency domain Zadoff-Chu generator with application to LTE and LTE-A systems. In: 2014 International Telecommunications Symposium (ITS), pp. 1–5 (2014)
3. Yang, F., He, Z., Wang, X., Huang, J.: GPP-based random access preamble detection in TD-LTE. In: IEEE Communications and Networking in China (CHINA-COM), pp. 802–806 (2012)
4. Kim, S., Joo, K., Lim, Y.: A delay-robust random access preamble detection algorithm for LTE system. In: IEEE Radio and Wireless Symposium (RWS), pp. 75–78 (2012)
5. Hu, X., Yihui, L., Xiaogang, L.: Research and implementation of PRACH signal detection in LTE-A system. Appl. Electron. Technol. 42(6), 74–76 (2016)
6. Cao, A., Xiao, P., Tafazolli, R.: Frequency offset estimation based on PRACH preambles in LTE. In: IEEE Wireless Communications Systems (ISWCS), pp. 22–26 (2014)
7. Hao, S.: Research on LTE random access detection technology. Xidian University, pp. 50–76 (2015)
8. Li, T., Wang, W., Peng, T.: An improved preamble detection method for LTE PRACH in high-speed railway scenario. In: 2015 10th International Conference on Communications and Networking in China (ChinaCom), Shanghai, pp. 544–549 (2015)
9. Leyva-Mayorga, I., Tello-Oquendo, L., Pla, V., Martinez-Bauset, J., Casares-Giner, V.: On the accurate performance evaluation of the LTE-A random access procedure and the access class barring scheme. In: IEEE Transactions on Wireless Communications, pp. 7785–7799 (2017)
10. Wang, Q., Ren, G., Wu, J.: A multiuser detection algorithm for random access procedure with the presence of carrier frequency offsets in LTE systems. In: IEEE Transactions on Communications, pp. 3299–3312 (2015)

Cryptographic Algorithm Invocation in IPsec: Guaranteeing the Communication Security in the Southbound Interface of SDN Networks

Deqiang Wang, Wan Tang[✉], Ximin Yang, and Wei Feng

College of Computer Science, South-Central Univ. for Nationalities,
Wuhan 430074, China
tangwan@scuec.edu.cn

Abstract. Due to the static configuration of IPsec cryptographic algorithms, the invocation of these algorithms cannot be dynamically self-adaptable to the traffic fluctuation of software-defined networking (SDN) southbound communication. In this paper, an invocation mechanism, based on the Free-to-Add (FTA) scheme, is proposed to optimize the invocation mode of cryptographic algorithms in traditional IPsec. To balance the link security and communication performance, a feedback-based scheduling approach is designed for the controller of IPsec-applied SDN to replace flexibly and switch synchronously the IPsec cryptographic algorithms in use according to the real-time network status. The feedback information is applied to decide which appropriate algorithm(s) should be employed for the cryptographic process in a special application scenario. The validity and effectiveness of the proposed invocation mechanism are verified and evaluated on a small-scale SDN/OpenFlow platform with the deployed IPsec security gateway. The results show that the FTA-based mechanism invokes IPsec encryption algorithms consistently with the requirement for communication security in the SDN southbound interface, and the impact of the IPsec cryptographic process on the network performance will be reduced even if the network traffic fluctuates markedly.

Keywords: Communication security · Software-defined networking (SDN)
IPsec · Algorithm invocation · Southbound interface (SBI)

1 Introduction

The software-defined networking (SDN) paradigm decouples the control plane from the underlying data plane and introduces network programmability and other features to promote network flexibility, adapting to the constantly changing network condition and facilitating the network's verification and deployment. However, due to less consideration of security issues in the initial design period of the SDN architecture, some new features introduced into SDN provide more convenience for the network management, but some new types of security threats consequently emerge [1–3].

The OpenFlow protocol is a widely adopted communication standard for the southbound interface (SBI) in SDN networks. While the control plane communicates with the data plane using the OpenFlow-supported instructions, the feature of

© ICST Institute for Computer Sciences, Social Informatics and Telecommunications Engineering 2019
Published by Springer Nature Switzerland AG 2019. All Rights Reserved
X. Liu et al. (Eds.): ChinaCom 2018, LNICST 262, pp. 583–592, 2019.
https://doi.org/10.1007/978-3-030-06161-6_57

separation between these two planes makes it insecure for the control flows when passing through exterior network links. OpenFlow thereby cooperates with the transport layer security (TLS) protocol to secure the communication between the SDN controllers and the switches (i.e. the SBI communication) [4]. Nevertheless, the TLS is too complicated in the verification and too fragile in defense of man-in-the-middle (MITM) attacks to guarantee the security, and it becomes optional instead of mandatory for OpenFlow [5]. Without the security protection of TLS, the TCP-based SBI communication is vulnerable to the tapping and forgery of control information, which makes the network more insecure and unreliable.

In recent years, some schemes have been proposed to enhance the SBI security [6–8]. These new controllers achieve better and more comprehensive security than general SDN controllers and reduce the risk from SBI. However, some risks still cannot be eliminated by the new controllers while exchanging control messages with switches, for instance MITM attacks, which exploit the flaw in the TLS protocol, and the risks of tapping and forging control messages when using TCP connections [9].

Internet protocol security (IPsec) is introduced to guarantee the security in the southbound interface of the controller and maintain secure communication between the controller and the switches in the SDN network [10, 11]. IPsec, originally developed for IPv6, can ensure the communication security in the Internet layer and does not require extra support from the controller. Meanwhile, as part of IPv6, it is in line with the current network evolution trend.

The IPsec protocol is mature in architecture but rigid in the invocation of cryptographic algorithms. Besides, the demand for customized algorithms or more algorithms supported by IPsec is urgent in various application scenarios, along with the mounting importance of network security. However, less research focuses on the flexibility of the subsequent invocation of IPsec algorithms. The rigid invocation of cryptographic algorithms in IPsec makes it hard to meet the diversified security demands of networks [12].

Furthermore, the invocation of these algorithms should consider the trade-off between the link security offered by IPsec and the communication performance of the SBI in SDN networks [11]. When IPsec is adopted to secure the communication between the SDN controllers and the switches, it is troublesome for the user to add the customized algorithm to the switches, because the vendors have limited the modification of switches, and device addition or upgrading thereby results in large costs [13]. Additionally, IPsec encryption/decryption will increase the performance consumption, even though there are certain performance requirements of SBI communication between the controller and the switches. When the traffic fluctuates, the consumption forms a bottleneck problem of communication and may amplify the variation in traffic and communication performance.

To address the above-discussed issue of IPsec algorithm invocation in SDN, a flexible FTA-based mechanism for IPsec encryption algorithm invocation will be studied in this paper. By striking a balance between IPsec-encrypted link security and communication performance, a scheme is proposed to ensure that the SDN SBI communication security has little impact on the transmission of control messages.

The rest of the paper is organized as follows. Section 2 provides an overall view of the proposed invocation mechanism of IPsec cryptographic algorithms. Section 3 demonstrates the proposal with an experiment and compares it with the native scheme in IPsec in terms of network performance. Finally, the conclusion of this paper is drawn in Sect. 4.

2 Invocation Mechanism for IPsec Cryptographic Algorithms

2.1 IPsec in SDN Architecture

Due to the separation of control and data planes, SDN controllers and switches are in different network locations. Controllers are usually high-performance hosts or servers, so the deployment of IPsec is straightforward and convenient. For most OpenFlow switches (i.e. Juniper EX4550), the vendors tend to limit their modification. That is to say, it is difficult to implement some users' customized demands, for example special security demands, locally.

Taking the above into consideration, adding a computer card or development board, for example Raspberry Pi, to OpenFlow switches can build an IPsec secure gateway. The open architecture of IPsec facilitates the addition of a new or customized cryptographic algorithm and is helpful for building a communication system with stronger closure and higher security. Moreover, IPsec can guarantee secure communication between controllers and OpenFlow switches with the IPsec gateways located in the switches. The added computer card or development board will enable optional and easily managed operation without exerting a negative impact on the configurations of OpenFlow switches, system running, data forwarding and so on.

2.2 Free-to-Add Invocation of Cryptographic Algorithms

Retaining the basic IPsec workflow, in our preliminary work, we proposed a mechanism (shown in Fig. 1) [13] titled Free-to-Add (FTA), providing flexible cryptographic algorithm addition and invocation for IPsec in SDN networks. Compared with the method in native IPsec, FTA makes the algorithm switching more adaptable and flexible, avoiding the need to rebuild the IPsec security association (SA) or modify the configuration file and restart IPsec. Besides, FTA applies a layer of user encapsulation for a special encryption algorithm to make attacks on encryption algorithms more costly and difficult.

2.3 Feedback-Based Algorithm Scheduling

In this section, we will discuss the method for scheduling IPsec cryptographic algorithms in FTA considering the requirements of both communication performance and link security.

Impact on Network Performance and System Resources. Primitively, to simplify the analysis of the impact that the cryptographic algorithm exerts on the network performance, we make three assumptions as follows:

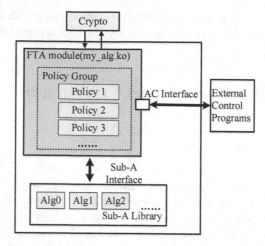

Fig. 1. Invocation process of encryption algorithms in FTA-based IPsec (AC: algorithm control, Sub-A: sub-algorithm)

Assumption 1. There are multiple algorithms with similar security strength and different system resource requirements, and the network performance can be optimized by applying these algorithms.

Assumption 2. The link bandwidth is not considered as a system resource that is required in IPsec encryption/decryption or communication.

Assumption 3. The variation of network delay is only associated with the queuing delay and processing delay of encryption/decryption, and other delays remain unchanged.

If the IPsec encryption/decryption processes quickly with few resources, the processing delay of encryption/decryption and the queuing delay of forwarding data in the link may decrease. Thus, we know that:

- IPsec encryption/decryption may degrade the communication performance. The stronger the encryption is, the greater the resource consumption and the more processing time it takes, leading to a greater impact on the performance. The changes in system resource consumption are proportional to the strength of the IPsec encryption/decryption, and the communication performance is inversely proportional to the strength.
- When the data traffic fluctuates, the offset of packet size distribution can result in throughput reduction and an increase in network latency. At this point, we can switch to a more appropriate encryption algorithm or policy to reduce the consumption of system resources and the network latency to ensure throughput. As a consequence, the impact of the IPsec encryption/decryption on the communication performance can be reduced under a guaranteed security level.

Feedback-Based Adjustment Model. The security capability of an encryption algorithm is positively related to the system consumption. Thus, an algorithm with higher security capability corresponds to lower communication performance and a higher level of security. Due to this feature, a less-consuming algorithm should be chosen to meet a high demand for communication performance and a higher-level algorithm for a higher security demand.

Therefore, we design a feedback-based adjustment model for the security algorithms, as shown in Fig. 2, according to the trade-off between IPsec encryption and communication performance. The state of an IPsec-secured link S is denoted by a triplet (T, D, E), where T, D and E denote the throughput, network latency and level of link security, respectively. Each cryptographic algorithm or policy corresponds to a level of security, and the level of link security E limits the other two factors T and D when the traffic load and the distribution of the packet size do not change. The security level of the network link should be adjusted according to the current T, D and security demand.

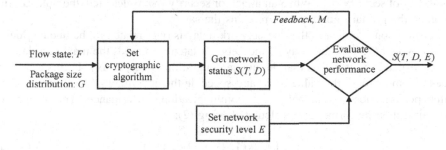

Fig. 2. Adjustment model according to the feedback

In the encryption process, an algorithm can be substituted by another algorithm with a higher security level when the security demand increases. Having set the levels of security, all the algorithms form an algorithm set A (a_1, a_2, ..., a_n), $n \geq 2$, in which an algorithm with a higher index means that it can provide higher security capability. A median algorithm a_j ($j = [n/2]$) is selected as the initial algorithm for the SBI communication.

The algorithm to be used will be switched according to the feedback of the later communication performance evaluation. The simple adjustment model maintains a balance between encryption security and communication performance, in which the key is to adjust the proper encryption algorithm for use according to the demands of both security and performance.

Scheduling IPsec Cryptographic Algorithms. In the adjustment model presented in the previous section, the feedback information indicates which appropriate algorithm should be employed for the cryptographic process. An appropriate encryption algorithm is one that can meet the demands of communication performance and link security in a balanced way. Here, we present an evaluation model to find the trade-off between communication performance and link security:

$$M = \frac{w_e}{w_s}, \tag{1}$$

where M, w_e and w_s denote the degree of balance, weight of the link security and weight of the communication performance, respectively. For instance, in a certain scenario in which the network performance is the top-priority factor, a higher-level encryption algorithm is needed in IPsec if the value of M increases and exceeds the effective range.

The weight of link security w_e is adjusted according to the network security status. Its value will increase when the risk of network security rises. With the security level of IPsec encryption algorithm elevating, the security risk keeps declining, and the security weight w_e is also reduced to the initial value 0. In terms of the weight of communication performance w_s, it is related to the link throughput, the network latency and the possibility of security risk. When an attack or security risk is detected, the link security becomes the primary goal, and w_s remains the same.

Assumption 3 indicates that the network delay is only related to the queuing delay and encryption/decryption delay. If the network latency exceeds the maximum latency of the normal link, it means that the queue congestion is very heavy. In this case, it is necessary to increase the value of w_s and schedule the encryption algorithm providing better performance to guarantee the communication performance. The weight of communication performance is calculated using (2):

$$w_s = \begin{cases} \max(1, \frac{D_n}{D_{\max}}) & (w_e = 1) \\ 1 & (w_e \neq 1) \end{cases}, \tag{2}$$

where D_n denotes the link latency obtained from the n-th sampling and D_{max} is the maximum latency when the network status is normal.

Since the processing performance of each encryption algorithm is not continuous and the link latency of the sampled network may change constantly, we set the algorithm switch criterion according to (3), where p_j is the processing performance of the j-th encryption algorithm in the algorithm set A. Note that the switch criterion is a value range instead of an exact value and the security level of the candidate algorithms for scheduling should meet the security demand.

$$\begin{cases} if\ M \geq \frac{p_{j-1}}{p_j}; & \text{switch to the algorithm with a higher security level, until it is the inital algorithm} \\ if\ M \leq \frac{p_{j+1}}{p_j}; & \text{switch to the algorithm providing higher process performance,} \\ & \text{until it is the algorithm with the lowest security level} \end{cases}$$

$$\tag{3}$$

Hence, the IPsec encryption algorithm used for IPsec and SBI communication may be replaced by another more appropriate algorithm when the link security and communication performance change to keep the balance between the two factors.

3   Experiment and Verification

3.1   Validity of the FTA Mechanism

In our work, a small-scale testbed is built to verify the validity of the FTA-based mechanism. The topology is shown in Fig. 3, and the configuration information is given in Table 1. In this testbed, the communication between the SDN controller and the OpenFlow switches is the SBI communication of SDN, and two Raspberry Pis act as the IPsec security gateways that ultimately use IPsec to secure the communication between the SDN controller and the OpenFlow switches. The performances of FTA-based IPsec and native IPsec are compared using three cryptographic algorithms.

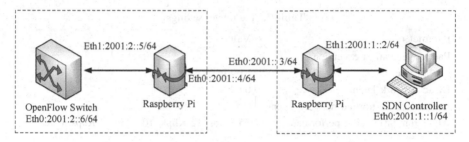

Fig. 3. Topology of the SDN testbed with IPsec gateways

Table 1. Configuration of the simulation software and hardware

SDN device	Equipment	Operating system	Hardware	Software
OpenFlow switch	Raspberry Pi 3B	Rappbian Stretch Lite	ARM Cortex-A53 1.2 GHz	OpenvSwitch 2.3.0
IPsec security gateway	Raspberry Pi 3B		Quad Core 1G RAM USB 2.0	StrongSwan5.4.0
SDN controller	DELL Inspiron 14 7000	Ubuntu 16.04.3	i5-4200H@2.8 GHz 8 G RAM, USB 2.0	OpenDaylight Beryllium SR4

From the results presented in Table 2, it can be seen that the network latencies and link throughputs are approximate in these two cases and that their variations are in the normal range. In addition, the actual bandwidths tested by *Iperf3* in both mechanisms are almost the same. In short, the use of the FTA-based mechanism has little impact on the network performance.

Table 2. Performance comparisons between native IPsec and FTA-based IPsec

Item	Method	Cryptographic algorithm		
		AES128	DES	3DES
Latency (ms)	Native IPsec	2.63	2,62	2,62
	FTA-based IPsec	2.75	2.53	2.62
Request/response per second	Native IPsec	378	382	381
	FTA-based IPsec	363	394	381
Test bandwidth (Mbps)	Native IPsec	64.1	55.7	34.8
	FTA-based IPsec	63.6	53	34.7

Table 3. Parameter settings

Parameter	Value
Packet generation exception	10
Packet size	$(0 \sim 1480$ Byte)
Basic network latency	0.02 s
Upper limit of normal network latency	0.4 s
Algorithm processing performance	15 Kbps, 12 Kbps, 10 Kbps, 8 Kbps, 6 Kbps

3.2 Verification of Feedback-Based Scheduling

To evaluate the availability of the feedback-based scheduling scheme, in this section, simulations are carried out using Matlab, and the network latency and link throughput are the performance metrics.

The settings of the simulation parameters are given in Table 3, in which the distribution of data packets and the sizes of the packets follow Poisson distribution and average distribution, respectively.

From Fig. 4(a), it is obvious that the link with the native scheduling scheme achieves an average network latency of 0.78 s and the distribution of latency is scattered with great variation. The results of the case of applying the feedback-based scheduling scheme are presented in Fig. 4 (b). The average network latency of the feedback-based case is 0.51 s. When detecting a latency exceeding the normal range of values, the algorithm will be substituted by another one to keep the peak value of latency less than 0.6 s. The results shown in Fig. 5 indicate that applying the feedback-based scheme can provide a more stable network latency with concentrated distribution and fluctuation within a narrow range.

In Fig. 5 (a), we can see that the link throughput in the native scheduling case is likely to be limited to about 10 Kbps, that is, the maximum processing performance of initial algorithms. This phenomenon is caused by the reduced processing performance in encryption/decryption nodes and results in heavy network congestion and packet loss. On the contrary, when the links apply the proposed feedback-based scheme to schedule the IPsec cryptographic algorithms, the processing performance and link throughput are improved, and the network congestion is thereby mitigated with higher throughput

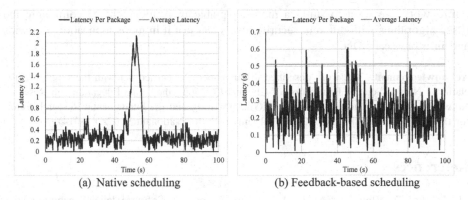

(a) Native scheduling   (b) Feedback-based scheduling

Fig. 4. Variations of network latency using different scheduling schemes for IPsec cryptographic algorithms

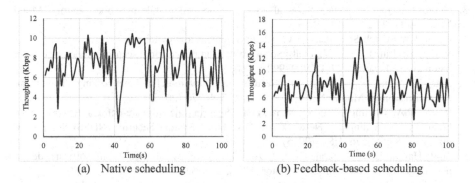

(a)   Native scheduling   (b) Feedback-based scheduling

Fig. 5. Variations of link throughput using different algorithm scheduling schemes for IPsec cryptographic algorithms

shown in Fig. 5 (b). In summary, the feedback-based scheduling for IPsec cryptographic algorithms can effectively switch the encryption algorithms and provides the SBI communication with a good balance between link security and communication performance.

4   Conclusion

In our work, an FTA-based mechanism considering the network feedback has been proposed to simplify the process of IPsec cryptographic algorithm addition and switching and ensure flexible and available invocation of IPsec cryptographic algorithms. The experimental results show that the feedback-based scheduling scheme enables the mechanism of algorithm invocation to invoke and switch the IPsec encryption algorithms, providing a good trade-off between the level of link security and the demand for communication performance in the SDN southbound interface. In future work, we want to study a more adaptable invocation scheme combining the IPsec algorithms with different security levels and test it in more application scenarios and in a physical SDN.

Acknowledgement. The work described in this paper was carried out with the support of the National Natural Science Foundation of China (61772562), the China Education and Research Network (CERNT) Innovation Project (NGII20150106), and the Fundamental Research Funds for the Central Universities, South-Central University for Nationalities (CZY18014).

References

1. Kreutz, D., Ramos, F.M.V., Veríssimo, P., et al.: Software-defined networking: a comprehensive survey. Proc. IEEE **103**(1), 14–76 (2015)
2. Liyanage, M., Ahmed, I., Okwuibe, J., et al.: Enhancing security of software defined mobile networks. IEEE Access **5**, 9422–9438 (2017)
3. Chen, M., Qian, Y., Mao, S., et al.: Software-defined mobile networks security. Mobile Netw. Appl. **21**(5), 1–15 (2016)
4. OpenFlow Switch Specification v1.5.1. http://OpenFlowSwitch.org
5. Wang, M., Liu, J., Chen, J., et al.: Software defined networking: security model, threats and mechanism. J. Softw. **27**(4), 969–992 (2016)
6. Shu, Z., Wan, J., Li, D., et al.: Security in software-defined networking: Threats and countermeasures. Mobile Netw. Appl. **21**(5), 764–776 (2013)
7. Porras, P.A., Cheung, S., Fong, M.W., et al.: Securing the software-defined network control layer. In: Proceedings of Network and Distributed System Security (NDSS) Symposium, pp. 1–15. San Diego, USA (2015)
8. Scott-Hayward, S., Natarajan, S., Sezer, S.: A survey of security in software defined networks. IEEE Commun. Surv. Tutor. **18**(1), 623–654 (2016)
9. Ferguson, A.D., Guha, A., Liang, C., et al.: Participatory networking: an API for application control of SDNs. In: Proceedings of ACM Special Interest Group on Data Communication (SIGCOMM), pp. 327–338. Hong Kong, China (2013)
10. Huang, X., Yu, R., Kang, J., et al.: Software defined networking for energy harvesting Internet of Things. IEEE Internet Things J. **5**(3), 1389–1399 (2018)
11. Marin-Lopez, R., Lopez-Millan, G.: Software-defined networking (SDN)-based IPsec flow protection. I2NSF Internet-Draft. July 2018. draft-ietf-i2nsf-sdn-ipsec-flow-protection-02. https://datatracker.ietf.org/doc/draft-ietf-i2nsf-sdn-ipsec-flow-protection/
12. Al-Khatib, A.A., Hassan, R.: Impact of IPSec protocol on the performance of network real-time applications: A review. Int. J. Netw. Secur. **19**, 800–808 (2017)
13. Yang, X., Wang, D., Feng, W., Wu, J., Tang, W.: Cryptographic algorithm invocation based on software-defined everything in IPsec. Wirel. Commun. Mobile Comput. (WCMC) **2018** (2018). https://doi.org/10.1155/2018/8728424. Article ID 8728424

HetWN Selection Scheme Based on Bipartite Graph Multiple Matching

Xiaoqian Wang[1](✉), Xin Su[2], and Bei Liu[2]

[1] Broadband Wireless Access Laboratory, Chongqing University of Posts
and Telecommunications, Chongqing, China
wangxiaoqian2017@mail.tsinghua.edu.cn
[2] Beijing National Research Center for Information Science
and Technology, Beijing, China

Abstract. Next generation communication networks will be a heterogeneous wireless networks (HetWN) based on 5G. Studying the reasonable allocation of new traffics under the new scenario of 5G is helpful to make full use of the network resources. In this paper, we propose a HetWN selection algorithm based on bipartite graph multiple matching. Firstly, we use the AHP-GRA method to calculate the user's preference for network and the network's preference for user. After these two preferences are traded off as the weights of edges in bipartite graph, we can extend the bipartite graph to a bipartite graph network. The minimum cost maximum flow algorithm is used to obtain the optimal matching result. Simulations show that our scheme can balance the traffic dynamically And it is a tradeoff between user side decision and network side decision.

Keywords: Heterogeneous wireless network · Bipartite graph
Minimum cost and maximum flow

1 Introduction

With the development of wireless communication technology, the future mobile communication networks will not be a single well-functional networks but a heterogeneous networks in which multiple wireless access technologies coexist. The performance in different networks, such as network throughput, coverage area and minimum delay will have a huge difference. So there is no wireless access technology that can satisfy all kinds of traffic needs [1]. It can be foreseen that the next-generation wireless communication networks will be a heterogeneous network consists of 5G networks and 4G networks (LTE-A and WIMAX2). In addition, the development of smart terminal such as mobile phone makes it possible for smart network selection. The reasonable selection results should

Supported by the National S&T Major Project (No.2018ZX03001011), Science and Technology Program of Beijing (No.D171100006317002) and the Ministry of Education and China Mobile Joint Scientific Research Fund (Grant No.MCM20160105)

satisfy the QoS requirements of different traffics and network operators' interests as much as possible at the same time.

Existing HetWN selection algorithms are mainly based on 3G or 4G network background and traffic background (Session class, streaming media class, interaction class, and background class). This article extends the network types and traffic types to 5G new scenario. The related works can be roughly classified into the following categories: Analytic Hierarchy Process (AHP) and its improvements, game theory methods and other methods. Reference [2] introduces the AHP method and the bankruptcy game model. References [3–5] are improvements to AHP algorithm. The former introduces an ordered weighted average operator to improve the performance of network selection handover. Reference [4] combines Grey Relational Analysis (GRA) and AHP, this paper is an improvement based on it. The concept of fuzzy logic is used in [5,6] to make AHP judgement matrix more suitable and reasonable. In addition, game theory tool is very suitable for analyzing the problem which contains a resource competing relationships. Game theory method includes non-cooperative game and cooperative game. Reference [7] considers that all users share the total network rate and therefore establishes a non-cooperative game model for each participant (user) regarding their respective data rate. References [8,9] propose an evolutionary game model. The proposed algorithm converges faster than non-cooperative game and Q-learning algorithm. Reference [10] proposes a multi-user TOPSIS-based matching game which improves the utilization of network and reduces the network blocking rate.

The rest of paper is organized as follows: In Sect. 2, a generic heterogeneous network model is established. In addition, 5G network parameters and new traffic types that may appear in 5G new scenario are also analyzed. In Sect. 3, a bipartite graph network model is established, and a minimum cost maximum flow algorithm is used to obtain the optimal matching results. In the last section, simulation analysis is carried out.

2   System Model

We consider a HetWN environment which consists of 5G, LTE-A, and WIMAX2 based on IEEE 802.16 m, as shown in Fig. 1. In addition, we assume that parameters such as per connection rate and average delay of network will not change before the user number reaches the upper limit of network capacity, that is, the network can provide service stably when the network capacity does not reach the upper limit. This article mainly analyzes the network selection of users in zone 3. The attributes and parameters of these three networks are listed in Table 1.

We have also summarized the related works on 5G new traffics, which the Global Mobile Suppliers Association briefly outlined in 2015. The impact of user mobility and energy efficiency must be considered in 5G new scenario. This paper takes transportation traffic, industrial automation and utility traffic, health traffic, virtual reality (VR) and augmented reality (AR) traffic and smart city traffic into consideration. The QoS requirement parameters are listed in Table 2.

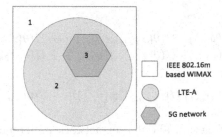

Fig. 1. The 5G heterogeneous wireless network system.

Transport traffic: There are multiple wireless applications that require low latency such as vehicle-to-vehicle (V2V), vehicle-to-infrastructure (V2I) and various intelligent transportation systems. They need a lower latency than the existing LTE networks can provide. For example, the expected delay for anti-collision system is 5 ms, and the reliability is 99.999%.

- *Industrial automation and utility traffic: Wireless sensors in industrial automation and robotics usually require secure, ultra-reliable communications and must have a low power consumption. In the public utilities, for example, many countries are developing smart grids. The low latency is necessary to protect power grid.*
- *Health traffic: The concept of mobile health applications has been developed for many years, such as personal health records and fitness data, wearable activity tracking and smart phone-based applications. In addition, mobile services can provide remote diagnosis for nursing staff.*
- *VR and AR traffic: VR and AR require a large amount of data. When the head show and other displayers are wirelessly connected, they must support low latency and high reliable data transmission.*
- *Smart city traffic: There are many applications of smart city in multiple fields as traffic, public management and others.*

3 Multiple Matching Algorithm Based on Bipartite Graph Networks

3.1 Bipartite Graph Networks Model

Assuming that there are M users and N alternative access networks in area 3 as shown in Fig. 1, denoted as $X = \{x_1, x_2, \ldots x_m\}$, $Y = \{y_1, y_2, \ldots y_n\}$, respectively ($N = 3$ in this paper). User i's preference for all access networks is denoted as $\theta(x_i)$. $\varphi(y_j)$ is the preference of access network j to all users. Therefore, the network selection model can be simplified as a bipartite graph model. The weight of edge in bipartite graph indicates the degree of matching between user and network. The weight matrix U is written as follows:

$$U = \begin{pmatrix} U(\theta(x_{11}), \varphi(y_{11})) & U(\theta(x_{12}), \varphi(y_{21})) & \cdots & U(\theta(x_{1n}), \varphi(y_{n1})) \\ U(\theta(x_{21}), \varphi(y_{12})) & U(\theta(x_{22}), \varphi(y_{22})) & \cdots & U(\theta(x_{2n}), \varphi(y_{n2})) \\ \vdots & \vdots & \ddots & \vdots \\ U(\theta(x_{m1}), \varphi(y_{1m})) & U(\theta(x_{m2}), \varphi(y_{2m})) & \cdots & U(\theta(x_{mn}), \varphi(y_{nm})) \end{pmatrix}_{M \times N} \tag{1}$$

Where U is a utility function about user-network preferences and network-user preferences.

This paper selects the minimum cost maximum flow algorithm to solve the bipartite graph multiple matching problem. The bipartite graph is firstly extended to a network as shown in Fig. 2. An aggregation node t and a source node s is added at the network side and user side, respectively. Each edge includes 2 elements. The former means the capacity of each edge, while the latter means the price of this edge. The edge capacity of source node to each user and users to networks are set to 1. The edge capacity of each network to aggregation node is the upper limit of each network's capacity. Only the edges between the users and networks have a price. This price depends on the remaining capacity of network. As the remaining capacity decreases, the price rises. The price of other edges are 0.

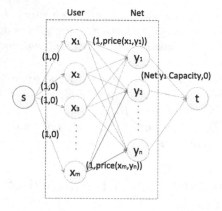

Fig. 2. Expanded bipartite graph networks.

3.2   Weight Decision of Bipartite Graph Using AHP and GRA

The weight of bipartite is determined by the utility function which consists of 2 preferences as described above. The price should be inversely proportional to weight in each corresponding edge. The greater the weight, the smaller the corresponding price. The user's preference for network mainly depends on user's traffic type, and the network always prefers to the user who can provide the highest price, so all networks have the same preference for each user.

The throughput per connection(1), average network delay(2), network supported mobility(3), network packet loss rate(4), network jitter(5), network energy efficiency(6) and price(7) are considered as the network parameters and different traffic's QoS requirement parameters. These 2 preferences are firstly calculated by AHP and then correlated by GRA.

Step 1: Parameter normalization. The parameters in Table 1 can be divided into two categories, namely the profit type (the bigger the better) and the cost type (the smaller the better). For users, profitable parameters include the throughput per link and network supported mobility, while the rest are cost-type parameters. For networks, all parameters are cost type except the user available payment. The normalization of profit types and cost types are as written as follows respectively:

$$a_{ij} = \frac{b_{ij} - \min\limits_{i} b_{ij}}{\max\limits_{i} b_{ij} - \min\limits_{i} b_{ij}} \tag{2}$$

$$a_{ij} = \frac{\max\limits_{i} b_{ij} - b_{ij}}{\max\limits_{i} b_{ij} - \min\limits_{i} b_{ij}} \tag{3}$$

b_{ij} is the original parameter in Table 1, a_{ij} is the normalized one.

Step 2: Construct the judgement matrix using the 1–9 ranking scheme.

Step 3: Calculation the network parameter's weight through judgement matrix C.

$$W_i = \prod_{j=1}^{np} c_{ij} (i = 1, 2, ...np) \tag{4}$$

$$\bar{W}_i = \sqrt[np]{W_i} \tag{5}$$

$$weight_i = \frac{\bar{W}_i}{\sum\limits_{i=1}^{np} \bar{W}_i} \tag{6}$$

$Weight = (weight_1, weight_2, \ldots weight_{np})$ is the parameter weight which decides the network selection results. np is the number of attribute parameters.

Step 4. Consistency test. The consistency index and average random consistency index are denoted as C_I and R_I, respectively. The check will pass if $C_I < 0.1$.

Step 5. Calculation of network selection weight. Simple additive weighting (SAW) scheme is used to decide the network's weight.

Step 6. Calculation of grey relational matrix. ρ (usually equals to 0.5.) is the correlation coefficient and a_{0j} is the optimal reference sequence.

Algorithm 1 Network Optimal Selection Algorithm Based on Bipartite Graph

---

Initialization:
1: There are M users and N alternative access networks. User's traffic is randomly distributed.
2: The price and capacity information for each edge in this bipartite graph network.
3: **while** 1 **do**
4:    Randomly select the minimum value in P.
5:    If the minimum value is P_{ij}, then allocate user i to network j and update P and edge capacity.
6:    If the edge capacity is 0, then delete the edge between user i and network j. $P_{ij} = nan$.
7:    If all users have been assigned, break out.
8: **end while**
Output:
9: Output the matching results.

$$R = \frac{\min_i \min_j |a_{0j} - a_{ij}| + \rho \max_i \max_j |a_{0j} - a_{ij}|}{|a_{0j} - a_{ij}| + \rho \max_i \max_j |a_{0j} - a_{ij}|} \tag{7}$$

Step 7. Calculation of correlation network's weight. Finally, according to the correlation matrix R obtained by GRA and network's weight obtained by AHP, the preferences of users with different types of traffic for all networks are obtained as:

$$\theta = R \cdot F^T \tag{8}$$

Similarly, we can also get the network's preference for users (φ).

Step 8. Calculation of utility function and price of edge in bipartite graph network. The utility function consists of 2 preferences. And a compromise factor is set to combine these 2 preferences which is denoted as α. The price can be calculated by the following formula:

$$P_{ij} = \frac{1}{\alpha\theta(x_{ij}) + (1-\alpha)\varphi(y_{ji})} + (\max_i \max_j P_{ij} - \min_i \min_j P_{ij})(NC_j - RC_j)^\gamma \tag{9}$$

NC_j is the network j's capacity and RC_j is its remaining capacity. γ is a number between 1 and 2.

3.3   Proposed Heterogeneous Wireless Networks Selection Algorithm

In this paper, the minimum cost maximum flow algorithm is used to solve this optimal matching problem. Firstly, the price and capacity of each edge is initialized by the analysis we have talked above. Then we find out all the paths from s to t. The average cost per path is equal to the total cost divided by the maximum capacity in this path. Since the capacity of edge connected to s is 1,

P_{ij} is the average cost per path. Then we find out the minimum value in P. If there are multiple paths with same price at the same time, randomly selection should be taken to ensure the fairness. After the user has been assigned to the corresponding network, the capacity and price of each edge in this bipartite graph networks will update. The edge will be deleted if its capacity is 0. The loop is ended until all users have been assigned. The specific algorithm refers to the Algorithm 1.

4   Simulation Results

In this simulation, we mainly study a NetWn which consists of 3 wireless network: 5G, LTE-A and WIMAX2 based on IEEE 802.16m. There are many users randomly distributed in area 3 as shown in Fig. 1. The network's parameters and traffic QoS requirement parameters are listed in Tables 1 and 2. The total user number is 20 to 500. The capacity of each network is set to 400. Due to the randomness of user traffic, we used 1000 Monte Carlo simulations per cycle.

Table 1. The attribute parameters of 5G, LTE-A and WIMAX2 network.

Network parameters	5G	LTE-A	WIMAX2 (Low mobility)	WIMAX2 (High mobility)
Throughput per connection (Gbps)	1	0.1	1	0.1
Network average latency (ms)	1	10	10	10
Network supported mobility (km/h)	500	350	6	120
Network packet lost (%)	0.001	0.003	0.002	0.002
Jitter (ms)	1	3	2	2
Energy Efficiency (1e-7 J/bit)	1	100	200	200
Price ($)	5	3	4	4

Due to the user in different moving conditions, the network parameters of WIMAX2 change a lot. So this paper considers the performance of WIMAX2 in low mobility scenario and high mobility scenario respectively. Figure 3 shows the weight of network selection under different traffics. Figure 4 shows the ratio of users which select different networks to the total number of users. In Fig. 4, the number of users who choose 5G network is the most. Because 5G network can provide the best service so that most of traffics are more suitable to choose 5G. With the total user number decreases, the ratio of user who selects 5G is increasing. With the total number of user increases, the result of network selection tends to be balanced, which shows that theproposed algorithm has the

Table 2. The required QoS attribute parameters of 5G new traffics.

QoS requirement attributes	Transport	Industrial automation and utility	Health	VR&AR	Smart city
Throughput per connection (Gbps)	0.0001	0.0001	0.05	0.5	0.05
Network average latency (ms)	5	10	10	1	50
Network supported mobility (km/h)	350	1	1	6	60
Network packet lost (%)	0.001	0.003	0.002	0.002	0.004
Jitter (ms)	1	3	3	2	5
Energy Efficiency (1e-7 J/bit)	300	10	200	500	500
Price ($)	10	7	12	10	5

function of balancing the traffic load while reasonably allocating services. This is due to an update of price in bipartite graph network.

The effect of compromise factor α on network selection results is shown in Fig. 5. When $\alpha = 0$, the network selection only considers the network's preference for user. Due to the prices of 3 networks are not much different, the proportion of users who choose different networks is not much different. When the compromise factor is 1, the network selection only considers the user's preference for network, which is the case in [4].

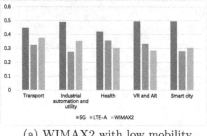
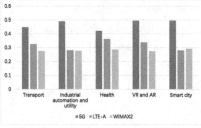

(a) WIMAX2 with low mobility   (b) WIMAX2 with high mobility

Fig. 3. The weight of network selection.

Figure 6 shows the average price of access network which is equal to the total system price divided by total system throughput in high mobility scenario. It also shows that our scheme is a tradeoff between AHP-GRA network selection at user side and network side. In addition, we propose a modified random network selection scheme: If the traffic can only be carried by 5G, then only the 5G network is selected, otherwise it is randomly allocated to these 3 networks. Simulation shows that our scheme has a lower average price than modified random selection scheme.

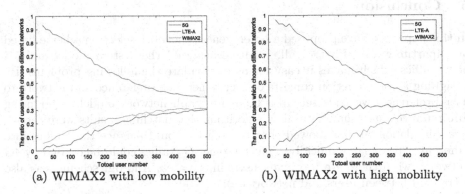

(a) WIMAX2 with low mobility   (b) WIMAX2 with high mobility

Fig. 4. Matching results ($\alpha = 0.5, \gamma = 1.4$).

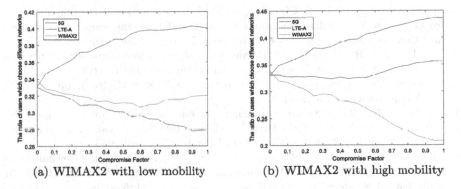

(a) WIMAX2 with low mobility   (b) WIMAX2 with high mobility

Fig. 5. Effect of compromise factor on matching results ($\gamma = 1.4$, user number is 500).

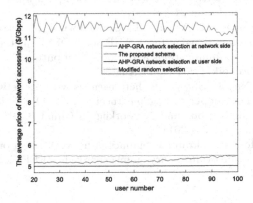

Fig. 6. The average price of networks with different schemes ($\gamma = 1.4$).

5 Conclusion

In this paper, we have proposed a heterogeneous network selection scheme based on bipartite graph. First of all, we have analyzed the system model and 5G new traffics, which makes it easy for us to calculate the following problems. By designing a price function consists of user to network preference and network to user preference, we have set up a bipartite graph network model. After using minimum cost and maximum flow algorithm, the matching results were carried out. Simulation results showed that our scheme could balance the traffics into different networks dynamically. That is, some 5G traffics could be carried by 4G networks when there are too many users in one area. And our scheme was also a tradeoff between users and networks.

References

1. Sgora, A., Vergados, D.D., Chatzimisios, P.: An access network selection algorithm for heterogeneous wireless environments. In: The IEEE symposium on Computers and Communications, pp. 890–892. Riccione, Italy, 22 June 2010
2. Liu, B., Tian, H., Wang, B.: AHP and game theory based approach for network selection in heterogeneous wireless networks. In: 2014 IEEE 11th Annual IEEE Consumer Communications and Networking Conference (CCNC), pp. 501–506. Las Vegas, USA, 10 Jan 2014
3. Preethi, G.A., Chandrasekar, C.: A network selection algorithm based on AHP-OWA methods. In: 6th Joint IFIP Wireless and Mobile Networking Conference (WMNC), pp. 1–4. Dubai, United Arab Emirates, 23 April 2013
4. Zhang, P., Zhou, W., Xie, B., Song, J.: A novel network selection mechanism in an integrated WLAN and UMTS environment using AHP and modified GRA. In: 2010 2nd IEEE International Conference on Networking Infrastructure and Digital Content, pp. 1–6. Beijing, China, 24 Sept 2010
5. Goyal, R.K., Kaushal, S.: Effect of utility based functions on fuzzy-AHP based network selection in heterogeneous wireless networks. In: 2015 2nd International Conference on Recent Advance in Engineering and Computational Sciences (RAECS), pp. 1–5. Chandigarh, India, 21 Dec 2015
6. Brajkovic, E., Sjekavica, T., Volaric, T.: Optimal wireless network selection following students online habits using fuzzy AHP and TOPSIS methods. In: 2015 International Wireless Communications and Mobile Computing Conference (IWCMC), pp. 397–402. Dubrovnik, Croatia, 24 Aug 2015
7. Cui, Y., Xu, Y., Xu, R., Sha, X.: A heterogenous wireless network selection algorithm based on non-cooperative game theory. In: 2011 6th International ICST Conference Communications and Networking in China (CHINACOM), pp. 720–724. Harbin, China, 17 Aug 2011
8. Niyato, Dusit, Hossain, Ekram: Dynamics of network selection in heterogeneous wireless networks: an evolutionary game approach. IEEE Trans. Veh. Technol. **58**(4), 2008–2017 (2009)

9. Zhu, K., Niyato, D., Wang, P.: Network selection in heterogeneous wireless networks: evolution with incomplete information. In: 2010 IEEE Wireless Communications and Networking Conference, pp. 1–6. Sydney, NSW, Australia, 18 April 2010

10. Xiulan, Yu., Zeng, Cheng: Matching game network selection algorithm based on TOPSIS. J. Chongqing Univ. Post Telecommun. (Natural Science Edition) **28**(4), 451–455 (2016)

Hybrid Deep Neural Network - Hidden Markov Model Based Network Traffic Classification

Xincheng Tan and Yi Xie[✉]

School of Data and Computer Science, Sun Yat-sen University, Guangzhou, China
tanxch3@mail2.sysu.edu.cn, xieyi5@mail.sysu.edu.cn

Abstract. Traffic classification has been well studied in the past two decades, due to its importance for network management and security defense. However, most of existing work in this area only focuses on protocol identification of network traffic instead of content classification. In this paper, we present a new scheme to distinguish the content type for network traffic. The proposed scheme is based on two simple network-layer features that include relative packet arrival time and packet size. We utilize a new model that combines deep neural network and hidden Markov model to describe the network traffic behavior generated by a given content type. For a given model, deep neural network calculates the posterior probabilities of each hidden state based on given traffic feature sequence; while the hidden Markov model profiles the time-varying dynamic process of the traffic features. We derive the parameter learning algorithm for the proposed model and conduct experiments by using real-world network traffic. Our results show that the proposed approach is able to improve the accuracy of conventional GMM-HMM from 77.66% to 96.11%.

Keywords: Network traffic · Content classification · Hidden Markov model · Deep neural network

1 Introduction

Internet traffic classification is the basis of effective network management. Such as Quality of Services (QoS), network planning and provisioning and network security. Currently, the common approach is to categorize traffic into different types of protocols or applications mainly by three methods: port-based, payload-based, and flow-based. However, with the development of network techniques, the design of network application and protocol is more sophisticated. Which allows one single protocol/application can carry various kinds of communication

Supported by the Natural Science Foundation of Guangdong Province, China (No.2018A03031303), the Fundamental Research Funds for the Central Universities (No.17lgjc26) and the Natural Science Foundation of China(No.U1636118).

contents [6] (e.g. HTTP can supply text, game, video and audio streaming). Just only identifying the protocol/application [1,3,8–10,15] type is not sufficient for effectively monitoring of network flows.On the other hand, traditional port-based and payload-based methods will meet certain limitation when the applications use dynamic port and encrypted transmission.

The goal of this paper is to classify Internet traffic into different content types. The core idea of the program is that the network flows generated by same type of content have similar behavior characteristics evolving over time. Using these characteristics can distinguish the content of network flow. Thus, we propose a new model that combines deep neural network and hidden Markov model to describe the behavior of network traffic generated by a given content type. For a given model, HMM profiles the time-varying dynamic process of the traffic features for specific content type; while DNN calculates the posterior probabilities of each hidden state based on contextual observation values. There are several reasons for using DNN-HMM. First, HMM is a simple and effective model to describe the time series data, which is suitable for modeling the behavior of network flow. However, HMM has two limitations: (i) the independent assumption of observation values; (ii) the ability of processing complex observation values is not efficient enough. The prominent performance of DNN in classification problems can not only solve the context-related problems, but also deal with complex observation values. Therefore, the advantages of DNN can be used to make up for the inherent defects of HMM. We perform experiments on real-world traffic, including six most common content types: audio, game, image, live video, radio, and video of demand (VOD). Results show that our approach has a significant improvement compared to conventional GMM-HMM.

The major contributions of this paper are summarized as follows.: (i) we use a DNN-HMM hybrid structure, the DNN is used to replace conventional probability distribution function (i.e. Gamma, Gauss, Poisson) of HMM state; (ii) we use packet arrival time (instead of inter-arrival time) and packet size as observation values, and multivariate mixed Gaussian distribution is used as the pdf of the observable vector; (iii) we propose a more fine-grained classification method based content type of traffic transferred.

The rest of the paper is organized as follows. In Sect. 2 we review related work about network traffic classification. Section 3 introduces the basic idea of DNN-HMM. Section 4 shows the experiment result of proposed model. Section 5 analyzes several key elements for the model performance improvement. Finally, Sect. 6 draws conclusions and outlines the future work.

2   Related Work

Recently, a number of traffic classification techniques have been studied, which can be categorized into three kinds: port-based, payload-based, and flow-based. However, the port-based approach is rather inaccurate due to dynamic port allocation, and some applications override the well-known port in order to bypassing the firewall [7]. The payload-based [4] approach cannot deal with the encrypted traffic and there is privacy concern with inspecting the payload.

The flow-based method was developed to address these problems. This method classifies traffic based on the flow statistics, e.g. flow duration, flow size and number of packets. Such methods usually use machine learning techniques to build profile patterns for specific traffic type, and do not affected by encryption due to not require to access to the payload. The authors of [9,10] provided surveys of techniques for traffic classification using classic ML. Although the classic ML methods can achieve high accuracy under some conditions, it's difficult to extract proper features for classification and calculate so many features are time consuming. An alternative is to use the raw packet-level information, such as the packet size (PS) and inter-arrival time (IAT).

Wright et al. [15] are the pioneers that applied HMM for traffic classification, For each traffic type, they built two HMMs by utilizing the PS and IAT respectively. Alberto et al. [1,3] proposed an improved model on basis of the previous work. They used a two-dimensional observation values trying to take account of the joint characteristics of PS and IAT, where PS and IAT obey Gamma distribution. However, the authors assumed that the PS and IAT are statistically independent, which contradicts our analysis of real-world traffic.

Recent years, deep learning [5] has been the new trend of ML, which has achieved successful applications in the fields of image and speech recognition. The performance of classic ML techniques heavily rely on handcraft features, therefore some researchers make applications of deep learning for IP traffic classification by using the raw traffic data. In [14], the authors used the consecutive payload bytes of TCP session as the input of DNN to identify protocols. Wang et al. [13] proposed an end-to-end scheme to classify encrypted traffic based on convolution neural network. Chen et al. [2] converted the early sequence of packets into image, and then classified the traffic according to the way of processing image by CNN. However, the approaches mentioned above do not consider the temporal correlation of flow features. Thus in this paper, our proposed method tries to combine the advantages of both HMM and DNN, and classify traffic from another point of view based on content.

3   Proposed Method

3.1   Traffic Pattern Characterization

Figure 1 displays the dynamic model of network flow. We can divide the time-varying dynamic process of the traffic flow into two parts: one is the observable layer, used to describe the changes of external flow shape and characteristics over time, such as the packet size and arrival time, i.e. the observation values; the other one is the state layer, used to describe the internal state changes of network flow generation mechanism or working mode over time. The internal states transition of network flow represents the change of flow pattern over time, and determines the measurements index of external shape and characteristics. However, in a practical application, the internal state of the network flow is difficult to measure directly, it can only be inferred and estimated by the measurement indicators of the external characteristics.

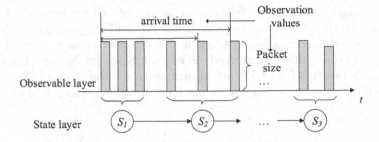

Fig. 1. The dynamic model of network flow

In general, network flows of the same type have relatively fixed change pattern, thus the behavior characteristic for a specific type of network flows can be described by external measurement indicators and internal states together, and used as the basis for identification. More specifically, for a given content type of network flow c, $\overrightarrow{O}_t^c$ denotes measurement vector at time t, Q_t^c is the corresponding internal state. To simplify the complexity of quantitative modeling, we assume that the current state Q_t^c is only related to the previous state Q_{t-1}^c, and independent of $Q_{1:t-2}^c$. In addition, the observation value $\overrightarrow{O}_t^c$ is only related to the current state Q_t^c, and independent of $Q_{1:t-1}^c$, $Q_{t+1:T}^c$, $\overrightarrow{O}_{1:t-1}^c$, $\overrightarrow{O}_{t+1:T}^c$. Thus, this work proposes to use HMM to profile the interaction between external measurement index and internal state of the network flow.

3.2 GMM-HMM

To built a DNN-HMM, firstly we need to train a baseline GMM-HMM. $O_T = (o_1, o_2, ..., o_T)$ is a sequence of observations generated by single HMM, T is the length of sequence. $o_t = (x_t, y_t)'$ is a continuous bi-dimensional observable at time t, $q_t \in \{s_1, s_2, ..., s_Q\}$ is the corresponding state of o_t, where Q denotes the number of states for HMM, x_t denotes the packet size (PS) at time t, y_t denotes $10 \log_{10}(AT/1\mu s)$. $\lambda = \{A, B, \pi\}$ denotes the set of parameters characterizing the HMM model, the details are given as follows:

1. $A = \{a_{ij}\}$ denotes the state transition probability matrix.

$$a_{ij} = \Pr(q_{t+1} = s_j | q_t = s_j), 1 \leqslant i, j \leqslant Q \tag{1}$$

2. $B = \{b_i(o_t)\}$ denotes the observation probabilities, where $b_i(o_t)$ represents the probability of observation vector o_t at state s_i.

$$b_i(o_t) = \Pr(o_t | q_t = s_i) \tag{2}$$

3. $\pi = \{\pi_i\}$ denotes the initial state distribution.

$$\pi_i = \Pr(q_1 = s_i), 1 \leqslant i \leqslant Q \tag{3}$$

We calculate the cross-correlation coefficient of packet arrival times and packet sizes within the same session, if the AT and PS is statistically independent, the cross-correlation coefficient will be very close to zero. Figure 2 shows that, in the traffic we analyzed, all content types exhibit significant cross-correlation between AT and PS, which demonstrates the AT and PS are clear related within same session. Thus we use joint distribution as pdf of observable vector, i.e.:

$$\Pr(\boldsymbol{o}_t|s_t = i) = \sum_{m=1}^{M} \frac{c_{i,m}}{2\pi|\boldsymbol{\Sigma}_{i,m}|^{1/2}} \exp[-\frac{1}{2}(\boldsymbol{o}_t - \mu_{i,m})^T \boldsymbol{\Sigma}_{i,m}^{-1}(\boldsymbol{o}_t - \mu_{i,m})] \quad (4)$$

$c_{i,m}$ is the weight of Gaussian mixture, and $\sum_1^M c_{i,m} = 1$ of certain state i. $\mu_{i,m} \in R^2$ is the mean vector of Gaussian, $\boldsymbol{\Sigma}_{i,m}^{-1} \in R^{2 \times 2}$ is the covariance matrix.

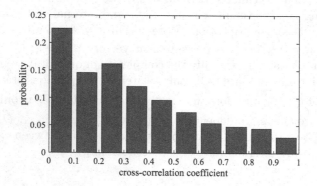

Fig. 2. Cross-correlation coefficient between AT and PS

3.3   DNN-HMM

Once the GMM-HMM is available, we replace the GMM-HMM model's Gaussian mixtures with a DNN and compute the HMM's state emission likelihoods $\Pr(\boldsymbol{o}_t|q_t)$ by converting state posteriors probability $\Pr(q_t|\boldsymbol{o}_t)$ obtained from the DNN. Compared to conventional GMM model, there are two advantage of DNN. Firstly DNN overcomes the complexity of the traditional analysis process and the difficulty of selecting the appropriate model function. Secondly, DNN can use the information of adjacent observation, which partially alleviates the violation to the observation independence assumption made in HMMs [16]. Figure 3 illustrates the architecture of the classifier. The details of the system are shown as follows:

1. The training procedure of DNN-HMM mainly includes three steps:
   (a) For each type of traffic $c \in \{1, 2, ..., C\}$, we train a GMM-HMM with parameters λ_c using the observation sequences of class c, where λ_c is easily estimated by Baum-Welch algorithm [11].

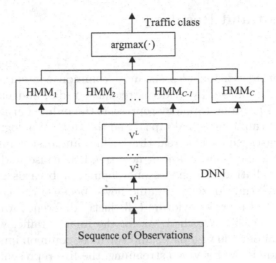

Fig. 3. The architecture of classifier

(b) For each observation sequence O_T^c of class c, using the corresponding model λ_c to obtain the optimal state sequence Q_T^* through the Viterbi algorithm [11]. The states of all HMMs are merged into one set $S = \bigcup S^c$, where S^c is the state space of model c. Then mapping the states space S of the HMMs to the output labels of DNN.

$$Q_T^* = \underset{q_1,q_2,\dots,q_T \in S^c}{\arg\max} \ \Pr(q_1, q_2, \dots, q_T | O_T^c, \lambda_c) \tag{5}$$

(c) Using the pairs of observable values and corresponding states (o_t, q_t) to train the DNN. Here, we use a regular Feedforward Neural Network (FNN), the more details about DNN can be seen in [14].

2. Classify procedure: For a new traffic flow sequence O, firstly it is input into the DNN, and DNN outputs the posterior probability $\Pr(q_t = s|o_t)$. Then we compute its probability $\Pr(O|\lambda_c)$ of being generated by each of the HMMs, since calculating $\Pr(O|\lambda_c)$ requires the likelihood $\Pr(o_t|q_t = s)$ instead of the posterior probability, converted it as follow:

$$\Pr(o_t|q_t = s) = \Pr(q_t = s|o_t)\Pr(o_t)/\Pr(s) \tag{6}$$

Here, $\Pr(s)$ is the prior probability of state s in the training set. $Pr(o_t)$ is independent of the state, thus it can be ignored. Finally, the observation sequence O is assigned to the class that generates it with the highest probability

$$c^* = \underset{1 \le c \le C}{\arg\max} \Pr(O|\lambda_c) \tag{7}$$

Where $\Pr(O|\lambda_c)$ is calculated by using the Forward-backward algorithm [11].

4   Experiment and Results

4.1   Dataset

The traffic we considered includes six most common content types of traffic, namely audio, game, image, live video streaming, radio, and video of demand, shown in Table 1. The audio topology represents the online music, including several audio formats (mp3, m4a and mp4), and the radio typology represents the web radio broadcast, which is a real-time online audio streaming. As regards game traffic, web game is becoming a new trend, because it does not require download and installation, we choose one popular web cards game: Fight the Landlord. We only study dozens of game flows, because the traffic of game is strongly dependent of user behavior and distinctly different form other types of traffic, so it can be easily classified. As for the image traffic, we analyzed the image types from about 100 web pages and two most common image types JPEG and PNG were chosen. As for video streaming, the live represents the live video streaming, and the VOD represents the video that allows the viewer to select content and view it at a time of his own choosing, the more detailed difference between live streaming and VOD can be found in [12].

In this paper, we only model the traffic from server to client, because we want to infer the type of content the user is browsing from the flow patterns. And we do not take account into packets with empty payload, like TCP control packets (SYN, ACK, Keep-Alive). All the sessions consisting of less than 5 packets are omitted from our analysis.

We collected the traffic on my PC by accessing the well-known websites and labeled it manually, the collection process lasted for multiple time periods. The dataset is separated into two portions: a training set and a testing set.

Table 1. The details of dataset

Type of traffic	Training set	Testing set
Audio	148	122
Game	32	30
Image	2697	1148
Live	241	210
Radio	221	101
VOD	227	135

4.2   Parameters of the Model

Each content typology generates a HMM model. For simplicity, we set each HMM has the same number of Q states, and each state has same number of M gauss mixtures. We try to keep the values for Q and M as small as possible in order to obtain lower computational complexity, and at same time provide sufficient accuracy in modeling all typologies of traffic. We make comprehensive comparison of the model performance for different values of Q and M, and then choose values of $Q = 3$ and $M = 2$ as the model parameters.

After the GMM-HMM training finished, we can obtain the training data for DNN. The dataset is divided into three subsets: 75% as the training set, 15% as the validation set, and 15% as the testing set. The validation set is used for checking the DNN parameters.

4.3 Result

We use Accuracy to evaluate the overall performance of a classifier, and use Precision and Recall to evaluate the performance for each class of traffic.

Table 2. Classification accuracy of GMM-HMM and DNN-HMM

Model	Training set	Testing set
GMM-HMM	89.46%	77.66%
DNN-HMM(4×20)	98.21%	96.11%

The overall classification accuracy of different system is compared in Table 2. The baseline GMM-HMM was trained on the training set, and achieved accuracy of 77.66% on the testing set. A DNN with four hidden layers each of which has 20 neurons was trained on basis of the GMM-HMM model. We can see that the accuracy of DNN-HMM is 8.75% higher than that of the GMM-HMM on the training set. Moreover, when the hybrid model is applied to the testing set, the classification accuracy increases from GMM-HMM's 77.66% to DNN-HMM's 96.11% - a 23% relative improvement, which is a very high gain.

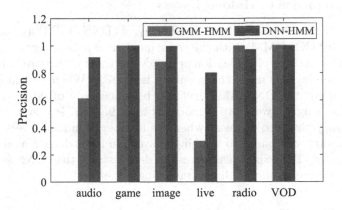

Fig. 4. The precision comparision of GMM-HMM and DNN-HMM

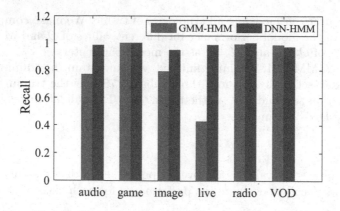

Fig. 5. The recall comparision of GMM-HMM and DNN-HMM

Figures 4 and 5 show the precision and recall comparision of six classes of traffic. As seen in Fig. 4, the precision of audio, image, and live in DNN-HMM is higher than GMM-HMM, and two classes is approximately equal, the radio class has a inferior precision of DNN-HMM. As seen in Fig. 5, the DNN-HMM can achieve more than 90% recall for each traffic type. Three typologies of traffic (audio, image and live) have a significant increase compared to GMM-HMM, and radio only has a slightly increase because it already has a very high recall. Both of the models obtain high recall for game, radio and VOD. In summary, the DNN-HMM achieves better performance than the GMM-HMM on the task of content classification of network traffic.

5   Discussion and Analysis

5.1   The Influence of Hidden Layers

Table 3 lists the results for multiple hidden layers (DNN-HMM) and corresponding single layer (NN-HMM) with the same number of parameters on testing set. One can notice that the classification accuracy increases with more hidden layers. When only one layer is used, the accuracy is 93.07%. When four hidden layers (4×20) are used, the DNN-HMM get the best accuracy of 96.11%, while the corresponding single layer (1×63) model obtains 95.25%. However, the classification accuracy starts to decrease when the number of hidden layers more than five. In summary, compared to the single layer model, a deeper model has better performance. The experimental results demonstrate that DNN with deeper structure has more strong nonlinear modeling power.

5.2   The Size of Contextual Window

Table 4 shows the accuracy comparison for whether using the information of adjacent packets. It can be clearly seen that regardless of whether shallow or deep networks are used, using information of adjacent packets significantly increases the classification accuracy. The performance of using five consecutive packets information is better than three packets, and three packets better than only one packet. Moreover, a deep network improves performance more than the single hidden layer network when using adjacent packets information, which can achieve high accuracy of 96.11%, while the corresponding network with single layer achieves 95.25%. Note that if only one single packets information is used, the DNN-HMM is even worse than GMM-HMM (77.15% versus 77.66%).

Table 3. Accuracy for multiple hidden layers versus single layer

L×N	DNN-HMM	1×N	NN-HMM
1×20	93.07%		
2×20	95.07%	1×34	94.16%
3×20	95.65%	1×50	94.67%
4×20	96.11%	1×63	95.25%
5×20	94.50%	1×77	94.33%

Table 4. Comparision for different size of window

Model type	1	3	5
NN-HMM (1×63)	76.98%	94.27%	95.25%
DNN-HMM(4×20)	77.15%	94.79%	96.11%

6   Conclusion

In this work, we propose a Deep Neural Network-Hidden Markov Model (DNN-HMM) for accurate traffic classification based on packet-level properties (packet arrival time and packet size). The classification scheme takes advantage of DNN's strong representation learning power and HMM's dynamic modeling ability. Moreover, we classify traffic into different content categories, including streaming audio/video, game, and image. We place our classifier in a real-world trace environment, the experiment results show that the proposed technique can achieve a significant performance improvement compared to the conventional GMM-HMM. Our results suggest that two key factors contributing most to the performance improvement of DNN-HMM are: (1) using deep neural networks with sufficient depth; (2) using the information of multiple consecutive packets. We will expand our method by considering more content typologies of traffic and real-time classification for the future work.

References

1. Alberto, D., Antonio, P., Pierluigi, S., et al.: An hmm approach to internet traffic modeling. In: proceeding of IEEE GLOBECOM (2006)
2. Chen, Z., He, K., Li, J., Geng, Y.: Seq2img: A sequence-to-image based approach towards ip traffic classification using convolutional neural networks. 2017 IEEE International Conference on Big Data (Big Data), pp. 1271–1276. IEEE (2017). https://doi.org/10.1109/bigdata.2017.8258054
3. Dainotti, A., De Donato, W., Pescape, A., Rossi, P.S.: Classification of network traffic via packet-level hidden markov models. In: Global Telecommunications Conference, 2008. IEEE GLOBECOM 2008. IEEE. pp. 1–5. IEEE (2008). https://doi.org/10.1109/glocom.2008.ecp.412
4. Finsterbusch, M., Richter, C., Rocha, E., Muller, J.A., Hanssgen, K.: A survey of payload-based traffic classification approaches. IEEE Commun. Surv. Tutor. **16**(2), 1135–1156 (2014). https://doi.org/10.1109/surv.2013.100613.00161
5. LeCun, Y., Bengio, Y., Hinton, G.: Deep learning. Nature **521**(7553), 436 (2015). https://doi.org/10.1038/nature14539
6. Li, W., Moore, A.W., Canini, M.: Classifying http traffic in the new age. In: ACM SIGCOMM 2008 Conference (2008)
7. Madhukar, A., Williamson, C.: A longitudinal study of p2p traffic classification. In: IEEE International Symposium on Modeling, Analysis, and Simulation. pp. 179–188 (2006). https://doi.org/10.1109/mascots.2006.6
8. Mu, X., Wu, W.: A parallelized network traffic classification based on hidden markov model. In: 2011 International Conference on Cyber-Enabled Distributed Computing and Knowledge Discovery (CyberC), pp. 107–112. IEEE (2011)
9. Nguyen, T.T.T., Armitage, G.: A survey of techniques for internet traffic classification using machine learning. IEEE Press (2008)
10. Perera, P., Tian, Y.C., Fidge, C., Kelly, W.: A comparison of supervised machine learning algorithms for classification of communications network traffic. In: International Conference on Neural Information Processing, pp. 445–454. Springer, Berlin (2017). https://doi.org/10.1007/978-3-319-70087-8_47
11. Rabiner, L.R.: A tutorial on hidden markov models and selected applications in speech recognition. Proc. IEEE **77**(2), 257–286 (1989). https://doi.org/10.1109/5.18626
12. Thang, T.C., Le, H.T., Nguyen, H.X., Pham, A.T., Kang, J.W., Ro, Y.M.: Adaptive video streaming over http with dynamic resource estimation. J. Commun. Netw. **15**(6), 635–644 (2013)
13. Wang, W., Zhu, M., Wang, J., Zeng, X., Yang, Z.: End-to-end encrypted traffic classification with one-dimensional convolution neural networks. In: 2017 IEEE International Conference on Intelligence and Security Informatics (ISI), pp. 43–48. IEEE (2017). https://doi.org/10.1109/isi.2017.8004872
14. Wang, Z.: The applications of deep learning on traffic identification. BlackHat USA (2015)
15. Wright, C., Monrose, F., Masson, G.M.: Hmm profiles for network traffic classification. In: Proceedings of the 2004 ACM workshop on Visualization and data mining for computer security. pp. 9–15. ACM (2004). https://doi.org/10.1145/1029208.1029211
16. Yu, D., Deng, L.: Deep neural network-hidden markov model hybrid systems. In: Automatic Speech Recognition, pp. 99–116. Springer, Berlin (2015). https://doi.org/10.1007/978-1-4471-5779-3_6

A New Clustering Protocol for Vehicular Networks in ITS

Mengmeng Liu⬤, Xuming Zeng, Mengyao Tao, and Zhuo Cheng⁽⊠⁾

China University of Geosciences, No. 388 Lumo Road, Wuhan, China
{1227664675,1404347104}@qq.com
{zengxvming,chengzhuo}@cug.edu.cn

Abstract. The dissemination of information is the main application of vehicular networks. The cost of network data transmission is significantly increased and the overall performance of the network routing protocol is deteriorated obviously in the vehicle networks without cluster. In order to ensure the reliability and timeliness of information transmission in VANET, it is necessary to cluster the network in most cases. The existed clustering protocols cannot solve the problem of the stability of the cluster due to the high dynamic mobility in VANET. In this work, we improve a clustering protocol based on the optimization of number of cluster heads, the distance from cluster head to cluster members, and the relative velocity of vehicles within the cluster. The optimized cost function contains three weighting factors for adjusting the specific gravity. The weighting factors can be decided with different demand. Simulation results show that our improved clustering protocol has very good performance for the stability of the cluster. It can provide a good support for the future network routing protocol.

Keywords: Cluster · VANET · Channel allocation

1 Introduction

A vehicular ad hoc network (VANET) is the fusion of the Internet, mobile communication network and the Internet of things, which has a key role in the intelligent transportation systems. In VANET, the idea of clustering is investigated to achieve some of the benefits of an infrastructure-based network without the need for physical infrastructure. Clustering algorithms are proposed to make a hierarchical network structure form in a distributed manner throughout the network. The so-called clustering is to make associating mobile nodes into groups according to some rule set, which is vital for efficient resource consumption and

Supported by CUG2018JM16, Fundamental Research Funds for the Central Universities and SJ-201816, Experimental technology research Funds.

X. Liu et al. (Eds.): ChinaCom 2018, LNICST 262, pp. 615–624, 2019.
https://doi.org/10.1007/978-3-030-06161-6_60

load balancing in large scale networks. Each group is called a cluster which is composed of cluster head (CH) to mediate between the cluster and cluster member (CM). Clustering routing algorithm is based on cluster to achieve node management and data forwarding, which improves routing discovery overhead, broadcast storm and network throughput. The existing clustering algorithms can be classified into four categories: mobility-based, weighted-based, DEA-based or ACS-based clustering, and Location-based clustering protocol.

1.1   Related Work

With the development of the vehicular ad hoc network, the improved protocols from the traditional MANET are difficult to meet VANET?s requirements. There are many routing protocols from different aspects. From the point of clustering protocol, it can be classified into four categories: mobility-based, weighted-based, DEA-based or ACS-based, and Location-based clustering protocol.

(1) Mobility-based clustering protocol: The protocols in this section take relative mobility between a node and its potential cluster heads as cluster head selection parameter, which can be more likely to be able to maintain a stable communication link with small relative velocities when nodes are moving together. Such as [1–4], these algorithms accurately identify nodes showing similar mobility patterns and group them in one cluster.

(2) Weighted-based clustering protocol: In Weighted-based clustering protocol [5–7], cluster head selection for its neighbours is to calculate an index quantifying its fitness. The index, such as the degree of link stability, connectivity, node uptime etc., is a weighted sum of various network metrics. In those protocols, the index representing a node's suitability is based on its entire neighbourhood relationship. When node's connectivity is quite marginal, it may be identified as a best cluster head candidate.

(3) DEA-based or ACS-based clustering protocol: Many DEA-based clustering have been studies in [8–11]. The combination of DEA (Data Envelopment Analysis) and clustering technique is categorized into two methods. The first method is a two-stage algorithm for clustering data by using DEA and a clustering technique such as k-means [9,10]. The second method is to use DEA results for clustering the data [8,11]. ACS-based (Ant Colony System) clustering algorithm is introduced in [12,13].

(4) Location-based clustering protocol: In the designing of routing protocols, considering geographic location information can control the number of cluster heads, reduce the communication cost of inter-cluster and enhance the efficiency of the routing forwarding [14–16]. Reference [15] uses node velocity estimation, introduction of virtual network central node, early warning for cluster head failure and inter-cluster load balancing to form a stable and reasonable clustering structure and achieve the purpose of being easy for frequency planning. Reference [16] is based on Firefly Algorithm and clustering techniques to improve the routing performance under different mobility structure.

1.2   Contribution of This Paper

In this work, a new clustering protocol is proposed to improve the stability and aggregation of network clustering in VANET. Our contribution includes:

(1) The proposed clustering method is related to density. It can be better applied to the different vehicle density environments. The performance of the clustering can well adapt to changes in vehicle density.

(2) The relative velocity of the vehicle is added to the cost function. It can improve the stability of the cluster and provide great stability for network routing protocols over the proposed clustering method.

(3) The weighting factors provide flexible scalability for network routing with different requirements.

2   Network Modeling and Problem Statement

2.1   Network Modeling

In this section, the network model is introduced. The assumptions of considered scenarios are as follows.

(1) We assume that the vehicle is uniformly distributed and its inter-vehicle space is $V_s = N/L$, where N is the total number of nodes and L is the length of highway.

(2) The multi-lane highway scenario is considered as a one-dimensional model.

(3) Each vehicle is equipped with a global positioning system (GPS) to know its own position at any given time.

(4) Each vehicle broadcasts beacon packet every 100 ms which includes its location information.

The straight highway scenario is shown in Fig. 1. There are n vehicles distributed the straight highway with the length of L , denoted as x_i, $i \in \{1, 2, \cdots, n\}$. Each vehicle periodically broadcasts a beacon packet. The probability of p determines that the vehicle can be chosen as a cluster head.

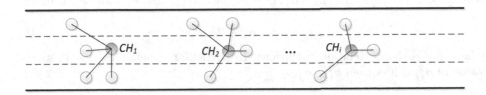

Fig. 1. Demonstration of a clustering scenario.

2.2   Problem Statement

In this work, our problem is to design a clustering protocol that aims to minimize the cost function and lets the nearby nodes divide into the one cluster as much as possible. A better cluster protocol can provide a higher reliability and lower delay for message dissemination.

Let's denote by $C(p)$ the cost function, p is the probability that a node can be chosen as a cluster head. The cost function consists of two parts: the average distance between CH and the distance of a CM to its CH. Therefore, we have:

$$C(p) = e_1 D_{CH}(p) + e_2 D_{CM}(p) + e_3 V(p) \qquad (1)$$

$$C = minC(p), \ for \ p_{opt} \in (0,1) \qquad (2)$$

where p_{opt} is the optimal value of p, e_1, e_2 and e_3 are the weighting factors.

2.3   Distribution of Relative Velocity

The previous studies show that the velocity of the vehicle traveling on the road is subject to the normal distribution $N(\mu, \sigma^2)$ [17]. Since we only focus on a unidirectional highway scenario, the velocity can be calculated as a scalar. Its probability density function (PDF) can be represented as:

$$f(v) = \frac{1}{\sqrt{2\pi}\sigma} e^{-(v-\mu)^2/2\sigma^2} \qquad (3)$$

where v is the vehicle's velocity. In reality, there is a correlation between the velocity of the vehicles. We denote v_i and v_j as any two vehicles' velocities which are jointly normally distributed random variables, then $v_i - v_j$ is still normally distributed. The PDFs of v_i and v_j are $f(v_i) \sim N(\mu_i, \sigma_i^2)$ and $f(v_j) \sim N(\mu_j, \sigma_j^2)$, respectively. So $v_i - v_j$ forms a multivariate normal distribution which can be represented as

$$f(v_{ij}) = \frac{1}{\sqrt{2\pi}\sigma_{ij}} e^{-(v_{ij}-\mu_{ij})^2/2\sigma_{ij}^2} \qquad (4)$$

where $v_{ij} = v_i - v_j$, $\mu_{ij} = \mu_i - \mu_j$. Due to the correlation, σ_{ij} can be calculated as

$$\sigma_{ij}^2 = \sigma_i^2 + \sigma_j^2 + 2\rho_{ij}\sigma_i\sigma_j \qquad (5)$$

where ρ_{ij} is the correlation coefficient. In particular, whenever $\rho_{ij} < 0$, so the value of σ_{ij}^2 is less then $\sigma_i^2 + \sigma_j^2$.

2.4   Clustering Method

To find the most stable vehicles as the cluster heads (CHs), a clustering method is proposed in this section. We assume that the distance to the source of vehicle x is uniformly distributed over the straight highway scenario with length L [18]. p is the probability that a node will be chosen as a CH. N_{CH} represents the average number of CHs. N_{CM} denotes as the average number of CM that will be merged into a CH. Therefore, we have:

$$N_{CH} = np \tag{6}$$

$$N_{CM} = \frac{n}{np} - 1 = \frac{1}{p} - 1 \tag{7}$$

The requirements for clustering are as follows: (i) to have less number of CHs; (ii) to let each CM within a CH be as close to its CH as possible; (iii) to let the velocities of each cluster?s CMs and CH have less variance. According to all of these requirements, a combined cost function is designed that can measure the cost of the number of CHs, the distance of a CM to its CH, and the variance of relative velocity between a CM to its CH. Therefore, the cost function $C(p)$ is:

$$C(p) = e_1 N_{CH} + e_2 \sum_{k=1}^{N_{CH}} N_{CM} E\{|x_{CH} - x_{CM}|^2\} + e_3 \sum_{k=1}^{N_{CH}} E\{|v_i - \mu|^2\} \tag{8}$$

where e_1, e_2, and e_3 are weighting factors to trade off these three requirements. Due to the uniform distribution of x, we can know that $E\{|x_{CH}|\} = L/2$ and $E\{|x_{CM} - x_{CH}|\} = L/(4np)$. As we know, the variance σ_c^2 of velocity of cluster can be calculated as

$$\sigma_c^2 = E\{|v_i - \mu_c|^2\} \tag{9}$$

where μ_c is the average velocity of all the CMs of a CH in a cluster. Its value can be determined by the speed limit of the highway. Then.

$$C(p) = e_1 np + e_2 np \left(\frac{1}{p} - 1\right) \frac{\Delta d^2}{16n^2 p^2} + e_3 np \sigma_c^2 \tag{10}$$

where $e_1 + e_2 = 1$, e_3 is selected by the value of σ_c^2. Let $dC(p)/dp = 0$, we find:

$$p^3 + w_1 p = w_0 \tag{11}$$

where

$$w_1 = e_2 L^2 / (16n^2(e_1 + e_3\sigma_c^2)) \tag{12}$$

$$w_0 = 2w1 \tag{13}$$

Note that the polynomial discriminant $D = (w_1/3)^3 + (w_0/2)^2 > 0$, one root is real and the other two are complex conjugates. In terms of the cubic formula, we have:

$$p_{opt} = \frac{w_1}{3\tau} - \tau \tag{14}$$

where

$$\tau = \sqrt[3]{-w_0/2 + \sqrt{w_0^2/4 + w_1^3/27}} \tag{15}$$

It can be verified that $d^2 C(p)/d^2 p|_{p=p_{opt}} > 0$, which means that $C(p)$ is indeed minimized at p_{opt}, which is the optimal probability for a node to be chosen as a CH.

2.5   Estimation of the Number of Vehicles

The number of vehicles n must be estimated in real time due to the cluster needs to update dynamically with the change of vehicle density. We have assumed that the vehicle is uniformly distributed. So each node can estimate the number of vehicles within the highway in length of L according to transmission range R and the number of vehicles within its range. Let's denote Ψ_t the total number of nodes within a node's transmission range and n_t the total number of nodes within the length of L. We have:

$$n_t = \frac{\Psi_t L}{R}. \tag{16}$$

We can get the value of Ψ_t easily with the help of beacon messages.

Assume that the measurement value of Ψ_t is denoted as $\Psi_t(i)$ during the i-th cluster updating cycle. In terms of Eq. 16, we have:

$$\hat{n}(i) = \frac{\Psi_t(i)L}{R}. \tag{17}$$

2.6   Clustering Protocol

All vehicles in the target area need to periodically broadcast beacon packets to initialize and update clusters. At the beginning, all the vehicles are in the status: $STANDALONE$. After the establishment of the cluster, the cluster head status is CH, the status of cluster member is CM. The process of clustering is described as below:

(1) All nodes broadcast a beacon packet to the whole nodes within its transmission range, which includes its node ID, GPS coordinates, and direction of travel.

(2) After received the beacon packet, each node stores all the information of the neighbor node and forming a the neighbor information table. Each node estimates the number of nodes $\hat{n}(i)$ in the entire network by using the neighbor information table.

(3) Each node will calculate its own optimized p_{opt} value by Eq. 14 in the updating process of clustering with the estimated value of σ_c^2 and $\hat{n}$. This node generates a random probability p between 0 and 1. If the p_{opt} is greater than the generated random probability p, this node turns to the cluster head.

(4) Each node can decide whether or not to enter the updating process by checking the change of the number of nodes within its neighbor table. If the change value $\hat{n}(i) - \hat{n}(i+1)$ is greater than the threshold β_{th}, it will activate the clustering updating process.

(5) If the time since last cluster updating is longer than a predefined constant, activate the cluster updating process.

(6) If a node does not added to any cluster all the time, it remains in the $STANDALONE$ status until the next updating process.

In this protocol, both the number of clusters and the cluster heads are adjusted dynamically with the change of n and σ_c^2.

3   Validation and Simulations

3.1   Simulation Configurations

The simulations are conducted to verify the proposed clustering protocol. The simulator we used is MATLAB version 8.6.

In order to present the performance of the proposed clustering protocol, we design the vehicle that is uniformly distributed in the straight highway with length $L = 2$km. The number of vehicles is $n = 50 \sim 450$. The transmission range is 300m. The range of e_1 is $0 \sim 1$. As a weight factor, it can adjust the weight of the distance between the cluster head and the number of cluster heads. In the analysis of the performance of the clustering protocol, we set up 1000 times runnings to get all of results, and then take its average value. It can obtain the most reliable simulation results. All simulation configuration parameters are shown in the Table 1.

Table 1. Simulation parameters and values

Parameter	Value
Highway length	2 km
Number of lanes	3
Number of vehicles	50~450
Minimum speed	80 km/h
Maximum speed	120 km/h
Transmission range	500 m

3.2   Simulation Results

In order to show the effect of the clustering algorithm, we use MATLAB to describe the distribution of the vehicle, and then use the proposed clustering protocol to get the CHs. CMs are added to the cluster by the principle of nearest joining. Finally, all the CMs and CHs in the same cluster can be connected directly by lines, so that the readers can directly see the effect of the vehicle network clustering. Figure 2(a) shows the vehicle distribution with no clustering, Fig. 2(b) shows the distribution of vehicles after clustering. As it can be seen from the Fig. 2, our proposed clustering protocol has a very good performance. The adjacent the vehicles are divided into one cluster. It is helpful to reduce the cost of data transmission, including reducing the network delay and improving the reliability of transmission. In order to obtain the performance of the proposed clustering protocol in different network size, the numbers of vehicles we choose are $n = 50, 150, 250, 350, 450$, and the range of weighting factor e_1 is $0.98 \sim 1$. We set $e_3 = 0$ that means the relative velocity of the vehicles is not

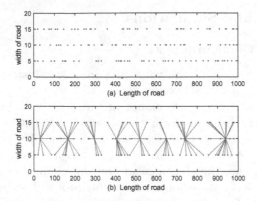

Fig. 2. a Vehicle distribution without clustering. **b** Vehicle distribution with clustering.

considered. After given the values of n and e_1, we can calculate the value of the optimized probability p_{opt} according to Eq. 14. Each node determines whether or not becoming a cluster head by comparing the value of p_{opt} with the random probability. After the CH is determined, other nodes join the cluster by judging the distance from cluster head. The average numbers of CHs in different network size of n are obtained by 1000 simulations. As it can be seen from Fig. 3, when the network size is small, the weighting factor e_1 has small effect on the number of CHs. When the network size is large, the number of CHs is obviously decreasing with increasing of the value e_1. This is in line with the requirements of clustering in the actual vehicle environment, it can not only guarantee the number of clusters in the sparse vehicle network, but also reduce the size of the actual cluster network in dense vehicle network environment. In order to

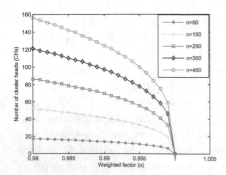

Fig. 3. For the different number of clusters, the values of the weighting factor e_1 have only minor differences when $e_3 = 0$.

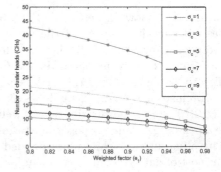

Fig. 4. The number of clusters VS the values of σ_c with $e_3 = 100$.

show the relative velocity of the vehicle on the results of clustering, we choose $\sigma_c = 1, 3, 5, 7, 9$ and $e_3 = 100$. In Fig. 4, as the variance σ_c of the relative velocity of the vehicle within the cluster increases, the number of clusters decreases. In reality, when the relative velocity of the vehicles in the network becomes large, the relative distance of the vehicle will increase. It leads to the requirement for a fast updating of the cluster. By reducing the number of clusters to get a better network topology stability is a good choice.

4 Discussion and Conclusions

In this work, we develop a new clustering protocol for vehicular networks. This clustering protocol is a dynamic clustering protocol. By optimizing the cost function, it obtains the optimal probability of CH. The cost function contains the weight factors, which can be used to adjust the proportion of the two part of the number of cluster heads, the distance of CMs to its CH, and the variance of relative velocity between CMs to its CH. It can achieve the actual needs in different network conditions. The clustering protocol we proposed is adaptive to the different vehicle density and the high dynamic mobility. It ensures that the performance of clustering does not face a significant decline for network routing due to increased vehicle density and the change of velocity.

Our future work would be that we need to find a good method to estimate the number of vehicles in the actual vehicle network and verify the performance and applicability of the clustering protocol by using the actual vehicle traffic data.

References

1. Rawashdeh, Z.Y., Mahmud, S.M.: Toward strongley connected clustering structure in vehicular ad hoc networks. In: 2009 IEEE 70th Vehicle Technology Conference Fall (VTC-Fall), vol. 1–4, pp. 1783–1787. Anchorage, AK, USA (2009)
2. Rawashdeh, Z.Y., Mahmud, S.M.: A novel algorithm to form stable clusters in vehicular ad hoc networks on highways. Eurasip J. Wirel. Commun. Netw. **2012**, 15 (2012). https://doi.org/10.1186/1687-1499-2012-15
3. Daknou, E., Thaalbi, M., Tabbane, N.: Clustering enhancement for VANETs in highway scenarios. In: 2015 5th International Conference on Communications and Networking (COMNET), pp. 1–5 (2015)
4. Kwon, J.H., Kwon, C., Kim, E.J.: Neighbor mobility-based clustering scheme for vehicular ad hoc networks. In: International Conference Platform Technology Service (PlatCon), pp. 31–32. ICT Platform Soc, Jeju, SOUTH KOREA (2015)
5. Daeinabi, A., Rahbar, A.G.P., Khademzadeh, A.: VWCA: an efficient clustering algorithm in vehicular ad hoc networks. J. Netw. Comput. Appl. **34**(1), 207–222 (2011)
6. Vodopivec, S., Beter, J., Kos, A.: A multihoming clustering algorithm for vehicular ad hoc networks. Int. J. Distrib. Sens. Netw. **10**(3), 107085 (2014). https://doi.org/10.1155/2014/107085

7. Hadded, M., Zagrouba, R., Laouiti, A., Muhlethaler, P., Saidane, L.A.: A multi-objective genetic algorithm-based adaptive weighted clustering protocol in VANET. In: 2015 IEEE Congress on Evolutionary Computation (CEC), pp. 994–1002. Sendai, Japan (2015)
8. Thanassoulis, E.: A data envelopment analysis approach to clustering operating units for resource allocation purposes. Omega-Int. J. Manag. Sci. **24**(4), 463–476 (1996)
9. Bojnec, S., Latruffe, L.: Measures of farm business efficiency. Ind. Manag. Data Syst. **108**(1–2), 258–270 (2008)
10. Sharma, M.J., Yu, S.J.: Performance based stratification and clustering for benchmarking of container terminals. Expert. Syst. Appl. **36**(3), 5016–5022 (2009)
11. Po, R.W., Guh, Y.Y., Yang, M.S.: A new clustering approach using data envelopment analysis. Eur. J. Oper. Res. **199**(1), 276–284 (2009)
12. Yang, X.B., Sun, J.G., Huang, D.: A new clustering method based on ant colony algorithm. In: 4th World Congress on Intelligent Control and Automation, pp. 2222–2226., Shanghai, China (2002)
13. Fathian, M., Jafarian-Moghaddam, A.R.: New clustering algorithms for vehicular ad-hoc network in a highway communication environment. Wirel. Netw. **21**(8), 2765–2780 (2015)
14. Edwards, R.M., Edwards, A.: Cluster-based location routing algorithm for inter-vehicle communication. In: IEEE 60th Vehicular Technology Conference, vol. 2, pp. 914–918. Los Angeles, CA (2004)
15. Liu, Z., Ma, Z.X., Shi, R.: Location information based clustering algorithm. Appl. Res. Comput. **28**(12), 4691–4694 (2011)
16. Sachdev, A., Mehta, K., Malik, L.: Design of protocol for cluster based routing in VANET using fire fly algorithm. In: 2016 IEEE International Conference on Engineering and Technology (ICETECH), pp. 490–495. Coimbatore, India (2016)
17. Wang, X., Wang, C., Cui, G., Yang, Q.: Practical link duration prediction model in vehicular ad hoc networks. Int. J. Distrib. Sens. Netw. **11**(3), 216934 (2015). https://doi.org/10.1155/2015/216934
18. Yu, M., David, G.D.: A new traffic aggregation technique based on Markov modulated Poisson processes. In: IEEE Global Telecommunications Conference (GLOBECOM 05), vol. 1–6, pp. 1726–1731. St Louis, MO (2005)

Accelerated Matrix Inversion Approximation-Based Graph Signal Reconstruction

Qian Dang, Yongchao Wang$^{(\boxtimes)}$, and Fen Wang

Collaborative Innovation Center of Information Sensing and Understanding, State Key Laboratory of ISN, Xidian University, Xi'an 710071, Shaanxi, China
qdang@stu.xidian.edu.cn;ychwang@mail.xidian.edu.cn

Abstract. Graph signal processing (GSP) is an emerging field which studies signals lived on graphs, like collected signals in a sensor network. One important research point in this area is graph signal reconstruction, i.e., recovering the original graph signal from its partial collections. Matrix inverse approximation (MIA)-based reconstruction has been proven more robust to large noise than the conventional least square recovery. However, this strategy requires the K-th eigenvalue of Laplacian operator $\mathcal{L}$. In this paper, we propose an efficient strategy for approximating the K-th eigenvalue in this GSP filed. After that, the MIA reconstruction method is modified by this proposed substitution, and thereby accelerated. Consequently, we apply this modified strategy into artificial graph signal recovery and real-world semi-supervised learning field. Experimental results demonstrate that the proposed strategy outperforms some existed graph reconstruction methods and is comparable to the MIA reconstruction with lower numerical complexity.

Keywords: Graph signal processing · Graph reconstruction Semi-supervised learning

1 Introduction

With massive production of irregularly structured signals, graph signal processing (GSP) becomes an overwhelming research filed, which intends to extend classical discrete signal processing tools into graph signal domain [1,2]. Recently developed GSP technologies, like graph-based filtering, sampling and reconstruction on graphs, have been applied into various real-life data analysis, such as transportation network monitoring, semi-supervised learning and recommendation systems [3–5].

Graph signal reconstruction attempts to recovery a smooth graph signal from its partial observed samples (noiseless or corrupted). A noiseless bandlimited

This work is supported by the National Science Foundation of China under Grant 61771356 and in part by the 111 project of China under Grant B08038.

graph signal can be perfectly recovered by its samples if the sample size is larger than its bandwidth [6]. Authors in [7,8] attempted to find out the analogous Nyquist-Shannon sampling theorem for graph signals and proposed some instructive graph sampling and reconstruction strategies. Least square (LS) reconstruction was investigated in [9] for designing a sampling condition for unique recovery. However, the classical LS method requires full eigen-pair decomposition of the graph Laplacian operator and large matrix inversion. Authors in [10] proposed an iterative least squares reconstruction (ILSR) algorithm based on *projection on convex sets theorem* for detouring above complex computations. A generalized ILSR algorithm was proposed in [11] based on frame theory, which has faster convergence rate compared to conventional ILSR algorithm. Authors in [12] designed a robust graph signal recovery strategy via truncated Neumann series, termed as matrix inverse approximation (MIA) reconstruction, which approximates the LS solution but without full eigen-decomposition or matrix inversion. Another approach for recovering a graph signal is regularization, which makes use of the inherited smoothness property of graph signals [13].

In this paper, we focus on designing an efficient strategy for accelerating the existed MIA reconstruction which needs the K-th eigenvalue of Laplacian operator $\mathcal{L}$. We propose an approximation method for computing this K-th eigenvalue in the GSP filed. Then, the MIA reconstruction method is improved by using this proposed substitution. In the sequence, we evaluate the performance of this method by applying it to artificial graph signal recovery and real-world semi-supervised learning field. Simulation results show that the proposed strategy is superior to some existing methods and is comparable to the MIA reconstruction with lower numerical complexity.

The remainder of paper is organized as follows. We first provide the preliminary background and MIA-based reconstruction strategy in Sect. 2 and propose the accelerated MIA graph signal reconstruction algorithm based on fast eigenvalue approximation in Sect. 3. Section 4 presents simulation results. Conclusions are presented in Sect. 5.

2   Notation and Background

A graph is represented as $\mathcal{G} = \{\mathcal{V}, \mathcal{E}, \mathbf{W}\}$, where $\mathcal{V}$ and $\mathcal{E}$ denote the set of nodes and edges, and weight $\mathbf{W}(i,j) = w_{i,j}$ on edge $(i,j) \in \mathcal{E}$ represents the similarity between node i and node j. The degree matrix $\mathbf{D}$ is defined by $\mathbf{D} = \text{diag}(\mathbf{d})$ in which $d_i = \Sigma_j w_{i,j}$. Then, the normalized graph Laplacian matrix can be written as $\mathcal{L} = \mathbf{D}^{-1/2}\mathbf{L}\mathbf{D}^{-1/2}$, which is adopted as the variation operator in this paper [14]. $\mathcal{L}$ is a symmetric and positive semi-definite matrix whose eigenvalues and eigenvectors are $0 = \lambda_1 \leq \lambda_2 \leq ... \leq \lambda_N \leq 2$ and $\mathbf{U} = \{\boldsymbol{u}_1, \boldsymbol{u}_2, ..., \boldsymbol{u}_N\}$ respectively. A graph signal is a function $f : \mathcal{V} \to \mathbb{R}$, which can also be represented as a vector $\mathbf{f} \in \mathbb{R}^N$, where each element represents the function value on its corresponding node. Analogous graph Fourier transform (GFT) of a signal $\mathbf{f}$ is defined as $\hat{\mathbf{f}} = \mathbf{U}^T\mathbf{f}$ and the inverse GFT is $\mathbf{f} = \mathbf{U}\hat{\mathbf{f}}$. A signal is called bandlimited when there exists a number $K \in \{1, ..., N\}$ so that its GFT satisfies

$\hat{\mathbf{f}}_i = 0$, for all $i > K$, and the smallest K is called *bandwidth* of a graph signal $\mathbf{f}$. Graph signals with bandwidth at most K are called K-bandlimited (K-BL) graph signals and can be expressed as $\mathbf{f} = \mathbf{U}_K \hat{\mathbf{f}}_K$, where $\mathbf{U}_K$ means the first K columns of $\mathbf{U}$. We use $\mathcal{S}^c$ to denote the complementary set of $\mathcal{S}$. A restriction of a matrix $\mathbf{A}$ to rows in set $\mathcal{S}_1$ and columns in set $\mathcal{S}_2$ is denoted by the sub-matrix $\mathbf{A}_{\mathcal{S}_1 \mathcal{S}_2}$. $\mathbf{A}_{\mathcal{S}\mathcal{S}}$ is abbreviated as $\mathbf{A}_{\mathcal{S}}$. Moreover, we use $\mathcal{S}_u$ to represent a uniqueness set [15] in this paper, and $\mathcal{S}_r$ to represent a random sampling set. $\mathbf{I}$ is a unit matrix whose dimension is determined by context.

Graph sampling is defined as a linear mapping $\mathbf{f}_{\mathcal{S}} = \mathbf{C}\mathbf{f}$, in which the sampling operator $\mathbf{C} \in \mathbb{F}^{M \times N}$ is [16]

$$\mathbf{C}_{ij} = \begin{cases} 1, & j = \mathcal{S}_i \\ 0, & \text{otherwise} \end{cases} \tag{1}$$

where $\mathcal{S}_i$ is the i-th sampling index, $\mathbb{F}^{M \times N}$ is the set of sampling operators corresponding to all sampling set $\mathcal{S}$ such that $|\mathcal{S}| = M$.

2.1 Matrix Inversion Approximation-Based Reconstruction Strategy

A sampled K-BL graph signal can be written as $\mathbf{f}_{\mathcal{S}} = \mathbf{C}\mathbf{U}_K \hat{\mathbf{f}}_K$ based on previously introduced notations. In noiseless condition, if $\text{rank}(\mathbf{C}\mathbf{U}_K) = K$, a unique and perfect reconstruction $\tilde{\mathbf{f}}$ can be obtained via the LS solution [17]

$$\tilde{\mathbf{f}} = \mathbf{U}_K (\mathbf{C}\mathbf{U}_K)^\dagger \mathbf{f}_{\mathcal{S}} \tag{2}$$

where $(\cdot)^\dagger$ denotes the pseudo-inverse operator.

According to proposition 1 in [12], this LS solution is equal to

$$\begin{aligned} \tilde{\mathbf{f}} &= \mathbf{U}_K [(\mathbf{C}\mathbf{U}_K)^T \mathbf{C}\mathbf{U}_K]^{-1} (\mathbf{C}\mathbf{U}_K)^T \mathbf{f}_{\mathcal{S}} \\ &= \mathbf{U}_K \sum_{l=0}^{\infty} [\mathbf{I} - (\mathbf{C}\mathbf{U}_K)^T (\mathbf{C}\mathbf{U}_K)]^l \mathbf{U}_K^T \mathbf{C}^T \mathbf{f}_{\mathcal{S}} \end{aligned} \tag{3}$$

After a series of derivations of (3), authors in [12] proposed the MIA reconstruction strategy which requires neither full eigen-decomposition nor matrix inversion:

$$\tilde{\mathbf{f}} = \mathbf{T}_{\mathcal{V}\mathcal{S}} \tilde{\mathbf{\Gamma}} \mathbf{f}_{\mathcal{S}} \tag{4}$$

where $\mathbf{T} = \mathbf{U}_K \mathbf{U}_K^T$ and $\tilde{\mathbf{\Gamma}} = \sum_{l=0}^{L} (\mathbf{I}_{\mathcal{S}} - \mathbf{T}_{\mathcal{S}})^l$.

The ideal low-pass graph filter $\mathbf{T}$ has a kernel function as follows

$$h(\lambda) = \begin{cases} 1, \lambda \le \lambda_K \\ 0, \lambda > \lambda_K \end{cases} \tag{5}$$

$h(\lambda)$ can be approximated by a truncated Chebyshev polynomial, thereby $\mathbf{T}$ will be approached by $\mathbf{T}^{\text{Ploy}} = \sum_{i=1}^{N} \left(\sum_{j=0}^{p} \beta_j \lambda_i^j \right) \boldsymbol{u}_i \boldsymbol{u}_i^T = \sum_{j=0}^{p} \beta_j \mathcal{L}^j$ without the information of $\mathbf{U}_K$. It is clear that λ_K is required for realizing the approximation of $\mathbf{T}$, since it is the cut-off frequency of this ideal low-pass filter. Actually, the complexity for calculating λ_K is $\mathcal{O}(N^3)$ in general. If Locally Optimal Block Prec-conditioned Conjugate Gradient (LOBPCG) method [18] is adopted for obtaining λ_K, its complexity is $\mathcal{O}\left((|\mathcal{E}|\,M + TM^3)\,T_1\right)$ [14]. Authors in [12] stated that λ_K can be computed via a series of fast algorithms, under which its complexity will be $\mathcal{O}(RN)$, where $K < R \ll N$. In order to reduce the complexity for obtaining λ_K and simplify the caculation steps, we propose an efficient strategy in this GSP field to approximate λ_K.

3   Accelerated MIA Graph Signal Reconstruction Based on Fast Eigenvalue Approximation

In this section, we propose a simple method to calculate λ_K of $\mathcal{L}$ in the GSP field based on the characteristics of cut-off frequency introduced in [19]. After that, we use the approximated eigenvalue to modify the conventional MIA reconstruction strategy.

3.1   Simple Strategy for Approximating λ_K of $\mathcal{L}$ in the GSP Field

A set $\mathcal{S}_u \subset \mathcal{V}$ is called a uniqueness set for the space $PW_\omega(G)$, any signal $\mathbf{f} \in PW_\omega(G)$ can be perfectly reconstructed from its noiseless samples $\mathbf{f}_{\mathcal{S}_u}$ [15]. Based on this definition and the work achieved in [8], the cut-off frequency of $\mathcal{S}_u$ can be estimated as

$$\omega_c(\mathcal{S}_u) = \lim_{j \to \infty} \Omega_j(\mathcal{S}_u) \approx \Omega_k(\mathcal{S}_u) \triangleq (\sigma_{1,k})^{1/k} \tag{6}$$

where $\Omega_k(\mathcal{S}_u)$ denotes an estimator of the ideal cut-off frequency and $\sigma_{1,k}$ denotes the smallest eigenvalue of the submatrix $(\mathcal{L}^k)_{\mathcal{S}_u^c}$. It is obvious that $\Omega_k(\mathcal{S}_u)$ tends to provide a better estimate with larger k, while the complexity will correspondingly increase. Moreover, as claimed in [19], the exact $\omega_c(\mathcal{S}_u)$ actually can be obtained by characterizing the uniqueness set in a different way, and the following theorem is presented in that paper.

Theorem 1. ([19]) *For a graph $\mathcal{G}$ with normalized Laplacian $\mathcal{L}$ with eigenvalues $0 = \lambda_1 \le \lambda_2 \le ... \le \lambda_N$ and corresponding eigenvectors $\mathbf{u}_1, ..., \mathbf{u}_N$, the cut-off frequency of a subset of nodes $\mathcal{S}_u$ is given by*

$$\omega_c(\mathcal{S}_u) = \max\{\lambda_i : \dim \mathcal{N}[\mathbf{u}_1, ..., \mathbf{u}_i, \mathbf{e}_j : j \in \mathcal{S}_u^c] = 0\} \tag{7}$$

Hence $\mathcal{S}_u$ is a uniqueness set for $PW_\omega(\mathcal{G})$ if and only if $\omega \le \omega_c(\mathcal{S}_u)$.

Let $\mathcal{S}_u = \{1, ..., N\} - \{j_1, ..., j_{N-K}\}$, then $\{\mathbf{e}_j : j \in \mathcal{S}_u^c\} = \{\mathbf{e}_{j_1}, ..., \mathbf{e}_{j_{N-K}}\}$. According to Steinitz exchange lemma, combined with the property that $\mathbf{u}_1, ..., \mathbf{u}_K$ are linearly independent, we can always find $\mathbf{e}_{j_1}, ..., \mathbf{e}_{j_{N-K}}$ from standard basis of $\mathbb{R}^N$ to make $\{\mathbf{u}_1, ..., \mathbf{u}_K, \mathbf{e}_{j_1}, ..., \mathbf{e}_{j_{N-K}}\}$ a basis for $\mathbb{R}^N$. Based on theorem that for a matrix $\mathbf{A} \in \mathbb{R}^{M \times N}$, $\text{rank}(\mathbf{A}) + \dim\mathcal{N}(\mathbf{A}) = \mathrm{N}$, when $i = K$, we can derive

$$\dim\mathcal{N}[\boldsymbol{u}_1, ..., \boldsymbol{u}_i, \boldsymbol{e}_j : j \in \mathcal{S}_u^c]$$
$$= K + (N - K) - \text{rank}(\boldsymbol{u}_1, ..., \boldsymbol{u}_K, \boldsymbol{e}_{j_1}, ..., \boldsymbol{e}_{j_{N-K}})$$
$$= 0$$

This actually means $\omega_c(\mathcal{S}_u) \geq \lambda_K$ based on Theorem 1. In the same way, when $i = K + 1$, we will have

$$\dim\mathcal{N}[\boldsymbol{u}_1, ..., \boldsymbol{u}_i, \boldsymbol{e}_j : j \in \mathcal{S}_u^c]$$
$$= K + 1 + (N - K) - \text{rank}(\boldsymbol{u}_1, ..., \boldsymbol{u}_{K+1}, \boldsymbol{e}_{j_1}, ..., \boldsymbol{e}_{j_{N-K}})$$
$$= 1 \neq 0$$

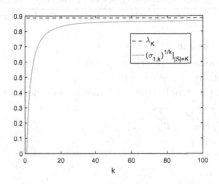

Fig. 1. Experimental result for evaluating equation (9), the estimated cut-off frequency of a random set $\mathcal{S}_r$ with $|\mathcal{S}_r| = 50$ and the exact solution λ_{50} of L.

which implies $\omega_c(\mathcal{S}_u) < \lambda_{K+1}$.

From Theorem 1, we know $\omega_c(\mathcal{S}_u)$ must be one eigenvalue of $\mathcal{L}$. Moreover, we have clarified that $\omega_c(\mathcal{S}_u) \geq \lambda_K$ and $\omega_c(\mathcal{S}_u) < \lambda_{K+1}$, thus leading to an important investigation in the GSP field that $\omega_c(\mathcal{S}_u) = \lambda_K|_{|\mathcal{S}_u|=K}$. In fact, it is empirically illustrated that picking randomly $\boldsymbol{e}_{j_1}, ..., \boldsymbol{e}_{j_{N-K}}$ will always lead $\{\boldsymbol{u}_1, ..., \boldsymbol{u}_K, \boldsymbol{e}_{j_1}, ..., \boldsymbol{e}_{j_{N-K}}\}$ to be a basis of $\mathbb{R}^N$, which implies that a random sampling set $\mathcal{S}_r = \{1, .., N\} - \{j_1, ..., j_{N-K}\}$ can also get this conclusion. Therefore, the exact solution of the cutoff frequency of a random set $\mathcal{S}_r$ with $|\mathcal{S}_r| = K$ can be solved by

$$\omega_c(\mathcal{S}_r) = \lambda_K|_{|\mathcal{S}_r|=K} \tag{8}$$

Equations (6) and (8) actually present an excellent method to compute λ_K in this GSP field. After combining (6) and (8), we propose the following corollary:

Corollary 1. *For a graph $\mathcal{G}$ with normalized Laplacian $\mathcal{L}$ whose eigenvalues are $0 = \lambda_1 \leq \lambda_2 \leq ... \leq \lambda_N$, its K-th eigenvalue λ_K can be approximate by*

$$\lambda_K \approx (\sigma_{1,k})^{1/k}|_{|\mathcal{S}_r|=K} \tag{9}$$

where $\mathcal{S}_r \subset \mathcal{V}$ is randomly selected set with $|\mathcal{S}_r| = K$, $\sigma_{1,k}$ denotes the smallest eigenvalue of the reduced matrix $(\mathcal{L}^k)_{\mathcal{S}_r^c}$. k is an estimation parameter which controls the trade-off between estimation accuracy and computational complexity, as exhibited in (6).

In this paragraph, we perform a numerical experiment to illustrate Corollary 1. We randomly generate a weighted graph with 500 nodes, where the connection probability between each pair is 0.3 and the weights on edges are randomly and independently generated from 0 to 1. Then we randomly select a sample set $\mathcal{S}_r$ such that $|\mathcal{S}_r| = 50$. With the increase of k, the approximate and exact solutions of the eigenvalue converge to a same value. As shown in Fig. 1, the larger k will make the $(\sigma_{1,k})^{1/k}$ closer to λ_{50}.

Algorithm 1 Outline of the proposed A-MIA reconstruction algorithm

---

Input: Graph variance operator $\mathcal{L}$, observed signal $\mathbf{f}_S$, bandwidth K, parameters L and k

Output: Reconstructed graph signal $\tilde{\mathbf{f}}$

1: Randomly select K nodes to constitute a set $\mathcal{S}_r$
2: Approximate λ_K of $\mathcal{L}$ by $\lambda_K = (\sigma_{1,k})^{1/k}|_{|\mathcal{S}_r|=K}$
3: Calculate the truncated Chebyshev polynomial coefficients of $h_a(\lambda)$ in (11) and then compute $\mathbf{T}_a^{\text{Ploy}} = \sum_{j=0}^{p} \beta_j \mathcal{L}^j$
4: Compute $\tilde{\boldsymbol{\Gamma}}_a = \sum_{l=0}^{L} \left[\mathbf{I}_S - (\mathbf{T}_a^{\text{Ploy}})_S \right]^l$
5: Return $\tilde{\mathbf{f}} = (\mathbf{T}_a^{\text{Ploy}})_{\mathcal{V}S} \tilde{\boldsymbol{\Gamma}}_a \mathbf{f}_S$

3.2 A-MIA Reconstruction Algorithm and Complexity Analysis

As we discussed in Sect. 2.1, $\lambda_K(\mathcal{L})$ is required to approximate the ideal low-pass filter $\mathbf{T}$. As discovered in Corollary 1, we can estimate $\lambda_K(\mathcal{L})$ efficiently by Eq. (9). According to (5) and (9), the kernel function can be approximated as

$$h_a(\lambda) = \begin{cases} 1, \lambda \leq (\sigma_{1,k})^{1/k}|_{|\mathcal{S}_r|=K} \\ \\ 0, \lambda > (\sigma_{1,k})^{1/k}|_{|\mathcal{S}_r|=K} \end{cases} \tag{10}$$

where $h_a(\lambda)$ denotes accelerated estimation of the original $h(\lambda)$.

With this new kernel function $h_a(\lambda)$, we can compute its corresponding Chebyshev matrix polynomial $\mathbf{T}_a^{\text{Poly}}$ more efficiently, since the calculation of λ_K is much simpler than the exact computation. Combining (4) and (10), we formally propose the accelerated MIA (A-MIA) reconstruction and illustrate the details of this modified MIA strategy in Algorithm 1.

$$\tilde{\mathbf{f}} = (\mathbf{T}_a^{\text{Ploy}})_{\mathcal{V}S} \tilde{\boldsymbol{\Gamma}}_a \mathbf{f}_S \tag{11}$$

where $\tilde{\boldsymbol{\Gamma}}_a = \sum_{l=0}^{L} [\mathbf{I}_\mathcal{S} - (\mathbf{T}_a^{\text{Ploy}})_\mathcal{S}]^l$.

If the bandlimited graph signal is recovered by LS method, as we can see from Eq. (2), the eigenvector matrix $\mathbf{U}$ is required whose computation complexity is $\mathcal{O}(N^3)$ in general. While the conventional ILSR method doesn't involve eigen-pair decomposition, it is essentially an iterative algorithm whose complexity and performance depends on the required steps for convergence. Both the MIA and the A-MIA algorithm just need to compute λ_K of $\mathcal{L}$. The computational complexity of λ_K in the MIA method is $\mathcal{O}(RN)$ via a series of fast algorithms, where $K < R \ll N$. As described in subsection D in section IV of paper [14], the complexity of computing $(\sigma_{1,k})^{1/k}|_{|\mathcal{S}_r|=K}$, that is $\lambda_{\min}\left[(\mathcal{L}^k)_{\mathcal{S}_r^c}\right]$, is $\mathcal{O}(kN)$ from the Rayleigh quotient perspective, where $k < K$ in general. In experiments, we set $k = 20$ and $K = 50$, which means the A-MIA algorithm is theoretically faster than the MIA method. Hence, we can safely claim that we propose a faster and simplified way for getting λ_K. The performance of this estimator compared to the exact one will be demonstrated in experiments.

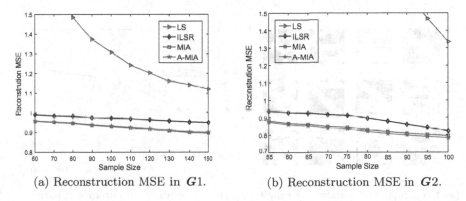

(a) Reconstruction MSE in $\boldsymbol{G}1$.   (b) Reconstruction MSE in $\boldsymbol{G}2$.

Fig. 2. Simulation results for different reconstruction strategies where graph signals are all sampled randomly.

4 Experiments

In this section, we conduct some experiments to evaluate the efficiency and performance of the proposed reconstruction strategy. All experiments were performed in Matlab R2016a, running on a PC with Intel Pentium(R) 2.9 GHZ CPU and 8 GB RAM.

4.1 Artificial Graphs and Graph Signals

For artificial data-related simulations, we use the following models [14]:

Artificial graphs: ($G1$) Erdös-Renyi random graph (unweighted) with 1000 nodes and connection probability 0.01; ($G2$) Unweighted Watts-Strogatz 'small world' model [20] with 1000 nodes, degree 8 and rewriting probability $\rho = 0.1$.

Artificial signals: The true signal is noise-free and approximately bandlimited with an exponentially decaying spectrum. The spectrum GFT coefficients are randomly generated from $\mathcal{N}(1, 0.5^2)$, followed by using the following filter to rescale $h(\lambda) = \begin{cases} 1, & \lambda \leq \lambda_K \\ e^{-4(\lambda - \lambda_K)}, & \lambda > \lambda_K \end{cases}$, where we choose $K = 50$.

Other Parameters: L is set to 10 and k is fixed at 20. The Chebyshev function in the GSP-toolbox package [21] is adopted to realize the Chebyshev polynomials approximation, where p $= 10$ and $\alpha = 8$. Random sampling is used for all reconstruction strategies.

As we analyzed in Sect. 3.2, the complexity of the proposed A-MIA algorithm is theoretically lower than the conventional MIA algorithm. Moreover, experimental results depicted in Fig. 2a, b demonstrate that the A-MIA algorithm is superior to the LS and ILSR algorithm and achieves almost the same performance as the MIA algorithm with lower complexity in both $G1$ and $G2$.

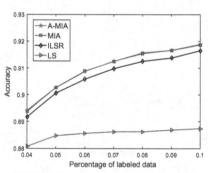

(a) Pixel images of handwritten digits.   (b) Classification accuracy comparison.

Fig. 3. Performance comparison for different reconstruction strategies on the handwritten digits dataset.

4.2   Application in the Semi-supervised Learning Field

We apply the proposed algorithm to a classification task on the USPS handwritten digits dataset [22]. This dataset consists of 1100 pixel images of size 16×16 for each digit 0 to 9. We randomly select 100 samples from dataset for each digit to form a subset which will consist of 1000 feature vectors of dimension 256. For each instance, these selected feature vectors are used to constructed a symmetrized κ-nearest neighbor graph via Gaussian kernel weighting function

$$\mathbf{W}(i,j) = \begin{cases} \exp\left(-\frac{[\text{dist}(i,j)]^2}{2\sigma^2}\right), & \text{if dist}(i,j) \leq \kappa \\ 0, & \text{otherwise} \end{cases} , \text{ where dist}(i,j) = \|\mathbf{f}_i - \mathbf{f}_j\|_2^2,$$

and $\mathbf{f}_i$ is the feature vector composed of pixel intensity values of the i-th image.

We fix parameter $\sigma = 1$ and $\kappa = 10$. The bandwidth K of the graph signal is approximately 50. We adopt the spectral proxies sampling algorithm [8] to select the nodes to label, based on which different recovery strategies are evaluated via reconstruction accuracy.

Figure 3a shows the pixel images of different handwritten digits. Figure 3b compares the classification accuracy of different reconstruction methods in terms of the percentages of labeled data. It depicts that our proposed method has higher accuracy than the conventional LS and ILSR algorithm and achieves almost the same performance as the original MIA reconstruction with theoretically lower complexity. Both the MIA and A-MIA reconstruction algorithms utilize an approximate low-pass filter $\mathbf{T}^{\text{Ploy}}$ which has a slowly decaying spectral kernel. Therefore, they can catch more information of approximately bandlimited graph signals, thus leading to superior performance when applied into real-world tasks.

5   Conclusion

In this paper, we propose an efficient method for approximating the K-th eigenvalue of the Laplacian operator in the GSP filed, and modify the conventional MIA reconstruction strategy by the approximate eigenvalue. Compared with some existed methods, the modified strategy can achieve better performance in both artificial datasets and real-world semi-supervised tasks with lower complexity.

References

1. Shuman, D., Narang, S., Frossard, P., Ortega, A., Vandergheynst, P.: The emerging field of signal processing on graphs: extending high-dimensional data analysis to networks and other irregular domains. IEEE Signal Process. Mag. **30**(3), 83–98 (2013)
2. Sandryhaila, A., Moura, J.: Big data analysis with signal processing on graphs: representation and processing of massive data sets with irregular structure. IEEE Signal Process. Mag. **31**(5), 80–90 (2014)
3. Sakiyama, A., Tanaka, Y., Tanaka, T., Ortega, A.: Efficient sensor position selection using graph signal sampling theory. In: Acoustics, Speech and Signal Processing, pp. 6225–6229 (2016)
4. Gadde, A., Anis, A., Ortega, A.: Active semi-supervised learning using sampling theory for graph signals. In: Proceedings of 20th ACM SIGKDD International Conference on Knowledge Discovery and Data Mining, pp. 492–501 (2014)
5. Chen, S., Sandryhaila, A., Moura, J., et al.: Adaptive graph filtering: multiresolution classification on graphs. In: Proceedings of 2013 IEEE Global Conference on Signal and Information Processing (Global SIP), pp. 427–430. IEEE (2013)
6. Chen, S., Sandryhaila, A., Kovačević, J.: Sampling theory for graph signals. In: 2015 IEEE International Conference on Acoustics, Speech and Signal Processing (ICASSP) (2015)
7. Pesenson, I.: Sampling in paley-wiener spaces on combinatorial graphs. Trans. Am. Math. Soc. **360**(10), 5603–5627 (2008)

8. Anis, A., Gadde, A., Ortega, A.: Towards a sampling theorem for signals on arbitrary graphs. In: 2014 IEEE International Conference on Acoustics, Speech and Signal Processing (ICASSP), pp. 3864–3868 (2014)
9. Narang, S.K., Gadde, A., Ortega, A.: Signal processing techniques for interpolation in graph structured data. In: 2013 IEEE International Conference on Acoustics, Speech and Signal Processing (ICASSP) (2013)
10. Narang, S.K., Gadde, A., Sanou, E., et al.: Localized iterative methods for interpolation in graph structured data. In: Global Conference on Signal and Information Processing (GlobalSIP), pp. 491–494. IEEE (2013)
11. Wang, X., Liu, P., Gu, Y.: Local-set-based graph signal reconstruction. IEEE Trans. Signal Process. **63**(9), 2432–2444 (2015)
12. Wang, F., Wang, Y., Cheung, G.: A-optimal sampling and robust reconstruction for graph signals via truncated neumann series. IEEE Signal Process. Lett. **25**(5), 680–684 (2018)
13. Chen, S., Sandryhaila, A., Moura, J., Kovačević, J.: Signal recovery on graphs: variation minimization. IEEE Trans. Signal Process. **63**(17), 4609–4624 (2015)
14. Anis, A., Gadde, A., Ortega, A.: Efficient sampling set selection for bandlimited graph signals using graph spectral proxies. IEEE Trans. Signal Process. **64**(14), 3775–3789 (2016)
15. Pesenson, I.: Sampling in paley-wiener spaces on combinatorial graphs. Trans. Am. Math. Soc. **360**(10), 5603–5627 (2008)
16. Chen, S., Sandryhaila, A., Kovačević, J.: In: (ICASSP) (ed.), pp. 3392–3396 (2015)
17. Chen, S., Varma, R., Sandryhaila, A., Kovačević, J.: Discrete signal processing on graphs: sampling theory. IEEE Trans. Signal Process. **63**(24), 6510–6523 (2015)
18. Knyazev, A.: Toward the optimal preconditioned eigensolver: locally optimal block preconditioned conjugate gradient method. Siam J. Sci. Comput. **23**(2), 517–541 (2001)
19. Shomorony, H., Avestimehr, A.S.: Sampling large data on graphs. In: 2014 IEEE Global Conference on Signal and Information Processing (GlobalSIP) (2014)
20. Watts, D.J., Strogatz, S.H.: Collective dynamics of small-worldnetworks. Nature **393**(6684), 440–442 (1998)
21. The Graph Signal Processing Toolbox. https://epfl-lts2.github.io/gspbox-html
22. USPS Handwritten Digits Data. https://cs.nyu.edu/~roweis/data.html

A Method of Interference Co-processing in Software-Defined Mobile Radio Networks

RenGui Gao and Dong Zhang[✉]

College of Mathematics and Computer Science Fuzhou University, Fuzhou University,
Fuzhou, Fujian, China
gaorengui186@gmail.com, zhangdong@fzu.edu.cn

Abstract. The intensive network technology that is one of the key technologies of 5G is the main means to solve the explosive growth of data traffic demands in the future. However, with the implementation of network-inte-nsive deployments, serious interference problems are generated. The software-defined network (SDN) allows the separation of the control plane from forward plane, which provides the flexibility of dynamic network programming. Combining SDN with 5G is an effective method to copy with the interference management. This paper proposes an SDN-based mobile wireless network architecture to solve the problem of interference coordination in mobile wireless networks. Through the advantages of the SDN controller, the underlying wireless network topology information is centralized to the control layer that is running the optimization algorithm of resource allocation, which solves the problem of interference coordination. We introduces an Integer Programming to sovle the problem of formulation. A Tabu heuristic algorithm is used to solve the problem of interference coordination. The experimental results show that compared with the algorithm of non interference coordination, the proposed algorithm evidently reduces the interference value of the whole network.

Keywords: Mobile wireless network · Software defined networking
5G · Interference coordination

1 Introduction

With the rapid growth of mobile network services and user data traffic demands, the 5G has been proposed to satisfy the users' the variety of network QoS requirements and the flexible network Service-Level Agreement (SLA) [1]. In order to ensure low latency, high reliability and continuous wide-area coverage (providing 10 Gb/s user experience rate on the premise of guaranteeing user mobility and service continuity) for the 5G communication systems, the deployment of mobile wireless base station (BS) nodes will be more intensive, which results

© ICST Institute for Computer Sciences, Social Informatics and Telecommunications Engineering 2019
Published by Springer Nature Switzerland AG 2019. All Rights Reserved
X. Liu et al. (Eds.): ChinaCom 2018, LNICST 262, pp. 635–644, 2019.
https://doi.org/10.1007/978-3-030-06161-6_62

in a significant increase in the interference of wireless channel [2]. The Mobile Radio Interference (MRI) is an important factor in limiting the network coverage and data transmission in radio communication. In Mobile Wireless Networks (MWN), interference that produces low Signal-to-Interference-plus-Noise Ratio (SINR) seriously affects the efficiency of data transmission between the users, which brings about poor Channel Quality Identifier (CQI). The problem has been studied for decades, but it has been satisfactorily dealt so far. Therefore, MRI is an urgent problem to be solved. At the same time, network operators need to constantly update the system to cope with the explosive growth of new network services, mobile traffic and network application development, which ensure the user's demand for network services. However, the current network architecture and the control functions of assimilation can not meet the requirements of the 5G network services. In the meantime, the network control and management tools of modern telecommunication system should possess extensibility, flexibility and reconfiguration capabilities.

Software Defined Networking (SDN) [3] by the Open Networking Forum seem to be key technology enablers in the direction of meeting requirements such as greater flexiblity and new business models. The high level abstraction of network control and management from the underlying network equipments is actualized by the separation of the data planes and the control. A friendly and programmable northbound interface is defined in a software mechanism to centralize control of the underlying network equipments for achieving network traffic scheduling, which evidently improves the network programmability, flexibility and innovation. The mobile wireless network (MWN) combined with the SDN will augment flexibility, extensibility and automatic configuration ability [4,5]. This combination can be applied to wireless resource scheduling, mobile management and cooperative interference processing (CIP) required by the 5G mobile communication network. The interference minimization algorithm of single BS is constructed by [6], but it does't take into account the problem of multi BSs interference coordination based on MWN.

In this paper, with the problem of CIP in multi BSs and multiuser (MBSMU), we optimize the allocation of the wireless network resources (WNR), which achieve the lowest interference value in the whole MWN. The contribution of this paper is divided into three points. The contributions of this paper are as follows: Firstly, we define a set of MWN-related parameters (such as the transmission power, Modulation and Coding Scheme (MCS) and Resource Block (RB), etc.) that can be exposed to the SDN controller, then the algorithm runs by the SDN controller, which optimize the WNR to reduce the interference of the MWN and improve the efficiency of data transmission. Secondly, we construct an integer programming model for interference co-processing under multi BS and multiuser, which will be solved by the Tabu heuristic algorithm.

Our roadmap for the rest of the paper is as follows. We first describe the related work in Sect. 2. Section 3 we present the system design and architecture. The CIP's model is introduced on Sect. 4. In Section 5, we use Tabu to solve that problem. Section 6 contains results from comparing the performance of Tabu ICP and not Tabu ICP, and Sect. 7 summarizes the paper.

2   Related Work

In the traditional MWN, its resources management is performed in a distributed fashion [5], where each BS has its own decision on network resources. At the same time, 5G MWN will be composed of ultra dense deployments of BSs [7]. The researchers have already identified new issues that interference management has become highly complex. At present, some researches are devoted to combining SDN with MWN to solve that problem. SoftRAN [8] proposes that the BSs be abstracted as a virtual large base station (i.e., to be abstract as a logical centralized control plane), which can implement the abstraction representation of wireless resources (i.e., space, time and frequency slots). RadioVisor [9] proposes that the 3-dimensional wireless resources should be dynamically divided based on the traffic between virtual operators. The sliced radio resources take interference into account. But no a practical scheme is mentioned in the above paper to solve the radio resource control problem.

SoftMobile [10] is the SDN architecture for building heterogeneous mobile networks. In order to solve the interference management problem, the author constructs the development API of the programmable network to realize the combination of the SDN concept and the MWN. However, the papers do not provide specific resource allocation algorithms.

In the literature [11], the concept of a virtual cell (Vcell) is proposed, which is designed to overcome the lack of flexibility and scalability in traditional wireless networks. Representation by means of parameterization of mobile wireless resources such as time, frequency, space and power, the operator can design the resource scheduling algorithm in the controller to optimize the resource allocation. However, the resource allocation algorithm based on interference management has not been actualized.

The paper [6] studies the interference model in the wireless self-organizing network and proposes a graph-based interference model that is ingenious and easy to interfere with the construction of graphs. The authors use the resources of mobile wireless networks, such as transmission links, resource blocks and MCSs as the parameters that are used to calculate the weighted interference collision graph as the input of optimal algorithm scheduling. As the same time, the Integer programming method is put forward with the objective function of minimizing. However, in this paper, only the calculation of the minimum interference generated by all links under a single base station and multi-user is considered, and the calculation method of global interference is not considered in the case of MBSMU.

3   System Design

From the perspective of mobile wireless core network, SDN is viewed as a breakthrough in solving the problems of lack of scalability, inflexibility of deployment, and complex management in existing architectures. As shown in Fig. 1, the key element of the SDN MWN is the open interface between the control layer and

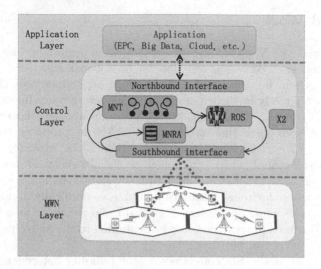

Fig. 1. Software-defined mobile radio access network architecture.

the data layer entity (BSs) and the programmability of the external application
to the network entity (BSs). The main strategy of this architecture is the logical
decoupling of the network from the software-based controller, which can implement network functions through the applications program that is to request and
manipalate the services provided by the network. The architecture of this paper
is designed as three layers: the MWN layer, the control layer and the application
layer.

MWN Layer. In SDN mobile wireless networks, user devices (such as cell-
phones, self-driving cars) exchange data through radio with BSs. The wireless
resources (such as RB, MCR and power) of the BSs will be reported to the controller through the southbound protocol to be aggregated and unified managed
by the controller. When in the coverage of the BSs signal, the access of the user
equipment will trigger the control information from the BSs to the control layer.
Through a series of analysis and calculation, the controller feed back resources
allocation policy to the BSs that finally completes the establishment of the data
transmission channel with the user equipment.

Control Layer. The optimal allocation of wireless network resources under
multi BSs multi-user environment is proposed in this paper to minimize the
total interference in the global network. In order to achieve the above goal,
the optimal allocation of mobile wireless network resources is achieved through
three control modules such as MNRA (Mobile Network Resource Abstration),
MNT (Mobile Network Topology) and ROS (Resource Optimization Scheduler).
Firstly, the MNRA module is an abstract representation of the BSs resources
reported from the Southbound interface, for example, the frequency, power and
MCS usage of the BSs are parameterized as the input data of the optimization
algorithm. This module ensures the optimization algorithm to make accurate

decisions by updating data in real time. Secondly, MNT is a visual module for the underlying network topology. The module with SDN network topology module of the main difference is that MNT transforms quickly, with the movement of the user equipment, which directly affect the optimization algorithm, so the importance of this module is self-evident. Finally, the ROS module is the location of the optimization algorithm in this article. The input of ROS is provided by MNRA and MNT that obtains the transmission distance between BS nodes and user devices. Similarly, the usage of base station node resources is generated through MNRA. The tabu heuristic algorithm is proposed in this paper to solve the problem of optimal resource allocation. In 5G, the amount of user equipments is regionally increasing. In the architecture, to meet the problem of different number of equipment in different regions, we should realize the function of the cluster deployment draw on the traditional X2 interface of LTE. In the control layer, the X2 module can be used to realize the message communication between the control layer and the control layer.

Application Layer. The purpose of separating the application layer from the MWN is to realize flexible control of the network through the unified interface provided by the control layer. By introducing evolved packet core networks (EPCs) [12], cloud computing, big data and other technologies, the innovation of network can be improved. In the LTE architecture, EPCs is respectively composed of the mobility management entity (MME), serving gateway (SGW), and packet data network gateway (PGW).In our architecture, the MME system in the application layer implements the network topology management. When the user moves at a high speed, the MME system need to constantly update the topology. The SGW and PGW are used to implement route forwarding of data packets. The special function of the PGW is to allocate IP addresses, packet filtering and QoS services for the user equipment. The cloud computing and big data in the application layer can provide a deep analysis of the movement direction of user equipments, and the combine the entire mobile network BSs topology to pre-transmit data processing to avoid loss of data packets due to mobility of user equipment.

4   The Interference Coordination Problem

We use the three parameters of the resource blocks, MCSs and power in the MWN to select the transmission channel. We define the network parameters as follows:

$BSs\ M = \{M_1, ..., M_m, ..., M_M\}$,

users per BS $U = \{UE_1, ..., UE_u, ..., UE_U\}$,

$K = \{1, ..., k, ..., K\}$,

$MCSs\ R = \{1, ..., r, ..., R\}$.

In the optimal allocation of mobile wireless network resources discussed in [13], the wireless transmission path of the BS m to the user u can be allocated to the block of the R block, and the method of calculating the interference value produced by the MCS of the K is as follows:

$$\omega_{m,u}^{k,r} = \frac{P_{m,u}^k * \chi_{m,u}}{\rho_r} - \sigma^2 \tag{1}$$

The formulation $\chi_{m,u}$ represents the channel gain of base station m to user u. σ represents the noise density. $P_{m,u}^k$ represents the power value assigned by the base station when the m BS uses the k resource block to communicate with user u. With multiple BSs and multiple users, the problem of optimizing resource allocation for minimizing total interference in MWN can be translated into the following integer linear programming problem:

$$min \sum_{m=1}^{M} \sum_{u=1}^{U} \sum_{k=1}^{K} \sum_{r=1}^{R} \omega_{m,u}^{k,r} * \xi_{m,u}^{k,r} \tag{2}$$

subject to:

$$\sum_{u=1}^{U} \sum_{r=1}^{R} \xi_{m,u}^{k,r} \leq 1 \qquad \forall k \tag{3}$$

$$\sum_{r=1}^{R} \rho_{u,r} \leq 1 \qquad \forall u \tag{4}$$

$$\xi_u^{k,r} \leq \rho_{u,r} \qquad \forall u,k,r \tag{5}$$

$$\sum_{m=1}^{M} \xi_{m,u}^{k,r} = 1 \qquad \forall u \tag{6}$$

$$\sum_{k=1}^{K} \sum_{r=1}^{R} TP_{m,u}^{k,r} * \xi_{m,u}^{k,r} \geq TP_{m,u}^{req} \qquad \forall m,u \tag{7}$$

$$\rho_{u,r} \in \{0,1\} \qquad \forall u,r \tag{8}$$

$$\xi_{m,u}^{k,r} \in \{0,1\} \qquad \forall m,u,k,r \tag{9}$$

Equations (3)–(9) represent the constraint functions for the optimization problem expressed in (2) where the expression for $\omega_{m,u}^{k,r}$ is given in Eq. (1). In this case, $\xi_{m,u}^{k,r}$ (9) is a decision binary variable that is equal to 1 if user u of macrocells m uses MCS r in RB k, or 0 otherwise. Similarly, $\rho_{u,r}$ (8) is a decision binary variable that is equal to 1 if user u make use of MCS r, or 0 otherwise. Constraint (3) makes sure that RB k is only assigned to at most one user u, and Constraints (4) and (5) together guarantee that each user is allocated to at most one MCS. Constraint (7) make sure that each link achieves its throughput demands $TP_{m,u}^{req}$. Equation (6) represents that user u can only belong to base station m.

Algorithm 1 Tabu search algorithm

1: $\pi := \pi_0$, $f_\pi := f(\pi_0)$
2: $\pi_{best} := \pi$, $f_{best} := f_\pi$
3: $T := \phi$
4: $t := 0$
5: *push* π_{best} *into* T
6: **while** $t < time$ **do**
7:     $t := t + 1$
8:     $\pi_{best}^{neigh} := NULL$
9:     $f_{best}^{neigh} := +\infty$
10:    **for each** $\pi' \in N(\pi)$ **do**
11:        **if** $\pi' \notin T$ **then**
12:            $f_\pi := f(\pi')$
13:            **if** $f_{\pi'} < f_{best}$ **then**
14:                $\pi_{best} := \pi'$; $f_{best} := f_{\pi'}$
15:                $\pi_{best}^{neigh} := \pi'$; $f_{best}^{neigh} := f_{\pi'}$
16:                $break$;
17:            **end if**
18:            **if** $f_{\pi'} < f_{best}^{neigh}$ **then**
19:                $\pi_{best}^{neigh} := \pi'$; $f_{best}^{neigh} := f_{\pi'}$
20:            **end if**
21:        **end if**
22:    **end for**
23:    **if** $f_{best}^{ncigh} < f_{best}$ **then**
24:        $\pi := \pi_{best}^{neigh}$; $\pi_{best} := \pi$; $f_{best} := f_{best}^{ncigh}$
25:    **end if**
26:    push π_{best} into T
27:    **if** $T.size() > tabulength$ **then**
28:        remove first element from T
29:    **end if**
30: **end while**

5   Algorithm

In this section, we use a heuristic algorithm to solve the problem of resource allocation in MWN with the minimum system interference in the case of multi BSs and multi users. The Tabu search Algorithm 1 solves all the local optimal solutions by using the evaluation function (i.e., the function f in the algorithm), which is solved by sequential iteration with TIME of the Tabu step to procure the global optimal solution. At the same time, in order to avoid the search process falling into a dead circle, the solution is marked by the Tabu table T. The input of the algorithm is π_0 and the output of the operation is π_{best}. Initially, an initial solution π_0 is defined, and the initial value f_π is obtained by solving the definition f, and the T is empty at the same time. In the algorithm, we define π_{best}^{neigh} and f_{best}^{neigh} record the local optimal solution and its function value in the search process, so that we can easily filter the solution in circulation. In each cycle, the possibility of searching the optimal solution π is carried out in the domain of solution. When the length of the Tabu table is larger than that of the pre defined tabulength, the Tabu table needs to be cleaned to remove the preferences in the table.

6   Performance Evaluation

In this paper, the Tabu algorithm is written through JAVA program code. We use the GT-ITM tool to generate the MNT. The MNT is configured to have 15 to 55 users. To emphasize interference of the number of users, The number of BSs in our experiment is fixed to 5. We proposes three algorithms (i.e., Tabu algorithm for non-interference co-processing (Tabu not ICP), Tabu algorithm for interference co-processing (Tabu ICP), and non-Tabu algorithm for interference co-processing (ICP not Tabu)) to compare the results.

Figure 2 shows that as the number of users increases with the constant BSs, the interference value of the whole network increases significantly, which is sufficient to affect the normal communication of the user. Through the comparison of Tabu not ICP and Tabu ICP algorithm, we can conclude that the interference coordination algorithm greatly reduces the interference value of the whole network. At the same time, Through the ICP not Tabu algorithm and the Tabu ICP algorithm, the Tabu genetic algorithm can effectively reduce the interference value in the case of the same consideration of interference coordination.

In Fig. 3, we verify the power consumption of the algorithm in the whole network. It is obvious that the power consumption of the three algorithms is very close, which is because the basic power values need to be guaranteed that each user is able to send data successfully. The power nonlinearly increases with the increase of the number of users. Tabu ICP reduce the interference value in the prophase, which brings about the data can be sent with less power. However at the end, it is equal to the other two algorithms because the ICP involves several BSs working together, which results in the power consumption is doubled at this stage.

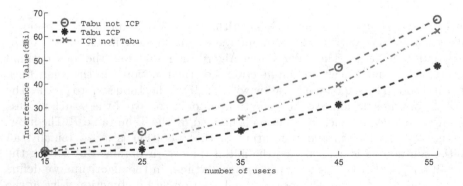

Fig. 2. Interference value

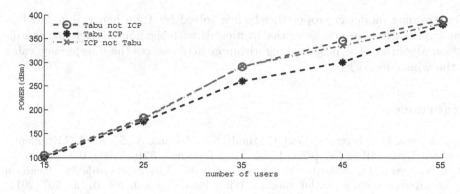

Fig. 3. Power usage

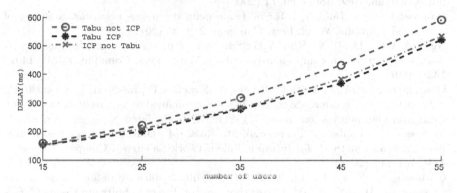

Fig. 4. Delay

The delay shown in Fig. 4 represents the time required for the algorithm to allocate the wireless resources to the user successfully. As we can see in the figure, compared to the other two algorithms, the delay of the algorithm Tabu not ICP is relatively large, which is mainly due to the utilization of the resources of the incomplete network. It only considers whether the resources can be allocated on a single BS. It will be waiting for adequate resources, which leads to the increase of time.

7 Conclusion

This paper mainly addresses the problem of resource allocation in MWN under multi-BSs and multi-users in order to minimize global network interference. With summarizing and expounding the current SDN wireless network, we propose a mobile wireless architecture based on SDN that is applicable to this paper. The three modules (MNRA, MNT and ROS) contained in the control layer are used as the modules of the input and loading algorithm to solve interference coordination problem. In order to solve the above resource allocation problem, an integer programming model is proposed, which is solved by Tabu heuristic algorithm. The experimental results show that compared with the non-interference coordination algorithm, the proposed algorithm greatly reduces the interference value of the whole network.

References

1. Andrews, J.G., Buzzi, S., Wan, C., Hanly, S.V., Lozano, A., Soong, A.C.K., Zhang, J.C.: What will 5g be? IEEE J. Sel. Areas Commun. **32**(6), 1065–1082 (2014)
2. Gavrilovska, L., Rakovic, V., Atanasovski, V.: Visions towards 5g: technical requirements and potential enablers. Wirel. Pers. Commun. **87**(3), 731–757 (2016)
3. Dave, T.: Openflow: enabling innovation in campus networks. Acm Sigcomm Comput. Commun. Rev. **38**(2), 69–74 (2008)
4. Rangisetti, A.K., Tamma, B.R.: Software defined wireless networks: a survey of issues and solutions. Wirel. Pers. Commun. **2**, 1–35 (2017)
5. Nguyen, V.G., Do, T.X., Kim, Y.H.: Sdn and virtualization-based lte mobile network architectures: a comprehensive survey. Wirel. Pers. Commun. **86**(3), 1401–1438 (2016)
6. Gebremariam, A.A., Goratti, L., Riggio, R., Siracusa, D., Rasheed, T., Granelli, F.: A framework for interference control in software-defined mobile radio networks. In: Consumer Communications and NETWORKING Conference, pp. 892–897 (2015)
7. Piovesan, N., Gambin, A.F., Miozzo, M., Rossi, M., Dini, P.: Energy sustainable paradigms and methods for future mobile networks: a survey. Comput. Commun. **119**, 101–117 (2018)
8. Gudipati, A., Perry, D., Li, E., Katti, S.: Softran:software defined radio access network. In: ACM SIGCOMM Workshop on Hot Topics in Software Defined NETWORKING, pp. 25–30 (2013)
9. Gudipati, A., Li, E., Katti, S.: Radiovisor: a slicing plane for radio access networks. In: The Workshop on Hot Topics in Software Defined NETWORKING, pp. 237–238 (2014)
10. Chen, T., Zhang, H., Chen, X., Tirkkonen, O.: Softmobile: control evolution for future heterogeneous mobile networks. IEEE Wirel. Commun. **21**(6), 70–78 (2015)
11. Riggio, R., Gomez, K., Goratti, L., Fedrizzi, R.: V-cell: going beyond the cell abstraction in 5g mobile networks. In: IEEE Network Operations and Management Symposium, pp. 1–5 (2014)
12. Khairi, S., Bellafkih, M., Raouyane, B.: Qos management sdn-based for lte/epc with qoe evaluation: Ims use case. In: Fourth International Conference on Software Defined Systems, pp. 125–130 (2017)
13. Lopez-Perez, D., Ladanyi, A., Juttner, A., Rivano H.: Optimization method for the joint allocation of modulation schemes, coding rates, resource blocks and power in self-organizing lte networks. In: INFOCOM, 2011 Proceedings IEEE, pp. 111–115 (2011)

Detecting Network Events by Analyzing Dynamic Behavior of Distributed Network

Haishou Ma[1], Yi Xie[2(✉)], and Zhen Wang[1]

[1] School of Electronics and Information Technology, Sun Yat-sen University,
Guangzhou 510006, China
[2] School of Data and Computer Science, Sun Yat-sen University, Guangzhou 510006,
China
xieyi5@mail.sysu.edu.cn

Abstract. Detecting network events has become a prevalent task in various network scenarios, which is essential for network management. Although a number of studies have been conducted to solve this problem, few of them concern about the universality issue. This paper proposes a General Network Behavior Analysis Approach (GNB2A) to address this issue. First, a modeling approach is proposed based on hidden Markov random field. Markovianity is introduced to model the spatio-temporal context of distributed network and stochastic interaction among interconnected and time-continuous events. Second, an expectation maximum algorithm is derived to estimate parameters of the model, and a maximum a posteriori criterion is utilized to detect network events. Finally, GNB2A is applied to three network scenarios. Experiments demonstrate the generality and practicability of GNB2A.

Keywords: Behavior analysis · Event detection · Network modeling

1 Introduction

Detecting network events has emerged as a common task in various network scenarios. Wireless Sensor Network (WSN) is applied to different fields including monitoring environment [4] and tracking targets [2], the collected sensed data is often analyzed to find interesting events. The WannaCry ransomware strikes across the globe and worm propagates throughout the mobile communication network lead to a critical problem of detecting malicious events. These diverse network scenarios carry out a uniform task of detecting network events. It watches what's happening to network, identifies the nature of network events, which is of great importance of network management.

This work is supported by the Natural Science Foundation of Guangdong Province, China (No. 2018A030313303), the Fundamental Research Funds for the Central Universities (No. 17lgjc26) and the Natural Science Foundation of China (No. U1636118).

© ICST Institute for Computer Sciences, Social Informatics and Telecommunications Engineering 2019
Published by Springer Nature Switzerland AG 2019. All Rights Reserved
X. Liu et al. (Eds.): ChinaCom 2018, LNICST 262, pp. 645–655, 2019.
https://doi.org/10.1007/978-3-030-06161-6_63

In order to detect network events, a lot of studies have been conducted [1]. These works mostly focus on detecting network events at a single node of network, such as at the border of network [6]. Due to the dynamic characteristics of network event, for example, a worm outbreaks in computer network [8], a single node's information may not be sufficient to identify events clearly. To overcome the limitation of single node's detection scheme, data fusion is considered in some researches, which collecting a collection of nodes' data to detect events [3,5]. However, the literatures review only considers the data features of network events, the correlation characteristics between network events and time continuity of network events are rarely applied. Moreover, the works mentioned in the literatures are applied to a specific network scenario instead of a general approach to deal with the uniform problem of detecting network events.

To solve limitations of the literatures review, this work is motivated to propose GNB2A to detect network events, which utilizes the correlation between network events and time continuity of network events. In GNB2A, a network is divided into three layers: topology layer is the actual network topology, event layer denotes the network events occur in network and behavior layer describes the external behavior of network nodes that is driven by underlying network events. Since the network events can not be measured directly, but the external behavior feature of network is measurable, therefore, event layer can be inferred from the behavior layer, the detecting task is map to an inference problem, and the goal is to infer the current event types of network nodes through measurable behavior feature. A two-layer mathematical model is introduced to model this inference problem, an observable random field denotes the measurable feature of behavior layer, and an unobservable Markov random field that describes the underlying event layer. In this work, an expectation maximum (EM) algorithm is applied to estimate parameters of model from the training data, and a maximum a posteriori (MAP) criterion is utilized to infer the event layer.

Contributions of this work include:

- Proposing GNB2A by leveraging hidden Markov random field, a spaio-temporal context is introduced to model the correlation of network events.
- Deriving algorithms for parameter estimation and network events detection.
- Evaluating the performance of GNB2A by designing three network scenarios in a simulated environment.

The rest of this paper is organized as follows. Section 2 describes GNB2A, the modeling approach and algorithms of this work are provided. In Sect. 3 experiments are designed to validate GNB2A. Finally, Sect. 4 includes the conclusion to this paper.

2   The Proposed Approach

2.1   General Network Behavior Analysis Approach

In most communication networks, a network node's behavior depends on the event it is currently encountering. For instance, a host may forward a large number of packets to a specific destination when it is involved in a DDoS attack. Here, "forward a large number of packets to a specific destination" is a network node's behavior, while the "DDoS attack" is the event that is affecting the node. This phenomenon is common in many network scenarios. A three-layer model is used to describe the relationship of (Node, Event, Behavior). Extending to distributed scenarios, it forms a general model, as shown in Fig. 1. Hence in GNB2A, a network is divided into three layers: topology layer, event layer and behavior layer. Topology layer is network topology consists of network nodes connected by links. Event layer denotes network events occur in network, the event occurs in a node also called "hidden state" in the following, since it cannot be observed by measurement directly. Behavior layer describes the external behavior of network which is driven by the underlying network events, the behavior also called "observation" in the following since it can be measured directly.

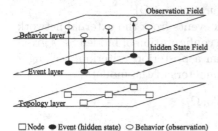

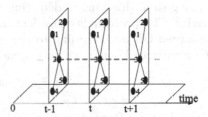

Fig. 1. The framework of GNB2A.

Fig. 2. Spatio-temporal context of network.

Due to the connectivity of network and the time continuity of events, an event E occurred on a node N is not only affected by N's neighbors, but also closely related to N's previous moment. To simply the modeling, this work only consider the impact of one-hop nodes, which is shown in Fig. 2, at time t, node 3 is influenced by its neighboring nodes' $\{1, 2, 4, 5\}$ events and previous event at time $t - 1$. This interdependent phenomenon is called "spatio-temporal context" in this paper.

Based on the above modeling approach in GNB2A, the event layer and behavior layer of a network can be considered as a double-layer random field: a hidden State Field (SF) and an Observation Field (OF), respectively, as shown in Fig. 1. Hidden SF consists of the events of each node in network, OF represents the measurements of external behavior feature of each node in network. Thus, detecting network events from network external behavior can be mapped to infer the hidden SF from a measured OF. In the follows, how to model the relationship between hidden SF and OF and how to infer hidden SF based on a measured OF are introduced in detail.

2.2   Formulation of the Model

In network layer with N nodes, $\mathbb{N}$ denotes the set of nodes, $x_{t,n} \in \mathbb{E}$ denotes the $n^{th}(n \in \mathbb{N})$ node appearing at the tth time slot, where $\mathbb{E}$ includes the collection of all $x_{t,n}$ for $\forall(t, n)$, and $|\mathbb{E}| = |\mathbb{N}| \times T$, where T denotes the number of time slot. Let $S_{t,n}$ denote the random variable of hidden state of $x_{t,n}$, i.e., the type of network event, and $s_{t,n} \in \mathbb{S}$ is an instance of $S_{t,n}$, where $\mathbb{S}$ is the set of all possible states. Then $S = \{S_{t,n} | \forall t \in [1, T], \forall n \in \mathbb{N}\}$ is a family of random variables defined on the set $\mathbb{E}$. Thus S can be used to describe the hidden SF, and $s \in \mathcal{S}$ denotes a configuration of S, where $\mathcal{S}$ is the set of all possible configurations of hidden SF. Use similar expressions, Let $O_{t,n}$ denote the random variable of observation of $x_{t,n}$, i.e., the behavior feature of network node, and $o_{t,n} \in \mathbb{O}$ is an instance of $o_{t,n}$, where $\mathbb{O}$ is the set of all possible observations. Then $O = \{O_{t,n} | \forall t \in [1, T], \forall n \in \mathbb{N}\}$ is a family of random variables defined on the set $\mathbb{E}$. Thus O can be used to describe the OF, and $o \in \mathcal{O}$ denotes a configuration of O, where $\mathcal{O}$ is the set of all possible configurations of OF.

The goal of this work is detecting network events from the external network behavior. This problem is equivalent to infer a configuration of hidden SF s given a measured OF o. According to the maximum a posteriori criterion, seeking $\hat{s}$ given o satisfies Eq. (1), where Ω denotes parameters of the model.

$$\hat{s} = \arg\max_{s \in \mathcal{S}}\{\Pr[s|o, \Omega]\} \tag{1}$$

Based on the Bayes theorem in Eq. (2):

$$\Pr[s|o, \Omega] \propto \Pr[o|s, \Omega] \cdot \Pr[s|\Omega], \tag{2}$$

The likelihood probability $\Pr[o|s, \Omega]$ in Eq. (2) describes the relationship between OF and hidden SF. In this work the conditional independent assumption is adopted to make the model solvable and tractable. Hence, $\Pr[o|s, \Omega]$ can be calculated by Eq. (3):

$$\Pr[o|s, \Omega] = \prod_{(t,n)} \Pr[o_{t,n}|s_{t,n}, \Omega]. \tag{3}$$

For the typical Gaussian distribution, the random variables of OF have the following probability density functions:

$$a_m(o_{t,n}) = \Pr[O_{t,n} = k|S_{t,n} = m, \theta_m] = \frac{1}{\sqrt{2\pi\sigma_m^2}}\exp(-\frac{(k - \mu_m)^2}{2\sigma_m^2}), k \in \mathbb{O}, m \in \mathbb{S}, \tag{4}$$

with the parameters $\theta_m = (\mu_m, \sigma_m)$, and k, m are the values of observation and hidden state, respectively. Prior probability $\Pr[s|\Omega]$ in Eq. (2) describes the interaction of hidden states between network nodes. Pseudolikelihood and first-order Markovianity are utilized to simplified the joint probability:

$$\Pr[S = s|\Omega] \simeq \prod_{(t,n)} \Pr[s_{t,n}|s_{\mathbb{N}_{t,n}^S}, s_{\mathbb{N}_{t,n}^T}, \lambda], \tag{5}$$

where $\mathbb{N}_{t,n}^S$ denotes the spatial neighboring nodes of node $x_{t,n}$, $\mathbb{N}_{t,n}^T$ denotes the corresponding one-hop temporal neighbor of node $x_{t,n}$. Based on the Hammersley-Clifford theorem, the partial probability can be written as

$$b_{t,n}(m) = \Pr[s_{t,n}|s_{\mathbb{N}_{t,n}^S}, s_{\mathbb{N}_{t,n}^T}, \lambda] = \frac{1}{Z_{t,n}(\lambda)} \exp(-U_{t,n}(m|\lambda)), m \in \mathbb{S}, \quad (6)$$

where $Z_{t,n}(\lambda) = \sum_{m \in \mathbb{S}} \exp(-U_{t,n}(m))$ and $U_{t,n}(m)$ are the marginal partition function and marginal energy function, respectively. And $U_{t,n}(m)$ has the form

$$U_{t,n}(m) = \varepsilon_{t,n} \sum_{n' \in \{\mathbb{N}_{t,n}^S, \mathbb{N}_{t,n}^T\}} V_{t,n}(m, s_{n'}), \quad (7)$$

where $\varepsilon_{t,n} = 1/(|\mathbb{N}_{t,n}^S| + |\mathbb{N}_{t,n}^T|)$ denotes the normalized factor of node energy, $|\mathbb{N}_{t,n}^S|$ and $|\mathbb{N}_{t,n}^T|$ are the number of spatial and temporal neighbor of node $x_{t,n}$, respectively. Potential function $V_{t,n}(m, s_{n'})$ defined by

$$V_{t,n}(m, s_{n'}) = \begin{cases} 0, & (s_{n'} = m) \\ \beta, & (s_{n'} \neq m) \end{cases}, n' \in \{\mathbb{N}_{t,n}^S, \mathbb{N}_{t,n}^T\}, \quad (8)$$

where β denotes the parameter correspond to the pairwise interactions between two nodes.

2.3   Algorithm

Infer hidden SF according to the MAP criterion satisfies Eq. (1), the pseudocode is shown in **Algorithm 1**.

Algorithm 1 Infer Hidden SF Algorithm

1: **function** Infer_Hidden_SF(o, Ω)
2: **Initialize** : $s^{(0)}$;
3: **for all** $x_{t,n} \in \mathbb{E}$ **do**
4:     **for all** $m \in \mathbb{S}$ **do**
5:         $a_m(o_{t,n}) = \Pr[O_{t,n} = k|S_{t,n} = m, \theta_m]$;
6:         $b_{t,n}(m) = \Pr[S_{t,n} = m|s_{\mathbb{N}_{t,n}^S}, s_{\mathbb{N}_{t,n}^T}, \lambda]$;
7:         $\xi_{t,n}(m) = a_m(o_{t,n})b_{t,n}(m)$;
8:     **end for**
9:     $s_{t,n} \leftarrow \arg\max_{m \in \mathbb{S}}\xi_{t,n}(m)$;
10: **end for**
11: $\forall x_{t,n} \in \mathbb{E} : \hat{s}_{t,n} \leftarrow s_{t,n}$;
12: **return** $\hat{s}$;
13: **end function**

Input of the algorithm are the observation of network behavior o and parameter of model Ω, output is the hidden state of the network. In the initialization process (2nd line), $s^{(0)}$ can be obtained by prior knowledge on OF and hidden SF. For every node in network (3rd line), algorithm traverses all the potential hidden states (4th line), then choose the state that has maximum probability as infer result (9th line). Note that the probability of a potential hidden state includes two parts based on Eq. (2), the first part is likelihood probability in Eq. (4), denoted by $a_m(o_{t,n})$ in algorithm (5th line), and the second part is partial prior probability in Eq. (6), denoted by $b_{t,n}(m)$ in algorithm (6th line).

Similar to most machine learning-based applications, parameter learning is required before using model. This work uses EM algorithm to estimate parameters. The core of EM algorithm is Q function, which is defined by

$$Q(\Omega|\Omega^{(i)}) = E_s\{\ln \Pr[o, s|\Omega]|o, \Omega^{(i)}\}, \tag{9}$$

where $\Omega^{(i)}$ and Ω denote the parameter sets obtained in the ith iteration and to be estimated in the $(i+1)^{th}$ iteration, respectively.

A computable form of Q function of this work is shown in Eq. (10), where $Q_A(\mu_m^{(i+1)}, \sigma_m^{(i+1)})$ and $Q_B(\beta^{(i+1)})$ denote the first term and the second term on the right-side of the second equal sign, respectively. Then the model's parameters can be estimated by maximizing $Q_A(\mu_m^{(i+1)}, \sigma_m^{(i+1)})$ and $Q_B(\beta^{(i+1)})$ independently since they are not related.

$$
\begin{aligned}
Q(\Omega|\Omega^{(i)}) &= E_s\{\ln \Pr[o, s|\Omega]|o, \Omega^{(i)}\} \\
&= \sum_{m \in \mathbb{S}} \sum_{t,n} \Pr[S_{t,n} = m|O_{t,n} = k, \Omega^{(i)}] \cdot \ln \Pr[O_{t,n} = k|S_{t,n} = m, \Omega] + \\
&\quad \sum_{m \in \mathbb{S}} \sum_{t,n} \Pr[S_{t,n} = m|O_{t,n} = k, \Omega^{(i)}] \cdot \ln \Pr[S_{t,n} = m|\Omega] \\
&= Q_A(\mu_m^{(i+1)}, \sigma_m^{(i+1)}) + Q_B(\beta^{(i+1)})
\end{aligned} \tag{10}
$$

In this work, parameter β is obtained by empirical approach, and the parameters $\Omega = \{\mu_m, \sigma_m\}, m \in \mathbb{S}$ can be estimated by by maximizing Q_A, and obtained by the following equation:

$$
\begin{cases}
\mu_m^{(i+1)} = \dfrac{\sum_{t,n} \Pr^{(i)}[s_{t,n}|o_{t,n}]o_{t,n}}{\sum_{t,n} \Pr^{(i)}[s_{t,n}|o_{t,n}]} \\
(\sigma_m^{(i+1)})^2 = \dfrac{\sum_{t,n} \Pr^{(i)}[s_{t,n}|o_{t,n}](o_{t,n} - \mu_m^{(i+1)})^2}{\sum_{t,n} \Pr^{(i)}[s_{t,n}|o_{t,n}]}.
\end{cases} \tag{11}
$$

Algorithm 2 Parameter Estimation Algorithm

1: **function** Parameter_Estimation(o)
2: **Initialize** : $i \leftarrow 0, \Omega^{(i)} \leftarrow \{\mu_m^{(i)}, \sigma_m^{(i)}, \forall m \in \mathbb{S}\}, \mathcal{L}^{(i)} \leftarrow 0, C_{em}$;
3: **repeat**
4:      $s^{(i)} \leftarrow$ Infer_Hidden_SF($o, \Omega^{(i)}$);
5:      $\{\hat{\mu}_m, \hat{\sigma}_m\} \leftarrow$ Update($Q_A, \Omega^{(i)}, s^{(i)}$), $\forall m \in \mathbb{S}$;
6:      $i \leftarrow i + 1$;
7:      $\Omega^{(i)} \leftarrow \{\hat{\mu}_m, \hat{\sigma}_m, \forall m \in \mathbb{S}\}$;
8:      $\mathcal{L}^{(i)} \leftarrow \sum_m \sum_{t,n} \ln \xi_{t,n}(m|s^{i-1}, \Omega^{(i)})$;
9: **until** $|\mathcal{L}^{(i)} - \mathcal{L}^{(i-1)}| \leq C_{em}$
10: $\hat{\Omega} \leftarrow \Omega^{(i)}$;
11: **return** $\hat{\Omega}$;
12: **end function**

Pseudocode of Parameter Estimation Algorithm is shown in **Algorithm 2**. Input of the algorithm is historical observation of network nodes, i.e. the training data of the model, output are the parameters of model. Parameters $\{\hat{\mu}_m, \hat{\sigma}_m\}$ update (5th line) based on Eq. (11). C_{em} denotes the given convergence condition for the iteration of algorithm, which measures the fitting degree of the model to the training data. To control the iteration process of algorithm, let $\mathcal{L} = \sum_s \ln \Pr[o, s|\Omega]$ denote the overall logarithmic likelihood.

3   Experiment

In this section, three network scenarios are designed to validate GNB2A. The scenarios are selected to demonstrate GNB2A is suitable from a wireless physical network to a logical connected network.

In WSN scenario, GNB2A is applied to detect events in the environment from the unreliable sensed data. The gradual depletion of the sensors' energy or sensors are compromised by attacker, the information transmitted by the sensors is inevitably subject to a degree of unreliability and error. In this experiment, WSN monitoring environmental variables (temperature here), 1000 sensor nodes are randomly distributed in the environment, and the base temperature data originates from the Intel Berkeley Research Lab[1]. A Gaussian noise is randomly added to the temperature data to model external disturbance. Neighboring nodes of node k consist of all nodes that are within a distance d from node k.

Events in this scenario defined as {high temperature, medium temperature, low temperature}, i.e., the hidden states of nodes. The observation of a node is the sensed data, due to the unreliability of sensed data, the real event may not be perceived directly from the threshold based approach, as shown in Fig. 3(a), three states are mixture in the environment. When applied GNB2A, the events are detected, and the environment are divided into three temperature regions, detection result is shown in Fig. 3(b). In this work, Accurate Rate and Macro

---

[1] http://db.csail.mit.edu/labdata/labdata.html.

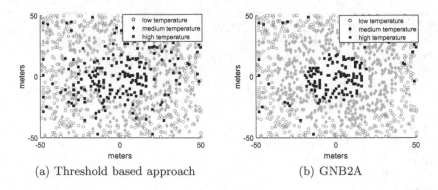

(a) Threshold based approach   (b) GNB2A

Fig. 3. Comparison of two approaches for detecting temperature events.

F1 are selected to evaluate the performance, Accurate Rate is the fraction of all correctly estimated state instances to all instances, Macro F1 is the arithmetic mean value of F1, the higher evaluation metric represents the better performance. Performance comparison is shown in the Table 1, it indicates that GNB2A outperforms the threshold based approach in WSN scenario.

Table 1. Performance comparison

Scenario	Approach	Accurate rate	Macro F1
WSN	Threshold based	72.6%	71.1%
	GNB2A	92.7%	91.6%
Internet	Kmeans	90.1%	89.2%
	GNB2A	96.7%	96.0%
SN	Kmeans	89.6%	89.6%
	GNB2A	96.6%	96.6%

In training process, training data originates from the historical sensed data. The parameters are convergent after 9 or 10 iterations, as shown in Fig. 4, it indicates algorithm converges quickly. And the parameters $\Omega = \{\mu_m, \sigma_m\}, m \in \mathbb{S}$ in this scenario describe the attributed of event. From a statistical point of view, observations can be expressed as a Gaussian mixed model, three Gaussian distribution represent three types of temperature event, as shown in Fig. 5. Due to the unreliability of sensed data, the sensed data of different temperature event are overlapping, it can not be distinguished accurately by a threshold based approach. By utilizing spatio-temporal context of network, GNB2A can achieve better performance in detecting events.

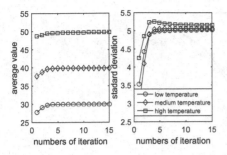

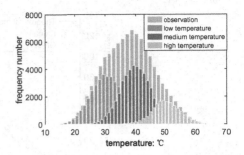

Fig. 4. Convergence of parameters.

Fig. 5. Distribution of observations.

In Internet scenario, GNB2A is applied to detect network state during a DDoS attack. DDoS attack is simulated in MATLAB. A small world network with 128 nodes and 257 links is generated, some nodes are chosen as botnet, sending packets to the victim simulates a SYN flood attack. The observation of a node in this scenario is entropy of destination IP address and the arrival rate of packets. Network events define as abnormal status of the traffic passing the network node, three discrete states are used to describe the abnormal status of traffic during a DDoS attack, i.e. $\{S_1, S_2, S_3,\}$, S_1 denotes the low abnormality, and S_3 denotes high abnormality.

The training data comes from synthetic traffic data, convergence of parameters and distribution of observations are similar to WSN scenario. Table 1 shows the performance comparison, it indicates that GNB2A outperforms Kmeans approach. GNB2A can detect the abnormality of network nodes effectively. The reason for the performance gain is that GNB2A combines spatial and temporal neighbors' states, which gets more information to detect network events.

In Social Network (SN) scenario, GNB2A is applied to detect the spammers. Considering a Short Message Service (SMS) worm propagates in social network, when a user gets infected (spammer), it sends SMS spam to others, and meanwhile the worm propagates via user's contact list. In this scenario, a scale free network with 128 nodes and 253 links is built, SMS worm propagation is modeled by Susceptible-Infected (SI) model and a hierarchical infection probability is used. The features of user according to the previous analysis [7], observation of "average SMS text length" is selected, the hidden states of a user are defined as {spammer, non-spammer}. Apply GNB2A, the worm outbreak is discovered by inferring from users' behavior.

Convergence of parameters and distribution of observations are similar to WSN scenario. In test process, when applied Kmeans approach, it may not be sufficient to estimate whether a user is spammer or not based on the user's behavior feature. While GNB2A combines user's previous and neighboring users' information to detect spammers, it gets more information and therefore better performance, Table 1 shows the performance comparison. GNB2A can tract the

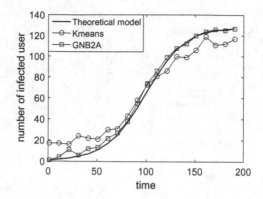

Fig. 6. The propagation of social network worm.

dynamic propagation of worm, as shown in Fig. 6, GNB2A is approximate to theoretical SI model in comparison to Kmeans approach. The result shows that GNB2A outperforms the Kmeans approach.

4 Conclusion

This work focuses on detecting network events, GNB2A is proposed to solve this problem, a hidden Markov random field is introduced to model the relationship between behavior layer and event layer of network, correlation characteristics between network events and time continuity of events are modeled by first-order Markovianity. An expectation maximum algorithm is derived to estimate parameters, and a maximum a posteriori criterion is utilized to detect network events. Experimental results demonstrate the generality and practicability of GNB2A.

References

1. Buczak, A.L., Guven, E.: A survey of data mining and machine learning methods for cyber security intrusion detection. IEEE Commun. Surv. Tutor. **18**(2), 1153–1176 (2016). https://doi.org/10.1109/COMST.2015.2494502
2. Demigha, O., Hidouci, W.K., Ahmed, T.: On energy efficiency in collaborative target tracking in wireless sensor network: a review. IEEE Commun. Surv. Tutor. **15**(3), 1210–1222 (2013). https://doi.org/10.1109/SURV.2012.042512.00030
3. Khaleghi, B., Khamis, A., Karray, F.O., Razavi, S.N.: Multisensor data fusion: a review of the state-of-the-art. Inf. Fusion **14**(1), 28–44 (2013). https://doi.org/10.1016/j.inffus.2011.08.001, http://www.sciencedirect.com/science/article/pii/S1566253511000558
4. Othman, M.F., Shazali, K.: Wireless sensor network applications: a study in environment monitoring system. Procedia Eng. **41**, 1204–1210 (2012)

5. Ramaki, A.A., Amini, M., Atani, R.E.: Rteca: real time episode correlation algorithm for multi-step attack scenarios detection. Comput. Secur. **49**, 206–219 (2015). https://doi.org/10.1016/j.cose.2014.10.006, http://www.sciencedirect.com/science/article/pii/S0167404814001527
6. Wu, S., Liu, S., Lin, W., Zhao, X., Chen, S.: Detecting remote access trojans through external control at area network borders. In: 2017 ACM/IEEE Symposium on Architectures for Networking and Communications Systems (ANCS), pp. 131–141 (2017). https://doi.org/10.1109/ANCS.2017.27
7. Xu, Q., Xiang, E.W., Yang, Q., Du, J., Zhong, J.: SMS spam detection using non-content features. IEEE Intell. Syst. **27**(6), 44–51 (2012). https://doi.org/10.1109/MIS.2012.3
8. Zhou, C.V., Leckie, C., Karunasekera, S.: A survey of coordinated attacks and collaborative intrusion detection. Comput. Secur. **29**(1), 124–140 (2010). https://doi.org/10.1016/j.cose.2009.06.008, http://www.sciencedirect.com/science/article/pii/S016740480900073X

Misbehavior Constraint MAC Protocol (MC-MAC) for Wireless Networks

Yupeng Ma, Yonggang Li$^{(\boxtimes)}$, Zhizhong Zhang, and Haixing Li

School of Communication and Information Engineering, Chongqing University
of Posts and Telecommunications, Chongqing, China
lyg@cqupt.edu.cn, {ma-yupeng, li_haixing}@foxmail.com

Abstract. The IEEE 802.11 protocol assumes that all wireless network nodes will abide by the protocol and cooperate well with it. However, in order to obtaining more channel resources or destroying network performance, some selfish nodes will be in misbehaviors when they are in the certain condition wireless contention-sharing channels, such as, Backoff Value Manipulation is a kind of misbehavior. And for this kind of misbehavior, this paper proposes a Misbehavior Constraint MAC protocol (MC-MAC), which can detect and penalize the backoff value manipulation, and it includes a new backoff value generating function with penalty function and a reputation model. Simulation experiments shows that the MC-MAC protocol has a significant inhibitory effect on misbehavior and can improve system throughput.

Keywords: IEEE 802.11 · Medium access control · Misbehavior constraint · Backoff value manipulation

1 Introduction

The IEEE 802.11 Medium Access Control (MAC) protocol, which at the basis of the Distributed Coordination Function (DCF) mechanism, is the most commonly used MAC protocols in current wireless network. When nodes access to wireless channel resources, they must follow the fairness and trust in the certain wireless sharing channels of a distribution network condition. However, there are some nodes will be in misbehavior that do not comply with the wireless network protocol rules. In addition, due to the great programmability of the network adapter (mobile base station), it is much easier for bad nodes to change the parameters of MAC protocols and achieve selfish or malicious purposes.

Nowadays many researches are focusing on MAC layer misbehaviors. The research in [1, 2] analyzes the greedy receivers misbehavior. And this misbehavior

This work is supported by the Defence Advance Research Foundation of China under Grants 61401310105 and the Chongqing Research Program of Basic Research Frontier Technology (No. cstc2017jcyjA1246).

is mainly reflected in the traffic that received by selfish nodes is much larger than sending. In research [2], the writer determines the scope of influence of greedy receiving nodes and quantifies the harm of greedy receiver misbehavior by using simulations and tests. The result is that the greedy receiving nodes will cause the nodes which affected by them to receive none traffic.

RTS/CTS (Request to Send/Clear to Send) DOS attack is also a kind of misbehaviors. The principle of RTS/CTS DOS attack is that making competing nodes set a longer Network Allocation Vector (NAV) by tampering with the duration field of the RTS/CTS control frame. The research [3,4] is on simulation analysis of this misbehavior. They found that as long as the NAV duration field is set to the maximum value and the rate of attacking nodes reach 30 frames per second, the normal node cannot access the channel.

Backoff value manipulation [5] as a common misbehavior, to obtain more channel resources, it mainly accesses the channel earlier by selecting a smaller backoff value. This misbehavior will not only reduce system performance, but also can lead to denial of service attacks [6] and result in good nodes cannot communicate properly. The research [7] classifies the backoff value manipulation as continuous misbehavior and intermittent misbehavior. It respectively evaluates and quantifies the harm to the network. After simulation analysis, they found that intermittent misbehavior will easily evade misbehavior detection, but when the size of the network becomes larger, this type of misbehavior will cause little harm to the network. But no matter how the size of the network changes, the continuous misbehavior can cause serious damage to the network.

There are many studies on misbehavior detection [8–10], but only few paper have studied how to suppress misbehavior [11]. Based on the dangerous and continuous of the misbehavior of backoff value manipulation, this paper proposes a MC-MAC protocol at the basis of CSMA/CA protocol. The MC-MAC protocol can detect and penalize the backoff value manipulation, and it specifically includes a new backoff value generating function with a penalty function and a reputation model.

The rest of this paper is organized as follows. We present the details of MC-MAC protocol in Sect. 2. The protocol implementation details and simulation results are discussed in Sect. 3. Finally, Sect. 4 draws the conclusion.

2 Proposed Misbehavior Constraint MAC Protocol(MC-MAC)

In the IEEE 802.11 DCF mechanism, when the channel is busy, a node should randomly select a backoff time in the range of $[0, CW]$ (Contention Window) if wants to send data, and wait until the backoff time goes back to zero before sending the control packet RTS. And it can win the channel if its random backoff time is shorter than the others. The misbehavior of reducing the backoff time is that the selfish node can access the channel earlier with a shorter backoff time than the normal node and preempt resources, then affect the throughput of the normal node and the entire system. In order to limiting the misbehavior nodes in wireless network, it is necessary to detect the misbehavior of the node, and then punish the selfish node to ensure the fairness of the communication environment. The detection mechanism of MAC-MAC protocol is implemented by modifying the message exchange mechanism of CSMA/CA. Next, calculating the Trust Value (TV) based on the detected performance of the node. Then the penalty level is graded according to the Trust Value (TV) of the node. Finally, the receiver calculates the penalty backoff value based on the $penalty backoff generating function$ proposed in this paper for the sender, and it will be used as the nodes to calculate the penalty backoff value at next time. Next, we will introduce MC-MAC in three parts.

2.1 Detection Mechanism Based on CSMA/CA

MC-MAC protocol detection mechanism is completed by modifying the CSMA/CA message exchange mechanism. The proposed modification can ensure that receiver R can assign a backoff value to sender S through RTS packet and Acknowledgement (ACK) packet. Therefore, R could verify whether the actual backoff time of S deviates from the backoff time allocated by R. This detection mechanism needs to modify the packet headers of the RTS, CTS and ACK packets. And the proposed modifications make the communication between the nodes more transparent. Figure 1 illustrates the message exchange mechanism of MC-MAC protocol and the related packet header changes.

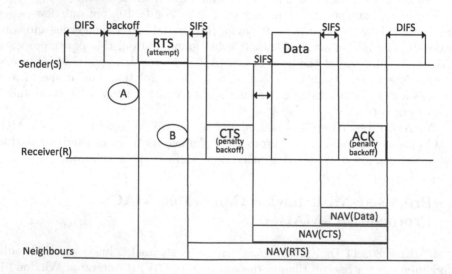

Fig. 1. Detection mechanism based on CSMA/CA

(1) The Sender S generate a backoff value according to the penalty backoff generation function (The following will introduce) during the first communication, but all subsequent transmission S should use the backoff value $(B_{exp} = penalty backoff)$ assigned by the receiver R. In point A of Fig. 1, S

sends an RTS packet to the R, and the number of retransmissions ($attempt$) is added to the packet header of the RTS.

(2) At point B, R receives the RTS packet, then extracts the number of retransmissions ($attempt$), and uses a monitoring function to detect the actual waiting backoff time B_{act}. The actual waiting backoff time B_{act} is equal to the interval which the receiver sends an ACK and receives the next RTS from the same sender.

(3) Receiver R calculates backoff value ($penaltybackoff$) for next transmission according to B_{act}, B_{exp} and $attempt$. Then add $penaltybackoff$ to the packet header of CTS and ACK return to S.

2.2 Trust Value and Statues

The MC-MAC introduces the Trust Value (TV) in order to score the performance of the nodes and then grades nodes according to the score. Changes in the communications environment may affect the protocol's judgment about node performance. However, the grading process is dynamically changing. So the judgment about the performance of a node depends on the multiple communications of the node. Therefore it turns out that grading improves the fault tolerance of the protocol. Equations (1), (2) demonstrate how to calculate the Trust Value (TV) by the receiver (R).

Firstly, the misbehavior factor (Mf) is obtained by Eq. (1). The Mf represents the ratio between receiver reported deviation $B_{exp}\alpha - B_{act}$ to the receivers expected backoff value B_{exp}. The parameter α can be adjusted according to the channel conditions to reduce the error of the judgment. However, when the smaller α is used, the protocol will miss some misbehavior. Therefore, this paper choose a reasonably large α for simulation. Equation (2) shows how to calculate the TV. The initial value of the TV is 100% for each node. Then update the TV according to each node's performance when each communication is completed. Table 1 shows the four grades of penalty level (PL), and the parameter PL is introduced. The PL is divided into four levels according to the TV. The protocol will perform corresponding operations on the nodes according to these four levels.

$$Mf = \frac{B_{exp} \times \alpha - B_{act}}{B_{exp}} \tag{1}$$

$$TV = TV - TV \times Mf \tag{2}$$

Table 1. Trust value and statues

Range of Trust Value	Status
$100 \geq T_v \geq 80$	$PL--, \min(PL) = 1$
$79 \geq T_v \geq 60$	$PL = PL + 1$
$59 \geq T_v \geq 40$	$PL = PL + 2$
$39 \geq T_v \geq 0$	Notifying the upper layer protocol

2.3   Penalty Backoff Generation Function

The penalty backoff generation function proposed in this paper can not only double the contention window value like the IEEE 802.11 BEB (Binary Exponential Backoff Algorithm) after a collision, but also generate a punitive backoff value for selfish node. Such a generating function can prevent the selfish node from selecting a smaller backoff value and not doubling the CW value after a collision. Penalty backoff generation function as shown in Eq. (3).

$$penaltybackoff = f(backoff, senderid, y) * 2^{y-1} * CW_{\min} \qquad (3)$$

backoff in Eq. (3) is the backoff value previously assigned to sender by receiver, senderid is the identifier of sender. Equation (4) shows that y is equal to the maximum value of the number of retransmissions attempt and penalty level PL. Sender retransmission may occur when there are nodes competing for the channel. So the attempt of sender may be greater than PL. In order to ease the channel conflict the receiver needs to use the attempt number to calculate a new backoff value (penaltybackoff) for the sender. But for the selfish node, the penalty level PL will be bigger than attempt, therefore the receiver will generate a punitive backoff value for the sender. The initial values of attempt and PL are equal to 1. CW_{min} is the node's minimum contention window $CW_{min} = 31$.

$$y = \max(attempt, PL) \qquad (4)$$

Function f uses a classical uniformly distributed random number method-linear congruential [12]. It can generate a uniform random number between 0 and 1, And Function f can be ensure that the sender will choose different backoff value after collisions [11]. Function f as shown in Eq. (5).

$$f(backoff, senderid, y) = ((aX + c) \bmod (CW_{\min} + 1))/CW_{\min} \qquad (5)$$

where $a = 5$, $c = 2 * y + 1$ and $X = (backoff + senderid) \bmod (CW_{\min} + 1)$.

3   Simulations Result Analysis

Actually NS2 network simulator is used to simulate MC-MAC protocol to evaluate if the MC-MAC protocol can restrain misbehavior. The simulation was processed at Wi-Fi environment. There are 8 senders and one receiver (AP). Simulation configuration as shown in Table 2. The traffic type is a CBR (constant bit rate) and rate 2 Mbps, wireless channel bandwidth is also 2 Mbps, packet size is 512 bytes.

Misbehavior Model. This paper adopts a dangerous continuous misbehavior model, and analyzes it in [6]. The continuous misbehavior model means that the selfish nodes always have a fixed selfish strategy. This model has a parameter which called misbehavior percentage (MP) to indicate the degree of misbehavior. For example, if the MP of a selfish node is 60%, then this node just needs to wait for 40% of the backoff value B_{exp} allocated by the receiver. As shown in Eq. (6). The larger the MP is, the smaller the actual backoff value of the selfish node is.

$$B_{act} = B_{exp} \times (1 - MP) \tag{6}$$

Table 2. Simulation Parameters

Parameters	Description
Traffic type	CBR
Packet size	512 bytes
Link bandwidth	2 Mbps
Transmission range	250 m
Number of total nodes	9
Number of misbehavior nodes	1
Routing protocol	DSR
Access method	RTS/CTS-DATA-ACK
Misbehavior percentage (MP)	(1%–100%)
Simulation time	60 s

In this section, we will compare the average throughput of good nodes, misbehavior node throughput, and system throughput for the IEEE 802.11 protocol and the proposed protocol, respectively.

3.1 Performance of MC-MAC Protocol Without Misbehavior

First we test the performance of MC-MAC protocol without misbehavior. The purpose of this test is to evaluate the effect of the occasional misjudgment of the MC-MAC. Therefore, we compare the node average throughput for the MC-MAC protocol with the IEEE 802.11 protocol under different network sizes.

In this simulation, we set the number of nodes from 1 to 60. It should be noted that all nodes are good. Other parameters are unchanged according to the settings in Table 2. Figure 2 shows the average throughput of the nodes for the MC-MAC protocol (red) and the IEEE 802.11 protocol (black) when there is no selfish node. It can be seen from the Fig. 2 that the two curves are in the same trend and almost coincide. This shows that in the absence of selfish nodes, the average throughput of the nodes for the MC-MAC protocol is almost same as the average throughput of the nodes for the IEEE 802.11 protocol. It means that there is little misjudgment of the MC-MAC protocol and the proposed protocol will not reduce the throughput of the network.

3.2 Performance of MC-MAC Protocol with Misbehavior

Figure 3 shows the average throughput of good nodes for the MC-MAC protocol (black) and IEEE 802.11 protocol (blue) under different misbehavior percentage

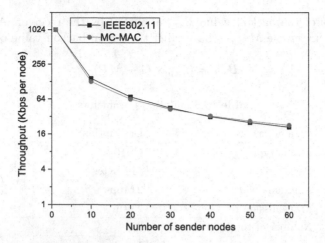

Fig. 2. Throughput of nodes without misbehavior

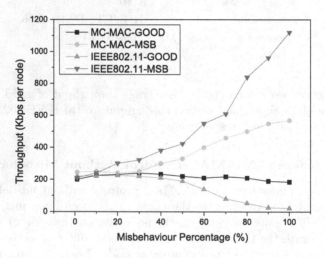

Fig. 3. Throughput of nodes with misbehavior

(MP). The figure also shows the throughput of misbehavior node on the MC-MAC protocol (red) and IEEE 802.11 protocol (pink). According to Fig. 3, when the MP of misbehavior node is from 1% to 100%, the throughput of selfish nodes for both protocols are increase (red and pink), but the MC-MAC protocol (red) is lower. In particular, since the MP reached 60%, the throughput of selfish nodes for the IEEE 802.11 protocol (pink) increased drastically, and the throughput of selfish nodes for the MC-MAC protocol (red) was not that drastic. Also, when the MP of the selfish node increases, the throughput of the good node for the MC-MAC protocol (black) decreases a little, but the average throughput of good nodes for the IEEE802.11 protocol (blue) drops almost to 0. Therefore, it can be concluded that the MC-MAC protocol can keep the average throughput of good nodes within a normal range in networks when competing misbehavior nodes, and the proposed protocol could reduce the throughput of selfish nodes.

Figure 4 plots the system throughput for the MC-MAC protocol (red) and IEEE 802.11 protocol (black) under different misbehavior percentage MP. From the figure, it can be seen that as the misbehavior percentage MP increases, the system throughput for the IEEE 802.11 protocol (black) decreased greatly, especially after the MP reaches 50%. The MC-MAC protocol (red) has a little change in system throughput as the MP increases. As a result, the MC-MAC protocol is more resilient in wireless networks with misbehavior nodes.

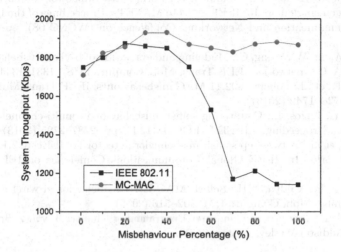

Fig. 4. System throughput

4   Conclusion

This paper proposes the MC-MAC protocol, which is implemented by modifying the IEEE 802.11 MAC protocol. The protocol can suppress the misbehavior that backoff value manipulation, and ensure the fairness and quality of the communication. The Simulation has proved that the protocol can effectively maintain the throughput of good nodes and maintain the throughput of the system in networks when competing misbehavior nodes.

References

1. Diwanji, H., Shah, J.: Effect of MAC layer protocol in building trust and reputation scheme in mobile ad hoc network. In: 2013 Nirma University International Conference on Engineering (NUiCONE), p. 1C3 (2013)
2. Han M.K. Qiu, L.: Greedy receivers in IEEE 802.11 hotspots: impacts and detection. IEEE Trans. Dependable Secur. Comput. **7**(4), 410–423 (2010)

3. Nagarjun, P., Kumar, V., Kumar, C., Ravi, A.: Simulation and analysis of RTS/CTS DoS attack variants in 802.11 networks. In: 2013 International Conference on Pattern Recognition, Informatics and Mobile Engineering, pp. 258–263 (2013)
4. Alocious, C., Xiao, H., Christianson, B.: Analysis of dos attacks at mac layer in mobile adhoc networks. In: Wireless Communications and Mobile Computing Conference (IWCMC), 2015 International, pp. 811–816 (2015)
5. Kyasanur, P., Vaidya, N.H.: Detection and handling of MAC layer misbehavior in wireless networks. In: Proceedings of the IEEE International Conference on Dependable Systems and Networks (DSN 03), pp. 173–182 (2003)
6. Szott, S., Natkaniec, M., Canonico, R., Pach, A.R.: Impact of contention window cheating on single-hop. In: IEEE 802.11e MANETs, Proceedings of the IEEE Wireless Communication and Networking Conference on (WCNC 08), pp. 1356–1361 (2008)
7. Lu, Z., Wang, W., Wang, C.: Modeling and evaluation of backoff misbehaving nodes in CSMA/CA networks. IEEE Trans. Mob. Comput. 11(8), 1331–1344 (2012)
8. Patras, P., et al.: Policing 802.11 MAC misbehaviours. IEEE Trans. Mob. Comput. 15(7), 1728–1742 (2013)
9. Zhang, Y., Lazos, L.: Countering selfish misbehavior in multi-channel MAC protocols. In: Proceedings - IEEE INFOCOM12.11, pp. 2787–2795 (2013)
10. Cao, X., et al.: A two-step selfish misbehavior detector for IEEE 802.11-based Ad Hoc networks. In: IEEE Global Communications Conference on IEEE, pp. 1–6 (2015)
11. Kyasanur, P., Vaidya, N.H.: Selfish MAC layer misbehavior in wireless networks. IEEE Trans. Mob. Comput. 4(5), 502–516 (2005)
12. Knuth, D.E.: The Art of Computer Programming, chapter 3, vol. 2, 3rd edn., pp. 10–17. Addison-Wesley, Boston (2000)

Blind Channel Estimation of Doubly Selective Fading Channels

Jinfeng Tian, Ting Zhou, Tianheng Xu, Honglin Hu, and Mingqi Li[✉]

Shanghai Advanced Research Institute (SARI), Chinese Academy of Science (CAS),
Beijing, China
limq@sari.ac.cn

Abstract. Blind channel identification methods based on second-order statistics (SOS), have attracted much attention in the literature. However, these estimators suffer from the phase ambiguity problem, until additional diversity can be exploited. In this paper, with the aid of the cyclic prefix (CP) induced periodicity, a channel identification algorithm based on the time varying autocorrelation function (TVAF) is proposed for doubly selective fading channels in Orthogonal Frequency Division Multiplexing (OFDM) systems. The closed-form expression for time-varying channel identification is derived within the restricted support set of time index. Particularly, the CP-induced TVAF components and their corresponding channel spread correlation elements implicitly carry rich channel information and are not perturbed by additive noise. These advantageous peaks can be employed to address the phase uncertainty problem, offering an alternative way of increasing the rank of signal matrix to achieve complementary diversity. Simulation results demonstrate the proposed method can provide distinctly higher accurate of channel estimation over the classical scheme.

Keywords: Channel estimation · Doubly selective fading channels
Time-varying autocorrelation function · Subspace

1 Introduction

The cyclic prefix orthogonal frequency division multiplex (CP-OFDM) technique, which is well known for its ability to resist inter-symbol-interference (ISI) in multicarrier communications, has been widely adopted in modern wireless communication systems. In practical OFDM systems, reliable channel estimation

This work is supported by Shanghai Excellent Academic Leader Program (No. 18XD1404100), Shanghai Technical Standard Project (No. 18DZ2203900), the Key Project of Shanghai Municipality of Science and Technology Commission (No. 17511104902), the Rising Star Program of Shanghai Municipality of Science and Technology Commission (No. 17QA1403800), and the program under Grant 6141A01091601.

X. Liu et al. (Eds.): ChinaCom 2018, LNICST 262, pp. 665–674, 2019.
https://doi.org/10.1007/978-3-030-06161-6_65

is an indispensable process to ensure coherent detection and plays a major impact on the whole system performance. Without use of training symbols, blind channel identification methods are well motivated for high bandwidth efficiency applications. Moreover, when the training sequence is not available or contaminated by interferences, blind channel estimation also plays a useful role. •

Most blind channel estimation algorithms are based on higher-order statistics (HOS) to identify the non-minimum phase channel [1]. If additional diversity is available, the channel identification issue can be settled with the sole help of second-order statistics (SOS). Subspace algorithm is one of the most popular SOS-based channel estimation methods for its robustness against noise. The additional diversity of channel, to enable the subspace-based methods workable, can be obtained by resorting to oversampling [2], multiple sensors [3,4], precoding [5], and predefined linear structure [6] etc.

Due to the practical requirement of mobility, there has been an increasing interest in wireless transmissions over time varying (TV) multipath channels. This time and frequency (doubly) selective fading effect makes channel identification more challenging. In order to reduce the number of unknown channel parameters, basis expansion model (BEM) is often applied to approximate the doubly-selective fading channel. In [7], a classical time-varying autocorrelation function (TVAF) based method is proposed to estimate the BEM coefficients of a TV single-input single-output channel via the subspace solution. The time varying nature of the autocorrelation of the received signal comes from the effect of time-variant channels. It was shown in [3] that the linear independence condition required in [7] does not hold for complex exponentials based BEM model. With the aid of multiple receive antennas, a subspace-based channel estimator associated with arbitrary basis functions is proposed over doubly selective fading channels [8,9]. In [10], a two-step subspace-based estimation method, by introducing splitting factor and permutation operation, is analyzed under time-varying single-input multiple-output (SIMO) channels. Overall, these improved estimators are designed with restriction on the antenna number or with the help of additional operation added to the system.

The motivation of this paper is to investigate blind channel estimation over doubly selective fading channels, without adding any other restriction to a CP-OFDM system. Not only the limitation of application scenarios can be relaxed, but the newly achievable diversity can also be integrated to other possible methods to further improve estimation performance. The standard subspace-based estimators assume that the transmitted signal is stationary [7–10]. Under such a premise, the time varying autocorrelations of the received signal are just exploited partially and some of the used correlations are corrupted by noise, which limit the estimation performance. Rather than stationary assumption, cyclostationary signal, which is a more realistic one, possesses extra information due to its hidden periodicity [11]. Based on the CP-induced cyclostationarity, we have extended the cyclostationary analysis method to BEM modeled doubly-selective fading channels [12]. This provides a more comprehensive view on the cyclostationarity at the receiver, thus additional channel diversity can be exploited for channel identification.

In this paper, we focus on a time-variant SOS based blind identification app-roach for doubly selective fading channels in OFDM systems. By decoupling the complicated effect of multiple paths in the TVAF of the received CP-OFDM sig-nal, the closed-form expression for blind identification of a doubly selective chan-nel is derived, which is an extension of the traditional TVAF-based time-varying channel identification method. With the use of the CP-induced time-varying autocorrelation components and their corresponding channel-spread correlation elements, the effect of additive noise can be canceled. Furthermore, a new param-eter is therefore introduced in the proposed estimator which increases the rank of the signal matrix, enabling substantial performance improvement.

The rest of the paper is organized as follows. Section 2 describes the sys-tem model. Section 3 reviews the TVAF of the received CP-OFDM signal over doubly-selective fading channels. In Sect. 4, a subspace-based time-varying chan-nel identification approach is proposed by exploiting the TV correlations con-tributed by the CP and the channel. Then the analysis of the simulation results is presented in Sect. 5. Finally, we conclude our work in Sect. 6.

2   System Model

Consider an OFDM system with CP, the discrete-time baseband equivalent transmitted signal can be written as

$$s(n) = \frac{1}{\sqrt{N}} \sum_{m=-\infty}^{+\infty} \sum_{k=0}^{N-1} d_{m,k} g(n - mM) e^{\frac{j2\pi k(n-mM)}{N}}, \tag{1}$$

where N is the fast Fourier transform (FFT) size. $d_{m,k}$ denotes the complex data symbol transmitted on the kth subcarrier in the mth OFDM symbol. We assume that $d_{m,k}$ is zero-mean and independent of each other such that $E\left\{d_{m,k}d^*_{m',k'}\right\} = \delta(m - m')\delta(k - k')$, where $E(\cdot)$, $\delta(\cdot)$, and superscript $(\cdot)^*$ stand for the mathematical expectation, the Delta function, and the complex conjugation, respectively. $g(n)$ is an M-length rectangular window. M is the length of an OFDM symbol with CP, i.e. $M = N + N_g$. N_g denotes the length of CP.

Then the transmitted signal passes through a doubly selective fading channel with additive white Gaussian noise (AWGN). Let us define $h(n, l)$ as the channel impulse response (CIR) at lag l and instant n. At the OFDM receiver, the discrete-time received signal can be expressed as

$$r(n) = z(n) + v(n) = \sum_{l=0}^{L_h} h(n, l)s(n - l) + v(n), \tag{2}$$

where $v(n)$ is a zero-mean white noise with variance σ_v^2. L_h denotes the maximum discrete delay spread of the channel. In order to eliminate ISI, N_g is set to be larger than L_h.

The doubly-selective fading channel is usually modeled as the BEM. Each channel tap in this model is represented as the weighted sum of a few complex exponential basis functions. According to [12,13], the BEM can be applied for a burst of K OFDM symbols. Considering that the sampling period at a receiver is equal to that at a transmitter, we have the discrete-time baseband equivalent channel model in a burst as

$$h(n,l) = \sum_{q=-Q/2}^{Q/2} h_q(l)e^{j\frac{2\pi}{KM}qn}, n = 0, \cdots, KM - 1 \qquad (3)$$

where Q denotes the discrete Doppler spread. $h_q(l)$, where $q \in [-Q/2, Q/2]$, are the channel parameters for the lth channel tap ($l \in [0, L_h - 1]$), which remain invariant per burst and vary independently from burst to burst.

3   Time-Varying Autocorrelation Function Over Doubly-Selective Fading Channels

For a cyclostationary signal, its autocorrelation function is not time-invariant, but time-dependent and periodic in time. The TVAF of a zero mean complex cyclostationary signal $s(n)$ is defined as

$$c_s(n, \tau) = E\left\{s(n)s^*(n + \tau)\right\}, \qquad (4)$$

where τ is an integer lag parameter. By substitution of (1) into (4), we have the result of $c_s(n, \tau)$ with

$$c_s(n, \tau) = \Gamma_N(\tau) \sum_{m=-\infty}^{\infty} g(n - mM)g(n + \tau - mM), \qquad (5)$$

where $\Gamma_N(\tau) = \frac{1}{N}\sum_{k=0}^{N-1} e^{-j2\pi k\tau/N}$. From (5), we can observe that $c_s(n, \tau)$ is M-periodic in n for each value of τ, i.e., $c_s(n, \tau) = c_s(n + M, \tau)$.

Using (2) and (3), we have derived the TVAF of the received OFDM signal $r(n)$ based on BEM model in the previous work in [12], which is given as

$$c_r(n, \tau) = \sum_{l=0}^{L_h} \sum_{\xi=\tau+l-L_h}^{\tau+l} \sum_{q=-Q/2}^{Q/2} \sum_{q'=-Q/2}^{Q/2} h_q(l)h_{q'}^*(l + \tau - \xi)$$
$$\times c_s(n - l, \xi)e^{j2\pi\frac{qn}{KM}}e^{-j2\pi\frac{q'(n+\tau)}{KM}} + c_v(\tau), \qquad (6)$$

where $c_v(\tau) = \sigma_v^2\delta(\tau)$. Since $c_s(n, \tau) = c_s(n + M, \tau)$, we have $c_r(n, \tau) = c_r(n + KM, \tau)$ for every τ. This signifies that $r(n)$ is a cyclostationary random process with cyclostationary period KM.

Figure 1(a), (b) separately illustrate the TVAF of the transmitted signal and received signal in CP-OFDM systems, where $N = 32$ and $N_g = 8$. It is shown that, in Fig. 1(a), all the nonzero correlation peaks have the value of 1 and appear at the three cross sections with $(\tau = 0, \pm N)$ in the correlation function plane. The components at $\tau = \pm N$ characterize the correlations caused by the CP, where the time varying characteristic is in a ladder manner. The set of correlations at $\tau = 0$ interprets the correlations induced by signal itself which is invariant in time n. In Fig. 1(b), the TVAF of the received OFDM signal over a doubly-selective fading channel is described, in which $K = 2$. The time varying behavior of the channel make the correlation peaks varying like a sinusoid in terms of time n. Due to the multipath delay effect, the correlation function is spread with respect to the lag parameter dimension. As AWGN $v(n)$ is stationary, $c_v(\tau)$ only has values of σ_v^2 when $\tau = 0$, which means the stationary noise only disturbs the information on $c_r(n, 0)$.

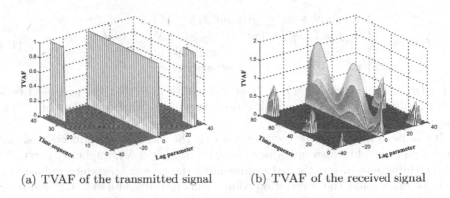

(a) TVAF of the transmitted signal   (b) TVAF of the received signal

Fig. 1. TVAF of CP-OFDM signals.

4   Proposed Blind Estimation Method Based on TVAF

With known parameters of L_h and Q as in [3,7], the goal of channel identification in this paper is to estimate the time-invariant coefficients $\{h_q(l)\}$. In this section, the estimation of the BEM coefficients is developed in two steps. In the first step, the correlations of the time-invariant coefficients are estimated by exploiting the CP-induced correlations and corresponding channel-spread correlations. In the second step, subspace method is applied to obtain the expansion coefficients.

4.1   Channel Estimator Based on TVAF

Substituting (5) into $c_s(n - l, \xi)$ in (6), we have

$$c_s(n - l, \xi) = \begin{cases} 1 & for \ N \le ((n - l) \ mod \ M) \le M - 1 \ and \ \xi = -N \\ 1 & for \ 0 \le ((n - l) \ mod \ M) \le M - N - 1 \ and \ \xi = N \\ 1 & for \ \xi = 0 \\ 0 & otherwise, \end{cases} \quad (7)$$

where mod stands for the modulus operator. To decouple the complicated effect of l of $c_s(n - l, \xi)$ in (6), we use the restricted support region of n for channel identification, where the non-zero values of $c_s(n - l, \xi)$ equal 1 for different l at a given ξ. Consequently, the TVAF of $r(n)$ can be derived as

$$c_r(n, \tau) = \sum_{q=-Q/2}^{Q/2} \sum_{q'=-Q/2}^{Q/2} R_h(\tau - \xi; q, q') f(n, \xi) b_q(n) b_{q'}^*(n + \tau) + c_v(\tau), \quad (8)$$

where

$$R_h(\tau - \xi; q, q') = \sum_{l=0}^{L_h} h_q(l) h_{q'}^*(l + \tau - \xi), \quad (9)$$

$$f(n, \xi) = \begin{cases} 1 \; for \; N + L_h \leq (n \; mod \; M) \leq M - 1 \; and \; \xi = -N \\ 1 \; \; for \; L_h \leq (n \; mod \; M) \leq M - N - 1 \; and \; \xi = N \\ 1 \qquad\qquad\qquad for \; \xi = 0 \\ 0 \qquad\qquad\qquad otherwise, \end{cases} \quad (10)$$

and $b_q(n) = \exp(j2\pi qn/KM)$. It is worthwhile to note that the proposed estimator can be reduced to the classical estimator in [7], when the effect of $v(n)$ is ignored and $\xi = 0$.

It can be seen that, at the receiver, not only the correlations induced by signal itself (i.e., $\xi = 0$) but also introduced by CP (i.e., $\xi = \pm N$), implicitly carry the channel information. For the sake of avoiding the noise uncertainty induced by $v(n)$, we just exploit the correlation components introduced by the CP and their channel-spread peaks for channel estimation. Thus, in the following, two values of ξ, i.e., $-N$ and N, are considered. Accordingly, the contribution of stationary noise is therefore canceled out in (8), because the values of $c_v(\tau)$ are nonzero only for $\tau = 0$ (ξ, at this moment, equals 0).

4.2   Recovering the Channel Correlations

The first step of identification of the expansion parameters $h_q(l)$ is to recover the correlations of the time-invariant coefficient pairs of channel $R_h(\tau - \xi; q, q')$. For the convenience of description, we denote $\gamma = \tau - \xi$. It can be easily found that $-L_h \leq \gamma \leq L_h$.

Then, the vector form of (8) can be written as

$$c_r(n, \xi + \gamma) = f(n, \xi) \phi(n, \xi, \gamma) \mathbf{R}_h(\gamma), \quad (11)$$

where

$$\phi(n, \xi, \gamma) = \left[b_{-\frac{Q}{2}}(n) b_{-\frac{Q}{2}}^*(n + \xi + \gamma), \cdots, b_{-\frac{Q}{2}}(n) b_{\frac{Q}{2}}^*(n + \xi + \gamma), \right.$$
$$\left. \cdots, b_{\frac{Q}{2}}(n) b_{-\frac{Q}{2}}^*(n + \xi + \gamma), \cdots, b_{\frac{Q}{2}}(n) b_{\frac{Q}{2}}^*(n + \xi + \gamma) \right], \quad (12)$$

$$\mathbf{R}_h(\gamma) = \left[R_h(\gamma, -\frac{Q}{2}, -\frac{Q}{2}), \cdots, R_h(\gamma, -\frac{Q}{2}, \frac{Q}{2}) \right.$$
$$\left. \cdots, R_h(\gamma, \frac{Q}{2}, -\frac{Q}{2}), \cdots, R_h(\gamma, \frac{Q}{2}, \frac{Q}{2}) \right]^T. \quad (13)$$

Define

$$\begin{bmatrix} n_{1_{i,1}}, \cdots, n_{1_{i,P}} \end{bmatrix} = [(i-1)M + N + L_h, \cdots, iM - 1]$$
$$\begin{bmatrix} n_{2_{i,1}}, \cdots, n_{2_{i,P}} \end{bmatrix} = [(i-1)M + L_h, \cdots, iM - N - 1], \tag{14}$$

where $P = N_g - L_h$. (11) can be further represented in a compact matrix form as

$$\mathbf{c}_r(\gamma) = \Phi(\gamma)\mathbf{R}_h(\gamma). \tag{15}$$

The components of $\mathbf{c}_r(\gamma)$ can be obtained from the instantaneous estimation of $\hat{c}_r(n,\tau) = r(n)r^*(n+\tau)$, given by

$$\mathbf{c}_r(\gamma) = [\mathbf{c}_{r_{1,-N}}^T, \cdots, \mathbf{c}_{r_{i,-N}}^T, \cdots, \mathbf{c}_{r_{K,-N}}^T,$$
$$\mathbf{c}_{r_{1,N}}^T, \cdots, \mathbf{c}_{r_{i,N}}^T, \cdots, \mathbf{c}_{r_{K,N}}^T]^T, \tag{16}$$

where

$$\mathbf{c}_{r_{i,-N}} = \begin{bmatrix} c_r\left(n_{1_{i,1}}, -N+\gamma\right), \cdots, c_r\left(n_{1_{i,P}}, -N+\gamma\right) \end{bmatrix}^T$$
$$\mathbf{c}_{r_{i,N}} = \begin{bmatrix} c_r\left(n_{2_{i,1}}, N+\gamma\right), \cdots, c_r\left(n_{2_{i,P}}, N+\gamma\right) \end{bmatrix}^T, \tag{17}$$

and $(\cdot)^T$ denotes transpose operation. The matrix $\Phi(\gamma)$, which is known a priori, has the following structure

$$\Phi(\gamma) = [\mathrm{A}_{1,-N,\gamma}^T, \cdots, \mathrm{A}_{i,-N,\gamma}^T, \cdots, \mathrm{A}_{K,-N,\gamma}^T,$$
$$\mathrm{B}_{1,N,\gamma}^T, \cdots, \mathrm{B}_{i,N,\gamma}^T, \cdots, \mathrm{B}_{K,N,\gamma}^T]^T, \tag{18}$$

where

$$\mathrm{A}_{i,-N,\gamma} = [\phi^T\left(n_{1_{i,1}}, -N, \gamma\right), \cdots, \phi^T\left(n_{1_{i,P}}, -N, \gamma\right)]^T$$
$$\mathrm{B}_{i,N,\gamma} = [\phi^T\left(n_{2_{i,1}}, N, \gamma\right), \cdots, \phi^T\left(n_{2_{i,P}}, N, \gamma\right)]^T. \tag{19}$$

The identification problem of $\mathbf{R}_h(\gamma)$, for every fixed γ, can be solved by the least squares (LS) method based on the $2KP \times 1$ vector $\mathbf{c}_r(\gamma)$ and the $[2KP] \times [(Q+1)(Q+1)]$ matrix $\Phi(\gamma)$. It has been verified that the use of instantaneous approximations for $c_r(n,\tau)$ is feasible for channel identification, as the number of equations is far greater than the unknown parameters [7]. It is the rank of $\Phi(\gamma)$ that is an important factor affecting the estimation performance. Since an additional variable ξ with value of N or $-N$ is introduced in (12), the linear independence of the columns in the matrix $\Phi(\gamma)$ can be largely increased compared to that in the classical TVAF-based method, which results in a significant improvement of the proposed estimator. In addition, the number of equations for estimating $R_h(\gamma; q, q')$ in the proposed method is reduced from KM to $2KP$. This decreases the computational complexity of the proposed method to a certain extent.

4.3   Identification of Channel Coefficients

The blind identification procedure is finally to estimate the $(Q+1)(L_h+1) \times 1$ vector $\mathbf{h}$ of the BEM coefficients

$$\mathbf{h} = \begin{bmatrix} h_{-Q/2}(0), \cdots, h_{-Q/2}(L_h), \cdots, h_{Q/2}(0), \cdots, h_{Q/2}(L_h) \end{bmatrix}^T. \tag{20}$$

According to [3], the parameters $R_h(\gamma; q, q')$ can be regarded as the output cross-correlation of a hypothetical SIMO system $y_q(n) = \sum_{l=0}^{L_h} h_q(l)w(n-l)$, where $w(n)$ is a common zero-mean white input with unit variance. Define the vectors $\mathbf{y}_q(n) = [y_q(n), \cdots, y_q(n-L)]^T$ for some order L and the vectors $\mathbf{Y}(n) = \left[\mathbf{y}_{-Q/2}^T(n), \cdots, \mathbf{y}_{Q/2}^T(n)\right]^T$. Based on the $[(Q+1)(L+1)] \times [(Q+1)(L+1)]$ correlation matrix of $\mathbf{Y}(n)$, as described in [7], we can uniquely identify $\{h_q(l)\}$ (up to a complex scalar factor) if $L \geq L_h$, by using the subspace solution.

5   Numerical Results

In this section, we present numerical comparisons between our proposed scheme and the classical TVAF-based subspace method. As illustrated in Fig. 2, the conventional TVAF-based estimator uses the correlation components at $\tau = 0 \pm l$ for time varying channel identification, while the proposed method exploits the correlation peaks at $\tau = \pm N \pm l$ within the restricted support region of n, where $0 \leq l \leq L_h$. In the experiments, the OFDM signal has 128 subcarriers and the length of the CP is $1/8$ of the useful symbol data. Subcarriers are modulated by 16QAM. The carrier frequency is 2.5 GHz. The OFDM symbol duration with CP is $102.86\mu s$. The BEM coefficients with $Q = 2$ and $L_h = 2$ are listed below.

$$\mathbf{h}^T = [0.1660 - 0.1722\mathrm{i}, 0.0101 + 0.1551\mathrm{i}, -0.3199 - 0.0863\mathrm{i},$$
$$0.0043 - 0.2809\mathrm{i}, 0.1423 - 0.1443\mathrm{i}, -0.1355 - 0.1699\mathrm{i},$$
$$0.3245 + 0.1537\mathrm{i}, -0.5881 - 0.0773\mathrm{i}, 0.2572 - 0.3079\mathrm{i}]$$

The channel coefficients are scaled so that the parameter vector $\mathbf{h}$ has unit norm. In addition, the order L adopted in the subspace identification process is set to L_h. Estimation is then carried out using Monte Carlo method with $N_i = 500$ runs. As a performance metric we use the normalized mean square error (NMSE), which is defined as $NMSE = \frac{1}{N_i} \sum_{i=1}^{N_i} E\left\{\left\|\hat{\mathbf{h}}^i - \mathbf{h}\right\|^2 / \|\mathbf{h}\|^2\right\}$. Before computing NMSE, the estimated parameter vector $\hat{\mathbf{h}}$ is scaled by $E\left\{\mathbf{h}/\hat{\mathbf{h}}\right\}$ to resolve the scaling ambiguity for simulation purpose.

In order to verify the validity of the LS estimates for $R_h(\gamma; q, q')$, the time varying signal power of the noise-free output data $z(n)$ is employed to evaluate the performance. The reconstructed signal power can be computed by $\hat{c}_{r,LS}(n, 0) = f(n, 0)\phi(n, 0, 0)\hat{R}_h(0)$, $0 \leq n \leq KM - 1$. From Fig. 2, it can be seen that the LS estimates for $R_h(\gamma; q, q')$ can be reliably recovered based on the instantaneous approximations for $c_r(n, \tau)$. Additionally, the reconstructed results of the proposed scheme are much closer to the accurate ones than those of the compared scheme. This deviation of the conventional method is mainly generated by using the noise-contaminated components at $c_r(n, 0)$ for estimation and by the fact of the column dependence in the matrix Φ.

Figure 3 illustrates the performance of NMSE as a function of SNR for $K = 10$ and $K = 20$. It can be observed that the considered methods both follow

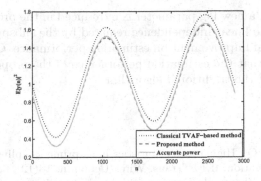

Fig. 2. Time-varying signal power ($SNR = 10\,\mathrm{dB}$ and $K = 20$).

a descending trend in NMSE with increasing SNR. Specifically, the proposed scheme outperforms the benchmark method. When the SNR is greater than or equal to 10 dB, the proposed estimator can achieve significant improvement in estimation performance. Since the effect of noise is minor at a higher SNR, these substantial performance gains of the proposed scheme are obtained mainly owing to the increased linear independence in the matrix Φ. As the number of symbols changes from 10 to 20, the NMSE performances of the two channel estimators are both enhanced while the superiority of the proposed estimator still maintains.

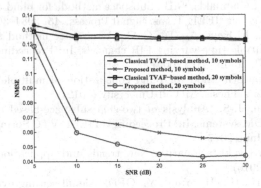

Fig. 3. NMSE versus SNR.

6 Conclusions

In this paper, a time-variant SOS based channel estimation method is proposed for doubly selective fading channels by using the inherent cyclostationrity of the transmitted signal. To address the phase ambiguity issue, the cyclostaionarity induced by the CP as well as the channel is exploited for channel identification.

As a result of this, a new lag parameter is introduced in the proposed estimator, which increases the linear independence required by the subspace method. This leads to substantial improvement on estimation performance. Computer simulation results show that the estimation performance of the proposed algorithm is superior to that of the traditional algorithm.

References

1. Wu, Q., Liang, Q.: Higher-order statistics in co-prime sampling with application to channel estimation. IEEE Trans. Wirel. Commun. **14**(12), 6608–6620 (2015)
2. Yu, C., Xie, L., Zhang, C.: Deterministic blind identification of IIR systems with output-switching operations. IEEE Trans. Signal Process. **62**, 1740–1749 (2014)
3. Giannakis, G.B., Tepedelenlioglu, C.: Basis expansion models and diversity techniques for blind identification and equalization of time-varying channels. Proc. IEEE **86**(10), 1969–1986 (1998)
4. Bonna, K., Spasojevic, P., Kanterakis, E.: Subspace-based SIMO blind channel identification: asymptotic performance comparison. In: Proceedings - IEEE Military Communications Conference, pp. 460–465 (2016)
5. Ghauch, H., Kim, T., Bengtsson, M., Skoglund, M.: Subspace estimation and decomposition for large millimeter-wave MIMO systems. IEEE J. Sel. Top. Signal Process. **10**(3), 528–542 (2016)
6. Mayyala, Q., Abed-Meraim, K., Zerguine, A.: Structure-based subspace method for multichannel blind system identification. IEEE Signal Process. Lett. **24**(8), 1183–1187 (2017)
7. Tsatsanis, M.K., Giannakis, G.B.: Subspace methods for blind estimation of time-varying FIR channels. IEEE Trans. Signal Process. **45**, 3084–3093 (1997)
8. Champagne, B., El-Keyi, A., Tu, C.-C.: A subspace method for the blind identification of multiple time-varying FIR channels. In: Proceedings of IEEE Global 2009, pp. 1–6. Honolulu, HI (2009)
9. Tian, Y.: Subspace method for blind equalization of multiple time-varying FIR channels, Master's Thesis. McGill University (2012)
10. Fang, S.-H., Lin, J.-S.: Analysis of two-step subspace-based channel estimation method for OFDM systems. In: Proceedings of IEEE VTC Spring, pp. 1–5. Sydney, Australia, 4–7 June 2017
11. Napolitano, A.: Cyclostationarity: New trends and applications. Signal Process. **120**, 385–408 (2016)
12. Tian, J., Guo, H., Hu, H., Yang, Y.: OFDM signal sensing over doubly-selective fading channels. In: Proceedings of IEEE GLOBECOM, pp. 1–5. Miami, USA, 7–9 Dec 2010
13. Tian, J., Jiang, Y., Hu, H.: Cyclostationarity-based frequency synchronization for OFDM systems over doubly-selective fading channels. Wirel. Pers. Commun. **66**(2), 461–472 (2012)

User Scheduling for Large-Scale MIMO Downlink System Over Correlated Rician Fading Channels

Tingting Sun[(⊠)], Xiao Li, and Xiqi Gao

National Mobile Communications Research Laboratory,
Southeast University, Nanjing, China
stt19931124@seu.edu.cn

Abstract. In this paper, we investigate the downlink transmission, especially the user scheduling algorithm for single-cell multiple-input multiple-output (MIMO) system under correlated Rician fading channels. Under the assumption of only statistical channel state information (CSI) at the base station (BS), the statistical beamforming transmission is derived by maximizing the lower bound of the average signal-to-leakage-plus-noise ratio (SLNR). Based on this beamforming transmission algorithm, three user scheduling algorithms are proposed exploiting only statistical CSI: (1) maximum SLNR: schedule the user with the maximum SLNR; (2) most dissimilar: schedule the user that is most dissimilar to the already selected users; (3) modified-treating interference as noise (TIN): treat the inter-user interference as uncorrelated noise to each user's useful signal and schedule the user with the largest signal-to-noise factor. Simulation results show that the proposed user scheduling algorithms perform well in achieving considerable sum rate.

Keywords: User scheduling · Rician fading · Downlink

1 Introduction

In recent years, massive multiple-input multiple-output (MIMO) has been recognized as one of the key techniques in future wireless communication systems [1] due to its channel-hardening effect and high potential in interference mitigation [2]. Scaling up the number of antennas in practice, however, faces several challenges. As point out in [1], the acquisition of the channel state information (CSI) at the BS is difficult. Most cellular systems today are in frequency-division duplexing (FDD) mode [3], where the BS gets access to the CSI through a feedback channel. Obviously, the instantaneous CSI feedback causes great overhead in the feedback link for transmission and scheduling algorithms as the number of antennas at the BS grows large. An alternative approach is to exploit the statistical information of the channel [4,5], which varies at a much slower rate

X. Liu et al. (Eds.): ChinaCom 2018, LNICST 262, pp. 675–687, 2019.
https://doi.org/10.1007/978-3-030-06161-6_66

and can be accurately obtained through long-term feedback. In [4], a two-stage precoding algorithm exploring both statistical and part of instantaneous CSI was proposed under Rayleigh fading channels. Moreover, the corresponding user scheduling algorithm was investigated in [6].

Note that most prior works on massive MIMO adopt the simple Rayleigh fading channel model, although this channel model assumption greatly simplifies the analysis, it can not capture the characteristics of the line-of-sight (LOS) component between the transmitter and the receiver. This is especially the case in millimeter-wave communication systems [7] in which LOS propagation dominates. Therefore it is of great importance to consider the more general fading channels which take into account the LOS conditions, i.e., Ricican fading channels [8]. The work in [10] investigates the downlink transmission and scheduling algorithm for full-dimension (FD) MIMO systems under uncorrelated Rician fading channels. In reality, correlation effects should be considered due to the space limitation of user equipments (UEs), the antenna configurations and the Doppler spread [11]. Under the assumption of the same correlation matrix for all users, [9] investigates the ergodic sum rate of downlink massive MIMO system with perfect CSI at BS.

Motivated by the above observations, in this paper, we consider a single-cell multiuser downlink transmission system under the more general correlated Rician fading channel. With only statistical CSI at BS, the statistical beamforming algorithm proposed under Rayleigh and uncorrelated Rician fading is extended to the correlated Rician fading channel by maximizing the lower bound on the average signal-to-leakage-plus-noise ratio (SLNR). Based on the beamforming algorithm, three user scheduling algorithms are proposed, which are the maximum SLNR, the most dissimilar, and the modified-treating interference as noise (TIN). Simulation results reveal that the proposed algorithms perform well in terms of the achievable sum rate.

2   System Model

We consider a single-cell MIMO downlink transmission system with L single-antenna user terminals. A uniform linear array with M antenna elements is employed at the BS. The distances between two neighboring antenna elements is $\lambda/2$, where λ is the wavelength of the carrier. The BS can serve at most U users.

2.1   Signal Model

Assume that a total of U_t $(U_t \leq U)$ users are scheduled. The received signal at user u can be expressed as

$$y_u = \sqrt{p_u}\mathbf{h}_u^T\mathbf{w}_u x_u + \sum\nolimits_{i=1,i\neq u}^{U_t} \sqrt{p_i}\mathbf{h}_u^T\mathbf{w}_i x_i + n_u, \tag{1}$$

where $\mathbf{h}_u^T \in \mathbb{C}^{1\times M}$ represents the channel vector between the BS and user u, $\mathbf{w}_i \in \mathbb{C}^{M\times 1}$ is the unit-norm beamforming vector of user i, x_i is the data symbol for user i satisfying $\mathrm{E}\{|x_i|^2\} = 1$, $n_u \sim CN(0,\sigma_u^2)$ is the complex additive white Gaussian noise, p_i is transmit power for user i with total power constraint $\sum_{i=1}^{U_t} p_i \leq P$. In this paper, we assume equal power allocation among the scheduled users, i.e., $p_i = P/U_t$.

2.2 Channel Model

We consider the correlated Rician fading channel model. Under this model, the channel vector $\mathbf{h}_u$ consists of a specular component corresponding to the LOS signal and a Rayleigh-distributed random component accounting for the diffused multipath signals. The channel vector $\mathbf{h}_u$ between the BS and user u can be written as

$$\mathbf{h}_u^T = \sqrt{\frac{K_u}{K_u+1}}\overline{\mathbf{h}}_u^T + \sqrt{\frac{1}{K_u+1}}\mathbf{h}_{w,u}^T \mathbf{R}_u^{1/2}, \tag{2}$$

where $\overline{\mathbf{h}}_u \in \mathbb{C}^{M \times 1}$ is the deterministic component, K_u is the ratio between the LOS and non-LOS channel power, the entries of $\mathbf{h}_{w,u} \in \mathbb{C}^{M \times 1}$ are independent and identically distributed (i.i.d.) complex Gaussian random variables (RVs) with zero mean and unit variance, and $\mathbf{R}_u \in \mathbb{C}^{M \times M}$ is the transmit channel correlation matrix. For the considered uniform linear array (ULA), the deterministic component $\overline{\mathbf{h}}_u$ of user u can be given by [12]

$$\overline{\mathbf{h}}_u = \left[1, e^{-j\pi \sin \theta_u}, \cdots, e^{-j\pi(M-1)\sin \theta_u}\right]^T, \tag{3}$$

where θ is the angle of departure (AoD) of user u. For the transmit correlation, we consider the one-ring scattering model [4] to determine $\mathbf{R}_u$.

We assume that each user has perfect effective CSI of its own, while the BS has only statistical CSI of all users, i.e., $K_u, \sigma_u^2, \overline{\mathbf{h}}_u, \mathbf{R}_u$, which are calculated by the user via a long-term statistics and are obtained by the BS through long-term feedback.

3 Downlink Transmission

Under the assumption of only statistical CSI at BS, it is difficult to get simple analytical optimization result directly maximizing the ergodic sum rate with respect to $\mathbf{w}_i$, $i = 1, \cdots, U_t$, due to the lack of analytical expression of the ergodic sum rate. Inspired by [14], here, we use the average SLNR metric. In the following, we first derive a lower bound of the average SLNR and then derive the beamforming transmission algorithm by maximizing the lower bound.

The SLNR of user u which measures the amount of power leaked from its beamforming direction to other users' channel direction, can be given by

$$\mathrm{SLNR}_u = \frac{\frac{P}{U_t}\left|\mathbf{h}_u^T \mathbf{w}_u\right|^2}{\sigma_u^2 + \frac{P}{U_t}\sum_{i=1,i \neq u}^{U_t}\left|\mathbf{h}_i^T \mathbf{w}_u\right|^2}. \tag{4}$$

Note that the numerator and denominator of (4) are independent. Based on Mullen's inequality [13], i.e., $E\{X/Y\} \geq E\{X\}/E\{Y\}$ if X and Y are independent random variables, we can obtain that

$$E\{\text{SLNR}_u\} \geq \text{SLNR}_u^L = \frac{\frac{P}{U_t}\mathbf{w}_u^\dagger E\{\mathbf{h}_u^*\mathbf{h}_u^T\}\mathbf{w}_u}{\sigma_u^2 + \frac{P}{U_t}\sum_{i=1,i\neq u}^{U_t}\mathbf{w}_u^\dagger E\{\mathbf{h}_i^*\mathbf{h}_i^T\}\mathbf{w}_u}. \tag{5}$$

From (1), we can get that

$$E\{\mathbf{h}_i^*\mathbf{h}_i^T\} = \frac{1}{K_i+1}\mathbf{R}_i + \frac{K_i}{K_i+1}\overline{\mathbf{h}}_i^*\overline{\mathbf{h}}_i^T, \qquad i = 1,\cdots,U_t. \tag{6}$$

Substituting (6) into (5), and after some manipulation, we have

$$\text{SLNR}_u^L = \frac{\mathbf{w}_u^\dagger\left(\frac{1}{K_u+1}\mathbf{R}_u + \frac{K_u}{K_u+1}\overline{\mathbf{h}}_u^*\overline{\mathbf{h}}_u^T\right)\mathbf{w}_u}{\sigma_u^2 + \frac{P}{U_t}\sum_{i=1,i\neq u}^{U_t}\mathbf{w}_u^\dagger\left(\frac{1}{K_i+1}\mathbf{R}_i + \frac{K_i}{K_i+1}\overline{\mathbf{h}}_i^*\overline{\mathbf{h}}_i^T\right)\mathbf{w}_u}. \tag{7}$$

Let us define

$$\mathbf{\Lambda}_u = \mathbf{F}_M^\dagger\mathbf{R}_u\mathbf{F}_M, \tag{8}$$

and

$$\mathbf{A}_u = \mathbf{F}_M^\dagger\overline{\mathbf{h}}_u^*\overline{\mathbf{h}}_u^T\mathbf{F}_M, \tag{9}$$

where $\mathbf{F}_M \in \mathbb{C}^{M\times M}$ is unitary DFT matrix, with the (i,j)-th element $[\mathbf{F}_M]_{i,j} = \frac{1}{\sqrt{M}}e^{-j\frac{2\pi}{M}(i-1)(j-1-M/2)}$, $i,j = 1,\cdots,M$.

According to [4], for ULA of large dimension, $\mathbf{\Lambda}_u$ becomes diagonal matrix with a few adjacent non-zero diagonal elements [14], i.e.,

$$\mathbf{\Lambda}_u = \text{diag}\left\{0,\cdots,0,\overbrace{\lambda_u^{(k_1)},\cdots,\lambda_u^{(k_2)}}^{\text{nonzero}},0,\cdots,0\right\}, \tag{10}$$

where $\lambda_u^{(k)}$, $k \in [k_1,k_2]$ denotes the k-th diagonal element of $\mathbf{\Lambda}_u$. In [10], it was shown that for ULA of large dimension, $\mathbf{A}_u$ is a diagonal matrix with only one non-zero diagonal element, i.e.,

$$\mathbf{A}_u = \text{diag}\left\{0,\cdots,0,a_u^{(j_0)},0,\cdots,0\right\}, \tag{11}$$

where $a_u^{(j_0)}$ denotes the j_0-th diagonal element of $\mathbf{A}_u$.

Based on these results, we have that when $M \to \infty$,

$$\text{SLNR}_u^L = \frac{\frac{P}{U_t}\mathbf{w}_u^\dagger\mathbf{F}_M\mathbf{\Omega}_u\mathbf{F}_M^\dagger\mathbf{w}_u}{\sigma_u^2 + \frac{P}{U_t}\sum_{i=1,i\neq u}^{U_t}\mathbf{w}_u^\dagger\mathbf{F}_M\mathbf{\Omega}_i\mathbf{F}_M^\dagger\mathbf{w}_u}. \tag{12}$$

where $\mathbf{\Omega}_i = \frac{1}{K_i+1}\mathbf{\Lambda}_i + \frac{K_i}{K_i+1}\mathbf{A}_i$, $i = 1,\cdots,U_t$. Then, when $M \to \infty$, SLNR_u^L can be upper bound as

$$\text{SLNR}_u^L \leq \frac{\frac{P}{U_t}\omega_u^{\max}}{\sigma_u^2 + \frac{P}{U_t}\sum_{i=1,i\neq u}^{U_t}\omega_i^{\min}}, \tag{13}$$

where $\omega_u^{\max}$ is the maximum diagonal element of $\boldsymbol{\Omega}_u$, and $\omega_i^{\min}$ is the minimum diagonal element of $\boldsymbol{\Omega}_i$. Assuming that the largest diagonal element of $\boldsymbol{\Omega}_u$ is the $\overline{m}_u$-th diagonal element. It can be seen that the upper bound in (13) can be achieved if and only if the beamforming vector is given by

$$\mathbf{w}_u = (\mathbf{F}_M)_{\overline{m}_u}, \tag{14}$$

and the $\overline{m}_u$-th diagonal element of $\boldsymbol{\Omega}_i$, $i \neq u$ is its minimum diagonal element which is zero, where $(\mathbf{X})_i$ denotes the i-th column of matrix $\mathbf{X}$. Therefore, to maximize the lower bound of the average SLNR, the BS should transmit to the user with beamforming vector (14) and the scheduled users should satisfy that the maximum diagonal element of $\boldsymbol{\Omega}_u$ is orthogonal to the non-zero diagonal element of $\boldsymbol{\Omega}_i$.

When $K_u = 0$, the channel becomes Raleigh fading channel. In this case, $\overline{m}_u$ in (14) reduces to the index of the largest diagonal element of $\boldsymbol{\Lambda}_u$. When $\boldsymbol{\Lambda}_u = \mathbf{I}_M$, the channel becomes uncorrelated Rician fading channel. In this case, $\overline{m}_u$ reduces to the index of the largest diagonal element of $\mathbf{A}_u$. These are consistent with result in [14] and [10] when 2D antenna array reduces to ULA. Therefore, in this paper, we employ the beamforming vector (14) for scheduled user u. However, the constraint that for $u \neq i$, the maximum diagonal element of $\boldsymbol{\Omega}_u$ should be orthogonal to the non-zero diagonal element of $\boldsymbol{\Omega}_i$ is too strict. In the next section, we will discuss the scheduling algorithm for the system we considered.

4   User Scheduling Algorithms

Based on the previous statistical beamforming algorithm, in this section, we focus on the user scheduling algorithm exploiting only statistical CSI under correlated Rician fading channel. Three user scheduling algorithms are proposed, which are the maximum SLNR, the most dissimilar and the modified-TIN.

4.1   Maximum SLNR

In the previous section, we get the statistical beamforming method by maximizing the SLNR lower bound. In the following user scheduling algorithm, we also consider the maximization of the SLNR lower bound. To achieve high sum rate, we would like that the SLNR of the selected user won't decrease much due to the power leaked to the candidate user's channel direction, while the candidate user could have high SLNR.

Based on the derived beamforming vector, and after some manipulation, we get

$$\mathrm{SLNR}_u^L = \frac{\omega_u^{(\overline{m}_u)}}{\sum_{i=1, i \neq u}^{U_t} \omega_i^{(\overline{m}_u)} + \frac{U_t \sigma_u^2}{P}}, \tag{15}$$

where $\omega_i^{(j)}$ denotes the j-th diagonal element of matrix $\boldsymbol{\Omega}_i$, $i = 1, \cdots, U_t$. Let's rewrite (15) as

$$\mathrm{SLNR}_u^L = \frac{1}{\sum_{i=1, i \neq u}^{U_t} r_{u,i}^{-1}}, \tag{16}$$

where

$$r_{u,i} = \frac{\omega_u^{(\overline{m}_u)}}{\omega_i^{(\overline{m}_u)} + \frac{\sigma_u^2 U_t}{P(U_t-1)}}. \tag{17}$$

It can be seen that lower bound (16) increases as $r_{u,i}$ increases. Define Q to be the set of already selected users and Ξ to be the set of unselected users. To make sure that the SLNR of the selected user u won't decrease much due to the candidate user i, and SLNR of the candidate user could be high, we would like that the minimum of $r_{i,u}$ and $r_{u,i}$ is as large as possible. Therefore, in this algorithm, we schedule the user with the largest

$$R_i = \min\left(\min_{u \in Q} r_{i,u}, \min_{u \in Q} r_{u,i}\right), \qquad i \in \Xi, \tag{18}$$

among the candidate users.

Based on the above analysis, we propose the following user scheduling algorithm. Firstly initialize Q to be empty. Select the user that can achieve the largest R_i, $i \in \Xi$. Since no user has been selected, that is to select the user with the largest $\omega_i^{(\overline{m}_i)}$. Add it into the set of selected users Q, and remove it from Ξ. Next, add the user with the largest R_i in Ξ into Q, and remove it from Ξ. Then, repeat the previous process for users in Ξ until there is no user left in Ξ or U_t users have been selected. The details of the proposed user scheduling algorithm are as follows:

Algorithm 1 Maximum SLNR

. Initialization:
(1) $N = 1$.
(2) Find user $\hat{s}_1$ such that $\hat{s}_1 = \arg\max_{i \in \Xi} \omega_i^{(\overline{m}_i)}$.
(3) Set $Q = \{\hat{s}_1\}$, $\Xi = \Xi \backslash \hat{s}_1$.
. Iteration:
(4) Increase N by 1.
(5) Find user $\hat{s}$ such that $\hat{s} = \arg\max_{i \in \Xi} R_i$.
(6) Set $Q = Q \cup \hat{s}$, $\Xi = \Xi \backslash \hat{s}$.
(7) If $\Xi = \varnothing$ or $N = U_t$, stop. Otherwise, go to 4) and repeat the iteration.

4.2   Most dissimilar

In Sect. 3, it was shown that user u and user i can perfectly be served simultaneously if $\omega_i^{(\overline{m}_u)} = 0$ and $\omega_u^{(\overline{m}_i)} = 0$. The worst case that user u and i can not be served simultaneously is that $\overline{m}_u = \overline{m}_i$.

Let us define

$$\mathbf{b}_{u,i} = \left[\omega_u^{(\overline{m}_u)}, \omega_u^{(\overline{m}_i)}\right]^T, \tag{19}$$

and

$$\mathbf{b}_{i,u} = \left[\omega_i^{(\overline{m}_u)}, \omega_i^{(\overline{m}_i)} \right]^T, \tag{20}$$

for user u and user i, respectively. Note that the cosine similarity between $\mathbf{b}_{u,i}$ and $\mathbf{b}_{i,u}$ is defined as

$$s_{u,i} = \frac{\mathbf{b}_{u,i}^T \mathbf{b}_{i,u}}{|\mathbf{b}_{u,i}||\mathbf{b}_{i,u}|}. \tag{21}$$

It can be seen that $0 \le s_{u,i} \le 1$, user u and user i can perfectly be served simultaneously when $s_{u,i} = 0$, and can not be served simultaneously at all when $s_{u,i} = 1$. Therefore, $s_{u,i}$ can be employed to measure whether user u and user i can be served simultaneously. The closer $s_{u,i}$ is to zero, the more dissimilar $\mathbf{b}_{u,i}$ and $\mathbf{b}_{i,u}$ are to each other, the more user u and user i can be served simultaneously. Therefore, the user with the smallest

$$S_u = \max_{i \in Q} s_{u,i}, \tag{22}$$

should be scheduled, i.e., the user which is most dissimilar to the already selected users.

Note that user u and user i are completely unsuitable to be served simultaneously when $\overline{m}_u = \overline{m}_i$. Therefore, to reduce the search complexity, there is no need to consider the user i, $i \in \Xi$, that satisfy $\overline{m}_i = \overline{m}_u$ for any user $u \in Q$.

Based on the above analysis, we propose the following user scheduling algorithm. Firstly initialize Q to be empty. Select the first user that can achieve the largest $\omega_i^{(\overline{m}_i)}$, $i \in \Xi$, since no user has been selected before. Add it into the set of select users Q and remove it from Ξ. Next, check the remaining user in set Ξ, and remove the user i in Ξ that does not satisfy $\overline{m}_i \ne \overline{m}_u$, for $\forall u \in Q$. By successively removing the users, the searching complexity for each selection is reduced. Then, add the user with the smallest S_i in Ξ into Q, and remove it from Ξ. Repeat the previous process for users in Ξ. The algorithm terminates when no more users in Ξ which indicates that the resulting set is maximal or U_t users have been selected. The details of the proposed user scheduling algorithm are as follows:

4.3 Modified-TIN

In this section, we propose a user scheduling algorithm by treating the inter-user interference as uncorrelated noise to each user's useful signal. Obviously, to achieve high sum rate, we would like that for each scheduled user, the useful signal power of it would be large, and the inter-user interference could be small.

Let us define

$$t_u = \frac{\frac{P}{U_t} |\mathbf{h}_u^T \mathbf{w}_u|^2}{\max_{i \in Q, i \ne u} \frac{P}{U_t} |\mathbf{h}_u^T \mathbf{w}_i|^2 \times \max_{i \in Q, i \ne u} \frac{P}{U_t} |\mathbf{h}_i^T \mathbf{w}_u|^2}, \tag{23}$$

for user u. It can be seen that t_u increases as the useful signal power of user u increases and decreases as the inter-user interference increases. Therefore, in

---

Algorithm 2 Most dissimilar

. Initialization:

(1) $N = 1$.

(2) Find user $\hat{l}_1$ such that $\hat{l}_1 = \arg\max\limits_{i \in \Xi} \omega_i^{(\overline{m}_i)}$.

(3) Set $Q = \{\hat{l}_1\}$, $\Xi = \Xi \backslash \hat{l}_1$.

. Iteration:

(4) Increase N by 1.

(5) Remove user i that satisfy $\overline{m}_i = \overline{m}_u$, for $\forall u \in Q$ from Ξ.

(6) Find user $\hat{l}$ such that $\hat{l} = \arg\min\limits_{i \in \Xi} S_i$.

(7) Set $Q = Q \cup \hat{l}$, $\Xi = \Xi \backslash \hat{l}$.

(8) If $\Xi = \varnothing$ or $N = U_t$, stop. Otherwise, go to 4) and repeat the iteration.

---

order to maximize the useful signal power and minimize the inter-user interference, we would like to schedule user u with the largest t_u.

Note that user scheduling based on t_u requires instantaneous CSI at BS, which leads to large amount of feedback overhead for FDD systems, especially when the number of users is large. Here, we would like to exploit only statistical CSI. Inspired by [15], we treat the inter-user interference as uncorrelated noise to the useful signal of each user, and use the following signal-to-noise factor

$$T_u = \frac{\frac{P}{U_t}\mathrm{E}\{|\mathbf{h}_u^T\mathbf{w}_u|^2\}}{\max\limits_{i \in Q, i \neq u} \frac{P}{U_t}\mathrm{E}\{|\mathbf{h}_u^T\mathbf{w}_i|^2\} \times \max\limits_{i \in Q, i \neq u} \frac{P}{U_t}\mathrm{E}\{|\mathbf{h}_i^T\mathbf{w}_u|^2\}}, \tag{24}$$

to design the user scheduling algorithm. Substituting (14) into (24), we have

$$T_u = \frac{\frac{P}{U_t}D_u}{\max\limits_{i \in Q, i \neq u} \frac{P}{U_t}I_{u,i} \times \max\limits_{i \in Q, i \neq u} \frac{P}{U_t}I_{i,u}}, \tag{25}$$

where $D_u = \omega_u^{(\overline{m}_u)}$, $I_{u,i} = \omega_u^{(\overline{m}_i)}$ and $I_{i,u} = \omega_i^{(\overline{m}_u)}$. To make sure that the quality of service of both selected user u and the candidate user i are acceptable, and also to reduce the searching complexity of the scheduling procedure, we would like that T_u and T_i are both above a certain threshold γ. Candidate users that do not satisfy this constraint are removed from the set of candidate users, and will not be considered in the rest part of the scheduling procedure. Then, schedule the user with the largest T_i among the remaining candidate users.

Based on the above analysis, we propose the following user scheduling algorithm. Firstly initialize Q to be empty. Since no user has been selected before, add the first user with the largest D_i, $i \in \Xi$ into Q and remove it from Ξ. Next, check all remaining users in set Ξ, and remove the user i that does not satisfy

$$T_i = \frac{\frac{P}{U_t}D_i}{\max\limits_{j \in Q} \frac{P}{U_t}I_{i,j} \times \max\limits_{j \in Q} \frac{P}{U_t}I_{j,i}} \geq \gamma, \tag{26}$$

or due to its addition, the already selected user u, $u \in Q$ can not satisfy

$$T_u = \frac{\frac{P}{U_t} D_u}{\frac{P}{U_t} \max\left(\max\limits_{j \in Q, j \neq u} I_{u,j}, I_{u,i}\right) \times \frac{P}{U_t} \max\left(\max\limits_{j \in Q, j \neq u} I_{j,u}, I_{i,u}\right)}. \tag{27}$$

Then, add the user with the largest T_i in Ξ into Q, and remove it from Ξ. Repeat the previous process until U_t users have been selected or no more users in Ξ. The details of the proposed user scheduling algorithm are as follows:

Algorithm 3 Modified-TIN

. Initialization:
(1) $N = 1$.
(2) Find user $\hat{k}_1$ such that $\hat{k}_1 = \arg\max\limits_{i \in \Xi} D_i$.

(3) Set $Q = \{\hat{k}_1\}$, $\Xi = \Xi \backslash \hat{k}_1$.
. Iteration:
(4) Increase N by 1.
(5) Remove user i that does not satisfy $T_i \geq \gamma$ and $T_u \geq \gamma$ for $\forall u \in Q$ from Ξ.
(6) Find user $\hat{k}$ such that $\hat{k} = \arg\max\limits_{i \in \Xi} T_i$.

(7) Set $Q = Q \cup \hat{k}$, $\Xi = \Xi \backslash \hat{k}$.
(8) If $\Xi = \varnothing$ or $N = U_t$, stop. Otherwise, go to 4) and repeat the iteration.

---

5   Simulation

In this section, we present numerical results to validate the performance of the proposed user scheduling algorithms. In all simulations, $M = 64$, $L = 100$, the noise level of all users are the same, i.e., $\sigma_u^2 = \sigma^2$. The BS can serve 16 users at most. The results are obtained by averaging over 500 user drops. Here, we take the user scheduling algorithm in [14] as a performance baseline which divides the users into 16 clusters, so that user u in i-th cluster satisfying $\overline{m}_u \in [4(i-1)+1, 4i]$ and the user with the largest $w_u^{(\overline{m}_u)}$ in each cluster is selected.

Figure 1 shows the average sum rate of the single-cell MIMO downlink transmission systems over correlated Rician fading channel under different user scheduling algorithms. In this figure, the Rician K-factor of each user is uniformly distributed in $[K_{\min}, K_{\max}]$, where $K_{\min} = -10$ dB, $K_{\max} = 10$ dB, the AoD of each user is distributed uniformly in $(-90°, 90°)$, and the angle spread of each user is uniformly distributed in $(5°, 15°)$. The threshold for the modified-TIN algorithm is $\gamma = 5$. Compared to the scheduling algorithm in [14], all the three proposed scheduling algorithms improve the average sum rate of the system significantly. It can be clearly observed from the figure that under the considered environment, the performance of the most dissimilar and the modified-TIN

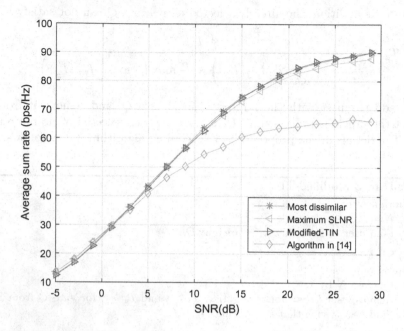

Fig. 1. Average sum rate of different user scheduling algorithms

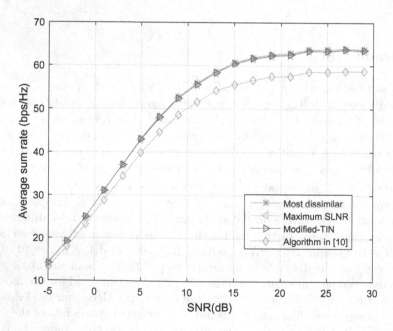

Fig. 2. Average sum rate of different user scheduling algorithms, $\mathbf{R}_i = \mathbf{I}_M$

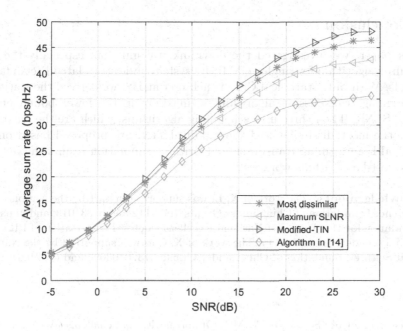

Fig. 3. Average sum rate of different user scheduling algorithms, $K_i = 0$

algorithm are almost the same, and is slightly superior to the performance of the maximum SLNR algorithm.

Figure 2 shows the average sum rate performance of the proposed scheduling algorithm when $\mathbf{R}_i = \mathbf{I}_M$, $i = 1, \cdots, L$, that is the channels of all the users are uncorrelated Rician fading. The Rician K-factor of each user is uniformly distributed in $[K_{\min}, K_{\max}]$, where $K_{\min} = -10$ dB, $K_{\max} = 10$ dB, the AoD of each user is distributed uniformly in $(-90°, 90°)$. The threshold for the modified-TIN algorithm is $\gamma = 5$. In this figure, the scheduling algorithm proposed in [10] for uncorrelated Rician fading channel is also presented. It can be seen that the performance of the proposed algorithms are almost the same, and all the proposed algorithms outperform the scheduling algorithm in [10].

Figure 3 shows the average sum rate of the single-cell MIMO downlink transmission systems when $K_i = 0$, $i = 1, \cdots, L$, that is the channel of each user is Rayleigh fading. In this figure, we assume that the AoD of each user is distributed uniformly in $(-90°, 90°)$, and the angle spread of each user is uniformly distributed in $(5°, 15°)$. The threshold for the modified-TIN algorithm is $\gamma = 5$. It can be seen that all the proposed algorithms still outperform the algorithm in [14] under this situation. And the modified-TIN performs best, while the maximum SLNR performs worst among the three proposed scheduling algorithm.

6   Conclusion

In this paper, we investigated the downlink transmission, especially the user scheduling algorithm for single-cell MIMO system under correlated Rician fading channels. With only statistical CSI of each user at BS, we derived the statistical beamforming transmission algorithm by maximizing the lower bound on the average SLNR. Then three user scheduling algorithms, which are the maximum SLNR, the most dissimilar and the modified-TIN, are proposed based on the statistical beamforming transmission algorithm. Simulation results showed that the proposed algorithms work well.

Acknowledgements. The work of X. Li was supported in part by the National Natural Science Foundation of China under Grants 61571112 and 61831013, and in part by A Foundation for the Author of National Excellent Doctoral Dissertation of PR China (FANEDD) under Grant 201446. The work of X. Gao was supported by the National Natural Science Foundation of China under Grants 61320106003 and 61521061.

References

1. Rusek, F., et al.: Scaling up MIMO: Opportunities and challenges with very large arrays. IEEE Signal Process. Mag. **30**, 40–60 (2013)
2. Marzetta, T.L.: Noncooperative cellular wireless with unlimited numbers of base station antennas. IEEE Trans. Wirel. Commun. **9**, 3590–3600 (2010)
3. Marzetta, T.L., Caire, G., Debbah, M., Chih-Lin, I., Mohammed, S.K.: Special issue on massive MIMO. J. Commun. Netw. **15**, 333–337 (2013)
4. Adhikary, A., Nam, J., Ahn, J., Caire, G.: Joint spatial division and multiplexing-The large-scale array regime. IEEE Trans. Inf. Theory **59**, 6441–6463 (2013)
5. Chen, J., Lau, V.K.N.: Two-tier precoding for FDD multi-cell massive MIMO time-varying interference networks. IEEE J. Sel. Areas Commun. **32**, 4418–4432 (2014)
6. Nam, J., Adhikary, A., Ahn, J.-Y., Caire, G.: Joint spatial division and multiplexing: Opportunistic beamforming, user grouping and simplified downlink scheduling. IEEE J. Sel. Top. Signal Process. **8**, 876–890 (2013)
7. Brady, J., Behdad, N., Sayeed, A.M.: Beamspace MIMO for millimeter-wave communications: System architecture, modeling, analysis, and measurements. IEEE Trans. Antennas Propagat. **61**, 3814–3827 (2013)
8. Kong, C., Zhong, C., Matthaiou, M., Zhang, Z.: Performance of downlink massive MIMO in Ricean fading channels with ZF precoder. Proc. IEEE **ICC**, 1776–1782 (2016)
9. Falconet H., Sanguinetti L., Kammoun A., Debbah M.: Asymptotic analysis of downlink MISO systems over Rician fading channels. In: Proceedings of IEEE ICASSP, pp. 3926–3930. Shanghai, China (2016)
10. Li, X., Jin, S., Suraweera, H.A., Hou, J., Gao, X.: Statistical 3-D beamforming for large-scale MIMO downlink systems over Rician fading channels. IEEE Trans. Commun. **64**, 1529–1543 (2016)
11. Taricco, G., Riegler, E.: On the ergodic capacity of correlated Rician fading MIMO channels with interference. IEEE Trans. Inf. Theory **57**, 4123–4137 (2011)
12. Han, Y., Zhang, H., Jin, S., Li, X.: Investigation of transmission schemes for millimeter-wave massive MU-MIMO systems. IEEE Syst. J. **11**, 72–93 (2017)

13. Mullen, K.: A note on the ratio of two independent random variables. Am. Stat. **21**, 30–31 (1967)
14. Li, X., Jin, S., Gao, X.: Three-dimensional beamforming for large-scale FD-MIMO systems exploiting statistical channel state information. IEEE Trans. Veh. Technol. **65**, 1529–1543 (2016)
15. Adhikary, A., Dhillon, H.S., Caire, G.: Massive-MIMO meets HetNet: Interference coordination through spatial blanking. IEEE J. Sel. Areas Commun. **33**, 1171–1186 (2015)

A Blind Detection Algorithm
for Modulation Order in NOMA Systems

Kai Cheng[(⊠)] [iD], Ningbo Zhang, and Guixia Kang

Key Laboratory of Universal Wireless Communications, Ministry of Education,
Beijing University of Posts and Telecommunications, Beijing , China
{chengkaibupt, nbzhang, gxkang}@bupt.edu.cn

Abstract. The blind detection algorithm for modulation order (MOD) of interference user in power-domain non-orthogonal multiple access (NO-MA) is studied by academics. Maximum likelihood method is the optimal approach, but with huge computational complexity. A sub-optimal approach based on max-log approximation is deduced which can reduce computational complexity, but with performance degradation. This paper investigates an improved blind detection algorithm for modulation order based on max-log likelihood approach in NOMA systems. Unlike the other two algorithms, the proposed algorithm takes the statistical characteristics of the received signal into consideration. The complexity analysis and link-level simulation results are provided to verify that the proposed method outperforms the max-log likelihood method with a little additional computational complexity, and it is a good trade-off between complexity and performance.

Keywords: NOMA · Blind detection · Modulation order

1 Introduction

With the rapid development of wireless communications, the number of users and the volume of services have exploded, putting higher demand on the system capacity of wireless networks. Subsequently, mobile communication technology is presently facing a new challenge, giving birth to the emergence of fifth-generation (5G) wireless communication. Since the exponential development of mobile Internet and the Internet of Things (IoT), one of the critical points that 5G needs to solve is high data-rate and capacity in applications.

Non-orthogonal multiple access (NOMA), as one of the candidate standards for next generation cell communication technology, has a series of advantages, such as higher spectral efficiency (SE) [8], higher sum channel capacity [7], smaller feedback requirement and lower transmission latency [5]. More importantly, with guaranteed user fairness assumption, the system throughput of NOMA can be significantly larger than orthogonal multiple access (OMA) [4].

© ICST Institute for Computer Sciences, Social Informatics and Telecommunications Engineering 2019
Published by Springer Nature Switzerland AG 2019. All Rights Reserved
X. Liu et al. (Eds.): ChinaCom 2018, LNICST 262, pp. 688–697, 2019.
https://doi.org/10.1007/978-3-030-06161-6_67

Generally, different NOMA solutions can be classified into two categories, power-domain NOMA and code-domain NOMA. This paper focuses on power-domain NOMA. According to the concept of NOMA, signals for multiple users are superposed in power domain and transmitted in the same time-frequency resources at transmitter. Multiuser detection (MUD) algorithms, such as successive interference cancellation (SIC) are utilized to detect desired signals at receiver.

Reference [1] shows that a significant performance gain can be achieved under the assumption that receiver has ideal interference parameters associated with undesired signals at the receiver end. The most common and easiest way to get interference parameters is broadcast signaling (higher layer signaling or dynamic signaling). Another method to obtain those dynamic interference parameters is blind detection (BD) [2]. Reference [2] shows the assistance information that is required for receivers to cancel superposition interference. Obviously, if dynamic interference parameters are transmitted through signaling, a large amount of signaling resources will be consumed and the number of UEs that BS can serve simultaneously will be reduced. Assume that N bits is needed to signal those parameters for one UE, when M UEs are superposed, the total consumed bits would be $M(M - 1)N/2$ bits. This is exactly opposite to the idea of designing the NOMA system. Therefore, blind detection or hybrid method would be the feasible solution in practice implementation.

Heunchul Lee et al. proposed blind detection algorithms based on max-log approximation for estimating the dynamic interference parameters TPR, RI, PMI, and MOD [6]. Alexei Davydov et al. proposed a blind maximum likelihood (ML) interference suppression receiver relying on direct estimation of the interfering signal parameters, such as transmission scheme, precoding vector, power boosting and modulation [3]. Maximum likelihood blind detection algorithm is the optimal solution but with high complexity, which can not be achieved in practice. To reduce complexity, max-log likelihood algorithm based on max-log approximation is deduced. However, This is done at the expense of performance for the reduction in complexity. And how to find the trade-off between performance and complexity has not been addressed so far. This paper focuses on blind detection algorithm for modulation order in power-domain NOMA.

The remainder of this paper is organized as follows: Sect. 2 presents system model, including multiuser superposition coding scheme. Section 3 proposes an improved blind detection algorithm for modulation order based on max-log likelihood algorithm and provides complexity analysis among different algorithms. In Sect. 4, we provide link-level simulation results to compare the performance of blind detection correct rate and link-level throughput between conventional algorithm and proposed algorithm. Finally, conclusions are made in Sect. 5.

2   System Model

This section presents system model. This paper considers a downlink single-cell scenario where consists one base station (BS) and N user equipments (UEs). The UEs are denoted as U_i, $i = 1 \cdots N$. The channel condition from BS to each

UE is denoted as h_i, $i = 1 \cdots N$. Assume that the channel conditions for every UE are sorted as

$$0 < |h_1|^2 \le |h_2|^2 \le \cdots \le |h_N|^2, \tag{1}$$

which means that the N-th user U_N holds the strongest channel condition and the first user U_1 holds the worst channel condition.

Based on the concept of NOMA, BS can serve more than one UEs on the same time-frequency resource simultaneously. And those UEs hold distinct channel conditions. Reference [4] has proved that NOMA with fixed power allocation (F-NOMA) can offer a larger sum rate than orthogonal multiple access (MA), and the performance gain of F-NOMA over conventional MA can be further enlarged by selecting users whose channel conditions are more distinctive. The NOMA scheme implements superposition coding (SC) in power domain at transmitter and decodes UE's signal with the help of SIC techniques at receiver. At transmitter, the UE with better channel condition would be assigned with a lower power ratio, and the UE with worse channel condition would be assigned with a higher power ratio. The portion of total power assigned to U_i is denoted as α_i, which satisfies $\sum_{i=1}^{s} \alpha_i = 1$, where s indicates the number of superposed signals on the same time-frequency resource. At receiver end, each UE needs to decode the signals of weaker UEs before decoding its own signal, i.e., U_i needs to decode signals of U_m, where $m < i$. The signals of weaker UEs would be reconstructed and subtracted from the received signal. U_i treats signals of U_n with $n > i$ as interference.

Without loss of generality, we choose a simple NOMA scenario with one BS and two UEs. The two UEs are marked as "Target UE" and "Interference UE", with channel condition h_2 and h_1, respectively. The channel conditions h_1 and h_2 satisfy $|h_2|^2 > |h_1|^2$, which indicates that "Target UE" has better channel condition than "Interference UE". Therefore, α, the portion of total power $\mathscr{P}$ allocated to target UE, is less than 0.5, which can be written as $\alpha < 0.5$.

Let us denote the K-dimensional complex signal vector transmitted from BS to user U_i as

$$\boldsymbol{x}^{(i)} = [x_1^{(i)}, x_2^{(i)}, \cdots, x_K^{(i)}]^T, \tag{2}$$

where $i = 1, 2$, $x_k^{(i)}$ denotes the k-th symbol for user U_i, K indicates the number of symbols for user U_i, and $(\cdot)^T$ denotes the transpose of a vector. Symbol $x_k^{(i)}$ is chosen from constellation set $\mathbb{C}^{(i)}$, whose cardinality is denoted by $|\mathbb{C}^{(i)}|$. Thus, the superposed signal to be transmitted to U_1 and U_2 can be written as

$$\boldsymbol{t} = \sqrt{\alpha \mathscr{P}} \boldsymbol{x}^{(2)} \oplus \sqrt{(1 - \alpha) \mathscr{P}} \boldsymbol{x}^{(1)}, \tag{3}$$

where $\boldsymbol{t}$ denotes the superposed signal to be transmitted, α represents the fraction of total power assigned to near user U_2, $\mathscr{P}$ denotes the total power used for transmission at transmitter, and the rules for $\oplus$ operation is shown in Fig. 5.1.2–2 in [2].

Let us define $\boldsymbol{r}^{(i)}$ as the received signal vector at the user U_i. Then, $\boldsymbol{r}^{(i)}$ can be written as

$$\boldsymbol{r}^{(i)} = \boldsymbol{H}^{(i)} \boldsymbol{t} + \boldsymbol{n}^{(i)}, \quad \text{for} \quad i = 1, 2, \tag{4}$$

where $\boldsymbol{H}^{(i)}$ denotes the channel matrix from BS to user U_i, $\boldsymbol{n}^{(i)}$ is the additive noise vector, whose elements are independent and identically-distributed (i.i.d.) complex Gaussian, $\mathbb{E}[|\boldsymbol{n}^{(i)}|^2] = \sigma_i^2$, where $\mathbb{E}[\cdot]$ denotes the expectation operator, and $|\cdot|$ represents the absolute value of a complex number.

The basic idea of non-orthogonal multiple access technology is to introduce interference information at the transmitter and simultaneously transmit the information of multiple users on the same time-frequency resource by superposition coding. Reference [2] describes candidate multiuser superposition transmission schemes, which can be categorized into three categories — Category 1, Category 2 and Category 3.

Because of the loss of power ratio in category 3 and the loss of gray mapping in category 1, this paper chooses category 2 as superposition coding scheme. An example of composite constellation of Category 2 is shown in [2]. With joint modulation mapping for target and interference UEs, gray mapping is kept for the label bits of the composite constellation. Moreover, receiver uses SIC technique to achieve the correct demodulation of the received signal.

3 Proposed Blind Detection Algorithm for Modulation Order and Complexity Analysis

This section investigates blind detection problem for estimating MOD interference parameter in NOMA systems. Note that by transmitting the downlink control information (DCI) through physical downlink control channel (PDCCH), the MOD parameter of target UE can be found explicitly. One way to get MOD parameter of interference UE is broadcast signaling. However, this method consumes too much unnecessary signaling. Another way to get MOD parameter of interference UE is blind detection, which will be presented in this section. Furthermore, complexity analysis will be presented in this section.

3.1 Proposed Algorithm for Blind Detection

It is well known that blind detection based on maximum likelihood (ML) estimation minimizes the error probability. Let $p(r_k^{(i)}|t_k)$ denote the conditional probability density function for $r_k^{(i)}$, given t_k, which is represented by

$$p(r_k^{(i)}|t_k) = \frac{1}{\sqrt{2\pi\sigma_i^2}} \exp(-\frac{||r_k^{(i)} - H_k^{(i,e)} * t_k||^2}{2\sigma_i^2}),$$ (5)

where $H_k^{(i,e)}$ is the k-th element in the effective channel matrix $\boldsymbol{H}^{(i,e)}$ and the superscript "e" is the abbreviation of "effective".

Assume that candidate modulation order set for interference user is a P-by-1 vector $\boldsymbol{M}^{\text{itf}}$ and the modulation order for target user is m^{tar}. P is the number of candidate modulation orders. Thus, the candidate composite modulation order for superposition coding is

$$M = M^{\text{itf}} + m^{\text{tar}}.$$ (6)

And p is one of elements in M. In case of a certain modulation order p, the maximum likelihood blind detection algorithm for interference modulation order is [6]

$$M_p = \frac{1}{K} \sum_{k=1}^{K} \frac{1}{|\mathbb{C}_p|} \sum_{t_k \in \mathbb{C}_p} \exp(-\frac{||r_k^{(i)} - H_k^{(i,e)} * t_k||^2}{\sigma_i^2}). \tag{7}$$

Constants in (5) is omitted in (7). Then, the ML detector performs an exhaustive search among all the possible constellation points corresponding to all the candidate composite modulation order p, and makes the best decision of p^{opt} which maximizes the metric

$$p^{\text{opt}} = \arg\max_p M_p. \tag{8}$$

Although ML detector is optimal, it is not practical, since it leads to prohibitive computational complexity. Thus, a suboptimal approach with reduced computational complexity to solve the optimal metric in (7), termed max-log likelihood, is derived. Max-log likelihood blind detection algorithm for interference modulation order can be described as [6]

$$M_p = \frac{1}{K} \sum_{k=1}^{K} \log \frac{1}{|\mathbb{C}_p|} \sum_{t_k \in \mathbb{C}_p} \exp(-\varpi_k^2/\sigma_i^2)$$

$$= \frac{1}{K} \sum_{k=1}^{K} -(\varpi_{k,\min}^2/\sigma_i^2 + \log|\mathbb{C}_p|) \tag{9}$$

$$+ \frac{1}{K} \sum_{k=1}^{K} \log(1 + \frac{\sum\limits_{\substack{t_k \in \mathbb{C}_p \\ t_k \neq t_{\min}}} \exp(-\varpi_k^2/\sigma_i^2)}{\exp(-\varpi_{k,\min}^2/\sigma_i^2)})$$

where

$$t_{\min} = \arg\min_{t_k \in \mathbb{C}_p} ||r_k^{(i)} - H_k^{(i,e)} * t_k||^2, \tag{10}$$

$$\varpi_k = ||r_k^{(i)} - H_k^{(i,e)} * t_k||, \tag{11}$$

$$\varpi_{k,\min} = ||r_k^{(i)} - H_k^{(i,e)} * t_{\min}||, \tag{12}$$

and the last term in (9) is omitted to reduce complexity [6]. Thus, max-log likelihood algorithm can be written as

$$M_p = \frac{1}{K} \sum_{k=1}^{K} \log \frac{1}{|\mathbb{C}_p|} \sum_{t_k \in \mathbb{C}_p} \exp(-\varpi_k^2/\sigma_i^2)$$

$$\approx \frac{1}{K} \sum_{k=1}^{K} -(\varpi_{k,\min}^2/\sigma_i^2 + \log|\mathbb{C}_p|). \tag{13}$$

Based on the above discussion, ML algorithm is optimal, but consumes too much time to get final result. Max-log likelihood algorithm consumes less time, but loses performance. And it demonstrates an interesting phenomenon that, in (7)–(13), the term $1/|\mathbb{C}_p|$ is a constant when p is given, which does not take the statistical characteristics of the received signal into consideration. Because of limited number of symbols and bandwidth constraints in a transmission process, the number of symbols can not be achieved statistically large. In another word, the joint modulated symbols at transmitter are non-uniform distribution. This results in the number of symbols, corresponding to each constellation point, is not exactly equal, but with slight deviations. Thus, the constant term $1/|\mathbb{C}_p|$ can not reflect actual characteristics of the signal, and the performance of blind detection degrades. The proposed method will take the features of signal into account, and can be written as

$$M_p = \frac{1}{K} \sum_{k=1}^{K} \log \xi_j \sum_{t_k \in \mathbb{C}_p} \exp(-\varpi_k^2/\sigma_i^2)$$

$$\approx \frac{1}{K} \sum_{k=1}^{K} -(\varpi_{k,\min}^2/\sigma_i^2 - \log \xi_j), \tag{14}$$

where

$$\xi_j = \frac{|\mathbb{S}_j|}{K}, \tag{15}$$

where $\mathbb{S}_j$ is the set of received symbols which have the minimum Euclidean norm to the j-th composite constellation point, and its cardinality is denoted by $|\mathbb{S}_j|$. j is the index of composite constellation symbol, which is in range $[1, \cdots, 2^p]$, where p denotes composite modulation order. K denotes the number of received symbols. The proposed method is termed as K-max-log likelihood algorithm.

3.2 Complexity Analysis

To further analyze the efficiency of the proposed K-max-log likelihood algorithm, we study the computational complexity of the proposed method. As described in Sect. 3.1, the proposed method is constructed using the max-log likelihood approach. Thus, as shown in Table 1, the computational complexity of maximum likelihood, max-log likelihood and K-max-log likelihood are compared associated with addition, subtraction, multiplication, division, exponent, logarithm and comparison. In Table 1, P denotes the number of modulation order candidates for interference user, K denotes the number of received symbols, and $|\mathbb{C}_p|$ denotes the number of constellation symbols in a certain composite modulation order p.

It is well known that multiplication, division, exponent and logarithm mathematical operations are more time-consuming. Both max-log likelihood and K-max-log likelihood reduce the number of such mathematical calculations, which can reduce computational complexity significantly. The term $\sum_{p=1}^{P} \mathbb{C}_p$ in K-max-log likelihood, compared to K, is pretty small. Therefore, with a little additional calculation, K-max-log likelihood algorithm almost has the same computational complexity as the max-log likelihood algorithm. And the proposed method has a much lower complexity than the maximum likelihood blind detection algorithm which requires exhaustive search and is impractical for real systems.

Table 1. Complexity comparison

	Max likelihood	Max-log likelihood	K-Max-log likelihood
Addition	$\sum_{p=1}^{P}(2K|\mathbb{C}_p|)+K$	KP	KP
Subtraction	$\sum_{p=1}^{P}2K|\mathbb{C}_p|$	$\sum_{p=1}^{P}(K|\mathbb{C}_p|+2K)$	$\sum_{p=1}^{P}(K|\mathbb{C}_p|+2K)$
Multiplication	$\sum_{p=1}^{P}3K|\mathbb{C}_p|$	$\sum_{p=1}^{P}(K+2K|\mathbb{C}_p|)$	$\sum_{p=1}^{P}(K+2K|\mathbb{C}_p|)$
Division	$\sum_{p=1}^{P}(K|\mathbb{C}_p|+K+1)$	$P(K+1)$	$P(K+1)$
Exponent	$\sum_{p=1}^{P}K|\mathbb{C}_p|$	0	0
Logarithm	0	0	$\sum_{p=1}^{P}|\mathbb{C}_p|$
Comparison	P	$P+\sum_{p=1}^{P}K|\mathbb{C}_p|$	$P+\sum_{p=1}^{P}K|\mathbb{C}_p|$

4   Performance Evaluation

In this section, we provide a series of link-level simulation results to verify the efficacy and accuracy of the proposed detection algorithm for estimating the MOD parameter of interference user. The candidate interference modulation set includes four kinds of modulation types, namely NONE, QPSK, 16QAM and 64QAM, where NONE means there is no interference user in current transmission process. Other simulation parameters are listed in Table 2. We use blind detection rate and link-level throughput, especially 70% throughput, as the measurements of the pros and cons of the proposed algorithm.

Table 2. Simulation parameters

Parameter	Value	Parameter	Value
Channel bandwidth	10 MHz	Carrier frequency	2 GHz
Sampling rates	15.36 MHz	TTI size/duration	14 OFDM Symbols/ms
CFI	2	Channel estimation	MMSE
Cyclic Prefix type	Normal	HARQ	Disabled
Number of FFT size	1024	CSI reporting mode	PUCCH 2-0
Fast fading	Rayleigh	No. of PRBs of PDSCH	50
Propagation channel	EPA	Receiver type	CWIC

First, we clarify the performance of blind detection rate of different algorithms. Figure 1 shows the blind detection correct rate of MOD parameter of interference user in NOMA transmission for 1-by-2 SIMO. The figure shows that K-max-log likelihood can always outperform max-log likelihood. On the other hand, the required SNR of K-max-log likelihood for achieving the correct rate of 100% is about 1.6–1.8 dB less than that of max-log likelihood algorithm. One thing should to be pointed out is that the number of simulation curves in Fig. 1 is less than the total number of simulation curves in Fig. 2a–d. The first reason is that there are no detection rate curves for "ideal" cases. The second reason is that the blind detection rate is affected by modulation order rather than transport block size.

Second, we clarify the performance of link-level throughput of max-log likelihood and K-max-log likelihood algorithm. Figure 2a illustrates the link-level throughput results of ideal, max-log likelihood and K-max-log likelihood method under certain conditions, where target user and interference user both use QPSK, but with different modulation coding schemes (MCSs). The "ideal" means that the MOD parameter of interference user is perfectly known through network signaling at the receiver. Figure 2b–d show the comparison of system throughput with those three methods under different conditions, which have already been shown in the corresponding captions and legends.

As shown in Fig. 2a–d, we can observe that the performance of proposed method always overcomes max-log likelihood's. In addition, the simulation results also show that target user throughput is significantly improved around 70% throughput point. Another noticeable feature is that the performance of the proposed method is very close to the ideal receiver around 70% throughput point in some simulation cases. Furthermore, the higher modulation order and the larger transport block size of the target user and the interference user is, the more performance degrades due to the failure of blind detection.

From the simulation results presented in this section, we can conclude that the proposed method can significantly improve blind detection rate and link-level throughput in NOMA systems, especially at 70% throughput point.

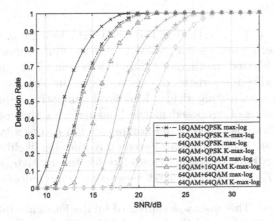

Fig. 1. The comparison of blind detection correct rate

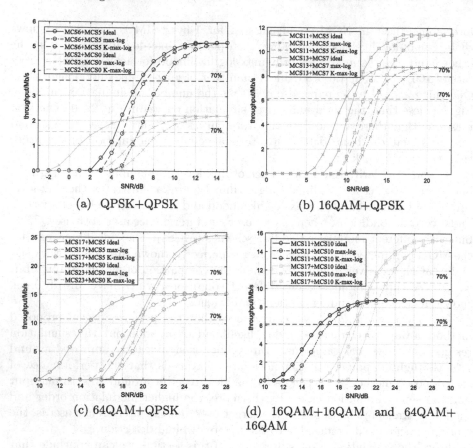

Fig. 2. Simulation results on different algorithms: link-level throughput

5   Conclusion

In this paper, we have investigated the blind detection problem of the MOD parameter for interference user in NOMA systems, and proposed an improved max-log-likelihood-based method for blind detection. The proposed method takes the statistical characteristics of received signal into consideration, which is not considered in max-log algorithm. Link-level simulation results have proved that the performance of proposed method outperforms that of max-log likelihood method with a little additional computational complexity. In conclusion, we have shown that the SIC receiver based on K-max-log likelihood blind detection algorithm can be a promising candidate for future high performance and low complexity UE devices in the next generation communication.

Acknowledgement. This work was supported by the Fundamental Research Funds for the Central Universities and the National Natural Science Foundation of China (61501056).

References

1. 3GPP: Study on network-assisted interference cancellation and suppression (naic) for lte. Technical report, 3rd Generation Partnership Project (3GPP) (2014)
2. 3GPP: Study on downlink multiuser superposition transmission (must) for lte. Technical report, 3rd Generation Partnership Project (3GPP) (2016)
3. Davydov, A., Morozov, G., Papathanassiou, A.: Blind maximum likelihood interference cancellation for lte-advanced systems. In: 2014 IEEE 79th Vehicular Technology Conference (VTC Spring), pp. 1–5 (2014). https://doi.org/10.1109/VTCSpring. 2014.7022846
4. Ding, Z., Fan, P., Poor, H.V.: Impact of user pairing on 5g nonorthogonal multiple-access downlink transmissions. IEEE Trans. Veh. Technol. **65**(8), 6010–6023 (2016). https://doi.org/10.1109/TVT.2015.2480766
5. Islam, S.M.R., Avazov, N., Dobre, O.A., Kwak, K.S.: Power-domain non-orthogonal multiple access (noma) in 5g systems: potentials and challenges. IEEE Commun. Surv. Tutor. **19**(2), 721–742 (2017). https://doi.org/10.1109/COMST.2016. 2621116. (Secondquarter 2017)
6. Lee, H., Lim, J.H., Cho, S., Kim, S.: Interference cancellation based on blindly-detected interference parameters for lte-advanced ue. In: 2015 IEEE International Conference on Communications (ICC). pp. 3143–3148 (2015). https://doi.org/10. 1109/ICC.2015.7248807
7. Liu, Y., Pan, G., Zhang, H., Song, M.: On the capacity comparison between mimo-noma and mimo-oma. IEEE Access **4**, 2123–2129 (2016). https://doi.org/10.1109/ ACCESS.2016.2563462
8. Saito, Y., Kishiyama, Y., Benjebbour, A., Nakamura, T., Li, A., Higuchi, K.: Non-orthogonal multiple access (noma) for cellular future radio access. In: 2013 IEEE 77th Vehicular Technology Conference (VTC Spring), pp. 1–5 (June 2013). https:// doi.org/10.1109/VTCSpring.2013.6692652

A Novel Non-WSSUS Statistical Model of Vehicle-Vehicle Radio Channel for the 5-GHz Band

Tao He[1]([✉]) [iD], Ye Jin[1], Weiting Fu[1], and Mingshuang Lian[2]

[1] Peking University, No.5 Yiheyuan Road, Haidian District, Beijing
People's Republic of China
hetao7@pku.edu.cn
[2] China University of Geosciences, No.29 XueYuan Road, Haidian District,
Beijing, People's Republic of China

Abstract. In recent years, with the dramatic development in intelligent transportation systems (ITS), vehicle-vehicle (V2V) radio channel models have drawn much attention. With the analysis of the preceding statistical models of V2V channel, it is obvious that the critical works in developing statistical channel models focus on two aspects, the modeling of the time-variant properties and the modeling of the severe multipath fading. In this paper, we discuss an innovative method to model the fading dispersive channels that do not satisfy the assumption of wide-sense stationary uncorrelated scattering (WSSUS). And the Weibull distribution is integrated to mimic the severe multipath fading of V2V radio channel. Moreover, based on the tapped-delay like (TDL) model, the non-WSSUS channel impulse response (CIR) function has been formulated. There are several statistical properties characterized to evaluate the performance of the proposed model, such as, Power delay profile (PDP), Temporal autocorrelation function (ACF), Local scattering function (LSF) and Power spectrum density (PSD). The simulation results demonstrate that the proposed model has a good performance in the characterization of the non-WSSUS V2V radio channel. Hence, the channel model presented will be beneficial in future V2V communications systems.

Keywords: V2V radio channel models · Non-stationary · Correlated scattering · Weibull fading · Statistical model

1 Introduction

V2V radio communication systems have recently drawn great attention for the potential to reduce traffic jams and accident rates [1]. Several attractive benefits of V2V radio communications are the intelligent perception of the road condition, the capacity for probing the dynamic weather and the awareness of the traffic condition, such as the traffic congestion, traffic distribution. For the advantages of V2V radio communication system, a recent standard for V2V communications in the 5-GHz Unlicensed National Information Infrastructure band has been developed, designing for extending the IEEE 802.11a application environment.

© ICST Institute for Computer Sciences, Social Informatics and Telecommunications Engineering 2019
Published by Springer Nature Switzerland AG 2019. All Rights Reserved
X. Liu et al. (Eds.): ChinaCom 2018, LNICST 262, pp. 698–708, 2019.
https://doi.org/10.1007/978-3-030-06161-6_68

However, the time-frequency selective fading features of such V2V channel are significantly different from the traditional cellular network channel, and thus requires distinct channel models [2]. Consequently, many V2V channel models have been proposed. And there are several statistical modeling works related to V2V radio channel, even though most of V2V channel models are the geometry-based stochastic models (GBSM) [3]. In [4], the authors take the Doppler spectrum shapes into account, reference [5] report the PDP and tap fading statistics. In [6], the authors have measured the features of the typical V2V channel rigorously based on the dedicated short-range communications (DSRC) standard. Furthermore, the implement of Rician or Rayleigh fading under the hypothesis of WSSUS has been presented in [7]. The work in [8] report the analysis of the multipath fading of V2V radio channel in the 5-GHz band. In addition, the description of severe multipath fading by means of taking the Weibull distribution into account.

Due to the complexity and diversity of the radio propagation environments, V2V radio channel tend to be non-WSSUS. According to the modeling of the non-stationarity, the work in [9] presents a basic non-stationary model algorithm named "birth and death" process, the multipath component is considered to be not static, it could appear or disappear after a generally short duration. In [10], the authors discuss that non-WSSUS can be divided into two part, one is the stationarity with respect to time, another is the correlation with respect to the scatterer, and the channel correlation function (CCF) is proposed to describe the non-WSSUS properties.

The remainder of this article is outlined as follows. In the next section, we give a specific description of the channel model. We then evaluate the performance of the channel model by analyzing the simulation results. Finally, conclusions are drawn in the final section.

2   Channel Model

In this section, we first describe the WSSUS CIR models. The modeling of the non-WSSUS is then addressed. Finally, the severe multipath fading is integrated to the presented channel model.

2.1   WSSUS CIR Models

In nearly all statistical models, the channel is modeled as a TDL model with a time-varying linear filter, and thus the impulse response of the filter is introduced to completely characterize the channel. However, the CIR can be defined as function $h(t, \tau)$, besides, the function is corresponding to the response of the channel at time t to an impulse input at time $t - \tau$. Furthermore, based on the hypothesis of WSSUS and Rayleigh fading distribution, the CIR function in the flat-fading channel can be expressed as

$$h(t, \tau) = \delta(\tau - \tau_0) \cdot \sum_{k=1}^{N} \alpha_k(t) e^{jw_{D,k}(t-\tau_0)-jw_c \cdot \tau_0} \tag{1}$$

where, at time t, $\alpha_k(t)$ represents the k th tap amplitude with respect to the Doppler spectrum, and the argument of the exponential term is the k th tap phase. According to the flat-fading channel, all the multipath component has same time delay τ_0. The δ-function is a Dirac delta, the carrier frequency is $w_c = 2\pi f_c$, and the term $w_{D,k} = 2\pi f_{D,k}$ represents the Doppler shift associated with the k th tap.

According to the practical propagation environments, V2V radio channel have numerous time-varying delay paths, so it will be frequency-selective rather than flat-fading. Besides, it is likely to have a Line-of-Sight (LOS) of V2V radio channel, so the distribution of the multipath fading will satisfy the Rician distribution. Consequently, the CIR function is given by

$$h(t,\tau) = \sum_{k=1}^{N} \left\{ \begin{array}{c} \sqrt{\frac{1}{K_k+1}}\alpha_k(t)e^{jw_{D,k}(t-\tau_k)} \\ + \sqrt{\frac{K_k}{K_k+1}}e^{j[w_{D,LOS,k}(t-\tau_k)+\varphi_{LOS,k}]} \end{array} \right\} e^{-jw_c\cdot\tau_k}\delta(\tau-\tau_k) \qquad (2)$$

where, K_k represents the Rician factor of the k th tap, however, it will be 0 when the LOS does not exist. And the term $w_{D,LOS,k} = 2\pi f_{D,los,k}$ represents the Doppler shift associated with the k th LOS, $\varphi_{LOS,k}$ represents the initial phase of the k th LOS.

2.2   Models for the Non-WSSUS

It is mentioned above that the non-WSSUS properties has two parts, one is the non-stationarity with respect to time, another is the correlated scatterer. Correspondingly, two algorithms are respectively presented to describe the two parts of the non-WSSUS.

Birth and Death Process. V2V radio propagation environments change frequently and rapidly because of the mobility and low transmitting and receiving antenna heights. With the time-variability of V2V channel, the number of multipath components and their strengths may not be static. In [9], An algorithm named the birth and death process is proposed to generate the time-vary multipath components. We incorporate this birth and death process into our developed channel models using persistence process $z_k(t)$.

For the persistence process, the Markov chain is frequently used to model it. And thus, we developed first-order two-state Markov chains specified by two matrices: the transition (TS) matrix and the steady-state (SS) matrix [11]. These matrices are expressed as follows:

$$TS = \begin{bmatrix} P_{00} & P_{01} \\ P_{10} & P_{11} \end{bmatrix} \quad SS = \begin{bmatrix} S_0 \\ S_1 \end{bmatrix} \qquad (3)$$

Here, the element P_{ij} in matrix TS is defined as the probability of converting the state i to state j, and each SS element S_i means that the steady-state probability of state i. Figure 1 shows sample persistence process associated with the third and fifth taps for the Urban-Antenna Inside Car (UIC) setting [9]. It is depicted that the third and fifth taps can appear or disappear after a short duration, so the non-stationarity in time domain can be perfectly modeled. In other words, the non-stationary CIR function is expressed as

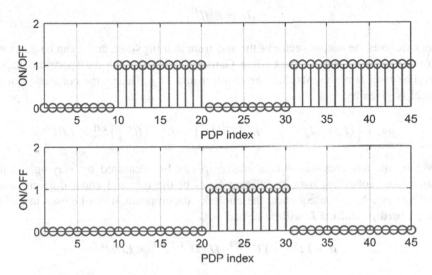

Fig. 1. Sample persistence processes $z_k(t)$ for taps 3 and 5 for the UIC case.

$$h(t,\tau) = \sum_{k=1}^{N} h'(t,\tau) \cdot z_k(t,\tau_k) \cdot \delta(\tau - \tau_k) \tag{4}$$

Here, $h'(t,\tau)$ represents the CIR of the WSSUS channel. The term $z_k(t,\tau_k)$ represents the k th tap persistence process.

Correlated Scatter Model. According to the non-WSSUS channel, the CCF can be computed as:

$$R_h(t,\tau;\Delta t,\Delta \tau) = E\{h(t,\tau + \Delta\tau)h^*(t - \Delta t,\tau)\} \tag{5}$$

Here, $\Delta t, \Delta \tau$ denote time lag and delay lag, respectively. $E\{\cdot\}$ denotes mathematical expectation, i.e., ensemble averaging, and it is reflected by the non-stationarity with respect to time and correlated with respect to delay.

In the former subsection, the birth and death process is introduced to model the non-stationarity with respect to time. Hence, we pay attention to the correlation with respect to delay. The delay-correlated matrix L is introduced to model it.

$$h(t,\tau) = \sum_{k=1}^{N} h''_{US}(t,\tau) \cdot L_k(\tau,\tau_k) = \boldsymbol{H}''_{US}(t,\tau) \cdot \boldsymbol{L} \tag{6}$$

where, $h''_{US}(t,\tau)$ represents the CIR of the US channel. The term $L_k(\tau,\tau_k)$ represents the delay-correlated vector with respect to the k th multipath component. Furthermore, the correlated with respect of delay is corresponding to the correlation of L. However, the

Doppler spectrum of the different taps is likely to be different, so the correlation matrix of the spectrum shaping filter output will be a diagonal matrix.

$$D = E[ff^H] \tag{7}$$

Here, f denotes the output vector of the spectrum shaping filter, and it can be generated by the transition of the complex white Gaussian noise through the spectrum shaping filter. The term D represents the correlation matrix of f. Hence, the correlation of the channel is given by

$$\rho = E\left\{ (Lf) \cdot (Lf)^H \right\} = E\{Lff^H L^H\} = L\{E\{ff^H\}\}L^H = LDL^H \tag{8}$$

However, the tap cross-correlation matrix ρ can be measured by varying antenna locations and collecting data at different times of the day and under different traffic conditions [9]. As a consequence, the cholesky decomposition can be used to resolve the measured ρ, and the L will be turned out.

$$\rho = TT^H = TD^{-1/2} \cdot D \cdot D^{-1/2}T^H = LDL^H \tag{9}$$

Note that T denotes the results of the cholesky decomposition of ρ. Above all, L is chosen to be

$$L = TD^{-1/2} \tag{10}$$

Besides, if all of the taps have the same Doppler spectrum, then we have $D = 1$ and $L = T$.

2.3 Non-WSSUS CIR Model with Severe Multipath Fading

Due to the severe delay disperse and the severe Doppler disperse in V2V channel, the multipath fading is likely to be severer than the Rayleigh fading. Furthermore, with the non-stationarity of V2V radio channel, the number of taps might be enormous and variable. The Weibull distribution [12] is used to model the severe multipath fading.

The Weibull model has two parameters, so it offers substantial flexibility. It can be given by

$$p_w(x) = \frac{\beta}{a^\beta} x^{\beta-1} exp\left[-\left(\frac{x}{a}\right)^\beta \right] \tag{11}$$

where β is the Weibull shape factor that is corresponding to the fading severity, $a = \sqrt{E(x^2)/\Gamma[(2/\beta)+1]}$ is a scale parameter, and Γ is the Gamma function. However, the Weibull distribution can be equal to the Rayleigh distribution when β equals 2. With the smaller β, the multipath fading will be severer.

Weibull distribution can be generated by Rayleigh distribution [13], we assume that W is a Weibull-distributed random variable, and there must be a Rayleigh-distributed random variable R that is submitted to the term $W = R^{2/\beta}$. Above all, the CIR function of the non-WSSUS V2V channel model is given by

$$h(t,\tau) = \sum_{m=1}^{N} \left\{ \sum_{k=1}^{N} \left\{ \begin{array}{c} \alpha_k(t)e^{jw_{D,k}(t-\tau_k)}e^{-jw_c\cdot\tau_k} \\ \times z_k(t,\tau_k) \end{array} \right\}^{\frac{2}{\beta_k}} \delta(\tau-\tau_k) \right\} \cdot L_m(\tau,\tau_m) \quad (12)$$

Here, τ_m denotes the m th multipath component, and the term $L_m(\tau,\tau_m)$ represents the m th row of the delay-correlated matrix L. Similarly, β_k denotes the Weibull shape factor of the k th multipath component. Furthermore, the proposed channel model is depicted in Fig. 2 followed.

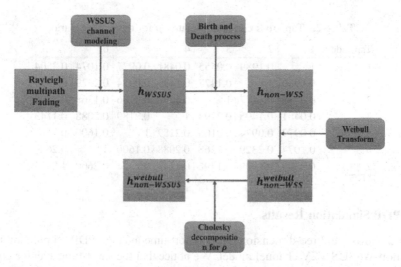

Fig. 2. Model structure for the proposed non-WSSUS V2V channel model.

3 Simulation Results and Analysis

In this section, we investigate the proposed V2V channel model in detail for several statistical properties based on the UIC scenario in [14]. And the following main parameters were chosen for the simulations: the carrier frequency f_c is set as 5.12 GHz, and the bandwidth for the simulations is equal to 10 MHz, the Doppler spectrum of all taps is considered as a same shape, such as, the U-shape. The following Table 1 gives the other parameters were chosen for our simulations.

Here, Energy is the measured mean-power of the taps. Weibull shape factor reflects the fading severity of the taps. And P_{00} and P_{11} are the Markov TS matrix diagonal elements, similarly, S_1 means that the steady-state probability of the 'birth' state. However, NA represents that the state of the tap does not exist. Furthermore, the tap cross-correlation matrix for simulations is given behind (Table 2).

Table 1. Channel parameters for the UIC scenario.

Tap Index	Energy	Weibull Shape Factor	$P_{00,k}$	$P_{11,k}$	S_1
1	0.756	2.49	NA	1.0000	1.0000
2	0.120	1.75	0.0769	0.9640	0.9625
3	0.051	1.68	0.3103	0.8993	0.8732
4	0.034	1.72	0.3280	0.8521	0.8199
5	0.019	1.65	0.5217	0.7963	0.7017
6	0.012	1.6	0.6429	0.7393	0.5764
7	0.006	1.69	0.6734	0.6686	0.4971

Table 2. Tap cross-correlation matrix ρ for the UIC scenario.

Tap index	1	2	3	4	5	6	7
1	1	0.1989	0.0555	0.0481	0.0977	0.1074	0.3504
2	0.1989	1	0.1477	0.1495	0.0974	0.2329	0.1999
3	0.0555	0.1477	1	0.2298	0.0106	0.1368	0.1496
4	0.0481	0.1495	0.2298	1	0.2189	0.2088	0.1143
5	0.0977	0.0974	0.0106	0.2189	1	0.1600	0
6	0.1074	0.2329	0.1368	0.2088	0.1600	1	0.2600
7	0.3504	0.1999	0.1496	0.1143	0	0.2600	1

3.1   PDP Simulation Results

Figure 3 shows the three-dimensional (3D) short-time average PDPs simulation results of the non-WSSUS V2V channel model. We notice that the short-time average changes with time t.

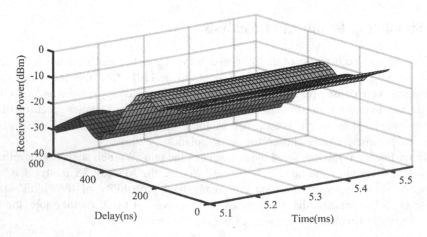

Fig. 3. Three-dimensional (3D) short-time averages PDPs of the non-WSSUS V2V channel model.

To emphasize these changes, five short-time PDPs, each over a duration of 20 μs, from 5.1 to 5.5 ms are depicted in Fig. 4. According to the varying time instants, it is obviously that the proposed channel models have the capacity for modeling the non-stationary delay disperse. In other words, the simulation results demonstrate the non-stationarity of V2V radio channel over the time period of 0.5 ms.

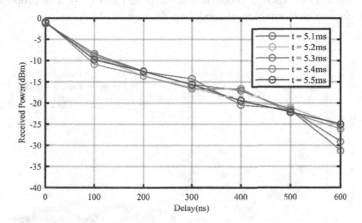

Fig. 4. Five short-time averages PDPs from 5.1 ms to 5.5 ms.

Figure 5 shows the comparison between the long-time PDP simulation results and the measured results for the UIC scenario. It is depicted in Fig. 5 that the simulation results have an approximate agreement with the measured results. However, the measured mean-power results are measured in a short duration, so the channel condition is considered as stationary. Due to the non-stationary properties, the simulation results have a little difference from the measured results. In other words, it is demonstrated that the proposed V2V channel model can perfectly mimic the delay disperse with the non-stationarity.

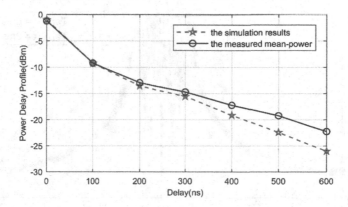

Fig. 5. Comparison between the PDP simulation results and the measured mean-power results for the UIC scenario.

3.2   Correlation Function Simulation Results

ACF simulation results. Figure 6 shows the absolute value of the time-variant ACF of different taps of the proposed V2V channel model at different time instants. And it is noticed that the absolute value of the ACF of different taps is different, besides, the simulation results of different time instants is different. The ACF is varying with time and delay, so the non-WSSUS properties are depicted in Fig. 6. The proposed model has an outstanding performance of modeling the non-WSSUS.

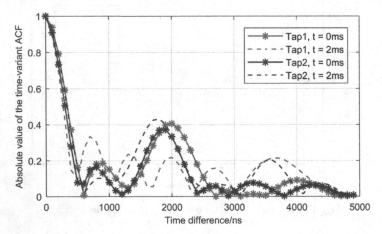

Fig. 6. Absolute value of the time-variant ACF of different taps of the proposed V2V channel model at different time instants for UIC scenario.

LSF simulation results. The absolute value of the time-variant LSF at different time instances is depicted in Fig. 7 above. It is obvious that the power of the effective scatterer varies with time instances. The simulation results show that the time-variant non-stationary properties of the scatterer are mimicked perfectly. Therefore, the proposed channel model is proved to be valuable.

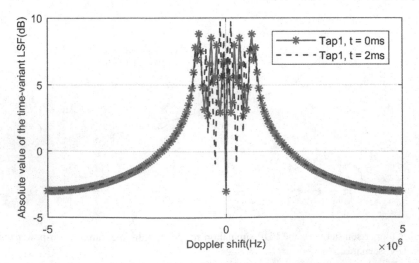

Fig. 7. Absolute value of the time-variant LSF of the first tap of the proposed V2V channel model at different time instants for UIC scenario.

3.3   PSD Simulation Results

Figure 8 shows that The PSD of the taps for the UIC scenario. We can easily notice that the PSD of the taps is similar to the U-shape. Due to the impact of the correlated scatterer, the interaction between each tap will be contributing. As a result, the PSD of the taps may not fit the standard U-shape, in other words, the spectrum distortion exists. However, the non-stationarity with respect to time also affects the PSD spectrum. It is obvious that the PSD of the taps is fluctuant rather than smooth. Furthermore, the simulation results turn out that the modeling of the Doppler disperse with the non-WSSUS is practical.

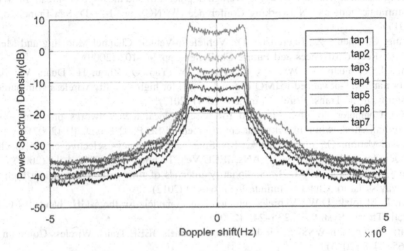

Fig. 8. The PSD simulation results of the taps for UIC scenario.

4   Conclusion

In this paper, we have proposed a novel non-WSSUS statistical model of V2V channel for the 5-GHz band. Several practical statistical models are integrated into the proposed model. Besides, the persistence process and delay-correlated matrix is adapted to characterize the non-WSSUS properties. Moreover, several statistical properties of the proposed model have been investigated. The PDP and PSD simulation results show that the proposed model describes the delay disperse and the Doppler disperse perfectly. According to the time-variant ACF and LSF behaviors, it is demonstrated that the proposed channel model could vividly mimic the non-WSSUS properties of V2V radio channel.

In conclusion, the novel model proposed in this paper can characterize the fluctuant and flexible V2V channel. As a result, the research of the V2V channel will be available for the future ITS construction. Our future work will further adapt the proposed model to more complex and practical scenario, such as the massive-MIMO scenario.

References

1. Borhani, A., Stuber, G.L., Patzold, M.: A random trajectory approach for the development of nonstationary channel models capturing different scales of fading. IEEE Trans. Veh. Technol. **66**(1), 2–14 (2017)
2. Jiang, D., Delgrossi, L.: EEE 802.11p: Towards an International Standard for Wireless Access in Vehicular Environments, pp. 2036–2040 (2008)
3. Dahech, W., Patzold, M., Gutierrez, C.A., Youssef, N.: A non-stationary mobile-to-mobile channel model allowing for velocity and trajectory variations of the mobile stations. IEEE Trans. Wirel. Commun. **16**(3), 1987–2000 (2017)
4. Patzold, M., Gutierrez, C.A., Youssef, N.: On the consistency of non-stationary multipath fading channels with respect to the average doppler shift and the doppler spread. In: Wireless Communications and Networking Conference (WCNC), pp. 1– 6.D. San Francisco, USA (2017)
5. Wang, C., Cheng, X., Laurenson, D.: Vehicle-to-Vehicle Channel Modeling and Measurements: Recent Advances and Future Challenges. pp. 96–103 (2009)
6. Ghazal, A., Yuan, Y., Wang, C-X., Zhang, Y., Yao, Q., Zhou, H., Duan, W.: A nonstationary IMT-advanced MIMO channel model for high-mobility wireless communication systems. IEEE Trans. Wirel. Commun. **16**(4) (2017)
7. IEEE Computer Society.: Standard for wireless local area networks providing wireless communications while in vehicular environment. IEEE P802.11p/D2.01 (2007)
8. Acosta-Marum, G., Ingram, M.A.: Six time- and frequency- selective empirical channel models for vehicular wireless LANs. IEEE Veh. Technol. Mag. **2**(4), 4–11 (2007)
9. Chen, B., Zhong, Z., Ai, B.: Stationarity intervals of time-variant channel in high speed railway scenario. China Commun. **9**(8), 64–70 (2012)
10. Sen, I., Matolak, D.W.: Vehicle–vehicle channel models for the 5-GHz band. IEEE Trans. Intell. Transp. Syst. **9**(2), 235–245 (2008)
11. Matz, G.: On non-WSSUS wireless fading channels. IEEE Trans. Wireless Commun. **4**(5), 2465–2478 (2005)
12. Zajic, A.G., Stuber, G.L.: Space-time correlated mobile-to-mobile channels: modelling and simulation. IEEE Trans. Veh. Technol. **57**(2), 715–726 (2008)
13. Sen, I., Matolak, D.W.: Vehicle-vehicle channel models for the 5-GHz band. IEEE Trans. Intell. Transp. Syst. **9**(2), 235–245 (2008)
14. Matolak., Sen, I., Xiong, W.: Channel modeling for V2V communications. In: Proceedings of the V2VCOM Workshop, San Jose (2006)

GPP-SDR Based GSM-R Air Interface Monitoring System and Its Big Data Interference Analysis

Xiang Chen[1,2(⊠)] and Zhongfa Li[1,2]

[1] School of Electronics and Information Technologies, Sun Yat-sen University,
Guangzhou 510006, China
chenxiang@mail.sysu.edu.cn
[2] Key Lab of EDA, Research Institute of Tsinghua University in Shenzhen,
Shenzhen 518075, China

Abstract. In the railway transportation industry, the monitoring of Global System for Mobile Communications for Railway (GSM-R) network is essential, which plays an important role in safety of the train. The traditional monitoring systems are mainly based on the A/Abis or PRI interfaces. Therefore, the traditional ways are difficult to monitor random interferences and faults occurred over wireless channels, which may causes the potential security menace. In this paper, we propose a GSM-R monitoring system based on the Um interface. Adopting the General Purpose Processor (GPP)-Software Defined Radio (SDR) framework, the GSM-R network can be monitored by full Um interface information, including spectrum, signaling and traffic information. We use the GPP-SDR based front-end processors to obtain the data from Um interface, which are transmitted to the center servers in a railway bureau data center. After receiving the original data, the C/S structure based center servers will process the data for users to monitor. The whole system design has been implemented and deployed in Guangzhou Railway Bureau, including Guang-Shen Line and Guang-Shen-Gang Line. Furthermore, a big data interference analysis framework is proposed based on the Um interface monitoring database, which has also been verified to successfully capture and classify traditional types of interferences in field tests.

Keywords: GSM-R monitoring · GPP-SDR framework · Big data analysis
C/S structure

1 Introduction

With the continuous promotion and development of high-speed railway, the Global System for Mobile Communications for Railway (GSM-R), as the transmission channel of automatic train control and detection information, is becoming more and more important and has been widely applied in railway operation around the world [1]. Since the International Union of Rail Ways (UIC) puts forward the system in 1992, the GSM-R developing into the most mature and widely used wireless communication system specially developed for railway application after more than 20 years of development.

© ICST Institute for Computer Sciences, Social Informatics and Telecommunications Engineering 2019
Published by Springer Nature Switzerland AG 2019. All Rights Reserved
X. Liu et al. (Eds.): ChinaCom 2018, LNICST 262, pp. 709–718, 2019.
https://doi.org/10.1007/978-3-030-06161-6_69

Since the GSM-R wireless communication system is carrying the important train control data, the safety and reliability of the GSM-R system will directly affect the security of the train operation. When the interferences occur, the train operation scheduling will face serious risk. If someone launches the radio interference signal in the GSM-R band, it will seriously affect the normal receiving and launching of the railway signal, which will not only lead to the decrease of railway operation efficiency, but also bring considerable hidden danger to railway safety operation, the personal safety of passengers and even national stability. Therefore, the GSM-R monitoring system, which can effectively monitor the signal and interference in the GSM-R network and analyze the useful signals in the GSM-R network under the condition of eliminating the interference, plays a vital role [2].

Because of the importance of the GSM-R monitoring system, how to better set up it has aroused wide attention. At present, the GSM-R monitoring system in China is mainly composed of some subsystems such as A interface monitoring, Abis interface monitoring, PRI interface monitoring and other managed and comprehensive analysis systems [3]. The locations of the different interfaces in the GSM-R system are shown as Fig. 1. This monitoring system has played an important role in the GSM-R wireless network optimization in China. Besides, it is becoming more and more accurate for the analysis of interference and failure. However, this system also has its insufficient aspects. When the interference and failure occur in the wireless channel between the MS (Mobile Station) and BTS (Base Transceiver Station), the existing means are difficult to find out the reasons.

As we know, the wireless channel is very fragile. Besides, at present, we are lack of wireless channel monitoring data. Therefore, it is still helpless for us to deal with some difficult faults, and the more accurate GSM-R monitoring system which can also monitor the wireless channel is becoming more and more urgent.

The traditional way to monitor the wireless channel is to test the channel along the railway by using a test cellphone or an instrument with an analytical function. However, the manual ways not only waste manpower, but also cannot continue to monitor the wireless channel. Accordingly, we propose a GSM-R monitoring method based on the GPP-SDR (GPP: General Purpose Processor, SDR: Software Defined Radio) framework which mainly uses the data from the Um interface usually called air interface. Based on this method, we have implemented the monitoring system through the combination of hardware and software, and we finally apply it to the actual project which can help us to monitor the GSM-R system automatically.

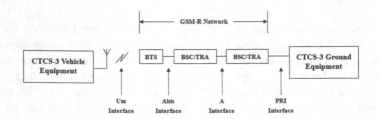

Fig. 1. The locations of the different interfaces in the GSM-R system

The structure of the paper is as follows: Sect. 2 describes the framework of our monitoring system; Sect. 3 provides the implementation method of our hardware and software platform; Sect. 4 introduces the application of our system; and Sect. 5 summarizes the paper.

2 The Framework of Our Monitoring System

In order to monitor the GSM-R network more intelligently, we have set up the monitoring system based on the GPP-SDR framework and big data analysis. Besides, we have further divided the servers by different functions.

2.1 The GPP-SDR Framework and Big Data Interference Analysis

Proposed in 1992, the Software Defined Radio (SDR) is radio broadcast communication technology [4]. Its wireless communication protocol is implemented by the software definition rather than the connection between the hardware which has made it more flexible for the wireless communication system to update itself. Besides, the SDR also brings many other benefits to us, such as reducing the space of the system and decreasing the cost. Different SDR devices may have different structures, however they are usually connected to the General Purpose Processor (GPP) which help to control the SDR devices [5]. With the continuous development of science and technology, the server has made great progress which makes it possible for the server to handle big data analysis just like the large railway data.

With the support of current science and technology, we propose a GSM-R monitoring system based on the GPP-SDR framework and big data analysis, as shown in Fig. 2. At the beginning, the front-end processor which communicates with the computer will collect the interface signaling, user service data and spectrum data from the Um interface through the control of the software rather than the connection of the hardware. Secondly all kinds of data will be transformed into JavaScript Object Notation (JSON) files and then transmitted to the server via Ethernet. Finally the server will carried out further parsing, processing and providing to users with the received data, so as to monitor the GSM-R system. In our system, a server can receive data from multiple front-end processors and processes them simultaneously.

2.2 The Data Server Structure

For the server, we use a hierarchical design architecture according to different functions, as shown in Fig. 3. In this structure, the DB-Servers only write the original data into the database and write the database two times after simple parsing. Besides, other big data analysis and query applications are distributed on the application servers (App-Servers) which obtain the data from DB-Server via the local area networks (LAN). The applications have their own databases on the application servers which store the data that the user needs. Therefore, the user can monitor the GSM-R system by observing the analysis results of different clients on the App-Server.

In order to make our system more robust and make it more convenient to update, we will further develop the Browser-Server (B/S) structure on the basis of the exiting Client-Server framework.

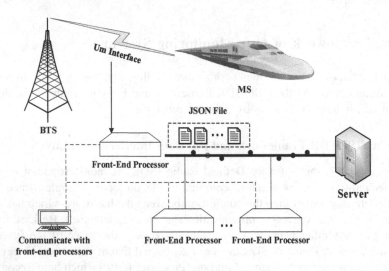

Fig. 2. The framework of the monitoring system

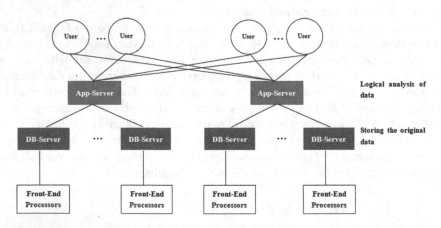

Fig. 3. The structure of the server

3   Hardware and Software Platform Implementation

In order to apply our system to the practical projects, we need hardware and software to realize our ideas. At present, we have implemented the framework mentioned above and applied it to some Chinese railways as the GSM-R monitoring system.

3.1 Hardware Implementation

The main hardware parts of out monitoring system are composed of the front-end processors and the servers. As shown in Fig. 3, multiple front-end processors will be connected with a DB-Server via Ethernet. Besides, the App-Server and DB-Server will communicate with each other through LAN.

The front-end processors mainly monitor the Um interface continuously and translate the received data into JSON file which will be transmitted to the DB-Server. The value of the monitoring frequency or other parameters can be modified by sending commands to the front-end professors from the host computers.

To monitor the GSM-R network, we need servers to analyze and deal with a large number of railway data obtained from the front-end processors. As shown in Fig. 3, different servers implement different functions. For example, the DB-Servers mainly simply process and store the original data. In addition, the App-Servers make logical analysis of data obtained from the databases in DB-Server.

3.2 Software Implementation

With the support of the above hardware, we have developed some software to handle and analyze the big data of the railway, so as to realize the monitoring of the GSM-R system.

(A) Database Design

Considering that we are going to deal with a large amount of data, we have designed the day table for the databases, which has a larger increase in the query speed compared with the month table. Besides, we use the MySQL as our databases management system.

In the database operation, we use the method which writes twice to the database to process original data, as shown in Fig. 4. At the first time, the spectrum data and signaling data will be separately inserted into spectrum databases and signaling databases after simple analysis. Afterwards, the signaling data will be decoded by the decoding program according to the protocols, which is multithreaded operation. After decoding, the decoded data will be reinserted into the signaling databases, which includes a lot of information for monitoring the GSM-R system.

(B) Intelligent Interference Analysis Software

For the purpose of monitoring the GSM-R system automatically, apart from the databases, we also need the intelligent interference analysis software. In our monitoring system, we mainly monitor the interference on the wireless channel via the intelligent interference analysis software. The implementation of the software is based on the big data obtained from the spectrum databases and signaling databases.

Since the spectrum data contains the amplitude values of different channels at different times in the GSM-R system which can directly reflect the work of the wireless channel, we have designed our intelligent interference analysis software mainly based on the big spectrum data and part of the signaling data, as shown in Fig. 5. In Fig. 5, the interference module mainly works as Fig. 6.

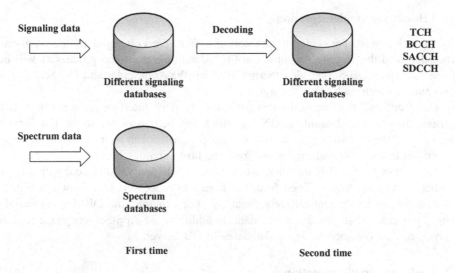

Fig. 4. The operation of the databases

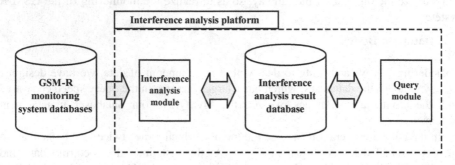

Fig. 5. The interference analysis platform

When analyzing the data of each front-end processor, we will firstly build a model based on the spectrum data selected from several days. Afterwards, we will automatically analyze the spectrum data based on the model and the signaling data every half an hour. The user can access the analysis data and restart the analysis when they need. The algorithm of processing a front-end processor is shown in Algorithm 1.

At present, machine learning has been developing rapidly and is applied in many fields. For our intelligent interference analysis system, we are going to take a machine learning approach, such as the decision tree algorithms, to optimize our methods to make the GSM-R monitoring more accurate.

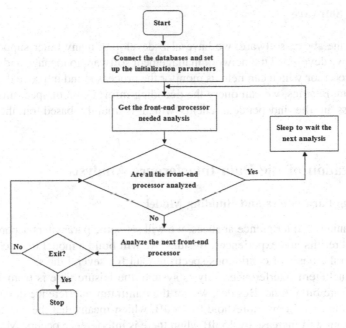

Fig. 6. The interference analysis module

Algorithm 1 Analyze the front-end processor
Get the begin time T_1 and the end time T_2.
Start with the processing time $t = T_1$, segment $= 30s$.
if not stop analysis **do**
   for $t < T_2$ **do**
         1. Spectrum data processing:
         Analyze the spectrum data based on the model values and measurement
         report.
         if have interference **do**
               Insert the information of the interference into the interfering
databases
         end
         2. Sweep frequency data processing:
         Analyze if there are interferences by using the sweep frequency data.
         if have interference outside the GSM-R system **do**
               Insert the information of the interference into the interfering
databases
         end if
         $t = t + segment$
      end for
   end if

(C) Other Software

Apart from the above software, we have also developed many other supporting software. We have developed the network management software to manage and control the front-end processor which can help us monitor the processor and improve the reliability of the system. Besides, we can query the decoding information of spectrum, signaling and business in the independent client, which is mainly based on the signaling databases.

4 Application of Big Data Interference Analysis

4.1 Setting Parameters and Building Model

At the beginning of interference analysis, we will set some parameters according to the previous test results and experiences. Besides, we will build a model for each front-end processor on the basis of continuous spectrum data for several days.

In our intelligent interference analysis system, the leisure time is from 1 to 5 a.m, while others are busy time. Besides, we set the minimum interference duration to one minute and the interference threshold to 15 dB, which means that the spectrum of the frequency point will increase by 15 dB when there is interference occurs. Moreover, we also set up the delay time of interference analysis for different base stations according to different traffic levels. In addition, the sleep time between two analyses is 30 min.

For each front-end processor, we choose two days of normal spectrum data, which need the field test to confirm whether there is interference in the data, to get the busy and leisure time model. The current selection strategy is to select the maximum data of busy or leisure time in the corresponding time period of two days at different frequency points as the model value of the different frequency.

After setting parameters and building model, we can start the application for continuous interference analysis.

4.2 The Application of Interference Analysis

On the basis of the above software and hardware, we implement the GSM-R monitoring system based on the GPP-SDR framework. Besides, our monitoring system has been applied into some Chinese railways, such as Wuhan-Guangzhou line and Guangzhou-Shenzhen line. Through the big data analysis, we have successfully monitored many interferences, which help to improve the reliability of the GSM-R system.

In June 2018, through the interference analysis platform, we discover an interference that occurred at a base station in Guang-Shen Line. The interference record is found through the software, as shown in Fig. 7 and Table 1. Figure 7 is the interference spectrum diagram and Table 1 is the detailed information about the interference.

According to the interference information, we can find that the interference happened at near 18 o'clock. Therefore, it was an interference in busy time which was defined from 1 to 5 a.m. From Fig. 7, we can also see that the interference is a narrowband interference. Since the channel 1019 is not contiguous to the TCH channel (1003 and 1016) and BCCH channel (1001) and the interference not satisfied with other specific conditions, the type of the interference was judged to other channel interference. In addition, this interference was most likely caused by the private amplifier, so the software determined the suspected source of the interference as the private amplifier.

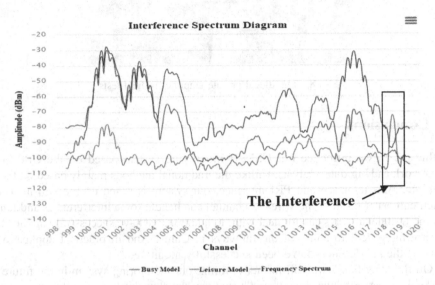

Fig. 7. The interference spectrum diagram

Table 1. Detailed information about the interference

Property	Value
Front-end processor number	450571032
Base station	ShiPai
Interference intensity	−69.226 dB
Interference bandwidth	0.2 MHz
Interference channel	1019
Strongest interference channel	1019
Type of interference	Other channel interference
Suspected source of interference	Private amplifier
Start-end time of interference	2018–06–10 18:28:53 ~ 18:30:42

In order to check this interference source, the tester used the spectrum analyzer to detect around the base station. As we expected, the result of the detection was roughly the same as that in Fig. 7. And then, the tester found the location of the most intense interference by changing the antenna position of the spectrum analyzer. Finally we found an illegal private amplifier on the roof of a residential building, as shown in Fig. 8. After we contacted the relevant department to deal with this illegal private amplifier, there was no similar interference in this base station.

Fig. 8. The illegal private amplifier in field tests

5   Conclusions

In this paper, we introduce our GSM-R monitoring system based on the GPP-SDR framework and big data analysis. Unlike the traditional methods mainly based on the A interface, Abis interface and PRI interface, our system is based on the Um interface which can better monitor the wireless channel and handle some interferences and faults that the traditional ways cannot deal with. In addition, we have successfully applied the monitoring system to some of the railways in China. And in practical applications, some of the interferences have been successfully monitored.

On the other hand, we will further optimize our monitoring system in the future to improve its performance. For example, we can develop the B/S structure to make it more flexible and use the machine learning methods to improve the efficiency and reliability of the big data processing.

Acknowledgement.   The work is supported by the NSFC (No. 61501527), Science, Technology and Innovation Commission of Shenzhen Municipality (No. JCYJ20170816151823313), State's Key Project of Research and Development Plan (No.2016YFE0122900-3).

References

1. Kastell, K., et al.: Improvments in railway communication via GSM-R. Veh. Technol. Conf. **6**, 3026–3030 (2006)
2. Wang, X., He, C.: Safety function design and application of CTCS on-board equipment in high-speed railway of China. In: International Conference on Electromagnetics in Advanced Applications, pp. 677–679 (2017)
3. Lu, W.: Introduction of GSM-R network interface detection technology for high speed railway. Public Commun. Sci. Technol. **7**(17), 63–64 (2015). (In Chinese)
4. Mitola, J: The software radio architecture. IEEE Commun. Mag. **33**(5), 26–38 (1995)
5. Schwall, M,, Jondral, F.K.: High-speed turbo equalization for gpp-based software defined radios. In: Military Communications Conference, Milcom 2013, pp. 1592–1596. IEEE (2013)

Shared Buffer-Based Reverse Scheduling for Onboard Clos-Network Switch

Wanli Chen[1,2], Kai Liu[3], Xiang Chen[1,2($\boxtimes$)], and Xiangming Kong[4]

[1] School of Electronics and Information Technology, Sun Yat-sen University,
Guangzhou 510006, Guangdong Province, China
[2] Key Lab of EDA, Research Institute of Tsinghua University in Shenzhen (RITS),
Shenzhen 518075, China
chenxiang@mail.sysu.edu.cn
[3] China Academy of Electronics and Information Technology, Beijing 100041, China
[4] Starway Communications Inc., Guangzhou 510663, Guangdong Province, China

Abstract. Onboard switching (OBS) is facing resource constraints and special requirements of hardware complexity and scheduling efficiency. By studying the existing OBS fabrics and scheduling algorithms, the Shared Buffer-based Reverse Scheduling (SB-REV) Algorithm is proposed, adopting the shared buffer in the input module (IM) and guiding the IM scheduling with the matching result of the central modules. Theoretical and experimental analysis shows that the SB-REV algorithm greatly improves the resource utilization and scheduling efficiency, while guaranteeing the cell delay and the throughput performance. The SB REV Algorithm is highly suitable for resource-constrained OBS environment.

Keywords: Onboard switching · Clos-network · Resource utilization · Scheduling efficiency

1 Introduction

Evolving from 1960s, the satellite communication has gained enormous attentions with the characteristics of large capacity, wide bandwidth, ubiquitous coverage, and the adaptability to multiple services [1]. Compared with the conventional bent-pipe forwarding technology, the onboard switching (OBS) technology only requires end-to-end transmission of one hop, which leads to higher security, lower latency, higher bandwidth utilization and less reliance on ground stations [2]. As the core of the OBS, the OBS fabrics determine the performance of throughput, cell delay, etc. Therefore, the main bottleneck of developing high-speed OBS lies in the OBS fabrics, with many challenges yet to be fulfilled.

In particular, the OBS of China are faced with even worse resource constraints. At present, the payload weight and power of China's largest satellite platform Dongfanghong No. 4 are only 700 kg, 8000 W, whereas that of the

X. Liu et al. (Eds.): ChinaCom 2018, LNICST 262, pp. 719–728, 2019.
https://doi.org/10.1007/978-3-030-06161-6_70

European satellite platform Alphabus are 2000 kg, 18000 W [3], far exceeding China. Since the resources are limited, it is necessary to fully consider the hardware complexity and enhance the resource utilization. On the other hand, as the bandwidth of the OBS gradually increases, it is vital to improve the speed of packet processing and the efficiency of scheduling algorithms. The Reverse Scheduling algorithm was proposed for great enhancement of scheduling efficiency, but with no buffer in the input modules (IM), the scheduling in IM is centralized, decreasing scalability. To solve the problem, crosspoint queue (CQ) can be adopted in IM. But the resource utilization of CQ is low, and thus the structure Shared Buffer was proposed individually, not deployed in the Clos network yet. In order to meet the above challenges, the Shared Buffer-based Reverse Scheduling (SB-REV) Algorithm is proposed, which greatly improves the resource utilization and the scheduling efficiency.

The outline of this paper is as follows. Section 2 presents the central features of the existing OBS fabrics and scheduling algorithms, and then leads to the shared buffer (SB) structure and the reverse scheduling (REV) algorithm. Section 3 analyzes the SB and the REV theoretically. Section 4 shows the performance of the SB and the REV with comprehensive experiment results. The conclusions are drawn in Sect. 5.

2   Related Works

As the switching capacity increases, the OBS fabrics evolve from Time-Division (TD) Switches, suitable for only small-capacity switching, to Space-Division (SD) Switches for large-scale OBS [4]. According to the uniqueness of the switching path, the SD switches can be further divided into single-path switches (such as Crossbar) and multi-path switches (such as Clos) [4]. When the switch size of Crossbar scales, the number of crosspoints added increases exceedingly, while the Clos network free of this problem [5]. The multi-path switch can establish multiple independent paths, and thus the cells of different input-output connections can be forwarded concurrently, with a capacity of up to 10 Tbps. In a Clos network $C(n, m, r)$ (where n, m, r are the number of input links of an input module, central modules and input modules respectively), if $m \geq 2n - 1$, the Clos network is strictly non-breaking [6]. Among multi-path switches, the Clos network is preferable and widely studied for the next-generation OBS, due to its higher reliability, scalability, and the feature of non-blocking.

The Crossbar switch is a key component of the Clos network. According to the existence of buffer inside of it, the Crossbar can be divided into Bufferless Crossbar (including input queue, IQ and output queue, OQ) and Buffered Crossbar (including crosspoint queue, CQ). Because the input and output ports are directly connected, the bufferless Crossbar requires a centralized scheduling algorithm to match the ports. When the switching scales, it becomes difficult for the bufferless Crossbar to meet the requirements of fast scheduling. On the contrary, in the buffered Crossbar (e.g. CQ), the input and output ports are separated. So the concurrent and distributed scheduling algorithm can be implemented, which reduces the complexity of scheduling and improves the scalability. However, the queue buffer size required for the CQ [7] is proportional to the squared number of ports n^2. To solve this problem, the current solution is adopting the SB [8] (as shown in Fig. 1). Note that each row of crosspoints share the same buffer. Contrasting with the CQ, the required buffer size of the SB is only proportional to the number of ports n, which can be conducive to resource utilizing. Yet the SB has not been deployed in the Clos network in previous literature.

As for the Clos network, it consists of three stages of Crossbar modules, referred to as input modules (IMs), central modules (CMs), and output modules (OMs) respectively. According to the memory deployment of each stage, it can be categorized into SSS-Clos, MSM-Clos, SMM-Clos, and MMM-Clos networks. In the SSS-Clos network, the port matching is actually an $N \times N$ bipartite graph matching problem (in a complexity of $O(N^2)$, where N is the switch size), degrading the scalability [9]. Both the SMM and MMM-Clos networks employ memory at the central module (CM), causing cells out-of-sequence at the receiver. To solve this problem, more buffers and feedbacks are needed, which hinders hardware implementation [10]. Free from the above problems, the MSM-Clos network only needs to solve the conflict at the output port of CM [11]. Herein we will concentrate on the MSM-Clos network.

In the MSM-Clos network, current scheduling algorithms are twofold: dynamic algorithms and quasi-static algorithms. The dynamic algorithm needs to perform real-time routing according to the arriving traffic and employs five-way handshake, which is complicated and time-consuming. The main dynamic algorithm is the Concurrent Round-Robin Dispatching (CRRD) algorithm [12]. In contrast, the quasi-static algorithm pre-configures the connection of CM and cannot adapt to the real-time traffic. To solve the above problems, Zhang et al. [13] proposed the Reverse Scheduling (REV) algorithm, by combining both the dynamic and the quasi-static algorithm. The REV guides the IM scheduling with the matching result of the CM, and cuts down the handshake times and the scheduling time. However in the IM, it's still using the bufferless Crossbar aforementioned, which needs centralized scheduling and lacks scalability.

In this article, we will combine the advantages of both the shared buffer and the reverse scheduling, propose the SB-REV algorithm, and prove its superiority by theoretical and experimental analysis.

3   The Shared Buffer-Based Reverse Scheduling Algorithm

3.1   The Shared Buffer Fabrics (SB)

In the crosspoint queue CQ_{ij}, only the cells from the input port I_i to the output port O_j can be buffered. By comparison, in the shared buffer SB_i as shown in Fig. 1, cells from the input port I_i to any output port $O_j(\forall j)$ can be buffered.

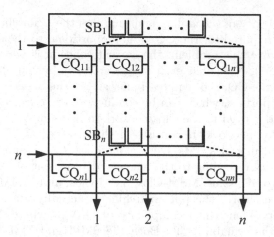

Fig. 1. The shared buffer fabrics

Accordingly we have the relationship between the queue length of the SB and the CQ $L_{SB} = sum(L_{CQ})$, where $L_{CQ}, L_{SB} \in [0, n]$. The required buffer size, denoted by L_{set}, is set according to the maximum queue length L_{max} when the queue length is not limited in the experiment. Consequently, we have

$$\frac{L_{set\_SB}}{L_{set\_CQ}} = \frac{n \times L_{max\_SB}}{n^2 \times L_{max\_CQ}} \in \left[\frac{1}{n}, 1\right], \tag{1}$$

where L_{max_SB} and L_{max_CQ} are the maximum queue lengths among n SBs and among n^2 CQs respectively. In other words, the required buffer size of the SB can be reduced up to $1/n$ that of the CQ. Note that the buffer utilization r is the ratio between the actually used buffer size and the required buffer size, i.e., $r = L_{act}/L_{set}$. If the SB and the CQ have the same L_{act} in the experiment, by Formula (1) we have

$$\frac{r_{SB}}{r_{CQ}} = \frac{\dfrac{L_{act\_SB}}{L_{set\_SB}}}{\dfrac{L_{act\_CQ}}{L_{set\_CQ}}} \in [1, n], \tag{2}$$

i.e., the buffer utilization of the SB can be improved up to n times that of the CQ.

3.2   The Reverse Scheduling Algorithm (REV)

In the MSM-Clos network, the concurrent round-robin dispatching (CRRD) algorithm and the reverse scheduling (REV) algorithm are compared.

(1) Hardware complexity
   The core component of the scheduling is the arbitration scheduler, so the hardware complexity of the scheduler represents the resource overhead.

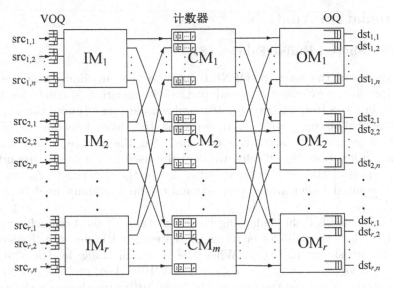

Fig. 2. The reverse scheduling algorithm in the MSM-Clos network

When adopting the CRRD, the hardware complexity of the r IMs is $O(nmr)$, so the complete hardware complexity is $O(mrN)$ [13]. When adopting the REV, as shown in Fig. 2, only the input port CI_{ji} in the CM needs a scheduler, so the hardware complexity of the m CMs is $O(r^2)$, and the complete hardware complexity is $O(mr^2)$. Consequently, the ratio of hardware complexity between the two algorithms is $\dfrac{O(\text{REV})}{O(\text{CRRD})} = \dfrac{O(mr^2)}{O(mrN)} = \dfrac{O(r)}{O(N)} = \dfrac{1}{n}$. That is, the resource overhead of the REV is much lower than that of the CRRD, which is conducive to solving the resource constraints of the OBS.

(2) Scheduling time

The scheduler is responsible for the decision of the handshakes or the matchings between the input and output ports. For a scheduler with n input numbers, the time complexity of the scheduler is $O(logn)$ [13]. When adopting the REV, the scheduling is only required in the CM. Assume that the Round-Robin (RR) algorithm is adopted in the CM to avoid starvation, then the scheduling time of the REV is proportional to $logr$, i.e., $t_{arb}(\text{REV}) = \alpha logr$, where α is a constant coefficient, determined by actual hardware performance.

As for the CRRD, it requires five-way handshake. Let $i(i \geq 1)$ be the iteration times in the IM, then the scheduling time of the CRRD is $t_{arb}(\text{CRRD}) = \alpha[i(logN + logm) + logr]$. So when $n = m = r$, we have [13]

$$\frac{t_{arb}(\text{REV})}{t_{arb}(\text{CRRD})} = \frac{1}{1 + ilog_r nmr} \stackrel{n=m=r}{=} \frac{1}{1 + 3i} = \frac{1}{4}. \tag{3}$$

4   Simulation Analysis

4.1   The Shared Buffer Fabrics (SB)

This paper uses the software OPNET to build the environment of the OBS simulation. In the experiment, we adopt the RR algorithm, simulate for 10 000 timeslots in a 16×16 Crossbar ($n = 16$), and compare the SB with the CQ. Here we define symbols $\eta_{L_{act}}, \eta_{L_{set}}, \eta_r$ to represent the ratios between the SB and the CQ in terms of the actually used buffer size, the required buffer size and the buffer utilization. Specifically, the actually used buffer size is proportional to the average of the actually used buffer size in the experiment $L_{act} \propto L_{avg}$, and the required buffer size is proportional to the maximum used buffer size $L_{set} \propto L_{max}$.

Under Bernoulli traffic, adopting the SB or the CQ can both achieve 100% throughput, but they differs in terms of the required buffer size. Experimental results are shown in Fig. 3. When the Bernoulli traffic load $\lambda \le 0.5$, we have $\eta_{L_{set}} = \eta_{L_{max}} = L_{max\_SB}/L_{max\_CQ} = 1/16 = 1/n$, and $\eta_{L_{act}} = \eta_{L_{avg}} = L_{avg\_SB}/L_{avg\_CQ} = 1$. So the ratio of the buffer utilization between the SB and the CQ is $\eta_r = \eta_{L_{act}}/\eta_{L_{set}} = 16 = n$.

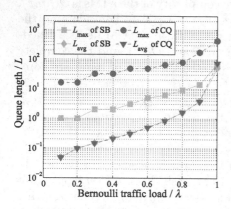

Fig. 3. Queue length of CQ and SB

Table 1. The throughput and cell delay of the iterative CRRD and REV algorithm

CRRD	Iterations	Cell delay	ρ_1	ρ_2
	1	389.44	0.99214	0.90800
	2	41.39	0.99982	0.97915
	3	40.80	1	0.97738
	4	40.80	1	0.97738
	5	40.80	1	0.97738
REV		46.24	N/A	0.99827

4.2   The Reverse Scheduling Algorithm (REV)

In the experiment, we simulate for 10 000 timeslots under the Bernoulli traffic ($\lambda = 1$) in the Clos network $C(8,8,8)$, and compare the REV with the $CRRD_i$ ($i = 1, 2, \ldots, 5$). Here we define the matching rate between the input and output ports in the IM as ρ_1, and the matching rate in the CM as ρ_2. Consequently we have the throughput of IM $\rho_{IM} = \rho_1 \times \rho_2$.

Experimental results are shown in Table 1. At the iterations of $i = 1, 2, 3$, we have the matching rate in the IM $\rho_1 = 99.214\%, 99.982\%, 100\%$, respectively. Iteration times higher than three cannot contribute to the matching rate and the cell delay, but further decreases the actual scheduling efficiency. On the contrary, the REV need no iteration to achieve a throughput of 99.8%, higher than that of the $CRRD_i (\forall i)$.

4.3 The Shared Buffer-Based Reverse Scheduling Algorithm

In the experiment, we simulate for 10 000 timeslots under the Bernoulli traffic ($\lambda = 1$) in the Clos network $C(8,8,8)$, and compare seven algorithms. In order to avoid the Head-of-Line (HoL) Blocking, the virtual output queue (VOQ) is adopted [14] in input stages. Here, the scheduling algorithms involving the CRRD iterate once.

According to the queuing strategy in the IM, seven algorithms can be divided into the following three groups:

(1) Bufferless Crossbar: the CRRD and the REV without buffer inside of the Crossbar;
(2) SB: the CRRD based on the SB and that with speedup (SB-CRRD and SB-$CRRD_S$) and the REV based on the SB and that with speedup (SB-REV and SB-REV_S);
(3) CQ: the CRRD based on the CQ (CQ-CRRD).

Among them, the second and the third groups of algorithms are based on Buffered Crossbar, which is conducive to concurrent and distributed scheduling and thus enhances scalability.

Specifically, there are two mechanisms of the RR scheduling in the SB: (1) The polling objects are all $n \times m$ cells in the buffer; (2) The polling objects are $n \times 1$ shared buffers (where n, m are the number of input and output ports of the IM respectively). The effect of adopting mechanism (1) is completely equivalent to that of not employing the SB, and thus mechanism (1) is not discussed specifically. When adopting mechanism (2), the polling number decreases to n, and the polling efficiency increases to m times of that without the SB. In this paper, when some shared buffer is successfully authorized, we only allow the oldest cell in it to be served first. Hereafter we adopt mechanism (2) in the SB by default.

As shown in Fig. 4a, the throughput of algorithms except CRRD, SB-CRRD, and SB-REV, reaches 100%. However, the throughput of SB-CRRD is even lower than that of CRRD. This is because we adopt mechanism (2) in the SB aforementioned, which causes the non-oldest cell in the buffer losing the right to poll, but meanwhile increases the RR scheduling efficiency up to m times. Different from the CQ, the cells in the SB can go to any output port. So as shown in Fig. 6, if some SB (SB_1) is not empty and some output port (O_2) is idle, then the throughput can be improved with a speedup. As shown in Fig. 7, the CQ is limited by the fixed output link rate and cannot increase the throughput with a speedup.

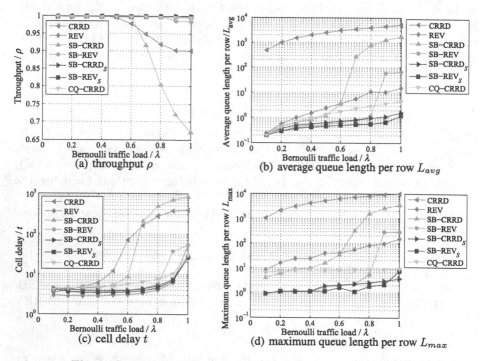

Fig. 4. Comparison of 7 algorithms in 4 different performances

As shown in Fig. 4a, SB-CRRD can achieve 100% throughput with a speedup, and as shown in Fig. 5, the speedup of SB-CRRD converges to $S = 1.208$. Similarly, the SB-REV algorithm has a throughput of 98.4% without speedup. And as shown in Fig. 5, the SB-REV needs a speedup of only $S = 1.05$ to outperform other algorithms in terms of throughput, cell delay and queue overhead.

As shown in Fig. 4a, c, the CQ-CRRD throughput is 100% and the cell delay performance is optimal. However, it can be seen from Fig. 4b, d that the queue overhead of the CQ-CRRD is large. Calculations show that the average required buffer size $\overline{L}_{set} = \dfrac{1}{10} \sum\limits_{\lambda=0.1}^{1} L_{max}(\lambda)$ of the CQ-CRRD is 4 times that of the SB-REV$_S$, much inferior to the SB-REV$_S$.

As can be seen from the four aspects in Fig. 4a, b, c, d, using the SB does not necessarily optimize the performance. However compared with the CRRD, adopting the REV can optimize the performance of all aspects. Among the seven algorithms, the SB-REV$_S$ outperforms others remarkably.

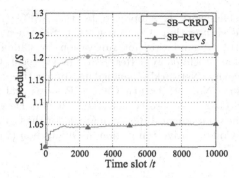

Fig. 5. Speedup performance when Bernoulli traffic load $\lambda = 1$

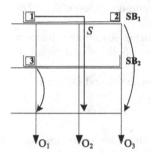

Fig. 6. SB improves throughput with speedup

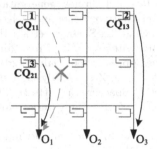

Fig. 7. CQ cannot speed up

5   Conclusions

This paper focuses on the problem of resource constraints of the OBS. Starting from the resource utilization and the scheduling efficiency, we study the shared buffer structure and the reverse scheduling algorithm, and propose the SB-REV algorithm. Theoretical analysis shows that the queue buffer size is reduced up to $1/n$, that the hardware complexity is reduced to $O(1/n)$, and that the scheduling time is reduced to $1/(1 + 3i)$ (where i is the iteration times of the compared algorithm CRRD).

This paper uses the software OPNET to build the environment of the OBS simulation. Experiment results show that: (1) By adopting the SB, the memory utilization is increased to n times that of the CQ; (2) By adopting the REV, no iteration is required, yielding a throughput ($\rho = 99.8\%$) higher than that of the CRRD$_i(\forall i)$. In addition, we also figure out the effects of adopting the SB and/or the REV from different points. Simulations show that the SB-REV algorithm only needs a speedup of $S \leq 1.05$ to outperform other algorithms in terms of cell delay and throughput. With lower hardware complexity and higher scheduling efficiency, the SB-REV algorithm is suitable for the resource-constrained onboard switching.

Acknowledgement. The work is supported by the NSFC (No. 61501527), Science, Technology and Innovation Commission of Shenzhen Municipality (No. JCYJ20170816151823313), Foundation for Innovation by China Academy of Electronics and Information Technology "Research on Protocol Oblivious Forwarding Technologies for Space Network", Guangdong Innovative and Entrepreneurial Research Team Program (No. 2013D014), China's Postdoctoral Science Foundation (No. 2017M620061), State's Key Project of Research and Development Plan (No. 2016YFE0122900-3), the Fundamental Research Funds for the Central Universities and 2016 Major Project of Collaborative Innovation in Guangzhou (No. 201604046008).

References

1. He, Y.Z.: Research on new genergation mobile satellite communication system[D]. Beijing University of Posts and Telecommunications (2015)
2. Wang, J., Qiao, L., Shao, S., et al.: High-performance routing search algorithm in satellite IP switches[C]. In: Proceedings of IEEE Computer Science and Network Technology, pp. 863–866. Dalian (2013)
3. Wang, M., Zhou, Z.C.: Analysis of the alphabus platform devel and design characteristics[J]. Spacecr. Eng. **19**(2), 99–105 (2010)
4. Chao, H.J., Liu, B.: High Performance Switches and Routers. Wiley, New York (2007)
5. Yang, W.X., et al.: Design and implementation of a multi-stage bufferless high radix router[J]. Comput. Eng. Sci. **39**(2), 245–251 (2017)
6. Tang, H.K.: Load balance technology of the Clos network[J]. Sci. Technol. Inf. **15**(8), 7–9 (2017)
7. Kleban, J., Suszynska, U.: Static dispatching with internal backpressure scheme for SMM Clos-network switches[C]. In: Computers and Communications, pp. 000654–000658. IEEE (2014)
8. Kornaros, G.: BCB: A Buffered CrossBar switch fabric utilizing shared memory[C]. In: Euromicro Conference on Digital System Design, pp. 180–188. IEEE Computer Society (2006)
9. Chao, H.J., Jing, Z., Liew, S.Y., et al.: Matching algorithms for three-stage bufferless Clos network switches[J]. IEEE Commun. Mag. **10**, 46–54 (2003)
10. Dong, Z., Rojas-Cessa, R., Oki, E.: Memory-memory-memory Clos-network packet switches with in-sequence service[C]. In: IEEE, International Conference on High PERFORMANCE Switching and Routing, pp. 121–125. IEEE (2011)
11. Gao, Y., Qiu, Z., Zhang, M., et al.: Distributed weight matching dispatching scheme in MSM Clos-network packet switches[J]. IEEE Commun. Lett. **17**(3), 580–583 (2013)
12. Oki, E., Jing, Z., Rojas-Cessa, R., et al.: Concurrent round-robin-based dispatching schemes for Clos-network switches. IEEE/ACM Trans. Netw. **10**(6), 830–844 (2002)
13. Zhang, M., Qiu, Z., Gao, Y., et al.: Reverse dispatching scheme for satellite Clos-network switches[J]. J. Xidian Univ. **40**(4), 96–101 (2013)
14. Kleban, J.: Packet dispatching using module matching in the modified MSM Clos-network switch[J]. Telecommun. Syst. **8**, 1–9 (2017)

Ergodic Capacity and Throughput Analysis of Two-Way Wireless Energy Harvesting Network with Decode-and-Forward Relay

Yingting Liu[1(✉)], Jianmei Shen[1], Hongwu Yang[1], Chunman Yan[1], and Li Cong[2]

[1] The College of Physics and Electronic Engineering, Northwest Normal University, No. 967 Anning East Road, Lanzhou, China
{liuyt2018,jianmeisl60,yancha02}@163.com,
yanghw@nwnu.edu.cn
[2] The Information and Communication Company, Jilin Electric Power Company Limited, Changchun, China
congli8462@163.com

Abstract. In this paper, we consider a wireless energy harvesting network, where two source nodes exchange information via a decode-and-forward (DF) relay node. The network adopts the time switching relaying (TSR) or power splitting relaying (PSR) protocols. In the TSR protocol, transmitting process is split into three time slots. In the first time slot, two source nodes send the signals to the relay node simultaneously and the relay node harvests energy from the radio frequency (RF) signals. In the second time slot, two source nodes send the information signals to the relay node simultaneously. In the third time slot, the relay node decodes the signals and then forwards the regenerated signal to two source nodes using all harvested energy. In the PSR protocol, every transmission frame is divided into two equal time duration slots. The energy constrained relay node splits the received power into two parts for energy harvesting (EH) and information processing in the first time slot, respectively, and forwards the reproduced information signal to the source nodes in the second time slot. We derive the analytical expressions of the ergodic capacity and ergodic throughput of the network both for the TSR and PSR protocols. Numerical results verify the theoretical analysis and exhibit the performance comparisons of two proposed schemes.

Keywords: Decode-and-forward · Wireless energy harvesting
Power splitting relaying · Time switching relaying · Ergodic capacity
Throughput

Yingting Liu: This work has been supported by the National Natural Science Foundation of China (Grant No. 61861041, 11664036 and 61741119), the high school science and technology innovation team project of Gansu (Grant No. 2017C-03).

X. Liu et al. (Eds.): ChinaCom 2018, LNICST 262, pp. 729–739, 2019.
https://doi.org/10.1007/978-3-030-06161-6_71

1 Introduction

Conventional wireless communication devices utilize the constant power supply such as batteries to support their operations. These devices need the human to change batteries periodically and this shortcoming limits the lifetime of wireless devices and causes the difficulty for maintaining them. In recent years, wireless harvesting energy has attracted more and more attention in the literature. Radio frequency (RF) can carry energy as well as information, so we can utilize this ability to accomplish simultaneous wireless information and power transfer (SWIPT). Utilizing the wireless harvesting energy technology, wireless devices such as wireless senor network have the infinite lifetime in the theory [1, 2].

In this field, a classical model consisted of three nodes which are one source node, one relay node and one destination node is well studied. In [3], the authors studied the outage probability and ergodic capacity of the above model. In [3], the system utilizes the power splitting relaying (PSR), the time switching relaying (TSR) or ideal relaying (IR) architectures to accomplish energy harvesting and information transmission, and the relay node uses the amplify-and-forward (AF) or decode-and-forward (DF) schemes to forward the received signal to the destination node. In [4], the authors studied the ergodic capacity of the system when the relay node uses the DF scheme. In [3, 4], all the analytical expressions evaluating the performance of the system have the integral forms, so the authors only use software tools to find the optimal parameters resulting in maximal ergodic capacity or minimal outage probability. On the basis of [3–6] studied the optimal power splitting factor which leads to maximal ergodic capacity or minimal outage probability. In [5], the system adopts the AF or DF schemes but not considering the direct link form the source node to the destination node. In [6], the system adopts the AF scheme and considers the direct link. Utilizing the high signal to noise ratio approximation, the authors obtain the closed-form solution of the optimal power splitting factor in [5, 6].

All above papers studied the one directional transmission only from the source node to the destination node. The authors in [7] studied the outage probability and ergodic capacity in the AF two-way channels and the authors in [8] studied the throughput of the system with a multiplicative relay node in the two-way channels. The authors in [9] split the whole transmission process into three time slots and derived the end to end throughput of the system. Simulation results verified the performance of the proposed scheme is superior to that of in [8]. The authors in [10–12] studied the ergodic outage probability of one-way log-normal fading channels.

To the best of our knowledge, the ergodic capacity and ergodic throughput of two-way DF network based on the TSR and PSR protocols considered in this paper have not been investigated in prior work.

The rest of this paper is organized as follows. Section 2 describes system model. In Sect. 3, we derive the ergodic capacity and ergodic throughput of the proposed TSR and PSR schemes. Numerical results are presented in Sect. 4. Finally, Sect. 5 concludes the paper.

2 System Model

As shown in Fig. 1, a wireless energy harvesting network consists of three nodes, which are two sources nodes, denoted by S1 and S2, and one energy constrained relay node which needs harvesting energy for its operation, denoted by R, respectively. R has no fixed power supply and needs harvesting energy for its operation. h and g are the channel coefficients between S1 and R and between S2 and R, respectively. The direct path between S1 and S2 is negligible, thus the information transmissions between two source nodes need a relay node [4]. We assume the channels are reciprocal and quasi-static Rayleigh block fading, so the channels remain constant during each block transmission time T. It is assumed that perfect channel state information (CSI) is available at all nodes. All nodes are equipped with a single antenna. It is assumed that the processing power required by the information decoding circuitry at the relay node is negligible as compared to the power used for signal transmission from R to S1 and S2.

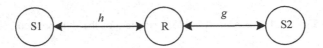

Fig. 1. System model.

The information transmission process in the TSR protocol as shown in Fig. 2, the transmission process is divided into three time slots. In the first time slot αT, S1 and S2 send the signals to R using the same powers. The second time slot $(1 - \alpha)T/2$ is used for information transmission form the source nodes to R, and the third slot $(1 - \alpha)T/2$ is used for information transmission form R to the source nodes. R consumes all the harvested energy when it forwards the information signal to the source nodes. α denotes the time fraction harvested energy from the source nodes and determines the ergodic capacity and ergodic throughput of the network, which is the key performance parameter of the network.

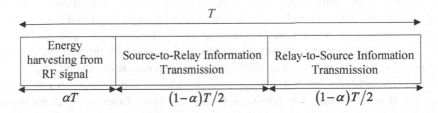

Fig. 2. The transmission block structure of the TSR protocol.

The information transmission process in the PSR protocol as shown in Fig. 3, the whole transmission block time is T, and the transmission process is divided into two equal time slots denoted by $T/2$. In the first time slot $T/2$, S1 and S2 send the signals to R using the same power simultaneously. P denotes the received signal power at R. The power splitter at R splits the received signal power P in $\rho : 1 - \rho$ proportion. The fraction ρP is used for EH, and the remaining fraction $(1 - \rho)P$ is used for information processing. R forwards the information signal to S1 and S2 using all the harvested energy in the second time slot $T/2$. The channels are reciprocal and quasi-static Rayleigh block fading among all nodes. It is assumed that perfect channel state information (CSI) is available at all nodes. All nodes are equipped with a single antenna. The processing power required by the information decoding circuitry at R is negligible compared to the power used for signal transmission [4, 7, 9].

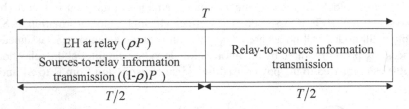

Fig. 3. The transmission block structure of the PSR protocol.

3 Analysis of Ergodic Capacity and Throughput

In this section, we analyze the performance of the proposed model for the TSR and PSR schemes.

3.1 Time Switching Relaying Protocol

The received signal at R in the first time slot $(1-\alpha)T/2$, $y_R^T(k)$ can be expressed as

$$y_R^T(k) = \sqrt{\frac{P_s}{d_1^m}} h s_1(k) + \sqrt{\frac{P_s}{d_2^m}} g s_2(k) + n_{a,R}(k) + n_{c,R}(k), \tag{1}$$

where P_s is the transmitting power of S1 and S2, d_1^m and d_2^m are the path losses from S1 and S2 to R, respectively, and m is the path loss exponent. $s_1(k)$ and $s_2(k)$ are the signals transmitted by S1 and S2. We assume that $s_1(k)$ and $s_2(k)$ have unit power. $n_{a,R}(k)$ and $n_{c,R}(k)$ denote the baseband noise signal received by the antenna at R and the noise due to RF band to baseband signal conversion, respectively. $n_{a,R}(k)$ and $n_{c,R}(k)$ are the additive white Gaussian noise (AWGN), which have zero-mean and different variances σ_n^2 and σ_c^2, respectively. Noting that in the formula (1), all the signals are expressed as the sampling signal forms.

On the basis of formula (1), the harvested energy E_R^T during the energy harvesting time slot αT can be expressed as

$$E_R^T = \eta \alpha T \left(\frac{P_s}{d_1^m} |h|^2 + \frac{P_s}{d_2^m} |g|^2 \right), \tag{2}$$

where $\eta \in (0,1)$ is the energy conversion efficiency which depends on the rectification process and the energy harvesting circuitry. In the formula (2), the noise energy is negligible because the noise energy is much smaller compared to the harvested energy in fact. Using (2), the signal-to-noise-ratio (SNR) of the signal transmitted by S1 and then received at R is derived as

$$\gamma_{S1,R}^T = \frac{P_s}{d_1^m \sigma^2} |h|^2. \tag{3}$$

In the formula (3), $\sigma^2 = \sigma_n^2 + \sigma_c^2$ is the variance of overall AWGN at R.

After the information transmission from the source nodes to R in the second time slot, R first decodes the received signals and then constructs the transmitted signal as $s_R(t) = s_1(t) \oplus s_2(t)$ applying physical-layer network coding (PNC), and finally sends the regenerated signal to S1 and S2 in the third time slot. The received signal at S2 in the third time slot can be expressed as

$$y_{S2}^T(k) = \sqrt{\frac{P_R^T}{d_2^m}} g_{SR}(k) + n_{a,S2}(k) + n_{c,S2}(k), \tag{4}$$

where $n_{a,S2}(k)$ and $n_{c,S2}(k)$ are the antenna and band conversion AWGNs at S2, having the zero mean and different variances σ_n^2 and σ_c^2, respectively. P_R^T denotes the transmitted power by R, which is given by

$$P_R^T = \frac{E_R^T}{(1-\alpha)T/2} = 2\eta\alpha \left(\frac{P_s}{d_1^m} |h|^2 + \frac{P_s}{d_2^m} |g|^2 \right) \Big/ (1-\alpha). \tag{5}$$

After several mathematical manipulations, γ_{S2}^T, the instantaneous SNR at S2, is given by

$$\gamma_{S2}^T = \frac{\left(d_2^m |h|^2 + d_1^m |g|^2 \right) |g|^2}{b_{11}}, \tag{6}$$

where $b_{11} = \frac{d_1^m d_2^{2m} \sigma^2 (1-\alpha)}{2\eta P_s \alpha}$. Because S2 knows $s_2(k)$ and CSI, S2 can recover the data transmitted by S1 via self-cancelation [9]. In delay-tolerant transmission mode, the ergodic capacity at R considering the signal transmission direction from S1 to S2 can be expressed as [4]

$$C_{S1,R}^T = \int_{\gamma=0}^{\infty} f_{\gamma_{S1,R}}^T(\gamma) \log_2(1+\gamma) d\gamma = e^{\frac{a}{\lambda_h}} E_1 \left(\frac{a}{\lambda_h} \right) \Big/ \ln(2), \tag{7}$$

where $f^T_{\gamma_{S1,R}}(\gamma) = \frac{a}{\lambda_h}e^{-\frac{a\gamma}{\lambda_h}}$ is the probability density function (PDF) of $\gamma^T_{S1,R}$ in (3), $a = d^m_1\sigma^2/P_s$, and $E_1(x) = \int_x^\infty e^{-t}/t\,dt$ is the exponential integral. λ_h is the mean of exponential random variable $|h|^2$. We can derive the ergodic capacity at S2 as follows

$$C^T_{R,S2} = \int_{\gamma=0}^\infty f^T_{\gamma_{S2}}(\gamma)\log_2(1+\gamma)d\gamma, \tag{8}$$

where

$$f^T_{\gamma_{S2}}(\gamma) = \frac{b_{11}d^{-m}_1 e^{-\frac{\sqrt{b_{11}d^{-m}_1\gamma}}{\lambda_g}}}{2\sqrt{b_{11}d^{-m}_1\gamma}\lambda_g} - \frac{1}{\lambda_g}\frac{b_{11}d^{-m}_1 e^{-V_1\sqrt{b_{11}d^{-m}_1\gamma}-\frac{b_{11}d^{-m}_2\gamma}{\sqrt{b_{11}d^{-m}_1\gamma}\lambda_h}}}{2\sqrt{b_{11}d^{-m}_1\gamma}} + \frac{1}{\lambda_g}\int_0^{\sqrt{b_{11}d^{-m}_1\gamma}}\frac{b_{11}d^{-m}_2 e^{-V_1x-\frac{b_{11}d^{-m}_2\gamma}{x\lambda_h}}}{x\lambda_h}dx,$$

$$V^T_1 = \frac{1}{\lambda_g} - \frac{d^m_1}{d^m_2\lambda_h}. \tag{9}$$

$f^T_{\gamma_{S2}}(\gamma)$ is the PDF of γ^T_{S2} in (6), and λ_g is the mean of exponential random variable $|g|^2$. Due to the page limit, we omit the proof here. Noting that there is no closed-form expression for $f^T_{\gamma_{S2}}(\gamma)$, we can get the value of (8) by the way of numerical computation.

The ergodic capacity from S1 to S2 is given by

$$C^T_{S1,S2} = \min(C^T_{S1,R}, C^T_{R,S2}). \tag{10}$$

Similarly, the ergodic capacity from S2 to S1 is given by

$$C^T_{S2,S1} = \min(C^T_{S2,R}, C^T_{R,S1}). \tag{11}$$

The ergodic throughput of the network can be expressed as

$$C_{TSR} = \frac{(1-\alpha)/2T}{T}(C^T_{S1,S2} + C^T_{S2,S1}) = \frac{(1-\alpha)}{2}(C^T_{S1,S2} + C^T_{S2,S1}). \tag{12}$$

It seems intractable to get the optimal α that result in the maximal throughput. The optimal α can be done offline by software tools for the given system parameters

3.2 Power Splitting Relaying Protocol

The received signal at R in the first time slot can be expressed as

$$y^P_R(k) = \sqrt{\frac{P_s}{d^m_1}}hs_1(k) + \sqrt{\frac{P_s}{d^m_2}}gs_2(k) + n_{a,R}(k). \tag{13}$$

Based on formula (13), the harvested energy E_R^P in the first time slot can be expressed as

$$E_R^P = \frac{\eta \rho P_s T}{2} \left(\frac{|h|^2}{d_1^m} + \frac{|g|^2}{d_2^m} \right).$$ (14)

The signal for information processing in the first time slot at R is given by

$$y_R'^P(k) = \sqrt{1 - \rho} y_R^P(k) + n_{c,R}(k).$$ (15)

Using (15), the instantaneous SNR of the link from S1 to R is given by

$$\gamma_{S1,R}^P = \frac{(1-\rho)P_s|h|^2}{d_1^m \sigma_R^2},$$ (16)

where $\sigma_R^2 = (1 - \rho)\sigma_n^2 + \sigma_c^2$ is the variance of overall AWGN at R.

In the second time slot, R constructs the decoded signals as $s_R(k) = s_1(k) \oplus s_2(k)$ applying PNC, and finally sends the regenerated signal to S1 and S2. The received signal at S2 in the second time slot can be expressed as

$$y_{S2}^P(k) = \sqrt{\frac{P_R^P}{d_2^m}} g s_R(k) + n_{a,S2}(k) + n_{c,S2}(k).$$ (17)

P_R^P denotes the transmitted power by R, which is given by

$$P_R^P = \frac{E_R^P}{T/2} = \eta \rho P_s \left(\frac{|h|^2}{d_1^m} + \frac{|g|^2}{d_2^m} \right).$$ (18)

After several mathematical manipulations, the instantaneous SNR at S2 is given by

$$\gamma_{S2}^P = \frac{\left(d_2^m |h|^2 + d_1^m |g|^2 \right) |g|^2}{b_{12}},$$ (19)

where $b_{12} = \frac{d_1^m d_2^{2m} \sigma_{s2}^2}{\eta \rho P_s}$ and $\sigma_{s2}^2 = \sigma_n^2 + \sigma_c^2$. In delay-tolerant transmission mode, the ergodic capacity of the link from S1 to R can be expressed as [4]

$$C_{S1,R}^P = \int_{\gamma=0}^{\infty} f_{\gamma_{S1,R}^P}(\gamma) \log_2(1 + \gamma) d\gamma = e^{\frac{a_1}{\lambda_h}} E_1 \left(\frac{a_1}{\lambda_h} \right) \Big/ \ln(2),$$ (20)

where $f_{\gamma_{S1,R}^P}(\gamma) = \frac{a_1}{\lambda_h} e^{-\frac{a_1 \gamma}{\lambda_h}}$ is the probability density function (PDF) of $\gamma_{S1,R}^P$ in (16), $a_1 = d_1^m \sigma_R^2 / (1 - \rho) P_s$. The ergodic capacity of the link from R to S2 is given by

$$C_{R,S2}^{P} = \int_{\gamma=0}^{\infty} f_{\gamma_{S2}}^{P}(\gamma) \log_2(1+\gamma)d\gamma, \tag{21}$$

where

$$f_{\gamma_{S2}}^{P}(\gamma) = \frac{b_{12}d_1^{-m}e^{-\frac{\sqrt{b_{12}d_1^{-m}\gamma}}{\lambda_g}}}{2\sqrt{b_{12}d_1^{-m}\gamma}\lambda_g} - \frac{1}{\lambda_g}\frac{b_{12}d_1^{-m}e^{-V_1\sqrt{b_{12}d_1^{-m}\gamma}-\frac{b_{12}d_2^{-m}\gamma}{\sqrt{b_{12}d_1^{-m}\gamma}\lambda_h}}}{2\sqrt{b_{12}d_1^{-m}\gamma}} + \frac{1}{\lambda_g}\int_0^{\sqrt{b_{12}d_1^{-m}\gamma}} \frac{b_{12}d_2^{-m}e^{-V_1^{P}x-\frac{b_{12}d_2^{-m}\gamma}{x\lambda_h}}}{x\lambda_h}dx,$$

$$V_1^{P} = \frac{1}{\lambda_g} - \frac{d_1^{m}}{d_2^{m}\lambda_h},$$

$$\tag{22}$$

$f_{\gamma_{S2}}^{P}(\gamma)$ is the PDF of γ_{S2}^{P} in (19). Due to the page limit, we omit the deriving process of $f_{\gamma_{S2}}^{P}(\gamma)$ here.

The ergodic capacity of the link from S1 to S2 is given by

$$C_{S1,S2}^{P} = \min(C_{S1,R}^{P}, C_{R,S2}^{P}). \tag{23}$$

Similarly, the ergodic capacity of the link from S2 to S1 is given by

$$C_{S2,S1}^{P} = \min(C_{S2,R}^{P}, C_{R,S1}^{P}). \tag{24}$$

The ergodic throughput of the network can be expressed as

$$C_{PSR} = \frac{T/2}{T}(C_{S1,S2}^{P} + C_{S2,S1}^{P}) = \frac{1}{2}(C_{S1,S2}^{P} + C_{S2,S1}^{P}). \tag{25}$$

Noting that there is no closed-form expression for $f_{\gamma_{S2}}^{P}(\gamma)$, it seems intractable to get the analytical expression for optimal ρ that result in the optimal throughput. The optimal ρ can be done offline numerically for the given system parameters.

4 Numerical Results

The parameters selected in this paper are the same depicted in [4]. The distances d_1 and d_2 are normalized to unit value. It is assumed that $P_s = 1$ W, $m = 2.7$, and $\eta = 1$. λ_h and λ_g are set to 1.

We set that the baseband antenna noise variances at all nodes are equal to σ_n^2, and the conversion noise variances are equal to σ_c^2. Figures 4 and 5 respectively show the analytical and simulation based results of the ergodic throughput with respect to α and ρ. The analytical results of the TSR and PSR protocols are produced based on formulas (7)–(12) and (20)–(25), respectively, and the simulation results are obtained by averaging over 10^5 random Rayleigh fading channel realizations. The analytical results perfectly match with the simulation results. This verifies our analysis.

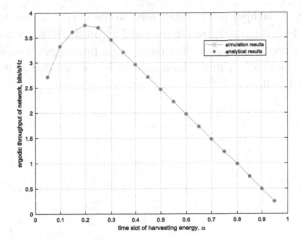

Fig. 4. The ergodic throughput comparisons of the simulation results and analytical results for the proposed TSR protocol with respect to α $(\sigma_n^2 = \sigma_c^2 = 0.01)$

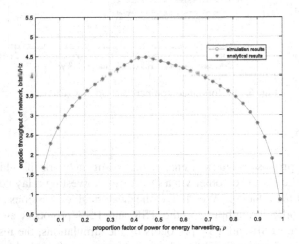

Fig. 5. The ergodic throughput comparisons of the simulation results and analytical results for the proposed RSR protocol with respect to ρ $(\sigma_n^2 = \sigma_c^2 = 0.01)$

The resolutions of ρ and α, which are the proportion factor used for EH in proposed PSR scheme and the time fraction used for EH in the TSR scheme, respectively, are both set to 0.01 when we find optimal solutions. We fix $\sigma_n^2 = 0.01$ or $\sigma_c^2 = 0.01$ and change the other noise power. As shown in Fig. 6, the maximal ergodic throughputs decrease with increasing the noise power both in the PSR and TSR protocols. When we fix σ_n^2 or fix σ_c^2, the performance of proposed PSR scheme is significantly better than that of the TSR scheme in a wide range of SNRs, and only at low SNR the performance of the TSR slightly outperforms that of the PSR when σ_n^2 fixed. In the TSR scheme, σ_n^2

and σ_c^2 affect the throughput in the same way, so the curves of the performance are overlap completely disregarded fixed σ_n^2 or fixed σ_c^2. In the PSR scheme, the optimal throughput with fixed σ_c^2 is higher than the optimal throughput with fixed σ_n^2 at the beginning but inferior to it with the increasing noise power, due to the effect of the proportion factor $1 - \rho$.

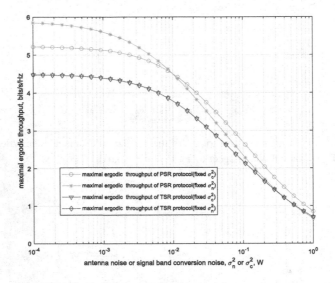

Fig. 6. Performance comparisons of two schemes

5   Conclusions

In the paper, we propose a wireless energy harvesting network, in which two source nodes communicate with each other via a DF energy harvesting relay node over quasi-static Rayleigh block fading. We derive the analytical expressions of the ergodic capacity and ergodic throughput for the TSR and PSR protocols and compare the performances of the PSR and TSR protocols by the simulations, the results verify the analytical results and reveal the PSR scheme achieves significantly higher throughput than the TSR scheme in a wide range of SNRs.

References

1. Liu, P., et al.: Energy harvesting noncoherent cooperative communications. IEEE Trans. Wireless Commun. **14**, 6722–6737 (2015)
2. Liu, P., et al.: Noncoherent relaying in energy harvesting communication systems. IEEE Trans. Wireless Commun. **14**, 6940–6954 (2015)
3. Nasir, A.A., et al.: Relaying protocols for wireless energy harvesting and information processing. IEEE Trans. Wireless Commun. **12**, 3622–3636 (2013)

4. Nasir, A.A., et al.: Throughput and ergodic capacity of wireless energy harvesting based DF relaying network. In: Proceedings of IEEE ICC, Sydney, NSW, Australia, pp. 4066–4071 (2014)
5. Ashraf, M.: Capacity maximizing adaptive power splitting protocol for cooperative energy harvesting communication systems. IEEE Commun. Lett. 22, 902–905 (2018)
6. Lee, H., et al.: Outage probability analysis and power splitter designs for SWIPT relaying systems with direct link. IEEE Commun. Lett. 21, 648–651 (2017)
7. Chen, Z., et al.: Wireless information and power transfer in two-way amplify-and-forward relaying channels. In: Proceedings of IEEE Global Conference on Signal and Information Processing (GlobalSIP), Atlanta, GA, USA, pp. 168–172 (2014)
8. Shah, S., et al.: Energy harvesting and information processing in two-way multiplicative relay networks. IET Electron. Lett. 52, 751–753 (2016)
9. Van, N., et al.: Three-step two-way decode and forward relay with energy harvesting. IEEE Commun. Lett. 21, 857–860 (2017)
10. Rabie, K.M., et al.: Half-duplex and full-duplex AF and DF relaying with energy harvesting in log-normal fading. IEEE Trans. Green Commun. Netw. 1, 468–480 (2017)
11. Rabie, K.M., et al.: Wireless power transfer in cooperative DF relaying networks with log-normal fading. In: Proceedings of IEEE Global Communications Conference (GLOBE-COM), Washington, DC, USA, December, pp. 1–6 (2016)
12. Rabie, K.M., et al.: Energy harvesting in cooperative AF relaying networks over log-normal fading channels. In: Proceedings of IEEE International Conference on Communications (ICC), pp. 1–7 (2016)

Author Index

Printed in the United States
By Bookmasters